In this series original contributions of physical research in all fields are contained, such as

— research monographs
— collections of papers to a single topic
— reports on congresses of exceptional interest for physical research.

This series is aimed at promoting quick information and communication between physicists of the various special branches.

In diese Reihe werden Originalbeiträge zu allen Gebieten der physikalischen Forschung aufgenommen wie

— Forschungsmonographien
— Sammlungen von Arbeiten zu einem speziellen Thema
— Berichte von Tagungen, die für die physikalische Forschung besonders aktuell sind.

Die Reihe soll die schnelle Information und gute Kommunikation zwischen den Physikern der verschiedenen Fachgebiete fördern.

Manuscripts
in English comprising at least 100 pages and not more than 500 pages can be admitted to this series. With respect to a quick publication the manuscripts are reproduced photomechanically. Authors who are interested in this series please turn directly to the Akademie-Verlag Berlin. Here you will get more detailed information about the form of the manuscripts and the modalities of publication.

Manuskripte
in englischer Sprache, die mindestens 100 Seiten und nicht mehr als 500 Seiten umfassen, können in diese Reihe aufgenommen werden. Im Interesse einer schnellen Publikation werden die Manuskripte auf fotomechanischem Wege reproduziert. Autoren, die an der Veröffentlichung entsprechender Arbeiten in dieser Reihe interessiert sind, wenden sich bitte direkt an den Akademie-Verlag Berlin. Sie erhalten dort genauere Informationen über die Gestaltung der Manuskripte und die Modalitäten der Veröffentlichung.

EPM '89

3rd International Conference
on Energy Pulse and Particle Beam
Modification of Materials

Physical Research

Volume 13

EPM '89

Energy Pulse and Particle Beam Modification of Materials

edited by
Karl Hohmuth
Edgar Richter

EPM '89

Energy Pulse and Particle Beam Modification of Materials

3rd International Conference held September 4—8, 1989
Dresden, GDR

edited by
Karl Hohmuth
Edgar Richter

With 468 Figures and 43 Tables

Akademie-Verlag Berlin

Editors:

Prof. Dr.-Ing. habil. Karl Hohmuth
Dr. sc. nat. Edgar Richter

Zentralinstitut für Kernforschung Rossendorf,
Akademie der Wissenschaften der DDR

Reproduction of the original authors' manuscripts

ISBN 3-05-500786-7

Erschienen im Akademie-Verlag Berlin, Leipziger Str. 3—4, Berlin, DDR-1086

Lizenznummer: 202 · 100/444/90
Gesamtherstellung: VEB Druckerei „Thomas Müntzer", Bad Langensalza, 5820
Lektorin: Heike Höpcke
LSV 1175
Bestellnummer: 764 095 3 (9287)

The international conference EPM'89 was the continuation of the conference EPM'84 and EPM'87.

The scientific programme covered

- implantation into silicon
- implantation and annealing of compound semiconductors
- transient heat treatment of semiconductors
- silicides
- implantation into metals
- surface modification of metals and other materials
- materials deposition
- silicon on insulators (SOI)
- diagnostics and ion beam equipments
- high temperature superconductors
- fundamentals

About 250 scientists from 16 countries attended the conference EPM'89.

In agreement with the proposal of the International Advisory Committee of this conference the continuation with EPM'91 is planned.

To the members of the International Advisory Committee, to the members of the Local Organizing Committee, to the invited speakers, to the chairmen of the sessions and all participants we would like to express our thanks for their active and successfull work.

Prof. Dr. K. Hohmuth
Chairman

Dr. E. Richter
Secretary

CONTENTS

INVITED PAPERS

SUBMITTED PAPERS

1. IMPLANTATION INTO SILICON

2. IMPLANTATION AND ANNEALING OF COMPOUND SEMICONDUCTORS

3. TRANSIENT HEAT TREATMENT OF SEMICONDUCTORS

4. SILICIDES

5. IMPLANTATION INTO METALS

6. SURFACE MODIFICATION OF METALS AND OTHER MATERIALS

7. MATERIALS DEPOSITION

8. SILICON ON INSULATORS

9. DIAGNOSTIC AND ION BEAM EQUIPMENTS

10. HIGH TEMPERATURE SUPERCONDUCTORS

11. FUNDAMENTALS

LATE PAPER

EXCIMER LASER ASSISTED DEPOSITION OF SiO_2-STRUCTURES ON SEMICONDUCTOR SUBSTRATES FROM SILICONORGANIC FILMS.

Ageev V.P., Dang Quoc Chung**, Konov V.I., Kuzmichov A.V.
Maslakov A.I.*, Velikov L.V.*, Zaroslov D.Yu.*

Introduction

At the present time a radiation produced local modification of solid surfaces is one of the most progressing field of laser applications. During last years an extended body of experimental and theoretical results has been presented in the literature, concerning with many aspects of laser induced crystallization of amorphous layers, synthesis of compound semiconductors, light enhanced surface oxidation, chemical etching etc. (e.g. /1-3/). Being of the great scientific interest these investigations of laser induced surface phenomena play also an important role in practice, for example, in modern micro- and optoelectronics, integrated optics, and others.

The significant place here belongs to development of new methods of laser controlled local film deposition. Till last few years the main attention here has been paid to deposition processes using gas phase or liquid metalloorganic precursors /1-3/. Recently a new approach to the problem of surface patterning began to develop. This approach is based on a laser produced decomposition of solid-state organometallic compounds (OMC), which put onto substrates in form of fine paste layers or amorphous films. So, a patterning of surfaces with metal or metal oxide features was achieved by using Cu- formiates /4/, Pd-acetates /6/, Au- and Pt-terpensulfids /5-7/, Fe- and Cr-abietates /8,9/ and other more complicated OMC- based compositions. In this work continuing the series of our investigations of laser produced deposition from solid state OMC we report novel results on surface patterning by dielectric structures of silicon oxide under the action of UV laser radiation on silicon-organic films. An interest to silicon organic materials arises mainly from the following. From one hand, silicon organic films can be easily transform into SiO_x layers by their annealing in the air (x=1-2, depending on annealing regime) at comparatively low temperatures T=250-650 C. From the other, these materials may have high resistance in plasma etching processes.

Institute of General Physics, USSR Academy of Sciences,
Vavilov str. 38, 117942 Moscow, USSR,
* Institute of Physics and Technology, USSR Academy of Sciences,
Krasikov str. 25a, 117218 Moscow, USSR.
** Present address: Institute of Physics, Ngiado-Tulien, Hanoi, Vietnam.

Experimental

Several modifications of oligomethylvinylcyclosiloxanes (OMVCS) were investigated. Oligomers were obtained by hydrolytic polycondensation of organochlorsilanes in organic solvents with the subsequent catalytic treatment by HCl to achieve the narrowing of molecule mass dictribution. Films of thicknesses H_0= 200 -1000 nm were spin-coated from solutions onto silica or silicon substrates. Remnants of solvent were removed by temperature pre-baking. This preliminary heat treatment also enables us to vary in some limits (before the decomposition would start) the average molecule mass of oligomer and to change optical properties of films.

The typical spectral transmittance of OMVCS (type II) film on quartz substrates are shown in Fig. 1. Note that initial films are transparent for visible radiation, but partial absorbing in the UV. The heat treatment at moderate temperatures give rises to some reduction of the UV- and increase of the visible light absorbance in films. This darkening makes it possible to use for further films processing many kinds of VIS/UV lasers. The final transformation of films to the SiO_2-state at temperatures of T = 640 C leads to a restoration of the transmittance to the quartz substrate level.

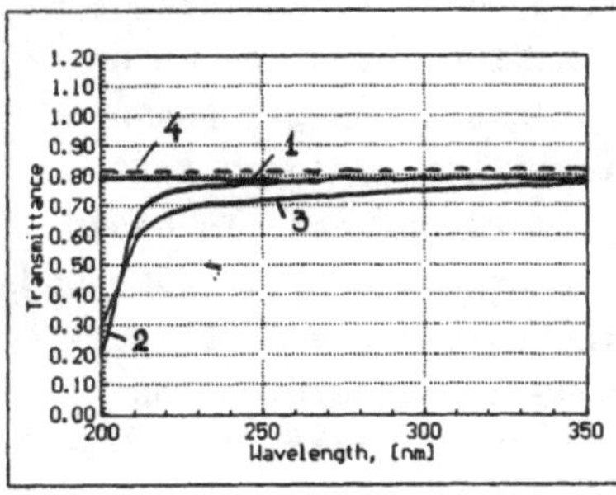

Fig.1 Transmittance spectra of 0.5 um thick OMVCS-II film on quartz substrate. 1-substrate; 2-initial film; 3-annealing (250 C,40 min); 4-annealing (640 C, 40 min)

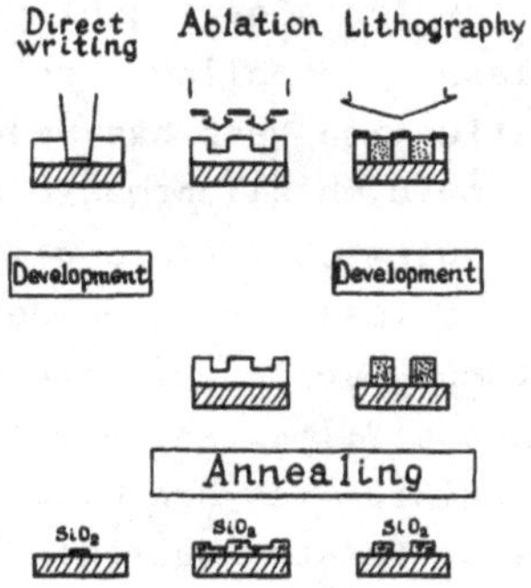

Fig.2

A films processing was performed by using pulse- repeated ArF excimer laser (model 1701 manufactured by Institute of General Physics of the USSR Academy of Sciences). generating 10-ns pulses of the UV radiation (wavelength =193 nm) at the rate till 100 pps.

Proceeding from properties of the testing materials it is expedient to consider different possible methods of a laser processing of films, including (see Fig.2) a total material removal via laser produced ablation, a direct one-step decomposition of the OMVCS to silicon-oxides of this or those stoichiometry, and (if it is possible) a laser lithography in films. In the former and latter cases residual silicon organic structures may be also transformed into the SiO_2 state by thermal annealing.

Results and Discussion

Experiments prove that all mentioned methods of the OMVCS films structurization and the SiO_x layer deposition can be really achieved at appropriate irradiation regimes. At the beginning consider the process of a multi-pulsed laser lithography in OMVCS films.

According to our analysis all sensitometric characteristics of OMVCS-based resists (contrast, sensitivity, resist type) appeared to be influenced to a great extend by the films preparation procedure and irradiation conditions: pulsed laser fluence (E) and exposure dose D= E N, where N is a number of successive laser shots.

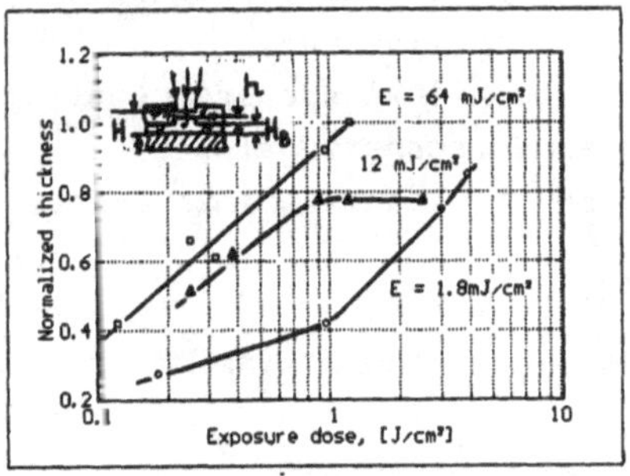

Fig.3 Characteristic curves of OMVCS-I resists.

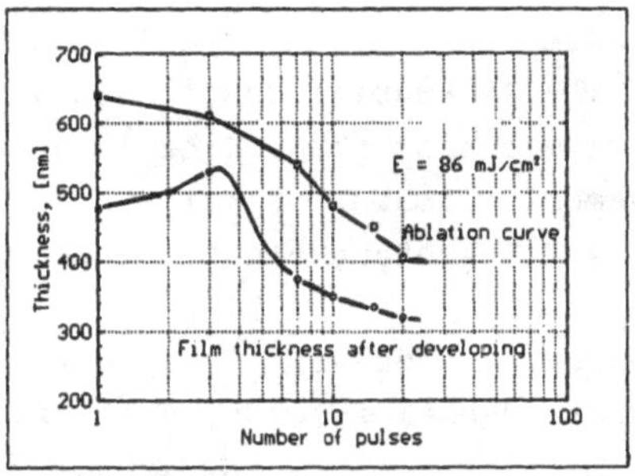

Fig.4 Thicknesses of exposed OMVCS-I films before and after development vs number of laser pulses.

In Fig.3 there are presented some characteristic curves of OMVCS-resist (of type I) taken in form of dose dependent normalized thicknesses $X = h/(H - H_b)$ of developed films (the methylisobutylketone: isopropanol (1:3) solution was used). Here, the partial insolubility of non irradiated (but pre-baked) films being responsible for an existence of typical background layer of thicknesses H_b (see inset in Fig.3) is taken into account. For "fresh" films the background thickness $H_b = 0$.

It follows from sensitometric curves, that one should make difference between three fluence-controlled regimes of the lithography, been described to the first time in our previous papers /8-9/ on the metal resinate films decomposition. At comparatively low fluences of the order of few tens of mJ/cm^2 or less the OMVCS-resist is always the negative one,i.e. normalized thicknesses of developed films increase monotonically up to levels of X 1 as the exposure dose rises (Fig.3). Variations of fluences practically do not change the contrast of resists, but make them more sensitive to the more intensive irradiation, i.e. the total irradiation dose needed for final material transformation into insolvable state diminishes.

The high -intensity films irradiation at fluences being above the material ablation thresholds provides another regime of the laser lithography. Under these conditions the material removal from exposed areas achieves during the laser pulse, and, additionally, at the stage of residual film development, but the latter is typical of a positive resist

The third (intermediate) regime appears to be also possible characterized by the effect of a resist sign conversion during exposure. It achieves at moderate fluences of E = 50-70 mJ/cm^2. The effect is illustrated in Fig.4, exhibiting thicknesses of exposed films before and after development as dependent on the number of laser pulses. As it is seen the film (here of type I) behaves initially as a negative resist, but under continuing irradiation it becomes the positive one.

The IR Fourier spectroscopy analysis has shown that appearance of negative resist properties is accompanied by some reconstruction of oligomer molecules. Firstly, a polycondensation of oligomers into longer chains takes place via hydroxyl groups:

$$-\overset{|}{\underset{|}{Si}}-OH + OH-\overset{|}{\underset{|}{Si}}- \longrightarrow -\overset{|}{\underset{|}{Si}}-O-\overset{|}{\underset{|}{Si}}- + H_2O$$

Secondly, a cross-linking process through the vinyl- and/or the methyl radical channel is under a progress:

$$-\overset{|}{\underset{|}{Si}}-CH=CH + CH=CH-\overset{|}{\underset{|}{Si}}- \longrightarrow -\overset{|}{\underset{|}{Si}}-\underset{\substack{CH_2\\CH_2\\CH_3}}{CH}-\overset{|}{\underset{|}{Si}}-$$

It is also important to underline here that considered laser produced reconstructions are qualitatively very similar to those ones, which took place during the pre-baking of OMVCS films. This fact as well as the substantial fluence-dependent behavior of characteristic curves allows us to suppose the negative resist properties to be rather of thermal than photolytic nature. It proves also by simple evaluations, which predict the pulsed increase of a films temperature above the polymerization level (T_p= 150 - 170 C) at fluences E 3-5 mJ/cm^2.

Studies of films exposed at high fluences point out to the destruction of initial oligomer chains into smaller ones dominating on the polymerization. To our opinion it provides both the sign conversion effect and positive resist properties.

Now let us to characterize the other two methods of the film structurization, namely the laser decomposition of OMVCS to silicon oxide phase, and the total material removal via a laser induced ablation. Phenomenologically, both processes can be controlled via the reduction of a film thickness during pulse -repeated irradiation.

Measurements performed under wide range of experimental conditions have shown both type of film processing to be of threshold character. So, the ablation takes place from the fist laser shot only at fluences E being above some critical values of $E_a = 30\ mJ/cm^2$ (for OMVCS-I), while the decomposition to SiO_x layer occurs at lower fluences $E_d = 15\ mJ/cm^2 \leq E < E_a$.

The existence of thresholds points first of all to the preferable thermal nature of process. It is also important to mention that multi-pulsed irradiation does not diminish the ablation threshold, which is abnormal for polymer materials (e.g. /10/). Curves of material losses from OMVCS-I films correspondent to SiO_x deposition as well as to ablation regimes are presented in Fig.5. One should pay attention to the principle in the kinetics of film etching in these cases. The deposition regime is characterized by the permanent reduction of the etch rate during the multi-pulsed irradiation, until the process would stop in a self-regulating manner at appropriated fluence-controlled thicknesses of silicon oxide films. Such behavior of the process can be easily understood in terms of the negative optical feedback between the UV laser radiation and OMVCS films, being decomposed towards the more transparent in the UV silicon oxides state. This is also in qualitative agreement with spectral transmittance measurements taken at different stages of films annealing presented above in Fig.1.

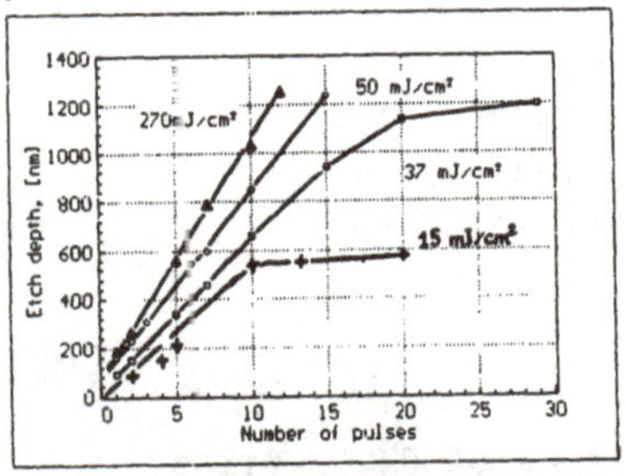

Fig.5 Thicknesses of OMVCS-I film vs exposure dose at different fluences.

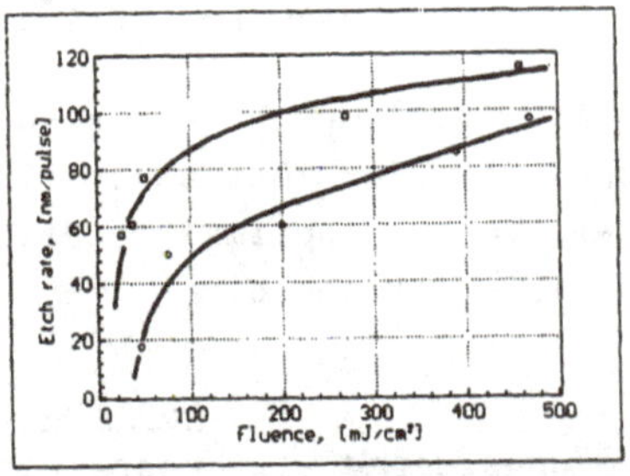

Fig.6 Dependencies of ablation rate of OMVCS films on fluence.

Analogously to discussed situation, the laser induced ablation may also be affected by a laser to films decoupling, which usually occurs at fluences near the ablation threshold (see curve for $E = 37\ mJ/cm^2$ in Fig.5). The nature of this effect lies obviously in the decomposition of deeper layers of the material, while upper ones are ablating. Nevertheless, this peculiarity disappeared quickly as fluence rising, and starting from approximately $E = 50\ mJ/cm^2$ the material removal achieves at constant rates in the every laser shot. It is important to note that under the action of more long-wave VIS/UV radiation the ablation kinetic features would expected to be quite different, because of the initially opposite feedback sign, that will be the matter of our foregoing investigations.

The fluence-dependent behavior of ablation rates derived from etching curves for OMVCS films of both types are presented in Fig.6. Films were characterized by practically the same thermodynamic properties, but had different values of UV absorption coefficients (k = 4.3 and 6.8 μm^{-1}, resp.). It follows from V(E)-curves, that the more the absorption coefficient the higher the efficiency of the etching. Additionally to these, analysis shows the curves to obey rather logarithmic than linear type of V versus E behavior.

Taken together both these facts are in a qualitative agreement with simple speculations, which consider the ablation process to start provided the laser energy supply per unit volume would reach some critical level of E^*. To prove this one must apply the Beer's law to describe the radiation energy input into material, and neglect heat-conductance energy losses to deeper layers of films. The latter is reasonable because of an extremely poor thermodifusivity of materials, $\varkappa < 10^{-3}$ cm^2/sec. so that $\varkappa\tau < 0.1$ ($\tau = 10^{-8}$ s - is the laser pulse duration). Then the etching rate (in units thickness per pulse) may by evaluated from the simple energy balance equation:
$(1-R)\ k\ E\ \exp\ (-k\ V) = E^*$, where R - is the light reflectance from the upper film surface, 1-R = 1. After trivial rearrangement one obtain:

$$V = k^{-1} \mathrm{Ln}\ (kE/E^*)$$

The derived scaling law has also those consequence that radiation energy expenditures for the total film ablation increase as the laser fluence rises, so that optimal conditions for the film processing are near the ablation threshold.

Conclusion.

We have considered all general characteristics of OMVCS films processing, enabling to form silicon organic as well as silicon oxide structures of required morphology. As an illustration of possibilities of methods Fig.7 shows the SEM of SiO_2 structures, which have been obtained

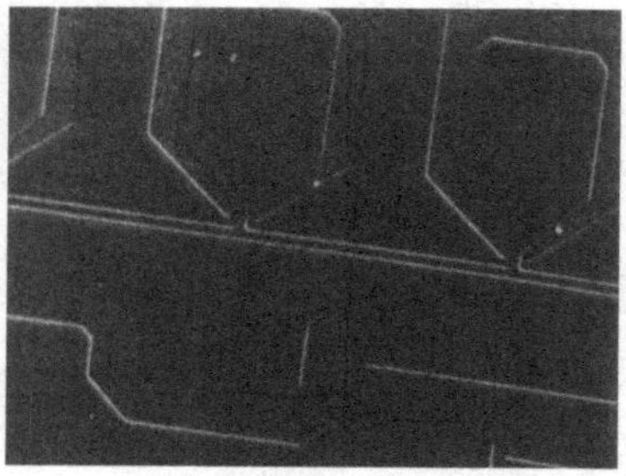

Fig.7 SEM of SiO_2 pattern obtained by UV laser lithography in OMVCS film

— 10 μm

by contact laser lithography in the 0.8 um thick film through a chromium-on-quartz mask, and the subsequent annealing. A submicron resolution is achieved. Special tests proved the deposited material to be characterized by electrical breakdown strength as high as 100 V/um, and pores surface densities less than 0.2 cm^{-2}. The quality of the structures material satisfies for main requirements of modern microtechnology /11/. Thus, a new way for surface patterning by SiO_2 films is propose, which may be the alternative to previously developed methods of the laser enhanced oxidation of silicon substrate surfaces or the chemical vapor deposition of SiO_2 layers.

References

(1) Bauerle D.(Ed.):Laser processing and diagnostics.Springer Series in Chemical Physics.Vol.39,Springer,Berlin,Heidelberg 1984.

(2) Metev S.(Ed.):Laser Assisted Modification and Synthesis of Materials,Sofia, University Publisher,Sofia 1985.

(3) Laude L.D.,Bauerle D., and Wautelet M. (Ed.):Interfaces under Laser Irradiation (NATO ASI Series.Series Applied Science),Martinus Nijhoff Publishers, Dordrecht, Boston, Lancaster, 1987.

(4) Gerassimov R.B.,Metev S.M., Savchenko S.K., Kotov G.A., and Veiko V.P.: Proc. of 12th Int.Conf.on Quant.Electron.,p.266 (1982).

(5) Yan R.Y., and Allen S.:SPIE 459, 71 (1984).

(6) Gross M.E. et al: Phys.Lett. 46, 1184 (1985); ibid. 47, 923 (1985); J.Appl.Phys. 57, 1139 (1985); ibid. 61, 1628 (1987); Mat.Res.Soc.Symp. Proc. Vol. 101, 89 (1988).

(7) Ageev V.P., Allen S., Konov V.I., Melnikov V.M. and Preisverk H.-P.: Preprint of the Gen.Phys.Inst. of the USSR Acad.Sci. No 143, Moscow 1986.

(8) Ageev V.P., Baranenkov I.V., Valiev K.A., Velikov L.V., Konov V.I., Maslakov A.I., Melnikov V.M., and Prokhorov A.M.: Dokl.Acad.Nauk SSSR 294, 1358 (1987) - in Russian.

(9) Ageev V.P., Konov V.I., Maslakov A.I., Melnikov V.M,and Velikov L.V.: in Proc. of Int.Conf. on Energy pulse and Particle Beam Modification of Materials (Dresden, DDR, 1987), p. 116.

(10)Mahan G.D., Cole H.S., Liu Y.S., and Philipp H.R.: Appl. Phys. Lett. 53, 2377 (1988).

(11)Wayne M. Moreau (Ed.): Semiconductor lithography, Plenum Press, New-York, 1988.

SYNTHESIS OF COMPOUNDS IN SEMICONDUCTORS BY THE IMPLANTATION OF REACTIVE IONS

P L F Hemment[1] and K J Reeson[1]

Abstract

Recent developments in the exploitation of ion beam synthesis are reviewed. The experimental conditions necessary to produce device worthy dielectric and "metallic" materials and the successful fabrication of MOS and bipolar circuits on silicon on insulator (SOI/SIMOX) substrates are reported. The achievement of a large reduction in the defect density in SIMOX substrates as a result of an improved understanding of the formation of these structures encourages optimism that ion beam synthesis will become an important enabling technology for the realisation of future generations of high performance microelectronic circuits.

1. Introduction

In this paper we will consider recent advances in the use of ion beams to synthesise compounds in silicon by the implantation of high doses of reactive ions[1][2]. The thin film structures that can be formed are being applied to advanced silicon devices and examples of the current use for the formation of high quality silicon on insulator (SOI) substrates and the prospects of forming high conductivity silicide layers will be described.

2. Background

The current interest in compound formation by ion beam synthesis (IBS) owes much to the need to develop a technology for the preparation of silicon on insulator (SOI) substrates to facilitate the fabrication of submicron CMOS circuits. The use of SOI substrates, in which the circuits are fabricated in a thin silicon layer above a dielectric, inhibits parasitic transistor action (latch-up) and enhances radiation tolerance and circuit speed. Thus, during the 1970's and early 1980's much effort in many laboratories around the world was devoted to the search for a viable SOI materials technology. Techniques researched included methods of improving silicon on sapphire (SOS), layer transfer, zone melt recrystallisation, porous silicon, heteroepitaxy and ion beam synthesis. Of these many techniques the use of O^+ ion implantation to form a buried layer of SiO_2 has been found to be the most versatile and controllable method. This situation owes much to the pioneering experiments of Izumi et al[3], who reported in 1978 the fabrication of high performance CMOS ring oscillators in an epitaxial silicon layer grown over a buried SiO_2 layer, formed by ion beam synthesis. This group had implanted 150 keV O^+ ions into 100 silicon and maintained the single crystal nature of the substrate by dynamic annealing during the implantation, utilising the kinetic energy of the ions to heat the substrate. The wafers were subsequently annealed at 1150°C for two hours, prior to growth of the epitaxial layer in which the devices were formed. Thus the initial recipe of (a) implantation, (b) annealing, and (c) epitaxial layer growth was established and this group proposed the acronym SIMOX (Separation by IMplanted OXygen). Subsequently it was shown that submicron CMOS circuits could be fabricated directly in the silicon overlayer without resort to silicon epitaxy[4] and that higher quality material results when background heating is used during implantation and anneal temperatures exceed 1300°C[1][8][11].

The success of SIMOX technology has stimulated additional studies of the synthesis of silicon compounds of importance in microelectronics including Si_3N_4, oxynitrides, SiC and, more recently, silicides including $CoSi_2$, $NiSi_2$, $TiSi_2$, $CrSi_2$, YSi_2 and $Co_xNi_{1-x}Si_2$[1].

---

[1] Department of Electronic and Electrical Engineering, University of Surrey, Guildford, Surrey, GU2 5XH, UK

3. Physical Aspects of Ion Beam Synthesis

Ion beam synthesis of microelectronic grade structures is a two stage process involving (i) implantation, and (ii) high temperature annealing. During stage (i) either a surface or buried layer may be formed, depending upon the combination of ion dose and energy. Formation of a buried layer is commonly accomplished by the implantation of a dose of reactive ions in the range 2 x 10^{17} cm^{-2} to 2 x 10^{18} cm^{-2} at an energy of typically 100 keV to 350 keV. The microstructure depends upon the substrate temperature, which is generally in the range 400°C to 600°C to inhibit amorphisation and to promote defect annealing in the silicon overlayer of SOI structures. The evolution of the synthesised structure also depends upon the diffusivity of the implanted species both in silicon and the compound. For doses above the critical value for compound formation (ϕ_C)[5], stoichiometry will be exceeded unless redistribution occurs, as seen in the O/Si system[5] and as has been reported for very low current density implantation of N^+ ions[6]. The full benefits of a high temperature anneal have been recognised since 1985[7][8], when it was shown that segregation of the implanted impurity, as well as annihilation of lattice defects, can be achieved. Typical anneal temperatures are 1300°C to 1405°C for O^+/Si[9], 1200°C for N^+/Si[1] and 1000°C to 1200°C for Co^+/Si[10]. The net mass transport is against the concentration gradient and, in the case of O/Si, the system relaxes by reducing the surface energy, during the dissolution of SiO_2 precipitates, until finally one planar precipitate remains, which is the buried layer[11].

Recipes have been reported for the formation of high quality SiO_2, Si_3N_4 and $CoSi_2$ layers buried within a single crystal silicon matrix. Over a limited dose range the thickness of these films scales linearly with dose, the structures have abrupt, although not necessarily planar, interfaces and contain threading dislocations as the dominant defect in the silicon overlayer. Dislocation densities as high as 10^9 cm^{-2} have been reported[1] and during recent years much work has been undertaken to eliminate these defects.

4. SIMOX

4.1 Status of SIMOX Technology

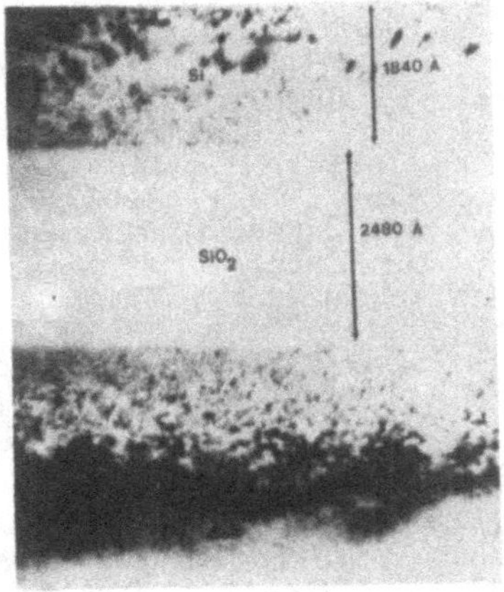

Figure 1: Cross-sectional TEM micrograph of an "as implanted" commercial SIMOX wafer[12].

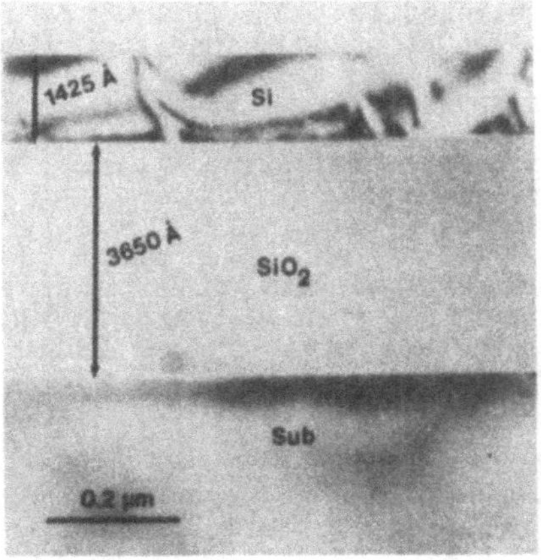

Figure 2: Cross-section of the sample shown in Figure 1 after annealing at 1405°C for 30 mins[12].

Figure 1, from Baumgart and van Ommen[12], shows a cross-sectional TEM micrograph of an "as implanted" SIMOX sample produced in an Eaton-Nova NV200 high current implanter (1.7 x 10^{18} O^+ cm^{-2}, 150 keV, T_i = 550°C, 730 Å screen oxide). A continuous layer of amorphous oxide is buried beneath a silicon single crystal overlayer of thickness 1840 Å. This layer is highly defective containing SiO_2 precipitates, extrinsic stacking faults, {311} defects, elongated interstitial dipoles, microtwins but very few dislocations. However dislocation loops are observed in wafers implanted at lower temperatures[12], when dislocation formation is the mechanism for point defect annihilation. The diameter of the oxide precipitates increases with depth and implantation temperature[11]. The oxide layer is found to be stoichiometric[1] although bond distortions, identified from IR and XPS measurements, suggest the co-existence of SiO_2 and SiO_x, where $x < 2$[13][14].

Figure 2 shows a TEM cross-section of a sample (as above) annealed at 1405°C for thirty minutes. Complete dissolution of the oxide precipitates has occurred and the released oxygen segregates at the buried layer which consequently grows in thickness and has abrupt, planar Si/SiO_2 interfaces. The dominant defects are threading dislocations ($\sim 10^9$ cm^{-2}) which span the silicon overlayer. Small geometry CMOS circuits fabricated in similar substrates[4] show all of the benefits predicted for SOI/CMOS. However, such a high density of dislocations is undesirable as they are a potential cause of circuit failure. Currently new processing recipes are being sought to inhibit dislocation formation.

4.2 Recent Developments

Point defects and internal stress are found to play an important role in the formation of dislocations[12]. Two processes generate excess point defects during O^+ implantation, these being displacement damage (Frenkel pair production) and internal oxidation. Jaussaud et al[15] have argued that the free silicon surface is the dominant sink for excess silicon interstitials (Si_{Int}) where annihilation may occur without additional defect formation. They propose that low dislocation material will be formed provided (i) the migration of Si_{Int} to the surface is not impeded, for example, by a high density of oxide precipitates, and (ii) the chemical state of the surface is conducive to annihilation, which may occur through internal epitaxial growth[16]. As the diffusion lengths of the point defects and, also, the size and density of oxide precipitates are a function of temperature, it is evident that, from the above arguments, the microstructure will depend upon both the beam current density (defect generation rate) and temperature. Several approaches have been used to reduce the dislocation density in the silicon overlayer. These include: implantation at a low defect introduction rate[17], sequential implantation and annealing[15][18], which generally results in the best quality SIMOX[18], and formation of voids[19][20]. Dislocation densities of less than 10^3 cm^{-2} are reported by authors.

Carbon Contamination - Large variations (4×10^{17} to 1×10^{19} cm^{-3}) in carbon contamination of SIMOX wafers have been measured[21] and the concentration has been correlated with the dislocation density after a 1300°C anneal. As carbon is known to form nucleation sites for precipitates and dislocations, it is important to maintain tight control of this contaminant. Dislocation densities below 10^6 cm^{-2} have been measured in conventional SIMOX substrates (1.6×10^{18} O^+ cm^{-2}, 150 keV, $T_i = 525$°C to 580°C, $T_a = 1300$°C) containing less than 2×10^{16} C cm^{-3}[21].

Novel Structures - The inherent versatility of IBS enables the dimensions and geometry of synthesised layers to be varied. For example implantation through a patterned masking layer enables non-planar oxide layers to be formed, which provide both lateral and vertical isolation of silicon device islands (total dielectric isolation[22]). The formation of thick oxide (0.75 µm) layers and multi-layer SIMOX structures has been reported by Namavar et al[23]. These workers used silicon epitaxy to increase the thickness of the silicon overlayer prior to implantation of a further dose of O^+ ions. Deep buried layers of SiO_2 have been prepared by 2 MeV to 5 MeV O^+ ion implantation[23][24] when, because of the reduced nuclear stopping cross-section in the near surface region, fewer defects are created and lower dislocation densities after annealing have been reported. Namavar et al[25] report a reduced density of threading dislocations in SIMOX substrates, after regrowth of an amorphous layer formed by the implantation of a low dose of Ge^+ ions to form a thin strained layer within the silicon overlayer. The interface between the strained layer and silicon causes the dislocations to change direction and inhibits their propagation to the surface. In contrast, Kamins et al[26] finds little interaction between the threading dislocation and a Ge/Si superlattice formed by MBE.

5. Other Materials

5.1 Wide Bandgap Materials

Achieving the goal of forming device worthy layers of Si_3N_4, oxynitrides and SiC by IBS is proving to be difficult. Buried Si_3N_4 structures are attractive for SOI substrates carrying the advantages over SIMOX of requiring a lower ion dose, lower anneal temperatures and providing a better diffusion barrier to impurities. However, they suffer from the disadvantages of nitride recrystallisation (α - Si_3N_4) forming a leaky dielectric[27], nitrogen pile-up[5] due to the low diffusivity of nitrogen in Si_3N_4 and anomalous nitrogen doping in the silicon overlayer[28]. The density of threading dislocations is high, typical values being 10^8 to 10^9 cm^{-2}. Attempts to form amorphous Si_3N_4 by subsequent implantation[29] or inclusion of oxygen[30] have largely been unsuccessful. Formation of the high temperature phase (β - Si_3N_4) has been reported by Bussmann et al[31] following high temperature (>600°C) implantation.

Oxynitride layers have been prepared by sequential implantation[32][33] involving (O^+ + N^+) and (N^+ + O^+) and, also, by simultaneous implantation using NO^+ molecules[34]. After annealing, different microstructures are observed, but the materials are still highly defective, some containing crystalline inclusions and bubbles in the samples implanted with the highest doses[28]. The lowest defect density was found after annealing at 1350°C[28].

SiC is a refractory, wide bandgap, cubic semiconductor with potential applications in microelectronic and optoelectronic circuits. IBS offers the advantage of preparation of pure layers under clean conditions, which is difficult to achieve by conventional deposition techniques. Layers of β-SiC have been formed after multiple energy implantation (total dose 1.9 x 10^{18} cm^{-2}) at 860°C[35]. Although the layers have an epitaxial relationship with the (100) silicon substrate, they are not continuous. Reeson et al[36] find that during a high temperature anneal (1300°C to 1405°C), the carbon partially redistributes and that a thin layer of β-SiC grows at the lower amorphous carbide/Si interface through a topotactic transformation. It is speculated that this layer can be used as a seed during subsequent epitaxy by conventional means.

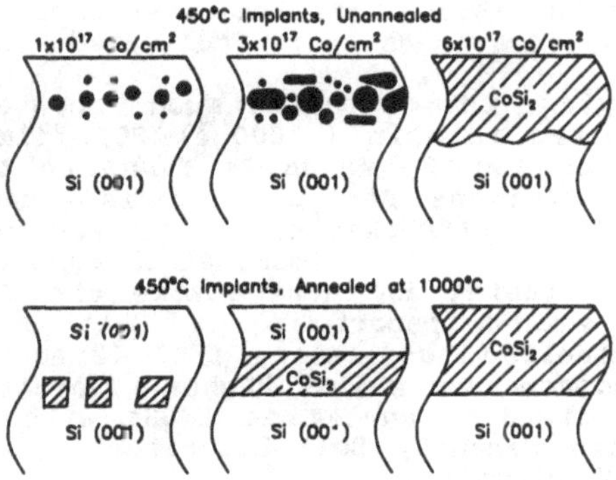

Figure 3: Schematic of $CoSi_2$ formation as a function of dose and annealing[45].

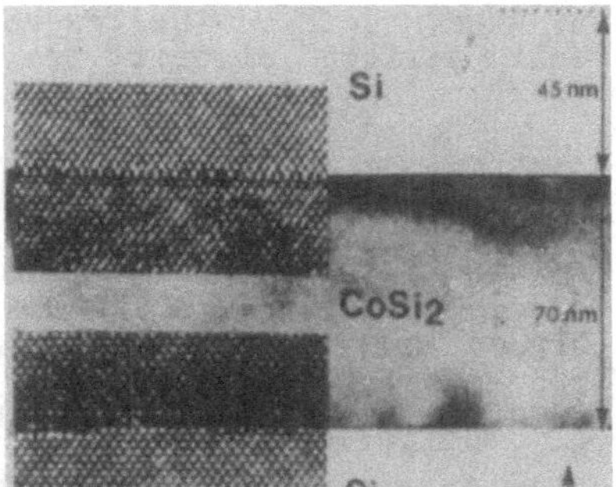

Figure 4: Cross-sectional TEM micrograph of a buried single crystal layer of $CoSi_2$ formed by implantation of 2 x 10^{17} Co^+ cm^{-2} at 170 keV and annealing at 1000°C for 30 mins. The inserts are HREM images of the interfaces showing coherence of the lattices[46].

5.2 Silicides

Currently cobalt disilicide is attracting most interest because of its high conductivity, close lattice match with silicon (1.2% mismatch) and fcc structure which enables epitaxy to be achieved on a number of crystal planes. Early experiments (1987) by White et al[37], who implanted a dose of 3×10^{17} Co^+ cm^{-2} at 200 keV into (111) and (100) silicon at about 350°C, followed by an anneal at 1000°C (30 minutes), showed that Co redistribution to form a buried layer of crystalline, stoichiometric $CoSi_2$ could be achieved. The silicide layer was 1100 Å thick with a silicon overlayer of only 600 Å, due to $\bar{R}_p$ and the high sputter rate of Si by Co^+ ions[38]. The electrical conductivity (12 μΩ cm at 300 K) is a factor of two better than the best UHV deposited layers, maybe due to high purity and the presence of excess metal atoms[40]. Brice and Barbour[38] report a temperature dependent sputtering coefficient for 350 keV Co^+ of 1.1 and 1.8 atoms per ion at 100°C and 450°C, respectively, which severely restricts the layer thickness that can be achieved.

More recent studies[39][10][45] of the evolution of hetero-epitaxial structures have elucidated the formation of the buried layers (Figure 3) through the growth of aligned (A-type) or twinned (B-type) $CoSi_2$ precipitates[39]. Reeson et al[10] find that during a high temperature anneal (4×10^{17} Co^+ cm^{-2}, 350 keV, $T_i \approx 550$°C) the A-type precipitates grow preferentially and that redistribution of the implanted Co occurs through progressive dissolution of the smallest precipitates, located within the wings of the depth distribution. The resulting $CoSi_2$/Si interfaces are atomically abrupt (Figure 4[39]) although pronounced steps (≈150 Å) have been reported, due to the coalescence of discrete precipitates[10]. Dislocations are observed in the silicon overlayer[10].

In the context of IBS, complete segregation of Co, O and N can all be achieved within an accessible temperature/time window.

6. Device Applications

Circuit Performance - The first demonstration of a large CMOS/SIMOX memory circuit (1 K) was by NTT in 1982[41]. This was followed by Texas Instruments who reported a 4 K memory in 1984 and a 16 K circuit in 1986, the latter achieving a 56% yield of working circuits[1], which was higher than the yield of bulk silicon control circuits. The company is now developing 64 K and larger memories. Circuits predating 1987 have been fabricated on thick SOI films annealed at low (1150°C to 1200°C) temperatures. Despite the high defect densities and non-ideal interfaces the circuits showed all the predicted benefits of SOI; however, at that time the high cost of substrates and the reluctance of bulk silicon circuit designers to accept change, frustrated the expectations of the SOI community.

The status of SOI is now changing, thanks to the much improved performance of CMOS circuits fabricated in thin (<1000 Å) SOI films annealed at temperatures in excess of 1300°C, which is found to be superior to the most optimistic predictions for circuits in thick (>3000 Å) SOI films[1]. This is due to the different electric field and charge distributions in fully depleted layers of silicon, which permits higher drive currents, more rapid switching and more stable circuit operation. For example, Hewlett-Packard report a 1.4 Gbit s^{-1} multiplexer and 1.8 GHz voltage-controlled oscillator, LETI (France) have developed a 16 bit 30 MHz microprocessor whilst Hughes Laboratories have achieved a 2:1 frequency divider running at 2 GHz with a minimum stage delay of 19 ps[1]. More recently Bell Laboratories[42] report 6.2 GHz operation of CMOS/SIMOX circuits.

The Challenges - The benefits of using SIMOX substrates rather than bulk silicon for CMOS circuits are undeniable and attention is now being given to bipolar applications[43] and the formation of optical waveguides[23]. Future applications for SIMOX may include multilayer $Si/SiO_2/Si/SiO_2/Si$ structures to combine high speed SOI logic with buried waveguide interconnects and incorporating optically active devices based upon heteroepitaxial III-V and Si/Ge alloy semi-conductors.

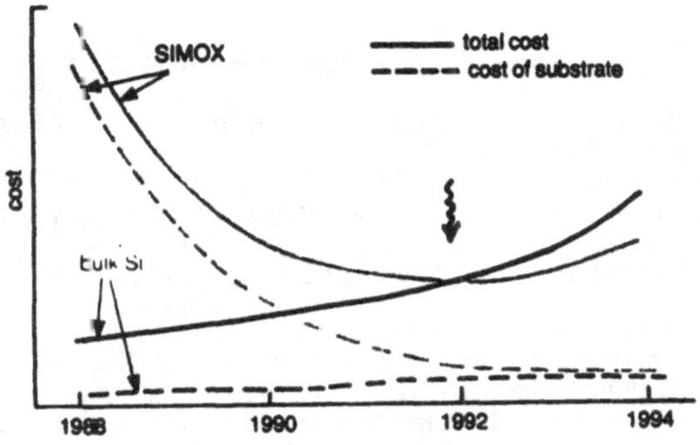

Figure 5: Projected costs of substrates and wafer processing as a function of calendar year[44].

Figure 5 illustrates the potential commercial benefits of SIMOX technology by comparing the projected costs of processing small geometry CMOS circuits on bulk silicon and SIMOX. The cross over in 1992 is predicted to be the watershed for the widespread use of SIMOX technology[44]. This augers well for the further exploitation of IBS as an enabling technology for advanced circuits which may incorporate heteroepitaxial $CoSi_2$ as a buried collector in small geometry, high density bipolar circuits and, more speculatively, for metal base transistors.

7. Concluding Remarks

The exploitation of IBS for the preparation of new materials and structures has been reviewed. Currently SIMOX is the front running application, thanks to the successful fabrication of demonstrator circuits on commercially produced substrates. There are exciting new prospects for IBS, provided that further development of high current ion implantation equipment, particularly for Co^+ ions, is undertaken to support new basic research.

Acknowledgements

The authors wish to thank colleagues for helpful discussions and supplying preprints. They are indebted to Karen Bray for preparing this manuscript.

References

[1] Sturm J C, C K Chen, L Pfieffer and P L F Hemment (Ed): Mat Res Soc Symp Proc 107 (1987).

[2] Wilson I H: Ion Beam Modification of Materials, Chapter 7 (Ed) P Mazzoldi and G Arnold, Elsevier Science Publishers (1987).

[3] Izumi K, M Doken and H Ariyoshi: Electron Letts 14 (1978) 593-594.

[4] Davis J R, K J Reeson, P L F Hemment and C D Marsh: IEEE EDL-8 7 (1987) 291.

[5] Hemment P L F: Mat Res Soc Symp Proc 53 (1986) 207-221, (Ed) A Chiang, M W Geis and L Pfeiffer.

[6] Namavar F: private communication.

[7] Hemment P L F, R F Peart, M F Yao, K G Stephens, R J Chater, J A Kilner, D Meekison, G R Booker and R P Arrowsmith: Appl Phys Letts 46 10 (1985) 925.

[8] Jaussaud C, J Stoemenos, J Margail, M Dupuy, B Blanchard and M Bruel: Appl Phys Letts 46 (1985) 1064.

[9] Celler G K, P L F Hemment, K W West and J M Gibson: Appl Phys Letts 48 (1986) 532.

[10] Reeson K J, A de Veirman, R M Gwilliam, C Jeynes, B J Sealy, J Landuyt, U Bussmann, J K N Lindner and E H te Kaat: Spring Meeting Mat Res Soc, San Diego, April 1989.

[11] Jaussaud C, J Margail, J Stoemenos and M Bruel: Mat Res Soc Symp Proc 107 (1987) 17-28.

[12] Baumgart H and A H Van Ommen: Mat Sci and Eng B2 (1989) 111-122.

[13] Hemment P L F, E A Maydell-Ondrusz, J E Castle, R Paynter, M C Wilson, R G Booker, J A Kilner and R P Arrowsmith: Thin Solid

Films 128 (1985).
[14] Harbeke G, E F Steigmeier, P L F Hemment, K J Reeson: Semicond Sci Tech 2 (1987) 687-690.
[15] Jaussaud C, J Stoemenos, J Margail, A M Papou and M Bruel: Spring Meeting Mat Res Soc, San Diego, April 1989.
[16] Stoemenos J: Solid State Phenomena 1 & 2 (1988) 115-132.
[17] Van Ommen A H, B H Koek and M P A Viegers: Appl Phys Letts 49 (17) (1986) 1062.
[18] Hill D, P Fraundorf and G Fraundorf: J Appl Phys 63 (10) (1988) 4933-4936.
[19] Maszara W P: J Appl Phys 64 (1) (1988) 123-128.
[20] El-Ghor M K, S J Pennycook, T P Sjoreen, C W White and J Narayan: Mat Res Soc Symp Proc 107 (1988) 79-85.
[21] Jastrzebski L, J T McGinn, P Zamzucchi and B Cordts: submitted J Electrochem Soc, January 1989.
[22] Hemment P L F, A K Robinson, K J Reeson, J R Davis, J A Kilner, R J Chater and J Stoemenos: Mat Res Soc Symp 107 (1988) 87-92.
[23] Namavar F, E Cortesi, R A Soref and P Sioshansi: Spring Meeting Mat Res Soc, San Diego, April 1989.
[24] Grob J J, A Grob, P Thevenin, P Siffert, A Golanski and C d'Anterroches: Mat Sci and Eng B2 1-3 (1989) 123-129.
[25] Namavar F, B Buchanan, E Cortesi and P Sioshansi: Spring meeting Mat Res Soc, San Diego, April 1989.
[26] Kamins T I, K L Wang, J Park, G E Davis: J Appl Phys 65 (4) (1989).
[27] Kwor R, R J Matson, M M Al-Jassim, S Polchlopek, P L F Hemment and K J Reeson: J Electrochem Soc 136 3 (1989) 876-878.
[28] Skorupa W, R Grotzschel and H Bartsch: Phys Stat Sol (a) 112 2 (1989) 661-666.
[29] Skorupa W, K Wollschlager, U Kreissig, R Grolzschel and H Bartsch: Nucl Inst Meths B19/20 (1989) 285.
[30] Skorupa W, K Wollschlager, R Grotzschel, J Schoneich, E Hentschel, R Kotte, F Stary, H Bartsch and G Gotz: Nucl Inst Meths B32 (1988) 440.
[31] Bussmann U, F H J Meerbach and E H te Kaat: Mat Res Soc Symp Proc 107 (1988) 479-482.
[32] Nesbit L, G Slusser, R Frenette and R Halback: J Electrochem Soc 133 (1986) 1186.
[33] Skorupa W, K Wollschlager, R Grotzschel, J Schoneich, E Hentschel, R Kotte, F Stary, H Bartsch and G Gotz: Nucl Inst Meths B32 (1988) 440.
[34] Reeson K J, P L F Hemment, D Meekison, C D Marsh, G R Booker, R J Chater, J A Kilner and J R Davis: Nucl Inst Meths B32 (1988) 427.
[35] Martin P, B Daudin, M Dupuy, A Ermolieff, M Olivier, A M Papou and G Rolland: submitted J Appl Phys, June 1989.
[36] Reeson K J, J Stoemenos, P L F Hemment: to be published.
[37] White A E, K T Short, R C Dynes, J Garno and J M Gibson: Appl Phys Letts 50 (1987) 95.
[38] Brice D K and J C Barbour: Nucl Inst Meths B36 (1989) 431-438.
[39] Bulle-Lieuwma C W T, A H Van Ommen and L J Van Ijzendoorn: Appl Phys Letts 54 (3) (1989) 244-246.
[40] White A E, K T Short, R C Dynes, J M Gibson and R Hull: Mat Res Soc Symp Proc 107 (1988) 3-15.
[41] Izumi K, Y Omura, M Ishikawa and E Sano: Symp VLSI Tech Dig 10 (1982).
[42] Celler G K, H-I Cong, R L Field, S J Hillenius, A Kamgar, W S Lindenberger, L E Trimble and J C Sturm: IEEE SOS/SOI Workshop, Navada, October 1989.
[43] Platteter D G and Cheek T F: IEEE NS, submitted for publication, December 1988.
[44] Colinge J P: Fifth Inter Workshop on Future Electron Devices, Miyagi-Zao, May 1988.
[45] Barbour J C, S T Picraux and B L Doyle: Mat Res Soc Symp Proc 107 (1988) 269-274.
[46] Van Ommen A H: Nucl Inst and Meths B39 (1989) 194-202.

SURFACE MODIFICATION OF SOLID CARBON BY ION IMPLANTATION

Masaya IWAKI

RIKEN (The Institute of Physical and Chemical Research)
Hirosawa 2-1, Wako-shi, Saitama-ken 351-01. JAPAN

1. Ion Beam Processing and Carbon

Ion implantation is a versatile technique for the formation of new surface layers with high- and multi-functional properties by means of a non thermal equilibrium process such as a room temperature process[1]. This feature means that it is one of the highest technologies for improving surface physical and chemical properties of materials as one of the surface finishing treatments[2]. Ion implantation in metals has been used for the improvement of the lifetime of industrial tools, and the surface layer modification of ceramics and polymers by ion implantation has been gradually activated to obtain desirable characteristics for electrical conductivity, wear resistance, chemical reaction and so on[3].

Solid carbon, which has structures such as diamond, graphite and amorphous, is a suitable material for many fields of materials science and engineering. These materials have a variety of electrical conductivity, thermal conductivity, mechanical properties and chemical properties. Figure 1 gives an outline of ion beam processing for carbon thin films. Many studies have been made of the formation of thin films of diamond or graphite like carbon, in which one of the most important subjects is how the formation of diamond or graphite and the transition of their phases can be controlled by an ion beam. The other is impurity doping in carbon thin films. Normally, it is difficult for any elements to be injected into solid carbon except for intercalation process in graphitic carbon. The ion implantation process is able to introduce every element into any type of solid carbon, although it induces radiation damage simultaneously.

Surface layer modification of solid carbon has widely spread from electronic devices fabrication to surface finishing of mechanical engineering components. It has been reported that the electric conductivity of ion implanted diamond depends on the energy and fluence[4,5], implanted species[6], and a target temperature [6,7]. The electrochemical reaction of diamond and glassy carbon is influenced by implanted species[5,8]. Mechanical properties of diamonds are improved by boron or nitrogen ion implantation[9,10]. The wear resistance of glassy carbon was also improved by ion implantation, depending on implanted species and energy[11-13]. Wettability of glassy carbon was changed stably by ion implantation rather than sputtering of a low energy ion bombardment[14].

In this report, I introduce electrical, electrochemical and mechanical properties of ion implanted diamond and glassy carbon. Their properties depending on implantation conditions such as ion species, energies, fluences and target temperatures will be discussed with respect to composition, structure and chemical bonding states in ion implanted layers.

2. Diamond

Diamond suffers radiation damage by ion beam bombardment to have disordered structures, which it is difficult to recover by thermal annealing. Surface layers of diamond damaged by ion implantation have electrical conductivity, which is dependent on ion species, energies, fluences and target temperatures. Figure 2 shows the sheet resistivity of argon implanted diamond as a funtion of fluence at various target temperatures[7]. It is found that the sheet resistivity decreases, as the fluence increases. This decrease in resistivity is according to the increase in radiation damage[4,5]. At the high fluence implantation, however, the sheet resistivity is dependent on the target temperature; argon implantation at relatively low temperatures causes it to gradually decrease and hot ion implantation leads it to a constant value. The difference of sheet resistivity for higher fluence implantation reaches two or three orders of magnitude.

The critical target temperature during ion implantation causing a large difference of the sheet resistivity is near room temperature ranging from 15°C to 35°C. Low temperature implantation in the specimen implanted at the high temperature caused the sheet resistivity to become higher[15]. This result suggests that the target temperature at the final stage of implantation is very important for determination of resistivity. The cause of sudden change in sheet resistivity may be due to the structure induced by the relations between radiation damage by ion implantation and the chemical stability at the substrate temperature of the final stage of ion implantation. Laser Raman spectroscopy of implanted diamond showed the difference of structure: Surface layers of specimens implanted at low tempera-

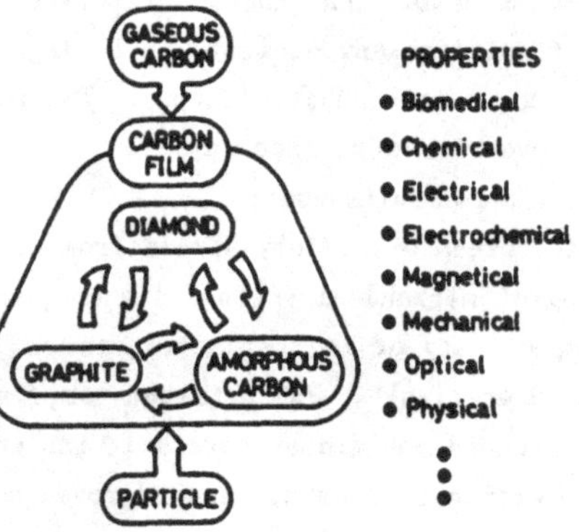

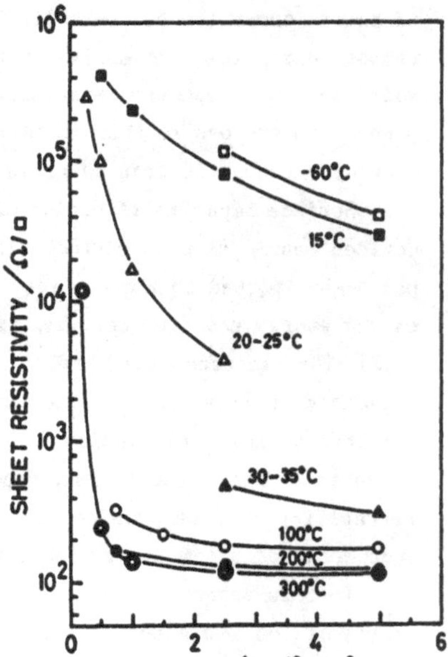

Fig. 1 Out line of ion beam processing for carbon thin films (above). Fig. 2 Sheet resistivity of argon implanted diamond at various target temperatures as a function of dose (right).

tures included large amounts of amorphous carbon in disordered graphite, which is easily produced by high temperature implantation. It is considered that the sheet resistivity is decided by the conflict of amorphous carbon and disordered graphite formed by the substrate temperature and radiation damage during ion implantation.

The difference of sheet resistivity due to ion species may be caused by the chemical bonding states of implanted species. For instance, titanium implantation did not show the target temperature dependence of sheet resistivity[6]. Argon implantation will introduce only the radiation damage, but implanted titanium will introduce not only the radiation damage but also the formation of new compounds such as titanium carbides made by bonding of implanted titanium with carbon.

Carbon electrodes in aqueous solution are often used as sensing electrodes in electrochemical measurements. Glassy carbon and graphite electrodes are chemically and electrochemically stable for usual solvents and supporting electrolytes, which is an advantage for a sensing electrode.

In order to compare the fundamental properties of an ion implanted diamond electrode with that of the usual carbon electrode, voltammograms were measured in typical supporting electrolyte solutions. Figure 3 shows cyclic voltammograms for diamond electrode/0.1M Na_2SO_4 aqueous solution systems[5]. The electrochemically-inert potential region is the widest and the residual current is the lowest for the zinc implanted diamond electrode in comparison with argon implanted diamond and glassy carbon. The wide stable potential region and low residual current, which are advantages for a sensing electrode, are obtained by ion implantation with metal ions.

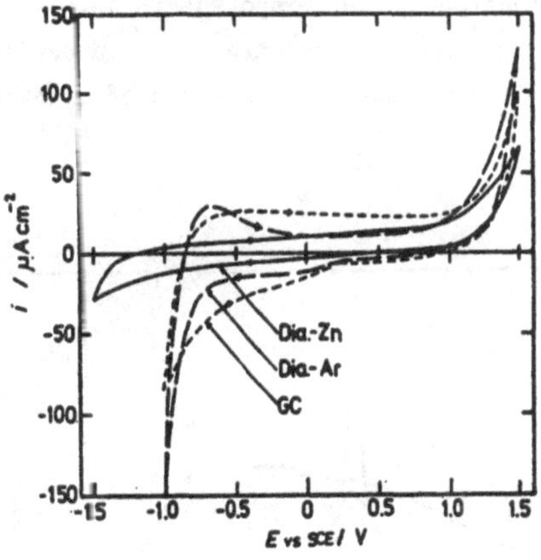

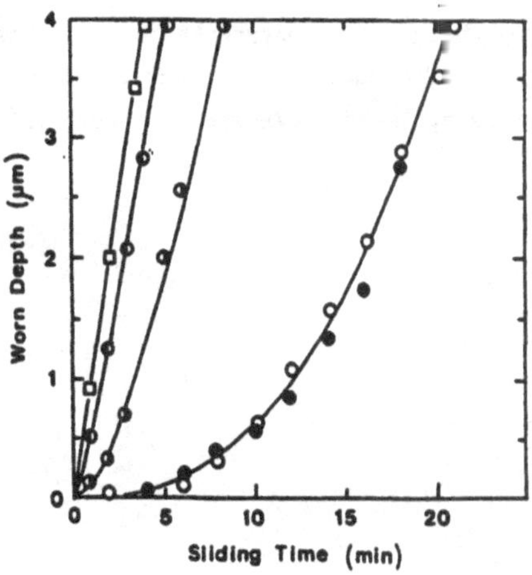

Fig. 3 Cyclic voltammograms for argon implanted diamond, zinc implanted diamond and unimplanted glassy carbon in 0.1M Na_2SO_4 aqueous solution (left).

Fig. 4 Relationship between worn depth and sliding time for 150 keV nitrogen implanted glassy carbon with fluences of $5x10^{14}$(◐), $1x10^{15}$(◑), $1x10^{16}$(o) and $5x10^{16}$(●) ions/cm^2 at about 25°C. ▢ indicates the pristine glassy carbon (right).

3. Glassy Carbon

In the case of glassy carbon, metal ion implantation also caused the wide stable potential region and low residual current for cyclic voltammograms[8]. No change in depth profiles of implanted elements was observed before and after electrochemical measurements. The result supports that the implanted elements in solid carbon is electrochemically stable, which is useful for a sensing electrode.

Wear properties have been investigated, using a conventional mechanical-polishing machine[13]. Figure 4 shows the relationship between the worn depth of nitrogen implanted glassy carbon and sliding time as a function of fluence. Nitrogen ion implantation was performed with the fluence ranging from $5x10^{14}$ to $5x10^{16}$ ions/cm^2 at an energy of 150 keV. The target temperature during ion implantation was near room temperature(25°C) by using a low ion beam current density. The sliding time was corresponding to the sliding distance. The worn depth of a pristine glassy carbon is directly proportional to a sliding time. This result supports that the experimental condition is appropriate for evaluating the wear property of glassy carbon.

It is found that the wear property is dependent on the fluence of nitrogen. At the low fluence of $5x10^{14}$ ions/cm^2, the wear property of glassy carbon was not a little improved by ion implantation. As the fluence increased, the duration of a low worn depth with respect to a sliding time became longer. At the highest fluence of $5x10^{16}$ ions/cm^2, the duration of the low wear rate was saturated. The results suggest that ion implantation causes the wear resistance to improve at the appropriate fluence.

The saturation shows that the improvement of wear resistance may be caused by the change in structure rather than the formation of new compounds in this experiment. The surface layer structure of ion implanted glassy carbon was investigated by using laser Raman spectroscopy. Figure 5 shows the Raman spectra of pristine

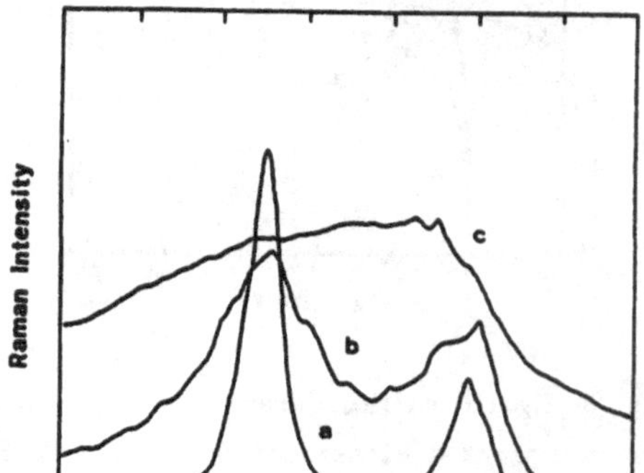

Fig. 5 Laser Raman spectra for a pristine glassy carbon (a), and for specimens implanted with $1x10^{15}$(b) and $1x10^{16}$(c) ions/cm^2 at about 25°C.

and implanted glassy carbon. Ion implantation was carried out with the fluence of $1x10^{15}$ and $1x10^{16}$ $ions/cm^2$ at an energy of 150 keV at near room temperature. The two peaks can be seen in the spectrum of a pristine glassy carbon. The peak of 1600 cm^{-1} is corresponding to graphite structure and the other at 1350 cm^{-1} is corresponding to long-range disordered structure of graphite, named the disordered graphite[8].

Ion implantation destroyed the graphitic structure to form a new structure. Finally, a high fluence ion implantation revealed a new peak at 1550 cm^{-1}. This peak can be always seen in the case of ion implantation into diamond or polymer with a relatively high fluence[7,16] and deposition of carbon thin films[17], at which the peak has been considered due to the formation of amorphous phase. The consideration will be applied to this experiment, and the Raman spectra suggest that nitrogen ion implantation with a high fluence causes the disordered graphite structure to be amorphous.

The relationship between the wear property and structure effected by ion implantation suggests that the improvement of the wear resistance of glassy carbon by nitrogen ion implantation may be caused by the amorphization of implanted layers. We already reported that the structure and electrical conductivity of ion implanted diamond is dependent on the target temperature[7]; when we implanted ions in diamond at the target temperature lower than room temperature, the structure of implanted layers became amorphous and the sheet resistivity was relatively high. When we implanted ions in diamond at higher temperatures than room temperature the structure of implanted layers was like a disordered graphite, which showed a lower sheet resistivity. It is considered that the dependence of target temperature on structure and sheet resistivity for implanted diamond is applicable to ion implantation modification such as tribological properties of glassy carbon.

4. Summary

A study has been made of electrical, electrochemical and tribological properties for ion implanted diamond and glassy carbon. The results show that ion implanted diamond has electrical conductivity and a wide stable potential region and low residual current in cyclic voltammograms, and implanted glassy carbon with electrochemical properties similar to those for implanted diamond has high wear resistance. These properties are dependent on ion species, fluences and target temperatures. In particular, room temperature is a critical temperature for the transition of amorphous and graphite, which is sensitive for surface properties. It is concluded that ion implantation at appropriate conditions is useful for the fabrication of surface layers of solid carbon with multi-functional properties.

Acknowlegement

I would record my acknowledgment to Dr. K. Takahashi for discussion, Mr. S. Sato of OPELECS Co. for his assistance of electrical measurements of implanted diamond and Mr. A. Sekiguchi of Toho University for his assistance of wear tests and laser Raman spectroscopy of implanted glassy carbon.

References

[1] Metastable Materials Formation by Ion Implantation, eds. S.T. Picraux, and W.J. Choyke, (North-Holland, New York, 1982).

[2] M. Iwaki, Nucl. Instr. and Meth., B37/38 (1989) 661.

[3] Proc. of 6th Intern. Conf. on Ion Beam Modification of Materials, Nucl. Instr. and Meth., B39 (1989).

[4] R. Kalish, T. Bernstein, B. Shapiro and A. Talmi, Radiat. Eff., 52 (1980) 153.

[5] M. Iwaki, S. Sato, K. Takahashi and H. Sakairi, Nucl. Instr. and Meth., 209/210 (1983) 1129.

[6] S. Sato, M. Iwaki and H. Sakairi, Nucl. Instr. and Meth., B19/20 (1987) 822.

[7] S. Sato and M. Iwaki, Nucl. Instr. and Meth., B32 (1988) 145.

[8] K. Takahashi, K. Yoshida and M. Iwaki, Nucl. Instr. and Meth., B7/8 (1985) 526.

[9] N.E.W. Hartley, p.295 in ref. [1].

[10] K. Kanazawa, K. Chijiiwa and M. Iwaki, Bull. Japan Soc. of Prec. Engg., 21 (1987) 63.

[11] M.J. Kenny, J.T.A. Pollock and L.S. Wielunski, p. 704 in ref. [3]

[12] K. Yoshida, K. Takahashi, K. Okuno, G. Katagiri, M. Iwaki and A. Ishitani, Appl. Phys. Lett., 52 (1988) 1046.

[13] M. Iwaki, K. Takahashi, K. Yoshida and Y. Okabe, p. 700 in ref. [3].

[14] M. Iwaki, Mater. Sci. Eng., (to be published).

[15] S. Sato and M. Iwaki, J. of the Japan Soc. of Powder and Powder Metallurgy, 35 (1989) 688 (in Japanese).

[16] K. Yoshida and M. Iwaki, Nucl. Instr. and Meth., B19/20 (1987) 878.

[17] J.S. Tsang, J.E. Demuth, P.N. Sanda and J.R. Kirtley, Chem. Phys. Lett., 76 (1980) 54.

ION BEAM INDUCED EPITAXIAL CRYSTALLIZATION AND PLANAR AMORPHIZATION OF SILICON

Jan Linnros[1]

ABSTRACT

An amorphous layer in silicon can be recrystallized by energetic ion irradiation at temperatures as low as 150 °C. The layer regrows epitaxially with a rate proportional to the atomic displacement rate at the amorphous/crystalline interface. At high ion dose rates and/or low temperatures the interface movement direction can be reversed and the amorphous layer grows into the crystal, the interface remaining planar. This reversible ion induced interface movement will be reviewed and proposed models will be discussed.

I. INTRODUCTION

Amorphous silicon in contact with crystalline silicon is thermodynamically unstable and crystallizes at temperatures above ≈ 450 °C. For an amorphous layer, e.g. formed by a high dose ion implantation, the amorphous/crystalline (α/c) interface is parallel to the surface and the layer crystallizes by a solid phase epitaxial growth (SPEG) process towards the surface. The rate of regrowth is strongly temperature dependent and is given by an Arrhenius expression with a single activation energy of ≈ 2.7 eV, even up to the melting point of Si. Figure 1 shows data by Olson (1983) for undoped Si.

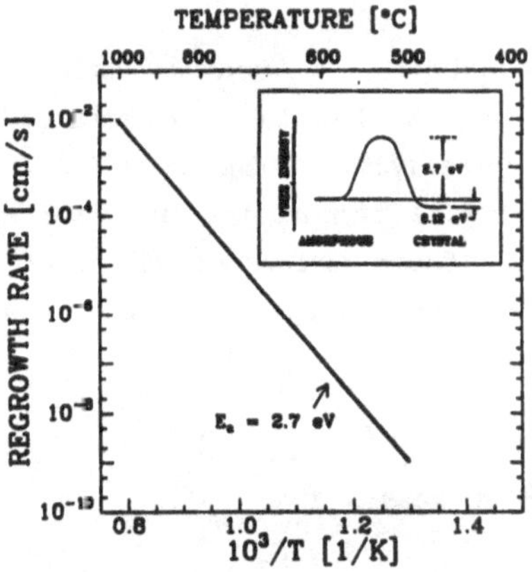

FIG. 1. Temperature dependence of the rate of regrowth during thermal epitaxial regrowth of an amorphous layer in (100) Si (data are from Olson 1983). The inset depicts a free energy diagram of the process where atoms in the amorphous phase are converted to the crystalline phase.

---

(1) Swedish Institute of Microelectronics, P.O. Box 1084, 164 21 Kista, SWEDEN

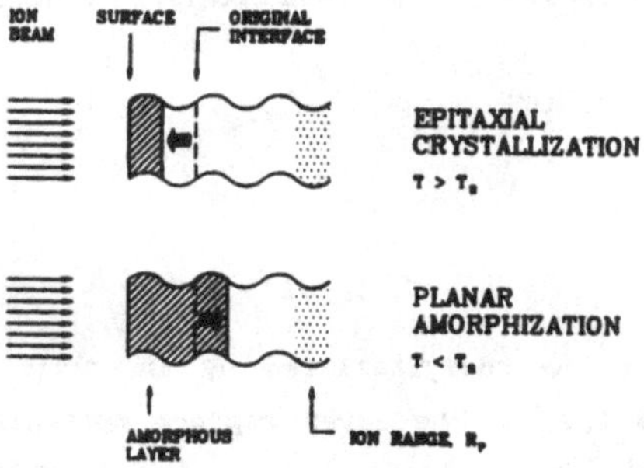

FIG. 2. Schematic of α/c interface motion during ion irradiation. Upper part displays epitaxial crystallization (IBIEC) (higher temperatures) and lower part planar, layer by layer, amorphization (at temperatures slightly below T_R, the reversal temperature).

The inset depicts a free energy diagram of the process: the 2.7 eV barrier and the 0.12 eV/atom (Donovan 1985) energy release upon crystallization which provides the driving force of the process.

Epitaxial crystallization of amorphous layers may also be induced by ion irradiation at temperatures as low as 150 °C (Golecki 1979, Nakata 1982, Gerasimenko 1974, Holmén 1975a,b, Elliman 1987b, Linnros 1984, 1985b, Williams 1985), for which the thermal SPEG rate is negligible. Local beam heating effects can usually be excluded since the rate of crystallization decreases with ion current. The effect is generally attributed to mobile defects in proximity of the α/c interface. The defects are formed during collision cascades of energetic ions passing through the interface, see Fig. 2. The amount of regrowth during ion beam induced epitaxial crystallization (IBIEC) is primarily related to ion dose instead of time as for thermal SPEG. The regrowth velocity depends critically on irradiation conditions as discussed below. Apart from Si IBIEC has been demonstrated for a number of semiconductor materials such as Ge, GaAs, InP.

Recently, it was shown that by decreasing the temperature during IBIEC the regrowth velocity not only decreased but the direction of movement of the α/c interface could also be reversed (Linnros 1985b, 1988b, Leiberich 1987). Thus, the amorphous layer would grow into the crystal the interface still remaining sharp and planar, see Fig. 2. The paper reviews experiments on ion beam induced epitaxial crystallization/planar amorphization and proposed models for the underlying mechanisms.

II. EXPERIMENTAL CONSIDERATIONS AND ANALYSIS TECHNIQUES

Amorphous layers for the study of beam induced crystallization are generally formed by implanting silicon ions at room temperature or at LN_2 temperature. For careful studies of regrowth rates crystallization is usually initiated by a thermal anneal to produce a sharp and stable amorphous/crystalline interface. Epitaxial crystallization or planar amorphization is subsequently induced by irradiation with energetic ions, usually in the MeV range. The dose rate, or rather the beam current, is kept low to avoid uncontrollable beam heating effects.

The amount of regrowth is traditionally extracted from spectra of Rutherford backscattering (RBS) using the channeling technique. To follow the regrowth in situ, time resolved reflectivity (TRR) measurements have recently been adopted. This allows the velocity of the advancing α/c-interface to be calculated at any position during the regrowth. Transmission electron microscopy (TEM) is used to characterize the defect structure in recrystallized regions.

III. EPITAXIAL CRYSTALLIZATION REGIME

Dependence on Dose and Temperature

The dose dependence of beam induced crystallization is illustrated by the backscattering spectra of Fig. 3. The substrate is silicon on sapphire, SOS, which crystallization rate is the same as for bulk Si. The sequence of spectra demonstrate epitaxial regrowth of the amorphous layer (of initial thickness ≈ 1600 Å) towards the surface similar to thermal SPEG. At a dose of $6 \cdot 10^{16}$ Si

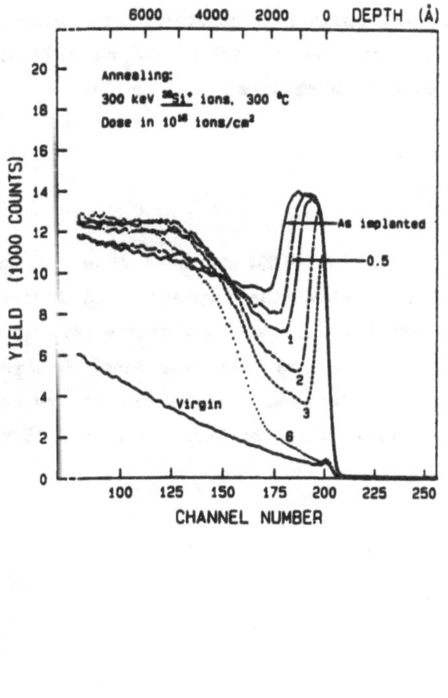

FIG. 3. Backscattering spectra of SOS samples with an initial ≈ 1600 Å thick amorphous layer subsequently recrystallized with progressively higher doses of 300 keV Si ions at 300 °C (from Linnros 1985a).

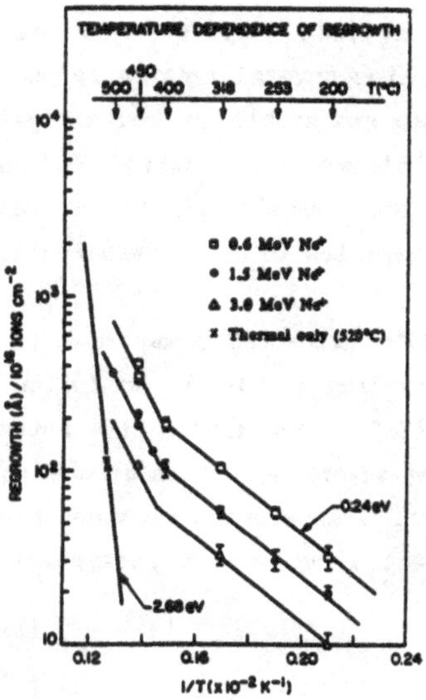

FIG. 4. Temperature dependence of the regrowth rate of ion induced crystallization at three different Ne bombardment energies (from Williams 1985).

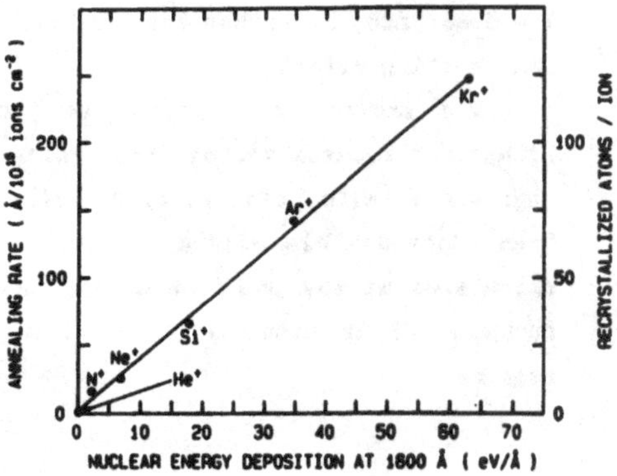

FIG. 5. Dependence of the regrowth rate on nuclear energy deposition at the a/c interface during IBIEC (from Linnros 1985a).

ions/cm^2 the complete layer has recrystallized although a damage peak has emerged at the range of the irradiating ions, i.e. at ≈ 5000 Å. For MeV ions this damage moves to larger depths which separates it from the regrown layer.

The temperature dependence of IBIEC is displayed in Fig. 4 by the data of Williams (1985). Regrowth rates have been extracted from RBS spectra following irradiation by Ne ions at three different energies. The included thermal annealing rate has been scaled to the appropriate irradiation time. Below 400 ^{0}C, the data exhibit an Arrhenius behavior characterized by an activation energy of ≈ 0.24 eV. This value is clearly much lower than the 2.7 eV activation energy of thermal SPEG.

Dependence on Nuclear Energy Deposition

The data in Fig. 4 also reveal that the IBIEC regrowth rate increases with decreasing ion energy. For different ions, the regrowth rate increases with ion mass. Thus, the rate of regrowth has the same qualitative behavior as the energy deposited in nuclear collisions by the retarding ions. In fact a proportionality between the IBIEC rate and the amount of nuclear energy deposition *at the interface* has been demonstrated (Linnros 1985a, Williams 1985b),

see Fig. 5. The right hand scale of Fig. 5 shows that, for heavy ions, more than hundred target atoms can be moved to crystal lattice positions by a single ion passing the α/c interface.

The role of nuclear energy deposition at the α/c interface can be further elucidated by comparing similar experiments using electron beams. Miyao (1986) found an enhancement of the SPE growth rate (see also Matthews 1973, Lulli 1987) at electron energies above the threshold energy for Si atom displacement (≈ 0.2 MeV). Therefore, we can conclude that the number of atomic displacements in proximity to the α/c interface provides the driving force for IBIEC.

Influence of Substrate Orientation and Impurity Atoms

The rate of thermal SPEG varies considerably with orientation of the α/c interface with respect to the substrate lattice: Regrowth on (100) substrates is almost 25 times faster than on (111) (Csepregi 1978). For IBIEC however, a less dramatic dependence is observed being only a factor of ≈ 2 - 4 slower in the <111> direction compared to the (100) direction (La Ferla 1987, Maher 1987). The effect of dopants and other impurities on the regrowth rate are also different for IBIEC and thermal SPEG, respectively. Dopants like P and B at high concentrations enhance thermal regrowth whereas impurities like inert gas atoms retards and eventually blocks the regrowth.

The effect of dopants on IBIEC is illustrated in Fig. 6. The data by Priolo (1988) are normalized to ion dose times nuclear energy deposition at the α/c interface. For IBIEC a smaller dopant enhancement effect is observed. At equal concentration of B and P dopants, however, the IBIEC rate is still enhanced whereas for thermal annealing the regrowth rate reverts to that for

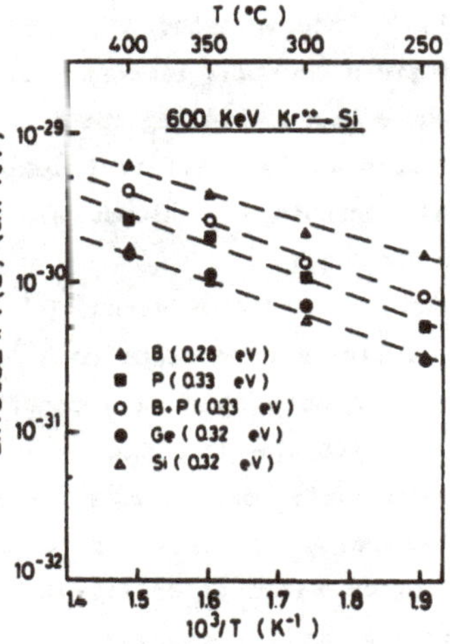

FIG. 6. Temperature dependence of the IBIEC regrowth rate, scaled to nuclear energy deposition at the α/c interface, for amorphous Si doped with B, P, B and P, Ge or Si (self implanted). All doses were $1\cdot10^{20}$ cm^{-2} (from Priolo 1988).

intrinsic Si. Such a compensational dependence has been attributed to a Fermi level dependence of thermal SPEG. In the IBIEC case, however, the fractions of electrically active B and P atoms are low and may not be equal due to a high residual defect density. Therefore, the effect of the Fermi level on IBIEC is not clear. Interestingly, recent measurements by Priolo (1989) seems to indicate a regrowth enhancement proportional to the logarithm of the dopant concentration. For oxygen Priolo (1989) found a retarding effect although at much higher impurity concentration than in the thermal SPEG case. Also for amorphous layers containing high concentrations of gold atoms IBIEC is much less affected compared to thermal SPEG (Elliman 1987a, Poate 1988, 1989).

Electrical activation of dopants following beam induced crystallization is generally poor. Alestig (1987) reported only ≈ 3 % electrically active As atoms for a $10^{15}/cm^2$ implanted As dose in SOS. Mobility was only slightly decreased as compared to thermally annealed samples. It was speculated that the poor electrical activation during IBIEC resulted from the formation of As defect complexes. This was supported by the angular dependence of the backscattering yield which showed that a large fraction of the As atoms were displaced from substitutional lattice positions.

IV. CRYSTALLIZATION/AMORPHIZATION REVERSAL

Recently, it was discovered that if ion irradiation was extended to still lower temperatures (c.f. Fig. 4) the IBIEC rate decreased faster than predicted by the Arrhenius behavior and in fact could be reversed resulting in a planar movement of the interface deeper into the crystal. This means that the amorphous layer grows deeper into the crystal at the expense of crystalline material in a layer by layer mechanism ("reversed epitaxy"), see lower part of Fig. 2. Furthermore, the same reversal could be accomplished by, instead, increasing the dose rate of the ion beam at a given constant temperature. The effect is illustrated in the inset of Fig. 7 where backscattering spectra have been taken after irradiation with 1.5 MeV Xe ions at two different temperatures, 271 and 240 ^{0}C, representing epitaxial regrowth and planar amorphization, respectively (Linnros 1988b).

Figure 7 shows the extracted interface displacement rates versus T^{-1} with ion dose rate as parameter. The data exhibit a linear dependence on T^{-1} and crosses the zero line which suggest that, for each dose rate, a temperature (the reversal temperature T_R) can be found for which the interface does not move upon ion irradiation. Figure 8 is an Arrhenius plot of the corresponding ion dose rate versus $1/T_R$ for different ions where T_R is derived from plots similar to the one in Fig. 7. As defect production rates generally increase with ion dose rate and defect annealing rates are controlled by temperature

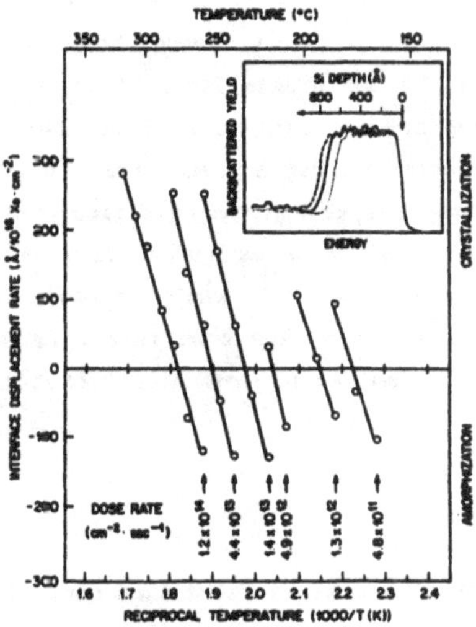

FIG. 7. Displacement of the α/c interface of a 750 Å thick amorphous layer induced by 1.5 MeV Xe irradiation versus 1/T. Data (from Linnros 1988) are displayed with dose rate as parameter. The inset shows examples of high resolution backscattering spectra illustrating IBIEC (dotted curve) at 271 °C and planar amorphization (dashed curve) at 240 °C, respectively. The full drawn line represents the initial amorphous layer.

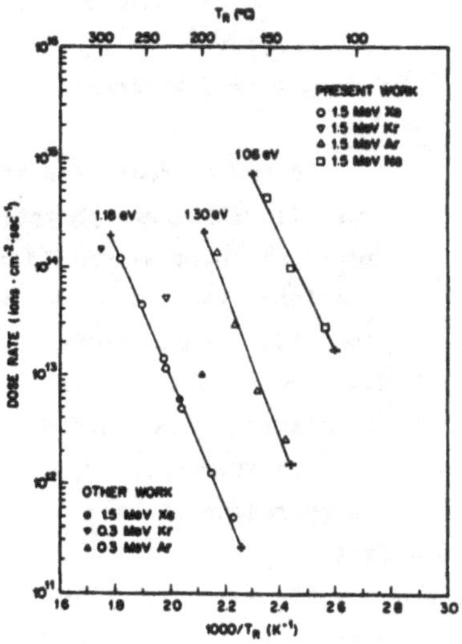

FIG. 8. Arrhenius plot of the reversal temperature for a given dose rate (from Linnros 1988). Data for Xe (filled symbol) is from Leiberich 1987 and for 0.3 MeV Ar and Kr from Linnros (1985b).

the curves fitted to the data can be interpreted as a dynamic equilibrium under which the defect production rate precisely balances annealing processes at the α/c interface. For beam parameters and temperatures to the left of the line annealing processes become relatively more efficient and epitaxial crystallization results whereas for parameters to the right of the line defectproduction overweighs and planar amorphization results. The deduced activation energies from the three curves in Fig. 8 may be averaged to about 1.2 eV, a value which coincides with the one reported for divacancy annealing (Vook 1969).

Scaling with the Square of the Density of Displaced Atoms

As stated above, the IBIEC rate scales with the density of displaced atoms at the α/c interface. The data for different ions in Fig. 8, however, do not scale with the density of displaced atoms but may, instead, be scaled with the *square* of this quantity (Linnros 1988b). Thus, by plotting the number of displaced atoms by a single ion, squared, times dose rate versus $1/T_R$ the data in Fig. 8 may be reasonably well described by one single line. Since the data do not, however, scale with the *total* displacement rate (i.e. dose rate times the number of displaced atoms per ion) squared we can conclude that this quadratic dependence on defect density only applies within one collision cascade. This result implies a second order defect production mechanism such as e.g. the formation of divacancies from single vacancies. This would, indeed, occur on a much shorter time scale than cascade overlap. The dose rate dependence would then arise from the accumulation of these second order type defects due to overlap of individual cascades.

Pulsed Beam Experiments

To reveal characteristic times of defect annealing rates that control the crystallization/amorphization processes pulsed beam experiments may provide important information (Linnros 1987, 1988a). Such experiments introduces two additional variables and can be performed either by varying (1) the pulse-on time, (2) the pulse-off time or (3) the pulse repetition frequency at constant duty cycle. This last mode is exemplified by the data in Fig. 9 for 1.5 MeV Xe irradiation (non-scanned, defocused beam) at 250 °C. The transition between the high frequency regime where crystallization is observed and the low frequency regime where amorphization is observed occurs at a frequency of a few Herz.

If the transition frequency is indicative of a characteristic defect annealing time one would expect this frequency to vary largely with temperature. Instead, the data by Linnros (1988a) show that the transition frequency is virtually temperature independent but strongly dependent on peak beam current. One may, therefore, conclude that the transition frequency must be related to a cascade overlap effect, see next section.

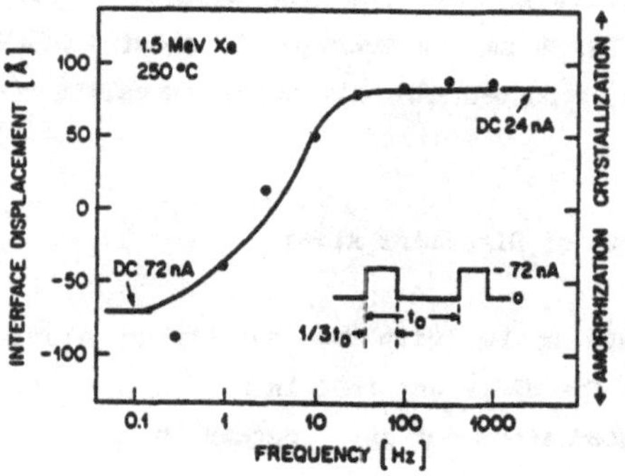

FIG. 9. Frequency dependence of the interface displacement for a pulsed Xe beam (from Linnros 1988) with a duty cycle of 0.33. The peak current of 72 nA (corresponding to $5\cdot10^{13}$ ions/cm^{2}s) and the average current of 24 nA is held constant. The full drawn line is a fit by Jackson's (1988) model to the data.

V. MODELS FOR ION INDUCED INTERFACE MOVEMENT

Simple Point Defect Models

Various models have been proposed for the IBIEC mechanism (Williams 1985, Chaki 1989, Linnros 1984, Heera 1989, Holland 1989). These generally assume that point defects arriving at the α/c interface convert atoms from the amorphous to the crystalline phase, tentatively on a one by one ratio. In particular a vacancy annealing mechanism have been put forward as the activation energy of the neutral vacancy is close to that of IBIEC. Diffusion of vacancies, and also interstitials, at a few hundred ^{0}C is, however, very fast and an annealing mechanism involving these defects is contradicted by the observed localized dependence of the regrowth rate on displacement density at the interface. To resolve this apparent discrepancy a very fast annihilation of simple point defects must be assumed, possibly before next cascade arrives in the same area. Defects of higher orders, such as the divacancy or other defect complexes, anneal at higher temperatures and may ,therefore, be candidates for possible IBIEC mechanisms.

An important issue for any defect model is from which of the two phases (amorphous or crystalline) defects responsible for the IBIEC mechanism originate. For this purpose experiments with channeled beams have been performed. The notion was that defect production within the crystal would be largely reduced for a channeled beam whereas in the amorphous phase the defect production rate would remain the same. Indeed, a channeling effect was found, thus, indicative of point defects within the crystal being responsible for the IBIEC mechanism. The IBIEC rate was, however, only reduced by a factor $\approx$ 2 and may as well be accounted for by a simple reduction in the cascade density at the interface due to isotropic cascade spreading. Lulli (1988), using electron beams, claims defects are coming from both sides of the interface and that the diffusion length of such defects is $\approx$ 40 nm.

The Model by Jackson

Recently, Jackson (1988) proposed a single defect model for IBIEC involving a detailed mathematical treatment which fitted experimental data surprisingly well. The assumptions of the model are: (1) Each ion converts a small volume of crystal at the α/c interface to the amorphous phase. (2) It also creates mobile defects that move around in the interface region with a certain jump frequency which temperature dependence is given by the 1.2 eV activation energy. (3) The concentration of this defect decays by binary recombination. (4) Each defect jump converts a fraction of an atom to the crystalline phase,

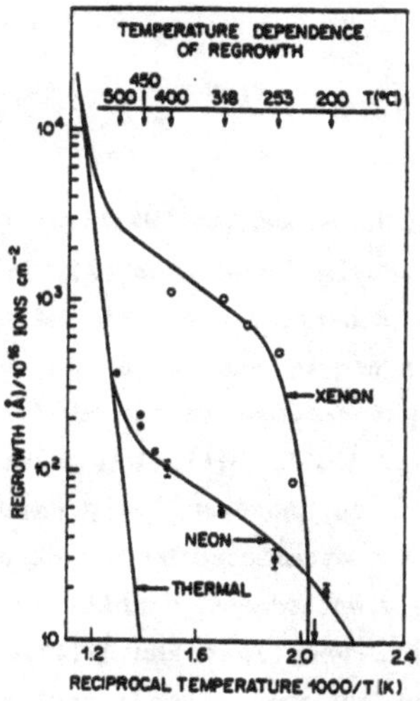

FIG. 10. Temperature dependence of the IBIEC rate and the thermal SPEG rate (Jackson 1988). The data are from Elliman (1987) and the curves were calculated by the Jackson model to fit the data.

i.e. the amount of crystallization is proportionate to the total number of jumps. Jackson suggests the defect to be a dangling bond in the amorphous phase.

The model accounts for ion induced crystallization and the reversal to planar amorphization as well as beam pulsing experiments. The temperature independence of the pulsing experiments follows as the defect decay rate is determined by annihilation of two defects which only depends on the rate of cascade overlap. One of the most striking consequences of the model is the single activation energy whereby the activation energy for the regrowth (Fig. 4) is only illusive when data are plotted in a logarithmic graph. This is shown in Fig. 10 where the full drawn lines have been calculated from the model. Thermal SPEG has also been included in the model assuming thermal generation of the same defect.

VI. CONCLUSIONS

The experimental data on ion beam induced epitaxial crystallization is extensive and its dependence on beam and target parameters is well characterized, particularly for Si but also for other semiconductors. This allows the regrowth rate to be predicted fairly well for new beam/target combinations. At lower temperatures or at higher ion dose rates defect production at the amorphous/crystalline interface out balances regrowth mechanisms and a reversal of the interface motion is seen. In this unique regime the α/c interface remains planar while the amorphous phase grows into the crystal. Scaling laws have been found where the amount of interface motion is linearly dependent on the number of atomic displacements at the α/c interface while a quadratic dependence on this density is observed for the scaling of dose rates. Models of IBIEC assume annealing mechanisms initiated by simple point defects or structural defects produced by the beam either at the crystalline side or at the amorphous side of the interface. In one of these models a detailed mathematical treatment has been formulated which fits experimental data very well. A detailed understanding of defect mechanisms at the interface and participating defects is still lacking. Comparison between IBIEC and thermal annealing may, however, provide important insights on interface mechanisms beneficial for understanding both processes.

ACKNOWLEDGEMENTS

Collaboration and discussions with G. Holmén, B. Svensson, W.L. Brown, R.G. Elliman, J.S. Williams, F. Priolo, J.M. Poate, G. Alestig, D.M. Maher through the course of the work on beam induced crystallization is gratefully acknowledged.

REFERENCES

Alestig, G., G. Holmén and J. Linnros: J. Appl. Phys. 62 (1987) 409
Atwater, H.A., C.V. Thompson and H.I. Smith: Mat. Res. Soc. Symp. Proc. 100 (1988)
Chaki, T.K.: Phil. Mag. Lett. 59 (1989) 223
Csepregi, L., E.F. Kennedy, J.W. Mayer and T.W. Sigmon: J. Appl. Phys. 49 (1978) 3906
Donovan, E.P., F. Spaepen, D. Turnbull, J.M. Poate and D.C. Jacobson: J. Appl. Phys. 57 (1985) 1795
Elliman, R.G., D.C. Jacobson, J. Linnros and J.M. Poate: Appl. Phys. Lett. 51 (1987) 314
Elliman, R.G., J.S. Williams, W.L. Brown, A. Leiberich, D.M. Maher and R.V. Knoell: Nucl. Instrum. Meth. B19/20 (1987) 435
Gerasimenko, N.N., A.V. Dvurechenskii, S.I. Romanov and L.S. Smirnov: Sov. Phys. Semicond. 7 (1974) 1461
Golecki, I., G.E. Chapman, S.S. Lau, B.Y. Tsaur and J.W. Mayer: Phys. Lett. 71A (1979) 267
Heera, V.: submitted to Phys. Stat. Sol.
Holland, O.W.: Appl. Phys. Lett 54 (1989) 320
Holmén, G., P. Högberg and A. Burén: Rad. Effects 24 (1975) 39
Holmén, G., S. Peterström, A. Burén and E. Bogh: Rad. Effects 24 (1975) 45
Jackson, K.A.: J. Mater. Res. 3 (1988) 1218
La Ferla, A., S. Cannavo, G. Ferla, S.U. Campisano, E. Rimini and M. Servidori: Nucl. Instrum. Meth. B19/20 (1987) 470
Leiberich, A., D.M. Maher, R.V. Knoell and W.L. Brown: Nucl. Instrum. Meth. B19/20 (1987) 457
Linnros, J., B. Svensson and G. Holmén: Phys. Rev. B30 (1984) 3629
Linnros, J., G. Holmén and B. Svensson: Phys. Rev. B32 (1985) 2770
Linnros, J.: Ph.D. thesis, Chalmers Univ. of Tech., Göteborg, Sweden (1985)
Linnros, J.and G. Holmén: J. Appl. Phys. 59 (1986) 1513
Linnros, J. and G. Holmén, J. Appl. Phys. 62 (1987) 4737
Linnros, J.L., W.L. Brown and R.G. Elliman: Mat. Res. Soc. Symp. Proc. Vol 100 (1988) 369

Linnros, J.L., R.G. Elliman and W.L. Brown: J. Mater. Res. 3 (1988) 1208
Lulli, G., P.G. Merli and M. Vittori Antisari: Phys. Rev. B36 (1987) 8038
Lulli, G., P.G. Merli and M. Vittori Antisari: Mat. Res. Soc. Symp. Proc. Vol. 100 (1988) 375
Maher, D.M., R.G. Elliman, J. Linnros, J.S. Williams, R.V. Knoell and W.L. Brown: Mat. Res. Soc. Symp. Proc. Vol. 93 (1987) 87
Matthews, M.D. and S.J. Ashby: Philos. Mag. 27 (1973) 1313
Miyao, M., A. Polman, W. Sinke and F.W. Saris: Appl. Phys. Lett. 48 (1986) 1132
Nakata, J. and K Kajiyama: Appl. Phys. Lett. 40 (1982) 686
Olson, G.L., S.A. Kokorowski, J.A. Roth and L.D. Hess: Mat. Res. Soc. Symp. Proc. Vol 13 (1983) 141
Poate, J.M., J. Linnros, F. Priolo, D.C. Jacobson, J.L. Batstone and M.O. Thompson: Phys. Rev. Lett. 60 (1988) 1322
Poate, J.M., D.C. Jacobson, F. Priolo and M.O. Thompson: Mat. Res. Soc. Symp. Proc. Vol 128 (1989) 533
Priolo, F., A. La Ferla and E. Rimini: J. Mater. Res. 3 (1988) 1212
Priolo, F., C. Spinella, A. La Ferla and E. Rimini: Proc. of Eur. Mat. Res. Soc. 1989, to be published
Vook, F.L. and H.J. Stein: Rad. Effects 2 (1969) 23
Williams, J.S., R.G. Elliman, W.L. Brown and T.E. Seidel: Phys. Rev. Lett. 55 (1985) 1482
Williams, J.S., R.G. Elliman, W.L. Brown and T.E. Seidel: Mat. Res. Soc. Symp. Proc. Vol 37 (1985) 127

ION BEAM MIXING: AMORPHIZATION, ICOSAHEDRAL PHASE AND FRACTAL PATTERN FORMATION

Bai-Xin Liu 1)

ABSTRACT

A phenomenological model was proposed by the author for predicting the glass forming ability(GFA) of the binary metal systems. Two parameters were found to be important in determining the GFA: (i) MPAR,i.e. the maximum possible amorphization range of a system, defined as 100% minus the maximum solid solubilities obtained from the equilibrium phase diagram, and (ii) the heat of formation(H_F) calculated by Miedema's model. A map was constructed by these parameters and accordindly the systems were classified into three catagories of hardly, possible and readily glass forming systems. Results from 56 systems so far obtained confirmed the proposed qualifications. Using ion mixing method, an icosahedral phase was formed, for the first time, in an equilibrium immiscible Fe-Cu system. Further study showed that this phase could also be formed by high temperature steady state inter-diffusion in thin films. When ion mixing was conducted to higher doses than required for forming a uniform mixed amorphous alloy. The additional dose stimulated anomalous structural phase transformation. The precipitated crystalline phases organized themselves in fractal patterns in some selected alloy films. It was found that the formation of those fractals was through a process of cluster-diffusion-limited-aggregation(CDLA).

I. INTRODUCTION

Since early 70s, research on ion-metal interactions has been focusing on two directions,namely on modifying some properties of materials for practical applications(G.K.Hubler 1987; Sioshansi 1987;J.J.Zhang et al. 1986) and on studying some metallurgical issues,such as alloy phase formation and transformation(C.M.Preece and J.K.Hirvonen 1979; B.X.Liu 1986; B.M.Paine and B.X.Liu 1989). Experimentally, two different schemes have been employed to conduct the investigations. One is direct ion implantation and the other is ion beam mixing, especially using multilayered films arrangement. In ion mixing, an ion beam is used to trigger the atomic collision for inducing inter-mixing between materials A and B. Using alternately deposited A-B multilayered films, ion mixing is capable of producing A_xB_{1-x} alloy of any desired composition by simply adjusting the relative thicknesses of the metals. The alloying process in this scheme is completed through a first step of atomic

--1) Dept. of Materials Science,Tsinghua University,Beijing,P.R.CHINA.

collision,which is of far-from-equilibrium in nature, and the second step of relaxation, in which an effective cooling speed as high as 10^{13} -10^{14} K/sec can be achieved. Apparently, this scheme is a powerful means for forming metastable phases,especially amorphous alloys(metallic glass)(B.X.Liu et al. 1983; B.X.Liu 1985; B.X.Liu, E.Ma et al. 1987; B.X.Liu 1987). Recently, this method has also been used to study the formation of quasicrystals(B.X.Liu,G.A.Cheng and C.H.Shang 1987, L.J. Huang et al. 1988; C.H.Shang et al. 1988), since the first discovery of this new metastable state of matter in 1984. An essntial point of ion mixing performed by author's group is to form the quasicrystals directly without post-thermal annealing required in other methods.

It is well known that amorphous alloy is in an unstable state and will transform into a lower energy state by crystallization upon thermal annealing. The precipitation of the crystalline phase is through a thermally excited rearrangement of atoms in the amorphous matrix. It has been shown earlier by the author that this kind of atomic rearrangement could also be stimulated by ion irradiation,e.g. the demixing effect observed in the Au-Si system(B.X.Liu et al. 1982). This phenomenon was observed when the multilayered alloy films were irradiated to higher doses than required to form a uniformly mixed amorphous phase. As the ion dose can be adjusted and controled easily, ion beam method is able to drive the amorphous state to a critical turning point of phase transition. Following this idea, the author's group has recently studied, for the first time, the formation of fractal patterns in some systems(B.X.Liu, L.J.Hunag et al. 1987; L.J.Huang et al. 1988; L.J. Huang, B.X.Liu et al. 1988),adding a novel method for experimental study of fractal issue, which is currently a very important subject in the field of condensed matter as well as of materials science.

This paper is attempted to present a brief review of the recent progress obtained by author's group on ion mixing induced amorphization, icosahedral phase and fractal pattern formation.

II. PREDICTION OF GLASS FORMING ABILITY

Table 1 lists the results of amorphous alloy formation studied by ion beam mixing. Based on the experimental results, two parameters were found of importance in determining the GFA: MPAR ,namely the maximum possible amorphization range, which equals 100%(as the whole composition range) minus the maximum solid solubilities, shown by a representative phase diagram in Figure 1. Obviously, the larger the MPAR, the broader the possible amorphization range will be. The second parameter is the heat of formation(H_F) calculated by Miedema's model(A.R.Miedema 1976). It has been shown that a negative H_F was favored for amorphous

Table 1. Metallic Glass Formaton by Ion Beam Mixing

System	Combination of MPAR & H_F	Glass Formed
Hardly Glass Forming(HGF) Systems		
Au-Cu;Au-Fe;Au-Ni;Mo-Nb; Ti-Zr.	MPAR less than 20 % H_F is also unfavored	No glass has been formed
Readily Glass Forming(RGF) Systems		
Al-Au;Al-Co;Al-Mo;Al-Nb; Al-Ni;Al-Pd;Al-Pt;Al-Ta; Al-Ti;Au-Ti;Co-Gd;Co-Mo; Co-Nb;Co-Tb;Co-Ti;Co-Zr; Cu-Zr;Er-Ni;Fe-Gd;Fe-Tb; Fe-Ti;Fe-Zr;Hf-Ni;Mo-Ni; Nb-Ni;Ni-Ti;Ru-Ti;Ru-Zr.	MPAR greater than 65 % H_F is of negative	Several glasses have been formed in many systems. Or,at least one glass has been formed.
Possible Glass Forming(PGF) Systems		
Ag-Co*;Ag-Cr;Ag-Cu;Ag-Nb*; Ag-Ni;Al-Fe;Al-Mn;Au-Co; Au-Ir**;Au-Os**;Au-Ru;Au-V; Au-W; Co-Cu;Cr-Cu*;Cu-Fe; Cu-Os*;Cu-Ta;Cu-W*;Fe-Mo; Fe-Nd;Fe-W; Mo-Ru.	either MPAR or H_F is not in the range for a RGF system defined above. (MPAR should be $>$20%)	Amorphization observed in narrow composition range,except the systems with *.

Note: Detailed data can be found in Ref.(B.X.Liu 1987),except for the systems with **,which can be found in Ref.(E.Peiner et al. 1988).

alloy formation,while a positive H_F limited amorphization to occur in a narrow composition range(B.X.Liu,E.Ma 1987). However, a positive H_F, even greater than +10 kj/mol,did not prohibit amorphization like J.A. Alonso predicted(J.A.Alonso et al.1986). A map was therefore proposed by the author to pursue further(see Figure 2.). One sees (i)when MPAR is less than 20 %, no amorphization has been observed. These systems are named hardly glass forming(HGF) systems. (ii) when MPAR is greater than 65 %, and at the same time H_F is negative, amorphous alloy can be formed in wide composition range. These systems are named readily glass forming systems(RGF). (iii)when MPAR is greater than 20 %,glass can be formed. If either MPAR or H_F is not in the range of a RGF system, amorphization can only occur in a narrow composition range.These systems are named possible glass forming(PGF) systems.

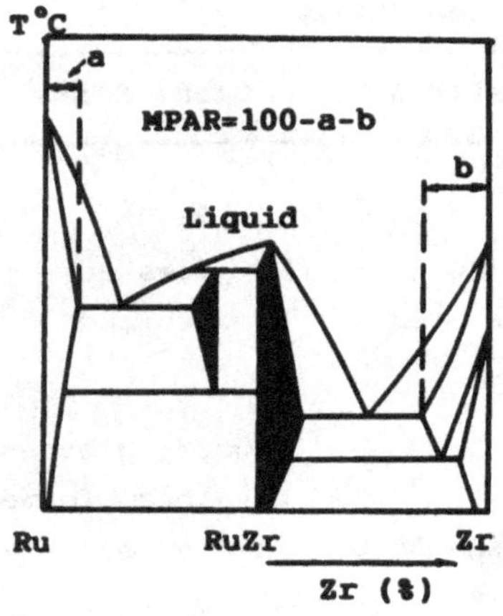

Fig.1. A representative phase diagram showing the definition of MPAR(Ru-Zr system).

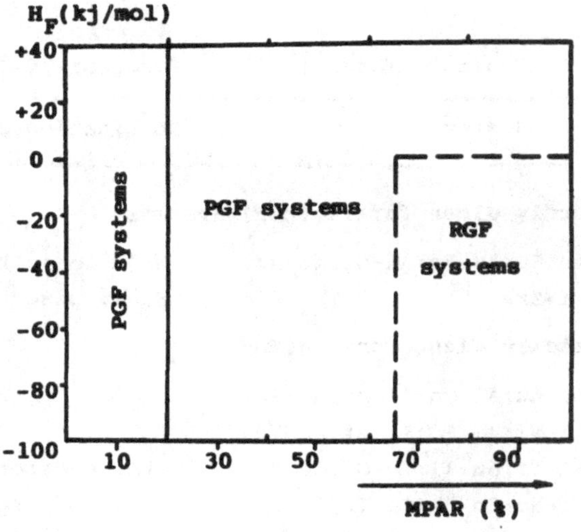

Fig.2. Possibility and readiness of glass formation by ion beam mixing in binary metal systems.

III. ICOSAHEDRAL PHASE AND FRACTAL PATTERN FORMATION

An icosahedral phase was formed by ion mixing in an $Fe_{60}Cu_{40}$ multilayered film,that had been irradiated by 300 keV xenon ions at RT to a dose of $1x10^{16}$ Xe/cm^2. Figure 3. is a TEM SAD pattern showing a dense set of diffraction rings. The diffraction lines could not be identified to be from the constituent metals. Moreover, there exists no any complicated compounds in this immiscible system providing such dense diffraction pattern. These results suggested that a quasicrystalline Fe-Cu phase was formed, with the composition of Fe-enriched. Another experiment was conducted by a Fe-rich multilayered film annealed at a temperature of 800°C. The outcome of high temperature annealing was the formation of icosahedral phase having similar pattern shown in Fig.3 (C.H. Shang and B.X.Liu 1989). All these results indicated that this Fe-Cu icosahedral phase was formed through a crystal-quasicrystal phase transformation,which was different from those reported being from an amorphous-quasicrystal phase transformation.Our results also suggested that quasicrystalline phase might be a common state existing in many metal systems, which of course needs much more studies.

Fractal patterns have been observed in some Ni-based alloy systems. In the Ni-Mo system,which has been studied by author rather extensively, a $Ni_{55}Mo_{45}$ multilayered film was irradiated at LN_2 temperature by 200 keV xenon ions to eight doses ranging from $1x10^{14}$ to $1x10^{16}$ Xe/cm^2. It was found that a dose of $1x10^{15}$ Xe/cm^2 could induce uniform mixing and

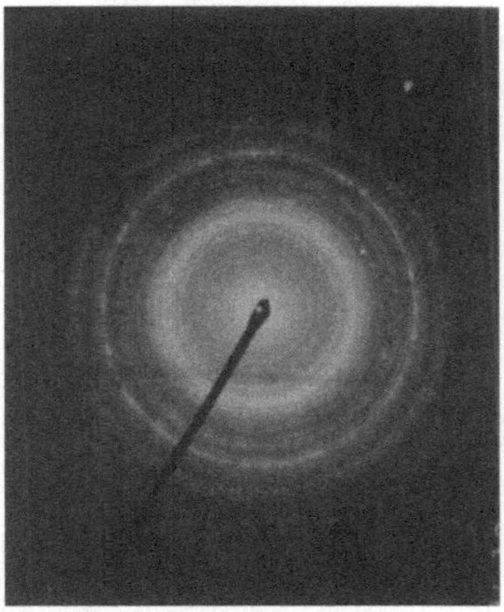

Fig.3. SAD diffraction pattern of an Fe-rich incosahedral phase formed by ion mixing.The indexing was completed by our QCII-program.

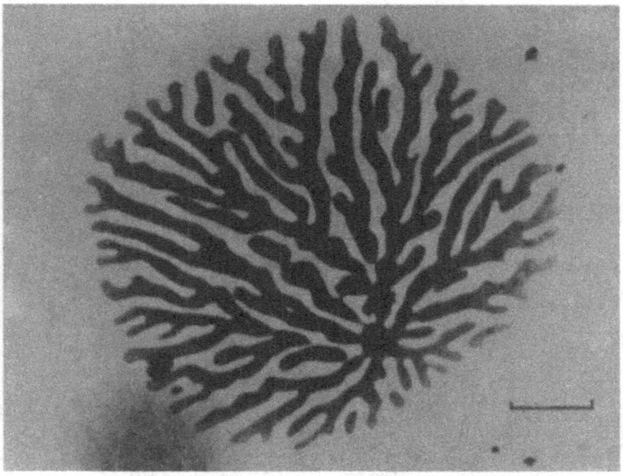

Fig.4. Fractal pattern observed in a $Ni_{55}Mo_{45}$ multilayered film after ion mixing to a dose of $7x10^{15}$ Xe/cm^2.The marker bar indicates 2 μm.

amorphization, and that a dose of $7x10^{15}$ Xe/cm^2 stimulated the previous formed amorphous phase to experience an anomalous structural phase transition,resulting in the formation of fractal patterns(Figure 4). The fractal pattern was composed of many branches, of which many single crystals of hcp structure were identified by micro-beam TEM SAD analysis. The dimension of the fractal was measured by first dividing the pattern into concentric disks with different radii R, and then counting the occupied pixels N by image processing computer with a resolution of 512x512 pixels. From the double log plots of N vs R, a straight line was figured out and the dimension,i.e. the slop,was determined to be 1.72±0.02. The formation of the fractals was most likely a multinucleation process,starting with some large clusters(precipitated crystals) from the amorphous matrix. The large clusters were stationary and distributed randomly in the film. It was followed by an attractive growth of small clusters. The small clusters were mobile,though limited in thin solid film, and gathered all precipitates to form the observed fractal patterns. More studies,both experimentally and theoretically, are now currently underway, and our computer simulation results performed by accumulating 2000 particles were in good agreement with the observations.

In conclusion, prediction of glass forming ability is now available by simply calculating two intrinsic parameters of the system. Ion mixing has been showed that it is a valuable means to study two currently very important issues, namely fractals and quasicrystals. Especially,our results indicated that some fractal patterns and quasicrystals were formed only by ion beam method, but not otherwise.

The author acknowledges the financial aids from the National Natural Science Foundation of China and the International Atomic Energy Agency(Contract No.4731/R1/RB) and many helps from his colleagues.

REFERENCES

J.A.Alonso and L.M.Lopez, Mater. Lett., 4 (1986) 765.

L.J.Huang,J.R.Ding,H-D.Li and B.X.Liu,J.Appl. Phys.,63 (1988) 2879.

L.J.Huang,B.X.Liu and H-D.Li, Il Nuovo Cimento 10D (1988) 1117.

L.J.Huang et al., J.Physics F:Metal Phys. 18 (1988) L69.

G.K.Hubler, NRL Memorandum Report 5928(1987).

B.X.Liu et al. Appl. Phys. Lett., 42 (1983) 45, and Nucl. Instrum. and Meth. 209/210 (1983) 229.

B.X.Liu et al., MRS Symposia Proc. Vol.7 (1982) 65.

B.X.Liu, Nucl. Instrum. and Meth. B7/8 (1985) 547.

B.X.Liu, Review Article,phys. stat. sol. (a) 94 (1986) 11.

B.X.Liu, E.Ma et al., Nucl. Instrum. and Meth. B19/20 (1987) 682.

B.X.Liu, Mater. Lett., 5 (1987) 322.

B.X.Liu,G.A.Cheng and C.H.Shang, Philos. Mag. Lett.,55 (1987) 265.

B.X.Liu, L.J.Huang et al., Phys. Rev. Lett.,59 (1987) 745.

A.R.Miedema, Philips Techn. Rev., 36 (1976) 217.

B.M.Paine and B.X.Liu, Review Chapter 6, in ION BEAM ASSISTED FILM GROWTH,EDITED BY T.Itoh(Elsevier, Amsterdam, 1989) 153-210.

C.M.Preece and J.K.Hirvonen, ION IMPLANTATION METALLURGY (Metallurg. Soc. AIME, New York, 1979).

E,Peiner and K.Kopitzki, Nucl. Instrum. and Meth. B34 (1988) 173.

C.H.Shang and B.X.Liu, J. Physics F: Metal Phys., 18 (1988) L169.

C.H.Shang and B.X.Liu, Mater. Lett., in press.

P.Sioshansi, Nucl. Instrum. and Meth. B19/20 (1987) 204.

J.J.Zhang,X.Z.Zhang, Z.T.Guo and H-D.Li, MRS Symposia Proc. Vol.55 (1986) 229.

NITRIDE SYNTHESIS WITH UV-PULSED LASERS: APPLICATIONS IN MICROELECTRONICS AND METALLURGY

V.Craciun*, A.V.Drigo***, G.Leggieri**, A.Luches**, M.Martino*, I.N.Mihailescu*, I.Ursu*

1. INTRODUCTION

The nitrides of refractory metals and semiconductors have properties with promising applications in various high technologies. Such compounds combine the advantages of a good hardness, high melting point, and good chemical stability. Besides, the nitrides of the refractory metals exhibit a very good electrical conductivity (a reason for which these are sometimes described as "metallic compounds"), while the nitrides of silicon and germanium have good bulk dielectrical characteristics - superior to the oxides of the two semiconductors.

Laser technologies were applied for the synthesis of such nitrides with characteristics such as high purity, strict spatial and temporal control of the temperature, the use of a controlled atmosphere during processing, a.s.o. - fully compatible with the requirements of state-of-the-art science and technology of materials (especially for VLSI microelectronics).

The titanium nitride was intensely studied for use as hard protective coating and more recently for important prospective applications in microelectronics, e.g. in diffusion barriers [1-3], gate electrodes [4], local connections [5], etc. The molybdenum nitride also exhibits good wear-preventive properties, in particular in connection with hard steel or copper alloys [6].

Silicon nitride is widely used in solid-state electronics for memory cells, MNOS (metal-silicon nitride-silicon dioxide-semiconductor) transistors and other devices [7]. The silicon oxinitride could provide an alternative, when faced with the difficulties entailed in obtaining thin SiO_2 layers (of less than 15 nm in thickness) - thus enabling one to obtain thin insulators of good quality [8,9].

The use of a germanium nitride or germanium oxinitride layer, regardless of thickness, is of interest as either a replacement or capping layer for GeO_2. As it is known, GeO_2 is hygroscopic and whenever used it has to be protected in order to avoid damage by either dissolution in water [10] or a chemical reaction with the ambient moisture in the atmosphere.

This paper reports new experimental results concerning the surface synthesis of nitrides/oxinitrides of Ti, Mo and Si, Ge, respectively, as a result of multipulse excimer laser (λ = 308 nm) irradiation of Ti, Mo, Si, and Ge in ambient NH_3 or N_2.

2. EXPERIMENTAL

The targets were made of high purity Si, Ge, Ti or Mo samples. Si samples were <100> - oriented single-crystalline wafers while Ge samples were <111> - oriented single-crystalline wafers. Ti samples were high purity (99.9 percent) foils of three different thicknesses - 0.025, 0.14 and 3 mm, respectively. Mo samples were high-purity foils of 0.5 mm thickness. Prior to irradiation, samples were carefully cleaned with alcohol and acethone, then rinsed in deionized water and dried in a nitrogen jet.

The irradiations were performed in a vacuum enclosure which had been first evacuated to 10^{-3} torr and then filled with high purity (99.9 percent) NH_3 or N_2. In order to avoid contamination from surrounding air, the working pressure in the irradiation enclosure was fixed at a slightly superatmospheric level (i.e. higher than 800 torr).

The irradiations were performed using a home-made XeCl* excimer laser (λ = 308 nm). The laser pulse duration was of $\tau_{FWHM} \simeq$ 20 ns. The laser radiation was focused on the surface of samples in spots of S_s ~ 0.1-0.2 cm^2 area.

The fluence with every incident laser pulse (of 300-800 mJ/cm^2) was slightly surpassing the threshold of optical breakdown of the gas in front of the solid sample. This kind of plasma is known to develop in the ambient gas

* Central Institute of Physics, Bucharest, ROMANIA
** University of Lecce, Department of Physics, Lecce, ITALY
*** University of Lecce, Department of Materials Science, Lecce, ITALY

starting from avalanche centers initially expelled from the sample surface under the action of the earliest part of the incident laser pulse. That is why the initiation threshold of this kind of breakdown plasma is only slightly depending on the target nature exhibiting, on the other hand, a strong dependence on the condition of its surface.

Sequences of 5, 10, 25, 50, 100 or 250 consecutive laser pulses were directed at a chosen location on the sample surface. To facilitate post-irradiation investigations, some continuous irradiation traces were formed on the sample surface - by a smooth careful translation of the sample after a multipulse irradiation sequence had been completed.

After irradiation the samples were examined by optical and electron (SEM) microscopy and then analyzed by Rutherford Backscattering Spectroscopy (RBS) with 2 MeV He^+ particles.

The quantitative measurement of the nitrogen and oxygen content on the irradiated samples was performed by using the $^{14}N(d,p_0)^{15}N$ and $^{16}O(d,p_0)^{17}O$ reactions with 0.61 MeV deuteron beams. The reaction products were recorded by using a 300 mm^2 surface barrier detector. It was covered by a 13 µm mylar foil in order to stop the backscattered particles and placed at 150° from the beam. The absolute calibration of ^{14}N and ^{16}O was performed by comparison with Ta_2O_5/Ta and Si_3N_4/Si calibrated standards [5,11].

X-ray investigations were performed using a SIEMENS KRYSTALLOFLEX IV diffractometer while in microhardness determinations a PMT 3 instrument was used.

X-ray photon spectroscopy (XPS) studies were conducted with an ESCALAB MARK II - VG apparatus. The line K-α of Al was used and the binding energy was determined - wherefrom information was inferred on the chemical bonds of the element(s) under study.

Sheet resistivity determinations were performed by the four-point probe method.

3. RESULTS

3.1. Nitridation of Ti and Mo

After a few tens of consecutive laser pulses, the colour of the laser-irradiated areas on the surface of Ti samples turns from bright grey to pale, and then intense, yellow. The yellow colour is particularly apparent at the superposition border between two neighbouring irradiation areas either of which is subjected to large numbers (100,250) of consecutive laser pulses. In this case, the yellow colour usually gets a typical "bronze" touch.

We note that according to Ref.12, Ti_2N is bright yellow and single-crystalline TiN is golden yellow. Also, according to the same reference, the titanium nitrides colour depends on the defects (covering any real surface) - in particular a quite rough surface has a typical "bronze"-yellow colour. This last piece of information is to be corroborated with the SEM evidence exhibiting a rather rough aspect of the compound growing on the Ti samples surface* as a result of multipulse laser irradiation in either NH_3 or N_2 (see e.g. Fig.1). The same is true for the Mo samples.

These observations got support from the results of the microhardness determinations and X-ray diffraction investigations.

We thus mention the typical case of a 0.14 mm thickness Ti sample, where a significant microhardness increase was observed in an area submitted to the action of 250 consecutive laser pulses in ambient NH_3 from ~300 HV (1 HV = 1 kg/mm^2) - the base material - up to 500-600 HV. Moreover, in the overlap zone between two neighbouring laser spots - which is showing an intense "bronze"-yellow colour - a microhardness value as high as 700 HV was found - in very good agreement with the values indicated in Refs. [6] as corresponding to thin TiN films.

It is to be mentioned here that the Ti samples of 0.025 nm in thickness where nitridation effects apparently different were observed - as resulting from the results of the nuclear reactions investigations - a different evolution of the surface microhardness was observed too: i.e. an increase in the surface material microhardness of 60 HV to ~80 HV; also a different structure was evidenced by X-ray investigations (a difference was thus evidenced between the positions of the X-ray diffraction maxima in the case of the samples of 0.025 mm and 0.14 mm thickness, respectively).

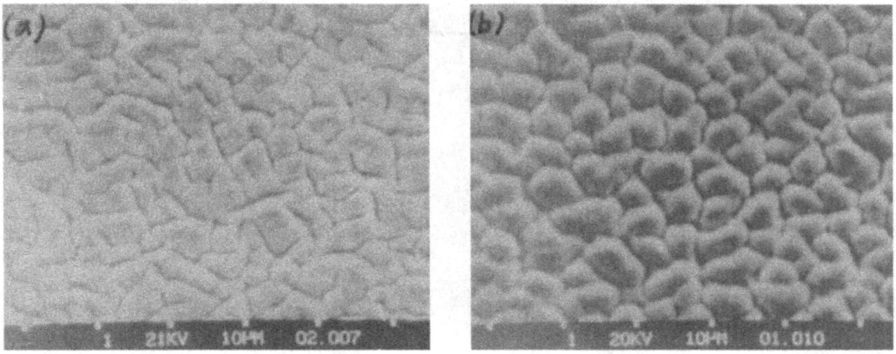

Fig.1 - SEM micrographs within the irradiation areas on the surface of a 0.14 mm thickness Ti sample submitted to the action of 100 (a) and 250 (b) consecutive laser pulses respectively in a NH_3 atmosphere; the fluence with every incident laser pulse was of ≈ 720 mJ/cm^2.

A microhardness increase was also observed, of the areas submitted to multipulse laser irradiation on the surface of Mo samples. Thus after only 15 consecutive laser pulses of 300 mJ/cm^2 fluence, the surface microhardness increased from 22-24 HV to 40 HV.

The X-ray diffraction analyses confirmed the presence of TiN and Ti_2N. We note - see e.g.ref.[6] - that the titanium nitride usually consists of a mixture of TiN, Ti_2N and a solid solution of nitrogen dissolved in bulk titanium.

Also $Mo_{2.3}N$, MoN and Mo_2N peaks were identified in the X-ray spectra of the Mo samples submited to multipulse excimer laser irradiation in ambient NH_3.

It is also to be mentioned that both titanium nitride and molybdenum nitride peaks are quite high, pointing to the fine dispersion of the two nitrides forming on the areas subject to multipulse laser irradiation.

The amplitude of the nitride peaks exhibit a tendency to increase with the number of consecutive laser pulses directed to the same irradiation area on the sample surface.

An estimation of the thickness of the nitride layer can be made based upon the data obtained by nuclear reactions studies of the areas submitted to multipulse laser irradiation in either NH_3 or N_2 ambient atmosphere. For this purpose one can use the information in Ref.13 that a nitrified Ti layer of 1 cm^2 area and $\simeq 1000$ Å thickness contains about 1.25×10^{17} nitrogen atoms. The thickness of the surface nitrified layer can be accordingly evaluated from

$$x(A) = \frac{\Delta n}{1.25\times10^{14}}, \quad \Delta n = n_f - n_i \qquad (1)$$

where n_i is the initial density of nitrogen atoms (i.e. on virgin surface) while n_f stands for the nitrogen atoms density at the end of the multipulse laser irradiation.

A typical example is given in Fig.2, corresponding to a Ti sample of 0.14 mm thickness submitted to multipulse excimer laser irradiation in ambient NH_3. Inspecting this figure one notices that

(i) Variation Δn of the concentration density of nitrogen atoms per unit surface submitted to multipulse excimer laser irradiation and, correspondingly, the x thickness of the surface nitrified layer - as inferred from the nuclear reactions data with the aid of Eq.(1) - both increase with the number of consecutive laser pulses applied to the same irradiation site.

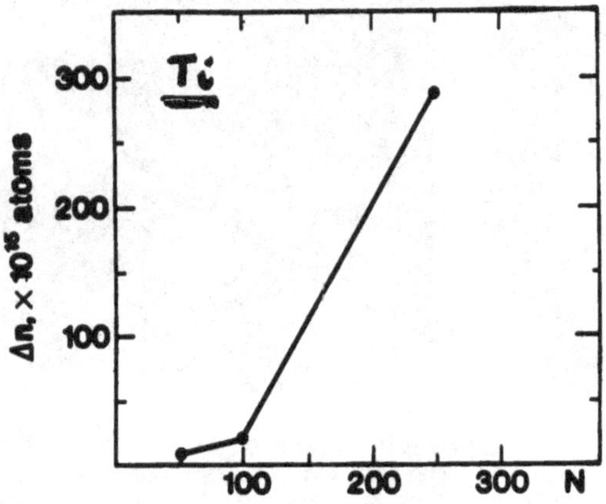

Fig.2 - The dependence on the number of consecutive laser pulses directed to the same irradiation site of both the variation (increase), Δn of the surface density of nitrogen atoms - as inferred from nuclear reactions data - and of the thickness of the nitrified layer, x - as estimated from these data with the aid of Eq.(1) - in case of a Ti sample submitted to multipulse excimer laser irradiation in a NH_3 atmosphere.

The increasing tendency is accentuated after the first hundred consecutive laser pulses, and one has good reasons to expect a further important increase in the two quantities, Δn and x, even after N > 250 pulses.

(ii) The thickness of the nitrified layer exceeds 2000 Å and, most probably is further rapidly increased upon extending the laser irradiation.

Further support to these data and complementary information were provided by the study of the XPS spectra of the areas submitted to multipulse laser treatment. Typical portions of some parts of interest in the XPS spectra of 0.025 mm Ti samples submitted to many (250) conscutive laser pulses are given in Fig.3.

We note that the quite high peak visible in Fig.3a at 458.5 eV corresponds to TiO_2 while the peak at 454.8 eV in the same figure is most probably a convolution between the Ti peak in TiC at 454.5 eV and the Ti peak in TiN at 455.2 eV, respectively [14-17]. Also, the N-1s peak at 396.6 eV visible in Fig.3b is significantly displaced as against the standard position of this peak with free nitrogen - which stands at 399 eV [18] -, pointing again to a Ti-N bond.

Similar evidence was found in XPS spectra of the other Ti samples submitted to multipulse excimer laser irradiation in ambient NH_3 - supporting in all cases the existence of Ti-N and Ti-O bonds and elliminating the possibility of a direct N-O bond.

3.2. Nitridation of Si and Ge

It is first to be emphasized that a perfect (mirror) condition of the areas submitted to multipulse excimer laser treatment was evidenced in this case by both optical and electron microscopy investigations. In particular this also serves as a proof of the quite uniform intensity distribution - of a "top hat" type - within the spot of the incident laser radiation on the sample surface.

Some typical SEM micrographs of areas on the surface of Si and Ge wafers submitted to multipulse excimer laser irradiation in ambient NH_3 are reproduced in Fig.4. One notices the high planeity degree along with the absence of any cracks and/or microcraters which are characterizing in this case the aspect of the laser-treated areas - particularly in comparison with the quite rough appearance of the multipulse laser irradiation areas on the Ti samples surface.

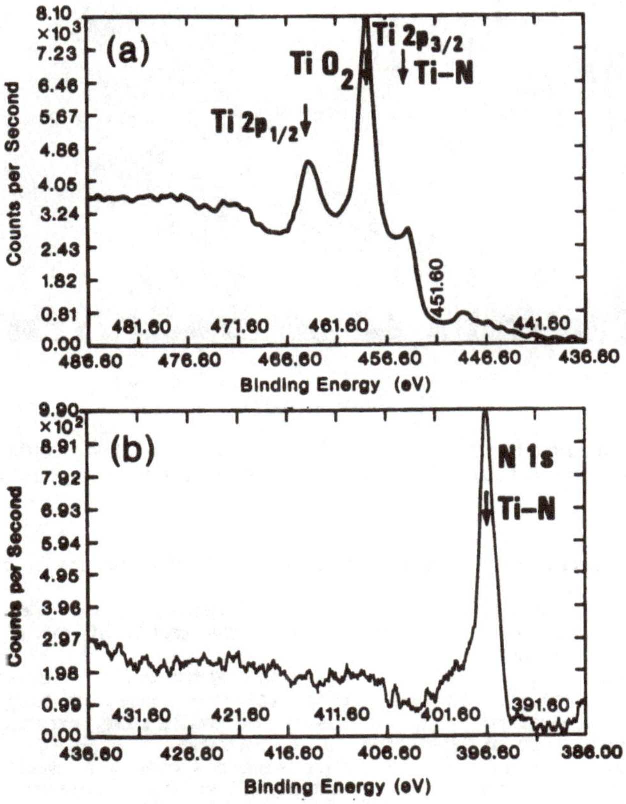

Fig.3 - Portions of the XPS - spectra of the top surface layer of a 0.025 mm thickness Ti sample submitted to the action of 250 consecutive laser pulses in ambient NH_3. The regions of the Ti-2p peak (a) and of the N-1s peak (b) are represented.

We have not reached yet any definite explanation for the quite large difference between the aspect of the laser-treated areas on the surface of the metallic and semiconductor samples, respectively. We believe however that this can be a consequence of the independent or the collective action of the following factors:

(i) The initial condition of the semiconductor surfaces - which were submitted to a high processing - compared to the metallic (Ti) samples which were only carefully cleaned.

(ii) The different kinetics of the nitridation process under laser irradiation on the semiconductor and the metallic (Ti) surfaces, respectively. We estimate that the elucidation of this aspect deserves further investigation.

In case of Si samples submitted to multipulse (250) excimer laser irradiation in a NH_3 ambient atmosphere, the XPS analyses have given the following experimental evidence: it was first noted that the Si-2p peak (Fig.5a) exhibits two maxima, at 102.8 eV and 99 eV. According to Refs.19-21, these maxima should be in connection with the Si-N and the Si-Si bonds, respectively. Also, the position of the N-1s peak at 398.2 eV (Fig.5b) corresponds to the Si-N bond - most probably in a Si_3N_4 compound [20]. On the other hand, the position of the O-1s peak at 532.8 eV (Fig.5c) is to be associated with a Si-O bond in SiO_2 [10,20].

We may conclude that the multipulse excimer laser irradiation resulted in this case in the formation of an oxinitride compound SiO_xN_y where both N and O are bound to a Si atom and not in-between. As a further confirmation of this hypothesis we note that, according to the Ref.10, a direct N-O bond is

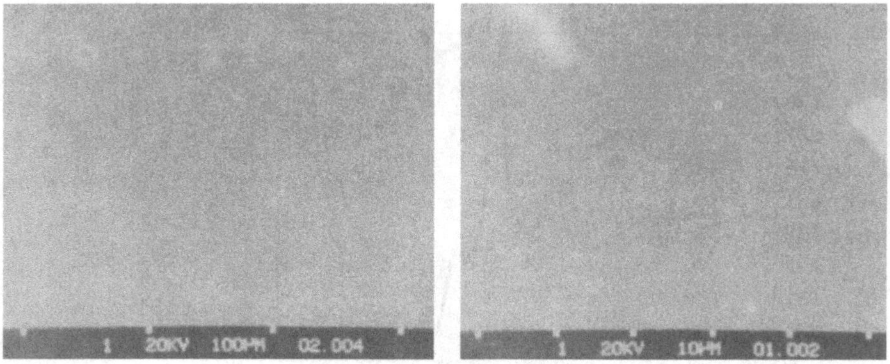

(a) (b)

Fig.4 - SEM micrographs of surface areas of a Si (a) and a Ge (b) wafers which were submitted to the action of 250 consecutive excimer laser pulses in a NH_3 atmosphere.

evident from the displacement of the N-1s peak to 402 eV - and not at 398.2 eV as visible from Fig.5b.

As for the process leading to the synthesis of the SiO_xN_y oxinitride we think, at this stage, that this is the result of the action of two - possibly simultaneous - phenomena:

(1) The non-complete cleaning of the Si wafers before submission to multipulse laser treatment, had left traces, possibly films, of SiO_2 native oxide which were converted - as a result of multipulse excimer laser irradiation in NH_3 -, into an oxinitride SiO_xN_y compound.

(2) Oxygen traces in the irradiation chamber - even very small ones - were sufficient to interfere in the interaction process and result in the final synthesis of a compound including oxygen besides Si and N. The large sensitivity of the laser-conducted nitridation process of the refractory metals (Ti, Zr, Hf) to the oxygen traces in the ambient nitrogen atmosphere [22,23] was also evident in this work upon multipulse laser irradiation of Ti in NH_3. Further experiments will have to be devised and performed in order to discriminate between the two mechanisms.

XPS analyses have shown that an oxinitride compound - of the GeO_xN_y type - is also forming as an effect of the multipulse excimer laser irradiation of germanium wafers in either NH_3 or N_2. This is convincingly illustrated by the sequence of XPS spectra regions given in Fig.6 corresponding to a Ge wafer submitted to the action of 250 consecutive laser pulses in ambient NH_3.

From Fig.6a we note that the binding energy of the Ge-3d peak at 33.45 eV is placed between the energy corresponding to the Ge-N bond - of 32.4 eV - and that of the Ge-O bond (most probably in GeO_2) - of 33.7 eV. Like in the case of Si, Ge appears to be also bound to both O and N - an identical observation is reported in Ref.10. Also the O-1s peak at 532.8 eV which is visible in Fig.6b is to be associated to the Ge-O bond (this peak corresponding precisely to the same bond was reported in Ref.10 at 532.7 eV). Finally, we observe that the N-1s peak (Fig.6c) consists of two maxima. The first maximum at 398.84 eV is to be associated again to the Ge-N bond (a value of 398.8 eV was reported in Ref.10 for this bond). The second maximum at 406.5 eV belongs most probably to an impurity in the base material.

Concluding these data - which are similar to those recorded by XPS analyses of other multipulse laser-irradiated Ge wafers - one may say that Ge is related to both N and O while no direct bond was observed between N and O atoms.

As expected, the growth of the nitrified surface layer also influences the surface electrical properties of the semiconductor samples.

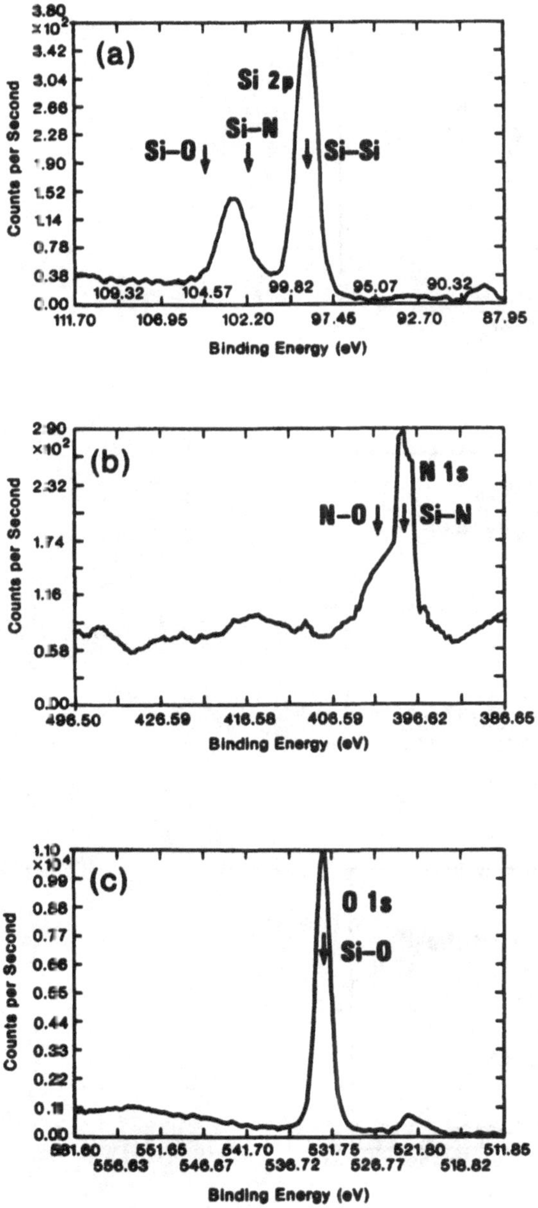

Fig.5 - Portions of XPS spectra corresponding to multipulse excimer laser irradiated areas on the surface of a Si wafer submitted to the action of 250 consecutive laser pulses in ambient NH_3. The regions corresponding to the Si-2p (a), N-1s (b) and O-1s (c) peaks respectively, are represented.

The increase in the sheet resistance - i.e. the transition to a pronounced insulator behaviour - is more significant in the case of Si samples. We thus observed an increase in the breakdown voltage of the structure forming on the surface of Si samples, from 2-3 V after the application of 25 consecutive laser pulses in ambient NH_3, to 15-20 V after the application of 100 pulses and to already 40 V after the application of 250 pulses to the same

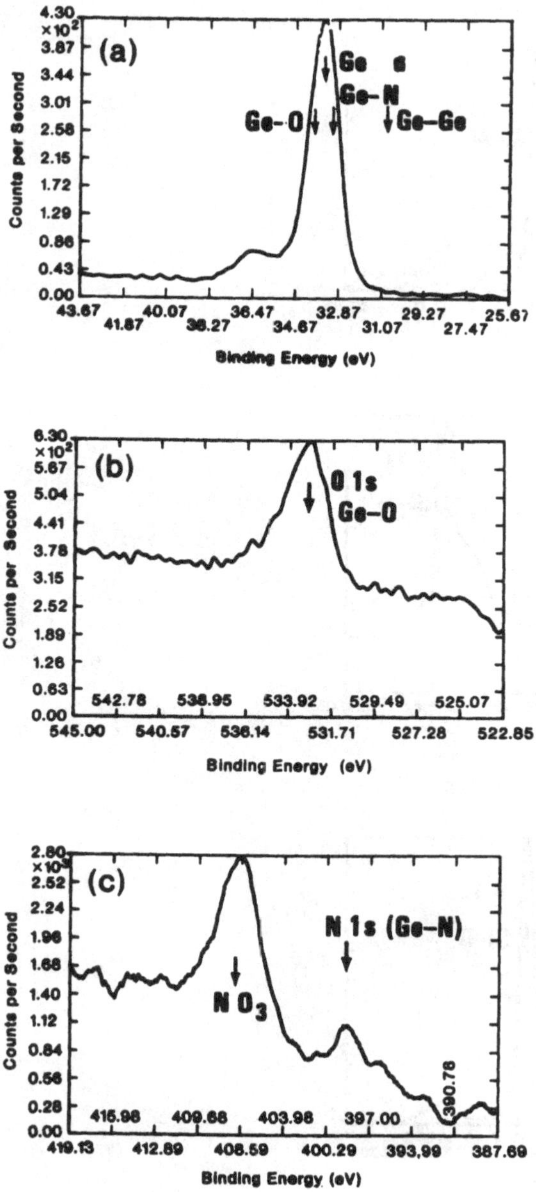

Fig.6 - Portions of XPS spectra corresponding to multipulse excimer laser irradiated areas on the surface of a Ge wafer submitted to the action of 250 consecutive laser pulses in ambient NH_3. The regions corresponding to the Ge-3d (a), O-1s (b) and O-1s (c) peaks are represented.

irradiation area. In this latter case, the leakage current was less than 10 nA for 35 V bias, which makes the forming structure already consistent with the requirements of integrated circuits processing.

The increase was less spectacular in case of multipulse excimer laser irradiation of Ge samples in ambient NH_3. We thus observed only a twofold increase in the sheet resistance - from an initial value of 2 Ω.cm - after the application of 50 consecutive laser pulses, which became a 40 -time increase after the application of 250 consecutive laser pulses.

However, according to the evidenced trend, we expect a further significant increase in the sheet resistance and the breakdown voltage when prolonging the multipulse excimer laser irradiation - to thousands of consecutive laser pulses applied to the same irradiation site.

4. CONCLUSIONS

We have demonstrated that the multipulse excimer laser irradiation of both metallic (Ti,Mo) and semiconductor (Si,Ge) samples in either NH_3 or N_2 atmosphere results in the synthesis of a surface nitrified layer very adherent to the bulk support. The thickness increases and the general properties improve with the number of consecutive laser pulses directed to the same irradiation site on the sample surface. In particular, the excimer laser nitridation of the two semiconductors proved to fulfil the basic requirements in order to make it consistent with VLSI microelectronics. We note e.g. that the rf plasma nitridation requires a quite large thermal energy budget for VLSI - i.e. the heating to 950°C for a few hours resulting in a dopant redistribution (as described in Ref.9). This makes the laser alternative fully consistent from the technical requirements point of view, and also economically more advantageous.

REFERENCES

(1) M.Wittmer, J.Vac.Sci.Technol. A2, 273 (1984).
(2) A.Armigliato, M.Finetti, J.Garrido, S.Guerri, P.Ostoja, and A.Scoryoni, J.Vac.Sci.Technol. A3, 2237 (1985).
(3) V.Craciun, I.N.Mihailescu, I.Ursu, G.Leggieri, F.Craciunoiu, A.Corici, G.Leggieri, A.Luches, V.Nassisi, and M.Martino, Appl.Phys.Lett. 52, 1225 (1988).
(4) M.Wittmer, J.Vac.Sci.Technol. A3, 1797 (1985).
(5) T.Tang, Ch.Ch.Wei, R.Haken, T.Holloway, Ch.F.Wan, and M.Douglas, IEDM Tech.Dig., 590 (1985).
(6) J.E.Sundgren and H.T.G.Hentzell, J.Vac.Sci.Technol. A4, 2259 (1985); A.Walker, J.Folkes, W.M.Steen, and D.R.F.West, Surf.Eng. 1, 23 (1985).
(7) N.A.Assadulayev, N.B.Brandt, S.M.Chudinov, S.N.Koslov and I.Ciric, J.Appl. Phys. 61, 4566 (1987).
(8) J.Nulman and J.P.Krusins, Appl.Phys.Lett. 47, 148 (1985).
(9) Paihung Pan, J.Appl.Phys. 61, 284 (1987); P.Debenest, K.Barla, A.Straboni, and B.Vuillermoz, Appl.Surf.Sci. 36, 196 (1989).
(10) D.J.Hyrnes and J.J.Rosenberg, J.Electrochem.Soc. 135, 961 (1988).
(11) M.Berti and A.V.Drigo, Nucl.Instr.and Meth. 201, 473 (1982).
(12) M.Savvides and B.Window, J.Appl.Phys. 64, 225 (1988).
(13) N.Laidani, J.Perriere, D.Lincot, A.Gicquel, and J.Amonroux, Appl.Surf.Sci. 36, 520 (1988).
(14) R.F.Hochman, A.Erdemir, F.J.Dolan, and R.L.Thom, J.Vac.Sci.Technol. A3, 2348 (1985).
(15) T.Brat, N.Parikh, N.S.Tsai, A.K.Sinha, J.Poole, and C.Wickersham Jr., J.Vac.Sci.Technol. B5, 1741 (1987).
(16) M.Repeta, L.Dignard-Bailey, J.F.Currie, J.L.Brebner, and K.Barla, J.Appl. Phys. 63, 2796 (1988).
(17) A.E.Morgan, E.K.Broadbent, K.N.Ritz, D.K.Sadana, and B.J.Burrow, J.Appl.Phys. 64, 344 (1988).
(18) F.Fujimoto, Y.Nakane, M.Saton, F.Komori, K.Ogata, and Y.Andoh, Nucl.Instr. and Meth. B19/20, 791 (1987).
(19) L.Kubler, R.Haug, E.K.Hlil, D.Bolmont, and G.Gewinner, J.Vac.Sci.Technol. A4, 2323 (1986).
(20) Yang-Sun Jo, J.A.Schultz, Shinichi Tachi, S.Contarini, and J.Wayne Rabelais, J.Appl.Phys. 60, 2564 (1986).
(21) S.Aritome, M.Morita,T.Tanaka, and M.Hirose, Appl.Phys.Lett. 51, 981 (1987)
(22) I.Ursu, I.N.Mihailescu, A.M.Prokhorov, and V.I.Konov, (in) Proceedings of the European Materials Research Society (E-MRS'86) Conference, Strassbourg, 1986, edited by D.Bauerle, K.L.Kompa, and L.D.Laude, Editions de Physique (Les Ulies, France, 1986), p.226.
(23) I.Ursu, I.N.Mihailescu, L.C.Nistor, V.S.Teodorescu, A.M.Prokhorov, V.I.Konov, and S.A.Uglov, J.Phys.D.: Appl.Phys. 20, 1519 (1987).

CHARACTERIZATION OF POLYCRYSTALLINE α-Al_2O_3 IMPLANTED WITH ZIRCONIUM, COPPER OR IRON

N. MONCOFFRE, C. DONNET, G. MAREST, J. TOUSSET
Institut de Physique Nucléaire de Lyon,
IN2P3-CNRS/Université Claude Bernard
43, Bd du 11 Novembre 1918, 69622 VILLEURBANNE Cedex, FRANCE

Copper, zirconium and iron implanted polycrystalline α-Al_2O_3 are studied at fluences ranging from 1 to 3 10^{17} ions cm^{-2} and at 110 and 190 keV implantation energies. By using XPS it is shown that after Zr implantation the 4 chemical states (ZrO_2, Al_2O_3, Al, Zr) are observed whereas for Cu, only Al_2O_3 and Cu are detected in the implanted layer. CEMS technique allows to determine the presence of Fe^0, Fe^{4+} and two Fe^{2+} ($FeAl_2O_3$ and Fe^{2+} vacancy complexes) charge states. RBS and SEM techniques are used to follow the implanted element depth distributions and the precipitate sizes (ZrO_2, $CuAl_2O_4$ and $(FeAl)_2O_3$ in particular) versus air annealings between 200 and 1600°C. Significant losses of the implanted species during the annealings occur from 700, 1400 and 1600°C respectively for Cu, Fe and Zr.

Introduction

Relatively little work has been reported on changes in the properties of implanted industrial polycrystalline ceramics compared to single crystals (1, 9, 13). Our purpose is to perform a fine physico-chemical characterization of Zr, Cu or Fe implanted polycrystalline α-Al_2O_3 before and after thermal annealings. The correlation is made between the oxidizing behavior of the implanted species and the nature of the chemical compounds detected at the surface. Indeed zirconia and alumina are two refractory oxides which have nearly similar Gibbs energies of formation. In contrast copper oxides are known to be much less stable whereas iron oxides have an intermediate thermodynamical stability.

Several complementary techniques are used to study this surface chemical evolution : Rutherford Backscattering Spectrometry (RBS) which is well adapted to determine the implanted depth distributions. The chemical state evolution is followed either thanks to X-Ray Photoelectron Spectroscopy (XPS) for implanted zirconium and copper samples or with Conversion Electron Mössbauer Spectroscopy (CEMS) for the iron ones. Further information is obtained with Scanning Electron Microscopy (SEM).

1. Experimental procedure

Thick α-Al_2O_3 samples were polished (Ra less than 0.02 µm) and annealed at 800°C in air for one hour before implantation. Specimen were implanted at room temperature (RT) with 110 and 190 keV ^{63}Cu, ^{57}Fe or ^{90}Zr ions at fluences ranging from 1 to 3 10^{17} ions cm^{-2}. Thermal annealings were performed in air for one hour from 200 to 1600°C.

RBS analysis were untertaken using an incident beam of 1.2 MeV He^+ ions. The experimental RBS spectra were compared to simulated spectra calculated on the basis of several successive sub-surface layers in which the different atomic concentrations are the adjustable parameters. This method allows us to obtain the implanted species concentrations (at cm^{-3}) versus depth (nm).

2. Results and discussion

2.1 As implanted samples

*RBS experimental spectra performed on Zr, Cu and Fe implanted samples are displayed in **Figure 1**. For a 10^{17} at cm^{-2} fluence at 110 keV, the values of the experimental mean ranges (37, 50 and 55 nm) and of the average atomic concentrations of the distributions (11, 13 and 13 %) for Zr, Cu and Fe respectively are in good agreement with TRIM code results which take into account the sputtering yield.*

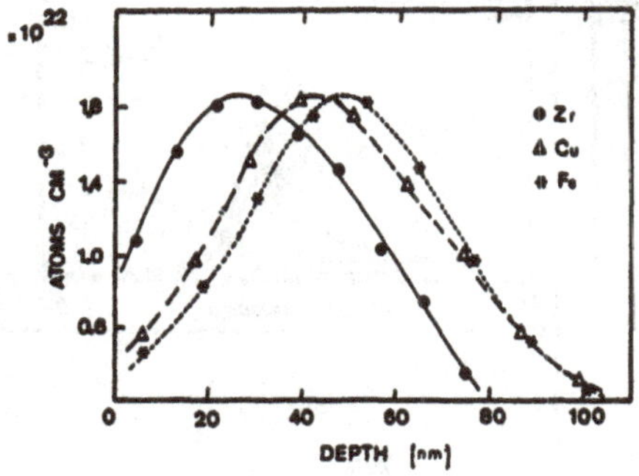

Figure 1 :
Experimental RBS distributions of as-implanted Zr, Cu and Fe into Al_2O_3 (ϕ = 10^{17} ions cm^{-2}, E = 110 keV).

*The chemical state of the different elements in the as-implanted layer was deduced either using XPS (for Cu and Zr implantation) or using CEMS (for Fe implantation). These results are presented in **Figure 2A, B, C**.*

*Implanted Zr shares into several components (Zr^{4+}, Zr^{3+}, Zr^0) corresponding to the oxidized and metallic states of Zr **(3)**. Moreover in addition to the Al^{3+} signal of the Al_2O_3 matrix, a peak of the Al^0 state is detected. Similar XPS experiments performed on the Cu implanted Al_2O_3 do not put in evidence any oxidized form of Cu. Only the Cu^0 and Al^{3+} peaks are detected. CEMS experiments on the Fe implanted alumina display three iron chemical states **(4,8)** : i) Fe^0 corresponding to small metallic precipitates. Such small γ-Fe precipitates have been already identified by TEM in implanted Al_2O_3 single crystals **(10)***

ii) Fe^{4+} due to a metastable iron charge state located in distorbed FeO_6 octahedra (2, 5, 7)
iii) two ferrous ion signals (Fe_I^{2+} and Fe_{II}^{2+}) attributed to the formation of the spinel compound $FeAl_2O_4$ (12) and vacancy configurations associated with iron (11). These results are significant of the oxidizing capacity of the 3 implanted species compared to alumina.

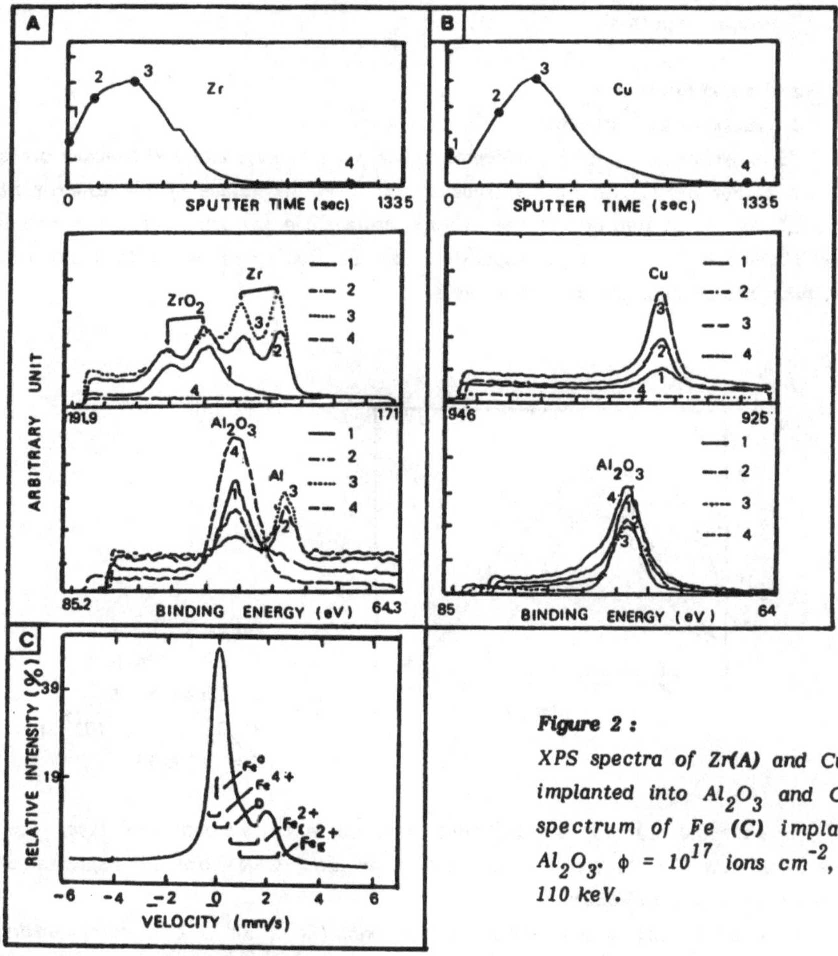

Figure 2 :
XPS spectra of Zr(A) and Cu (B) implanted into Al_2O_3 and CEMS spectrum of Fe (C) implanted Al_2O_3. $\phi = 10^{17}$ ions cm^{-2}, E = 110 keV.

2.2 *Post-implantation annealings*

*The RBS results related to the annealed samples between 400 and 1600°C are presented in **Figure 3A, B** and **C** for Zr, Cu and Fe implantations respectively. The Mössbauer spectra of* the Fe implanted Al_2O_3 annealed at 800, 1200 and 1400°C are displayed in **Figure 4. Figure 5** gives some typical SEM micrographs of Zr, Cu and Fe implanted surfaces annealed at various temperatures. In fact RBS profiles have to be correlated with SEM observations since the classical RBS techniques does not provide any lateral resolution. Indeed **Figure 3** indicates that the distribution profiles broaden versus annealing temperature whatever the implanted element. This tendancy is not significant of a depth diffusion process but to some progressive surface phase precipitation which is well observed by SEM. We have already shown **(3)** on Zr implanted samples annealed from 700 to 1600°C that ZrO_2 precipitate growth occurs at the surface as the annealing temperature is increased. These precipitates are observed by SEM and the coupling of the two techniques (RBS and SEM) allows us to follow the grain size evolution as a function of the annealing temperature. In addition Grazing angle X-Ray Diffraction (GXRD) results **(14)** show that zirconia has the tetragonal structure up to 1000°C which then becomes monoclinic. This last structure is the stable one at RT but requires a minimum grain size of 100 nm which is reached only from 1200°C. It is worth pointing out that at 1600°C the big ZrO_2 nodules are distributed upon the surface very unhomogeneously and that a 60 % Zr loss occurs. Moreover thermal annealing experiments performed in the electron microscope confirm the Zr oxidation into small ZrO_2 precipitates (≈ 20 nm) from 800°C **(3)**.

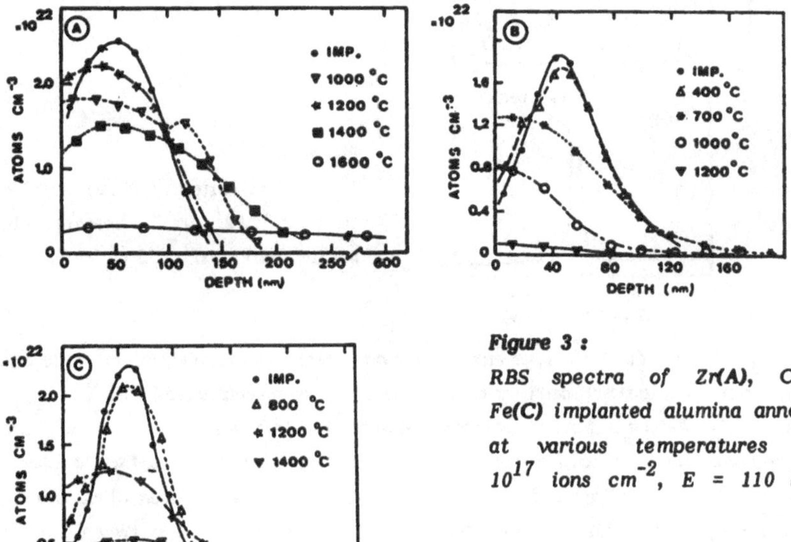

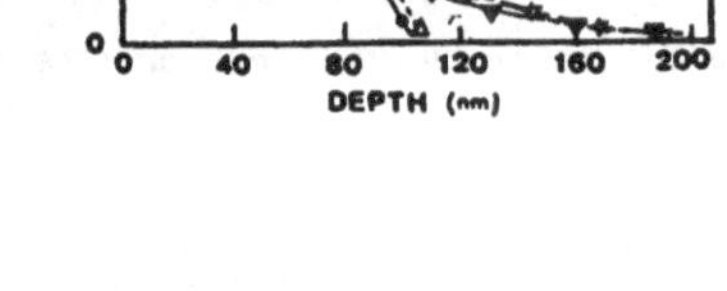

Figure 3 :
RBS spectra of Zr(A), Cu(B), Fe(C) implanted alumina annealed at various temperatures (ϕ = 10^{17} ions cm^{-2}, E = 110 keV).

A similar evolution is observed on Cu implanted Al_2O_3 at lower temperatures. From 400°C a slight Cu migration toward the surface is noted by RBS ***(Figure 3).*** *Between 500 and 700°C, $CuAl_2O_4$ precipitates are identified by GXRD* ***(14). Figure 5A*** *displays their homogeneous distribution on the surface at 500°C. From 700°C a progressive Cu loss is noticed on RBS spectra correlated with a copper oxide precipitate release well observed on SEM micrographs (craters). A 1200°C anneal leads to an almost complete Cu loss. In the case of Fe implantation, up to 800°C RBS does not indicate any change in the iron distribution profiles. In contrast, CEMS, which is a very powerful technique to investigate the local iron environment, detects some evolution in the iron chemical state from 200°C* ***(8).***

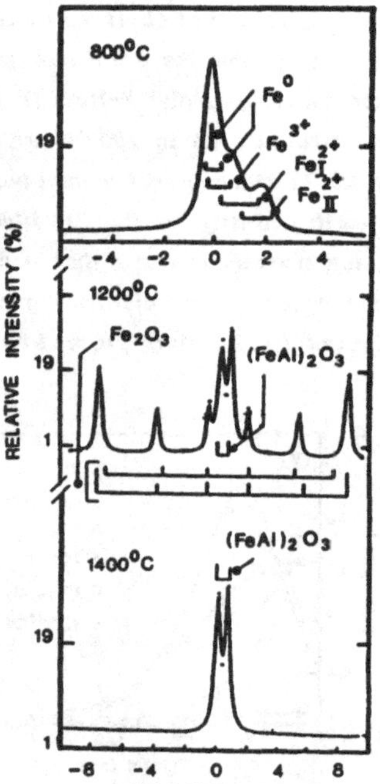

Figure 4 :

Thermal evolution of CEMS spectra corresponding to a 10^{17} Fe^+ cm^{-2} implanted alumina annealed at 800, 1200 and 1400°C.

At 400°C, the Fe^{4+} ions completely disappear. The number of metallic clusters increases significantly together with the formation of a mixed oxide $(Fe_xAl_{1-x})_2O_3$ containing sufficiently few Fe^{3+} ions to be paramagnetic ***(6).*** *This mixed oxide phase increases at 800°C whereas at this temperature a slight iron migration toward the surface is observed by RBS. It can be associated with the formation of small precipitates located along grain boundaries and well identified on SEM micrographs. In the temperature range between 1000 and 1200°C all the iron contributions initially created by implantation totally disappear. Iron atoms form two populations of $(Fe_xAl_{1-x})_2O_3$ precipitates : i) the magnetic one (sextets in the CEMS* *spectrum in **Figure 4**) corresponds to large hematite-like precipitates ($x > 0.9$) detected by SEM along grain boundaries (**Figure 5**). These precipitates could be initiated by a Fe enrichment of the mixed phase detected at 800°C ii) a paramagnetic component which could be due to the mixed oxide distributed on the whole surface of alumina grains. At 1400°C a total vanishing of the biggest magnetic phases located on grain boundaries happens. Simultaneously the CEMS sextet disappears and only the mixed oxide signal remains. This outdiffusion corresponds to an iron loss of 30 %. At 1600°C this loss of iron continues and reaches 50 %.*

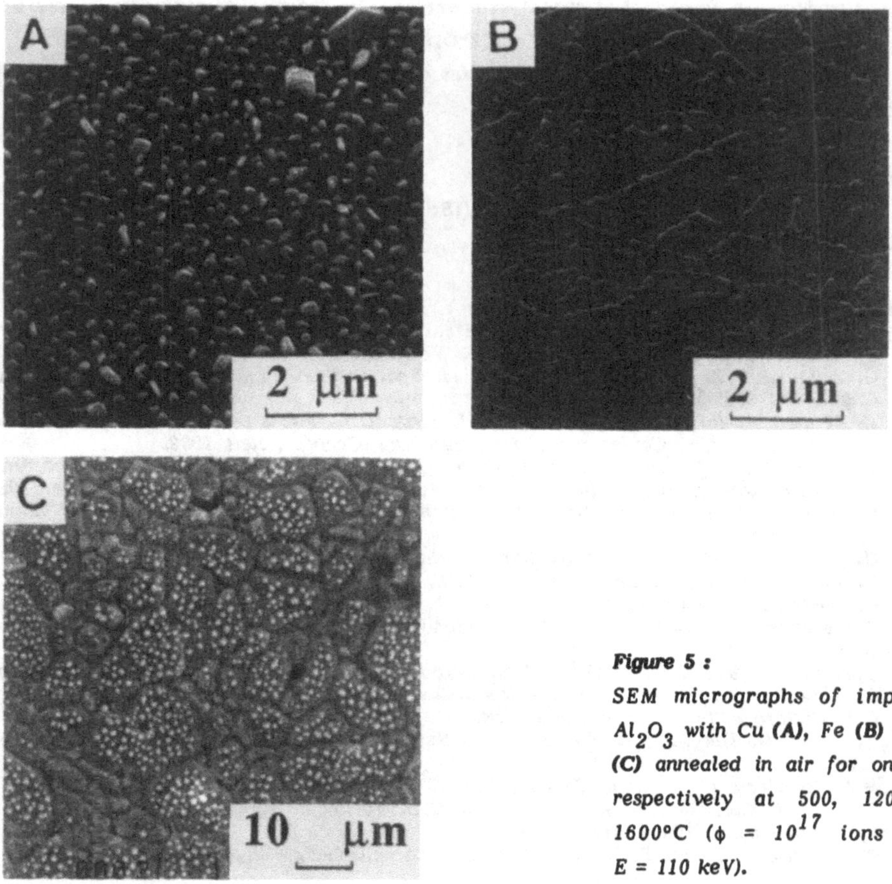

Figure 5 :
*SEM micrographs of implanted Al_2O_3 with Cu **(A)**, Fe **(B)** and Zr **(C)** annealed in air for one hour respectively at 500, 1200 and 1600°C ($\phi = 10^{17}$ ions cm^{-2}, $E = 110$ keV).*

Conclusion

Several complementary techniques (RBS, CEMS, SEM, XPS) have been used to compare the behavior of three different implanted species (Zr, Cu, Fe) into alumina. Significant differences between these elements have been put in evidence in regard to thermal treatments in air. In fact thermal evolution and phase precipitation occur in various temperature ranges :

- *Cu is hardly oxidizable thus explaining the metallic form of Cu in the as-implanted specimen. The* $CuAl_2O_4$ *spinel is only detected between 500 and 700°C*
- *Fe slightly more oxidizable than Cu gives raise to the* $(Fe, Al)_2O_3$ *mixed oxide from 400°C whereas the spinel* $FeAl_2O_4$ *is detected directly after implantation as well as the metallic iron*
- *The oxide* ZrO_2 *is already detected in the as-implanted sample. It puts forward the competition between Zr and Al in regard with oxygen. The thermal Zr evolution is consistent with a phase segregation between* Al_2O_3 *and* ZrO_2.

Significant losses of the implanted species during the annealings occur from 700, 1400 and 1600°C respectively for Cu, Fe and Zr.

Acknowledgements

The authors are gratelul to A. PLANTIER who performed the implantations and to G. FUCHS for helpful discussions.

References

(1) B.R. Appleton, H. Naramoto, C.W. White, O.W. Holland, C.J. McHargue, G. Farlow, J. Narayan, J.M. Williams : Nucl. Instr. & Meth. **B1**(1984) 164.

(2) G. Demazeau, Z. Li-Ming, L. Fournier, M. Pouchard, P. Hagenmuller : J. Solid State Chem. **72** (1988) 31.

(3) C. Donnet, H. Jaffrezic, N. Moncoffre, J. Tousset, G. Fuchs presented at the Radiation Effects in Insulators Conference (REI5), Hamilton (Canada), June 1989.

(4) C. Donnet, H. Jaffrezic, G. Marest, N. Moncoffre, J. Tousset, 1st European Conference on Accelerators in Applied Research and Technology, Frankfurt (FRG), September 1989.

(5) L. Fournes, Y. Potin, J.C. Grenier, G. Demazeau, M. Pouchard : Solid State Comm. **62** (1987) 239.

(6) C. Janot, H. Gilberg : Bull. Soc. Franç. Minéral. Cristallogr. **93** (1970) 213.

(7) G. Marest, Surface and Interface Analysis **12** (1988) 509.

(8) G. Marest, C. Donnet, N. Moncoffre, J. Tousset presented at the International Conference on Applications of the Mössbauer Effect, Budapest (Hungary), September 1989.

(9) C.J. McHargue, G.C. Farlow, P.S. Sklad, C.W. White, A. Perez, N. Kornilios, G. Marest : Nucl. Instr. & Meth. in Nucl. Res. **B19/20** (1987) 813.

(10) C.J. McHargue, Private Communication.

(11) A. Perez, G. Marest, B.D. Sawicka, J.A. Sawicki, T. Tyliszezak : Phys. Rev. **28** (1983) 1227.

(12) M.J. Rossiter : J. Phys. and Chem. Solids **26** (1965) 775.

(13) P.S. Sklad, C.J. McHargue, C.W. White, G.C. Farlow : High Tech. Ceramics, edited by P. Vincenzini, (1987) 1073.

(14) D. Tréheux, D. Juvet, B. Pain, S. Ramos, to be published.

PLASMA FORMATION AND FILM GROWTH BY LASER-INDUCED DEPOSITION TECHNIQUE (LPVD)

W.Pompe[1) , H.J. Dietze[2)

The application of laser evaporation of solid targets as a plasma source for thin film deposition was coming in the centre of interest in 1988. Using this method, it was possible to prepare single crystalline superconducting films of Y-Ba-Cu-O at low substrate temperatures and in short times.
During the last ten years the LPVD was studied in various laboratories. The growing application of this technique has induced also a lot of questions on possible further developments, that means the question of further progress of appropriate lasers, of deposition at lower substrate temperature in order to open the way for the deposition of HT_c films on silicon as well as GaAs wafers, for instance. To give an answer to these questions, it seems useful to elucidate also the physical background of this new technique from the point of view of interaction of laser radiation with a solid target.

1. Plasma formation

The essence of the LPVD process, i.e. the vaporization of a solid substance by means of an incident laser beam, can be visualized by a plot temperature vs. power density in the simplest theoretical approach (Fig. 1). At very low power densities the target is only heated, without evaporation worth mentioning. With larger power density the surface temperature becomes so high that most of absorbed energy is used up in evaporation. Thus, increasing power density yields an increasing amount of vapor with only a slight increase in temperature. First such a model was developed by Krokhin in 1964 (1).
Of course, the transition point between heating up and intensive evaporation depends on the pulse time.The evaporation is different near the critical point. Then the highly dense vapor cloud explodes, that means the evaporation proceeds by ablation. A further increase of laser power density is connected with a sharper rise of the vapor temperature. It seems reasonable to assume that in the vicinity of the critical point the vaporization of the material looses its character as a thermally activated process.

---

1) Zentralinstitut für Festkörperphysik und Werkstofforschung, Helmholtzstr. 20, Dresden, 8027, DDR

2) Zentralinstitut für Isotopen- und Strahlenforschung, Permoserstr. 15, Leipzig, 7050, DDR

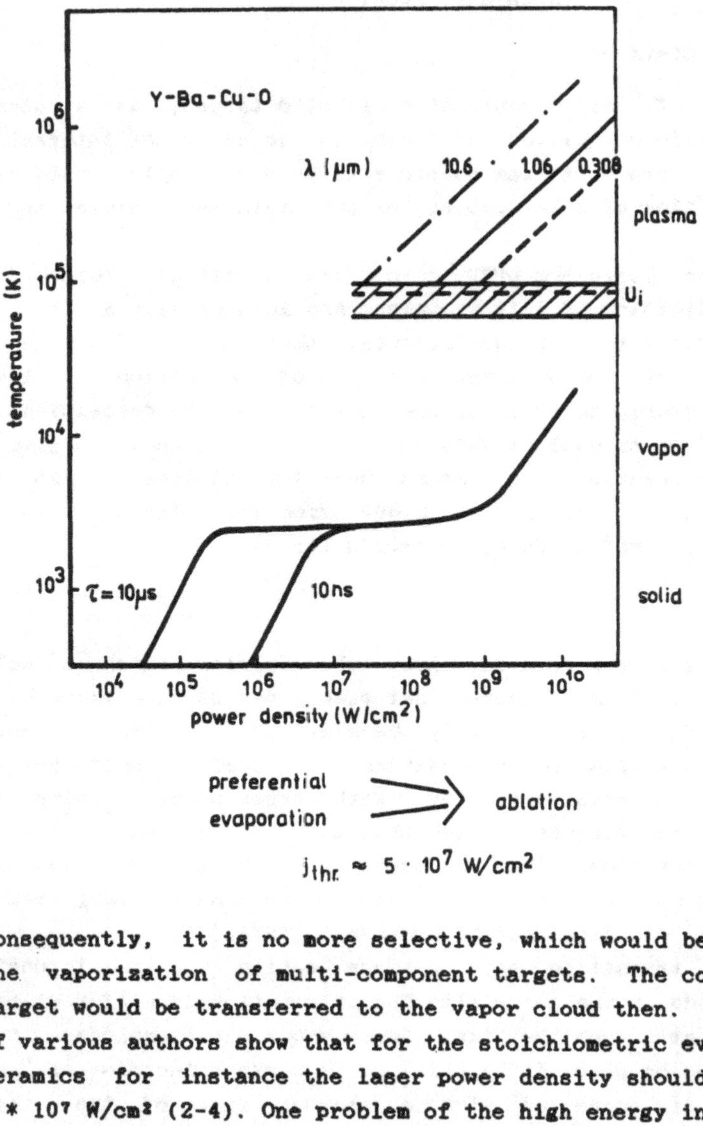

Fig. 1 Change of temperature with laser power density

Consequently, it is no more selective, which would be advantageous for the vaporization of multi-component targets. The composition of the target would be transferred to the vapor cloud then. Experimental data of various authors show that for the stoichiometric evaporation of HTc-ceramics for instance the laser power density should be larger than $5 * 10^7$ W/cm² (2-4). One problem of the high energy input in the target arises if also deeper zones of the target are heated. Then, together with the evaporation, melt instabilities or microcracks can be generated. That is the reason for the ejection of larger droplets from the target surface. According to the growth of the depth of the heat affected zone proportional $(\tau \cdot a)^{1/2}$ with a = thermal conductivity, τ = pulse duration, the droplet ejection is diminished for shorter laser pulses. With a pulse duration of 10 ns the depth is about 100 nm for Y-Ba-Cu-O ceramic. Furthermore, the penetration depth of the laser radiation can determine the zone formation. For longer penetration depths and short pulses the heat affected zone is governed chemical reactions. Applying multicomponent targets we can generate various compound clusters in this way. One interesting possibility is, for instance, the formation of clusters of carbides, oxides or nitrides with high evaporation temperatures. Then we can start with a target of a mixture of the components with lower evaporation temperatures. Also plasma chemical reactions with an additional gas can be used for film deposition. As examples the LPVD of GaAs in an As_2-atmosphere, of BN in a N_2-atmosphere, and of Y-Ba-Cu-O in an O_2-atmosphere can be mentioned. The high evaporation rate per laser pulse causes also a large deposition rate. Using a XeCl-laser with a power density of 10^9 W/cm² the thickness of the deposited film varies between tenth of the lattice constant up to few monolayers for characteristic distances (8). Due to the average flight time of particles between target and substrate the deposition rate changes between $10^{19} - 10^{22}$ $cm^{-2}s^{-1}$. Such high deposition rates offer the possibility to generate non-equilibrium conditions during the film growth. This makes it possible to realize for instance the layer growth mode of films, if at lower deposition rates an undesirable island growth dominates. In connection with the pulsed discontinuous evaporation process the LPVD yields the possibility to deposit very thin films or layered structures with high accuracy. One outstanding example is the generation of multilayer structures in the nm scale, which are interesting for X-ray mirrors. Thus W/C- or Ni/C-multilayers have been produced by LPVD. One problem arises if the vacuum pressure is not low enough. Then nondesired adsorbtion layers can grow in the time between pulses. For high quality structures the LPVD should therefore be performed in a UHV chamber.

2. Activated film growth

As we had seen, the photon flux transfers a larger amount of energy by the inverse Bremsstrahlung mechanism on the electrons and ions. At a power density of about 10^8 W/cm² the particle energy varies from a few eV to 10 eV, depending on the laser frequency. Experimentally we observe that a certain part of ions can extract more energy out of the plasma.

For a plasma with high electron density the ambipolar diffusion of light electrons and heavy ions is the reason that at larger distances from the laser spot we find an amount of ions with higher kinetic energy which move together with the fast electrons in front of the plasma cloud. Such a mechanism should be dominating at high power densities.

In the following table data of electron and ion energy of a carbon plasma are summerized. The measurements were performed with a plasma probe at larger distances (Table 1). We see the essential difference between the ion and electron energies. Furthermore the increase of the

by the absorption length. The latter phenomenon seems to be essential for the different interactions of Y-Ba-Cu-O ceramic with laser radiation. Whereas CO_2-laser irradiation is connected with the ejection of larger particles from the target, the XeCl-laser irradiation with larger absorption yields only a few small droplets. Because in the presence of larger particles there are more defects within the film the superconductive properties of such films are insufficient.
With increasing laser power density the ionization of the vapor will essentially depend on the laser frequency. For the Y-Ba-Cu-O system it can be estimated that electron collision will lead to ionization. If we assume that the well known model of laser-plasma interaction by inverse Bremsstrahlung for high power, elaborated by Krokhin (1), Puell (5) et al. can be extrapolated to this low limit case, then we get the following thresholds for growing ionization of Y-Ba-Cu-O vapor:

λ	10.6/um	1.06/um	0.308/um
$j_{ion.thr.}$ (W/cm²)	$8 * 10^6$	$8 * 10^7$	$2.5 * 10^8$

The remarkable shift of the ionization threshold with the laser frequency ($j_{ion.thr.} \sim 1/\lambda$) according to those theories seems to be a key point for the understanding of the preference of excimer lasers for deposition of stoichiometric films.
An essential assumption for this estimate is the hypothesis that the electron temperature and the ion temperature near the laser focus are the same for the given power densities. For the laser evaporation of carbon it was shown that this assumption is fulfilled very well (6).
A first summary of the evaporation scenario shows that for every material we can find an appropriate range of laser power density for intensive evaporation. This fact determines one outstanding feature of LPVD - the existence of a very high instanteneous evaporation rate which could be used also for very high deposition rates.
The high ablation rate and and the formation of a very dense vapor cloud at the first stage give rise to cluster formation. Therefore the vapor cloud can be used as an effective cluster source (7). The cluster size distribution varies with the power density in such a way that the largest clusters are formed at an intermediate power density. This is so because at low power the vapor is not dense enough for aggregation while at high power the temperature is so high that clusters would dissociate, and the expansion of the cloud is so quick that there is not much aggregation in the cooler stages. Obviously this dependence can be used if we are interested in an ion cluster beam deposition at intermediate power densities. At lower or higher power densities larger clusters are not generated, that means we work with an atomic or ion beam, respectively.
The formation of a high density plasma cloud can be used for plasma

photon-plasma interaction with decreasing laser frequency is visible. That can be understood in the model of inverse Bremsstrahlung mechanism.

Laser	λ/µm	τ/ns	Focus r/µm	j/GWcm^{-2}	$\bar{\varepsilon}_i$/eV	Te/eV	relation sp^3/sp^2	refractive index, n
CO_2 TEA	10.6	100	500	1	1000	27		2.0
Nd-Glass Nd-YAG	1.06	30	500	1	340	10	~25%	1.9
Excimer (XeCl)	0.308	30	250	1	100-150	2		1.55-1.75

Table 1 Dependence of characteristic parameters for LPVD of amorphous carbon films

Studying the evaporation of Ti and of Y-Ba-Cu-O ceramic with a Nd-pulse laser in a scanning microscope, Müller and Siemroth observed high electron temperatures and high kinetic energies of the various ions also at lower power density (9). The values are significantly larger than the data following from the inverse Bremsstrahlung model. The inspection of the target surface showed that at a laser power density of 10^8 W/cm^2 the plasma cloud interacts with the target surface by the formation of unipolar arcs. The localized power density of a single arc is in the order of 10^{10} W/cm^2. Therefore unipolar arcs emit high energetic electrons as well as ions. If we want to suppress the generation of higher energetic ions we have to avoid such instabilities. One promising way seems to be the application of excimer lasers. Then high evaporation rates are possible without strong plasma heating. A second way could be found by a gas dynamical or electrodynamical handling of the plasma beam in order to prevent higher energetic particles from the direct impingement on the film.

Particles with energies in the range of 20 to a few 100 eV can enhance various film growth processes. Various effects are discussed as for instance the change of the attachment probability, sputtering, preferential evaporation ,implantation, formation of metastable phases, anomalous diffusion, decrease of the temperature of epitaxial growth or crystallization. Besides the desired activating effect there is also a damaging effect. With particle energies larger than 100 eV sputtering leads to smaller or even negative growth rates. Thus, the question arises what is the optimum particle energy to support film growth. Recently we discussed this problem within the frame of a simple model

(10). We have estimated the influence of thermal spikes on thermally activated processes at the surface of a growing film. In thin film formation, migration and desorption of adatoms are governed by the surface diffusion coefficient D and the mean adatom stay time, respectively.
Analogous kinetic coefficients apply to cluster migration and to structural relaxations in the film. The impinging high-energetic particles give rise to a transient local heating near the surface. Therefore the surface processes are influenced by a fluctuating temperature field T_r (R, t).
The depth of the spike centre d below the surface and the random spike rate determine the random temperature at the surface. We have considered the influence of these fluctuations on an effective Arrhenius factor W.

$$W = \exp(-E/kT_{eff}) = \langle \exp(-E/k(T + T_r(R,t))) \rangle$$

The average with respect to surface position and time can be performed if the temperature field is known. The result of numerical calculations is shown in Fig. 2.

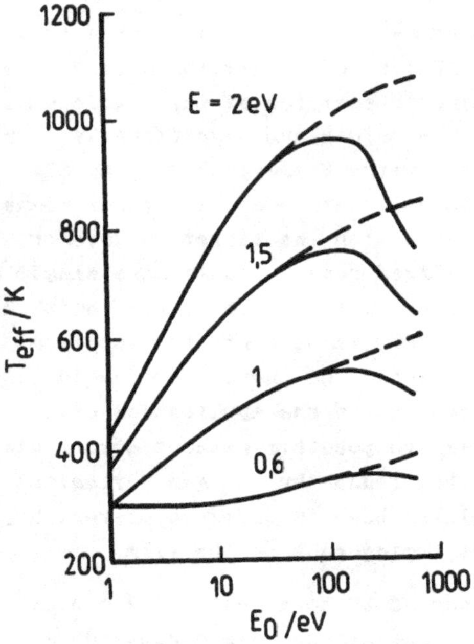

Fig. 2
Effective temperature versus energy E_0 of deposited particles for different activation energies (the dashed lines corresponds to heavy ion bombardment or small angle of incidence, $J_h = 10^{22}/m^2s$, T = 300 K)

Only rate processes with energies E > E_c are activated by spikes. The higher the particle flux, the stronger is the acceleration of rate processes. For $J_h > 10^{22}/m^2s$ ratios $E/kT_{eff} < 20$ can be achieved. The effect is especially pronounced at low substrate temperature and at particle energies between 20 .- 100 eV. Obviously the LPVD is characterized by such parameters. It is known from a number of examples that the film growth is activated by higher energetic particles during LPVD. For instance in the case of carbon our investigations have shown that the refractive index of the deposited films depends strongly on the wave length of the applied laser. Laser deposited carbon films have a remarkable content of sp^3-bonds. Together with this we find a high IR-transparency. It is assumed that the formation of these diamond like films is governed by the action of high energetic particles. Thermal spike effects as well as the selective etching of weaker bonds could be the reason for the formation of the sp^3-bonds.

Many authors had observed that in the LPVD of epitaxial films of semiconductors the substrate temperature can be lowered by 100 - 150 degrees below the value applied in the conventional thermal evaporation technique. As examples the following systems can be mentioned: Bi_2Te_3, GaAs, Ge, Si, InSb, CdTE, PbTe. The progress in this field has motivated Dubowski (1988) to introduce the term "pulsed laser evaporation and epitaxy" (PLEE) (11).

An outstanding example of the advantages of LPVD in epitaxial growth is the above mentioned deposition of highly oriented superconductor films. Starting with the LPVD of Y-Ba-Cu-O on $SrTiO_3$ at temperatures of 800 - 900°C, now first results have been published on the deposition of high quality films on silicon at a substrate temperature of 500°C without post thermal treatment.

We studied the LPVD of Y-BA-Cu-O films with a XeCl laser and of Bi-Sr-Ca-Cu-O films with a Nd-YAG laser. Whereas with the XeCl laser the transfer of the target stoichiometry to the substrate was possible, for the Nd-YAG-laser stoichiometry corrections of the target were necessary. The Y-Ba-Cu-O films were deposited on Si with a ZrO_2 buffer layer at 700° C, the Bi-Sr-Ca-Cu-O films were prepared on Si with a $SrTiO_3$ buffer layer at 450° C. Up to now a post annealing was yet necessary for the two systems. Short time heating up to 800°C and up to 830°C, respectively, followed by slow cooling in oxygen yield the desired properties. Also examples for deposition of superconducting films without buffer layer have been given. That is only possible if the substrate temperature is low enough, excluding in this way a strong interface reaction with Si. It seems to be not quite clear what is the optimum particle energy for activated film growth. Especially the Y-Ba-Cu-O films seem to be very sensitive with respect to a bombardment with higher energetic particles. The observed upper limit of the laser power density for growth of high quality films is probably caused by this fact.

3. Further development

It is obvious that the progress of LPVD is essentially connected with progress of further development of appropriate powerful pulse lasers. In conclusion of the argumentation above, lasers with a wave length smaller than 1/um seem to be especially promising for deposition with high rates as well as not too high particle energies. Lasers with larger wavelength have advantages for the generation of film structures if small scale implantation effects for the structure formation are to be applied.

In order to realize high average growth rates the repetition rate of evaporation should be increased. Progress in this way reduces the disturbing influence of a poor vacuum quality.

An other problem which has to be solved is the increase of the average depostion rate as well as the energetic efficiency of the method. We have found a possibility to increase the deposition rate by a combination of LPVD with a vacuum arc. Here we use the laser evaporation only for the repeated ignition of the vacuum arc (12). Deviating from the conventional vacuum arc technique we switch off the arc after about 2/us periodically. Thus larger droplets on the target can be avoided. The arc can switch on again by the next laser pulse. This technique was tested successfully for the deposition of Ti, C as well as Ti-C multilayers. We think that this modification of LPVD opens the way for the generation of thicker films with higher energetic efficiency.

In conclusion it can be stated that the growing knowledge on the peciularities of the laser evaporation offers further possibilities for the development of the LPVD . We think that in this way it will become a reliable novel deposition technique with increasing application during the next few years.

(1) Krokhin, O.: Laser Handbook, ed. by F. T. Arecchi, E. O. Schulz-Du Bois, North-Holland, vol. 2 (1972) 1371

(2) Roas, B., L. Schultz, G. Endres: Appl. Phys. Lett. 53 (1988)1557

(3) Gaponov, S. V., N. V. Iljin et. al.: Pisma ZETF 48 (1988) 155

(4) Dijkkamp, D., T. Venkatesan et. al.: Appl. Phys. Lett. 51 (1987) 619

(5) Puell, H.: Z. Naturforschung 25a (1970) 1807

(6) Völlmar, S., Th. Schlegel et. al.: EPM87, Physical Research, ed.by K. Hennig, Akademieverlag, vol. 8 (1988) 408

(7) Dietze, H.-J., S. Becker et. al.: ZFI-Mitteilungen vol.134 (1987), ZfI Leipzig, Acad. Scie. GDR

(8) Scheibe, H.-J., A. A. Gorbunov et. al.: Thin solid films (submitted 1989)

(9) Müller, H., P. Siemroth, K. Wetzig: EPM 89, Physical Research (paper 9.19)

(10) Bobeth, M., W. Pompe: Phys. sts. sol. (submitted 1989)

(11) Dubowski, J. J.: Chemtronics 3 (1988) 66

(12) Siemroth, P., H.-J. Scheibe et. al.: EPM89, Physical Research (paper 7.16)

RANGE PROFILE CALCULATIONS BY DIRECT NUMERICAL SOLUTION OF LINEARIZED BOLTZMANN TRANSPORT EQUATIONS

M. Posselt[1)]

1. Introduction

In the framework of the linear cascade theory (1) two alternative procedures are employed to describe the ballistic processes during ion implantation: (i) "static" BCA Monte Carlo simulations (e.g. TRIM (2), ACAT (3) etc.) and (ii) linearized Boltzmann transport equations (LBTE). The LBTE are linear partial integro-differential equations for the statistical distribution functions of the particles moving in ion-beam-induced collision cascades. The method of the LBTE is not so universal and versatile as the Monte Carlo simulations. But it yields ion range distributions, recoil range profiles etc. by the solution of equations, not by statistical experiments. Therefore, in many applications the computing time is essentially smaller than in the case of Monte Carlo simulations.

In the last decade several authors proposed methods for a direct calculation of the profiles by the LBTE. In contrast to the well-known LSS(WSS) theory (4)(5) which yields the statistical moments of the range and damage distributions these procedures can be applied to multi-layer targets frequently used in modern ion beam technologies. The most of the works consider a one-dimensional geometry.

In section 2 of the present work we elucidate the general form of the one-dimensional LBTE. An overview on the methods of solution known from the literature is given in section 3. In section 4 we develop a new efficient method to calculate the depth distributions of implanted ions and recoiled targets atoms by a direct numerical solution of the LBTE. We apply this procedure to characteristic examples of ion implantation into homogeneous and layered targets. The results are compared with profiles obtained from TRIM Monte Carlo simulations (section 5).

2. Linearized Boltzmann transport equations (LBTE)

In the case of a one-dimensional planar geometry with normal incident ions the statistical distribution function of the particles only depends on three variables: (i) the depth coordinate x, (ii) the energy E and (iii) the cosine of the angle θ between the direction of motion of the particles and the x-axis, $\eta = \cos\theta$. For the scalar flux $f_I(x,E,\eta)$ of the ions the LBTE is given by

$$\eta(\partial f_I/\partial x)= I_I(f_I) = I_{In1}(f_I)+I_{In2}(f_I)+I_{Ie}(f_I) \qquad (1)$$

1) Academy of Sciences of the GDR, Central Institute for Nuclear Research, Rossendorf, PF 19, DDR-8051 Dresden

I_I consists of three parts. The first part describes the elastic atomic scattering for energy transfers T above a certain threshold T_{min}

$$I_{In1}(f_I) = N \int_{T'>T_{min},\ E+T'=E'} d\sigma_{TI}(E'\to E,\eta'\to\eta)\ \ f_I(x,E',\eta')$$
$$- N \int_{T''>T_{min},\ E=E''+T''} d\sigma_{TI}(E\to E'',\eta\to\eta'')\ \ f_I(x,E,\eta) \qquad (2)$$

N and $d\sigma_{TI}$ are the particle density of the target atoms and the differential scattering cross-section for the collisions between the incident ions and the target atoms, respectively. For energy transfers below T_{min} the change of the direction of motion of the projectiles can be neglected. In this case we obtain

$$I_{In2}(f_I)=(\partial/\partial E)[N\ S_n(T_{min},E)\ f_I] \quad \text{with} \quad S_n(T_{min},E)=\int_{T'<T_{min},E+T'=E'} d\sigma_{TI}(E'\to E)\cdot T' \qquad (3)$$

Because of the divergence of $d\sigma_{TI}$ for energy transfers T -> 0 the nuclear scattering must be described by the two terms I_{In1} and I_{In2} with the cut-off T_{min}. The electronic stopping acts as a continuous braking force. $I_{Ie}(f_I)$ is given by

$$I_{Ie}(f_I) = (\partial/\partial E)[N\ S_e(E)\ f_I] \qquad (4)$$

S_e is the electronic stopping cross-section. The scalar flux f_I must obey the following boundary condition

$$\eta \geq 0:\ \ f_I(0,E,\eta)\sim\delta(\eta-1)\,\delta(E-E_0),\ \ \eta<0:\ \ f_I(\infty,E,\eta)=0 \qquad (5)$$

E_0 is the incidence energy of the ions. δ denotes the Dirac's delta function. For the scalar flux of the target atoms $f_T(x,E,\eta)$ we obtain the LBTE

$$\eta(\partial f_T/\partial x)=I_{IT}(f_I,f_T) \quad \text{with} \quad f_T(0,E,\eta)=0\ (\eta\geq 0),\ f_T(\infty,E,\eta)=0\ (\eta<0) \qquad (6)$$

I_{IT} is an operator similar to I_I. The part of I_{IT} which contains the scalar flux of the incident ions f_I corresponds to a source term. Using the solutions f_I and f_T of the LBTE we can determine the ion range distribution $F_I(x)$ and the recoil range profile $F_T(x)$ by

$$F_I(x) = \lim_{E\to E_C} \left(\int d\eta\ f_I(x,E,\eta)\right)/\left(\int dx\ \ d\eta\ f_I(x,E,\eta)\right)$$
$$F_T(x) = \lim_{E\to E_C} \left(\int d\eta\ f_T(x,E,\eta)\ (m_T/m_I)^{1/2}\right)/\left(\int dx\ \ d\eta\ f_I(x,E,\eta)\right) \qquad (7)$$

Below the cut-off energy E_C we assume the particles to be stopped. m_I and m_T are the masses of the ion and the target atom, respectively. Equations (1)-(7) can be easily generalized to multi-component and multi-layer targets. In the latter case the particle density and the differential scattering cross-section - as a function of the target atomic number and mass - show discontinuities at the interface between the two layers.

3. Methods of solution

In the last decade several methods for solving the one-dimensional LBTE have been proposed. In general these are either special approximation procedures which are employed to treat the LBTE in particular cases or numerical calculation schemes derived on a more or less intuitive way.

The bipartition model of Luo et.al.(6) stems from the electron transport theory. It is only applicable to the implantation of light ions into heavier targets. Luo et.al. calculated range profiles and reflection coefficients. If the energy transfer per collision and the probability of large angle deflections are small the pertubation expansion of Bardos (7)(8) can be employed. Bardos determined the range distributions for implantation energies above the maximum of the electronic stopping cross-section by a first order approximation. It can be shown analytically that the first order approximation of Bardos is rather similar to the single scattering approximation of Biersack (9) which have been recently applied to the calculation of range profiles for the MeV implantation. The method of Bardos was also used in the case of the implantation of light ions at lower energies (10). Here, the pertubation expansion is not valid in the whole depth region but only in certain depth intervals.

The one-dimensional LBTE is also employed to describe the sputtering of target atoms. Using an expansion with respect to Legendre polynomials Urbassek and Waldeer (11)(12) determined the energy and directional distributions of the sputtered particles. Such an expansion is justified since the sputtered particles are mainly low energy recoils with a nearly isotropic directional distribution.

Lizunov et.al.(13)(14)(15) applied methods of the neutron transport theory. Their procedure is probably equivalent to a direct numerical solution of the one-dimensional LBTE. Lizunov et.al. determined ion range distributions and range profiles of recoiled target atoms. Most of the examples considered dealt with problems of ion-solid interactions in nuclear physics.

Christel, Gibbons, Giles and Mylroie (16)(17)(18)(19) and Ashworth, Moulavi-Kakhi et.al.(20)(21) developed numerical schemes on a three-dimensional point grid formed by the variables x, E, η . Both procedures are not consequently derived from the one-dimensional LBTE. But they seem to be rather equivalent to a direct numerical solution of the LBTE. The main differences between the calculation schemes and a direct numerical solution concern (i) the treatment of the continuous energy loss described by the stopping cross-sections S_e and $S_n(T_{min},E)$ and (ii) the consideration of the angular scattering. The algorithms of Christel et.al. and Ashworth et.al. were successfully employed to determine depth profiles of implanted ions in many homogeneous and layered targets. The method of Christel et.al. yields not only ion range distributions but also the profiles of the recoil implantation. Modifications of the procedure of Christel et.al. concerning the treatment of the angular scattering and details of the numerical scheme were introduced by Burenkov et.al.(22)(23).

4. Direct numerical solution of the LBTE

Our aim was to develop a new method of solution which satisfies the following conditions. (i) It is based on a consequent derivation from the LBTE. (ii) The method is not limited to a special class of ion-target combinations or to a certain range of implantation energies. (iii) The procedure is applicable to multi-component and multi-layer targets. (iv) The method yields ion range distributions and profiles of recoil implantation. (v) The numerical solution of the LBTE for the scalar fluxes of the incident ions and the target recoils is performed directly using an explicit iteration scheme with finite differences.

In the following we want to discuss some details of the procedure. Let us begin with the LBTE (1) for the scalar flux f_I of the incident ions. As it is well known the differential scattering cross-section $d\sigma_{IT}$ in the first integral of (2) ("gain" term) can expressed by

$$d\sigma_{IT}(E'\to E,\eta'\to\eta) = d\sigma_{IT}(E'\to E)\,\delta(e'e - \cos\vartheta(E',E))\,d\eta'\,d\varphi'/4\pi \qquad (8)$$

The differential cross-section $d\sigma_{IT}(E'\to E)$ is given by

$$d\sigma_{IT}(E'\to E) = \pi a_s^2\, f(t'^{1/2})/(2t'^{3/2}),\quad t' = (C_E^2\, E'\, T')/\gamma,\quad E' = E+T' \qquad (9)$$

with

$$C_E = a_s m_T/(Z_I\, Z_T\, e^2\, (m_I+m_T)) \qquad \text{and} \qquad \gamma = 4\, m_I\, m_T/(m_I+m_T)^2 \qquad (10)$$

Z_I and Z_T are the atomic numbers of the projectile and the target, respectively. In dependence on the special interatomic potential different relations for the screening length a_s and the scattering function f are known. In the delta function of equation (8) e' and e are vectors characterizing the directions of motion of the projectile before and after collision, respectively. The scalar product $e'e$ must be equal to the cosine of the laboratory scattering angle given by

$$\cos\vartheta(E,E')=((m_I+m_T)/(2\,m_I))\,(E/E')^{1/2} + ((m_I-m_T)/(2\,m_I))\,(E'/E)^{1/2} \qquad (11)$$

The product $e'e$ does not only depend on the direction cosines η' of e' and η of e with respect to the x-axis but also on the azimuthal angle φ'. Since in the first integral of $I_{In1}(f_I)$ the scalar flux depends on η' but not on φ' one have to integrate the delta function with respect to φ'. We have succeeded in performing this integration on an analytical way by the solution of a problem of the spherical trigonometry. The result of this somewhat lengthy calculation is (24)

$$a(E,\eta,E',\eta') = \int_0^{2\pi} \delta(e'e - \cos\vartheta(E',E))\,d\varphi'/2\pi =$$

$$\frac{\Theta[\sin\vartheta\,(1-\eta'^2)^{1/2} + \eta - \eta'\cos\vartheta]\;\Theta[\sin\vartheta\,(1-\eta'^2)^{1/2} - \eta + \eta'\cos\vartheta]}{\pi\,[\sin^2\vartheta\,(1-\eta'^2) - (\eta - \eta'\cos\vartheta)^2]^{1/2}} \qquad (12)$$

Θ denotes the Heavyside s theta function. In the second integral of $I_{In1}(f_I)$ ("loss" term) the scalar flux is independent of the integration variables. Therefore, the integration of the corresponding delta function with respect to η'' and φ'' yields a constant which is equal to one in our notation. Finally, we obtain the following expression for $I_{In1}(f_I)$:

$$(C_X\, C_E^2/\gamma)\int_{T_{min}}^{\min[(\gamma/(1-\gamma))E,\,E_0-E]} dT'(2T'+E)\, f(t'^{1/2})/(2t'^{3/2}) \int_{-1}^{1} d\eta'\, a(E,\eta,E+T',\eta')\, f_I(x,E+T',\eta')$$

$$-(C_X\, C_E^2\, E/\gamma)\int_{T_{min}}^{\gamma E} dT''\, f(t''^{1/2})/(2t''^{3/2})\, f_I(x,E,\eta) \tag{13}$$

with $C_X = \pi a_s^2 \gamma N$.

The expression $I_{In2}(f_I)$ can be transformed into

$$I_{In2}(f_I) = (C_X/C_E)\, (\partial/\partial E)[(T_{min}/\gamma E)^{1/2}\ S_n(C_E(ET_{min}/\gamma)^{1/2})\, f_I(x,E,\eta)] \tag{14}$$

The function S_n is a reduced (dimensionless) nuclear stopping cross-section (2).

Similar to other methods for solving the LBTE we only consider the motion of particles with $\eta \geq 0$ ("forward" direction). Particles with $\eta < 0$ are assumed to be stopped. This approximation is justified in many cases. For a detailed discussion of this problem we refer to (1).

With the above approximation and $I_{In1}(f_I)$ from (13), $I_{In2}(f_I)$ from (14) and $I_{Ie}(f_I)$ from (4) we obtain a LBTE which further numerical treatment is evident.

The LBTE (6) for the scalar flux of the target recoils can be treated in a similar manner. Additionally, in $I_{ITn1}(f_I, f_T)$ we have to consider source terms which describe the generation of moving target atoms by the energy transfer from incident ions. The generalization to multi-component targets leads to further source terms characterizing the generation of target recoils of a given sort by moving target atoms of another type.

We have developed the computer program BOTE (BOltzmann Transport Equations) using an explicit iteration scheme with finite differences $\Delta x, \Delta E, \Delta\eta$. Details of this numerical procedure will be outlined in a forthcoming paper.

5. Results and discussion

Figs. 1, 2, 3 and 4 show ion range profiles obtained from the direct numerical solution of the LBTE by the program BOTE (curves) and from TRIM Monte Carlo simulations (histograms). For both procedures we have used similar physical inputs. In particular, the nuclear atomic scattering is described by the "universal" potential (2) and the electronic energy loss is given by the combination of the Lindhard-Scharff and the Bethe-Bloch formulae (2). The agreement of the range distributions calculated by BOTE with the histograms simulated by TRIM is good. In fig. 1 the range profile obtained by the algorithm of Christel et.al. is also shown. The input parameters have been chosen to be similar to those of BOTE and TRIM. The resulting range distribution agrees well with the profiles obtained by the other methods. In figs. 1 to 4 some experimental data from the literature are shown for comparison (25)(26)(27)(28). They are results of SIMS and RBS measurements. Figure 5 illustrates an example of recoil implantation. The agreement between the curves obtained by BOTE with the histograms calculated by TRIM and with the experimental data (29) is satisfactory. In all examples discussed above the program BOTE is about 10 to 100 times faster than corresponding TRIM simulations with 10,000 particle histories.

6. Conclusions

We have developed a new effective and general method to calculate ion range distributions and profiles of recoiled target atoms by direct numerical solution of the one-dimensional LBTE. Our method has two advantages: (i) It is based on a consequent derivation from the LBTE. (ii) It is not limited to special applications. The procedure is more than one order of magnitude faster than TRIM Monte Carlo simulations. A more detailed description of the numerical scheme and some more results will be given elsewhere.

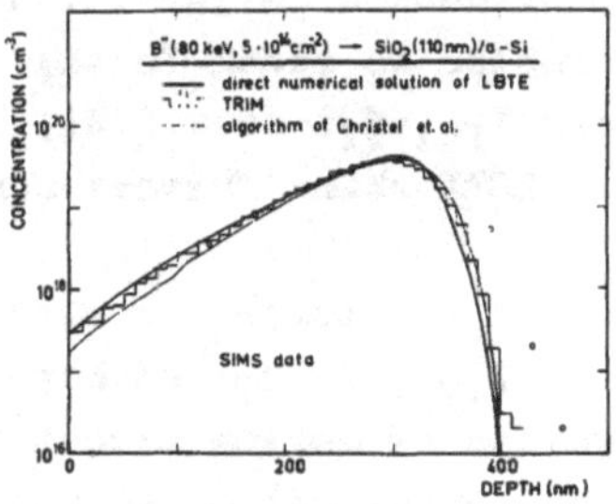

Fig. 1

Range profiles for 80 keV B^+ implantation into SiO_2/Si. Experimental data from (25).

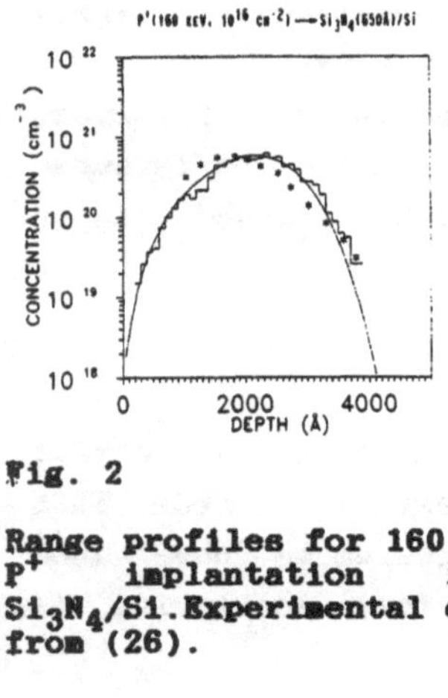

Fig. 2

Range profiles for 160 keV P^+ implantation into Si_3N_4/Si. Experimental data from (26).

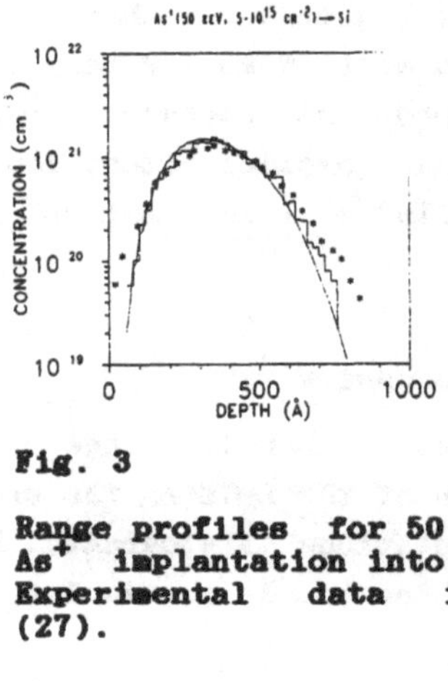

Fig. 3

Range profiles for 50 keV As^+ implantation into Si. Experimental data from (27).

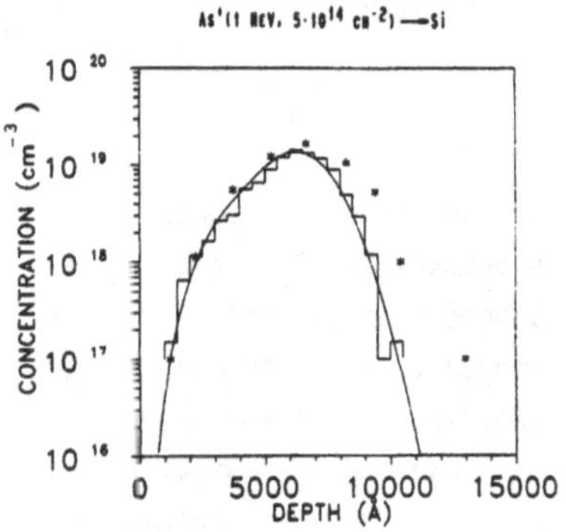

Fig. 4

Range profiles for 1 MeV As^+ implantation into Si. Experimental data from (28).

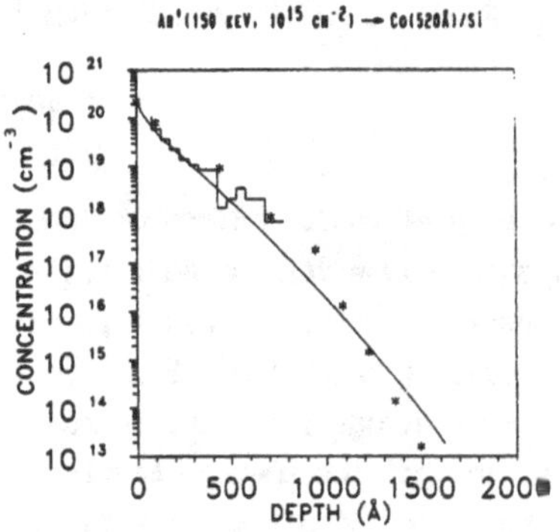

Fig. 5

Recoil range profiles (Co in Si) for 150 keV Ar^+ implantation into Co/Si. Exp. data from (29).

(1) Posselt, M.: phys. stat. sol. (a) 112 (1989) 263.
(2) Ziegler,J.F., J.P. Biersack and U. Littmark: The Stopping and Range of Ions in Solids, Pergamon Press, 1985.
(3) Takeuchi,W. and Y. Yamamura: Rad. Eff. 71 (1983) 53.
(4) Lindhard,J., M. Scharff and H.E. Schiott: Kong. Danske Vid. Selsk., mat-fys. Medd. 33 (1963) no.14.
(5) Winterbon,K.B., P. Sigmund and J.B. Sanders: Kong. Danske Vid. Selsk., mat.-fys. Medd. 37 (1970) no 14.
(6) Luo Zengming and Wang Shiming: Phys. Rev. B36 (1987) 1885.
(7) Bardos G.: Phys. Lett. A119 (1987) 415.
(8) Bardos G.: Rad. Eff. 105 (1988) 191.
(9) Biersack, J.P.: Nucl. Instr. and Meth. B35 (1988) 205.
(10) Bardos, G.: phys. stat. sol. (a) 112 (1989) 311.
(11) Urbassek, M.: Nucl. Instr. and Meth. B4 (1984) 356.
(12) Waldeer,K.T. and H. M. Urbassek: Nucl. Instr. and Meth. B18 (1987) 518.
(13) Lizunov,Yu.D. and A.I. Ryazanov: Rad. Eff. 60 (1982) 95.
(14) Lizunov,Yu.D. and A.I. Ryazanov: Poverchnost 1987, no.5, 121 (in Russian).
(15) Lizunov,Yu.D. and A.I. Ryazanov: Rad. Eff. 107 (1989) 185.
(16) Christel,L.A., J.F. Gibbons and S.W. Mylroie: J. Appl.Phys. 51 (1980) 6176.
(17) Christel,L.A., J.F. Gibbons and S.W. Mylroie: Nucl. Instr. and Meth. 182/183 (1981) 187.
(18) Christel,L.A. and J. F. Gibbons: J. Appl. Phys. 52 (1981) 5050.
(19) Giles,M.D. and J.F. Gibbons: Nucl. Instr. and Meth. 209/210 (1983)33.
(20) Ashworth,D.G., M. Moulavi-Kakhi and Kranti V. Anand: J. Phys. C17 (1984) 2449.
(21) Moulavi-Kakhi,M. and D. G. Ashworth: J. Phys. C18 (1985) 1135.
(22) Burenkov,A.F., F.F. Komarov, M.M. Temkin and G. Schlotzhauer: Rad. Eff. Lett. 86 (1983) 161.
(23) Burenkov,A.F., F.F. Komarov and M.M. Temkin: Poverchnost 1989, no.4, 15 (in Russian).
(24) Posselt, M.: Annual Report 1986 ZfK-621 (1986) 63.
(25) Posselt, M. and T. Feudel: this volume.
(26) Hirao,T., K. Inoue, S. Takayanagi and Y. Yaegashi, J. Appl. Phys. 50 (1979) 193.
(27) Nakata,J. and K. Kajiyama: Jap. J. Appl. Phys. 21 (1982) 1363.
(28) Tamura,M., N. Natsuaki, Y. Wada and E. Mitani: Nucl. Instr. and Meth. B21 (1987) 438.
(29) Kwok, H.L.: Solid State Phenomena 1/2 (1988) 195.

PHASE FORMATION BY ION BEAM MIXING OF METAL SYSTEMS

B. Rauschenbach*)

Ion beam mixing is an effective method of forming nonequilibrium phases and equilibrium compounds, i.e. a technique for the modification of surface properties. In a typical experiment, a composite target of a thin film of one material deposited on another will be bombarded with energetic heavy ions. Each ion slows down through a series of collisions with the target atoms alonge its path. In most ion mixing experiments the ion beam parameters are chosen so that the energy deposited in displacement has the maximum at the interface /1/. The degree of mixing, the final phase formed, depend on the relative importance of the contributing mechanisms. Experimental results indicate the strong role both of collisional effects and of thermochemical effects /2/.
In this paper we review recent observations of phase formation by ion beam mixing in metal-metal systems with emphasis on binary metal systems that exhibit distinct differences in thermochemical and ballistic behaviour.

1. Mixing mechanisms

Following a brief introduction to the mixing mechanisms is given from the standpoint of the ion induced phase formation. Several atomic mixing mechnisms are discussed. Basically, the mixing processes can be divided into collisional and thermochemical effects.
(i) collisional regime
In this regime, two processes can induce mixing, the recoil implantation and the cascade mixing. Both these processes are athermal and the mixing which results does not therefore depend on equilibrium thermodynamic considerations. Incoming ion with an initial kinetic energy of a few 100 keV travels through the target and transfers its energy in the first 10^{-13} s. The primary knock-on atom (PKA) starts a random walk, which may be described as a thermal-like process. The PKA may induce secondary collision and, if its energy is high enough, displacement cascades. The energy transferred to the PKA is about 1 keV and in the displacement cascade up to 100 eV, whereas the binding energy is about 1 eV and the energy transfer is finished after about 10^{-11} s (1000 K corresponds to a thermal energy of about 0.1 eV) /3/. This indicates that ballistic processes are very far off thermodynamical mechanisms.
(ii) thermochemical regime
The phase formation which result in this regime depend on many factors including the nature, distribution and mobility of radiation induced defects, the diffusion coefficients of the various elemental constituents, chemical driving forces and the availability of trapping sites such as free surfaces and bulk defects at which solute atoms can segregate /4/. According to the laws of thermodynamics a system tends towards an equilibrium state with a minimum of the Gibbs free energy, G=H-TS. The equilibrium diagram discribes the equilibrium states for each phase. This diagram is directly related to the free energy vs

----

*)Central Institute for Nuclear Research Rossendorf
GDR-8051 Dresden, P.O. Box 19

concentration diagram, (G,x)-diagram, which gives the free energy G at a given temperature for various phases as a function of the mole fraction x and can be used for predicting whether metastable phase will occur /5/.

2. Amorphous phase : Fe-Zr

We use the Fe-Zr system as an illustration. The phase diagram for this system is characterized by small ranges of solubility and complex crystal structures. Such systems are prime candidates for transformation to an amorphous phase and also the short period of the cooling time favor polymorphic transformation into amorphous structures. RBS spectra taken from an as-evaporated Fe-Zr multilayered film and after mixing with $1*10^{16}$ Xe^{+}-ions/cm^2 at room temperture (RT) are shown in Fig. 1. Spectra showing the dis-

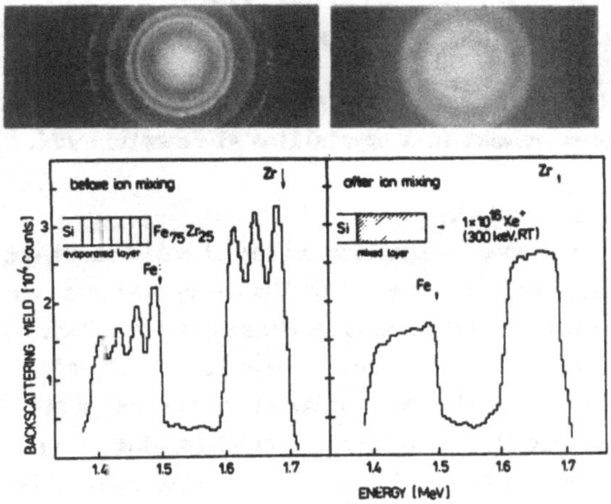

Fig.1:
RBS spectra from a Fe-Zr multilayer ($Fe_{75}Zr_{25}$) as-evaporated and after mixing with $1*10^{16}$ Xe^{+}-ions/cm^2 at RT and the corresponding electron diffraction patterns.

crete layered structure was smeared out. Electron diffraction patterns of uniformly mixed Fe-Zr layer demonstrate the transition of crystalline into the amorphous structure /6/. This result is in good coincidence with the "structure-difference" rule by LIU et al. /7/. It has been suggested that an amorphous binary alloy will be formed by mixing of multilayers in which the two constituent metals are of different crystalline structure when the composition is fixed near a deep eutectic. Our measurements and studies by BØTTINGER et al. /8/ demonstrate that the glass-forming range is always smaller with increased temperature. In the low temperature regime, the formation of a crystalline final state is suppressed by kinetic restictions, because those state require longer nucleation and growth times.

3. Quasicrystalline phase : Al-Fe

An amorphous phase can be processed by subsequent thermal anneals to form an other metastable structure. A specific example is the transition into quasicrystalline state. Multilayered Al-Fe samples ($Al_{80}Fe_{20}$) were mixed at RT to form the amorphous phase. The amorphous layer was than heated slowly ($<$ 5 °C/min) and annealed at about 200 °C for some minutes (5...30 min). Fig. 2 shows THEED patterns (a,b) taken from large grains (c) after mixing and annealing at 230 °C for 25 min. The diffraction patterns indicate

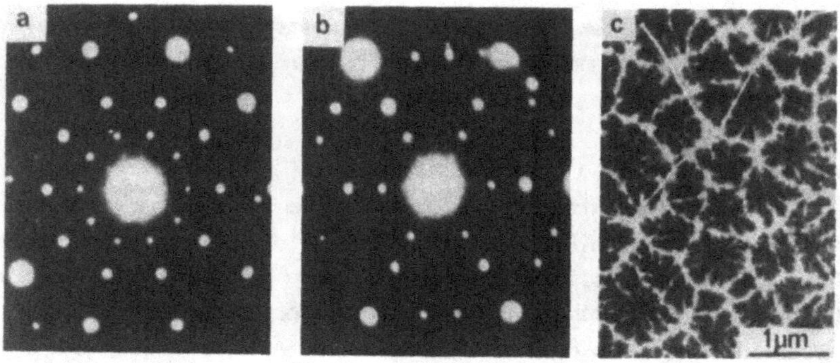

Fig. 2: Selected area diffraction patterns (a,b) and a HVEM micrograph (c) taken from mixed and annealed Al-Fe multilayered system. Electron diffraction patterns reveal the five-fold (a) and three-fold (b) symmetry.

the existence of a quasicrystalline phase dispersed in a crystalline Al-Fe matrix /9/.

4. Solid solution and ordered compound : Cu-Au, Cu-Ag

The equilibrium phase diagram of Cu-Au system shows a complete series of solid solutions and several ordered compounds. In contrast, the phase diagram of the Cu-Ag system shows that in thermal equilibrium there is vanishingly small mutual solid solubility of the components Ag and Cu at RT. All metals have f.c.c. structure. We have found that uniformly mixed Cu-Au layers were obtained by mixing with Xe^+- or Ar^+-ions and that a pronounced increase of the grain size is a result of ion bombardment /10/. For the Cu-Ag system, electron diffraction studies show an extension of the terminal solid solubility by ion mixing. According to PEINER and KOPITZKI /11/, a strong correlation between the ion induced solid solubility and the heat of formation ΔH_f is obvious, independ on miscibility in the solid state. The solibility decreases with increasing ΔH_f. Therefore, a complete solubility of Cu-Au ($\Delta H_f \approx -10$ kJ/mol) and a remarkable solubility was also observed for the Cu-Ag system by ion mixing although $\Delta H_f > 0$, ($\Delta H_f \approx +40$ kJ/mol).

Also the possibility of producing ordered compounds was investigated in the Cu-Au system. When mixed films with an initial composition of $Cu_{50}Au_{50}$ were annealed at a temperature below 350 ^oC, there appeared cross-like diffuse spots at some positions of superlattice reflection in the diffraction pattern, as seen in Fig. 3.

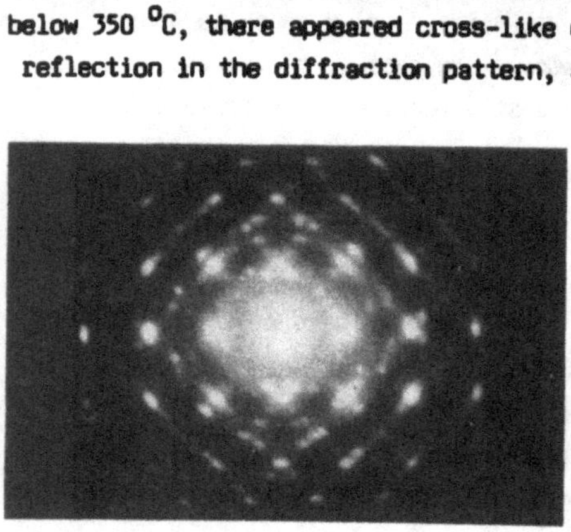

Fig. 3: The $L1_0$-superstructure by ion mixing of Cu-Au thin film. The electron diffraction pattern shows some positions of superlattice reflections (in general, a normal reflection point is surrounded by two up to four superlattice reflection points).

This figure shows an electron diffraction pattern of a $L1_0$-type ordered alloy CuAu I after annealing for 10 min at 250 °C /10/. According to HUNG et al. /12/, the formation of ordered alloys is favoured against the foramtion of solid solution at high temperatures. Likely, very small clusters are formed during ion mixing which are increased at post-annealing. The transformation to CuAu I is accompanied by twinning in order to relieve the tetragonality. Therefore loop formation or misfit dislocations were observed (see Ref. /10/).

5. Metastable crystalline phase: Fe-Cr, Al-Fe

Ion beam mixing is considered as a rapid quench process. If the cooling time is sufficient for precipitates to nucleate, metastable crystalline phase formation may be achieved. The time required for nucleation is strongly influenced by crystalline structure and composition of mixed layer. The preferred crystal structures are simple crystalline structures with only two atoms per unit cell, e.g. A2-type phases (Im3m, W-type) or B2-type phases (Pm3m, CsCl-type) /2/. Formation of the FeAl-phase (B2-type) /13/ and the FeCr-phase (A2-type) after mixing by high-energetic noble gas ions are two examples. Both phases are ordered cubic phases. Multilayered Fe-Cr or Al-Fe samples were mixed. When the ion mixing was performed at temperatures between 200 and 400 °C, metastable crystalline phases with simple cubic structure were formed due to thermally activated atomic diffusion at the elevated substrate temperature in the presence of radiation induced defects. An example is shown in Fig. 4. In this case

Fig. 4:
HVEM-micrograph of multilayered Fe-Cr film after mixing with $1*10^{16}$ Xe^+-ions/cm^2 at 250 °C.

Fe-Cr multilayered sample was irradiated with Xe^+-ions at 250 °C. This relatively low temperature is sufficient for the atoms to move toward "metastable" equilibrium positions. The transmission electron micrograph shows mixing induced FeCr needle-like precipitates (note the preferred orientation of the FeCr-precipitates). Formation of complex structures require higher atomic mobility, so that such structures are inhibited.

6. Complex phase : Y-Ba-Cu

The formation of complex crystalline phases with large unit cells are suppressed by kinetic restrictions. Only when the temperature of mixed system is sufficiently high and the transport over long distances is possible, phases expected from the equilibrium phase diagram form within the mixed layer. An example is the formation of the Y-Ba-Cu oxide high-superconducting phase with a complex structure (details see Ref. /14/ and this Proceedings).

Polycrystalline superconducting thin films were prepared by vapor deposition of alternate layers of Y, Ba, and Cu, followed by O^+- or Xe^+-ion beam induced mixing and short-time post-annealing at temperatures up to 800 ^oC. The samples were completely superconducting between 79 and 82 K with onsets as high as 110 K. Electron diffraction studies show the superconducting phase to crystallize in a orthorhombic structure. The microstructure is dominated by lamellar domains. These were indentified as twins, does terminate O-Cu-O in the inter-Ba Cu-O planes.

7. What are the metallurgical phases induced by ion beam mixing ?

In absence of delayed effects (i.e. radiation enhanced diffusion) and low temperature regime, the mobility of atoms during mixing is very small. In general, such quench processes leads to metastable phases. For predictions of formed phases the (G,x)-diagram should be known /5/. Fig. 5 represents a binary equilibrium phase diagram at constant pressure for a system A,B, which forms a single intermetallic γ. The solid solubility limits for the terminal solid solutions α and β are given. Also, two free energy-composition diagrams for the temperatures T_1 and T_2 are illustrated in Fig. 5. The

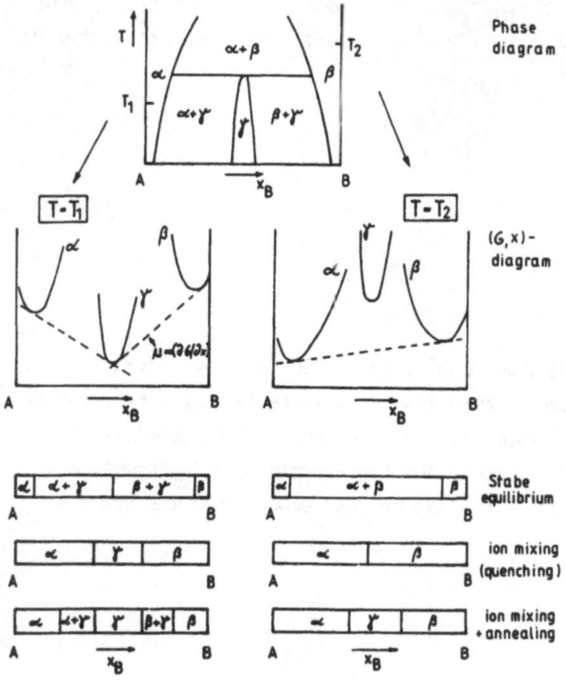

Fig. 5: Equilibrium phase diagram and (G,x)-diagram at T_1 and at T_2 for the system A-B and the ranges of existence of phases (bars) in the state of stable equilibrium, after mixing and after mixing and subsequent annealing.

shape of the curves in the (G,x)-diagram for the various phases depends on the temperature, G=G(T). It can be seen that the common tangent of the (G,x)-curves must be drawn to find the Gibbs free energy of a mixture in the range between $x_\alpha < x_B < x_\gamma$ and $x_\gamma < x_B < x_\beta$ or $x_\alpha < x_B < x_\beta$. However, of all the possible lines that connect points on the curves, the common tangent to the curves at the fraction x_α, x_β and x_γ represents the lowest Gibbs free energy is possible for intermediate fractions.
In the case of slow cooling from T_2 (state of stable equilibrium, first bar on the right side of Fig. 5), on either side of the two-phase region $\alpha + \beta$ there are single-phase regions α and β. The structure γ is not formed, because the (G,x)-curve lies above the common tangents of α and β. The middle bar in Fig. 5 shows the situation after ion mixing (quenching) at T_2. The boundary of regions α and β is defined by the point where the corresponding curves in the (G,x)-diagram intersect. The γ-phase is only formed by subsequent annealing.
The free energy diagram together with kinetic restictions during relaxation following cascade events should allow one to predict the preferred final states arising following ion mixing of metal-metal systems /15/.

References

/1/ R.S.Averback: Nucl. Instr. Meth. B 15 (1986) 675

/2/ L.S.Hung and J.W.Mayer: Nucl. Instr. Meth. B 7/8 (1985) 675

/3/ L.E.Rehn and P.R.Okomota: Nucl. Instr. Meth. B 39 (1989)

/4/ R.Kelly: this Proceedings

/5/ N.Saunders and A.P.Miodowink: CALPHAD 9 (1985) 289

/6/ B.Rauschenbach and V.Heera: in Ion Implantation (Ed. F.H.Wöhlbier), Trans Tech Publ., Aedermannsdorf 1988, p.143

/7/ B.X.Liu, W.L.Johnson, M.-A. Nicolet and S.S.Lau: Nucl. Instr. Meth. 209/210 (1983) 299

/8/ J.Bøttinger, K.Dyrbye, K.Pampus and R.Poulsen: Phil. Mag., to be published

/9/ B.Rauschenbach and V.Heera: J.Mater. Sci. Lett. 6 (1987) 401 and J. Less-Common Met. 145 (1988) 445

/10/ B.Rauschenbach: Appl. Phys. A 40 (1986) 47

/11/ E.Peiner and K.Kopitzki: Nucl. Instr. Meth. B 34 (1988) 173

/12/ L.S. Hung, M.Nastasi and J.W.Mayer, J. Appl. Phys. 56 (1984) 1420

/13/ B.Rauschenbach and K.Hohmuth: Nucl. Instr. Meth B 23 (1987) 323

/14/ B.Rauschenbach and K.Hohmuth: Z. Phys. B - Condensed Matter 74 (1989) 155

/15/ W.L.Johnson: Report SAND -83 -1230 (1983)

RAPID PHASE TRANSITIONS IN SILICON UNDER PULSED-LASER IRRADIATION

W.C. Sinke, A. Polman and P.A. Stolk*)

ABSTRACT

Different pulsed laser induced phase transformations in silicon are discussed and compared, with emphasis on random and epitaxial explosive crystallization of ion-implanted amorphous Si and on formation of amorphous Si from the melt.

I. INTRODUCTION

Rapid phase transitions in Si induced by pulsed-laser irradiation have been studied for almost 15 years now. In the early years of research on this subject there was a lively scientific discussion concerning the nature of the phase transformations involved. The question was roughly whether they are purely thermal in nature or influenced by the laser-generated electrons and holes. The present opinion is that most processes can be described well by simple thermal models, although there are indications that in a limited number of cases non-thermal effects may play a role (see for a review reference [1]). Nevertheless laser-induced phase transformations are still a subject of fundamental interest. The use of nano- or picosecond laser irradiation offers the possibility to study melting and solidification as well as solid-phase crystallization under conditions far from thermodynamic equilibrium. As yet there are no theories available to predict the behaviour of Si under these circumstances in an accurate way. This is both true for nucleation and for growth of phases.

In this contribution we will discuss different pulsed laser induced phase transformations in ion-implanted amorphized Si and single-crystal Si, leading to the formation of single-crystal Si (c-Si), large-grain polycrystalline Si (lgp-Si), fine-grain polycrystalline Si (fgp-Si) or amorphous Si (a-Si).

- First, epitaxial regrowth from the melt. This process can be used to remove ion-implantation damage in c-Si (pulsed-laser annealing). By complete melting of the damaged or amorphized region and subsequent solidification from c-Si substrate seed, it is possible to obtain c-Si without extended defects [2].
- Second, formation of a-Si from the melt. Liquid Si (l-Si) may solidify into a-Si at high velocities and/or large undercooling [3] or at low velocities when a seed for crystal growth is absent [4].
- Third, random explosive crystallization (EC) of a-Si. It is well-known that a-Si may be transformed into fgp-Si via a self-propagating process of melting and solidification [5-14]. Besides fgp-Si, usually a surface layer of lgp-Si is formed.
- Finally, epitaxial EC of a-Si. Recently, Polman and coworkers studied this process, which can be initiated in buried a-Si layers formed by high-energy ion implantation and which yields single-crystal Si [15-20].

The results from these different experiments will be compared and used to address the issues of ultra-rapid solidification and nucleation in Si.

*) *FOM-Institure for Atomic and Molecular Physics, Kruislaan 407, 1098 SJ, Amsterdam, The Netherlands. Telephone (+31)20946711. FAX (+31)206684106.*

II. EPITAXIAL REGROWTH FROM THE MELT

The first use of pulsed lasers in silicon processing was by Russian researchers for annealing of ion implantation damage [21]. For this purpose one generally uses nanosecond pulses from ruby lasers, Nd:YAG lasers or, more recently, excimer lasers. By irradiation at an energy density in the order of 1 J/cm^2 a damaged or amorphized surface layer of up to 1 μm thickness can be melted. If the melt thickness is sufficiently large, subsequent epitaxial crystal growth starts from the undamaged c-Si substrate and the resulting material is basically defect-free [2]. Typical regrowth velocities are several m/s. A schematic of this process is given in Fig.1. The process appears relatively simple but has been studied in detail because it offers a potential alternative to furnace annealing of implantation damage. In addition, it has been used to test various thermal models and their parameters to describe laser-induced melting and solidification [22,23].

LEGEND FIGURES 1-5

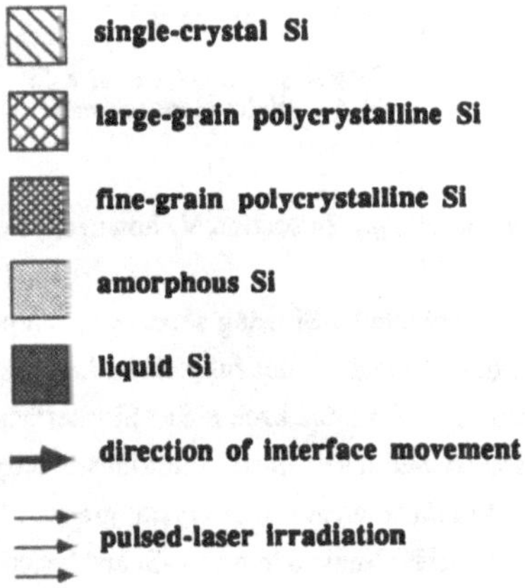

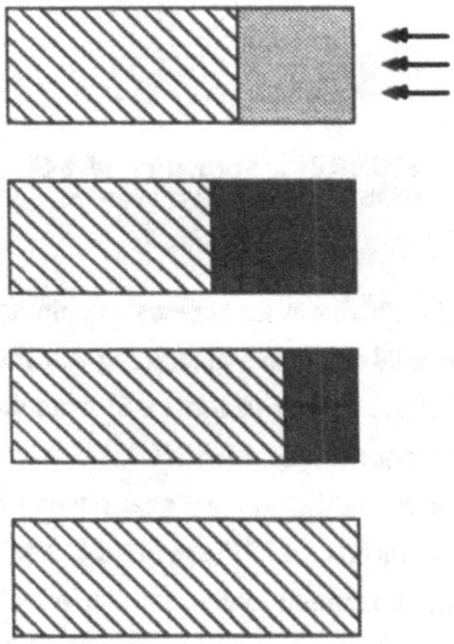

FIGURE 1. Pulsed-laser annealing of ion-implantation damage

III. AMORPHOUS REGROWTH FROM THE MELT

The situation described in the previous section becomes more complicated when the regrowth velocities exceed ≈10 m/s. This can be achieved for shallow melting of c-Si, by using picosecond pulses or UV pulses of < 10 ns [3,24-27]. It has been found that epitaxial regrowth breaks up to amorphous regrowth at interface velocities of ≈10-15 m/s, dependent on the orientation of the crystal [3], see Fig.2. Breakdown occurs at relatively low velocities for Si(111) and at high velocities for Si(100). It has been attributed to the fact that at high velocities the atomic rearrangements at the interface are too slow to permit (perfect) crystal growth. This results in defect formation and multiplication and ultimately in amorphous phase formation. We note that this kinetic arguement is not the only possible explanation for the occurrence of a maximum crystal growth velocity. This will be discussed in detail in section V.

a-Si can only be formed when the interface temperature is below the melting temperature of a-Si (T_{ma}), being approximately 200-250 K below that of c-Si (T_{mc}) [5,28]. Measurements of the velocity *vs.* undercooling relationship in the temperature range around T_{mc} (1685 K) indicate that an undercooling of ≈90 K is required to obtain an interface velocity of 6 m/s, or 15 K/(m/s) [28]. Extrapolation to larger undercooling then yields an undercooling of 15 m/s • 15 K/(m/s) = 225 K around the velocity at which a-Si forms (in case of melting of

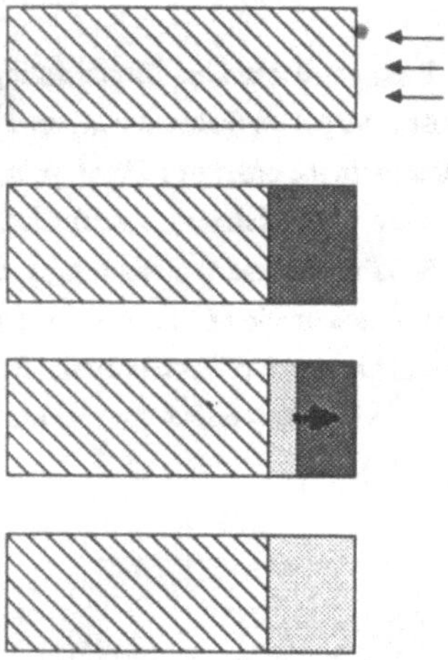

FIGURE 2. Formation of a-Si from the melt: fast growth.

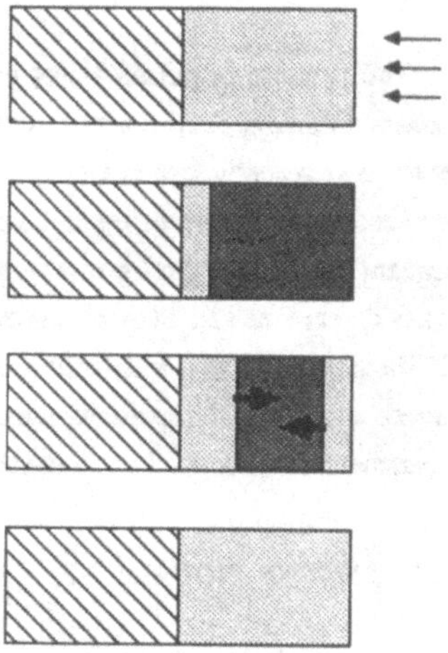

FIGURE 3. Formation of a-Si from the melt: slow growth.

Si(100)). This number appears to coincide with the estimated value of T_{ma}. In section V, however, we will show that this may be a misleading coincidence.

Another way to form a-Si from the melt is by melting ion-implanted a-Si using short (<10 ns) pulses from a frequency-doubled Nd:YAG laser or a ruby laser. When the a-Si layer is not fully melted, there is no seed for crystallization and amorphous regrowth is found to occur both from the back a-Si/l-Si interface and from the surface [4,29] (see Fig. 3). The velocity at which a-Si is formed under these conditions is only 1-3 m/s, which is much lower than the velocity of ≈15 m/s required to induce break-up of crystal growth. It has been argued that amorphous regrowth upon melting of a-Si has never been observed in pure a-Si and hence is an impurity-related process [30]. This statement is not in agreement with observations on Si-implanted a-Si [31]. Amorphous regrowth, however, is often found to evolve into EC [32], which may complicate its observation.

IV. RANDOM EXPLOSIVE CRYSTALLIZATION

Since the pioneering experiments by Takamori et al. [33] and Matsuda et al. [34] it is well-known that a-Ge and a-Si can be crystallized in a spontaneous way by applying a localized energy pulse. Once crystallization has been started, it is driven further by the latent heat effectively released upon transformation from the amorphous to the crystalline phase. It has been shown that in most, if not all, cases the transformation is mediated by melting of the amorphous phase [20]. Most experiments have been carried out on pulsed laser induced EC of ion-implanted a-Si [5-14]. This system is well-defined and very suitable for detailed micro-structural and time-resolved analysis. The general picture that has evolved is given in Fig.4. Upon irradiation with a low-energy laser pulse (pulselength typically >20 ns), a thin surface layer of a-Si is melted. This melt is highly undercooled with respect to the crystalline phase and will therefore tend to crystallize. Since no seed for crystallization is present, the material formed is polycrystalline (lgp-Si). The latent heat released upon this crystallization is sufficient to heat and melt underlying a-Si and hence a secondary, highly undercooled buried layer of l-Si is formed. This melt also crystallizes under the release of latent heat and causes further melting. In this way a self-sustained process of melting and crystallization is initiated. A thin buried layer of l-Si runs into the a-Si at a velocity of 8-15 m/s [5,9,10,12], transforming amorphous material into randomly-oriented fgp-Si. It is noted that this picture is somewhat simplified [12,14], but it is nevertheless very useful to discuss the basic processes involved.

The most difficult part in random EC is to understand and describe nucleation of the crystalline phase. This problem is discussed in detail in Ref. [20] and we will only address it briefly here. As far as formation of fgp-Si is concerned, it is essential to know where and when the nuclei are formed which eventually grow out to crystallites of 5-10 nm. At present it is not even clear whether they are formed in the a-Si in front of the buried l-Si layer [35], at the a-Si/l-Si interface [36] or in the l-Si layer itself [37] (although recent experiments [38] have shown that nucleation rates in l-Si are much too small to support the latter model). The nucleation leading to lgp-Si formation is not well understood either. Basically the same questions hold as for formation of fgp-Si, but a difference is the velocity at which the a-Si/l-Si interface moves into the a-Si, which is low (especially around the turning point of the surface melt) [39] in comparison to the 8-15 m/s found for fgp-Si formation.

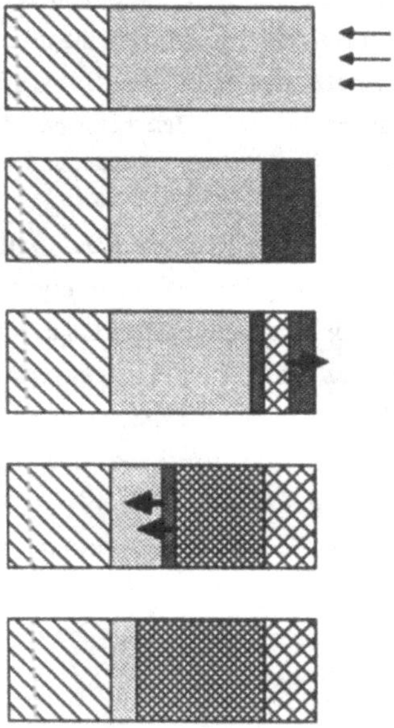

FIGURE 4. Random explosive crystallization of a-Si.

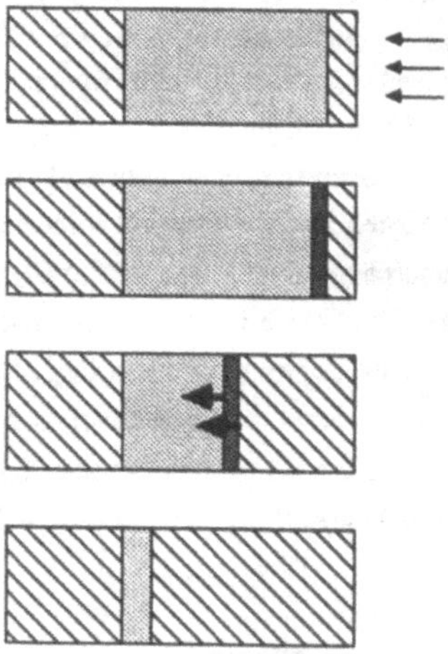

FIGURE 5. Epitaxial explosive crystallization of a-Si.

V. EPITAXIAL EXPLOSIVE CRYSTALLIZATION

Recently, Polman and coworkers have investigated pulsed laser induced EC of a-Si layers buried beneath a c-Si surface layer [15-20]. These structures can be prepared by high-energy ion implantation into c-Si. Since the c-Si top layer has a lower optical absorption coefficient and a higher melting temperature than the a-Si layer underneath, it is possible to irradiate and heat a-Si through the c-Si, which results in buried melting of a-Si at the a-Si/c-Si interface. This melt is highly undercooled with respect to the adjacent c-Si and immediately crystallizes with the c-Si as a seed. The latent heat released upon crystallization is sufficient to heat and melt underlying a-Si, which will also crystallize. In this way a self-sustained process is triggered, which is similar to that in a-Si surface layers. The major difference is that in this case the material formed is single-crystal, epitaxially aligned with the c-Si on top (see Fig.5).

It has been found that the c-Si formed by epitaxial EC contains twins, but is free of polycrystallites. This is remarkable, since the overall features of this type of EC are similar to those of random EC. Apparently the nucleation and growth mechanism leading to fgp-Si formation is suppressed under the conditions of epitaxial EC (see Ref. [20] for a discussion).

The velocity of epitaxial EC has been measured to be 15-16 m/s for Si(100) [15,16,19]. In contrast to the velocity of the l-Si layer in random EC, the velocity in epitaxial EC remains constant during propagation through the a-Si (over ≈350 nm), as was concluded from time-resolved optical reflectivity measurements. The temperature of the freezing l-Si/c-Si interface, on the other hand, is calculated to make an excursion over ≈70 K. Immediately after melting of the a-Si, the temperature at the l-Si/c-Si interface is approximately equal to T_{ma} (≈1460 K [5]). Due to the release of latent heat upon crystallization and to additional heating by the tail of the laser pulse the temperature at this interface rises. Note that the temperature at the a-Si/l-Si melt front is assumed to remain close to T_{ma}. After reaching a maximum of ≈1530 K it decreases again due to heat conduction into the c-Si substrate. Eventually the temperature of the melting and freezing interfaces drops below T_{ma} and EC is quenched. From the measured crystallization velocity and the corresponding calculated temperatures, it is possible to construct a part of the velocity *vs.* undercooling relationship in a temperature regime which is very difficult to explore in other experiments (≈155-205 K undercooling). It is found that the crystallization velocity saturates at 15-16 m/s for this large undercooling, in qualitative agreement with predictions [40,41]. Various models for crystal growth, however, differ considerably in their basic assumptions. Best agreement between theory and experiment is found for models in which diffusive motion in l-Si at the interface is the rate limiting factor [19,40,41].

An interesting consequence of the experiments described just before is that the crystallization velocity can never exceed 15-16 m/s, which coincides with the observed threshold velocity for break-up of crystal growth into amorphous growth (see section III). This implies that break-up is not necessarily related to the velocity of the interface (which is constant in the temperature regime of interest) as has been assumed, but may also be related to the degree of undercooling below T_{ma}. The observation of a plateau in the undercooling *vs.* velocity relationship shows that an extrapolation to large undercooling of data obtained at small undercooling is not a useful thing to do. The coincidence of T_{mc}-T_{ma} ≈ 225 K and the extrapolation 15 m/s • 15 K/(m/s) = 225 K (see section III) may just be good (or bad) luck. Note that the value of 15 K/(m/s) is not accurately known [42].

VI. CONCLUSIONS

We have discussed and compared different pulsed laser induced rapid phase transformations in silicon. As a result of intensive investigations, understanding of these processes has steadily improved over the past decade. Most processes can now be described well in phenomenological way. Important remaining questions to be solved mainly concern the fundamental mechanisms underlying the phase transformations, particularly random nucleation of the crystalline phase and solidification at very high (>10 m/s) velocities.

VII. ACKNOWLEDGEMENTS

The work described in this paper is part of the research program of the Foundation for Fundamental Research on Matter (FOM) and was financially supported by the Netherlands Organization for the Advancement of Research (NWO) and the Netherlands Technology Foundation (STW). The authors like to thank F.W. Saris for his stimulating interest throughout this work.

VIII. REFERENCES

[1] M. Wautelet, P. Quenon and A. Jadin, Semicond. Sci. Technol. **3** (1988) 54.
[2] R.T. Young, J. Narayan and R.F. Wood, Appl. Phys. Lett. **35** (1979) 447.
[3] A.G. Cullis, N.G. Chew, H.C. Webber and D.J. Smith, J. Cryst. Growth **68** (1984) 624.
[4] J.J.P. Bruines, R.P.M. van Hal, H.M.J. Boots, W. Sinke and F.W. Saris, Appl. Phys. Lett. **48** (1986) 1252.
[5] M.O. Thompson, G.J. Galvin, J.W. Mayer, P.S. Peercy, J.M. Poate, D.C. Jacobson, A.G. Cullis and N.G. Chew, Phys. Rev. Lett. **52**, 2360 (1984).
[6] W. Sinke and F.W. Saris, Phys. Rev. Lett. **53**, 2121 (1984).
[7] J. Narayan, C.W. White, O.W. Holland and M.J. Aziz, J. Appl. Phys. **56**, 1821 (1984).
[8] J. Narayan, S.J. Pennycook, D. Fathy and O.W. Holland, J. Vac. Sci. Technol. **A2** (1984) 1495.
[9] J.J.P. Bruines, R.P.M. van Hal, H.M.J. Boots, A. Polman and F.W. Saris, Appl. Phys. Lett. **49** (1986) 1160.
[10] D.H. Lowndes, G.E. Jellison, Jr., S.J. Pennycook, S.P. Withrow and D.N. Mashburn, Appl. Phys. Lett. **48**, 1389 (1986).
[11] K. Murakami, O. Eryu, K. Takita and K. Masuda, Phys. Rev. Lett. **59**, 2203 (1987).
[12] D.H. Lowndes, S.J. Pennycook, G.E. Jellison, Jr., S.P. Withrow and D.N. Mashburn, J. Mater. Res. **2**, 648 (1987).
[13] P.S. Peercy, J.Y. Tsao, S.R. Stiffler and M.O. Thompson, Appl. Phys. Lett. **52**, 203 (1988).
[14] D.H. Lowndes, S.J. Pennycook, R.F. Wood, G.E. Jellison, Jr. and S.P. Withrow in *Fundamentals of Beam-Solid Interactions and Transient Thermal Processing*, ed. by M.J. Aziz, L.E. Rehn and B. Stritzker (MRS Symp. Proc. **100**, MRS, Pittsburgh, 1988) p. 489.
[15] P.A. Stolk, A. Polman, W.C. Sinke, C.W.T. Bulle-Lieuwma and D.E.W. Vandenhoudt in *Ion Beam Processing of Advanced Electronic Materials*, edited by N. Cheung, J. Roberto and A. Marwick (MRS Symp. Proc. **147**, MRS, Pittsburgh, Pennsylvania, 1989) (in press).
[16] A. Polman, thesis, Utrecht State University (1989).
[17] A. Polman, D.J.W. Mous, P.A. Stolk, W.C. Sinke, C.W.T. Bulle-Lieuwma and D.E.W. Vandenhoudt, to be publ. in Appl. Phys. Lett.
[18] A. Polman, P.A. Stolk, D.J.W. Mous, W.C. Sinke, C.W.T. Bulle-Lieuwma and D.E.W. Vandenhoudt, submitted for publication.
[19] A. Polman, P.A. Stolk and W.C. Sinke, submitted for publication.
[20] W.C. Sinke, A. Polman, S. Roorda and W.C. Sinke, to be publ. in *Beam Processing and Laser Chemistry*, ed. by I.W. Boyd and E. Rimini (Proc.1989 E-MRS Meeting), also to be published as a special issue of Appl. Surf. Sci..
[21] N.N. Gerasimenko et al., Proc. Int. Conf. Ion Implantation, Budapest (1975).
[22] H.C. Webber, A.G. Cullis and N.G. Chew, Appl. Phys. Lett. **43** (1983) 669.
[23] R.F. Wood and G.E. Giles, Phys. Rev. **B23** (1981) 2923.
[24] P.L. Liu, R. Yen, N. Bloembergen and R.T. Hodgson, Appl. Phys. Lett. **34** (1979) 864.
[25] R. Tsu, R.T. Hodgson, Teh Yu Tan and J.E. Baglin, Phys. Rev. Lett. **42** (1979) 1356.
[26] M.O. Thompson, J.W. Mayer, A.G. Cullis, H.C. Webber, N.G. Chew, J.M. Poate and D.C. Jacobson, Phys. Rev. Lett. **50** (1983) 896.
[27] J.A. Yater and M.O. Thompson, in *Selected Topics in Electronic Materials*, ed. by B.R. Appleton, D.K. Biegelsen, W.L. Brown and J.A. Knapp (MRS Ext. Abstr. **18**, MRS, Pittsburgh, 1988), p.219.
[28] J.M. Poate, J. Cryst. Growth **79** (1986) 549.
[29] S.U. Campisano, D.C. Jacobson, J.M. Poate, A.G. Cullis and N.G. Chew, Appl. Phys. Lett. **46** (1985) 846.
[30] P.S. Peercy, J.Y. Tsao and M.O. Thompson, in Ref. [27], p.215.
[31] J.J.P. Bruines, thesis Eindhoven Tecnical University (1988).
[32] J.J.P. Bruines, R.P.M. van Hal, B.H. Koek, M.P.A. Viegers and H.M.J. Boots, Appl. Phys. Lett. **50** (1987) 507.
[33] T. Takamori, R. Messier and R. Roy, Appl. Phys. Lett. **20** (1972) 201.
[34] A. Matsuda, A. Mineo, T. Kurosa and M. Kikuchi, Solid State Commun. **13** (1973) 1165.
[35] S. Roorda and W.C. Sinke, Appl. Surf. Sci. **36** (1989) 588.
[36] J.Y. Tsao and P.S. Peercy, Phys. Rev. Lett. **58** (1987) 2782.
[37] R.F. Wood, D.H. Lowndes and J. Narayan, Appl. Phys. Lett. **44** (1984) 770.
[38] S.R. Stiffler, M.O. Thompson and P.S. Peercy, Phys. Rev. Lett. **60** (1988) 2519.
[39] R.F. Wood and G.A. Giles, Phys. Rev. **B34** (1986) 2606.
[40] P.M. Richards, Phys. Rev, **B38** (1988) 2727.
[41] M.H. Grabow, G.H. Gilmer and A.F. Bakker, in *Atomic Scale Calculations in Materials Science*, ed. by J. Tersoff, D. Vanderbilt and V. Vitek (MRS Symp. Proc. **141**, MRS, Pittsburgh, 1989), p.349 and references therein.
[42] P.S. Peercy and M.O. Thompson, in *Energy Beam-Solid Interactions and Transient Thermal Processing*, ed. by D.K. Biegelson, G.A. Rozgonyi and C.V. Shank (MRS Symp. Proc. **35**, MRS, Pittsburgh, 1985), p.53.

THE CURRENT STATUS OF ION IMPLANTATION PROCESS IN INDUSTRIAL APPLICATIONS

Piran Sioshansi*

ABSTRACT

Ion implantation, a process that is commonly used for doping of semiconductor devices, offers the potential to selectively modify surface properties of materials. The modification in properties is the result of changes in the composition or morphology of the near surface layers by the incoming ion beams. Extensive research in this field has demonstrated that hardness, coefficient of friction, wear resistance, corrosion resistance, surface energy, and other properties of the surface can be influenced by ion beams without adversely affecting bulk properties. The unique capabilities, outstanding features, and overall potential of the technology are described. Based on Spire Corporation's experience, successful applications of the ion implantation process are presented and the criteria for choosing promising candidates are outlined. The built-in quality control which characterizes the ion implantation process has created "niches" in the marketplace in applications where a high degree of reliability and reproducibility is essential. The technology is expected to have an increasing contribution in selective applications from aerospace to biomaterials.

A. INTRODUCTION

In the ion implantation process the energetic ions penetrate into the substrate surface and interact with the substrate material by changing the composition and structure of the near surface region. Changes in composition result in the formation of second phases (carbides, nitrides, oxides) and solution hardening. The metastable compositional and structural changes produced in the surface by ion implantation are frequently observed to be stable under operational conditions. The aim of this article is to review the successful applications of the ion implantation field.

B. APPLICATIONS IN BEARINGS AND GEARS

The ion implantation process has proved to be an attractive surface treatment for overall performance of high precision bearings and gears. Bearings and gears are good examples of components that have tight dimensional tolerance for which few other conventional surface modification procedures are suitable. Furthermore, the threat of delamination makes the application of coatings to these components highly risky. The preservation of dimensional integrity and surface finish in high precision bearings and gears and the absence of a sharp interface between the implanted layer and substrate material as a result of ion implantation removes the concern that exists with the delamination of hard coatings.

Spire uses a mass analyzed ion implantation system to separate different ions and allow surface treatment with a pure ion species. The beam is focused by means of magnetic quadrupole

---

* Spire Corporation, Patriots Park, Bedford, MA, 01730, USA

lenses and is raster scanned electrostatically to cover a large area inside the treatment chamber (Figure 1). A large area ion beam minimizes the temperature rise associated with the process. Various quality control procedures are built into the system to guarantee reliability and reproducibility of the process. Thus, ion implantation appears to be ideally suited for processing of sensitive aerospace components. Examples of bearing and gear components which have been ion implanted are presented in Table 1.

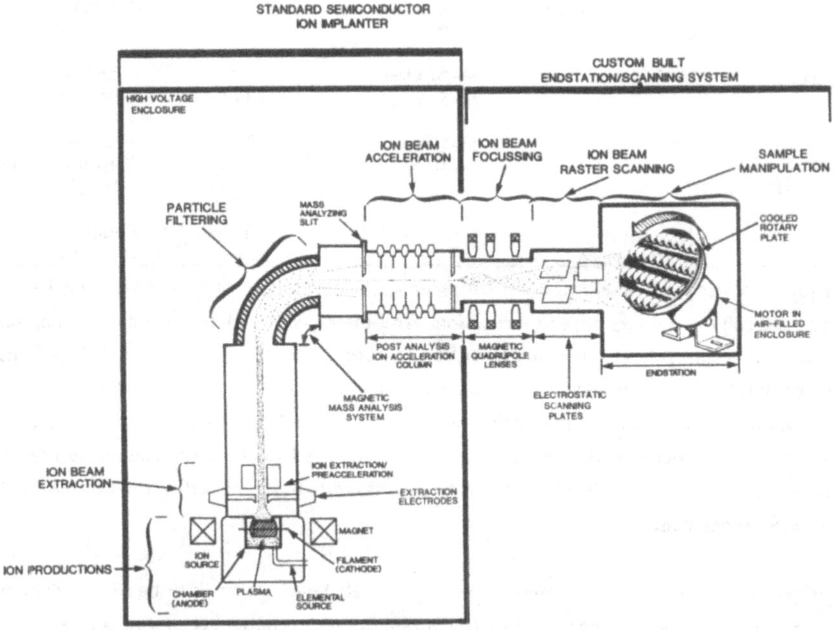

FIGURE 1. SPIRE'S MASS ANALYZED ION IMPLANTED SYSTEM.

The ion implantation of titanium ions followed by carbon has been used by a number of groups[1,2] for imparting wear resistance on bearing grade 52100 and 440C steel. Sioshansi and Au[3] reported a significant improvement in the wear resistance of mating surfaces sliding against a 52100 steel surface that had been ion implanted with titanium and carbon under lubricated conditions.

The titanium and carbon implanted surface not only showed a much reduced sliding wear but it manifested a reduced wear variability. Furthermore the low wear rate proved to be independent of material combination sliding against the 52100 titanium and carbon implanted surface. The result of this work supports an oxidation layer formed on the surface of the implanted component that minimizes the metal-to-metal contact between the sliding surfaces.

Recent work in high precision instrument bearings has demonstrated that the titanium and carbon ion implantation is an effective process for increasing the fretting wear[4] resistance of the 52100 and 440C alloys. The improved performance of the instrument bearings as a result of ion

TABLE I. SUCCESSFUL BEARING AND GEAR APPLICATIONS FOR THE ION IMPLANTATION PROCESS.

Component	End Use	Material	Ion Species	Benefits (Reference)
Turbopump Bearings	Space Shuttle Main Engine	440C stainless steel	Ti+C	Improved Wear Resistance (5)
Instrument Bearings	Navigational Guidance System (Gyroscope)	52100 steel 440C stainless steel	Ti+C	Improved Fretting Resistance (4)
Main Shaft Bearings	Jet Engine	M-50 steel M-50 NIL steel	Cr Ta	Improved Corrosion Resistance (6)
Main Shaft Bearings	Jet Engine	M-50 steel M-50 NIL steel	Ta	Improved Rolling Contact Fatigue Life (7)
Gears	Helicopter Transmission	9310 steel	Ta	Reduced Scuffing Wear (6)

implantation is attributed to elimination of asperities interlocking and cold welding between the balls and races under boundary layer lubrication. The instrument bearings successful application is a major breakthrough for the ion implantation process. Due to their small dimension, large quantities of ball bearings can readily be treated at low cost by the ion implantation process. The treatment can be applied to both the balls and races of the ball bearings. It is generally believed that a dissimilar treatment on the races from the one applied to the balls should be used in these applications. It is anticipated that ion implanted instrument bearings will be available in the near future for widespread use.

Perhaps the most dramatic improvement[5] of ball bearings to date has been demonstrated in laboratory tests of the bearings used in the fuel oxidizer turbopumps of the space shuttle main engine. Implantation of dissimilar ions on the races and balls show a significant reduction in corrosive wear of these bearings. Figure 2 shows the ball bearings for the space shuttle program as they are being loaded into fixtures for the ion implantation process.

Extensive research in corrosion protection of ferrous alloys by ion implantation has been performed at various laboratories. The U.S. Naval Research Laboratory (NRL) has investigated the corrosion resistance of bearings under laboratory and field conditions.[6] The NRL results clearly indicate that ion implantation of chromium, tantalum, and similar species is an effective process to impart corrosion resistance to M-50 steel components. When chromium is added in solid solution in the steel samples, a passive layer is formed that provides corrosion protection against chloride attack in aqueous conditions. Prototype quantities of main shaft bearings of gas turbines have been chromium ion implanted by NRL with satisfactory overall corrosion results.

Ion implantation of bearing and gear samples with tantalum has been performed at NRL[6] and independently at Spire. Tantalum is a particularly interesting ion species for the treatment of steel bearings and gears because tantalum can impart both corrosion and wear resistance to ferrous alloys. The NRL results have centered on evaluating the performance of tantalum ion implantation into 9310 steel gears in sliding and scuffing wear situations. The results show a significant

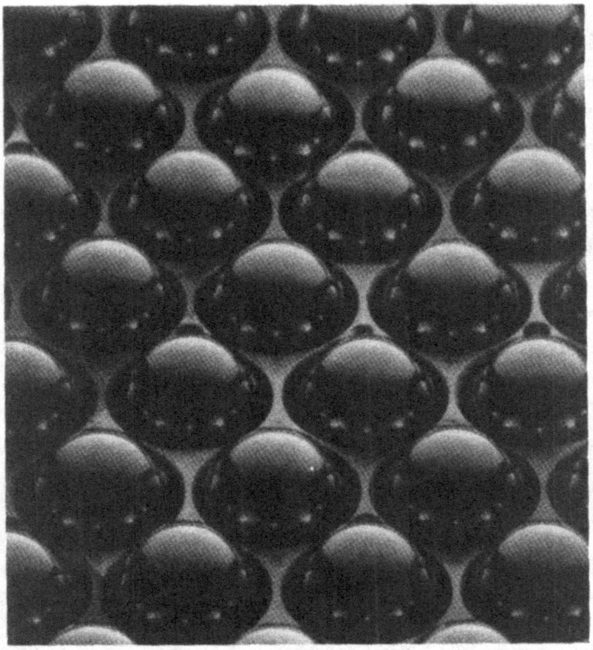

FIGURE 2. THE SPACE SHUTTLE BALL BEARINGS AS THEY ARE LOADED INTO FIXTURES FOR ION IMPLANTATION.

reduction in friction and wear of different steel components. Spire has ion implanted tantalum in gears of gas turbine compressors as well as helicopter engine transmission systems. Preliminary laboratory evaluations indicate that the tantalum ion implanted gears have performed significantly better than the conventional gears and in many instances the scuffing wear can be drastically reduced. Field results from the shaft and drive gears that received a dual treatment of titanium and carbon have proved very effective in the compressor section of the gas turbine.

C. APPLICATIONS IN HIGH PRECISION TOOLINGS

The ion implantation process has been used extensively for processing of industrial tools. The industrial tool applications is quite varied and have been reviewed by this author in a number of articles.[8,9,10] The most successful examples in this field are tools that are used for shaping, cutting, and piercing of materials such as plastics, papers, synthetic fibers, and soft tissues. These tools experience a mild adhesive or abrasive wear and in some cases the corrosive attack aggravates the wear process. Examples of toolings in this category include: high speed steel dies used for piercing high quality paper and cutting polymer sheets; 400 series stainless steel cutting blades used for surgical process. The surgical cutting blades are particularly noteworthy as they are believed to result in sharper cuts with less damage to the surrounding tissue. Roller cutters for impression cutting of plastics and papers is another successful application of ion implantation. Cutting blades and slitter knives used for cutting currency have been ion implanted for the U.S. government. Pins and perforating plates used for making stamps have also been treated by ion implantation. The most common treatment in these applications is nitrogen ion implantation. Alternatively, titanium and carbon implantation have been used on these tools with great success.

Another interesting application for the ion implantation process is treatment of hot rolling mill rolls for continuous forming of nonferrous rods. Figure 3 shows a pair of these rolling mill rolls. These rolls are made of tool steels and are subject to severe adhesive wear. Ion implantation with titanium and carbon (Spire uses the IONGUARD-2001 for titanium and carbon ion implantation) has consistently increased the life of the rolls by a factor of six (6X) and has produced a smoother and better surface finish on the rods. This rolling application and similar extrusion of soft metals are believed to be near market maturity.

Recently the die plates for forming compact disks have been ion implanted with titanium and carbon for field testing. These dies have a mirror surface finish that is, of course, not altered by the ion implantation process. Few other surface coating processes are acceptable for this application. The ion implantation is believed to enhance both the mold release and life performance of these molds. Various other molds for engineering plastics and thermosetic plastics have received the ion implantation process with satisfactory results. In particular, nickel coated fiber optical coupler molds show a distinct increase in hardness and wear resistance as a result of nitrogen ion implantation.

High speed steel punches and dies that are used for forming and punching hot-rolled steel and austenitic stainless steel show[11] a factor of 10-12 times increase in life when implanted by nitrogen ion beam. Similar dies with similar ion implantation treatment when used for processing of cold-rolled steel or martensitic stainless steel show no improvement in life. This particular case points at the significance of tracking the conditions under which the tools are used and explains

FIGURE 3. A PAIR OF ROLLING MILL ROLLS FOR FORMING OF NONFERROUS METAL RODS, TREATED WITH THE IONGUARD-2001 PROCESS.

why there are sometimes discrepancies in the open literature on performance of ion implanted tools.

D. CONCLUSIONS

Ion implantation can significantly influence surface properties such as corrosion and wear resistance of materials. The mechanism of changes induced by ion implantation is varied. In many applications the improved surface properties are of a physical nature such as amorphization, homogenization or changes in surface morphology. In other cases the improved properties are due to chemical changes such as formation of nitrides and carbides that contribute to precipitation hardening. In many instances the improved wear and corrosion resistance occur simultaneously. This is of particular interest in bearings and gears as well as biomaterials applications.

Ion implantation processing offers a high degree of reliability and reproducibility. This feature of ion implantation is crucial for a wide range of sophisticated high precision applications requiring the highest possible process yield where no failure can be tolerated. The ion implantation process (Spire Corporation uses the trademark IONGUARD for proprietary ion beam parameters) has been chosen for surface treatment of titanium based artificial joints such as knees, hips, shoulders, and fingers by several major U.S. orthopedic manufacturers. Ion implantation has also been specified for treatment of bearings used in the space shuttle program. A large variety of toolings for stamping, shaping, cutting, and piercing applications are treated for evaluation by the ion beam process. As more consistent field results become available, it is anticipated that ion implantation would play a more important role in processing of a larger variety of high precision, high technology, high value added applications.

E. REFERENCES

(1) I.L. Singer in Ion Implantation and Ion Beam Processing of Materials, eds., G.K. Hubler, O.W. Holland, C.R. Clayton, and C.W. White, North Holland, New York, p. 585, (1984).

(2) L.E. Pope, F.G. Yost, D.M. Follstaedt, S.T. Picraux, and J.A. Knapp, in Ion Implantation and Ion Beam Processing of Materials, eds., G.K. Hubler, O.W. Holland, C.R. Clayton, and C.W. White, North Holland, New York, p. 661 (1984).

(3) P. Sioshansi and J.J. Au, "Improvements in Sliding Wear of Bearing-Grade Steel Implanted with Titanium," Materials and Science and Engineering 69, 161-166 (1985).

(4) R.A. Hanson, "The Effect of Advanced Materials Technologies on Wear in Instrument Bearings, 1987 Bearing Conference Proceedings, March 10-12, Orlando, Florida, sponsored by DOD/Instrument Bearing Workshop Group (1987).

(5) L. Ng. and Y. Naerheim, "Wear Resistant Material and Coatings for Cryogenic Bearings," 1987 Bearing Conference Proceedings, March 10-12, Orlando, Florida, sponsored by DOD/Instrument Bearing Workshop Group (1987).

(6) K.S. Grabowski, G.K. Hubler, E.T. Hodge, R.A. Jeffries, C.R. Clayton, Y.F. Wang, and G. Kuhlman, "Ion Beam Processing of Bearing Steels for Corrosion and Wear Resistance," NRL Memo Report, 5592, Naval Research Laboratory, Washington, DC, June 27, 1985.

(7) Eric Bamberger, General Electric Company, private communication, 1987.

(8) P. Sioshansi, "Surface Modification of Industrial Components by Ion Implantation," Materials Science and Engineering, 90, 373-383 (1987).

(9) P. Sioshansi, Thin Solid Films, 118, 61-71 (1984).

(10) P. Sioshansi, Materials Science and Engineering, 90, 373-383 (1987).

(11) R.B. Alexander, proceedings of the ASM's Conference on Ion Implantation and Plasma Assisted Processes for Industrial Applications 23-25 May 1988, Atlanta, GA, USA, to be published.

ION BEAM PROCESSING FOR SILICON-ON-INSULATOR

Wolfgang Skorupa[1]

0. INTRODUCTION

During the last few years ion beam processing penetrated very aggressively in many branches of advanced solid state technology. This holds also for the modern semiconductor technology in the area of VLSI (very large scale integration) and ULSI (ultra large scale integration).

SOI (silicon-on-insulator) is one of the most discussed candidates of this branch offering the possibility to produce integrated circuits of high packing density, low power consumtion and high speed.

Several techniques has been known for the production of SOI-structures using ion beam processing directly or as an efficient support, and will be reviewed in this paper:

SILICON-ON-SAPPHIRE (SPE-SOS)
(combined with Solid Phase Epitaxy)

FULL ISOLATION BY POROUS OXIDISED SILICON (HI-FIPOS)
(combined with hydrogen implantation)

LATERAL-ION BEAM INDUCED EPITAXIAL RECRYSTALLISATION (L-IBIEC)

SEED SELECTION THROUGH ION CHANNELING (SSIC)

SEPARATION BY IMPLANTED OXYGEN/NITROGEN (SIMOX/SIMNI)

MOLECULAR BEAM EPITAXY ON SOI (MBE-SOI)

This is not a state-of-the-art-review of the described techniques. The first aim of this study was to take advantage of the knowledge concerning the different possibilities of using ion beams for the production of SOI-structures.

1. SPE-SOS

Besides the Harris-process of Dielectric Isolation, the well-known SOS-technology is the only one being implemented in production lines. A SOS-substrate consists of a single crystalline sapphire wafer with a single crystalline silicon film heteroepitaxially grown upon. The main disadvantages of this type of substrate are (i) impurity outdiffusion from the sapphire substrate during the epitaxial growth taking place at high temperatures, (ii) formation of planar defects (stacking faults, twins) at the sapphire-silicon interface and in the silicon film due to lattice mismatch and different thermal expansion coefficients, and (iii) high substrate costs. Due to the latter fact electronic devices and circuits produced on such substrates are first of all used for space and military applications. The formation of planar defects during the heteroepitaxy was that problem being succesfully solved by ion beam processing (see Fig.1). A distinct reduction of the defect density of the silicon films can be reached by amorphisation of the interface region sapphire-silicon subsequently followed by SPE at 550-600°C /1/. Sometimes, an additional annealing step at 1000°C will be performed to further eliminate residual defects. It is clear, that during the amorphisation step the surface region of the silicon film must remain single crystalline to serve as a seed during SPE. A reduction of the defects in the surface region is possible by amorphising it in a second step and to repeat the SPE-regrowth using the lower part of the silicon film as a seed /2/. In this manner, a drastic lowering of the density of

[1]Academy of Sciences of the GDR, Central Institute for Nuclear Research, Rossendorf, PF 19, DDR-8051 Dresden, GDR

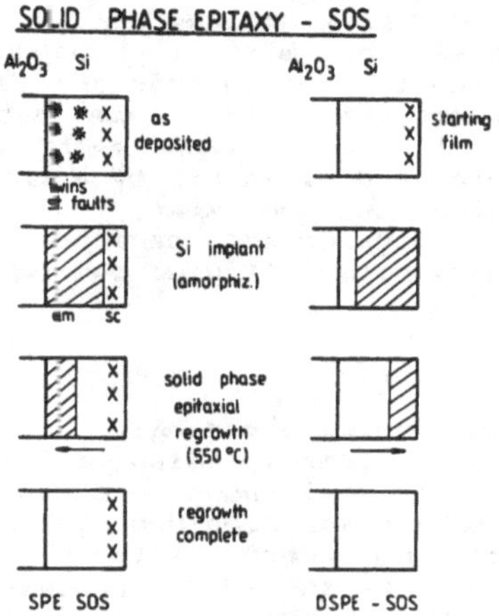

Fig.1: Schematic of the SPE-SOS process (after Golecki /1/)

planary defects occurs leading to higher minority carrier lifetimes and, hence, to a lower junction leakage in MOS and bipolar devices. Both, ion implantation and furnace annealing as used for amorphisation and SPE, respectively, are processes of high reproducibility and widely used in the microelectronic technology. Consequently, the described method of defect reduction is state-of-the-art of modern SOS-technologies.

2. HI-FIPOS

Although being one of the well known front running methods for the formation of SOI-structures the conventional FIPOS-process /3/ suffers from several disadvantages /4/. In this process n-type regions are formed at the surface of p-type silicon substrates. During the anodisation step necessary to form the porous silicon - only p-type material will be transformed - the current flow begins between the n-type regions and then spreads out in all directions within the bulk region. There, porous silicon with a higher density is formed because the density of porous silicon increases if the current density falls down. In this manner, material with mechanical stress is formed leading to dislocation densities up to $10^7 cm^{-2}$ within the silicon islands. On the other hand, due to current spreading thick porous layers are needed to isolate even small islands and, moreover, stalks are formed at the bottom side of the islands so that the dielectric strengh between the island and the silicon substrate is lowered.

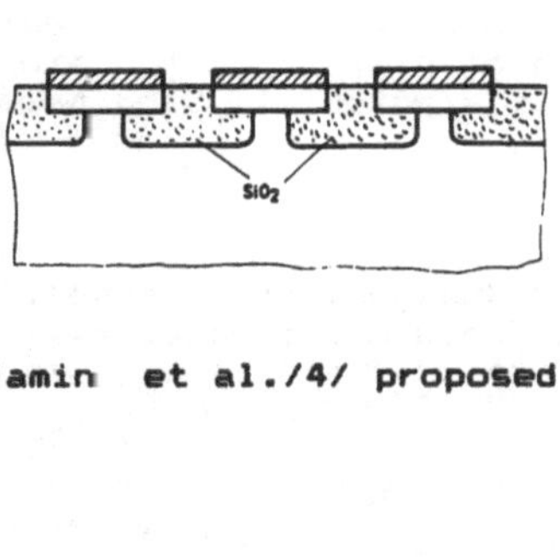

Fig 2: Schematic of the HI-FIPOS process (after Benjamin etal./4/)

Benjamin et al./4/ proposed the use of a buried n-type grid formed by ion beam processing to circumvent the above mentioned problems. This configuration shown in Fig.2 leads to a much more homogenous current flow and thus to the formation of porous silicon with a homogenous density. The stalks at the bottom side of the silicon islands are also avoided in this manner. It was demonstrated that using this method silicon islands with a larger area and with much less mechanical strength can be formed /4/. The buried n-type grid is formed by a masked hydrogen implantation. During the oxidation step necessary for the conversion of the porous silicon to silicon dioxide and performed at temperatures higher than 800° the implanted hydrogen diffuses out having no further influence on the SOI-structure.

3.L-IBIEC

Recently, the basic IBIEC-process has been devoted a strongly growing interest as being extensively reviewed and reported also at this conference /5/. Shortly spoken, IBIEC is the well known process of thermally induced solid phase epitaxy (TH-SPE) being combined with ion beam processing for the purpose of depositing energy into nuclear processes. In this manner the activation energy of the TH-SPE process of about 2.7 eV can be lowered up to about 0.3 eV. TH-SPE normally needs a temperature of about 550°C or higher to take place. This limit can be lowered below 200°C by using IBIEC with appropriate parameters. By further lowering of the temperature the inverse process of step-by-step interface amorphisation can be induced /5/.

All the work concerning the basic mechanisms of IBIEC was performed at amorphous silicon layers produced by ion implantation directly on or within single crystalline silicon wafers. During the last year, the succesfull epitaxial recrystallisation of an amorphous silicon layer deposited by chemical vapour deposition on a single crystalline silicon substrate was firstly demonstrated /6/. No special precautions against impurities and the natural silicon oxide layer were used in these experiments as necessary to recrystallise such layers by TH-SPE. This type of processing can be principally used to produce layer systems with steep doping gradients. On the other hand, such a vertical driven recrystallisation of a deposited layer from a single crystalline seed is an important prerequisite to drive a epitaxial growth front laterally on an insulator. In this way a SOI-structure can be formed which was already demonstrated for the case of TH-SPE /7/. Recently, our group has shown for the first time lateral oriented growth of a silicon layer by IBIEC to be possible /8/. The SOI-structure consisted of a thermally grown silicon dioxide layer (200 nm) with seeds defined by fotolithography and wet chemical etching and a silicon layer (400 nm) deposited upon. The ion irradiation was performed by 330 keV silicon ions with doses in the $10^{17}cm^{-2}$ range. The temperature during IBIEC was 400°C simply realised by the ion beam itself. It was demonstrated by Transmission Electron Microscopy both at plan view and cross sectional samples a texture-like lateral growth of the silicon layer. Of course, the quality of this material is not yet satisfactory. Further efforts are awaited from a better understanding of the phenomena taking place at the interface between the substrate and the deposited layer as well as from an appropriate design of the mask edge between the seed and the insulating layer.

Nevertheless, this technique offers the possibility to produce SOI-structures in the low temperature regime of solid phase epitaxy combined with the advantages of ion beam irradiation which allows the use of lower process temperatures as compared to TH-SPE as well as to perform the epitaxial growth only in selected regions of a wafer-both laterally and/or vertically. In this manner, L-IBIEC is a potential candidate for three dimensional integration.

4. SSIC

The topic SOI includes not only methods of producing single crystalline but also polycrystalline films on any type of insulating material. Such substrates are mainly used for the production of flat panel displays including thin film transistors and elated devices. Especially, it is necessary to prepare material with a high grain size to minimise the influence of grain boundaries on the electronic transport properties. An interesting idea to produce not only grains with higher size but also with nearly the same crystal orientation related to the surface was proposed by Reif and coworkers /9,10/. It was demonstrated that by a skilful utilise of the channeling phenomenon of ions in single crystalline material silicon films with the above-mentioned properties can be formed.

SEED SELECTION THROUGH ION CHANNELING

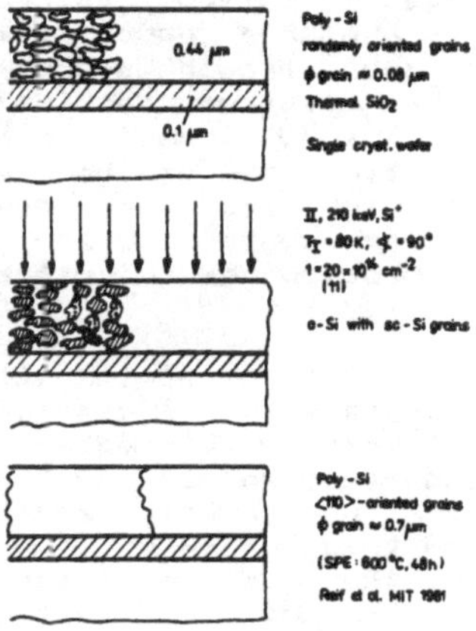

Fig.3: Schematic of the SSIC process (after Reif et al./9-11)

This process (see Fig.3) begins with low pressure-chemical vapour deposition of a polycrystalline silicon layer on a silicon wafer covered with a thermally grown silicon dioxide layer serving as an insulator. This leads to the formation of a silicon film with randomly oriented grains with a grain size of about 0.08 um. Then, ion implantation of silicon ions follows by performing the irradiation under a distinct angle. In this manner, grains oriented with their main crystal axis parallel to the ion beam will be scarcely damaged due to the ion channeling effect whereas all other grains not aligned to the ion beam will be amorphised. The ion beam irradiation is performed by silicon ions to exclude any impurity or doping effects. Subsequently, annealing at 600°C for several hours leads to recrystallisation of the irradiated silicon film by seeding at the remaining grains with uniform orientation. Using this method, polycrystalline silicon layers with (110)-oriented grains and a size of 0.8 um were produced /10/. The maximum angle deviation from the (100)-axis was 4°. It was demonstrated by the use of X-ray diffraction that an optimum irradiation dose exists where most of the grains are of the same orientation after the recrystallisation process. This is caused by the energy deposition into nuclear processes occuring during implantation of the silicon ions. At too low doses the energy deposition is not sufficient to render the misoriented grains amorphous. If the implanted dose is too high the nuclear energy deposition reaches the critical amount to amorphise also single crystalline silicon.

Thin film transistors produced in all three types of material did show the lowest threshold voltages and the highest channel mobilities for the silicon layers implanted with the optimum dose /10/. Moreover, it was observed that this effect is first of all related to the better surface properties of the (110)-oriented grains as compared to that of randomly oriented grains /11/. On the other hand, the grain boundary effect was demonstrated to be of less importance.

5. SIMOX/SIMNI

This technique comprising high dose implantation of oxygen and/or nitrogen is on a very high level concerning basic research and application tests /12,13/. Few years ago, the construction of high current implanters being able to deliver oxygen currents in the 100 mA-range has begun. At present, several laboratories and companies have installed such machines for the pilot production of SOI-substrates and, of course, integrated circuits upon them. Furthermore, the possibility to produce SOI-substrates with high quality silicon layers of a thickness of about 100 nm has SIMOX/SIMNI made a powerful candidate to produce ULSI devices with high performance /14/. Beyond the traditional speed, power consumtion and rad-hard advantages of SOI-devices, a MOS transistor in a thin film SOI-substrate shows a higher short channel threshold voltage stability, an improved subthreshold slope, an increased saturation current and reduced hot electron degradation. Additionally, the well known disadvantage of a SOI-MOS transistor -the "kink-effect"- can be avoided /15/. So, this technique can be really named the most promising future SOI-technology for producing devices with submicron dimensions. A more detailed review concerning ion beam synthesis is presented by P. Hemment in these proceedings. Further, we refer to some results of our own recent work.

During the last ten years, especially, the formation of buried silicon dioxide and silicon nitride layers was extensively investigated. Recently, the formation of silicon oxynitride layers by NO^+-implantation or the sequential implantation of oxygen and nitrogen has become of interest /16/. Both, interesting basic problems of ion beam synthesis but also new application aspects evolved from this research. On the other hand, most of the basic research to study the problems of ion beam synthesis was performed by methods like RBS, TEM and IRS. With the emergency of high current implanters leading to a technologically oriented research -the necessity to quickly inspect the properties of many substrates, both vertically and laterally, has become of great importance. A powerful method fulfilling this is Spectroscopic Ellipsometry (SE). This method firstly applied to SIMOX-substrates /17/ offers the possibility of inspecting the vertical microstructure, both, nondestructively and quickly on many lateral positions of a SOI-substrate. Using appropriate models developed by comparing the results of SE and TEM (at cross sectional samples) a fast evaluation of the properties is possible.

Firstly, we have used this method for the investigation of buried silicon oxynitride layers formed in single crystalline silicon by sequential implantation of oxygen and nitrogen. The results were compared with those got by TEM, RBS,etc./18/. The SE-TEM comparison has shown the same thickness values for the different depth regions. Moreover, by SE the oxygen/nitrogen content of these regions could be simulated. Considering the results of all applied methods it was concluded that implanting nitrogen before oxygen led to both, a higher quality of the microstructure and the silicon oxynitride compound, than the reverse sequence. This should be caused by a different stopping power against ions of silicon rich in oxygen or nitrogen /18/.

6. MBE-SOI

It may be somewhat bold to extend the topic of ion beam processing up to the Molecular Beam Epitaxy where most of the particles are not charged but are moved in form of a directed vapour beam. This aspect but also thinking on a possible replace of MBE by the Ion Cluster Beam-technique /19/ with its own advantages should satisfy the discussion of two interesting MBE-applications for SOI in this paper.

The first one joins immediately the methods to reduce the dislocation density of SIMOX-structures recently reported /12,13/. Adapting an experimental finding of compound semiconductor superlattices, Kamins et al./20/ tested the influence of Si-Ge-superlattice structures on the dislocation density of epitaxial silicon layers deposited above SIMOX-wafers. To describe shortly the mechanism, a strain near heteroepitaxial interfaces produces a shear force which is able to bend dislocations parallel to the superlattice rather than allowing them to propagate to the surface. In this manner, nearby dislocations then also interact and will be annihilated or the dislocation may terminate at the edge of the sample. Of course, it is necessary that the superlattice is properly designed.

But one must state that the results were not very encouraging. Although the bending of some dislocations could be observed by XTEM in such SIMOX-SiGe-structures most of the dislocations with a density of about $10^{9}cm^{-2}$ propagated to the surface. However, it must be claimed that the used SOI-structures were no optimised ones from the viewpoint of damage annealing. A 1250°C,6h-anneal step was used so that beyond dislocations also oxide precipitates still remain.

Whereas much more effective and also economic techniques are known to reduce the dislocation density of SIMOX-silicon films, e.g.the multiple implant-anneal-cycle /12,13/, the formation of a single crystalline compound semiconductor film on a SOI-structure could supply all the advantages of both heteroepitaxy and SOI and, may be, some additional ones. This should be of importance for the production of electronic and optoelectronic devices on the same wafer. First successful experiments to use MBE for the heteroepitaxial growth of GaAs on SOI-structures were reported by Das et al./21/ performed on SIMOX-structures. The growth of the GaAs layer was dominated by hillock formation at microscopic mechanically damaged regions of the silicon surface. The cause for this mechanical damage was not reported. However, between the hillocks there was already found an uniform layer growth. Microtwins, threading dislocations, twinning dislocations and antiphase domain boundaries were the predominant defects in these layers. Generally, this is a problem of heteroepitaxy. But on the other hand, it must be stated again that the quality of the SIMOX-structure was not the best one due to the used anneal step: 1200°C,3h. In this manner, a lot of defects still remain within the silicon top layer, especially precipitates. Undoubtly, more optimised SOI-substrates from the viewpoint of defect engineering should lead to much more encouraging results.

7. CONCLUSIONS

1. Nearly all types of ion-solid-interaction causing mass transport and energy deposition into nuclear and electronic processes leading to

-doping (HI-FIPOS),

-compound formation (SIMOX/SIMNI),

-materials deposition (MBE-SOI),

-damage/amorphisation (SPE-SOS, SSIC), and

-dynamical annealing (SIMOX/SIMNI, L-IBIEC)

were successively used to produce SOI-structures itself or to support efficiently other SOI-technologies. Moreover, the specialty of particle channeling in single crystalline matter was also applied with advantage. Only sputtering remains to be used in a constructive manner for SOI.

2. Several techniques reported above are still in the period of early beginning (L-IBIEC, MBE-SOI) whereas SPE-SOS is a well established technology. SIMOX/SIMNI is now a powerful motor to bring SOI to ULSI. Being full of optimism, one can only applaud to J.P.Colinge /22/ for his tentative estimation (Fig.4) of that time where the processing costs with Thin-film SOI substrates will be cheaper than that with conventional bulk wafers.

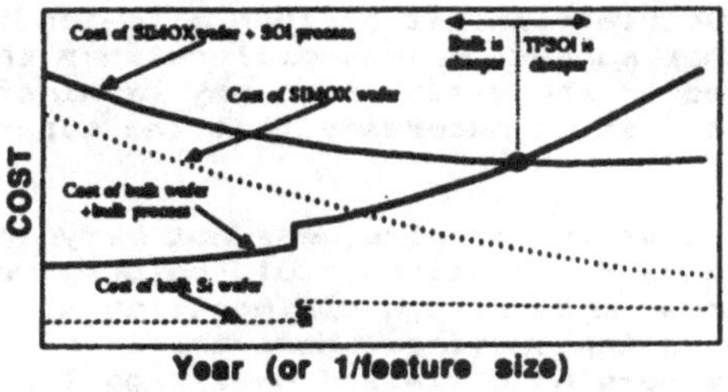

Fig.4: Projected cost evolution of ULSI processing using bulk and Thin film-SOI substrates basing on SIMOX (after J.P.Colinge /22/)

ACKNOWLEDGEMENTS

I would like to thank Drs. R.Grötzschel, V.Heera, K.Wollschläger, M.Voelskow (ZfK Rossendorf), H.Bartsch (IFE Halle), A.De Veirman (RUCA Antwerpen) and J.Vanhellemont (IMEC Leuven) for stimulating discussions on SIMOX/SIMNI and IBIEC.

REFERENCES

/1/ M.E.Roulet, P.Schwob,I.Golecki, M.A.Nicolet: Electron. Lett. 15 (1979)527.
/2/ T.Yoshii, S.Taguchi, T.Inoue, H.Tango: Jpn.J.Appl.Phys. 21(1982)21-1,175.
/3/ K.Imai, H.Unno: IEEE Trans.Electron.Dev. ED-31(1984)297.
/4/ J.D.Benjamin, J.M.Keen, A.G.Cullis, B.Innes, N.G.Chew: Appl.Phys.Lett.49(1986)716.
/5/ See the review paper by J.Linnros and the original papers by D.Hoehl et al. and V.Heera in these proceedings.
/6a/ A.La Ferla, E.Rimini, G.Ferla: Appl.Phys.Lett.52(1988)712
/6b/ W.Skorupa, M.Voelskow, J.Matthäi, P.Knothe: Electron.Lett. 24(1988)876.
/7/ For a recent review see: H.Ishiwara, T.Asano, K. Tsutsui, H.C.Lee, S.Furukawa: Mat.Res.Soc.Symp.Proc.107(1988)241.
/8/ M.Voelskow, W.Skorupa, K.Wollschläger, J.Matthäi, P.Knothe, K.-H.Heinig, H.Bartsch: E-MRS Mtg.,Strasbourg, May 30-June2,1989 and these proceedings.
/9/ R.Reif, J.E.Knott: Electron.Lett.17(1981)587.
/10/ K.T.-Y.Kung, R.Reif: J.Appl.Phys.62(1987)1503 and references therein.
/11/ K.T.-Y.Kung, R.Reif: J.Appl.Phys.63(1988)2131.
/12/ For state-of-the-art see: Mat.Res.Soc.Symp.Proc., 107(1988).
/13/ W.Skorupa, R.Grötzschel, H.Bartsch: phys.stat.sol.(a) 112(1989)661.
/14/ J.P.Colinge: Microelectronic Engineering 8(1988)127.
/15/ J.P.Colinge: IEEE Electron.Dev.Lett.9(1988)97.
/16/ W.Skorupa, K.Wollschläger, R.Grötzschel, J.Schöneich, E.Hentschel, R.Kotte,F.Stary, H.Bartsch, G.Goetz: Nucl. Instr.Meth.B32(1988)440 and references therein.
/17/ J.Vanhellemont, H.E.Maes, A. De Veirman: J.Appl.Phys. 65(1989)4454 and references therein.
/18/ W.Skorupa, M.Voelskow, R.Grötzschel, J.Albrecht, J.Vanhellemont, A.De Veirman: these proceedings.
/19/ I.Yamada: these proceedings.
/20/ T.I.Kamins, K.L.Wang, J.Park, G.E.Davis: J.Appl.Phys. 65(1989)1505.
/21/ K.Das, T.P.Humphreys, J.B.Posthill, N.Parikh, J.Tarn, N.El-Masry, S.M.Bedair, W.K.Chu, J.J.Wortman: Mat.Res.Soc.Symp. Proc.107(1988)513.
/22/ J.P.Colinge: Spring Mtg. of the Electrochem. Soc.,1989

THIN FILM HIGH TEMPERATURE SUPERCONDUCTORS

M. SUBOTOWICZ[1)]

1. Introduction

In connection with the progress in the high temperature superconductors (HTS) their possible applications are widely discussed, especially that of HTS thin films. The perspectives of the realistic zero-resistance electronics are very promising, as well as the other properties related to superconductivity. Now the superconductivity can be achieved at temperatures comparable to those of the semiconductor operation.

Some possibilities for device interconnection for complex, high-speed systems HTS have zero resistance whatever are the sizes of the connections.

They may be used in many high performance electronic systems, demanding high bandwidth and high interconnection density. Let a computer would work at 1 Gbit/s. Typical dimensions are about 10 cm because of the signal transit time in the connections. Here interconnections, high frequency and low dissipation are required. Superconductor films represent the only possible way to build such system. Also- for high bandwidth telecommunication switching, for high interconnection density, heavily loaded lines in computers a.s.o. Other applications are: infrared detectors, Josephson junctions based components (bolometric elements), Josephson switches in some picoseconds, SQUIDs composed of two Josephson junctions connected in a ring etc.

2. HTS - materials

There are following perovskite - type HTS - materials based on rare earth (RE), mainly yttrium, bismuth, thallium and other elements: LaBaCuO, LaSrCuO, YBaCuO, REBaCuO where RE means rare earth element. The critical temperature T_c in yttrium based HTS is equal up to 90 K. The bismuth or thallium based superconducting compounds may have even higher critical temperatures: BiSrCu($T_c \sim$ 20K), Bi(Sr,Ca)Cu ($T_c \sim 80 \div 120$ K), Tl(Ba,Ca)CuO($T_c \sim 80 \div 120$ K), BiPb(Sr,Ca)CuO($T_c \sim 50 \div 120$ K), TlPb(Sr,Ca)CuO($T_c \sim$ 130 K), $Pb_2Sr_2CaYCuO$($T_c \sim$ 50 K). All the above materials reveal the hole conductivity. There exist also some materials showing the electron conductivity, namely $Nd_2CuO_{4-x}F_x$ with T_e = 27 K, as well as $Nd_{1.85}Ce_{0.15}CuO_4$ and $Nd_2CuO_{3.6}F_{0.4}$ with T_e = 21K, /3/, /5/.
The copper - free superconducting compounds as $Ba_{1-x}K_xBiO_3$ with $T_c \sim$ 25 - 30K give also the evidence of the electron conductivity, /3/.

---

1) Experimental Physics Department, Institute of Physics, University M. Curie-Skłodowska, 20-031 Lublin, Poland.

In the paper /9/ was reported another copper - free compound, $La_{2-x}Ni_xO_4$ ($T_c \sim 70K$) where copper was replaced with ferromagnetic element nickel, Ni. Similar replacement was performed in other La-based superconductor, namely in $La_{2-x}Sr_xCuO$($T_c \sim 35K$), where Cu was replaced by Ni. The superconducting material was $La_{2-x}Sr_xN_iO_4$, /9/.

3. HTS thin film processing

Thin films are evaporated on the following single crystalline substrates: $SrTiO_3$, Al_2O_3, MgO, $ZrO_2(+Y_2O_3)$, BaF_2 and Si. Thin film deposition techniques are following: a) vacuum deposition techniques(UHV), b) atmosphere deposition techniques.

There are mainly the YBCO-type films and rare earth - free thin films: Bi-Sr-Ca-Cu-O and Tl-Sr(Ba)-Ca-Cu-O- type.

The following vacuum deposition techniques are used:

1) molecular beam epitaxy (MBE),
2) ion beam mixing,
3) electron beam coevaporation,
4) sputtering (dc, ac, magnetron),
5) ionic sputtering,
6) gas source vacuum deposition, 7) Thermal evaporation.

To the atmosphere deposition techniques belongs the metalloorganic vapour deposition (MOCVD).

Basic processes for the deposition of the perovskite - type thin films led to the following phases; /16/.

I amorphous phase, $T_s < T_{cr}$,
II polycrystalline phase, $T_s > T_{cr}$ or $T_s < T_{cr}$ with postannealing,
III single crystals, $T_s > T_e$ or $T_s < T_{cr}$ with postannealing,

where the temperatures mean: T_s - of the substrate, T_{cr} - crystallizing, T_e - epitaxial, T_a - annealing.

T_{cr} for the perovskite - type oxide is around 500 - 600^oC.

For polycrystalline phase is $T_s > T_{cr}$. This phase can be obtained from the amorphous phase at the temperature $T_a > T_{cr}$, and the single crystalline phase - when $T_s > T_e \gg T_{cr}$.

4. HTS materials where electricity is carried by electrons

In1989 was discovered the superconductivity in $Nd_{1.85}Ce_{0.15}CuO_4$ as the first case of a HTS with electrons as the charge carriers, /3/. All previously characterized copper oxide superconductors are hole-doped. It was shown that substitution of fluorine for oxygen in T´- phase Nd_2CuO_4 provides an alternation route to achieve formal reduction of copper, electron conductivity and T_c = 27 K. The result is unusual, because aniomic rather than cationic substitution gives rise to the superconductivity. Fluorinated $Nd_2CuO_{4-x}F_x$ after annealing in N_2 at T_a = 890°C for t_a = 16h was HTS. Onset temperature was 21K and T_c = 12K. All the fluorinated samples exhibited n-type conductivity at room temperature (Seebeck-effect measurements).

5. Cu - free superconductor (electrons as electricity carriers)

The superconducting perovskite $Ba_{1-x}K_xBiO_3$ has the highest transition temperature (T_c = 31K) aside from the copper oxide compounds /11/. Substitution of monovalent potassium for divalent barium in the parent insulator $BaBiO_3$ is analogous to the substitution of Sr for La in La_2CuO_4.

In $BaBiO_3$ optical spectrum /11/sharp dominant absorption peak at 2eV was observed. It is interpreted to be a Peierls insulator with an optical energy gap of about 2eV.

Measurements of Hall and Seebeck effects in thin films $Ba_{1-x}K_xBiO_3$ give negative coefficients at 300K (Hall) and 80 ÷ 350K (Seebeck). The charge carriers are electrons. Single conduction band is less than half-full as a result of potassium (K) doping. There is no lattice distortion in the superconducting phase of $Ba_{1-x}K_xBiO_3$. In this respect the normal state of metallic $Ba_{1-x}K_xBiO_3$ seems completly different from that in cooper oxide superconductor, in which the energy gap generated by the large Coulomb repulsive energy is preserved and the substitution of divalent for trivalent elements, for example Sr for La in La_2CuO_4, produces a hole in the lower split -off band.

6. Nickel for cooper in HTS

Z. Kąkol, J. Spałek and J.M. Honig of Indiana Center for Innovative Superconducting Technology at Purdue University believe that they have found superconducting material, where nickel substitutes the copper: $La_{2-x}Sr_xNiO_4$. If it is true then: 1) One could expect higher transition temperatures than the current record-holding 125K (or perhaps 130K) materials and 2) formation of a cubic structure with $T_c > 77K$, which would benefit many applications, especially those in the film electronics, /9/.

In the measurements performed so far the following results were obtained:

- an "enormous" anisotropy of the magnetic moments,
- susceptibility X = 10^{-3}emu/cm^3, a value corresponding to 1% of what would be expected for complete flux expulsion,
- four- point resistivity data indicate a crossover from semi- to superconducting properties at 40K with superconducting onset at 19K,
- this material appears to split into two phases below 70K, one is superconducting and the other- antiferromagnetic. The proportions of these

two phases changes with the thermal dycling: the superconducting phase is growing at the expense of the antiferromagnetic one. Deposition process consists of three stages: 1) mixing, 2) annealing for crystallization and sintering, 3) annealing for control of oxygen vacancies. For the deposition /16/ of the HTS thin films the process I is commonly used. After postannealing we obtain the polycrystalline phase. Single crystal films can be obtained in the processes I and II. The vacancies control can be conducted by the temperature in the process II or by oxygen bombardement on the growing film. The transition from the non-superconducting tetragonal phase into the superconducting orthorhombic phase occurs during annealing around 600-700°C.

7. Doping induced insulator - metal transition in HTS

Some recently discovered oxide superconductors are surprisingly sensitive to their doping amount in following cases (S.Uchida, 1989, Inf.Conf. on HTS in Dubna 28.06-01.07, p.19 Abstracts) a) $La_{2-x}Sr_xCuO_4$, b) $Ba_{1-x}K_xBiO_3$, c) $Nd_{2-x}Ce_xCuO_{4-y}$, /14/.

a) $La_{2-x}Sr_xCuO_4$ The substitution of La with Sr corresponds to the doping with holes in the antiferromagnetic insulator La_2CuO_4. The Sr composition can be varied over a wide range $0 \leqslant x \leqslant 0.35$ without making oxygen vacancies. At x=0.25 the superconductor to normal metal transition occurs. The remarkable change in the transport coefficient and magnetic susceptibility has been observed. The Hall coefficient changes sign from negative to positive. Beyond this composition the magnetic susceptibility exhibits a Curie-type temperature dependence.

b) $Ba_{1-x}K_xBiO_3$ This is the only Cu-free compound having $T_c > 25K$. An insulator to metal transition takes place at $x \sim 0.35$. Optical measurements in $Ba_{1-x}K_xBiO_3$ have show optical energy gap for $x < 0.35$. Its magnitude depends on x, decreases with increasing x and vanishes for $x > 0.35$. Explanation of this effect and transport properties in the metallic phase accepts the existence of charge-density-wave and its destruction by doping influences the insulator- metal transition.

c) $Nd_{2-x}Ce_xCuO_{4-y}$ The system N_2CuO_4- type (T′structure) consists of the sheet of isolated Cu-O squares. Contrary to the La-based 2-1-4 stechiometry (T structure) the T′compound cannot be doped with holes. Tetravalent Ce ions can be dissolved reducing the resistance of the antiferromagnetic insulator Nd_2CuO_4. The Hall coefficient indicates for electrons as the charge carriers in the Ce-doped compounds. After reducing the oxygen content in the Ce-doped compound the group (Y.Tokura, H.Takagi and S.Uchida, Nature 337,1989) achieved recently T_c=24K.

8. Tunneling measurements of the HTS-films

Typical methods of the measuring of the superconducting energy gap of HTS- materials are tunneling spectroscopy or low- temperature scanning tunneling microscope and infra- red reflectance techniques. These both /7/

methods were applied in the paper. The best tunneling data of Y-Ba-Cu-O correspond to a superconducting energy gap Δ of about 16 meV or $2\Delta/k_BT_c \approx 4.3$ for the tunneling tip oriented both parallel and perpendicular to CuO_2 planes. The infra- red measurements showed gaps of $\Delta \approx 9$meV, almost two times less than those from the tunneling measurements. The tunneling results are consistent with strong coupling superconductivity. For $La_{2-x}Sr_xCuO_4$ the authors /7/ received $2\Delta/k_BT_c \approx 1.3 \div 2.7$, well below of the BCS prediction of $2\Delta/k_B\ T_c = 3.53$.

The experimental curve shows excess conductance at zero-bias, asymmetry in the two opposite sweep directions, a back-ground conductance that rises nearly linearly with voltage and large additional structure at voltages higher than the gap voltage. This type of characteristic can be connected with granular structure of the HTS-material.

In the paper /15/were measured (I-V) tunnel characteristics in the temperature 4.2 and 53K of the new HTS-material: $Bi_4Ca_3Sr_3Cu_4O_{16+\delta}$ with $\delta = 0.32$. For Δ the authors obtained values Δ ranging between 15 and 23 meV, that corresponds to $2\Delta/k_BT_c$ values between 4.5 and 6.5, consistent with strong-coupling superconductivity. There were very short range of spatial variations.

In the paper /4/the polycrystalline samples of $Bi_2(Sr_{0.6}Ca_{0.4})Cu_2O_4$-HTS material investigated in the temperature range 8.2÷90K. The in the temperature 8.2K obtained Δ value was 6meV that corresponds to the $2\Delta/kT_c = 8.7$. The difference between the BCS and experimental result the authors explain by theory of resonance valence bonds of Surumurai.

For the Y-Ba-Cu-O thin film material was /13/obtained for the pressurized contacts the value $2\Delta = 31.6 \div 34.2$ meV and $2\Delta/kT_c = 5.4 \div 5.9$.

In the paper/10/ the contacts SIS (superconductor-isolator-superconductor) were investigated by the tunnel spectroscopy method. SIS-contact is extremely sensitive detector of the super-high-frequency EM-field, equal up to quantum limit $e/h\omega$, the crvogenic computer using SIS has the switching time 20 ns, the noise level of the mixing device built on the SIS-contact is equal 5×10^{-15}W/Hz$^{1/2}$ for the frequency 90 GHz. This corresponds to the noice level of the Schottky diode at the temperature 1K.

In the paper /1/were measured tunnel characteristics for $YBa_2Cu_3O_{7-x}$ in several phases. For the tetragonal phase (T_c=10K) they received $2\Delta = 30 \pm 0.3$ meV and $2\Delta/kT_c = 3.5 \pm 0.5$. On the SIS contacts of the single crystalline rhomboedric phase was observed non-stationary Josephson effect.

In the paper /2/the $YBa_2Cu_3O_{7-x}$ thin films on various substrates ($SrTiO_3$, MgO, sapphire) with T_{co} = 82; 75 and 72 K, resp. were investigated. There were found for the samples with T_{cc} = 68K and 93K the values of the ratio $2\Delta/kT_c = 5.4$ and 3.9, resp.

In my paper /8/ with the coworkers the $2\Delta/kT_c = 3.5 - 4.5$ was determined in the BiSrCaCuO-HTS-system, as well as the temperature dependence of the energy gap in the bulk material $Bi_{1-x}Pb_xSr_2Ca_2Cu_3O_{10-x}$ for x=0.2.

Both results correspond to the BCS theory.

9. Rare earth and rare earth-free HTS thin films

During the deposition, T_s satisfies the relation $T_e \lessapprox T_s \lessapprox T_t$, where T_t means the transition temperature from tetra-into ortho-rhombic phase. Low temperature synthesis of rare earth HTS thin films was classified in two cases: 1) $T_s > T_{cr}$ (500 - 700°C) followed by postannealing in O_2(Ta $\approx$ 400 ÷ 600°C). This case corresponds to the process II with "in situ annealing" in high vacuum, or when in oxygen- "ex situ deposition"
2) $T_s > T_{cr}$ without postannealing corresponds to the process III ("in situ deposition"), /16/.

Both processes of annealing were studied in sputtering and laser deposition: $T_s \approx 500$ - 700°C and $T_a \approx$ 400 - 500°C in O_2.
The rare earth HTS material has the general formula $REBa_2Cu_3O_{6+x}$ and that of rare earth- free is following: $(TlO)_2Ba_2Ca_{n-1}Cu_nO_{2n+2}$ for n=1,2,3. The most general formula is: $(AO)_mM_2Ca_{n-1}Cu_nO_{2n+2}$, where M may be Ba or Sr or their mixture. The cation (A) may be Tl, Bi, a mixture of Bi and Pb, Bi and Tl or Tl and Pb. The structures contain the CuO sheets, and the number of such copper sheets which have stacked consecutively is indicated by n. The number of consecutively stacked AO sheets is indicated by m, which may be 1 or 2 for A=Tl but only 2 for A=Bi. The environment of the A cation is highly distorted octahedral. They may be some substitution for Ca by Sr or Y. There is no n=0 phase, /12/.

10. Possible high temperature superconductivity at about 300 K of the oxidized atactic polypropylene

In the Russian "Letters to the Journal of the Experimental and Theoretical Physics", March 25, 1989 /6/ was publiehed the paper of N.S. Jenikolapian et al. on the possible HTS of the oxidized polypropylene at about 300 K. The measurements were performed in the Institute of the Synthetic Polymers of the Academy of Sciences of USSR.

The authors investigated the physical properties of the thin atactic polypropylene (PP) films 0.3 - 100 µm thick. The films deposited from the solution on the copper or indium substrate were oxidized in air at the room temperature during three years (PP-1) or in more accelerated oxidization form by thermal heating and treated by UV radiation (PP-2). They contained about 10% oxidized functional groups. Before oxidization the film had good dielectric and small diamagnetic properties; the susceptibility, $\kappa \sim 10^{-6}$.

The samples PP-1 0.3 µm thick in 60% have the resistivity smaller than $10^5 \Omega$, the rest 40% have the typical dielectric's resistivity higher than $10^9 \Omega$. The anomally small resistivity measured in 60% of points on the film surface cannot be distinguished from the point contact resistivity between metal and recording instrument. In some points the PP-1 film has zero resistivity, typical property of the superconductor.

When cooling the PP-1 film between two indium foils from the temperature 293 K ($R \approx 0.3$ GΩ) up to 77 K the resistivity does not change significantly in the temperature range 250 $\div$ 280 K and no change up to 77K. At the following heating from 77 K the resistivity slowly decreases and at the glazing temperature ($T_g \approx 240$ K) stepwise decreases up to the instrumental zero and does not change up to 280 K, when stepwise the resistivity returns to the value $\sim 10^8 \Omega$. This transition equal 280 K is reversible when realizing the scanning between 293 $\div$ 240 K. The reproducibility of the transition temperature 280 K is equal $\pm$3K. The transition width is so small that it was impossible to measure it. In the thin film samples with the accelerated aging (PP-2) the transition temperature was not determined because the indium electrode with growing current up to 0.5 A started to melt (indium melting point is equal 429 K) and the resistivity of the samples was practically equal zero. It means that the transition temperature is $T \gg 429$ K for the films PP-2.

About 10% of the samples attain the anomalous strong diamagnetic state, stable in the room temperature. This state can be destroyed stepwise at the magnetic field $B \geqslant B_{cr}$ with the spontaneous transition of the film into the ferromagnetic state. In connection with the zero resistivity this indicates that the superconductivity is possible in the room temperature. Heat is not emitted in the PP-2 even at the current density 3×10^6 A/cm^2 what confirms the superconducting state of the investigated thin films.

11. Ion beam formation of the HTS thin films

HTS films are formed by ion beam induced mixing of the multilayered Y-Ba-Cu systems by Rauschenbach and Hohmut (1988, 1989) at about 400°C and postannealing in flowing oxygen at temperatures up to 800°C. Favourable is mixing with oxygen against the xenon mixing. They have onsets between 110 and 120 K, and completely HTS at around 80 K.

In the paper of Kato et al. (1988) the effect of the hydrogen ion implantation on the superconductivity in Y-Ba-Cu-O was investigated . Both the temperatures of superconducting onset and zero resistance were lowered through H^+ ion implantation and the superconductivity disappeared by implantation of 1.5×10^{17} ions/cm^2. In several papers presented now during EPM-89 the role of ion beams in the formation of HTS is discussed, namely - the role of electron and ion beam pulses on the surface effects in yttrium HTS (Korenev et al.1989). Ion other paper (Nikonov et al., 1989) the enhancement of T_c in Y-Ba-Cu-O after proton irradiation was reported. Suvorov et al. (1989) investigated the HTS of Bi-Sr-Ca-Cu-O thin films obtained by ion beam sputtering. After implantation of 40-90 keV oxygen ions and annealing ($T_a = 850 \div 890$°C) they obtained $T_c < 68$ K.

Acknowledgement

This paper was sponsored by the Program CPBP 01.08 and 01.12 (Poland).

Compound	Film deposition	T_c(K)	I_c(A/cm^2)	Stochiometry	Substrate
BiSrCaCuO	r-f-planar magnetron sputtering, evaporation	80 120		2-2-1-2 2-2-2-3	MgO MgO
TlBaCaCuO			$1.2x10^5$(T= =77K) $8.5x10^6$(T= =4.2K)		
YBaCu	Ion beam mixing	~80		1-2-3	ZrO_2
$YBa_2Cu_3O_{7-x}$	Hydrogen implantation	no HTS at dose 1.5x $x10^{17}H^+cm^2$			
$YBaCu_2Oy$	7MeV electron irradiation	no HTS at dose 1x $10^{17}e/cm^2$		1-1-2	
YBCO	evaporation	80 69	$9x10^4$(4,2K) 10^6(77K) $3x10^3$(4.2K)		Si $SrTiO_3$ Si,buffer
YBCO	sputtering, evaporation, laser	60-85-91	$5x10^3$(77K)	1-1.9-2.6	ZrO_2,Si_3N_4
YBCO	evaporation				Al_2O,$SrTiO_3$ Si
$Bi_4Sr_3Ca_3Cu_4O_x$	"flash" evaporation	78-72		4-3-3-4	MgO
YBCO	pulsed laser (KrF excimer laser)	85 75			$SrTiO_3$ sapphire
$Y_1Ba_2Cu_3O_{7-x}$	laser writing SQUID			1-2-3	MgO
$DyBa_2Cu_3O_{7-x}$	MBE	89	$4.8x10^5$(4.2K) $3.9x10^4$(77K)		$SrTiO_3$
$Y_1Ba_2Cu_3O_{7-x}$	MBE	82	10^3(77K) $5x10^4$(70K) 10^6(4.2K)		
$Y_1Ba_2Cu_3O_{7-x}$	SQUID				
$Bi_2(Sr,Ca)_4\cdot Cu_3Oy$		110			MgO
BiSrCaCuO		107			MgO(100)
BiSrCaCuO	laser sputtering	120			MgO,ZrO_2, $SrTiO_3$
YBaCuO	UV-light irradiation				

Remark: The references to the above table may be obtained from the autho of the present paper.

References

(1) Aminoff B.A. et al., "Tunnel spectroscopy of the single crystals $YBa_2Cu_3O_{7-x}$", preprint of the Moscow State Univ., Phys. Fac. 1988, No. 27, p.1-5,

(2) Benatchka S. et al., "Tunneling spectroscopy in thin film superconducting junctions YBCO Pb", Solid State Commun., 68(1988)753,

(3) "Holes and electrons", Superconductivity News, 3(1989)1, No.2,

(4) Ikuta H. et al., "Tunneling spectroscopy of the 80K Superconducting Phase of the Bi-Sr-Ca-Cu-O system", Jap.J.Appl.Phys., 27(1988)LT038,

(5) James A.C.W., Zahurak S.M. and Murphy D.W., "Superconductivity at 27K in fluorine-doped Nd_2CuO_4", Nature, 338(1989) 240,

(6) Jenikołopian N.S. et al., "Wozmoshnaja swiechprowodimost, okislionnogo polipropilena w obłasti 300K", Letters to Zh.Eksp.Th.Physics, 49 (1989) 326,

(7) Kirtley J.R. et al., "Tunneling and infrared measurements of the energy gap in the high critical temperature superconductor Y-Ba-Cu-O", Phys. Rev. B, 35 (1987) 8846,

(8) Mazurek P. et al., "Properties of the BiSrCaCuO - HTS films prepared by the "flash" evaporation and tunneling characteristics", Proc. of the HTS-Conference in Wrocław (Poland), June, 1989,

(9) "Nickel for copper", Superconductivity News, 3(1989)1, No.2,

(10) Picirillo L. and V. Covelli, "SIS junctions: Theory and applications" Note recens. a notiz., 36 (1987) 107,

(11) Sato H. et al., "Optical study of the metal-insulator transition on $Ba_{1-x}K_xBiO_3$ thin films", Nature, 338 (1989) 241,

(12) Sleight A.W., "Chemistry of High-Temperature Superconductors", Science, 242 (1988) 1519,

(13) Svistunoff V.M. et al.,"Tunnel effect in the superconducting films Y-Ba-Cu-O", Phys. of Low Temp., 14 (1988) 1095,

(14) Uchida S., "Doping induced insulator-metal transition in high T_c oxide superconductors", Abstracts of the International Conference on HTS in Dubna, USSR, 28.06-01.07.1989, p. 19,

(15) Vieira S. et al., "Tunneling measurements of the energy gap in $Bi_4Ca_3Sr_3Cu_4O_{16+x}$", Phys. Rev. B, 38 (1988) 9295,

(16) Wasa K. et al., "Thin films processing for high T_c superconductors", Moriguchi, Osaka, Matsushita Electric Industrial, Osaka, 1988.

STIMULATED PHASE TRANSITION IN THE AMORPHOUS SILICON LAYER-SINGLE CRYSTAL SILICON SUBSTRATE SYSTEM

A. F. Vyatkin

Institute of Microelectronics Technology and High Purity Materials, USSR Academy of Sciences, 142432 Chernogolovka, Moscow District, USSR

ABSTRACT

A phenomenological model of the processes of thermal solid phase epitaxial growth (SPEG) and ion-beam or electron-beam induced epitaxial growth (IBIEG) is suggested. The model is based on the idea about the common nature of SPEG and IBIEG. The model takes into account all the processes occuring in amorphous layer, crystal phase and amorphous-crystalline interface (a-c interface).

It is known [4] that solid-phase epitaxial growth takes place in the amorphous silicon layer-single crystal silicon system (a-c system) at $T > 500^{\circ}C$. Solid-phase epitaxial growth also proceeds in the a-c system at $150^{\circ}C < T < 500^{\circ}C$ under ion-beam irradiation [26] or at room temperature under electron-beam irradiation [15]. We will denote the former process as SPEG and the two latter as IBIEG. The thermodynamic driving force of the processes is the difference in Gibbs free energy between a-Si and c-Si. This difference is negative for all temperatures, ΔG_{a-c} 0. However, under the equilibrium conditions visible crystal growth occurs only at $T > 500^{\circ}C$ because of kinetics limitations. In [26, 15] it is shown that ion-beam induced and electron-beam induced growth are athermal phenomena and both processes are proportional to nuclear deposited energy. SPEG and IBIEG exhibit different values of activation energy, 2.7 ± 0.35 eV and 0.3 ± 0.5 eV, respectively [4, 26, 29]. This enabled some authors to conclude that SPEG and IBIEG are of different nature and, hence, based on different atomic processes.

The goal of the present paper is to prove that SPEG and IBIEG have a common nature and to develop a phenomenological model based on this statement. The main atomic process of a-c transition should be a process when at the initial stage of the transition the atom belongs to the amorphous phase and at the final stage of the transition it belongs to the crystal phase. As mentioned above, appropriate kinetics conditions along with the energy conditions are required to provide such a-c transition. The atom migrates across the a-c interface when the kinetic processes of the amorphous and crystalline phases provide a proper condition for transition. For example, variation of the atomic structure of the phases adjacent to the interface is possible. This variation results in decreasing potential barrier height and, hence, increasing transition probability. Atomic structure variation could take place due to defect formation near the a-c interface and following the defect migration on the a-c interface. All these processes of defect formation and migration like the basic atomic process of a-c transition are characterized by their own energy parameters. Thus, a-c transition includes a sequence of several atomic processes. The main atomic process in this case could be diffusion jump of an atom across the a-c interface.

The opposite way of a–c transition is atomic motion due to the development of the collision cascade near the c–a interface during IBIEG. In this case the atom belonging to the amorphous phase is involved in the collision cascade induced by ion-beam irradiation. It is quite feasible that the atom stops its cascade motion on the crystal side of the a-c system. However, the reverse process of atom transition from the crystalline phase to the amorphous phase is equally probable. Therefore, the net effect of such transition is equal to zero. Hence, the diffusion jump across the a–c interface is likely to be the main IBIEG process This conclusion is confirmed by the experimental data. Almost all the experimental investigations of SPEG have been made using specimens with ion implanted amorphous layers. In this case nonuniform depth accumulation of damage takes place during ion irradiation [27] . This damage distribution is conserved in the material after ion beam bombardment of the sample because irradiation is generally carried out at room or liquid nitrogen temperature. This indicates that at least the initial stages of both SPEG and IBIEG occur under equal conditions. This is also confirmed by the experimental data of crystallization in an a–c system under the pulsed ion irradiation conditions [12]. In this case ion-beam induced crystallization proceeds for some time after the termination of the pulse action.

The thermodynamics laws permit the transition in the a–c system for all temperatures in one direction only, namely, from the amorphous state to the crystalline state (a–c transition). However the reverse transition from the crystal to the amorphous phase is feasible during IBIEG [12, 13] . Some authors call this process, step-by-step amorphization. At first sight the presence of the reverse transition points to the difference in nature between the SPEG and IBIEG processes, because for SPEG as an equilibrium process thermodynamics does not permit reverse transition. Nevertheless, the reverse c–a transition has been experimentally found under the thermal equilibrium conditions of the SPEG process [17]. This is apparently inconsistent with thermodynamics. When calculating energy balance it is necessary to take into account the free energy of the a–c interface. The a–c interface length decreases during reverse c–a transition [17] (Fig. 1). Here the free energy loss caused by c–a transition is compensated by the energy gain due to the decrease in the a–c interface length.

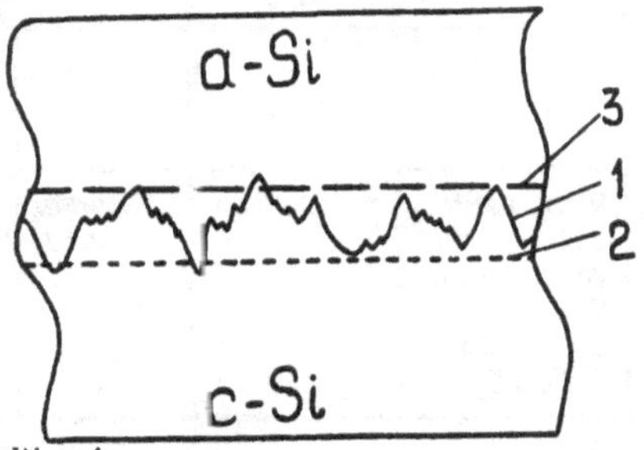

Fig 1
1–a–c interface initial position
2–a–c interface position after c–a transition
3–a–c interface position after a–c transition

It is clear that the a–c interface flattening through a–c transition results in a greater energy gain then does that caused by c–a transition. Probably, under the experimental conditions of [17] (structure of the amorphous and crystalline phases near the a–c interface, the number and mobility of point defects in this region) the a–c transition was prevented by the kinetics limitations.

Thus, it follows from the foregoing that in order to understand and describe the SPEG and IBIEG processes it is necessary that they should be considered from the point of veiw of their common nature taking into account the processes that occur in all the components of the a-c system, i.e. in the crystal, amorphous layer and the a-c interface. In other words, comprehensive knowledge is required about the atomic structure of the crystalline and amorphous phases near the interface, the a-c interface structure and the defect properties in the 3 components of the a-c system.

Sufficient experimental data are now available only on the crystalline part of the a-c system (e.g. see [9]). There are some data on the amorphous phase [10, 11], and none is available on a-c interface atomic structure and its defects. On the atomic scale a-c interface is unlikely to be sharp but has a thickness of several lattice parameters and is characterized by a transition structure [12].

Clearly the experimental data available are not sufficient for analytical description of SPEG and IBIEG. Therefore, only phenomenological description of the processes is possible.

A number of theoretical models have been developed, which describe different experimental results [1,6,13-16]. Yet none of them affords complete description of the process.

Table 1 lists the most characteristic experimental dependencies of the SPEG and IBIEG processes.

Table 1.

Method of the energy input into the a-c system Characteristic dependence	Thermal annealing	Rapid thermal annealing	Ion-beam irradiation	Electron-beam irradiation	Laser annealing
Temperature E_a(eV) dependence V_o	2.7±0.35 [4,11]	2.7±.2 V_o↑ [1]	.3±.05 [12]	-	V_o↑[19]
Orientation dependence $V_{Si(100)}: V_{Si(110)}: V_{Si(111)}$	[3] 24: 6: 1	-	[25] 2: -: 1	-	-
Concentration dependence $V_B: V_p: V_{Si}: V$ gas impurity	[24] 20:6:1:.1	[1] 4:(3.5-6.5):1:-	-	[8] 1:2:1: -	-
Pressure dependence $V_{(p=5\cdot10^4 atm)}: V_{(p=1atm)}$	[14] 10	-	-	-	-
Type of defects responsible for the process	Extended defects, vacancies dangling bonds	Interfacial defect sites	Divacancies vacancies interstitials [29]	Vacancies	Charged defects

Here: E_a-activation energy of the process

V-interface motion velocity, $V=V_o \exp(-\mathcal{E}a/kT)$

↑↓-arrows denote increasing or decreasing of the V_o value

The data of the last line were taken from the models which were developed in the experimental works referred to in the Table 1. Analysis of the most of the experimental data and the models developed bring us to the conclusion about dominating role of the defects in SPEG and IBIEG. However, it is not possible to inducate a single type of defects which are responsible for the whole processes. Unambiguous relationship has been found between activation energy of the processes and activation energy for migration or formation of point defects (e.g. see [6]).

However, up to date ehere have been no comprehensive studies of the influence of a given type of defects on the SPEG and IBIEG parameters. Experiments on the determination of the possible effects of interstitial migration across the a-c interface on its motion velocity have been made recently [2]. The data obtained did not reveal any such effect. The method of creating excess defects may have had a decisive influence on the results. Excess concentration of interstitials was ensured near the a-c interface but the atomic structure of the phases and defect mobility, in this region remained unaltered.

It might be well to point out another experimental result when a flux of excess defects onto the a-c interface (or from the a-c interface) did not affect its motion velocity [9]. In this experiment Si – SiO_2 interface in SIMOX material may have been a source of point defects. Clearly the origin of defects in this case is quite different from that of defects induced by ion-beam annealing [12]. We have recently found that Si – SiO_2 interface in SIMOX material can serve as a sink for defects produced by ion-beam irradiation [28] (Fig. 2). As seen from Fig. 2, at equal doses of irradiation of bulk silicon and the silicon layer in SIMOX material the amount of damage is different in these two materials and is less in SIMOX material.

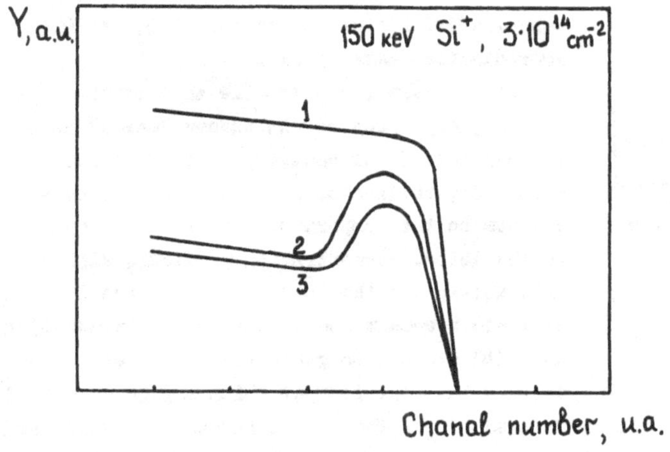

Fig. 2 RBS spectra of two silicon samples.
1. Random spectrum; 2. Aligned spectrum of bulk silicon specimen;
3. Aligned spectrum of the silicon layer in SIMOX material.

The data obtained in [2,9] point to the absence of an unambiguous influence of point defects on SPEG and IBIEG.

Summarizing all the experimental data of SPEG and IBIEG (but a small amount of them has been discussed in the foregoing) we can suggest a phenomenological model of SPEG and IBIEG. This model is based on a set of assumptions: 1) the main atomic process is common for SPEG and IBIEG; 2) SPEG and IBIEG are multistep processes; 3) SPEG and IBIEG include processes which occur in crystalline and amorphous phases and on the a-c interface; 4) orientation dependence is described by Narayan's model [18] ; 5) the concentration dependence is described by the model suggested in [16] . Within the frame work of the above assumtions the essence of the model is as follows.

Under the conditions of SPEG and IBIEG two defect fluxes are directed onto the a-c interface. These defects originate in the regions, situated near the a-c interface in the amorphous and crystalline phases. Let these fluxes be denoted by Q_1 and Q_2 respectively. The a-c interface has the property of transferring across itself atoms from one side of the interface into the other side. This points to diffusion jumps across the a-c interface. Atoms occupy equilibrium positions in the crystalline or amorphous phases as soon as diffusion jumps are over. A-c or c-a transition is feasible when there are proper conditions for diffusion jumps in the proximity of the atom. Such conditions are created due to the changes in the atomic structure of the phases near the a-c interface. Such changes are initiated by the fluxes excess defects directed onto the a-c interface. We assume that the process of atom migration across the a-c interface has a high rate and is not a limiting stage of the whole process. The direction of a-c interface motion, i.e. realization the crystallization or amorphization is determined by the Q_1-Q_2 flux ratio (Fig. 3). When the defect flux from the amorphous phase side exceeds that from the crystalline phase side, i.e. $Q_1 > Q_2$, the direction of interface motion coincides with the positive direction of the Z-axis and, hence, crystallization occurs. At $Q_2 > Q_1$ amorphization takes place.

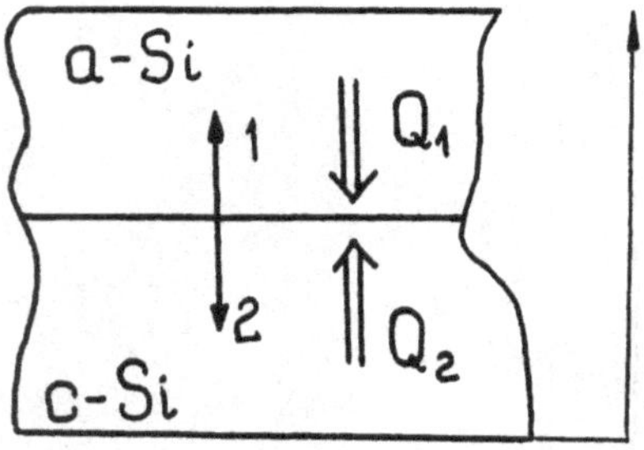

Fig. 3 Scheme of displacement a-c interface VS the Q_1-Q_2 flux ratio
1- Q_1 Q_2
2- Q_2 Q_1

This pattern resembles the zone melting process, as if one of the phases "melts" near the a-c interface. However, solidification into a crystalline or amorphous phase occurs not due to the temperature gradient but due to the defect concentration gradient. Within this assumption the role of temperature in SPEG and ion-beam and electron-beam irradiation in IBIEG amounts to generation of excess point defects the behaviour of which may be described by the model recently developed by Holland [10] . The set of equations describing the defect dynamics in the crystalline phase should be supplemented with an analogous set of equations describing the defect dynamics in the amorphous phase.

Taking into account this model, it is easy to explain the experiments on pulsed ion-beam irradiation and the inversion of crystallization into step-by-step amorphization 6,12 . It is clear that the inversion follows from the change in the Q_1-Q_2 flux ratio. Pulsed ion-beam annealing experiments and, particularly, the frequency dependence due to varying amplitude of the ion-beam pulse may also be accounted for by this model. This is illustrated by the diagrams of Fig. 4, 5 and 6. If, according to the model, the distance of the a-c interface motion is proportional to the diffusion flux of defect onto the a-c interface, it is easily seen that the narrow frequency band exists when the ratio of the distance of a-c interface motion ΔS and the total value of the interface shift S changes within visible limits. At low frequences is negligibly small. At high frequences it reaches the value 1/2 and becomes constant. This behaviour of the $\Delta S/S$ ratio agrees with the experimental findings [6].

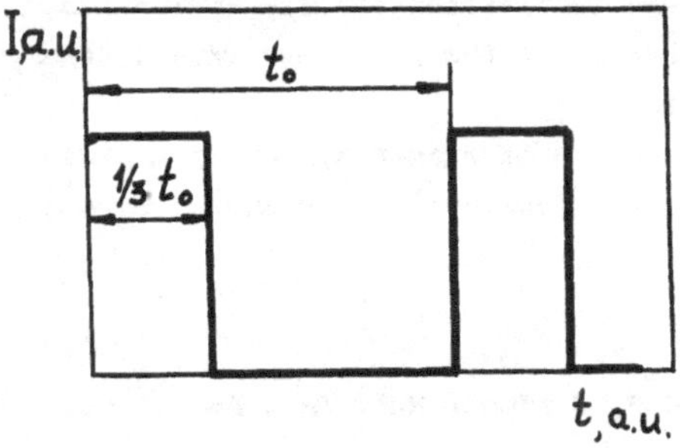

Fig. 4

Fig. 6 shows, that variation of ion current leads to variation of the $\Delta S/S$ ratio. In this case it is assumed that the defect flux is proportional to defect generation rate G and G in turn is proportional to dose rate I .

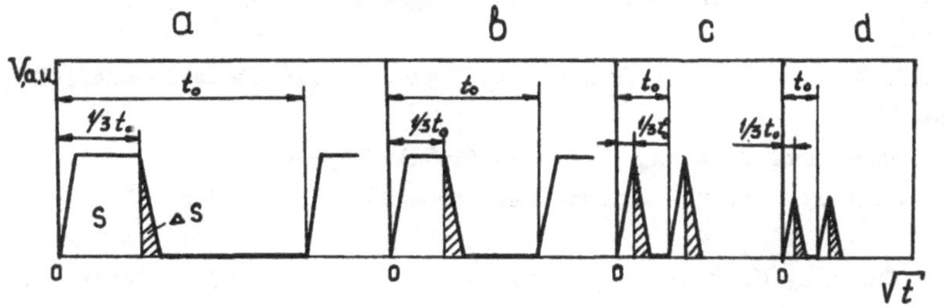

Fig. 5

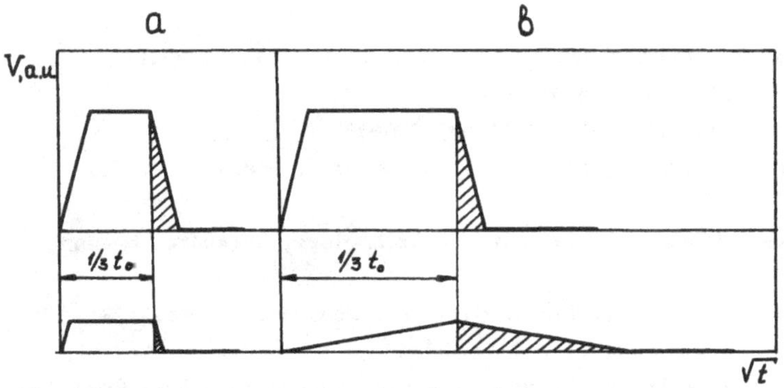

Fig. 6
Dependences of the velocity of a-c interface motion VS time; a,b,c, d-illustrate frequency variations $S=v \cdot t$ -distance of the a-c interface motion ΔS -shaded part - addition to the value S; ΔS is proportion to the value of the interface motion when ion irradiation is switched off.

It is also seen that to provide conditions for the shift of the frequency dependence in the direction of small frequences, as it is shown in 6 , it is necessary to assume that decrease in pulse amplitude is accompanied by a decrease in defect mobility. This results in the change of the curve slope of the V vs t dependence.

CONCLUSIONS

1. The essence of the SPEG and IBIEG processes can be understood taking into account all the processes occuring in the components of the a-c system: amorphous phase, crystal phase and a-c interface.

2. The difference in the number of parameters (activation energy, activation volume, etc.) results from the multistage mechanism of the processes rather than the difference in nature between SPEG and IBIEG.

3. Solid-phase epitaxial growth requires creation of excess defect of a given type which can stimulate variations of the atomic structure of the amorphous and crystalline phases near the a-c interface.

LITERATURE

1. Adekoya, W.O., M. Hage-Ali, J.C. Muller and P. Siffert: Mat. Res. Soc. Symp. Proc., 100 (1988) 447-452.
2. Buravlyov, A.V., A.G. Italyantsev, Z.Ya. Krasnobayev, V.N. Mordkovich and A.F. Vyatkin: Nuel, Instr. Meth., 39 (1989) 366-370.
3. Csepregi, L., E.F. Kennedy, J.W. Mayer and T.W. Sigmon: J. Appl. Phys., 49 (1978) 3906-3911.
4. Csepregi, L., R.P. Kullen, J.W. Mayer and T.W. Sigmon: Solid State Commun., 21 (1977) 1019-1024.
5. Drozd, R. and J. Washburn, J. Appl. Phys., 53 (1982) 397-402.
6. Elliman, R.G., J. Linnros and W.L. Broun: Mat. Res. Soc. Proc., 100 (1988) 363-367.
7. Feola La, A., S. Cannavo, G. Feola, S.U. Campisano, R. Rimini and M. Servidori: Nucl. Instr. Meth., B 19/20 (1986) 470-475.
8. Fahey, P.M., P.B. Griffin and J.D. Plummer: Reviews of Modern Physics, 61 (2) (1989) 289-294.
9. Hemment, P.L.F., E.A. Maydell-Ondrusz, K.G. Stephens and P.D. Scovell: Electronics Letters, 19/13 (1983) 483-485.
10. Holland, O.W.: Appl. Phys. Lett. 54 (4) (1989) 320-324.
11. Lietoila, A., A. Wakita, T.W. Sigmon and J.F. Gibbons: J. Appl. Phys., 53 (1982) 4399-4403.
12. Linnros, J.: PLD thesis Chalmers University of Technology, Göteborg, Sweden, 1985.
13. Linnros, J., R.G. Elliman and W.L. Brown: Mat. Res. Soc. Symp. Proc., 74 (1987) 477-478.
14. Lu, G.Q., E. Nygren, M.J. Aziz, D. Turnbull and C.W. White: Mat. Res. Soc. Symp. Proc., 100 (1988) 433-437.
15. Lulli, G., P.G. Merli and M.V. Antisari, Proc. Mat. Res. Soc., 100 (1988) 375-379.

16. Mosley, L.E., P.J. Germaine, M.A. Paesler and K. Zellama: J. Non-Cryst. Solids 59/60 (1983) 273-278.
17. Myers, E., G.A. Rozgonyi, D.K. Sadana, W. Maszara, J.J. Wortman and J. Narayan: Mat. Res. Soc. Symp. Proc., 35 (1985) 521-526.
18. Narayan, J.: J. Appl. Phys., 53 (1982) 8607-8612.
19. Olson, G.L., J.A. Roth, L.D. Hess and J. Narayan: K.I.K. Scientific Publishing Tokyo, (1985) 73-78.
20. Pantelides, S.T., Mat. Res. Soc. Symp. Proc., 100 (1988) 387-392.
21. Pantelides, S.T.: Phys. Rev. Lett., 57 (1986) 2979-2984.
22. Spaepen, F. and D. Turnbull: Laser-Solid Interaction and Laser Processing, edited by S.D. Ferris, H.J. Leamy and J.M. Poate (American Institute of Physics, New York, (1987) 73.
23. Sinke, W.C., T. Warabisako, M. Miyao, T. Tokuyama, S. Rooda and P.W. Saris: Mat. Res. Soc. Symp. Proc., 100 (1988) 549-554.
24. Suni, I., G. Goltz, M.G. Grimaldi, M.A. Nicolet and S.S. Lan: Appl. Phys. Lett, 40 (1982) 269-274.
25. Sundgren, J.E.,: Programme E-MES 1989 Conference, A-IV 1 Strasbourg, France.
26. Svensson, B., J. Linnros and G. Holmen: Nucl. Instrum. Meth., 209/210 (1983) 755-759.
27. Thornton, J., K.C. Paus, R.P. Webb, A. Albu-Yaron, G.R. Booker and I.H. Wilson, submitted to publication.
28. Vyatkin, A.F., P.L.F. Hemment, K. Reesson, A. Robinson: unpublished.
29. Williams, J.S. and R.G. Elliman: Phys. Rev. Lett., 51 (1983) 1069-1074.

IONIZED CLUSTER BEAM DEPOSITION
-RECENT PROGRESS IN RESEARCH AND APPLICATIONS-

Isao Yamada

I INTRODUCTION

About 17 years have passed since the ionized cluster beam (ICB) technique was initiated at Kyoto Universtiy [1]. Since then, we could see much progress with respect to both fundamentals and applications. During the past years, characteristics of ICB deposition in forming high quality thin films and in specific new application areas have become much more clear. Especially in the areas of optical coatings and VLSI metallization, the ICB technique becoming one of the most important technologies and cannot be replaced by any other techniques. This paper reviews present status of the ICB technique research and applications.

II RECENT PROGRESS IN FUNDAMENTAL STUDIES

a. Film formation processes

Computer simulation and experimental study of growth processes have been made to understand the effects of ICB bombardment in film growth processes. According to the simulation by Yamamura [2], it has been shown that the cluster bombardment enhances adatom migration during ICB deposition. Experimental verification of enhancement of surface migration has been made by Levenson et. al. [3]. Observations of migration distance as a function of acceleration voltage and of effects of the surface condition upon the migration distance have shown the characteristic behavior of the ICB deposition. Müller used a molecular dynamics technique to simulate ICB deposition in a two dimensional model [4]. ICB deposition mechanisms cannot be directly observed during the process of deposition. The computer molecular dynamics simulations can offer a detailed study of time evolution of cluster-substrate collective interaction. We expect better understanding of the characteristics of ICB deposition from the results.

Grain growth control during ICB deposition could be clearly observed in Al deposition onto an SiO_2 surface by Yahashi et.al. [5]. Much larger grains than those formed by vacuum evaporation could be formed in the ICB deposition. It has been also shown that in ICB deposition on the bottom of a V-groove substrate, Al grains larger than grains on the slopes of the V could be deposited. The results suggest that large grains which seem to be single crystal could be generated.

Ion Beam Engineering Experimental Laboratory, Kyoto University, Sakyo, Kyoto 606 Japan

Discovery of epitaxial growth at room temperature of aluminum thin films on Si(111) and Si(100) substrates by ICB technique has opened a new area of materials science in terms of epitaxial growth on large misfit substrates and a new approach to VLSI metallization which faces difficulties for further device miniaturization [6]. A remarkable fact concerning these results is that the epitaxial films could be formed at nearly room temperature and on a large mismatch (25%) substrate surface. The aluminum films exhibited a sharp interface with the silicon surface, are thermally stable and resist oxidation and electromigration. Recent ICB deposition of epitaxial Al films onto Si substrates has proceeded toward formation of nearly perfect single crystal films [7] and to analyses of the atomic structure at the large misfit interface. Atomic resolution observation and in-situ annealing tests were carried out with the collaboration of Westmacott and Dahmen of the National Center for Electron Microscopy, Lawrence Berkeley Laboratory, University of California. The results show that epitaxial Al films on Si(111) and Si(100) substrates have sharp interfaces and only one atomic layer forms the boundary between the aluminum film and the silicon substrate surface [8].

More recently, deposition of gold films onto silicon substrates at low temperatures by the ICB technique in an UHV environment has been studied in conjunction with in-situ XPS analysis [9]. The results suggest that the ion irradiation suppresses Au-Si mixing at the interface and evokes layer by layer growth over three-dimensional island formation. Sosnowski et. al. [10] studied interface characteristics of copper films on Si(111) substrates. The results suggest epitaxial growth in the early stages of film formation and a smooth surface with no silicon segregation in the film.

Unique characteristics of ICB deposition recently addressed are summarized as follows: (1) Extension of epitaxial growth to large lattice misfits, (Al epitaxy on Si, Al_2O_3, CaF_2 [11]), (2) Formation of stable interfaces, [8,9,10]), (3) Control of Au/Si [9] and Cu/Si interfaces [10]), (4) Control of crystalline system, (Change from anatase to rutile or vice versa in TiO_2 deposition [12]) and (5) Control of stoichiometry, (Oxides, nitrides, etc, in reactive deposition [13]).

b. Cluster formation

Research on metal cluster formation is one of the important subjects needed to support ICB technique. We have developed a cluster formation theory which assumes homogeneous nucleation in the expanding vapor from the crucible nozzle [14]. Computer simulations of this process indicate that for crucible pressures above a few Torr, clusters should be formed and that for higher pressures, the cluster size should increase. Using the computer simulation. formation mechanisms of large size clusters, such as effects of vapor pressure in the crucible upon cluster nucleation and growth can be examined. Since the computer simulations cannot simulate the exact expansion processes, the numerical data obtained using the simulation may not be quantitatively correct. However, large size cluster formation mechanisms could clearly be addressed. Knauer of Hughes Research Laboratories [15] has recently shown the existence of an alternative process by which clusters may be formed on the crucible walls by heterogeneous nucleation and then subsequently swept out. His calculations indicate that significant numbers of clusters may be formed at operating conditions which are presently obtainable.

Experimental verifications of cluster formation have been made by several other groups. Gspann of the University of Karlsruhe has used a specialized crucible nozzle [16]. Using cesium he obtained significant quantities of clusters in the size range of 1000 atoms per cluster. Shanks of Iowa State University reported silver clusters of up to 35 atoms observed by time of flight (TOF) method [17]. More recently Ag clusters of 650 atoms median value were obtained with a 0.5 mm nozzle. In our laboratory, TEM observations of clusters collected on cooled substrates have continued in order to develop a direct method to observe the cluster size. In recent experiments, special attention has been given to deposition of small amounts of material with reasonable control [18]. This was accomplished by introducing a rotating chopper to reduce the effective deposition. In order to suppress the surface migration of the deposited atoms, a carbon overcoating has been introduced during or after collecting the clusters. The results are quite consistent with previous observation. The size of the clusters is in a range of 1.5 nm diameter at typical operation conditions.

111 APPLICATIONS TO OPTICAL AND ELECTRON DEVICES

a. Optical applications

High quality ICB films are expected to have many new applications over optical range from X-ray to visible. Some recent results of an X-ray mirror with low stray scattering, an excimer laser mirror with very high damage threshold power and dielectric optical coatings will be introduced here.

Hashimoto et. al. [19] developed high X-ray reflection films with very smooth surfaces. Gold films 50 nm thick were deposited on polyimide substrates. Deposition area was 100 mm in diameter. The RMS height variation, which is a measure of the surface roughness, was 0.47 nm for the ICB deposited films, but for the film deposited by vacuum evaporation was 1.05 nm. The results show that stray scattering can be reduced one order of magnitude by the ICB films and an increase in relative integrated reflection of 10 - 30 % can be obtained. The films are expected to be used in space for X-ray telescope in a space rocket and therefore the films must exhibit high reliability. Strong adhesion, long lifetime, and high radiation damage resistance are required of the films.

Applications of high power excimer lasers which emit ultraviolet light at high intensity in a wavelength region of 193 nm (ArF) - 351 nm (XeCl) are rapidly developing in many fields. Accordingly, thin film coatings with high damage resistance on various substrate materials are required. However, presently available high reflectance metal coatings do not have sufficient damage thresholds to be used with pulsed lasers, particularly in the short wavelength region. Characteristics of aluminum single crystal films are being studied in order to apply them as high power excimer laser mirrors [20]. A 500 nm Al single crystal film was deposited onto a Si(111) substrate. After the deposition of the Al film, a MgF_2 protective overcoating was deposited in a separate chamber by electron beam evaporation. The MgF_2 film thickness was 70 nm. The damage thresholds of the Al films were measured using an acousto-optical Q-switched ArF excimer laser with a pulse duration of 20 ns and repetition at 100 Hz for 5 sec. The damage threshold of the films was 0.4 J/cm . The value is substantially higher than that of films deposited by vacuum evaporation. Angular dependence of the reflection has been measured over a rang of 200 - 800 nm. Incident angle of the beam was varied from 0 to 45 degrees from surface normal. No significant angular dependence was observed. This result suggests that the film could be used for a wide range of beam wavelengths at any incident angle. Careful examination of the damaged area has revealed that no melting of the coated Al and MgF_2 could be seen. The damage to the film is clearly seen as peel-off of the MgF_2 film. This indicates that the epitaxial Al film itself has even higher damage threshold. Improvement of the MgF_2 film coating is now being studied.

In dielectric coatings, formation of high density, smooth and pinhole-free films is an important issue. Flat and crack free CaF_2 films and high density TiO_2 and Al_3O_2 films can be formed by ICB deposition. In CaF_2 deposition onto Si(111) and Si(100) substrates, the large difference in thermal expansion coefficient between the film and substrate causes peel off and cracking of the deposited films. The effect of ion acceleration voltage was studied [21]. No cracks were observed on films deposited at higher acceleration voltages. Crystalline quality is also improved as the acceleration voltage is increased. In TiO_2, Al_2O_3 and other oxide depositions, increases in refractive index with increase of the acceleration voltage are generally characteristic of ICB deposition. In TiO_2 deposition, surface smoothness has been measured as a function of acceleration voltage [22]. Significant reduction of surface roughness could be obtained. These results show that ICB can form films with high density and extremely smooth surfaces.

b. Applications to electron devices

Two important achievements by which we could demonstrate unique characteristics of ICB deposition are formation of epitaxial aluminum thin films [11] and epitaxial Y-Ba-Cu-O thin films [23]. Those films could be used as contacts and as interconnects for VLSI or ULSI semiconductor devices. Metallization is presently the major factor limiting the density of device elements in high-speed semiconductor devices. Many problems asociated with small scale metallization have been recognized including electromigration, thermal in stability, poor step coverage, corrosion etc. These are influenced by the film composition and structure as well as by the film-substrate interface. These features are dependent not only on the materials used but also on theprocessing and film deposition technology.

Epitaxial Al film deposited by ICB has remarkably high thermal stability [11]. No voids or hillocks were found, and no appreciable Si interdiffusion was observed even after annealing at 550°C. Very encouraging data were also obtained for step coverage characteristics and for electromigration resistance. These properties of the ICB Al films may provide solutions to the various problems in the metallization for VLSI. In recent studies, Al epitaxy on GaAs, CaF_2 and sapphire substrates has also been performed. All-epitaxial metal/insulator/semiconductor structures for three-dimensional devices may be made by use of ICB technology.

Epitaxial Y-Ba-Cu-O thin films with high transition temperature have been formed by ICB [23]. The films were deposited at 630°C. Resulting film surfaces are smooth and diffraction analysis shows a fine streak pattern. No annealing was done after the depositions. Film Tc zero was at 82 K. The results show that this as-grown method allows synthesis of superconducting thin films at low substrate temperatures. Although further reduction of the deposition temperature to about 500°C and selection of substrate material are still necessary, the results indicate a near term possibility of using superconducting materials as interconnections.

lV CONCLUSIONS

Results of investigations concerning cluster beam generation mechanisms, fundamental film formation mechanisms and example ICB film applications have been reviewed. In many cases, ICB deposited films have been observed to differ markedly from films by other deposition process.

REFERENCES

[1] T.Takagi, I.Yamada, M.Kunori and S.Kobiyama, Proc. 2nd Int. Conf. on Ion Sources (Österreichiche Studiengesellshaft für Atomenergie, Vienna, 1972) p.790.

[2] Y.Yamamura, I.Yamada and T.Takagi, Nucl. Inst. Methods in Physics Research, B37/38 (1989) 902.

[3] L.L.Levenson, M.Asano, T.Tanaka, H.Usui, I.Yamada and T.Takagi, Journal of Vacuum Science and Technology A, Vol.6, No.3 (1988) 1552, and L.L.Levenson, B.Swartzlander, H.Usui, I.Yamada and T.Takagi, Materials Research Society 1988 Fall Meeting, November 1988, Boston, Paper No. A2.4.

[4] K-H.Müller, J. Appl. Phys., 61 (1987) 2516.

[5] A.Yahashi, L.L.Levenson, H.Usui and I.Yamada, European Materials Research Society 1989 Strasbourg Meeting, May, 1989.

[6] I.Yamada, H.Inokawa, and T.Takagi, J. Appl. Phys., 56 (1984) 2746.

[7] I.Yamada, Nucl. Inst. Methods in Physics Research, B37/38 (1989) 770.

[8] I.Yamada, European Materials Research Society 1989 Strasbourg Meeting, May, 1989.

[9] I.Yamada, H.Usui, H.Harumoto and T.Takagi, in Laser and particle-Beam Chemical Processing for Microelectronics, edited by D.J.Ehrlich, G.S.Higashi and M.M.Opryskc (Mat. Res. Soc. Symp. Proc., Vol.101, Pittsuburgh, PA 1988) pp. 195-200.

[10] M.Sosnowski and I.Yamada, Nucl. Inst. Methods in Physics Research, B37/38 (1989) 874.

[11] I.Yamada and T.Takagi, IEEE Trans., ED-34 (1987) 1018.

[12] K.Fukushima, I.Yamada and T.Takagi, J. Appl. Phys., 58 (1988) 4146.

[13] H.Hashimoto, L.L.Levenson, H.Usui, I.Yamada and T.Takagi, J. Appl. Phys., 63 (1988) 241.

[14] I.Yamada, H.Usui and T.Takagi, J. Phys. Chem., 91 (1987) 2463.

[15] W.Knauer and R.L.Poeschel, J. Vac. Sci. Technol., B6 (1989) 456.

[16] J.Gspann, Nucl. Inst. Methods in Physics Research, B37/38 (1989) 775.

[17] D.E.Turner, K.M.Lakin and H.R.Shanks, to be printed in Processing and Characterization of Materials Using Ion Beams, edited by L.E.Rehn, J.E.Greene and F.A.Smidt , (Mater. Res. Soc. Proc., Pittsburgh, PA 1989).

[18] H.Usui, M.Tanaka, I.Yamada and T.Takagi, Nucl. Inst. Methods in Physics Research, B37/38 (1989) 886.

[19] Y.Hashimoto, T.Maeyama and K.Machida, Spring Meeting of Japn. Inst. Metallurgy paper No 552 (1988).

[20] I.Yamada, to be printed in Optical Materials: Processing and Science, edited by G.Ortiz and D.B.Poker (Mater. Res. Soc. Proc.,Pittsuburgh, PA 1989).

[21] I.Yamada, L.L.Levenson, H.Usui and T.Takagi, in Materials Modification and Growth Using Ion Beams., edited by U.J.Gibson, A.E.White and P.P.Pronko (Mater. Res. Soc. Proc., Pittsburg, PA 1987) pp. 253-258.

[22] K.Fukushima and I.Yamada, European Materials Research Society 1989 Strasbourg Meeting, May, 1989.

[23] K.Yamanishi, Y.Kawagoe, S.Yasunaga, K.Imada and K.Satoh, presented at 1989 spring meeting. Materials Research Society, April 1989, San Diego, Paper No. M8.21.

STUDIES OF PRECIPITATE FORMATION IN OXYGEN-IMPLANTED LAYERS OF SILICON

Aleshin A. N., Malinin A. A., Mordkovich V. N.

Institute of Microelectronics Technology and High Purity Materials, USSR Academy of Sciences, 142432 Chernogolovka, Moscow District, USSR

The behaviour of oxygen in CZ silicon (oxygen content $5 \cdot 10^{17} - 2 \cdot 10^{18} cm^{-3}$) has recently received comprehensive investigation [1-3]. But the behaviour of oxygen implanted in silicon has get to be studied. It is of particular interest in view of the problem of fabricating an internal (built-in) getter by ion implantation of oxygen into a silicon wafer (such an attempt was made, for instance, in [4]). It is to be noted that the behaviour of implanted oxygen is likely to differ from that of growth oxygen on heat treatment of silicon wafers due to the formation of radiation defects in the ion-doped layer during implantation.

The present work was intended to investigate the state and behaviour of oxygen implanted in silicon by IR-spectroscopy and transmission electron microscopy (TEM) upon heat treatment of FZ silicon over a wide temperature range. The implantation doses were chosen so that the mean oxygen concentration in the layer was comparable to the content of growth oxygen in CZ silicon. An important aspect of the work was the study of the influence of another implanted impurity, namely, carbon, on the behaviour of implanted oxygen.

FZ <100> p-type silicon wafers were implanted with 350 keV oxygen ions at a dose of $3 \cdot 10^{14} cm^{-2}$. Prior to oxygen implantation some of the silicon wafers were implanted with 370 keV carbon ions at a dose of $2.5 \cdot 10^{14} cm^{-2}$. The implanted samples were annealed in vacuum ampoules in the temperature range 750-1000°C for 5.5-50 h. Some of the samples were subjected to two-stage annealing: 750°C, 9 h and 1050°C, 4 h.

Figs. 1 and 2 show the IR-spectra of samples subjected to different modes of heat treatment. The spectra illustrate the annealing-induced changes in the state of implanted oxygen. For the samples annealed at 750°C for 9 h the spectrum is a "wave" and its absorption peak (the left "semiwave") observed at a wave number of 1095 cm^{-1} is closest to the 1107 cm^{-1} absorption line corresponding to free interstitial oxygen in CZ silicon. With increasing annealing temperature (and annealing time for T=750°C), the absorption peak increases in magnitude and is shifted to the long-wave part of the spectrum approximating a wave number of 1080 cm^{-1}. Besides, the samples subjected to heat treatment at 1000°C and two-state annealing exhibit additional absorption peaks in the vicinity of the wave number 805 and 470 cm^{-1}. It has been noted [5,6] that such absorption peaks are typical of thermally fabricated amorphous SiO_2 films.

Electron-microscopic studies have revealed that heat treatment in the above-mentioned temperature range gives rise to small defects with a black-and-white deformation contrast up to 10-20 nm in size at a depth (Fig. 3) the value of which is close to that of the projective range oxygen ions.

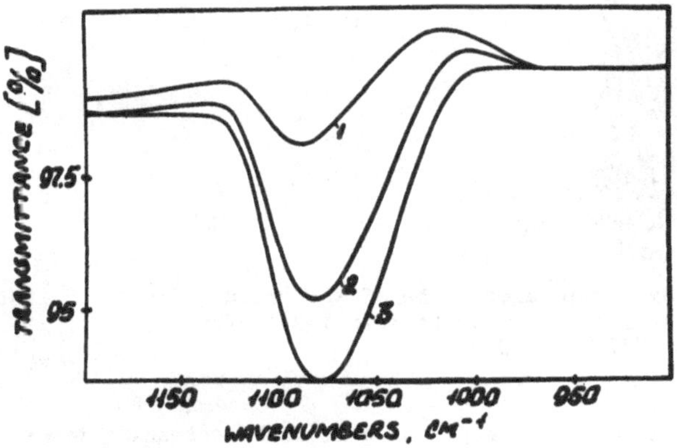

Fig. 1. IR-spectra of oxygen- and carbon-implanted samples after annealing at 750°C:
1) - 9 h, O^+-implantation; 2) - 50 h, O^+-implantation;
3) - 9 h, C^+- and O^+-implantation.

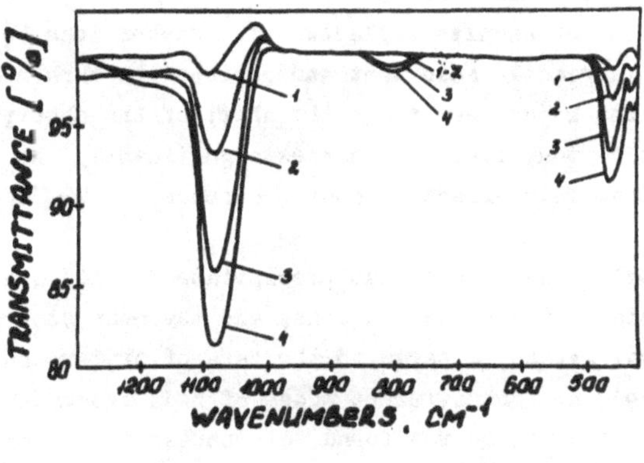

Fig. 2. IR-spectra of samples subjected to heat treatment in the 750-1000°C temperature range and two-stage annealing:
1) - 750°C, 9 h; 2) - 800°C, 12 h;
3) - 1000°C, 5.5 h; 4) -two-stage annealing.

The presence of these defects points to the formation of fine-dispersed oxygen precipitates in the oxygen-implanted layer during annealing. And the IR-spectroscopic data indecate that the oxygen precipitates are in the amorphous state.

Fig. 3. Electron-microscopic micrograph of the cross-section of an ion-implanted silicon layer after annealing at 750°C for 9 h.

The increase in magnitude of absorption peaks with increasing temperature or duration of annealing (at 750°C) shows that a part of implanted oxygen is in the optically inactive state which undergoes transformation into amorphous oxygen precipitates revealed by IR-spectroscopy. The shift of the oxygen absorption peak from 1095 cm^{-1} to 1080 cm^{-1} is indicative of the transformation of "loose" oxygen-silicon SiO_x clusters formed by ion implantation into compact amorphous SiO_2 precipitates. The oxygen-silicon clusters are seeds of heterogeneous precipitate nucleation.

Fig. 1 shows an IR-spectrum of samples implanted with carbon ions before oxygen ion implantation. It is readily seen that the presence of carbon ions in the oxygen-implanted silicon layer results in the shift of the absorption line to a more compact phase region, i.e. it enhances significantly the formation of SiO_2 precipitates. This effect is most pronounced at 750 and 800°C.

Electron-microscopic studies have shown that precipitate density does not increase with additional implantation of carbon ions, and the mean size of precipitates becomes even smaller, as compared to the case of oxygen ion implantation, amounting to 6-12 nm for different modes of heat treatment. Besides, preliminary carbon implantation was found to suppress formation of stacking faults during two-state annealing which is observed only on oxygen implantation. Hence, carbon ions have influence on implantation-induced point defects, mainly interstitial silicon atoms, which are involved in the formation of both precipitates and secondary defect structure. In precipitate formation carbon ions act as a catalyst modifying the point defect ambient so that the free energy of crystal is decreased. Decrease of free energy is an extra driving force of precipitation. It is well known [2-3] that carbon ions are regarded only as heterogeneous seeds for oxygen precipitates for CZ silicon containing growth oxygen and carbon.

REFERENCES

1. Batavin V. V.,Kristallografiya, 1970, v.15, p.125.
2. Kishino S., Matsushita Y., Kanamori M., Iizuka T.,Jap. J. Appl. Phys., 1982, v.21, N1, p.1.
3. Ogino M., Appl. Phys. Lett., 1982, v.41, p.847.
4. Weiner D. H., Simor Wang S., Drowley C. I.,Appl. Phys. Lett., 1987, v.50, p.986.
5. Cachard A., Roger J. A., Pivot J., Dupuy C.H.S.,Phys. Stat. Sol. (a), 1971, v.5, p.637.
6. Nagasima N.,Jap. J. Appl. Phys., 1970, v.9, p.879.

STRUCTURAL CHARACTERIZATION OF ION-IMPLANTED SILICON LAYERS BY MEANS OF CROSS-SECTIONAL TEM IMAGING

H. Bartsch[1)]

INTRODUCTION: The kinetics of the amorphous to crystalline (a/c) phase transition in highly doped ion-implanted a-silicon layers is influenced not only by the structure of the a/c interface region /1/ but also by precipitation processes during recrystallization /3, 6, 7/. This problem is of particular interest for creating very thin layers in shallow-junction processing. Examples are given of As^+- and BF_2^+-implanted (100)-silicon studied by cross-sectional(XTEM) imaging.

EXPERIMENTAL: High-voltage TEM (HVTEM) is applied using XTEM specimens prepared by the "face to face" method /4/. Images are taken in the bright-field (BF) as well as the dark-field (DF) mode. Specimens are recrystallized in situ in the vacuum of the TEM using an electric heater (VA). The results of the recrystallization experiments (T_a: 670...820K) are compared to those of conventional furnace-annealed (CFA) as well as rapid-thermally annealed (RTA) samples.

RESULTS: The recrystallization of BF_2^+- as well as P^+/As^+ double-implanted silicon layers (data see Figs. 1 and 3) is influenced by the precipitation of doping atoms or fluorine atoms, respectively, as recently described by Olson et al. /6, 7/. The recrystallization kinetics may be characterized by four stages:

1. Incubation regrowth: The irregular a/c phase boundary region recrystallizes at temperatures between 670 and 720K /1/ depending on the implanted dopant, with no further shift of the phase

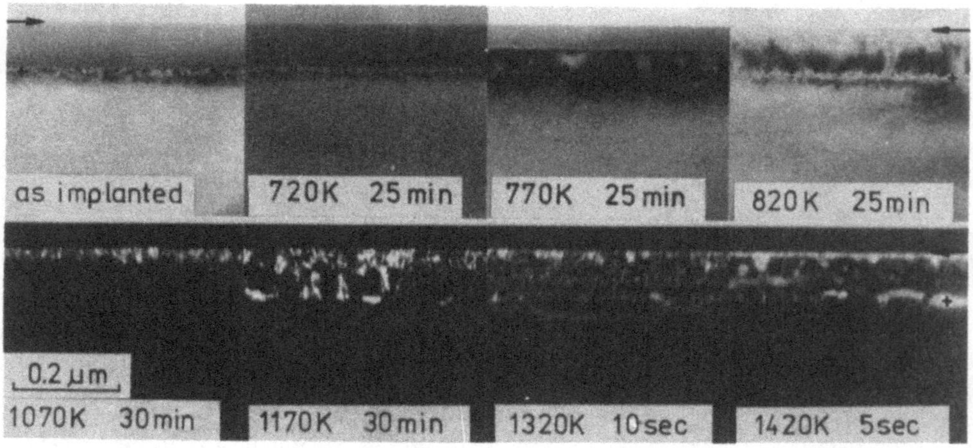

Figure 1: XTEM images of the recrystallization and defect formation in BF_2^+-implanted Si-layers (E=50keV, D=5x10^{15}cm^{-2}, T_i=300K). Upper row: in situ VA (BF); lower left: CFA in N_2; lower right: RTA in air; arrows: surface, crosses: end-of-range defects.

1) Institute of Solid State Physics and Electron Microscopy, Academy of Sciences of the GDR, Halle/S., Weinberg 2, GDR-4050

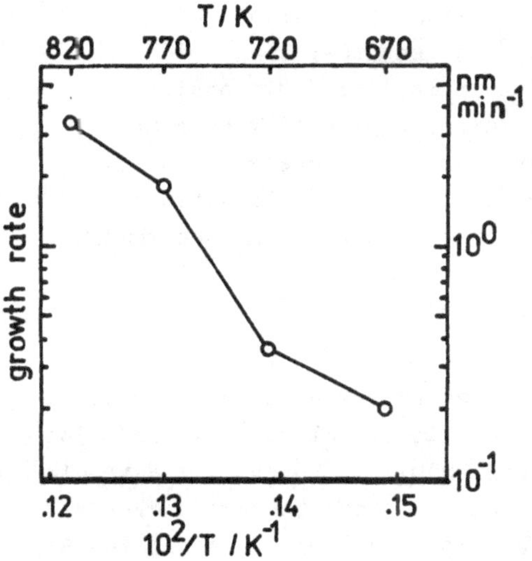

Figure 2: Growth rate vs. reverse temperature of the BF_2^+-implanted silicon layer shown in Fig. 1.

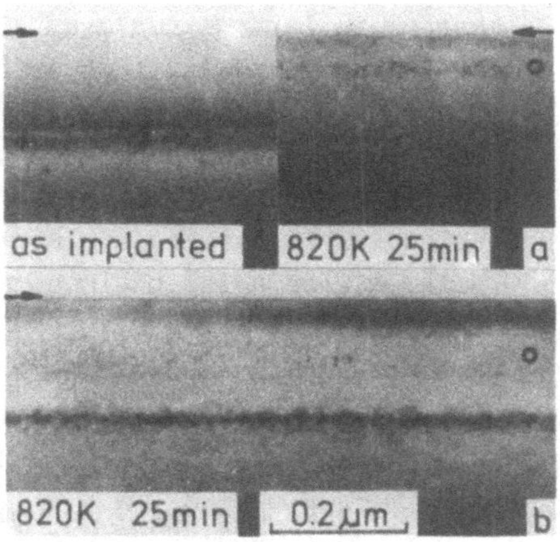

Figure 3: XTEM images (BF) of As-precipitation in VA Si-layers.
(a) P^+-Implantation: E=60keV, D=$4.5x10^{14}cm^{-2}$, T_i=77K + As^+-Implantation: E=60keV, D=$6x10^{15}cm^{-2}$, T_i=77K.
(b) As^+-Implantation: E=130keV, D=$6.5x10^{15}cm^{-2}$, T_i=300K.
arrows: surface, circles: precipitation layer, crosses: end-of-range defects.

boundary being observed. The growth rate (Fig. 2) is therefore extremely low (approximately 0.2...0.3nm min^{-1}).

2. Solid phase epitaxy (SPE): Up to a depth near the projected range (R_p) of the implanted ions undisturbed SPE growth is generally observed at temperatures between 720 and 770K. The end-of-range defects remain in the initial depth position, and unlike at high temperatures /2/ no extended defects are observed to form.

3. Precipitation: When the phase boundary reaches the depth near R_p precipitates occur and the growth rate is reduced as revealed by time-resolved reflectivity (TRR) measurements /6, 7/, or the growth is even stopped at this depth at lower temperatures. The BF_2^+-implanted layer (Fig. 1-770K) shows the interface to be roughened into very fine facets. As^+-implanted layers show small precipitates in a layer near R_{pAs}. The widths of the layers depend on the ion dose and the ion energy (Fig. 3).

4. Final growth: If the precipitates are small and the disturbance of the moving c-interface is low, a temperature rise causes the growth to continue up to the surface (e.g. As^+-implantation - Fig.3). No extenced defects are formed up to 770K; at 820K the formation of small dislocation loops is firstly observed, sometimes growing out to the surface. In the BF_2^+-implanted sample, however, microtwin lamellae have formed, originating in the former facets near R_p and growing up to the surface. The growth rate of the twin layer is comparable to the usual (100)-SPE growth.

Comparing the layer structures after VA with those after CFA and RTA (Fig. 1 - second row) displays three main features:

- end-of-range defects remain in their initial depth position
- gas bubbles always form in the central region of the layer
- microtwins in a surface-near layer are still present

Some structural features are modified by the thermal treatment:

- dislocation emission into the substrate (RTA-1320K) and to the surface
- increase of the bubble size (RTA-1420K)
- reduction of the twin density by RTA.

DISCUSSION: The precipitation of implanted atoms may locally reduce the SPE rate /6, 7/ and also the electrical activity /5/. The formation of large precipitates (or gas bubbles), however, may result in strong obstacles acting on the moving a/c phase boundary. Facets originate at these obstacles in the a/c interface which change the SPE growth completely into a twin-controlled mechanism. The twin formation overcomes the stresses caused by the obstacles with twin lamellae growing up to the surface at a growth rate comparable to the (100)-SPE growth. Microtwins, however, are of high thermal stability. Their concentration may solely be reduced by RTA above 1300K, possibly caused by the reduced segregation of fluorine. Furthermore, microtwins form a crystallographically inhomogeneously oriented surface. Diodes formed on such structures show insufficient electrical characteristics due to an irregular-shaped interface between the twins and the aluminium contact.

SUMMARY: Precipitation processes occurring at the a/c phase boundary during recrystallization play an increasing important role in shallow-junction processing. Large precipitates may introduce microtwins into the implanted layer even in (100)-SPE which are stable up to the high-temperature treatment.

ACKNOWLEDGEMENT: I would like to thank Mr. W. Einbrodt from the "Kombinat Mikroelektronik" for his kind engagement in arranging the implantations and different annealing cycles.

REFERENCES:

(1) Bartsch,H., D.Hoehl and P.Werner: phys.stat.sol.(a) 112 (1989) 715
(2) Bartsch,H., J.Heydenreich, D.Hoehl, G.Kästner, and P.Werner: Izv.Akad.Nauk SSSR, Fiz.Ser.Conf. (Svenigorod) 1987, 1600-1605
(3) Bartsch,H.: Proc.Int.Conf.on Sci. and Technol. of Defect Control in Semicond., Yokohama 1989. Ed: K.Sumino, in press
(4) Bravman,J.C. and R.Sinclair: J.Electron Microsc.Techn. 1(1984) 53
(5) Grob,J.J., S.Unamuno, A.Grob, M.Ajaka, A.Slaoui and R.Stuck: Nucl.Instr.Meth. B19/20 (1987) 501-506
(6) Olson,G.L., J.A.Roth, L.D.Hess and J.Narayan: MRS Symp. Proc. Vol.23: Energy Beam-Solid Interactions and Transient Thermal Processing, Boston Mass. 1983. Eds.: J.C.C.Fan, N.M.Johnson North-Holland N.Y., Amsterdam 1984, 375-382
(7) Olson,G.L., J.A.Roth, L.D.Hess and J.Narayan: In: Layered Structures and Interface Kinetics. Ed.: S.Furukawa, KTK Sci.Publ., Tokyo 1985 73-98

GENERATION AND ANNEALING OF POINT DEFECTS IN SILICON AFTER HIGH ENERGY IMPLANTATION

A. Mertens, J. Röhrich [1)]

Introduction

High energy ion implantation is a promising technique of semiconductor doping up to depths of some microns. Annealing is necessary to activate the dopands and to remove the implantation-produced defects. The annealing behaviour of these defects is from great interest, because low defect concentrations already influence the parameters of electronic devices (carrier life time).

In the present work deep levels generated in silicon by high energy ion implantation were electrically characterized by means of Deep Level Transient Spectroscopy (DLTS). Using non-doping elements it is possible to measure the defect concentration with high accurancy down to a limit of $10^{-4} * N_{D,A}$ ($N_{D,A}$ - doping density of the material). Defect level concentration profiles were measured by a double correlation method (DDLTS /2/) with a lock-in system.

Experiments

For the preparation of the samples n-type silicon (phosphorus-doped, $N_D = 1.2 * 10^{15}$ cm^{-3}) and p-type silicon (boron-doped, $N_A = 1.5 * 10^{15}$ cm^{-3}) were used. Random implantations were performed at room temperature using C^{++} and Si^{+++} ions in an energy range from 400 keV up to 990 keV. The values of the dose were chosen to $1 * 10^{8}$ cm^{-2} and $2 * 10^{10}$ cm^{-2}. After annealing and chemical cleaning Schottky barrier structures were prepared by vacuum evaporation of Au and Al onto the n- and p-type samples, respectively. Isochronal heat treatment of 30 min in N_2-atmosphere had been carried out at 400 °C, 600 °C and 800 °C.

Fig. 1 shows some DLT-spectra after carbon-implantation into n- silicon depending on the annealing temperature. The additionally represented minority carrier spectrum (as implanted state) was generated by intense light pulses from a GaAs luminiscence diode through the back contact of the sample. The defects E1 ($\Delta E = 0.17$ eV), E2 ($\Delta E = 0.23$ eV) and E4 ($\Delta E = 0.42$ eV) are attributed to the well known A-center (VO -/0), the double charged divacancy (V_2 --/-) and the divacancy (V_2 -/0) together with the E-center (PV -/0). Although it has been sometimes reported about the level E3 ($\Delta E = 0.31$ eV), a clear identification was not made.

---

1) Humboldt University, Department of Physics
Invalidenstraße 110, Berlin 1040, GDR

The defect H3 is believed to be the VOC-complex (0/+). The traps generated after annealing were not identified. The depth distributions of several defects in n-type silicon after carbon-implantation are plotted as a function of of the implantation parameters in Fig. 2. The maxima of the depth distributions were determined only for the low dose implantations, because the defect concentration must be less than the doping density. With respect to the data given by Burenkov et al. /1/ for the projected range of damage (R_pD) our values are somewhat lower. The depth distributions for the carbon-implantations with the higher dose depending on the annealing temperature are shown in Fig. 3a. After 600 °C annealing a reduction of the defect concentration was found. Additionally, a widening of the profile, which may be a result of diffusion, occurs. After 800 °C heat treatment, DDLTS was not applicable. The avarage concentration was estimated from the height of the DLTS-peak. The annealing characteristics of the deep levels in silicon-implanted n-Si is given in Fig. 3b. In contrast to Fig. 3a the defect tail is deeper. After 600 °C annealing only a low defect concentration in the range of the former maximum (R_pD) was found.

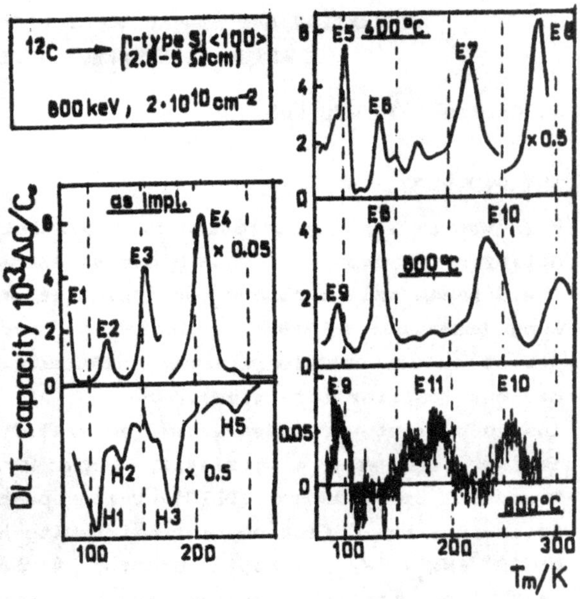

Fig. 1 DLT-spectra (e_n=10s) in dependence on annealing temperature

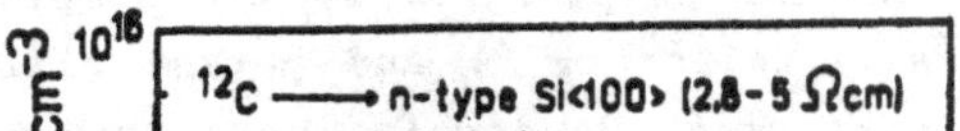

Fig. 2 Depth distributions of deep levels (+/E1, o/E4) dependent on ion energy and dose

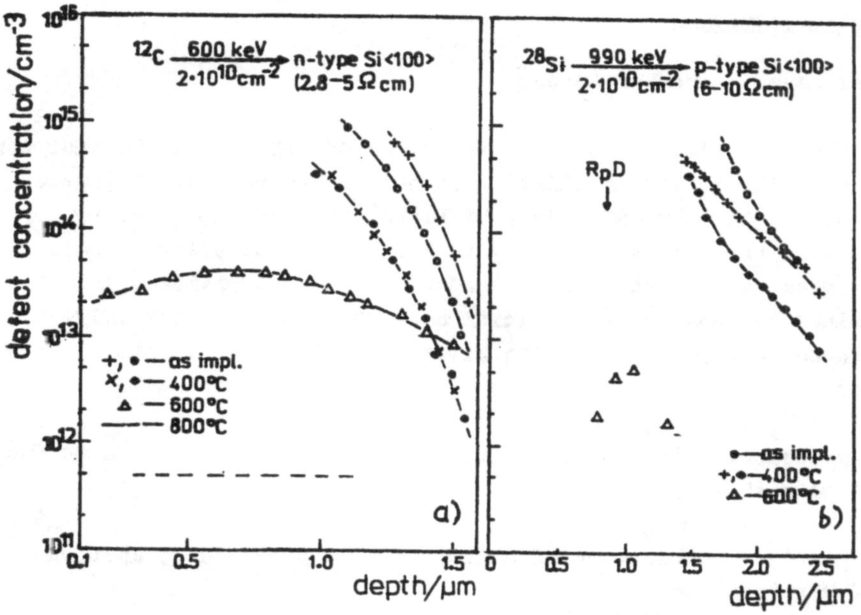

Fig. 3 Profiles of deep levels, Parameter: annealing temperature
C -implantation: +/E1, o/E4, x/E5, •/E8, Δ/E6
Si-implantation; o/H3, +/H8, •/H10, Δ/H10

Conclusions

In the as implanted-samples only vacancy-related deep levels were detected. They annealed out up to 400 °C but, a lot of other defects are generated and the overall concentration does not change very strongly. Their identification demands further investigations. After 600 °C annealing a reduction of the concentration, especially in p-type silicon was measured. The overall defect concentration for both, the n- and p-type material after 800 °C heat treatment was lower than $1*10^{12}$ cm^{-3}. To remove all defects it is necessary to go to higher annealing temperature. In this case very careful sample preparation and annealing is necessary [1].

References

(1) Burenkov, A. F., F. F. Komarov, M. A. Kumakhov, M. M. Temkin, Spatial energy distributions deposited in solids during atomic collision cascades (in Russian), Ed. by Energoatomizdat., Moscow 1985
(2) Lefèvre, H., M. Schulz: Appl. Phys. 12 (1977) 45.

1) We are most grateful to the team of Dr. Kerkow for carrying out implantation and annealing procedures.

IMPURITY EFFECTS AT AMORPHIZATION OF Al AND Ga DOPED ION-BOMBARDED SILICON

A.V.Dvurechenskii and V.P.Popov [1)]

The effects of simulation and retardation of crystalline-to-amorphous (C-A) transition at ion irradiation of heavily doped silicon layers (Si<B>, Si<P>, Si<B:P>, Si<B:As>, Si<B:Sb>) we have explained by the difference in the interaction of impurity atoms with point defects /1,2/.

It has been assumed that the retardation of C-A transition in Si<B> is explained by indirect recombination of the Frenkel pairs (Fig. 1) when B_s atoms are replaced in the sites by self-interstitials I (Watkins reaction)

$$B_s + I \rightarrow B_I \qquad (1)$$

and by subsequent returning transition of the B_I atom to the site due to interaction with vacancies

$$B_I + V \rightarrow B_s \qquad (2)$$

The indirect recombination of defects can be provided by another chain of reactions too

$$D_s + V \rightarrow (D_sV) \qquad (3)$$

$$(D_sV) + I \rightarrow D_s \qquad (4)$$

or

$$D_s + I \rightleftharpoons (D_sI) \qquad (5)$$

$$(D_sI) + V \rightarrow D_s \qquad (6)$$

where D_s is the impurity in the site of the lattice.

The indirect recombination at a high impurity and defect concentration and/or prolonged irradiation is determined by the energy barrier of complexes that are formed between them. The values of barriers are unknown for most reactions. Besides that, they are strongly dependent on the charge states of the components. The greatest cross section in p-Si is expected to is reaction (1) with acceptor impurity, while in n-Si reaction (3) with donor impurity must prevail. It is known from experiments the reaction (1) for Si<B> and Si<Al>, but in Si<Ga>only reaction (3) was revealed. The aim of this work has been to clarify the microscopic mechanism of C-A transition suppression by changing B_s to Al_s and Ga_s.

Heavily doped silicon layers with maximum Al and Ga concentration 1.5×10^{21} cm^{-3} have been produced by Al^+ (E_{Al}=50 keV) and Ga^+ (E_{Ga}=100 keV) ion implantation with dose of 2×10^{16} cm^{-3}. The annealing of implantation defects and impurity build-up in the sites was performed by pulsed liquid-phase recrystallization induced by XeCl-eximer laser (λ = 308 nm, τ =10 ns, E=0.3 J/cm^2). The quality of annealing is characterized by RBS spectra given in Fig.2.The minimum efficiency of RBS (χ_{min})= 8-9% is considerably higher than χ_{min} for the original Si crystal

1) Institute of Semiconductor Physics, 630090 Novosibirsk, USSR

χ_{min}=3-5%). At recrystallization a segregation of Al and Ga also occured. The highest concentration in sites of Ga atoms N_{Ga_s} =(6-8)x10^{20} cm^{-3} was observed at the depth of 40-60 nm. About 50 percent of impurity atoms did not occupy their regular sites in the lattice near the surface

The C-A transition was studied at Ne$^+$ ion irradiation (E=50 keV, j≃ 0.1 μA/cm^2). The peak of defect depth distribution R_D =70 nm at ion bombardment was corresponded to the depth of the peak of impurity concnetration in the substitutional site (Fig.3). The kinetics of defect accumulation (Fig.4) was found by procedures described in /1/.

The absence of reaction (1) in Si<Ga> is manifested by enhanced defect accumulation over the whole range of irradiation doses and by stimulation of the C-A transition (Fig.4). In Si<Al> samples the rate of defect accumulation is less than in Si<Ga> up to dose 3x10^{14} cm^{-2}. But unlike Si<B>, the retardation effect in C-A transition is absent. There are three possible reasons to explain these results. First, χ_{min} of heavily doped crystals is different from the original one due to high concentration of structural defects: twins, phase inclusions etc. These defects can be nucleation centres for the amorphous phase production. Second, Al_s concentration does not exceed 5x10^{20} cm^{-3} which, according to /2/, is the threshold for the retardation of the C-A transition in Si<B>. Third, the ratio of the cross sections of reactions (1) and (3) σ_1/σ_3 should be greater for B_s than for Al_s due to the short-acting potential of elastic forces. Fig. 1 illustrates the dominant interaction of self-interstitials with B_s atoms having covalent radius less than one of Si (r_B=0.088, r_{Si}=0.117 nm), whereas vacancies are trapped by impurity atoms streching the lattice ($r_{Al}\simeq r_{Ga}\simeq$ 0.126) nm). In order to detect the reason for C-A transition stimulation in Si<Al>, it is necessary to increase the concentration of the dopant in the lattice sites and to decrease the defect concnetration after recrystallization.

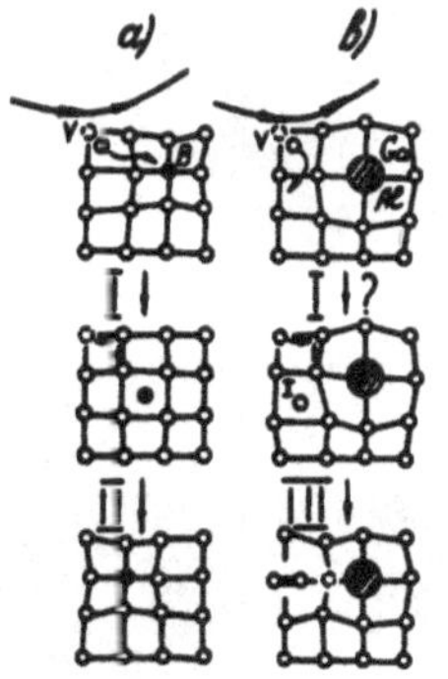

Fig.1. Scheme of point defect reactions with III group atoms. The arrow number corresponds to the reaction number in the text.

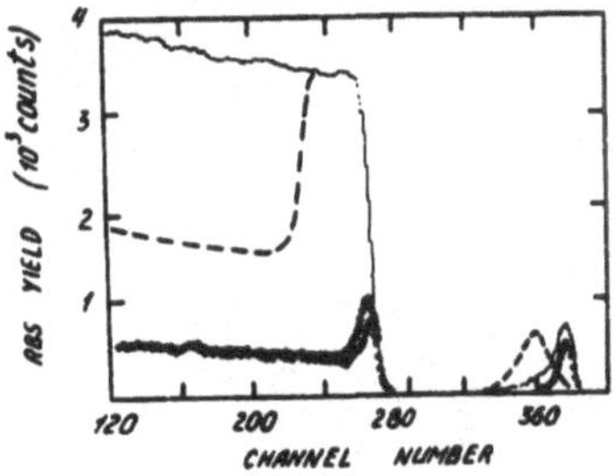

Fig.2. The RBS spectra of Si Al (o) and Si<Ga> samples. The dashed line shows spectra after Al, Ga implantation without annealing, the solid-the "random" spectrum of Si<Ga> after recrystallization. The He$^+$ ion energy is E= 1.5 MeV.

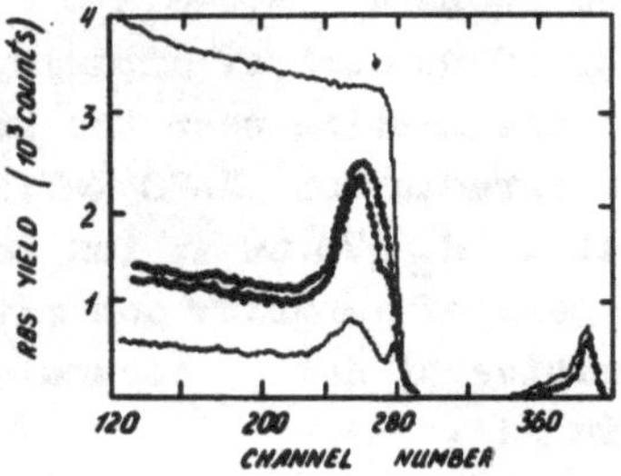

Fig.3. The same as in Fig.2 after Ne^+ irradiation (E= 50 keV, Φ =4x10^{14} cm^{-2}). The solid line at the bottom is the original Si after irradiation with Ne^+.

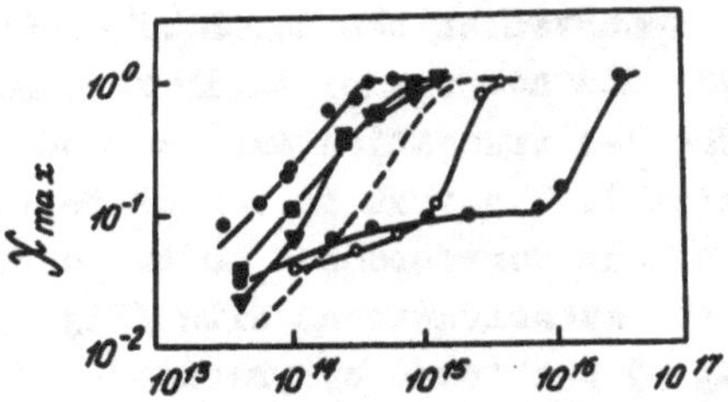

Fig.4. Normalized efficiency of RBS χ_{max} versus Ne^+ ion dose (E_{Ne}= 50 keV) in samples: ▼ -Si<Al>, ■ -Si<Ga>, dashed line-Si. E_{Ne}=80 keV: ○-Si<P>, ●-Si<B>, ◐-Si<B:P> /1,2/.

REFERENCES

(1) A.V.Dvurechenskii, R.Grötzschel, V.P.Popov. In: EPM 87, Academie-Verlag Berlin, 1988, p.44.

(2) A.V.Dvurechenskii, R.Grötzschel, V.P.Popov. In: Abstracts of Int. Conf. om ion implantatien in semiconductors and other materials. Lublin, Poland, 1988, p.68.

RAMAN SCATTERING IN HEAVILY DOPED SILICON LAYERS PREPARED WITH LASER ANNEALING

V.A.Geisler[1)], R.Grötzschel[2)], A.V.Dvurechenskii[1)], T.V.Kurochkina[1)] and Yu.A.Manzhosov[1)]

Raman study of heavily doped silicon layers obtained by ion doping and subsequent nanosecond laser pulse annealing have been performed. Ion doping was carried out by several impurities with different covalent radii: P, As, Sb, B. The ion doping parameters were chosen to form a layer of 0.1 μm thickness with carrier concentration $N_i \approx 8 \times 10^{20}$ cm^{-3}. To recover the crystal structure, a ruby laser pulsed anneal with 1-2 J/cm^2 energy in the spot of 4 mm-diameter has been used. The impurity depth distribution was studied with Auger-spectrometry and Rutherford backscattering. The Raman scattering spectra were taken in the geometry of backscattering at argon laser light excitation with a wave length of $\lambda_i = 488$ nm at T=295 K. The analysed thickness was 0.3 μm.

The obtained Raman spectra contain two components: one from the near-surface heavily doped layer and the other from the bulk material (ω = 520.5 cm^{-1}) (Fig.1). For impurities as As, P, they are well discernible and exhibited as two peaks; for Sb only asymmetrical broadening is observed. Such strong dependence of the spectrum on the dopant type at constant carrier concentration have been here interpreted in the framework of the existing ideas of the influence of free carriers and mechanical strains on the phonon crystal spectrum /3,2/.

We believe that in the spectrum formation three mechanisms participate: 1) electron-phonon interaction, 2) a change of force constant interaction induced by introduction of a large quantity of impurity with a different covalent radius; 3) mechanical strains arising in the implanted layer due to lattice parameter inconsistence between doped layer and substrate.

For n-type silicon, the Fermi level is shitted to the conduction band at the carrier concentration $N_i \approx 7 \times 10^{19}$ cm^{-3} /5/. The electron energy at transition from the occupied levels to the empty one in the band becomes comparable with the phonon energy, active in the Raman scattering. The interference of these excitations leads to the phonon frequence renormalization which is evidenced as a considerable broadening or shift of the Raman scattering spectrum to the low-frequency side. Using the estimation of /4/, we have determined that for carrier concentration of N=8x10^{20} cm^{-3} the frequence shift of the phonon $\Delta\omega_i$ is -5 cm^{-1} ("-" denotes the shift to the high-frequency side, "+" denotes the shift to the high-frequency side). In p-type silicon at carrier concentration above 1x10^{19} cm^{-3} the phonon energy becomes comparable with the energy of transition

1) Institute of Semiconductor Physics, 630090 Novosibirsk, USSR
2) Central Institute of Nuclear Studies, DDR

between the bands of light and heavy holes which also leads to the frequency renormalization $\Delta\omega_1 \approx -8$ cm^{-1} /2/.

The introduction of a great amount of impurity directly influences the phonon frequency and is estimated from the difference of masses and force constants of impurity ions and matrix. The simplified expression for the estimation of the maximum contribution of this mechanism is:

$$\frac{\Delta\omega_2}{\omega} = -\eta/2\left[\frac{M_i - M_{Si}}{(1-\eta)M_i + \eta M_{Si}} + 3\gamma\frac{\rho_i - \rho_{Si}}{\rho_{Si}}\right]$$

where $\eta = N_i/N_{Si}$, M_i, M_{Si} are masses of implanted ion and silicon respectively; ρ_i, ρ_{Si} are covalent radii; γ - is the Grünaisen parameter. The results of the estimation are given in Table 1.

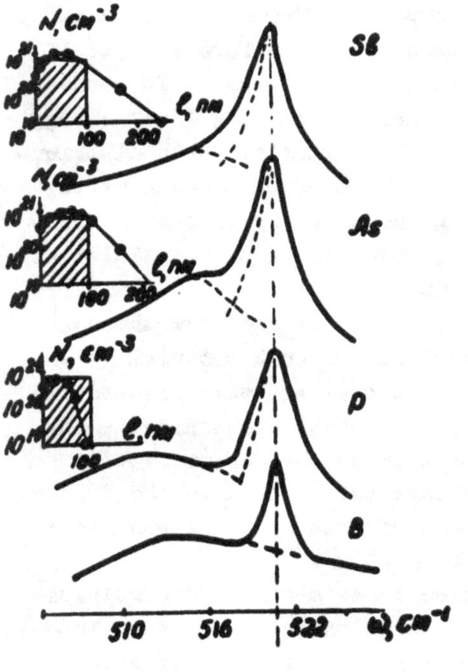

Fig.1. The Raman spectra of heavily doped silicon layers prepared with laser annealing.

Until now the investigators of heavily doped materials have been taking into consideration these two contributions. In our opinin, it is necessary to take account also of the built-in mechanical strains, since we are dealing with a heavily doped thin layer as an island with a changed lattice parameter rigidly fixed in the Si matrix. We have assumed the rectangular depth distribution of the doped impurity and the non-uniformity in the layer is neglected. By solving the problem of the symmetrical flat load applied to the thin layer, we obtain the expression for the phonon frequency change:

$$\Delta\omega_3 = [pS_{12} + q(S_{11} + S_{12})]\,\sigma/\omega \;; \quad \sigma = \frac{\varepsilon}{S_{11} + S_{12}}$$

where ε is uniaxial deformation, S_{11}, S_{12} are elastic constants of silicon; p and q are phenomenological coefficients /1/.

So the spectra of thin supersaturated silicon layers have been experimentally studied and theoretically analysed. The account taken of the three mechanisms have allowed us to describe the experimental data. One may to assume a possibility of solving the reverse problem, i.e., to determine the value of mechanical strains and carrier concentration from the Raman spectra.

Table 1.

ρ_{Si} = 1.17 Å			Phonon frequency shift, , cm^{-1}				
M_{Si}=28.09 g/mol			Estimation by 3 mechanisms			Theor.	Exper.
Atom of impurity	Covalent. radius, Å	Massa, g/mol					
Sb	1.36	121.76	- 5	- 5.2	+ 12.1	+ 1.9	- 5
As	1.18	74.91	- 5	- 2.5	+ 0.62	- 6.8	- 6
P	1.1	30.98	- 5	- 0.35	- 5.21	-10.5	-10
B	0.88	10.82	- 8	+ 9.2	- 18	-16.8	- 7

REFERENCES

(1) F.Cerdeira, et al. Phys.Rev.B, 5(1972)580.
(2) M.Cerderia, M.Cardona. Phys.Rev.B 5(1972)1440.
(3) M.Chandrasekhar, J.B.Ranuaci, M.Cardona. Phys.Rev.B 17(1978)1623.
(4) G.Contreras et al. Sol.State Commun 49(1984)303.
(5) M.Jouanne et al. Sol.State Commun 8(1975) 1047.

ELECTRONIC ENERGY LOSS OF IONS IN SILICON AND GERMANIUM

J. Hlávka

1. Introduction

The knowledge of specific energy loss of ions slowing down in a target during ion implantation is of importance mainly in calculations of ion distribution due to implantation. From this point of view any experimental result concerning ion energy loss is of importance. The total specific energy loss can be divided into three components: nuclear, electronic and charge- exchange. The aim of our paper is to demonstrate in which way it is possible to investigate electronic energy loss in semiconductors and to present preliminary results gained by the method.

2. Experiment

The method is based on the fact that excess electrons and holes are produced by implanted ions during their slowing down. These excess carriers cause a change of conductivity of the target sample which is called excess (radiation) conductivity. The excess conductivity is proportional to the excess carrier concentration which depends not only on the generation rate of excess carriers but on their recombination rate as well. The situation is complicated by the fact that it is difficult to appreciate the recombination rate of excess carriers which, moreover, is changed during implantation [1] . This problem is possible to overcome in the way as follows: excess carriers can be generated by a proper light (the proper absorption coefficient) in the region similar to that in which excess carriers are produced during implantation. Under this condition equal excess conductivity is observed in cases in which the generation rate is the same independent on the way of excess carrier generation [2] (either by ions or by light). In the case of light the number of generated excess carriers can be

1) Department of Solid State Physics, J.E. Purkyně University, Kotlařská 2, 611 37 Brno, Czechoslovakia

found on the basis of the number of absorbed photons. By comparison the excess conductivity caused by light (photoconductivity) with that caused by ions it is possible to evaluate the number of excess carriers produced by incident ions and to calculate the number of excess carriers produced by one ion. In such a way it is possible to investigate the excess carrier production as a function of ion energy, dose, direction of ion beam with respect to crystallographic axis, etc.

A simple apparatus was used to verify proposed idea. A surface ionization ion source [3] was used to create a beam of Na^+ ions. the beam was chopped with frequency of 37 Hz. In the same way a light beam incident on the sample was chopped. The excess conductivity (photoconductivity) was measured in a standard way [4] by the use of lock-in nanovoltmeter.

3. Experimental results

Some preliminary results are presented in Figs. 1., 2. where

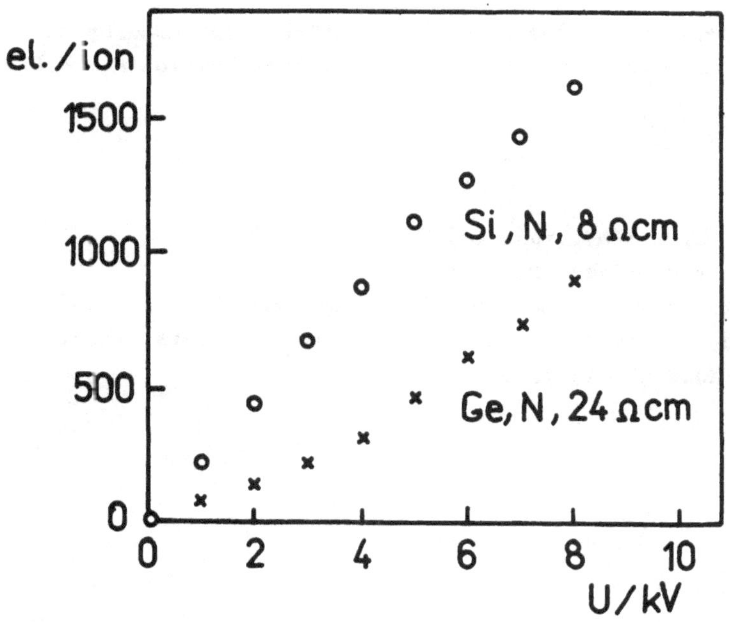

Fig. 1. Energy dependence of the number of electrons produced by one Na^+ ion in germanium and silicon (measured at doses greater than 10^{15} cm^{-2}).

the energy and dose dependences of the number of electrons produced by one Na^+ ion in germanium and silicon are shown. An experimental error of about 60% is estimated.

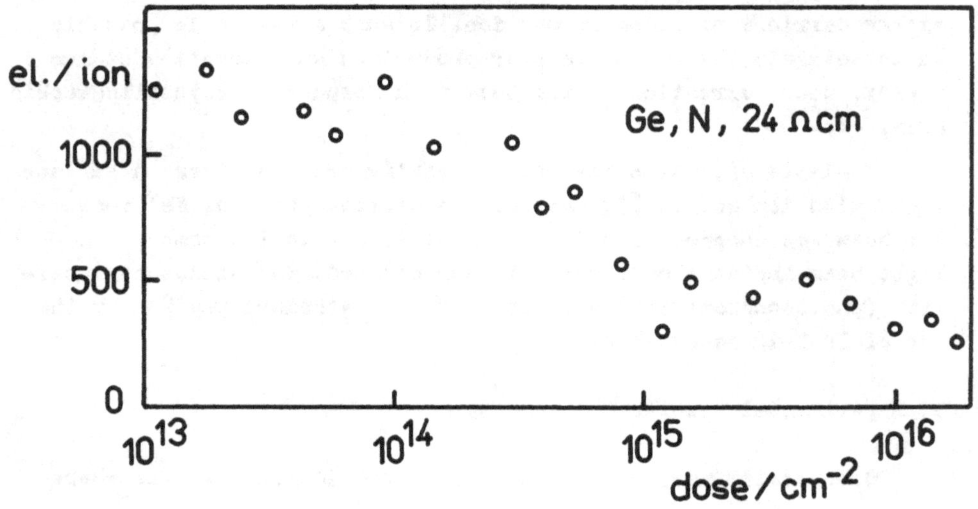

Fig. 2. Dose dependence of the number of electrons produced by one Na^+ ion in germanium (measured at accelerating voltage 4 kV),

References

[1] J. Hlávka, phys. stat. sol. (a) 94 (1986) 353
[2] M. Líbezný, unpublished results
[3] M.J. Zarabi and M. Satyam, Rev. Sci. Instrum. 42 (1971) 257
[4] S.M. Ryvkin, Photoelectric Effects in Semiconductors, Plenum Press, New York, 1964, p. 37

In situ RECRYSTALLIZATION OF IMPLANTED Si IN A HIGH-VOLTAGE ELECTRON MICROSCOPE

D. Hoehl[1], H. Bartsch[1]
V. Heera[2], W. Skorupa[2], M. Voelskow[2], K. Wollschläger[2]

<u>INTRODUCTION.</u> In the VLSI-technology low-temperature processes are important for removing the radiation damage/recrystallization of implanted amorphous layers. Radiation-induced Solid-Phase Epitaxy (SPE) can be used at temperatures below that for the thermally activated recrystallization. The Ion-Beam Induced Epitaxial Crystallization (IBIEC) is a well-investigated process /1/. Present investigations are concerned with the application of an electron beam to the same purpose /2-5/. Enabling irradiation and observation at the same time, an electron microscope is a suitable instrument to investigate in situ the regrowth behaviour at the amorphous/crystalline interface. It was the aim of our investigations to determine the activation energy (E_A) for the EBIEC-process.

<u>EXPERIMENTAL.</u> To produce a thick, homogeneous, amorphous layer, we implanted (100)-Cz-Si with Si^+-ions: 660 keV, 6 x 10^{15} cm^{-2} + 400 keV, 5 x 10^{15} cm^{-2} + 300 keV, 2 x 10^{15} cm^{-2} + 100 keV, 2 x 10^{15} cm^{-2}, subsequently preannealed at 450 °C for 30 min in N_2, only for smoothening the interface. Since the SPE-rate depends on the impurity ion species /6/ which is implanted, we used only Si^+ to exclude this influence. A high-voltage electron microscope was employed, always working at an electron energy of 1 MeV. The beam current densities were 10 - 15 A/cm^2 with the beam spot diameter being 3.5 µm. The regrowth at the a/c-interface was observed always on cross section specimens - Fig. 1. The specimens were irradiated at room temperature (RT) as well as at: T = 100°, 200°, 300°, 400° and 450 °C. According to an estimation by Lulli /2/ one can exclude an essential additional heating of the specimen by the intensive electron beam. Calculating the temperature rise by applying the formula of Hobbs /7/ is impossible, because of the complicated geometry of the specimens.

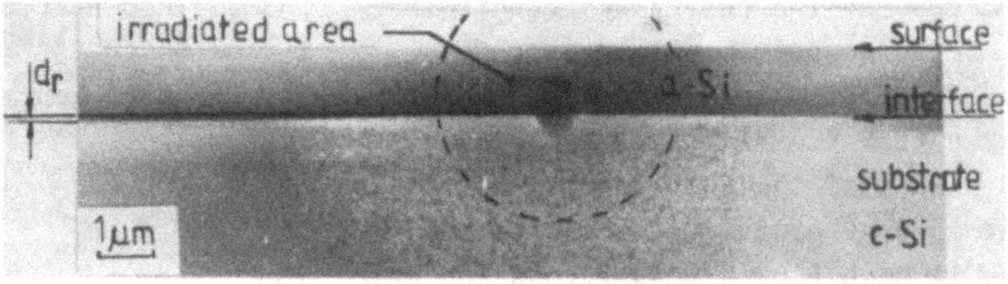

Fig. 1 Transmission electron microscope image of a cross section specimen
T = 200 °C, t = 157 min, i = 12 A/cm^2
d_r: thickness of the recrystallized layer

---

1 Institute of Solid State Physics and Electron Microscopy, Academy of Sciences of the GDR, Weinberg 2, Halle, DDR-4050

2 Central Institute of Nuclear Research, Academy of Sciences of the GDR, Rossendorf, PSF 19, Dresden, DDR-8051

RESULTS and DISCUSSION. The thickness (d_r) of the recrystallized layer can be measured from the electron microscope images - Fig. 1.
The measurements taken of the images every 30 min at a total irradiation time of 2.5 h are summarized in Fig. 2. Since the beam current was constant, the curves show the regrowth in dependence on the dose. The procedure was repeated at various temperatures revealing that the regrowth rate is not constant. At the beginning of the epitaxial process the regrowth is stronger. Then even no further regrowth is visible (T = RT ... 300 °C) or the rate decelerates (400°, 450 °C). The third region of the curves shows that the regrowth always decelerates, except for 450 °C, because here the SPE is strongly thermally activated. The total regrowth behaviour is comparable with IBIEC: with further irradiation the regrowth efficiency decreases /8/. The different regrowth stages can be explained by a polynucleation, i.e. a radiation-induced formation of crystalline nucleation centres in the amorphous layer, far away from the interface (at the temperatures used - $T < 450$ °C - thermally activated nucleation can be excluded). If the moving interface approaches the centres, it will slow down (second region of the curves) - this phenomenon was observed by Lulli /2/, too. After this, an interaction takes place between the epitaxial regrowth at the interface and the growth of the nuclei (third region).
This observation shows that only the first rise of the curves represents the real SPE-process. Thus the regrowth rates used for determining E_A are extrapolated from the first rise. The activation energy of the EBIEC-process can be determined by plotting the regrowth rate per unit dose (10^5 C/cm^2) versus the inverse temperature - Fig. 3.

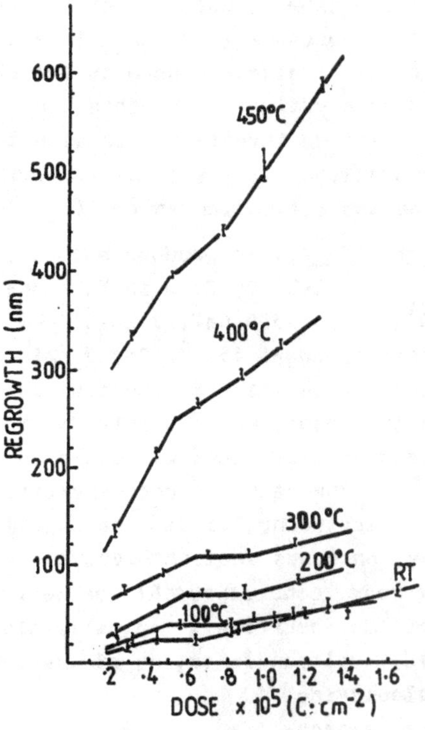

Fig. 2 Thickness of the recrystallized layer in dependence on the dose at various specimen temperatures

Using the formula $R = R_0 \exp(-E_A/kT)$

R: regrowth rate, T: temperature
k: Boltzmann constant,
E_A: activation energy

and the extrapolated values for R = 60 nm and 150 nm at T = RT and T = 300 °C, respectively, the activation energy was calculated to be 0.05 eV, which is essentially smaller than that for IBIEC (0.24 eV - /9/). Over the temperature range $T > 300$ °C, E_A was determined to be 0.6 eV (which is in

good agreement with the analogue value for IBIEC /1/).
These activation energies are very small in comparison to the thermal ones (2.7 eV), i.e. in general, the EBIEC process is athermal. Up to now, the microprocesses of the recrystallization at the interface have been explained in analogy to IBIEC. If the energy of the incoming electrons is high enough to displace a target atom, an interstitial and a vacancy are created. The vacancies diffuse to the interface, reconstructing the regular crystal lattice. Since up to now no E_A-values for EBIEC have been reported, only comparable values for IBIEC are considered (e.g. 1.2 eV - /9/), which is the dissociation energy of divacancies, leading to the hypothesis that the divacancies are the controlling factor of the transformation at the interface. But recent electron beam experiments at room temperature and at T = -171 °C (Corticelli /5/) have questioned this concept. E_A, determined for EBIEC, alone will not clarify this problem.

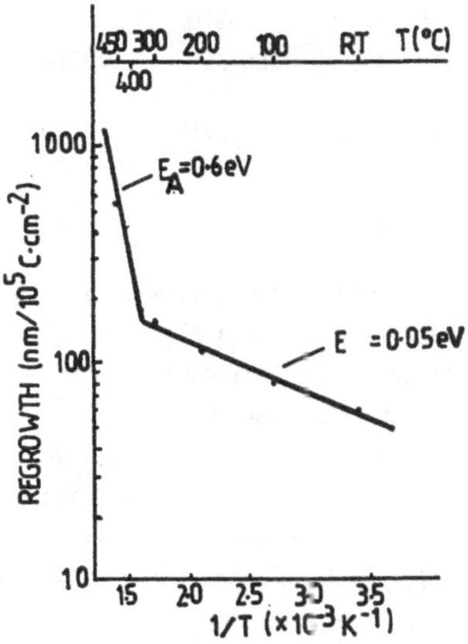

Fig. 3 Electron-beam induced crystallization as a function of the specimen temperature used for determining the activation energy E_A.

SUMMARY. The activation energy E_A = 0.05 eV (T = RT ... 300 °C) here first determined for EBIEC is surprisingly small. This evidences that EBIEC is practically athermal over this temperature range. Polynucleation should be considered a mechanism hindering the regrowth of thicker a-layers.

(1) Elliman, R.G., J.S. Williams, D.M. Maher, W.L. Brown Mater. Res. Soc. Proc. 51 (1986) 319
(2) Lulli, G., P.G. Merli, M. Vittori Antisari Phys. Rev. B 36 (1987) 8038
(3) Lulli, G., P.G. Merli, M. Vittori Antisari Mat. Res. Soc. Symp. Proc. 100 (1988) 375
(4) Lulli, G., P.G. Merli, M. Vittori Antisari, B. Jouffrey Proc. 9th Int. Congr. Electr. Microsc., Kyoto 1986, p. 985
(5) Corticelli, F., G. Lulli, P.G. Merli: Solid phase epitaxy of implanted silicon at liquid nitrogen and room temperature induced by electron irradiation in the electron microscope; submitted to Appl. Phys. Lett.
(6) Csepregi, L., E.F. Kennedy, T.J. Gallagher, J.W. Mayer, T.W. Sigmon J. Appl. Phys. 48 (1977) 4234
(7) Hobbs, L.W., in: Quantitative Electron Microscopy, ed. by J.N. Chapman, A.J. Craven, Proc. 25th Scottish Universities Summer School, Glasgow, 1983
(8) Kerkow, H., G. Kreysch et al., phys. stat. sol. (a) 94 (1986) 793
(9) Brown, W.L., R.G. Elliman et al. Inst. Phys. Conf. Ser. 87 (1987) 61

DEPENDENCE OF $Si-SiO_2$ INTERFACE STATE DENSITY ON ION IMPLANTATION PROCESS AND ITS EXPLOITATION FOR EVALUATING LOW DOSE ION IMPLANTATION:

Ing.Martin KEJHAR[1)]

INTRODUCTION

Uniformity of created layers is important parameter in semiconductor wafer fabrication. Ion implantation is modern method for creating differently doped layers. There are several methods for measuring ion implantation homogeneity (2). Electrical measurements are the simplest and most frequently used methods for this purpose. Unfortunately the methods based on measurement capacitance-voltage (C-V) characteristics of MOS structures, which are the most sensitive methods from the group of electrical one, have insufficient sensitivity for dose lower than $1.10^{15}m^{-2}$. Only the "Ion scan" and the "Thermal waves" methods have sufficient sensitivity for so low dose.

In this paper we discuss results which we have reached in investigation of ion implantation induced damage of silicon and we refer to possibility of exploitation the measurement of this damage for evaluating homogeneity of low dose implantation.

THEORY

It is well known that number of defects appears due to ion implantation. Concentration of this defects is many times greater than concentration of implanted ions. This defects are in semiconductor device fabrication removed by postimplantation heat treatment which aim is to reduce their concentration so they have no or only a little influence on electrical properties of fabricated devices. If we omit postimplantation heat treatment, electrical properties of fabricated devices will be considerably influenced by ion implantation induced defects. Change of electrical properties of structures related to defects will be much more greater than changes related to change of implanted dopant concentration in case of low dose ion implantation. Thus measurement of electrical characteristics of structures without annealing would be theoreticly sensitive tool for low dose ion implantation monitoring.

There are several structures and their characteristics which we can use for this measurements. In our work we devote attention to MOS capacitor because of by measuring its characteristics we can monitor both the defects in active layer of silicon and the damage of $Si-SiO_2$ interface due to ion implantation to wafer covered by SiO_2 layer. The second reason for this choice is that MOS capacitor is simple structure and its preparation is easy. The defects in silicon cause change of minority carrier lifetime and the damage of $Si-SiO_2$ interface appears as interface state density and surface generation velocity change.

1) Czech Technical University, Faculty of Electrical Engineering,K334, Suchbátarova 2, Prague, 166 27, Czechoslovakia.

EXPERIMENTS AND RESULTS

We have studied the influence of low dose ion implantation of phosphorus $^{31}P^+$ in N-type Si wafers with (100) orientation and $2.10^{21}m^{-3}$ bulk donor concentration as well as boron $^{11}B^+$ in P-type wafers with (100) orientation and $1.10^{21}m^{-3}$ acceptor concentration. We have studied dose range from $1.10^{14}m^{-2}$ to $4.10^{15}m^{-2}$. The thicknees of covering thermaly grown oxide has been 100nm. Test chips with MOS capacitors were created after ion implantation. No temperature treatment was applied after ion implantation with exception operations necessary for test chips preparation (Al sputtering, photolitography). Temperature during this operations did not exceed 150°C. Finished wafers was put to the first measurement. Annealing at 450°C twenty minutes, in N_2+H_2 was applied subsequently and then second measurement carried out.

Interface state density is determined from measured quasistatic and 1MHz high-frequency C-V caracteristics (1). The difference between interface state densities of implanted and nonimplanted MOS capacitors is considered as the measure of damage $Si-SiO_2$ interface due to ion implantation. Zerbst method (3) is used for lifetime and surface generation velocity measurements.

We find frommeasurements done before annealing that capacity is frequency dependent in region from 10kHz to 1 MHz and is approximately independent for frequency above 1MHz. Fig.1. showes obtained dependence of ion implantation induced interface state density on implanted dose of phosphorus. This dependence is approximately linear in dose range from $1.10^{14}m^{-2}$ to $1.10^{15}m^{-2}$ and can be expressed by equation

$$N_{SS} = 18.7\ D + 1.5\ 10^{15} \qquad (1)$$

where N_{SS} designates interface state density and D implanted dose. This dependence could be useful for ion implantation monitoring. Unfortunately only very slight dependence of interface state density on implanted dose is obtained for boron implanted sampes.

Properties of structures change substantially after annealing. Frequency dependence disappears in the whole measured range and interface state density decrease. Values of interface state density are in range from $2.10^{14}m^{-2}eV^{-1}$ to $4.10^{14}m^{-2}eV^{-1}$ on all samples after annealing. C-t characteristics change substantially too. Transient times prolong and dependence of life time on implanted dose appears. Fig.2. showes this dependence for boron implanted samnles. Lifetime on nonimplanted structures is 25 ms. The same dependence for phosphorus implanted samples is slight and lifetime decreases from 5 us to 2 us in observed dose range. Nonimplanted structures lifetime is 30 ms for N-type samples. We can infer from measured data that greater lifetime changes could appear if dose lower than $10^{14}m^{-2}$ was implanted.

CONCLUSION

Experiments indicate the possibility of exploitation measurements of implantation induced damage for monitoring low dose implantation. Strong dependence of interface state density on dose has been found for phosphorus. It turns out from experiments that lifetime measurement could be sensitive tool for monitoring very low dose implantation.

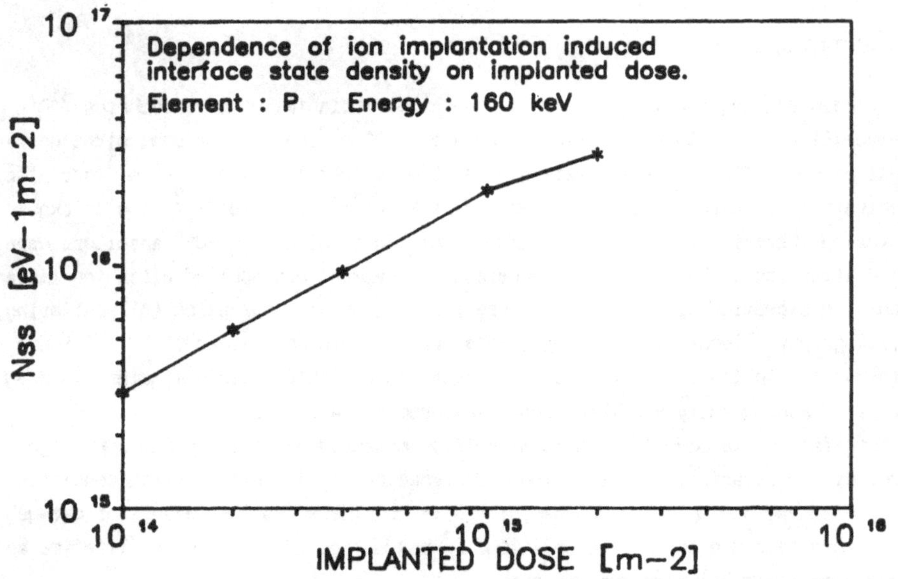

Fig.1. Dependence of ion implantation induced interface state density on implanted dose.

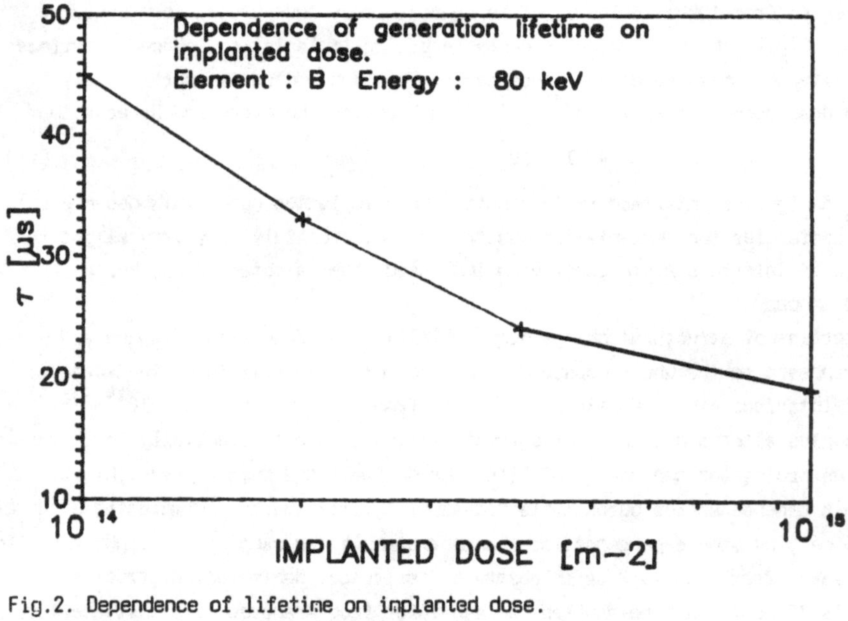

Fig.2. Dependence of lifetime on implanted dose.

ACKNOWLEDGEMENTS

The autor gratefully acknowledges helpful discussions with V Hašlar and wishes to thank P.Kalina and his coworkers for samples preparation.

(1) Nicollian,E.H.,Brews,J.R.:MOS Physics and technology, J.Wiley, New York 1982
(2) Ryssel,H.,Glawishnig,H.:Ion implantation techniques. Springer Verlag,Berlin 1982
(3) Zerbst,M.: Z. angev. Phys.,22(1966)30-36

INVESTIGATION OF THE INFLUENCE OF A HIGH OXYGEN OR NITROGEN CONTAMINATION ON THE ELECTRICAL ACTIVATION AND RESIDUAL DEFECT EVOLUTION OF ARSENIC IMPLANTED SILICON LAYERS

R. Kögler [1] and D. Baither [2]

The presence of a high oxygen or nitrogen concentration in Si can be caused by different process steps in microelectronics:

- the recoil implantation due to dopant implantation through oxide or nitride cap layers,
- the direct implantation of oxygen or nitrogen because of the formation of buried oxide or nitride layers,
- the gettering of oxygen or nitrogen in damaged layers of silicon.

It is well known that a high O-concentration strongly influences the residual defect evolution of an As-implanted Si-layer.
Recently it was also shown, that beside the defect evolution also the electrical activation of arsenic in Si is limited after RTA under the influence of a high O-concentration in the implanted Si-layer /01/.
The formation of small As-O clusters was proposed. They are assumed to be immobile and electrically inactive.
In this paper the electrical activation of arsenic after RTA is related to the observed residual defects.
Moreover, the investigation is extended to the influence of nitrogen at high concentrations in Si.

1. EXPERIMENTAL

A high contamination of Si with oxygen or nitrogen proceeds by As-implartation through a SiO_2-cap or Si_3N_4-cap, respectively.
An additional O- or N-implantation was also used.
After RTA the electrical activation of the arsenic fraction implanted into Si was determined by sheet resistance measurement and Hall-Effect measurement using the method of Van der Pauw.
TEM-results are related to the results of the electrical measurements.
As- and O- depth profiles are determined by SIMS.

2. RESULTS FOR THE EFFECT OF OXYGEN ON ACTIVATION OF As BY "RTA" AND THE RESIDUAL DEFECTS IN SILICON

The RTA-treatment of 1175°C, 1s was choosen.
Under these annealing conditions a nearly defect free silicon is reached for an As-implantation into virgin silicon. However, the presence of oxygen in Si decreases the velocity of the defect evolution. Therefore, under the same conditions, in this work a high density of residual defects is to see.
During a short RTA-impuls of about 1s the implanted oxygen depth profile is generally unchanged but the gettering of oxygen at the "end of range" defects is shown in Fig. 1.
In Tab. 1 values of the electrical activation of arsenic implanted into Si (below the SiO_2-cap) is shown for comparison with the residual defects in Fig. 2 and Fig. 3.

Tab. 1 As-activation (%) in Si with high O-contamination after RTA at 1175°C, 1s

As-dose / oxide thickness:	5nm	20nm	35nm	additional $10^{16}O^+/cm^2$ 5nm	20nm	35nm
1* $10^{15}As^+/cm^2$	78	79	78	41	40	31
5* $10^{15}As^+/cm^2$	62	59	53	40	29	26
1* $10^{16}As^+/cm^2$	46	40	34	35	25	22

1) Zentralinstitut für Kernforschung, Rossendorf
PF 19, Dresden 8051, GDR
2) Institut für Festkörperphysik und Elektronenmikroskopie
PF 250, Halle (Saale) 4010, GDR

Fig. 1 As- and O- depth profiles (SIMS) for the case of dual-implantation as a function of time after RTA-treatment at 1125°C

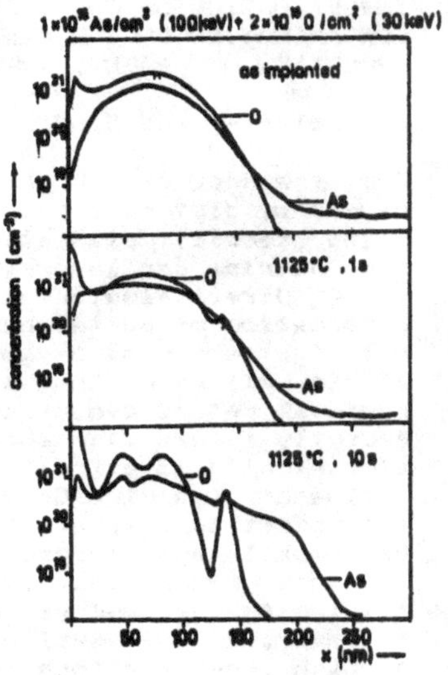

At a dose of $1*10^{15}As^{+}/cm^{2}$ no influence of the thickness of the SiO_2-cap can be observed neither for the arsenic activation nor for the residual defects.
In this case the implantation dose of As is too low to produce a high O-concentration by recoil implantation. An O-concentration lower than $5*10^{20}/cm^{3}$ does not influence the As-activation /01/.
However, the additional O-implantation ($c_{max}(O) = 1.42*10^{21}/cm^{3}$) causes a decrease of the activation by about 50 % as shown in Tab. 1.
At a higher dose of $5*10^{15}As^{+}/cm^{2}$ the influence of the thickness of the SiO_2-cap is clearly shown for the As-activation as well as for the residual defects.

In Fig. 2d perfect prismatic loops are seen and all of them exhibit b-vectors inclined to the surface. Consequently, they should be able to glide to the surface and leave the specimen.
Only a few of them have interact to form new segments with surface-parallel b-vectors.
The glide process is resticted by the high content of recoiled oxygen. Therefore in the case of the thicker oxide (Fig. 2e,f) the density of the segments with surface- parallel b-vector is high enough to form an irregular network.
At a dose of $1*10^{16}As^{+}/cm^{2}$ a sligth increase in the defect density in Fig. 2g confirms the described formation process of the defects. However, for a thicker oxide in Fig. 2i again single dislocation loops arise. A similar defect structure is obtained also for implantation through the 5nm thick oxide by an additional O-implantation.
Obviously, the oxygen reduces the growth rate of the dislocation loops. The irregular shape and the rough boundary of the loops indicate the disturbed climb process.
At a certaine oxygen content the creation of elongated defects is nearly compleatly suppressed.
Nevertheless, the activation of arsenic and the density of residual defects does not show the same tendency.
Obviously, the excess silicon availible for the formation of the residual defects is restricted in this case, namely by the formation of the As-O clusters, they cannot be seen by TEM.

3. RESULTS FOR THE EFFECT OF NITROGEN ON ACTIVATION OF As BY "RTA" AND THE RESIDUAL DEFECTS IN SILICON

The influence of a high N-concentration in Si was investigated by an As-implantation through a Si_3N_4-cap of the thickness of 60nm which is nearly equal R_p.

Tab. 2 Activation of As, implanted through Si_3N_4 in Si

As^{+} 100keV dose	--------> Si_3N_4 / Si		($d(Si_3N_4)$=60nm)	
	RTA: 1000°C, 1s		1125°C, 1s	
$1*10^{15}/cm^{2}$	36 %	(75 %)	37 %	(78 %)
$5*10^{15}/cm^{2}$	26 %	(64 %)	36 %	(75 %)
$1*10^{16}/cm^{2}$	17 %	(33 %)	33 %	(50 %)

Values for arsenic activation without Si_3N_4-cap in brackets.

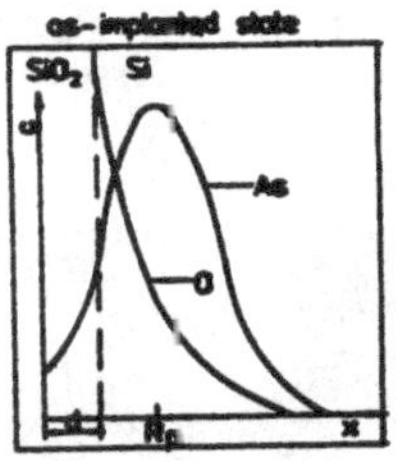

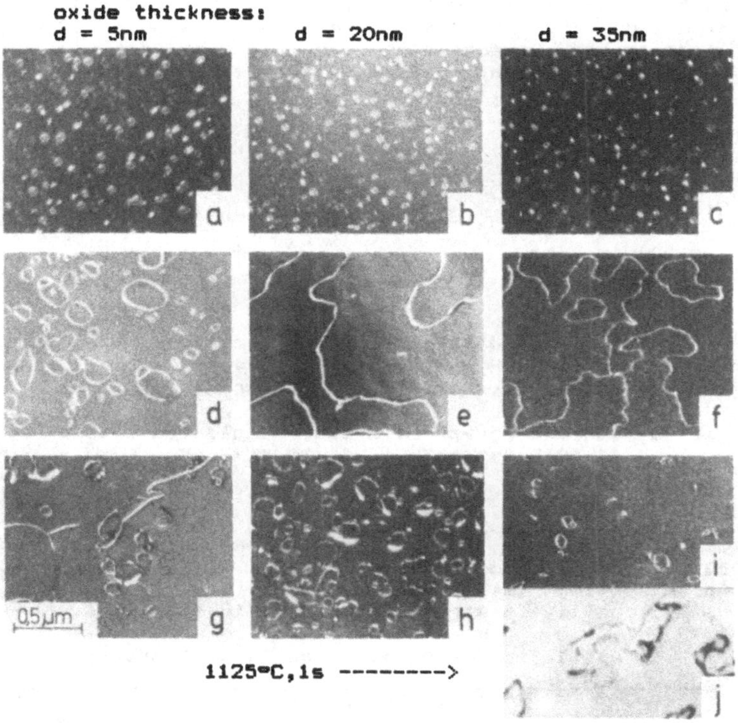

Fig. 2
TEM-micrographs
As^+ at a dose of:
a, b, c :
1* $10^{15}/cm^2$
d, e, f :
5* $10^{15}/cm^2$
g, h, i, j :
1* $10^{16}/cm^2$
implanted into Si
through SiO_2 with
different d

RTA: 1175°C, 1s

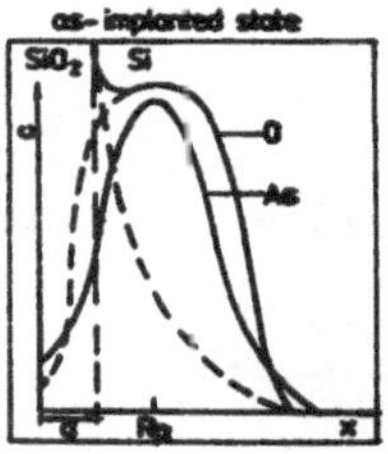

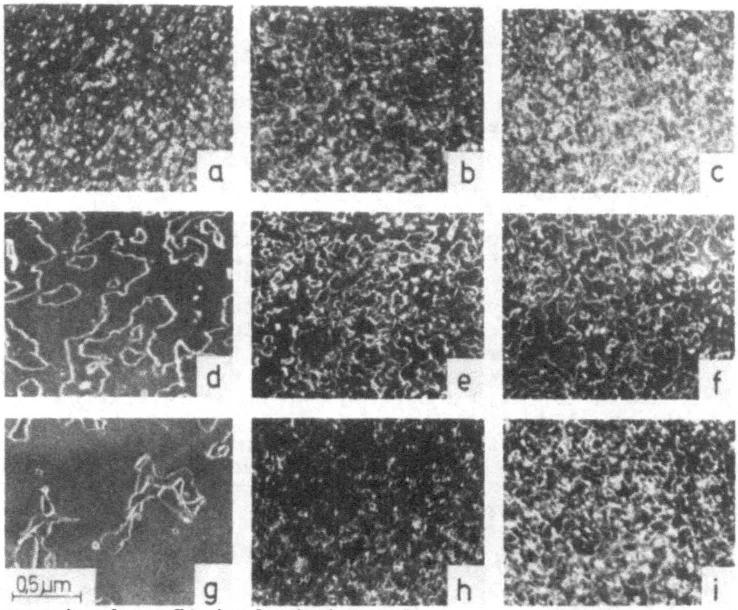

Fig. 3
TEM-micrographs
As^+ implanted
into Si through
SiO_2 under the
same conditions
as in Fig. 2,
but with
additional O^+-
implantation of
1* $10^{16}O^+/cm^2$

RTA: 1175°C, 1s

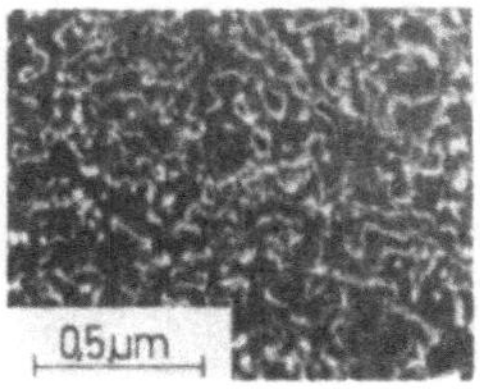

Fig. 4 TEM-micrograph of an O^+-implanted sample 5* $10^{15}O^+/cm^2$ after RTA at 1125°C, 1s.

Additional oxygen implantation (Fig. 3) at thicker oxide layers leads to a strong increase in the defect density. In this case defect structures occur like a superposition of defects obtained after As-implantation (Fig. 2) and O-implantation alone (Fig. 4).

In this case a very effective recoil implantation of nitrogen in Si is expected. The results of the As-activation are shown in Tab. 2. A decreased As-activation for the case of the presence of a high nitrogen concentration is clearly observed.

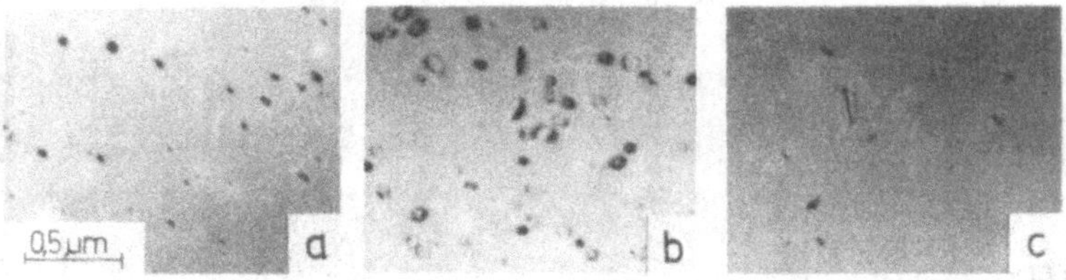

Fig. 5
TEM-micrographs of As^+ implanted into Si through Si_3N_4 d= 60nm
a 1* $10^{15}/cm^2$
b 5* $10^{15}/cm^2$
c 1* $10^{16}/cm^2$
after 1125°C, 1s

On the other hand the residual defect concentration in the specimens in Fig. 5 is very low. Qualitatively, it shows the same behaviour how in the case of implantation into virgin Si.

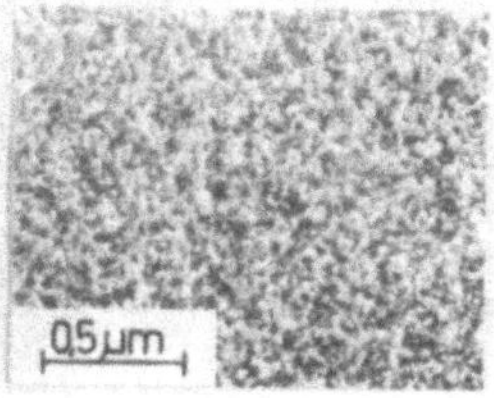

Fig. 6 TEM-micrograph of a dual- implanted sample
1* 10^{15} As^+/cm^2 + 5* $10^{15}N^+/cm^2$
after RTA at 1125°C, 1s

Contrary to this observation the implantation of 5* $10^{15}N^+/cm^2$ does not result in any change of As-activation (see Tab. 3), but the residual defect density is strongly enhanced in this case.

Tab. 3 Activation of As after dual- implantation of As^+ and N^+

RTA:	1* $10^{15}As^+/cm^2$ + 5* $10^{15}N^+/cm^2$		5* $10^{15}As^+/cm^2$ + 1* $10^{16}N^+/cm^2$	
1000°C, 1s	80 %	(75 %)	24 %	(64 %)
1075°C, 1s	---	---	35 %	(67 %)
1125°C, 1s	82 %	(78 %)	56 %	(75 %)

Values for As-activation without N^+-implantation in brackets.

4. CONCLUSIONS

1. A high oxygen or nitrogen concentration in Si reduced the electrical activation of As after a RTA-treatment.
2. The behaviour of the desity of residual defects in Si does not completely correspond to the behaviour of the As-activation.
Under certain circumstances the residual defect density is very low - nevertheless the arsenic is not electrically active.
3. This behaviour can be explained by the formation of small clusters (for instance As-O clusters), which are electrically inactive and not visible by TEM.
The formation of these clusters strongly decreases the excess silicon availible for the residual defect evolution during RTA.

5. ACKNOWLEDGMENT

The authors thank Dr. H. Syhre for SIMS measurements.

6. REFERENCE

/01/ Kögler R., E. Wieser, J. Albrecht und P. Knothe
phys. stat. sol. (a) 113, 321 (1989)

LONG-RANGE EFFECT IN ION IMPLANTATION OF SILICON AND ITS ASSOCIATION WITH MECHANICAL STRAIN IN CRYSTAL

Gorin S.N., Kalinushkin V.P., Kuzemchenko T.A., Manenkov A.A., Mikhailova G.N., Sokolov S.Yu.[1], Khaibullin I.B.[2]

Ion implantation is nowadays the standard doping technique in semiconductor technology. The space locality was so far considered an important advantage of this method, allowing to suppose that using the spectral masks one might carry out the selective doping with micron resolution. But the results published in the last decade gave rise to discussions and even to serious objections of specialists to possibility of such utilization of this technique. The case is that the new experimental evidence of a so-called long-range effect (LRE) in ion implantation were obtained. This phenomenon is displayed by changes in the semiconductor structure and physical properties arising in the region considerably exceeding the maximum implantation ion path in the crystal. In some papers /1,2/ new defects and their complexes formed during ion implantation at the depth from 10 to 100 μm under the implanted layer were reported. Some lately the lateral LRE was discovered /3,4/: changes of the material properties inside the semiconductor Si and Ge wafers arose in the area covered by mask during ion implantation at 5-10 mm from the boundary of the implanted region. Evidently that LRE may become the key limiting factor for further integration in microelectronics when advancing to submicron range. Therefore the task of systematic and profound studies of principle regularity and mechanism of this effect seems to be quite important.

For explanation of lateral LRE in /3,4/ we supposed that ion implantation give rise to mechanical strains in the regions covered by mask. Indeed,corresponding deformatins of the crystal lattice were observed in undermask part in silicon /4/ and germanium /5/. The appearing mechanical strains also stimulate accelerated migration of radiative defects to much longer distances and even themselves can rouse formation of new defects and their complexes thus contributing to changes of crystal structure and consequently to material properties.

In this paper in order to verify the propositions which seemed to be responsible for LRE the implantation have been performed at heating of the sample in range 20 to 500°C. As a result there appeared a possibility to control the processes of generation, migration, interaction and annealing of radiation defects. We used three methods for estimating of defect quantity variation after ion implantation: microwave photoconductivity /4/, low-angle light scattering at the wavelength 10.6 μm /3/ and X-ray topography. The applied microwave method for the charge carriers life-time measurement (at frequency 10 GGz) with laser

1) Institute of General Physics.Vavilov str.38, Moscow, USSR 117942
2) Kazan Phys.-Techn.Institute, Kazan, USSR 420029

photoexcitation of carriers at $\lambda = 1.06\ \mu m$ gives information about the e-h pairs life-time which in turn assosiated with the number of defects in crystals being centers of sticking and recombination, and also about the state of crystal surface /4,6/. The method of low-angle light scattering is sensitive to the presence of impurity clusters and allows to determine their concentration and size.

The initial Si sample was a commercial boron doped <111> oriented, 10 Ohm.cm plate of 100 mm diameter and 0.5 mm width. It was divided into 6 parts of an approximately equal size. Parameters of implantation by P^+ ions were as follows: energy-40 keV, dose-$6 \cdot 10^{15}/cm^2$, current density $j=1\ \mu A/cm^2$, temperature interval - 20-500°C. About half of the sample was covered by a mask - a steel plate of 1.5 mm width preventing the crystal from ion beam. Effective life-time of nonequilibrium carriers and the intensity of low-angle light scattering were measured before and after implantation in the same regions of crystal surface located at 5 mm from the mask boundary on both sides from it.

X-rays topograms of implanted samples were studied with the help of photomicrodensitometer. A part of the sample near the boundary of the mask diviation the implanted and unimplanted areas was monitored in the direction perpendicular to the boundary. The densitograms obtained allowed to measure the width of the boundary region containing strong crystal distortion.

Fig.1 shows the temperature dependence of effective life-time of nonequilibrium carriers τ_e/τ_o in implanted and neighbouring regions of the sample during implantation. At T=20°C and 5 mm far from the boundary in the masked region, the effective life-time of the carriers became about 1.5 times higher *) in comparison with the initial. Fig.1 and 2 also show the X-rays topography data: the higher the temperature during ion implantation, the less is the width of the damaged boundary area.

Analogous peculiarities were observed by light scattering (Fig.3): only at T=20°C the scattering intensity increased, and at $T \geq 200$°C no changes of scattering in the undermasked area observed.

The obtained results indicate a strong dependence of the LRE on the temperature of implantation. Let's try to understand, at least qualitively, the mechanism of this effect and processes causing the experimentally observed temperature dependence.

---

*) Unlike the earlier observed decrease of the nonequilibrium carrier life-time as a result of ion implantation /4/ in this case probably takes place the growth of observed life-time thanks to generation of sticking centers /6/.

It is known that irradiation of Si by P+ ions in mentioned regimes (40 keV, $6 \cdot 10^{15}/cm^2$) arises in the semiconductor surface layers the processes of generation, migration, accumulation and interaction of point radiative defects (vacancies, intersities) between themselves and with atoms of controllable and uncontrollable impurities. Besides, simultaneously take place recombination and annealing of point defects and their complexes. So at T < 150°C thermally activied complex formation and accumulation of radiative defects dominate in recombination and annealing, and this leads to a considerable disorder of crystal lattice and, finally, to complete amorphization of the surface layer. As a result of these phenomena strong mechanical strains appear in the implanted layer, which lead to accelerated migration of primary and secondary radiative defects into the depth, as well as to the area covered by mask causing the observed LRE.

The temperature growth till 200°C and higher keenly accelerates recombination and annealing processes of radiation defects and their complexes just during implantation. As a result, strongly decrease magnitudes of deformation fields and their gradients, thus removing the force, contributing to a rapid and deep penetration of various defects into the area under the mask and deeper layers of the sample, that is removing LRE. Besides, that the growth of irradiation temperature makes silicon more plastic, so mechanical strains become smaller and LRE decreases.

Basing upon this model and using an important property of LRE discovered earlier - its dependence on the crystal orientation /4/, enabled to make clear the question (to a certain degree), what sort of defects was mainly responsible for this effect.

In silicon, point and packing defects, dislocation loops, A-clasters are known to display increased mobility in the mechanical strain field of crystal. The fulfulled analysis showed that the character of the dislocation displacement anisotropy and mechanical strain magnitude indispensable for initiating of such displacement (10-100 kg/cm^2/7/) well agree with appropriate experimental data obtained while observing LRE.

The influence of dislocations upon the life-time of nonequilibrium carriers is well known: they can be at the same the centers of recombination and capture, as it is in our experiments. This is a additional argument in favour of the dislocational type of the defects, participating in the appearance of LRE of ion implantation. So the dislocational origin of defects, causing this effect agrees with experiments as well as with published data.

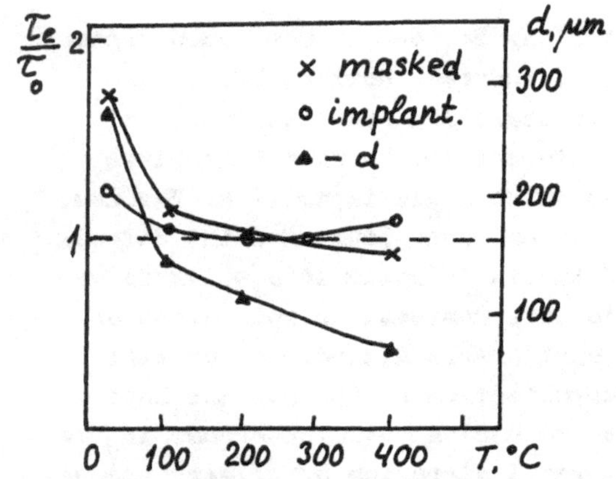

Fig.1. Dependence of effective life-time τ_e/τ_o of nonequilibrium carriers on the implantation temperature. τ_o =7.2 μs in the virgin material. Dashed line - τ_e/τ_o in the virgin plate.

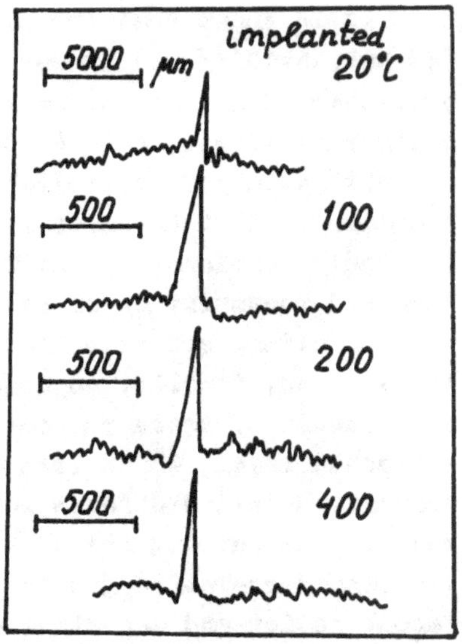

Fig.2. Densitograms of X-ray microtopograms in the region of boundary dividing irradiated and nonirradiated area in the sample.

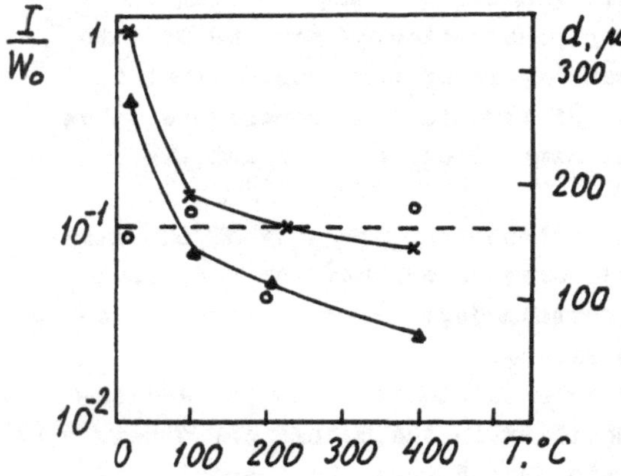

Fig.3. Dependence of scattered IR irradiation intensity on the temperature of sample heating during the implantation. Dashed line - the level of light scattering intensity in the virgin plate. Denotations are the same as in Fig.1

References

/1/ Vavilov V.S., G.N.Galkin, M.I.Guseva, L.M.Ershova, K.V.Kiseleva, V.V.Krasnopevtsev and Yu.N.Milutin: Radiation Effects in Semiconductors Ed. by F.L.Vook, N.Y., Plenum Press (1968) 346

/2/ Zorina G.V., Yu.S.Popov and V.L.Shargel: Ion implantation in semiconductors and other materials (in Russian) Vilnus (1985) 276

/3/ Kalinushkin V.P., A.A.Manenkov, G.N.Mikhailova, M.G.Ploppa, A.M.Prokhorov, A.S.Seferov, Yu.N.Chekhonadskii and I.B.Khaibullin: Microelectronica, 15 (1986) n.6, 528

/4/ Kuzemchenko T.A., A.A.Manenkov, G.N.Mikhailova and S.Yu.Sokolov: Phys.Lett.A 129 (1988) n.3, 180

/5/ Voronov V.V., V.P.Kalinushkin, I.B.Khaibullin, G.N.Mikhailova and A.S.Seferov: Kratkie soobschenia po fizike FIAN SSSR, 5 (1989) 40

/6/ Kuzemchenko T.A., S.Yu.Sokolov, S.G.Tikhodeev: Preprint IOFAN SSSR, Moscow, n.69 (1988) 17

/7/ Titov V.V. Preprint of Institute of Atomic Energy, Moscow, n.3774/11 (1983)

ACTIVATION OF PHOSPHORUS IN SURFACE LAYERS OF P^+-IMPLANTED SILICON UNDER RF PLASMA TREATMENT

Valakh M.Ya., Lysenko V.S., Nazarov A.N., Rudko G.Yu., Tkachenko A.S., Shakhraychuk N.I.[1)]

1. Introduction

Low temperature treatment in the plasma of rf discharge was proposed in paper [1] for the annealing of implantation-induced defects in Al-SiO_2-Si structures. This technique implies the sample temperatures less than 400-500°C even at maximum discharge powers and therefore it permits to avoid the drawbacks of high temperature annealing. Processes occuring in semiconductors at rf plasma treatment are investigated insufficiently, and the mechanism of dopant activation has not been investigated at all.

2.Experimental

The samples. We have investigated traditional Al-SiO_2-Si systems. The implanted layer was located under the dioxide layer. The implantation dose was equal to $5\cdot10^{13}$ion/cm^2. Some of the structures were isochronously annealed in nitrogen atmosphere in the temperature range 300-1000°C during 15 min. Another were treated in nitrogen rf plasma at various levels of discharge power during 15 min.

The electrode and the dioxide layer were chemically etched before optical measurements.

Experimental technique. The photoluminescence of bound exciton complexes was investigated at T=4,2 K. It gives the information about the amount of electrically active impurities in Si and the lattice disordering. Radiation of Ar^+ laser was used for the excitation. The luminescence sygnal was recorded using double grating monochromator, cooled photomultiplier and photon counting system.

3. Experimental results

Luminescence spectra of P^+-implanted silicon at various regimes of subsequent treatment are shown in Fig.1. The observed bands are the non-phonon boron NP(B) and phosphorus NP(P) bound exciton bands. It is seen that in the untreated implanted sample the intensity of phosphorus bound exciton is smaller than that of the boron bound exciton (curve 1 Fig.1b). Both thermal and rf plasma annealings cause the increase of phosphorus

----

1) Institute of semiconductors, Ukrainian Academy of Sciences
Prospekt Nauki 45, 252650 Kiev-28, USSR

bound exciton luminescence intensity.

Comparison of curve 3 (Fig.1a) and curve 4 (Fig.1b) shows that rf treatment at discharge power density of 1,6 W/cm^2 produces almost the same relative increase of phosphorus bound exciton intensity as the thermal annealing at 650°C does. Note that the temperature of the sample during rf treatment at 1,6 W/cm^2 does not exceed 400°C.

Further increase of discharge power density in exceed of 1,6 W/cm^2 leads to some decrease of phosphorus bound exciton luminescence intensity.

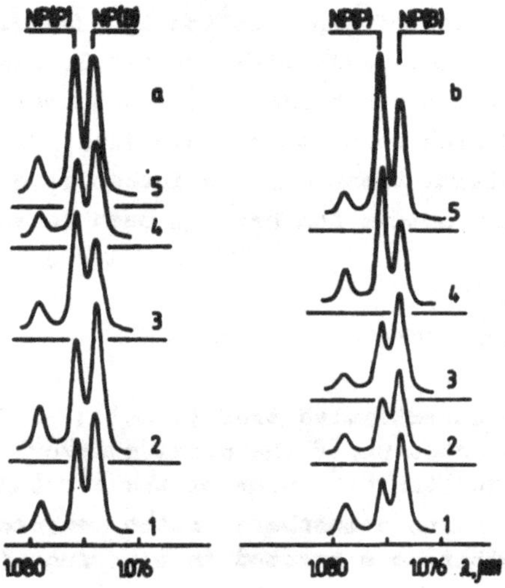

Fig.1. Non-phonon luminescence bands of excitons bound to boron NP(B) and phosphorus NP(P) for the samples treated in the rf discharge plasma (a) and thermally annealed samples (b). (a) The discharge power density is equal to: 1 - 1,0 W/cm^2, 2 - 1,3 W/cm^2, 3 - 1,6 W/cm^2, 4 - 2,0 W/cm^2, 5 - 2,3 W/cm^2. b) Annealing: 1- without annealing, 2 -T=300°C, 3 - T=500°C, 4 - T=650°C, 5 - 950°C.

In Fig.2a,b we present the dependences of normalized ratio I^* of phosphorus bound exciton intensity I_P to boron bound exciton intensity I_B on discharge power density and annealing temperature, correspondingly. The value I^* is equal to $I^* = (I_P/I_B)/(I_P/I_B)_{950^oC}$ where $(I_P/I_B)_{950^oC}$ is the ratio of phosphorus and boron bound exciton intensities in the sample annealed at 950°C. It is seen that even the largest of the examined discharge powers do not permit to attain the activation degree obtained at 950°C annealing.

We have also investigated the spectra of the samples which were treated in the rf plasma in two different ways. In the first case Al-SiO_2-Si structures were treated in such a way that the metal electrode (the "front" side) faced the discharge plasma. In the second case the semiconductor substrate (the "back" side) faced the plasma. It was shown that the increase of discharge power density in both above-mentioned cases had led to the increase of luminescence intensity of excitons bound to implanted phosphorus. But the "front" side treatment causes the higher rate of phosphorus bound exciton band increase than the "back" side one. Furthermore, exciton bands of "back" side treated samples have essentially larger halfwidths than those in "front" side treated samples.

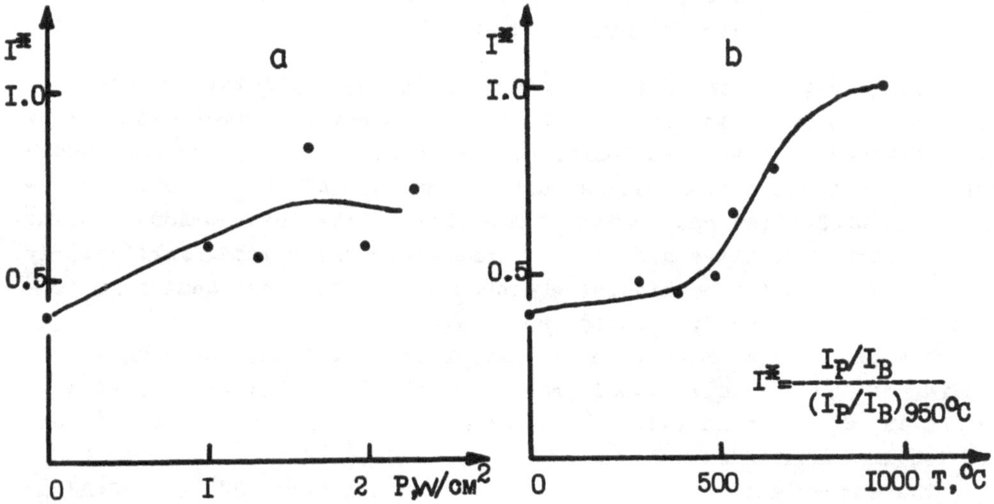

Fig.2. Normalized intensity ratio $I^* = (I_P/I_B)/(I_P/I_B)_{950^\circ C}$ versus discharge power density (a) and annealing temperature (b).

4. Conclusions

i) Both rf and thermal treatments cause the activation of the implanted phosphorus.

ii) The activation under the rf discharge treatment takes place at more low temperatures than that of traditional thermal annealing. Therefore the conclusion was made that there is some nonthermal mechanism of phosphorus activation which takes place during the rf treatment.

iii) Even the largest of the examined discharge powers do not permit to attain the level of activation obtained at 950°C annealing.

1. Lysenko V.S., M.M.Lokshin, A.N.Nazarov, T.E.Rudenko, A.S.Tkachenko: Zh. tekhn. Fis., Pisma 9(1983)796

CRYSTALLOGRAPHIC NATURE OF THE HEXAGONAL SILICON FORMED BY THE HIGH-INTENCITY ION IMPLANTATION

S.A.Petrov, F.F.Komarov, A.P.Novikov, V.Yu.Troitsky

Institute of Applied Physics Problems, Kurchatov str. 7, Minsk 220064, USSR

The investigation of the high-intensity ion implantation into silicon has shown that the local macroscopic heating dramaticaly changes the character of the radiation-induced processes. Many phenomena observed at this implantation regime such as the crystalline structure ordering, the nucleation and growth of new phases, the interaction of point defects with impurities are not yet completely understood. Particularly, it is known that the considerable amount of the residual damage is produced beyond the mean projacted range $\bar{R}_p$.

The aim of the present investigation is to get the basic understanding of the crystallographic nature of the hexagonal precipitates formed in (001) and in (111) silicon during the high-intensity As^+ and Ar^+ ion implantations.

The energy and the ion beam current density were 120 keV, 50 $\mu A/cm^2$ for As ions and 150 keV, 100 $\mu A/cm^2$ for Ar^+ ions. Transmission electron microscopy (TEM) and electron difraction (TED) were used in these studies. The precipitates of new phase were observed at depths ranging from Rp to 2Rp. The maximum temterature of the samples during implantation of arsenic ions were 880 K, and during argon irradiation the temperature reached 1100 K. The temperature was controlled by a pyrometer.

The preparation of the samples for the transmission electron microscopy have been made in a rotated glass in an solution $HF : HNO_3 = 1 : 7$. Such etching gives bevelled in samples, so that one can observe the depth depdence of the damage. The TEM - investigation have been made on a 100 kV electron microscope. The electron diffraction studies showed that the inclusions observed at depth larger than the mean projected range Rp have the hexagonal crystalline structure of wurtzit type with latice parameters: a=0.380 nm c=0.628 nm. This phase have been found on both (001) and (111)-samples /1,2/. As it is seen from fig. 1. the hexagonal inclusions in silicon after the As-implantation into (001)-samples during 10s and with the dose $D = 2.4 \times 10^{15}$ cm^{-2} have the shape of thin lamels. The further irradiation ($t = 40$ s, $D = 1.4 \times 10^{16}$ cm^{-2}) of the silicon monocrystals results in a growth of the new phase. The oval $\{0002\}$ type reflexes on the microdifraction pictures steaming from the hexagonal modification

Fig.1. Electron-diffraction photography of Si (001), implanted with As^+ (120 keV, current density 50 μA/cm^2).

Fig.2. Electron-diffraction photography of Si (111), implanted with P^+ (100 kev, current density 100 μA/cm^2).

have been transformed into round high intensity reflexes.

The analysis of the electron microscopy pictures shows that the near surface layer with the depth of 40 nm remains free from defect inclusions which are located at greater depths. Besides these defect inclusions there are pores observed near the crystal surface. The location of the pores let us suppose that their origin is due to generation of a large number of the recoil atoms at this depth. As a result of this process the considerable exsess concentration of the vacancies at small depths is created immediatelly affer cascade propagaticn.

For the (oo1) silicon the folowing nev phase // matrix orientations were determined: [0001] // [110] and [$\bar{1}$010] // [001] . For the (111) silicon these were: [0001] // [112] and [1$\bar{2}$1$\bar{2}$] // [1$\bar{2}$1] . The structure of the hexagonal pnase inclusions which are located at the depths larger then the mean progected range does mot depend on the type of the implanted ions.

(1) Komarov F.F., Kotov E.V., Novikov A.P., Petrov S.A. and Chit'ko V.I.: Radiation Effects 105 (1987) 79-84

(2) Komarov F.F., Novikov A.P. and Petrov S.A.: Journal of Technical Physics 58, 3 (1988) 548-551

APPLICATION OF ION IMPLANTATION TO CHEMICAL MICROSENSORS

Pham M.T., Hoffmann W., Hüller J., Schöneich J.

Akademie der Wissenschaften der DDR
Zentralinstitut für Kernforschung Rossendorf,
PF 19, Dresden, 8051, DDR

Problem and concept

Ion sensitive field effect transistors (ISFET) represent a new type of chemical microsensors being currently under intensive research and development /1/. The conversion of the chemical signal (ion concentration) into an electrical signal is accomplished by the ion selctive membrane (ISM) which represents the key component of an ion sensor. There are a few ways to achieve the ISM bound to the ISFET, Fig.1.

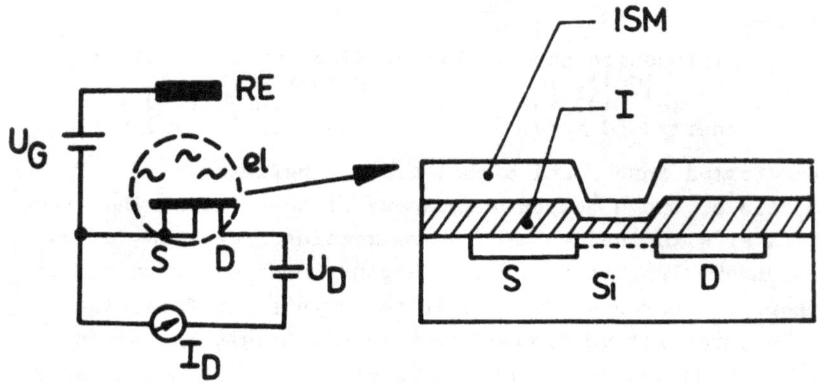

Fig.1 Schematic diagram of an ISFET
ISM : ion selective membrane
Si : Silicon
el : electrolyte
I : gate insulator
RE : reference electrode
S,D : source, drain

The results are not yet satisfactory considering the major requirements
- adherence (firmly attached ISM improves the selectivity and increases the livetime)
- coating in locally defined areas (for realizing multifunction and integrated sensors)
- technology (compatibility to IC technologies for a low cost production).

The method presented here is based on the synthesis of the ISM by implanting its components directly into the surface of the FET-gate insulator. Advances cover the mentioned problems. /2, 3, 4/.

Optimizing

The method was verified for alkali alumino silicate glasses as ISM for Na^+ and K^+ ions. /5/.

Glass films were prepared by implanting Na^+ or K^+ ions and Al^+ ions (or homologous ions such as B^+, Ga^+, In^+) in Si_3N_4 or SiO_2 followed by a thermal treatment, Fig.2. The treatment step is necessary for annealing the damages introduced by ion bombardment and for the glass formation.

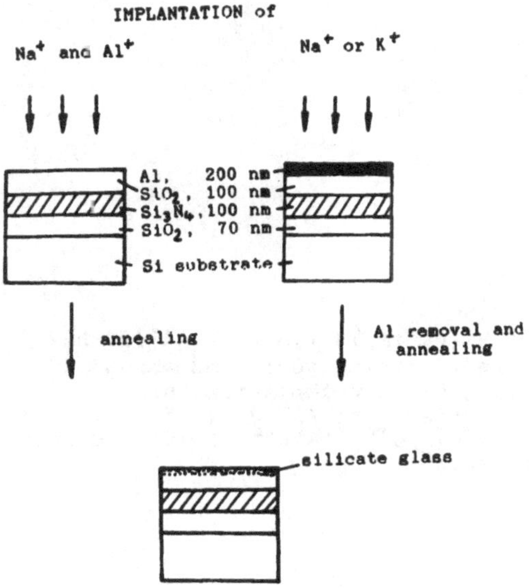

Fig. 2
Processing steps for preparing the silicate glass membrane

The parameters of the ion implantation (ion dose and energy) and the thermal treatment were optimized by capacitance measurements of the capacitor structure electrolyte/insulator/silicon (equivalent to the MOS capacitor). For example the K^+ ion sensitivity of the implanted Si_3N_4 surface is shown in Fig.3 by presenting changes of the flat band potential as a function of the electrolyte concentration.
The sensitivity to Na^+ and K^+ ions shows a strong dependence on implantation dose. By varying the implantation dose, a maximum for K^+ and Na^+ sensitivity of 50...55mV/decade concentration was found. With ion implantation the sensitivity to Na^+ and K^+ increases and that of H decreases.

- Low ion energy
- implantation through an interlayer of Al
- usage of a Si_3N_4 barrier layer
- short time annealing

were found to be effective with respect to avoiding the deterioration of the electrical characteristics of the field effect capacitor due to mobile ions and damages by ion bombardement.

Sensor realizing

Fig.4 shows the characteristics of an ISFET for Na^+ ions realized with ion implanted ISM. The measured sodium sensitivity compares well with results for optimized ISM. Na^+ and Al^+ were implanted into the SiO_2 layer (LP-CVD, 100 nm) on top of the ISFET gate using the photolithografic masking technique to define the areas to be implanted on the wafer.

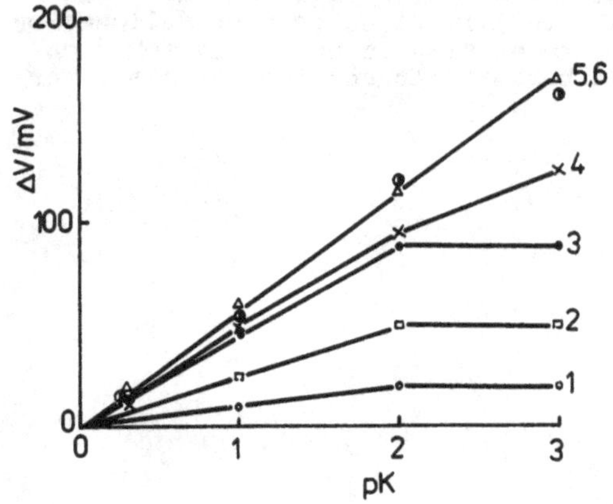

Fig. 3 K^+sensitivity at pH = 8 of Si_3N_4 implanted with Al^+ (20 keV, $5*10^{16}$ ions/cm^2) and K^+(30 keV, varying dose) and annealed (O_2, 900 °C,100 s) : 1. Si_3N_4 without implantation.
2. implanted with Al^+. 3, 4, 5, 6 : K dose : $5*10^{15}$, $1*10^{16}$, $5*10^{16}$ and $1*10^{17}$ ions/cm^2

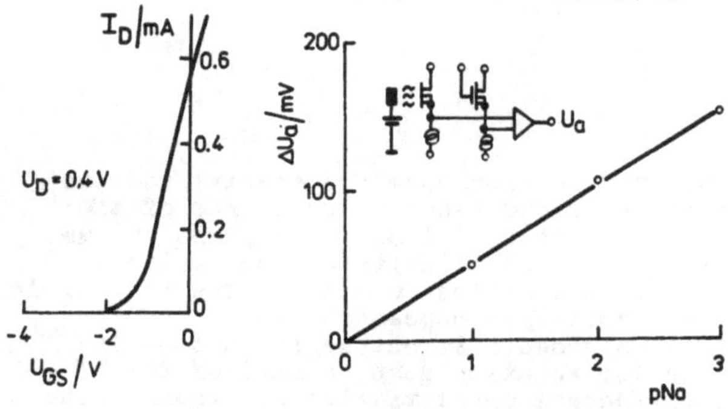

Fig. 4 Na^+ - sensitive FET with ion-implanted membrane : transfer characteristics and response to sodium concentration charges

Dual structure : ISFET-MISFET
Chip size : 1,5 x 5,9 mm^2
Gate area : 12 x 432 μm^2
implanted with Na^+(20 keV, $5*10^{16}$ ions/cm^2) and Al^+(20 keV, $2*10^{16}$ ions/cm^2) annealed at 1000 °C for 100 s

References

/1/ 2 nd Int.Sensor Meeting, Bordeaux 1986, Transducers' 87, Tokyo 1987, Eurosensors II, Enschede 1988
/2/ Y.Sanada, T.Akiyama, Y.Ujihira, E.Niki: Fresenius Z.Anal.Chem.: 3/2 (1982) 526
/3/ M.T.Pham, W.Hoffmann: - Sensors and Actuators 5 (1984) 217.
/4/ T.Ito, H.Inagaki, I.Igarashi: IEEE trans. on ED, ED-35 (1988) 56
/5/ G.Eisenmann: "Glaselectrodes for Hydrogen and other Ions" Dekker, New York 1967.

PULSE IMPLANTATION DOPING OF SILICON WITH BORON AND PHOSPHORUS

J.Piekoszewski[1], J.Białoskórski[1], A.Horodeński[1], J.Kaczanowski[1], J.Langner[1], C.Pochrybniak[1], A.Turos[1], Z.Werner[1], J Zaręba[1].

1. Introduction

Pulse Implantation Doping (PID) is an original process of instantaneous doping the semiconductor surface layer, recently developed at our Institute (2-7). In this process, a high-intensity pulse of plasma containing the dopant species is applied to the surface of the processed material and provides both, a dose necessary for doping and an energy necessary to melt its near-surface layer. In the liquid phase, a rapid diffusion of the dopant occurs; next, the layer recrystallizes epitaxially.

In our previous works we have demonstrated the feasibility of shallow p-n junction formation by forming p^+-type silicon layer doped with boron from BF_3 plasma. In this communication we report the first results of forming n^+-type silicon layer in PID process by using PF_5 plasma. Since it is our ultimate goal to apply PID to solar cell technology, we have attempted to produce n^+-p-p^+ photovoltaic structures using exclusively PID technique for photovoltaic junction and ohmic contact formation. For comparison, p^+-n-n^+ photovoltaic structures were also prepared.

2. Experimental

All samples were cut from 4-8 Ω·cm, (111) silicon of either n- or p-type. Plasma pulse irradiations were performed using JONOTRON '85 machine [1] generating pulses of BF_3 and PF_5 plasma of an energy density 2-5 $J \cdot cm^{-2}$ and duration of about 0.5 µs. The adopted procedure for n^+-p-p^+ photovoltaic structure preparation was as follows:

a) cleaning in hot $H_2SO_4 : H_2O_2$,
b) ohmic contact formation with 2 pulses of BF_3 plasma,
c) n^+-p front junction formation with 3 pulses of PF_5 plasma,
d) Ti-Pd-Ag vacuum metallization,
e) sintering at 550°C,
f) edge trimming by chemical etching,
g) vacuum evaporation of SiO_x AR coating.

The procedure for p^+-n-n^+ structure preparation differed in steps b) and c) as follows:

b*) ohmic contact formation by evaporation of 200 Å of Sb and plasma pulse alloying with 2 pulses of H_2 plasma,
c*) p^+-n junction formation with 1 to 12 pulses of BF_3 plasma.

---

1) Soltan Institute for Nuclear Studies, 05-400 Otwock, Poland.

3. Results and discussion

The results of measurements of photovoltaic properties of the prepared structures are shown in Table 1.

Table 1. Summary of the preparation conditions and photovoltaic properties of PID produced structures.

FRONT JUNCTION FORMATION					PHOTOVOLTAIC PARAMETERS			
SAMPLE No	STRUCTURE, DOPING GAS	E_{mean} $\left[\frac{J}{cm^2}\right]$	E_{max} $\left[\frac{J}{cm^2}\right]$	NUMBER of PULSES	V_{oc} [mV]	I_{sc} $\left[\frac{mA}{cm^2}\right]$	FF	η [%]
895		3.6		1	557	30.5	0.70	11.7
900		3.7		1	564	29.6	0.70	11.7
891	p^+-n	4.0		1	566	28.4	0.70	11.3
814		4.8		2	574	28.3	0.74	12.0
812	BF_3	4.3		3	574	29.3	0.72	12.0
813		4.8		5	574	26.3	0.76	11.4
811		4.2		12	578	23.3	0.76	10.2
2157			2.9	3	550	32.4	0.71	12.6
2150	n^+-p		3.2	3	555	31.3	0.71	12.3
2154			3.2	3	556	30.6	0.72	12.2
2152	PF_5		3.3	3	557	30.7	0.72	12.4
2159			3.6	3	551	30.7	0.73	12.3
2153			3.7	3	557	30.7	0.71	12.1

The following conclusions can be drawn from this table:

- For p^+-n-n^+ structures, the increase of the number of pulses leads to an increase of V_{oc} and FF and a decrease of I_{sc}. This result is in qualitative agreement with the results of Young et al. [8] on ion implanted and laser annealed samples which exhibited identical trends with an increase of implantation dose. We showed previously [7] that boron concentration in the sample grows with the number of pulses.
- For n^+-p-p^+ samples, higher I_{sc}, close FF, and slightly lower V_{oc} are observed as compared to the former ones. Howewer, the resultant efficiency falls in the higher range than that for the best p^+-n-n^+ samples obtained till now. An apparent decrease of I_{sc} with an increase of pulse energy can be interpreted as related to the junction depth which should grow with pulse energy.

An interesting observation is that crystallinity of the processed layers is more sensitive to pulse energy for PF_5 than for BF_3, being better for lower energy densities as it is shown in the Figs. 1. and 2.

4. Conclusions

We have demonstrated for the first time that it is possible to dope effectively silicon with phosphorus using pulses of PF_5 plasma without a need of any further treatment. Together with the previously acquired experience on doping with boron using BF_3 plasma pulses, this result makes it possible to fabricate photovoltaic n^+-p-p^+ and p^+-n-n^+ struc-

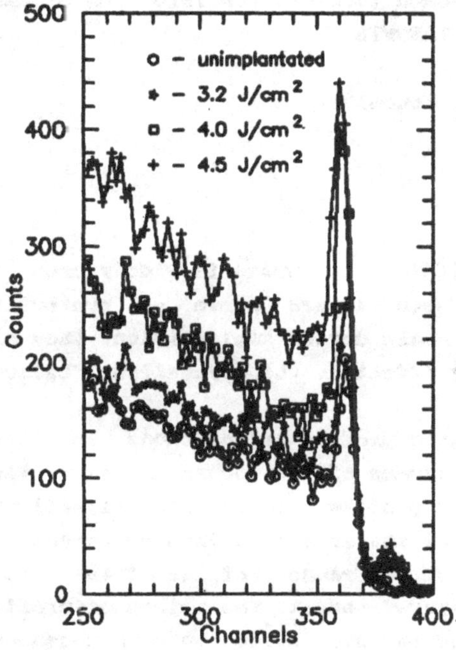

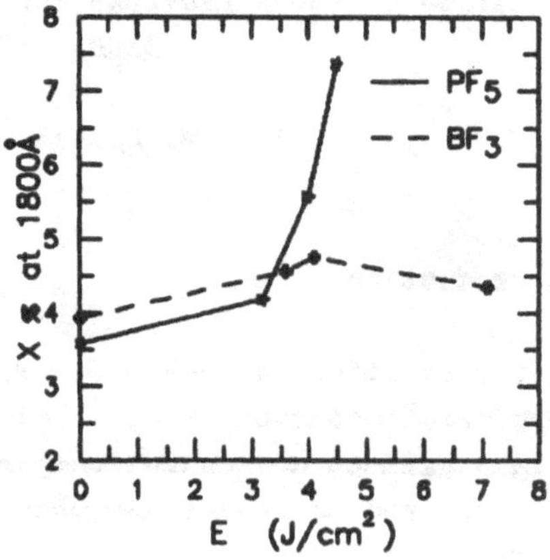

Fig. 2. Dechanneled fraction (at 1800Å) as function of pulse energy density for BF_3 and PF_5.

Fig. 1. Channeling spectra (2.0 MeV He^{++} ions) of p-type Si [111] procesed by PID with PF_5 for different pulse energy densities.

tures using exclusively PID technique. The n^+-p-p^+ structures recently obtained by PID with PF_5 plasma exhibit the parameters superior to those of p^+-n-n^+ structures obtained previously by PID with BF_3 plasma. PID produced photovoltaic junctions exhibit a slight growth of V_{oc} and a decrease of I_{sc} with the growth of doping concentration in agreement with the reported results on conventionally implanted and laser annealed junnctions.

References

[1] Langner,J., K.Czeus, E.Górski, M.Gryziński, A.Horodeński, J.Piekoszewski, C.Pochrybniak, Z.Werner, K.Wojtowicz Proc. EPM-87 Cof. Dresden (1987)

[2] Piekoszewski,J., M.Gryziński, J.Langner, Z.Werner: phys.stat.sol. (a) 67 (1981) K163-K166

[3] Piekoszewski,J., M.Gryziński, J.Langner, Z.Werner, G.C.Huth: J.Physique 43 (1982) 1353-1358

[4] Piekoszewski,J., Z.Werner, J.Langner, A.Jakubowski, C.Pochrybniak, A.Harasiewicz: Nucl.Instr.Meth. 209/210 (1983) 477-482

[5] Piekoszewski,J., Z.Werner, C.Pochrybniak, M.Gryziński, J.Langner, S.Gębalski, M.Klimkiewicz, A.Horodeński: Proc. EPM-84 Conf., Dresden (1984) 142-146

[6] Piekoszewski,J., Z.Werner, C.Pochrybniak, J.Langner, M.Gryziński, A.Horodeński: phys.stat.sol. (a) 112 (1989) 757-760

[7] Werner,Z., J.Piekoszewski, C.Pochrybniak, Z.Haratym, A.Sulik: . Proc. epm-87 Conf. Dresden (1987) 312-314

[8] Young,R.T., R.F.Wood, G.E.Jr.Jellison, W.H.Christe: Proc. 3rd E.C.Photovoltaic Solar Energy Conf., Cannes,(1983)703-707

SIMULATION OF RANGE PROFILES FOR THE BORON IMPLANTATION INTO SIO_2/SI AND SI_3N_4/SIO_2/SI TARGETS

M. Posselt[1), T. Feudel[2)

. Introduction

Ion implantation through SiO_2 or Si_3N_4/SiO_2 cap layers is widely used in modern semiconductor technologies. These layers serve as protective coatings against contaminations and as masks during implantation. They are also employed to reduce the channeling effect in the crystalline silicon substrate.

In the present work we consider the range profile calculations in these layered targets. For applications in programs of the technology simulation the simple and efficient method of profile combination (1)-(4) is favoured. Here, the range distribution of the ions in a layered target is obtained by an appropriate combination of the range profiles of these ions in the homogeneous components of the layered target. The method of profile combination is only valid for layered targets which have homogeneous components with similar atomic numbers and masses. We want to demonstrate that this method is applicable to the B^+ implantation into SiO_2/Si and Si_3N_4/SiO_2/Si targets. The range distributions simulated by this procedure are compared with range profiles obtained by TRIM Monte Carlo simulations (5) and with experimental data.

2. Calculation of range profiles in layered targets

Figs. 1a and 1b show the range profiles of 30, 80 and 120 keV B^+ implantations into SiO_2/a-Si and Si_3N_4/SiO_2/a-Si layered targets. They were calculated by TRIM and by range profile combination. The agreement between the results of both methods is good. This can be considered as a theoretical demonstration of the fact that the method of profile combination can be applied to the boron implantation into SiO_2/a-Si and into Si_3N_4/SiO_2/a-Si targets. In comparison with TRIM simulations with 10,000 ion histories the method of profile combination is about 20,000 times faster. In our example the input parameters of the method of profile combination were obtained as follows: In a first step TRIM simulations were performed to yield the range distributions of 30, 80 and 120 keV B^+ implantations into the homogeneous a-Si, SiO_2 and Si_3N_4 targets. Then the calculated histograms were fitted to a Pearson-IV

1) Academy of Sciences of the GDR, Central Institute for Nuclear Research, Rossendorf, PF 19, DDR-8051 Dresden
2) VEB Forschungszentrum Mikroelektronik Dresden, Betrieb des Kombinates VEB Carl Zeiss Jena, PF 34, Grenzstr. 28, DDR-8080 Dresden

function applying a least-squares fit. So, we got 4 range profile parameters for every implantation energy and for every component.

In general, the validity of the method of profile combination only depends on the difference between the atomic numbers and masses of the components in the layered target but not on the fact whether a component has an amorphous or a crystalline structure. Consequently, this procedure can be also used in the case of B^+ implantations into SiO_2/c-Si and Si_3N_4/SiO_2/c-Si. Since the TRIM program is only applicable to amorphous targets the boron profile in crystalline silicon cannot be obtained by this method. Therefore, we employ a semi-empirical approach used in the technology simulator TESIM-3 (6): The boron profiles in crystalline silicon are described by a Pearson-IV distribution with the parameters of the a-Si but modified by an exponential channeling tail. This tail is characterized by only one parameter. A similar procedure is also applied in the process simulator SUPREM II (7).

3. Comparison with experimental data

We performed 30, 80 and 120 keV B^+ implantations into SiO_2/a-Si, SiO_2/(100)-Si, and Si_3N_4/SiO_2/(100)-Si layered targets with a tilt angle of 7^0 (no rotation of the wafer). SiO_2, Si_3N_4 and a-Si were deposited by CVD on the (100)-Si substrate. After the implantation the SiO_2 and the Si_3N_4 layers were etched off. The layer thickness was determined by ellipsometric measurements. The boron profiles in a-Si and in (100)-Si were investigated by SIMS.

Fig. 1a shows the results of the B^+ implantations into a-Si through SiO_2. Except in the region of the depth tail the SIMS data agree satisfactorily with the theoretical profiles calculated in the last section. The agreement of the curves obtained by the method of profile combination with the experimental data can be further improved using the empirical range parameters of TESIM-3. This is demonstrated by the dashed curves in fig. 1a. The improvement is mainly caused by a slight modification of the parameters for a-Si. The range distributions for boron implantations into (100)-Si through SiO_2 and Si_3N_4/SiO_2 cap layers are given in figs. 2a and 2b, respectively. A satisfactory agreement between the profiles measured by SIMS and the range distributions calculated by the method of profile combination is obtained. As inputs of the simulation we used the empirical values for the range parameters and for the exponential channeling tail given in TESIM-3.

4. Conclusions

Our theoretical and experimental considerations have shown that the method of profile combination can be employed in the case of the boron implantation through SiO_2 and Si_3N_4/SiO_2 cap layers into a-Si and c-Si.

For the use in computer codes of the technology simulation this procedure has essential advantages. It is about 4 orders of magnitude faster than TRIM simulations.

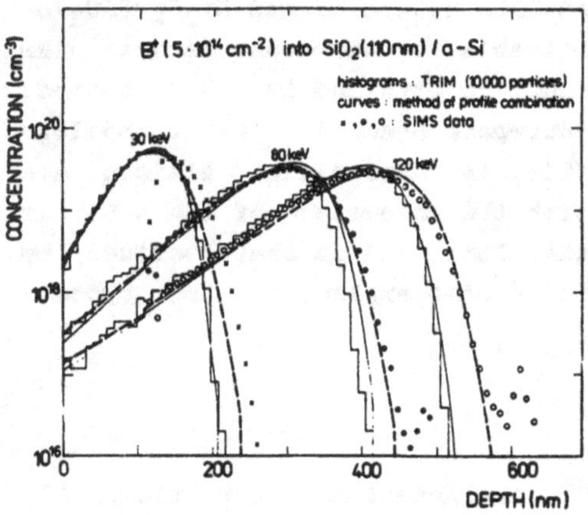

Fig. 1a

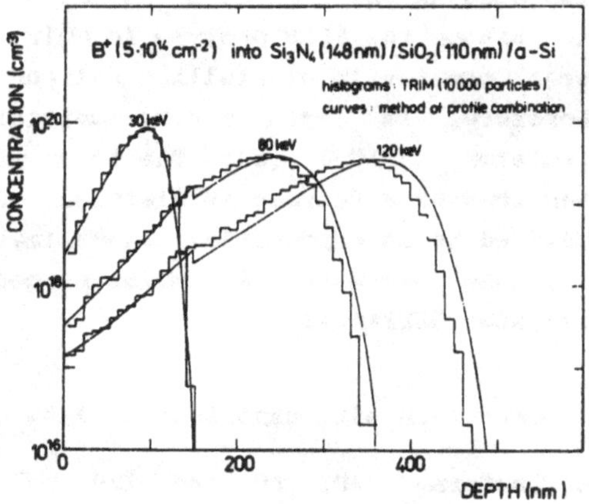

Fig. 1b

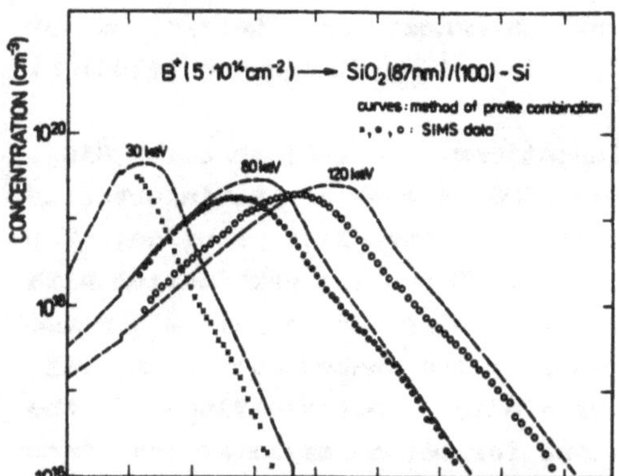

Fig. 2a

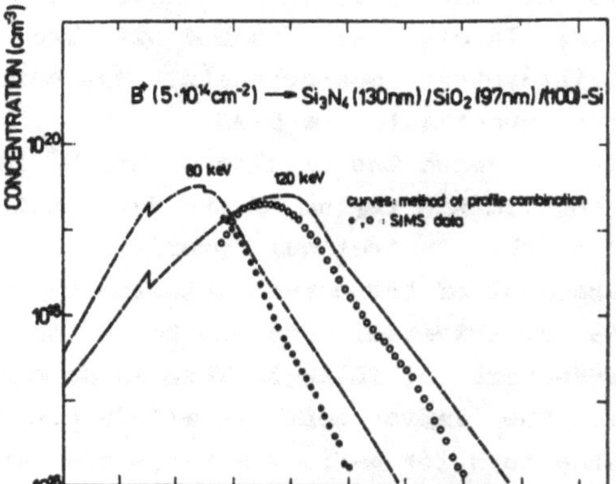

Fig. 2b

(1) Posselt, M.: phys. stat. sol. (a) 112 (1989) 263.
(2) Ishiwara, H., S. Furukawa, J. Yamada and M. Kawamura, in: Ion Implantation in Semiconductors (ed. S. Namba), Plenum Press, New York, 1975, p. 423.
(3) Ryssel, H., J. Lorenz and K. Hoffmann: Appl. Phys. A41 (1981) 201.
(4) Amaratunga, G.A.J., K. Sabine and A.G.R. Evans: IEEE Trans. on Electron Devices ED-32 (1985) 1889.
(5) Ziegler, J.F., J.P. Biersack and U. Littmark: The Stopping and Range of Ions in Solids, Pergamon Press, 1985.
(6) Feudel, T., S. Ulbricht, A. Hürrich, H.U. Jäger and R. Stephan: TESIM-3, VEB Forschungszentrum Mikroelektronik Dresden, Dresden 1988.
(7) Antoniadis, D.A., S.E. Hansen and R.W. Dutton: IEEE Trans. on Electron Devices ED-26 (1979) 490.

EPITAXIAL LATERAL OVERGROWTH OF AMORPHOUS CVD SILICON FILMS INDUCED BY ION IRRADIATION

M. Voelskow[1], W. Skorupa[1], K. Wollschläger[1], J. Matthäi[1], P. Knothe[1] and H. Bartsch[2]

1. INTRODUCTION

During the last years, lowering of the processing temperature in manufacturing integrated circuits has become of increasing importance to realise devices with a higher speed, lower power consumption and smaller geometrical dimensions. Among others, an interesting new method is to lower the temperature of the thermally induced solid phase epitaxy (SPE) by the use of ion beam processing. This process is called ion beam induced epitaxial crystallization (IBIEC). A lot of work has been performed to study basically IBIEC of amorphous silicon (a-Si) layers produced by ion implantation directly on or within monocrystalline silicon substrates /1,3,6/. For different practical purposes, like the production of low doped silicon layers on a highly doped substrate or the use of lateral SPE over insulating layers for the realization of silicon-on-insulator (SOI) structures, the recrystallization of deposited a-Si layers on a monocrystalline substrate is of great importance. Recently, this was proved to be possible by using preamorphization of the transition region between the deposited a-Si and the monocrystalline substrate /2,5/. Problems arising from the interfacial oxygen in such systems during conventional thermal recrystallization can be overcome by IBIEC as well studied by Priolo et al. /4/.

In this letter, a first test will be reported concerning epitaxial lateral overgrowth by IBIEC at 400 °C.

2. EXPERIMENTAL

To clarify the optimum regrowth conditions for the lateral IBIEC process, at first the process was studied at vertical structures. The a-Si layers with a thickness of 400 nm were deposited onto monocrystalline silicon wafers with (100)-orientation by LPCVD at 1 Torr and a temperature of 570 °C. Before the deposition step cleaning of the wafers was performed by boiling them in a H_2O_2/H_2SO_4-mixture followed by thermal oxidation at 1000 °C in dry oxygen. In the vertical IBIEC-experiments the SiO_2-layer

1) Academy of Sciences of the GDR, Central Institute for Nuclear Research Rossendorf, PF 19, DDR - 8051 Dresden, GDR

2) Academy of Sciences of the GDR, Institute for Solid State Physics and Electron Microscopy Halle, PF 250, DDR - 4010 Halle, GDR

with a thickness of about 100 nm grown during this step was etched away by buffered HF immediately before loading the wafers into the CVD reactor. After deposition, part of the wafers was implanted with 660 keV arsenic ions (R_p = 440 nm) at a dose of 2×10^{15} cm^{-2} and room temperature to amorphise the transition region /5/. For the lateral IBIEC-experiments the thermally grown silicon dioxide film was etched away only in windows. In such a manner, these windows in the SiO_2 film may act as seeds for epitaxial lateral overgrowth of the deposited amorphous CVD silicon film in order to produce SOI-structures. To eleminate all nucleation centers for crystallization in the deposited film, the whole silicon layer was additionally amorphised by implantation of arsenic atoms at three different energies (150 keV, 400 keV, 660 keV) at doses of 2×10^{15} cm^{-2}. Before the IBIEC stept the arsenic implanted wafers of both series were annealed by furnace at 450 °C for 30 min in dry nitrogen to eliminate a damaged but not fully amorphised region below the thickened a-Si layer.

IBIEC was performed with 330 keV silicon ions at a target temperature of 400 °C. Target heating was done by the ion beam itself as described earlier /6,7/.

Rutherford backscattering spectrometry (RBS) and channeling techniques with 1.7 MeV, $^4He^+$-ions were used to study the vertical crystallization. Transmission electron microscopy at plane view and cross sectional samples were used to investigate the lateral crystallization of the silicon film over the SiO_2 layer (SOI-structures).

3. RESULTS

In the vertical crystallization experiments without the additional amorphization step no epitaxial regrowth was found despite fact that the implantation of a silicon dose of 2×10^{17} cm^{-2} should be sufficient for a full regrowth of the whole layer as estimated from calculations taking into account the energy deposition into nuclear processes and IBIEC rates of implantation amorphised silicon layers /3,6/. This can be as ascribed to the properties of the transition region containing first of all a thin silicon oxide layer /4/.

The introduction of the additional step of amorphising the transition region by arsenic implantation led to much better results. As shown by channeling measurements, after implanting silicon doses of 6×10^{16} cm^{-2} and 1×10^{17} cm^{-2}, about half of the layer and the full layer were recrystallized, respectively.

Taking into account the results of this first series of experiments a high dose of 5×10^{17} cm^{-2} silicon ions was used to crystallise the SOI-samples. Fig. 1 shows a micrograph of a cross sectional sample. As clearly visible,

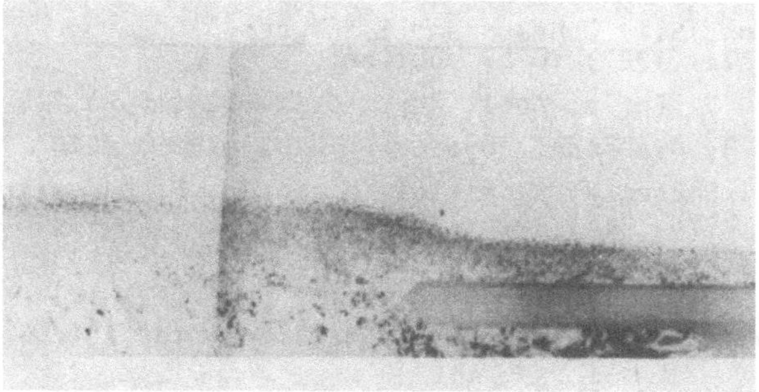

Fig. 1

Micrograph of a cross sectional SOI-sample after irradiation with Si ions at 400 °C (330 keV, 5×10^{17} cm^{-2})

starting form the edge of the SiO_2 film, the crystallization front was moved over the SiO_2 film over a distance of approximately 1.5 µm. Furthermore, no random nucleation in the amorphous film beyond the crystallization front was found (see left part of Fig. 1). However the crystallized part of the film didn't show the expected single crystalline growth but large grains with a strong (100)-texture.

Firstly, this may be caused by extended defects introduced into the crystallized silicon above the seed region which can act as nucleation centers during the lateral growth. Twins and loops were found in that region as visible in Fig. 1.

A second reason for the grain growth can be the nucleation at the imperfectly etched seed edges.

As shown in earlier experiments /6/ the high defect concentration is caused by the presence of arsenic atoms in the layer. Due to the high flux of point defects created by the silicon annealing atoms they tend to form point defect clusters and secondary defects like loops and twins.

4. CONCLUSIONS

Firstly, lateral overgrowth to produce SOI-structures was shown to be possible by IBIEC at 400 °C leading to a strong (100)-texture. Further experiments are necessary to avoid problems arising from the high defect concentration in the seed region and by imperfect seed edges.

Acknowledgement

We would like to acknowledge K.-H. Stegemann and I. Hilme for their experimental assistance.

REFERENCES

/1/ Elliman, R.G., J.S. Williams, S.T. Johnson and E. Nygren, Mat. Res. Soc. Symp. Proc. 74 (1987) to by published

/2/ La Ferla, A., E. Rimini and G. Ferla, Appl. Phys. Lett. 52 (1988) 712

/3/ Linnros, J., G. Holmen and B. Svennson, Phys. Rev. B 32 (1985) 2770

/4/ Priolo, F., A. La Ferla, C. Spinella, E. Rimini, G. Ferla, F. Baroetto and A. Licciardello, Appl. Phys. Lett. 53 (1988) 2605

/5/ Skorupa, W., M. Voelskow, J. Matthaei and P. Knothe, El. Lett. 24 (1988) 875

/6/ Skorupa, W., M. Voelskow and J. Matthaei, Phys. Stat. Sol. (a) 101 (1978) k101

/7/ Voelskow, M., W. Skorupa, J. Matthaei, K.-H. Heinig and H. Bartsch, Proc. E-MRS Mtg. Strasbourg 1987, Les Editions de Physique, Paris 1987, pp. 343

RAPID THERMAL ANNEALING OF ION IMPLANTED GaAs WITH A GRAPHITE STRIP HEATER

T. Bachmann, W. Wesch and K. Gärtner[1)]

1. Introduction

Selenium ion implantation has generated a great deal of interest for the fabrication of n-conducting layers in semiinsulating GaAs. The annealing of the damaged layers by unconventional rapid thermal methods in the solid phase regime (graphite strip heater, halogen lamps) has some advantages compared to conventional annealing. In this contribution we report detailed results about the generation of ion implantation damage and its removal by means of a graphite strip heater as a function of implantation temperature and ion fluence.

2. Experimental

<100>-oriented Cr-doped semiinsulating GaAs wafers were implanted in a non-channeling direction with various fluences ($5x10^{12}$ to $5x10^{14} cm^{-2}$) of 300 keV $^{80}Se^+$ ions at different temperatures (300 to 513 K). A low current density ($\leq 0.05 \mu A/cm^2$) was chosen in order to prevent additional heating of the samples by the ion beam. After that, annealing with a SiO_2-caplayer (thickness: 110 nm) produced by reactive sputtering has been carried out at a temperature of 1273 K and a time of 50 s (heating rate: 200 K/s) by means of a graphite strip heater described in [1]. To characterize the as-implanted as well as the annealed samples RBS-channeling measurements were performed using 1.4 MeV He^+ ions at a backscattering angle of 170^0. Additionally, some samples were examined by electron microscopy and low temperature channeling measurements to determine the nature of the defects occuring in the as-implanted layers. Electrical measurements were carried out using the van der Pauw method.

3. Results

3.1. RBS-measurements

Room temperature implants lead to strongly damaged layers (radiation damage peak visible in the backscattering spectra) for doses higher than $1x10^{13}cm^{-2}$. Amorphization has been achieved by implanting fluences $\geq 4x10^{13}cm^{-2}$. The situation changes for higher implantation temperatures which is demonstrated in fig.1. The minimum yield at the depth of 300 nm (behind the damage profile) is used as a characteristic measure of the

---

1) Friedrich-Schiller-Universität Jena, Sektion Physik, Max-Wien-Platz 1, Jena, DDR, 6900

integral defect concentration. It is depicted as a function of the dose for different implantation temperatures T_i. As it can be seen in fig.1 within the dose range investigated stronger damaged layers are obtained only for the lowest temperature T_i=333 K at fluences larger than $1x10^{14} cm^{-2}$. In all the other cases the samples are only weakly damaged (small enhancement of the minimum yield). TEM-micrographs (not depicted here) made for such samples show no dislocations or other extended defects. The diffraction patterns indicate the presence of a monocrystalline layer without any extra spots. Therefore,

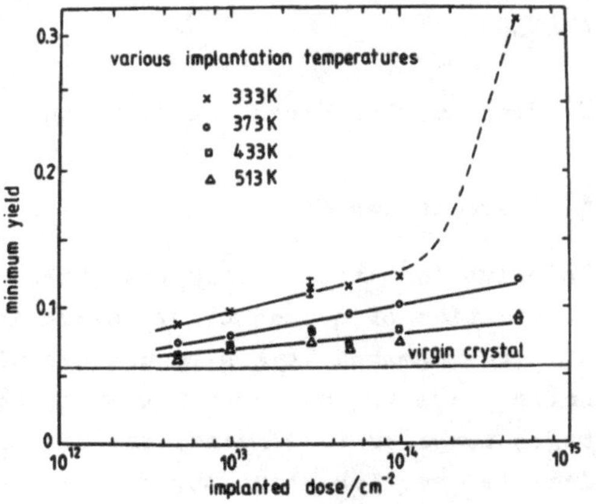

Fig.1. Minimum yield as a function of ion fluence and implantation temperature

the defects responsible for the small enhancement of the minimum yield can only be point defects or point defect complexes. To get more detailed information about these point defects the difference in minimum yield $\chi_{min}(z,T) - \chi_{min,perf.}(z,T)$ was measured as a function of the depth z for different temperatures during the backscattering experiments. Two examples of the results are given in fig.2 and 3 for 373 K implants with fluences of $1x10^{14}$ and $5x10^{14} cm^{-2}$, respectively. Previous investigations [3] of the temperature dependence of the minimum yield in similar cases have shown that the defects can be characterized by two discrete values of the distance r_a perpendicular to the string. The low value $r_{a,1}$=0.018 nm can be related to distortions of the lattice in the immediate neighbourhood of point defects or -complexes whereas the higher one $r_{a,2}$=0.065 nm represents for instance interstitials [3]. Therefore the analysis of the experimental data given in fig.2 and 3 is performed using the same values $r_{a,1}$ and $r_{a,2}$. It provides the total relative concentration of displaced atoms (point defects) and the percentages of the atoms which are displaced by $r_{a,1}$ and $r_{a,2}$, respectively.

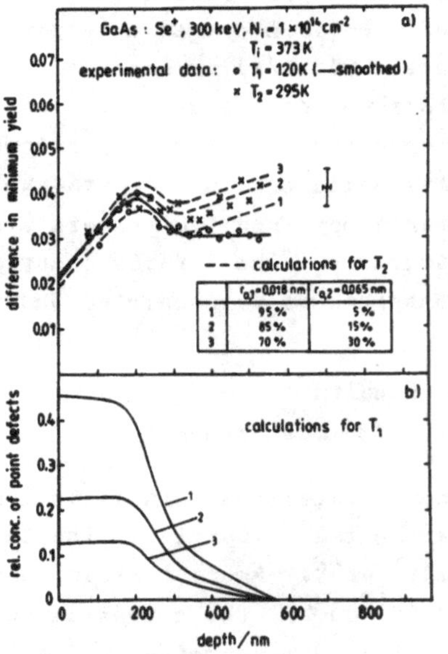

Fig.2. Difference in minimum yield (a) and relative total defect distribution (b) for $1x10^{14} Se^+ cm^{-2}$ implanted at 373 K

For the lower fluence the difference in minimum yield increases with increasing temperature (positive temperature dependence). Assuming different percentages for the two kinds of defects with the distances $r_{a,1}$ and $r_{a,2}$ (see table in fig.2a and 3a) in a first step the corresponding defect distributions were calculated from the difference of the minimum yield at 120 K. The second step provides the difference in minimum yield for the higher temperature of 295 K using the calculated defect distributions. As it can be seen in fig.2a the best fit to the experimental data is gained for the curve 2 characterized by the fact that 85 % of the total defect concentration depicted in fig.2b (curve 2) are related to defects with the low value of $r_{a,1}$. For the higher ion fluence of $5x10^{14} cm^{-2}$ the temperature dependence of the difference in minimum yield becomes negative connected with defects of low $r_{a,1}$ only (curves 1 in fig.3). The defect profiles in fig.2b and 3b are broadened and do not agree with the vacancy distribution obtained by Monte-Carlo-Simulation.

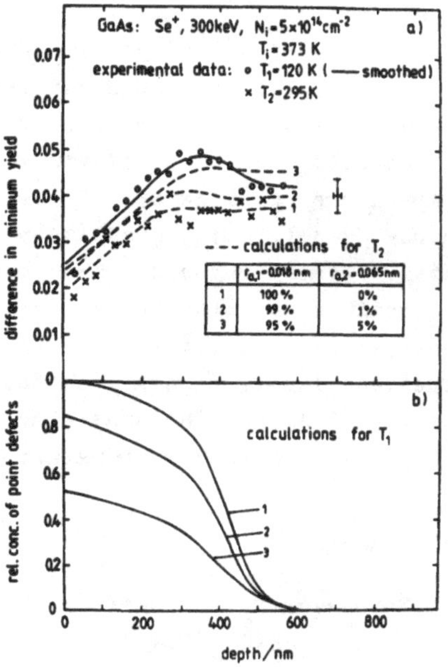

Fig.3. Difference in minimum yield (a) and relative total defect distribution (b) for $5x10^{14}$ Se^{+} cm^{-2} implanted at 373 K

3.2. Electrical measurements

Fig.4 shows the sheet carrier concentration and sheet mobility for annealed samples. The activation is different for different regions of the implantation temperature (dash-dotted-line for 333 K implants and dashed line for higher implantation temperatures). It is clearly seen that the activation threshold is shifted to lower doses if the implantation temperature exceeds 333 K. Room temperature implants are characterized by a missing activation.

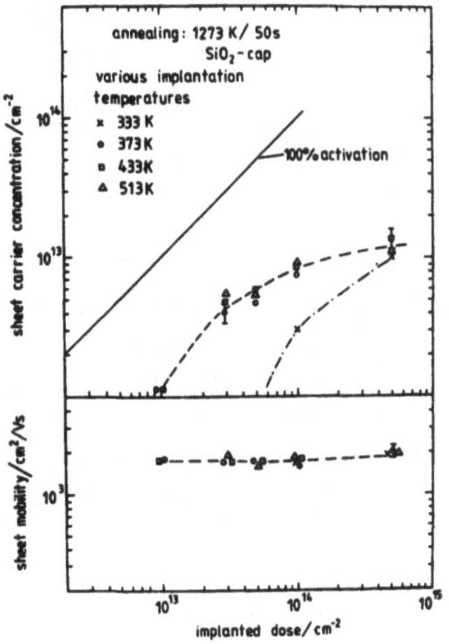

Fig.4. Electrical sheet properties

4. Discussion

The absence of an amorphized layer for 333 K implants even for the fluence of 5×10^{14} Se^+ cm^{-2} has not been reported in the literature. The strong recombination of Frenkel pairs created during ion implantation can be responsible for the observed weakly damaged layers. The analysis of the temperature dependence of the minimum yield indicates the presence of mainly weakly displaced atoms with $r_{a,1}$=0.018 nm related to distorted lattice atoms in the neighbourhood of defect clusters or -complexes. Such distortions can also derive from the existence of anti-site-defects as a consequence of incorrect recombinations. The electrical properties of the annealed layers show a much better activation for implantation temperatures $\geq$373 K, but the annealing temperature is not high enough to achieve activation for room temperature or low dose implantations.

References

[1] Bachmann, T., H. Bartsch: Nucl.Instr. and Meth B, in press

[2] Wesch, W., K. Gärtner, E. Wendler, G. Götz: Nucl. Instr. and Meth. B 15 (1986) 431-434

[3] Wesch, W., K. Gärtner, A. Jordanov, G. Götz: Nucl. Instr. and Meth. B, in press

PHYSICAL AND ELECTROPHYSICAL PROPERTIES OF $Ge_{2x}(GaAs)_{I-x}$ LAYERS INDUCED BY PULSE ELECTRON BEAM ANNEALING

V.I.BALJUBA, V.P.VORONKOV, A.P.VJATKIN
SIBIRIAN PHYSICAL-TECHNICAL INSTITUTE, TOMSK
N.I.LEBEDEVA,D.E.PROSKUROVSKY
INSTITUTE OF HIGH CURRENT ELECTRON, TOMSK

Earlier we reported about obtaining of the $Ge_{2x}(GaAs)_{I-x}$ solid solution by the pulse electron beam annealing [I]. The studies of the physical and electrophysical properties of this materials is of indubitable interest to their practical use in the microelectronics. In this report are presented the results of the investigation of the dependence of the $Ge_{2x}(GaAs)_{I-x}$ solid solution films electro physical properties on the electron beam annealing conditions.

The Ge-films got dusty on the GaAs single crystal falser were putted to the annealing. The beam pulse durations $8 \cdot 10^{-8}$ sec. with the maximum electron energy 50 KeV. The annealing energy density changed from 0,2 to 0,9 J/cm^2. The solid solution layers crystal structure was being studied by the slow electrons diffraction. Initial Ge-films had the polycrystal structure. Following the models pulse annealing we observed the system of the sharp individual reflexes, defining the cut-centre cubic crystal structure. The reflexes interstate changed in comparison with the reflexes system on the GaAs single crystal. The individual reflexes were enclosed by the more weak satellites. The analysis shows that the reflexes splitting was provoked by the double diffracion on the doublers. The electrographic investigations and also the component composition measurement data, obtained by the Auger-spectroscopy method, show that the solid solution single crystal films shapes as the result of the electron beam annealing action. Perfection of the structure and also solid solution composition depends on the Ge-got dusty layers thickness and on the electron beam energy density. The most perfect films were obtained with following conditions: the Ge--films thickness is 0,03 μ , the pulse energy density is 0,5 J/cm^2 and the ones durations is $8 \cdot 10^{-8}$ sec.

The charge carriers concentration and the ones mobility in the solid solution layers were determined on the Holl-effect measurement by the Van der Pao method.

After the electron beam annealing we obtained both the conductivity electron type layers and the hole ones.

The conductivity type bepends on the Ge-raised dust film thickness and the annealing energy density. The conductivity type conversion of the solid solution film was observed with the annealing energies 0,2 J/cm^2. The energy density, changing the conductivity sing, increases with the Ge-layer thickness increase.
The energy density, changing the conductivity sign, measured by the thermoelectromotive force sign disagrees with one measured by the Holl-tension. Such disagree is connected with the solid solution film composition uniformity on an area and with the used methods peculiarities. With the thick films measurement the Holl-electromotion force on the measurent part gives the conductivity mean value, while the thermoelectro-motive force is the more locai method.

Dependence of the charge carriers concentration in the solid solution film on the annealing energy density has a complicated character (fig.I).

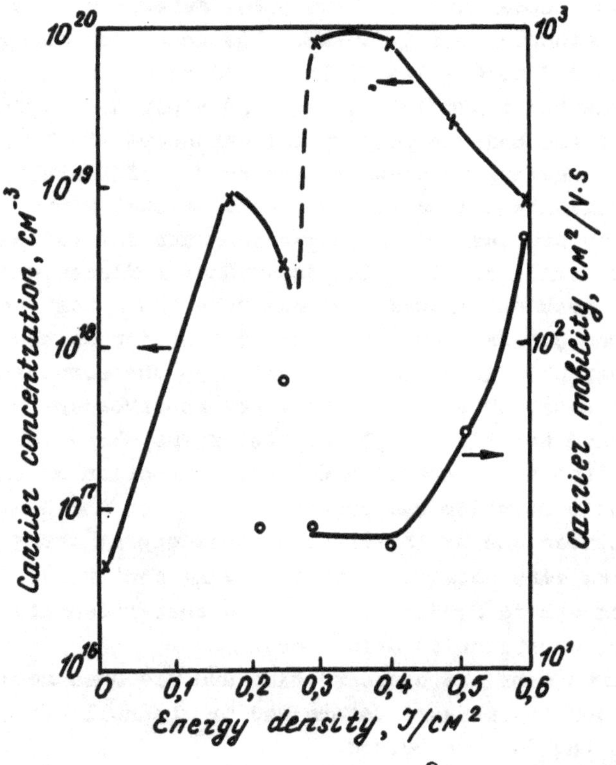

With the annealing energies below 0,2 J/cm^{2}the solid solution films have the conductivity hole type and with the annealing energies above 0,3 J/cm^2 they have the electron one. The electron concentration and the hole one changes along the curve with the maximum. The maximum electron concentration $2 \cdot 10^{20}$cm^{-3} is obtained with the annealing energies 0,4 J/cm^2. Change of the charge carriers conductivity type and the concentration one is the determined by the Ge-atoms concentration in the Ga and As lattice. The calculations show that the Ge-melting and the GaAs-surface layer one are the result of the anneeling energy 0,2 J/cm^2. The melted layer thickness increases with the annealing energy density increase. With the Ge-invariable quantity it leads to the layers enrichment by the GaAs. The element composition investigations conducted by the Auger-spectroscopy method show the Ge-maintenance decrease in the solid solution layers with the annealing energy increase. Thus, the control by the solid solution composition takes place as the result of the annealing energy density change. The solid solution layers with the Ge-large maintenance had the conductivity hole type, and the electron one with the layers impoverishment by the Ge because of the diamond - sphalerit phase change. The $Ge_x(GaAs)_{1-x}$ solid solution layers obtained by the pyrolytic synthesis method had the conductivity analogous dependence on the composition [2].

The electron mobility in the solid solution layers increases with the annealing energy increase and its absolute value is closed to the mobility in the GaAs-single crystals with the concentration corresponding value.

REFERENCES

I V.I.Baljuba, V.P.Voronkov, A.P.Vjatkin, D.E.Proskurovsky, N.I.Lebedeva, Proc of Int. Conf. on Energy Pulse Modification of semiconductors and related materials, Dresden, 1984.

2 Zh.I.Alferov et all,Fiz.Tekh.Poluprov. 1982, 16, N5, p.887-890

MODIFICATION OF ZnSe PROPERTIES BY HIGH CURRENT PULSED ELECTRON BEAMS

B.P. Dement'ev[1], M.B. Kotljarevsky[2],
D.I. Proskurovsky[1], G.E. Ozur[1], N.I. Yakubovich[2]

The paper deals with the analysis of high current electron beam irradiation (HCEBI) influence on the inherent defects structure of ZnSe crystals, being thus the continuation of the work [1]. A model of the defects formation is proposed in the paper based on the studies of the crystal luminescent and electric properties modification during HCEBI as well as on the data of microscopic and microanalysis investigations.

The initial substance used in our study were specially unalloyed monocrystals grown from melt with n-type conductivity, specific resistivity of which being 10^{11} Ohm·cm. The electron beams used for HCEBI had the under threshold electron energies E = 20 to 40 keV and current density of 0.8 to 2.5 kA/cm^2 at the pulse duration t_p = 30 to 100 ns[2]. Also were used the pulsed electron beams with E = 10 to 30 keV and duration of t_p = 0.1 to 1.0 μs [3]. Energy density of the beams varied from 0.2 to 1.4 J/cm^2.

In order to assess the defects creating action of the HCEBI and to identify the defects type we used photoluminescence analysis. Specific feature of the edge emission which is of donor-acceptor nature in ZnSe, is the existence of the photon recurrence with the interval equal to the energy of longitudinal optical phonon. The phononless band is stronger in the case of a more perfect crystal lattice. Intensity of the exciton band is also strongly dependent upon the defectivity of the crystal lattice [4].

At the energy densities of HCEBI W = 0.2 to 0.6 J/cm^2 when no liquid phase recrystallization in the irradiated layer occurs (according to data of scanning electron microscopy) intensities of the exciton and edge emission bands decrease, Fig. 1. The decrease of the photoluminescence intensity reveals the creation of defects during HCEBI including the appearance of nonradiative recombination centers. There occurs the shift of the phononless band maximum from 458.8 nm to 459.3 nm during this process amd also certain changes in the phonon energies take place due to the lattice deformations caused by the appearance of atomic defects in it. Electric measurements in these samples after their HCEBI treatment at W = 0.4 to 0.6 J/cm^2 revealed, from the data on thermoelectromotive force, the change of the conductivity type. The activation energies of the acceptor centers determined from the temperature behaviors of the samples electric conductivity were $E_{a1} = E_v + 1.1$ eV and $E_{a2} = E_v + 0.6$ eV. In ZnSe the electrically active centers with such energies are identified as the inherently defective centers (doubly charged vacancies of Zn and Se, respectively) [5].

1) Institute of High Current Electronics, Tomsk, USSR
2) Teaching Training Institute, Berdjansk, USSR

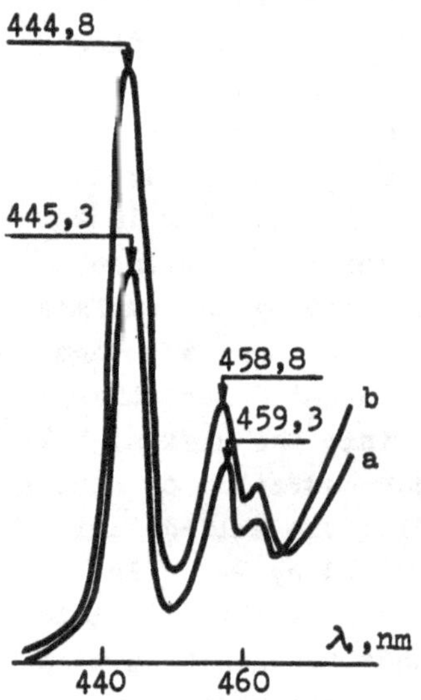

Fig.1. Photoluminescence spectra of ZnSe. T= 77 K, $\lambda_{excit.}$ = 337 nm. a) is for the initial crystal; b) for the crystal after HCEBI (E = 40 keV, t_p = 60 ns, W = 0.6 J/cm^2.

Investigation of the samples' surfaces modification at HCEBI using the scanning electron microscopy allowed the determination of the solid phase and liquid phase boundaries of the annealing by varying the pulse energy density [1]. On the surfaces of samples subjected to HCEBI at t_p = 90 ns and W = 0.5 J/cm^2 there appears condensed matter in the form of small balls (see Fig.2). This is more distinctly observed after ten HCEBI pulses. At the HCEBI by a single pulse at t_p = 500 ns the appearance of the condensed matter is observed starting with the energy W = 0.8 J/cm^2 (see Fig.2c). It was found from the data of X-ray spectral microanalysis that the condensed matter is Zn (Fig.3). Thus, the HCEBI of samples at medium energy densities results in the deviation of the irradiated layer stoichiometry towards the excess of metalloid (the metal escapes to the surface), what is confirmed by the data of electric measurements and of the microanalysis of the surface.

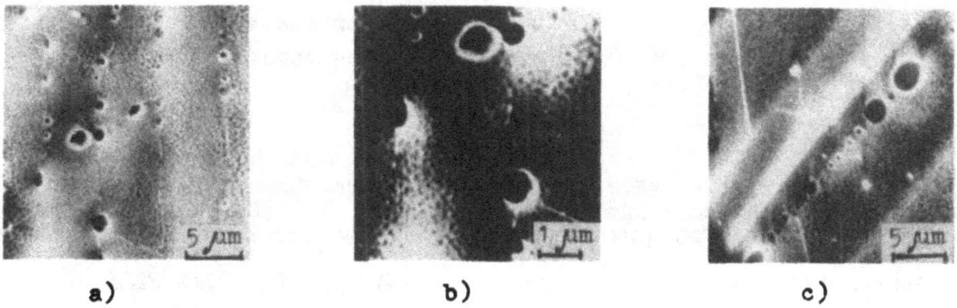

Fig.2. Modification of the ZnSe surface at HCEBI (scanning electron microscopy). a) and b) E = 40 keV, t_p = 90 ns, W = 0.5 J/cm^2, 10 pulses c) E = 30 keV, t_p = 500 ns, W = 0.8 J/cm^2, single pulse.

In the case of equilibrium heating in vacuum the conditions for formation of the defects provide, at the same time, the excess of metallic component and, as a consequence, the conductivity of samples becomes of n-type. The occurrence of the acceptor centers (Zn vacancies) in the irradiated layer can be explained by the following model. At the HCEBI there is generated the electron-hole plasma in a thin (5 to 10 μm in thickness) under surface layer just at the first nanosecond, which has almost stationary number density about 10^{17} to 10^{18} cm^{-3}. This plasma provides the sheet conductivity of the order of σ_{sh} = 10 to 100 Ohm^{-1}x cm^{-1}. The excess charge of the injected electrons is ejected to the surface during the Maxwellian time $t_M = \varepsilon\varepsilon_0/\sigma_{sh}$ 10^{-9} to 10^{-8} s. During

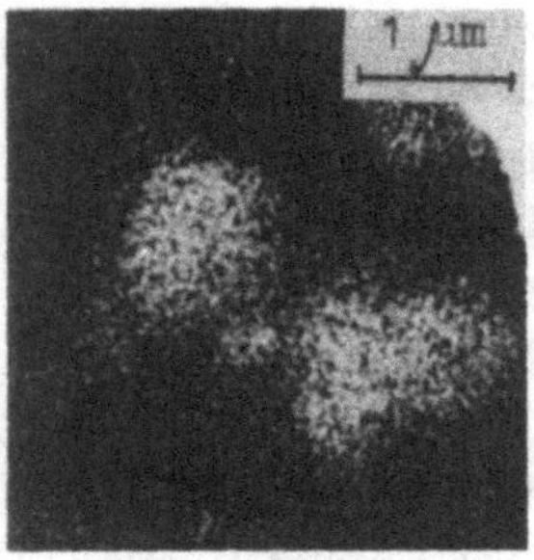

Fig. 3. The results of scanning X-ray microanalysis of the ZnSe surface after HCEBI ($E = 40$ keV, $t_p = 90$ ns, $W = 0.5$ J/cm^2, 10 pulses series. Zinc lines K_2 and K were recorded.

this same time the irradiated layer warmup starts ($T = 700$ to 1000 K). The shielding distance at such conductivity values is shorter than 10 nm, so one can surely assume that all the injected charge of electrons (10^{-6} to 10^{-3} C) is distributed over 10 to 100 atomic layers just under the surface. These electrons are captured by the surface Zn ions which thus become neutral and then they are desorbed. Thus created vacancies of the metal then diffuse into the crystal bulk, where now the excess concentration of metalloid is produced providing the self-defective p-type conductivity, guided by V_{Zn} value. This process is competited by the metalloid vaporization process what results in a self compensation of the self-defective p-type conductivity. For this reason, even insignificant changes in the irradiation conditions, i.e. in the heating kinetics, in the change accumulation or its leakage, can result in an essential change of the ratio of selfdefects in both sublattices. At the energy densities $W = 0.5$ to 0.6 J/cm^2 there can take place the generation of the Frenkel pairs by the oscillational-thermal mechanism. In this case the vacancies of Zn formed inside the crystal bulk remain at their places, while the interstitial Zn atoms capture electrons and diffuse towards the surface.

References

(1) Dement'ev, B.P. et al.: Izv. Vyssh. Uchebn. Zaved. Fiz. 3(1984) 126
(2) Koval, B.A. et al.: Pis'ma Zh. Tech. Fiz. 7 (1981) 1227-1230
(3) Ozur, G.E. and Proskurovsky, D.I. : Pis'ma Zh. Tech. Fiz. 14 (1988) 413-416
(4) Georgobiani, A.N. et al.: Kratk. Soobsh. po Fiz. FIAN SSSR 10 (1987) 12-14
(5) Georgobiani, A.N. et al.; Izv. Vyssh. Uchebn. Zaved. Fiz. 3 (1989) 51-56

AMORPHIZATION OF CD-IMPLANTED GAAS*)

J.Krynicki[1)], H.Rzewuski[1)], R.Groetzschel[2)] and A.Claverie[3)]

INTRODUCTION

Damage accumulation of postimplantation defects in a semiconductor crystal can often results in the formation of a continous amorphous layer. The critical ion dose ror the process depends on many parameters like ion mass, ion energy, target temperature and the density of a crystal.

The amorphization process in Si is the one which is most [1-3] clearly, although not completely, understood. Much less explored situation exists in the case of semiconducting compounds [4]. In general, it is believed that in the latter case formation of an amorphous layer takes place in a different manner.

At present, two basic models exist which describe the amorphization process under ion bombardement:

a. a model of homogeneous amorphization where the past of energy which is lost through elastic nuclear collisions and deposited within the crystal lattice plays a key role.

b. a model of overlapping amorphized zones produced by each ion impinging upon the crystal.

In this paper we present the results of investigations on the amorphization of GaAs under Cd ion implantation at liquid nitrogen and room temperatures.

The transformation processes conducing to the observed distribution have been discussed.

EXPERIMENTAL

Commercialy available (Hoboken, Holland) GaAs n-type <100> samples containing $2*10^{17}cm^{-3}$ Te atoms were used. The amorphization of Cd implanted samples in the wide range of ion energy 20-180 keV and ion doses $2*10^{13}$- $9*10^{13}cm^{-2}$ at LN_2 and RT has been studied. Current density of the ion beam was lower than 1 $\mu A/cm^2$ to suppress heating of the sample. The temperature of the target was monitored by a thermocouple and did not exceed 40^oC. The samples were

---

*) Partly supported by Institute of Physics Polish Acad. of Science.

1) Institute of Nuclear Chemistry and Technology, Dorodna 16, 03-195 Warszawa, Poland.

2) Zentralinstitut fuer Kernforschung, Rossendorf, GDR.

3) CEMES-LOE du CNRS, 29 rue J.Marving, 31400 Toulouse, France.

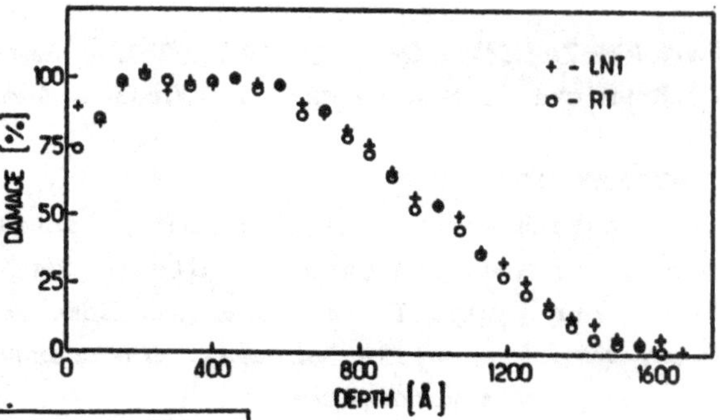

Fig. 1
Damage distributions of Cd - implanted GaAs for different implantation temperatures.
E = 120 keV, D = 9×10^{13} cm^{-2}

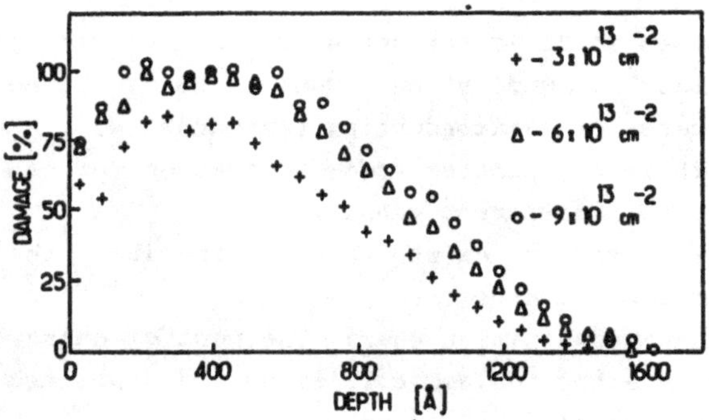

Fig. 2
Damage distributions in GaAs implanted with various doses of Cd - ions of 120 keV.

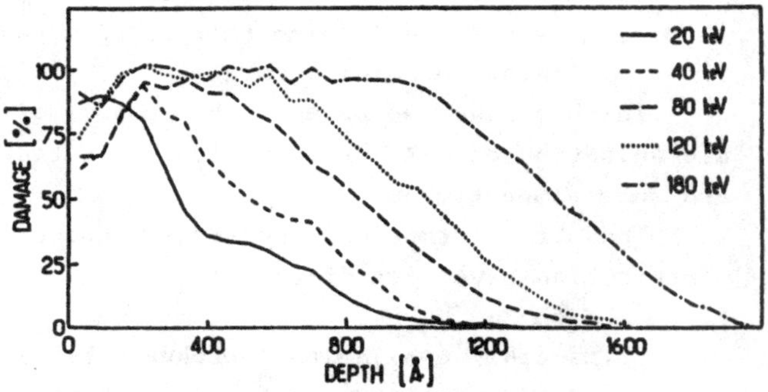

Fig. 3
Damage distribution profiles in GaAs implanted with Cd ions of various energies;
D = 9×10^{13} cm^{-2}.

Table 1
Critical doses and nuclear deposited energy versus energy of Cd ions implanted into GaAs

Energy (keV)	Critical dose (cm^{-2})	Deposited energy (keV/cm^3)
20	1×10^{14}	1.8×10^{21}
40	8×10^{13}	1.8×10^{21}
80	5.5×10^{13}	1.5×10^{21}
120	4×10^{13}	1.2×10^{21}
180	1.8×10^{13}	0.5×10^{21}

misoriented before implantation to avoid channelling.

Damage distribution were measured using 1.7 MeV $^{4}He^{+}$ Rutherford backscattering technique. The critical dose for amorphization and widths of amorphized layers for each Cd ion energy have been determined.

RESULTS AND DISCUSSION

Fig.1 presents the degree of amorphization of GaAs layers implanted with 120 keV and $9*10^{13}cm^{-2}$ Cd ions at 90 K and 300 K. In both cases the damage created is almost identical. Fig.2 presents the damage created with 120 keV Cd ions for different doses. Fig.3 shows that the dose of $9*10^{13}$ of Cd ions with 20 keV energy is not sufficient to create an amorphized layer. For this lowest energy used the maximum of damage distribution occurs close to the surface of sample in conformity with theoretical calculations (for 20 keV Cd ions the projected range $R_p \simeq 90$ Å). The observed distributions show an interesting feature of a very heavy damage extending to the bulk of a sample.

For higher energies the damage created extends much deeper than the expected range of Cd ions - Figs.2 and 3. In Table I are presented the critical doses obtained for each energy of Cd ions and deposited damage energy calculated by method described in ref.5. It is assumed that the critical doses corresponds to the damage measured by RBS exceeding 95%.

CONCLUSIONS

Damage distributions in GaAs created by Cd ions are independent of the implantation temperature between 90 K and 300 K.

The critical doses for various energies of Cd ions have been estimated. The observed dependence of a critical dose on the energy of implanting ions is explained by annihilation of defects near the surface of a sample which requires a much higher dose for amorphization in this area.

The experimental results and calculated deposited energy values do not confirm the validity of the model for homogeneous amorphization in GaAs by Cd ions implantation.

REFERENCES

1. F.F.Morehead, B.L.Crowder Rad. Eff. 6, 27, 1970.
2. J.R.Dennis, E.B.Hale J. Appl. Phys. 49, 1119, 1978.
3. S.Prussin, D.I.Margolose, R.N.Tauber J.Appl.Phys 57, 180, 1985.
4. D.K.Sadana Nucl. Instr. Meth. in Phys. Res. B 7/8, 375, 1985.
5. C.Vieu et al. Nucl. Instr. Meth. B28, 229, 1987.

Au-Ge/n-GaAs OHMIC CONTACTS BY RAPID THERMAL ANNEALING

SKRABKA Tomasz, KONG HYON Chol[1)]

Introduction

The fabrication of modern microwave GaAs devices requires low resistance and smooth surfaces ohmic contacts due to the effects of their small sizes. The most commonly used ohmic contacts to n-type GaAs are based on Au-Ge eutectic alloy metallization. The components of Au-Ge metallization are usually converted into liquid phase during formation orocedure and subsequently form the thin heavily doped semiconductor layer, which is responsible for contact resistance. The morphology and electrical properties of alloyed contacts strongly depend on annealing conditions i.e. the duration time, the annealing temperature and the ambient in which the process is performed (2,4).

Shortening of annealing time reduces the outdiffusion of As from GaAs lattice. Creation of the As vacancies cause Ge to occupy As sites resulting in p-type doping of GaAs surface layer, which may lead to high resistance ohmic contacts. Moreover, fast alloying and quenching are conductive to the formation of smooth contacts surfaces.

Thus, interest of RTA (Rapid Thermal Annealing) application to the formation of Au-Ge ohmic contacts is recently increasing (3).

Experimental

In our study the attempts of Rapid Thermal Annealing of the ohmic contacts by laser and halogen lamps were undertaken. The ⟨100⟩ tellurium bulky doped of $N_d = 5 \cdot 10^{23} m^{-3}$ wafers were used to carry out the experiments. Contact metallization consisted of sequentially evaporated Au-Ge alloy (88/12)(100 nm), Ni (30 nm) and Pd (50 nm) layers. With the purpose of measurement contact specific resistance ρ_c by Circular Transmission Line Model special structures were patterned using lift-off technique (5)

To compare among annealing techniques the samples prepared in one process were annealed in three different ways. One group of wafers was heated in the small RTA system, what was previously described (6).

Another samples were treated by normally incident, focused to 100 µm diameter, 7 W, CW CO_2 laser beam. Wafers were placed on movable holder. Scanning rate was down to 0.5 mm/s, which corresponds to 0.2s pulse duration.

Because of great difficulty in local temperature measurement on very small areas, we decided to estimate surface temperature theoretically. The most commonly spread solution of heat equation is given by (1):

---

1) Institute of Electron Technology, Technical University of Wrocław
Janiszewskiego 11/17, 50-372 Wrocław, Poland

$$T(0,t)=I_a \delta/(K\sqrt{\pi})$$ (1)

where: δ - heat diffusion length   V - molar volume
K - thermal conductivity C_p - specific heat
I_a - incident irradiance   t - pulse duration

Above formula is valid in case where the heat diffusion length equal $\delta=2\sqrt{KVt/C_p}$ is much greater than the optical absorption depth α^{-1}. In the case of Pd and wavelength 10.6 µm, α^{-1} equals 14 nm. Reflection coefficient was calculated on the base of following formula:

$$R=1-2\sqrt{\frac{2\varepsilon_0 \omega}{\sigma}}$$ (2)

and for Pd it was found to be 96%, and surface temperature was estimated to be 980°C.

At last the other wafers were routinely annealed in resistively heated quartz tube furnace at 450°C in 3 min. For comparison of these three processes the schematic time-temperature profiles are demonstrated in Fig. 1.

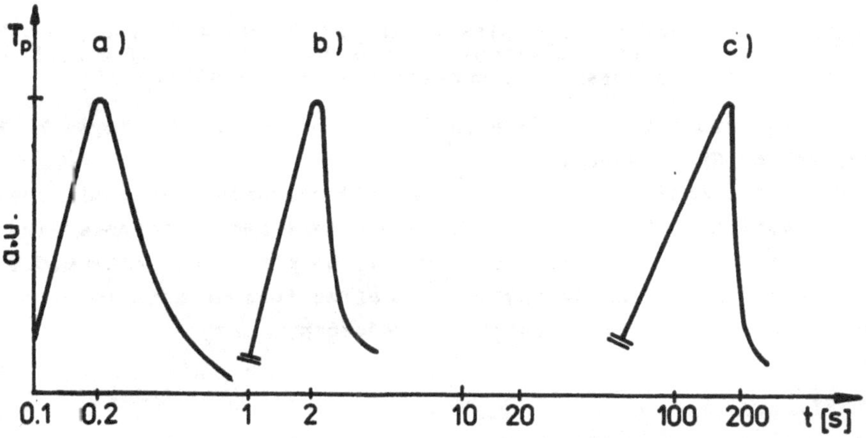

Fig.1. Schematic curves of annealing time duration in log scale vs. peak temperature T_p: a) laser annealing, b) halogen lamp annealing, c) conventional annealing.

Results and conclusions

The fabricated Au-Ge/Ni/Pd ohmic contacts formed by conventional and halogen lamps heatings exhibit linear and symmetrical I-U characteristics. Unfortunatelly, wafers annealed by laser showed nonohmic character even at slowest scanning rate (Fig.2). Such behavior may be caused by greater than estimated reflection coefficient or/and greater than expected lateral heat flow.
The average specific contact resistances were measured to be $2\cdot10^{-5}\Omega\cdot cm^2$, and $2\cdot10^{-6}\Omega\cdot cm^2$ for wafers conventionally and pulse annealed, respectively. One order of magnitude decrease of contact resistance resulted probably from reduction of arsenic loss during contacts formation.

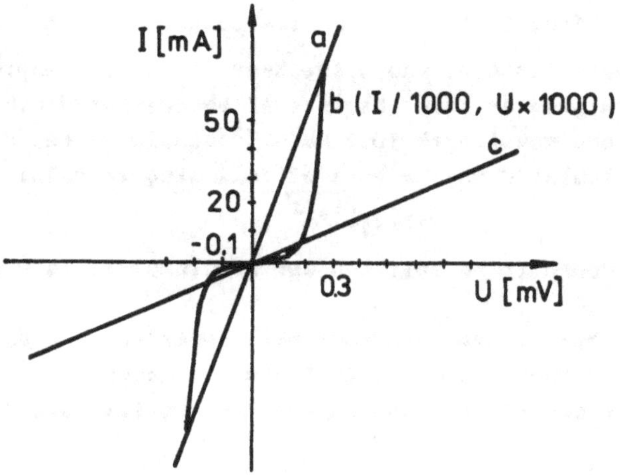

Fig.2. Current-Voltage characteristics of fabricated Au-Ge/Ni/Pd ohmic contacts: a) annealed by halogen lamp, b) annealed by CO_2 7 W laser, c) conventionally annealed.

Also great improvement in surface uniformity of contacts alloyed by RTA was revealed by SEM observation.

Obtained results confirmed usefulness of RTA technique for ohmic contacts fabrication. Halogen lamp annealing significantly improved the contact quality. Laser application seems to be promising process due to very short times of annealing, but requires further experiments with higher power density or shorter wavelength.

References

(1) Allmen v. M.: Laser-Beam Interactions with Materials, ch.3, Springer-Verlag 1987

(2) Howes, M.J., Morgan, D.V.: Gallium Arsenide, John Wiley Sons, 1985

(3) Perez, R.: Microelectr. Manuf. Testing 7, 21, (1986)

(4) Piotrowska, A., Guivarc'h, A. and Pelous, H.G.: Solid-St. Electr. 26, 179, (1983)

(5) Reeves, G.K.: Solid-St. Electr. 23, 487, (1980)

(6) Skrabka, T., Słaby, M.J., Mizera, B.: Phys. Stat. Sol.(a), 112, K105, (1989)

APPARATUS FOR INFRARED ANNEALING OF GaAs IN CONTROLLED ATMOSPHERE

Yurchenko V.A., Slobodich S.V., Teselkin A.V., Pivovar V.S., Chernenko S.V.[I)]

Recently a short-time exposure to the infrared (IR) radiation has become most widely used method for GaAs annealing. The method allows to improve the quality of the obtained ion-implanted structures, to extend the raw material thermostability range, to simplify the process and to raise the operation safety. The most commonly used apparatus design has turned out to be the one that has two halogen lamp unets symmetrically located on the both sides of quartz reactor being filled with the hydrogen-containing gas mixture. Analysis of the IR-annealing results and their comparison with theththermal annealing results reveal that process temperature comes out to be the main dominant factor when activating the implanted dopant. The factor of time is less important here. Each temperature value of GaAs annealing in also known to be conformed by the critical time that, if being exceeded will result in material dissociation (rise in the annealing temperature depresses this parameter). Besides, the semiconductor dissociation depends not only on the holding at the maximum temperature, but also on the temperature rise and fall periods.

The above-mentioned information has become the basis for the conclusion that for the semiconductor compound IR annealing should go out to the maximum temperature as fastly as possible. The apparatus shold also have the high rate of cooling. It was recommended to use more powerful IR-sources and more effective systems for lamps and reactor cooling (I).

The article deseribes the attempt to meet the above-said requirements. And what's more, the requirements were fulfilled only by changing the apparatus construction (without raising the power of the lamp unets), the possibility for carrying out the process in the controlled atmosphere being retained. Fig.I shows the layont of the new-made apparatus for the IR-annealing. For a comparison it also gives a convention variant.

The spicific features of the new-made apparatus include first of all the absence of the quartz reactor II. Lamp unets I,2 are placed in the hermetically sealed annealing chamber 3. Chamber walls 4 are constantly water-cooled and serve as the reflectors. Susceptor 5 for positioning the samples 6 is symmetrically located between the lamp unets. The characteristic feature of the apparatus is also the asymmetrical arrangement of the lamps and susceptor in the annealing chamber. Besides, gas is fed into the chamber through the upper wall (gas inlet 7), gas outlet 8

I) Scientific Research Institute of Radiomaterials, SU-220600 Minsk,USSR is located in the bottom of the chamber. Additional reflector 9 is placed below the bottom lamp unit. The temperature is controlled by the thermocouple IO fixed on the susceptor.

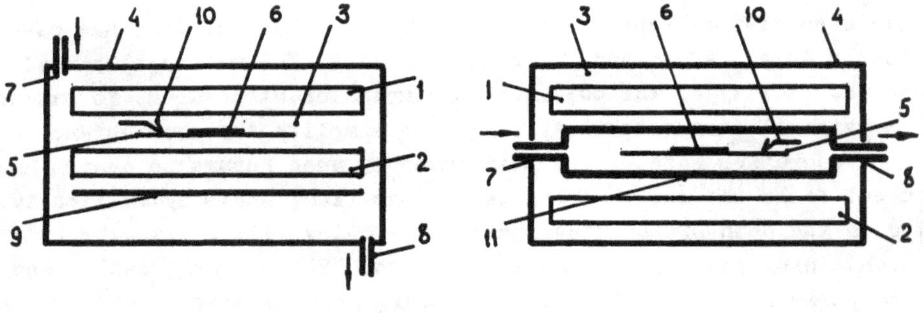

Fig.I. Apparatus for IR-annealing: a) new-made variant and b) conventional variant:
I,2 - halogen lamps unets; 3 - annealing chamber; 4 - chamber walls; 5 - susceptor; 6 - sample; 7 - gas inlet; 8 - gas outlet; 9 - additional reflector; IO - thermocouple; II - quartz reactor.

The removal of the quartz reactor from the apparatus has become the main factor for increasing the rate of sample heating and cooling: firstly, by diminishing the "thermal"lag" of the annealing zone (the mass of the reactor is mach more than that of sample and susceptor) and secondly, by providing the more compact arrangement of the radiating unets. The reduction of the distance between the lamps also resulted in the increase of the uniform heating area.

Fig.2 shows an example of a measured temperature cycle in the apparatus of the both types. Fig.3 presents the temperature distribution along the susceptor at the given working conditions. It's necessary to note that the problem of uniform heating of the structures at the high rates of temperature rise has been sufficiently reviewed, as wee see it, in literature. That's why it was not the subject of investigation in this work.

The possibility for carrying out the process in the hydrogen atmosphere was provided first of all by the hermetical sealing of the annealing chamber. Hydrogen is fed in a conventional manner after the operating volume has been purged by a neutral gas - nitrogen. The arrangement og the inlet and outlet connections in the upper and bottom parts of the chamber has improved the atmosphere - getting-ready efficiency at the minimum gas consumptioñ - hydrogen fills the chamber from the top pressing out the heavier gases downwards. During the process (heating stage) and high-temperature holding stage the hydrogen supply can be cut off at all. The location of the sample to be annealed in the up-

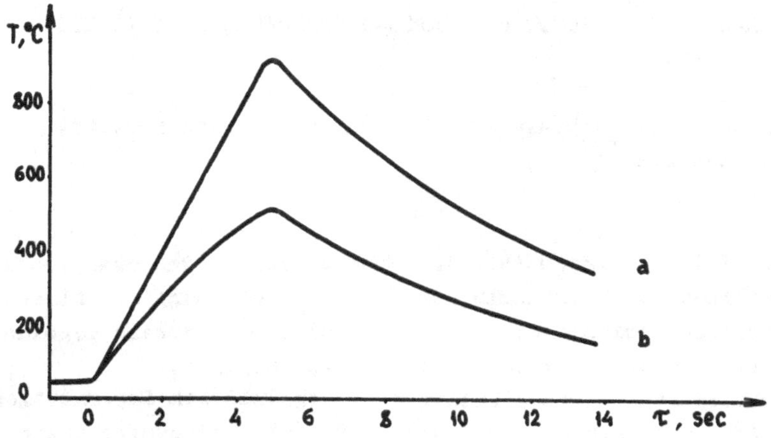

Fig.2. The curve of temperature changing versus time when the halogen lamps have been switched on for 5 second in the installation: a) without the reactor; b) with the reactor; U=2IO V.

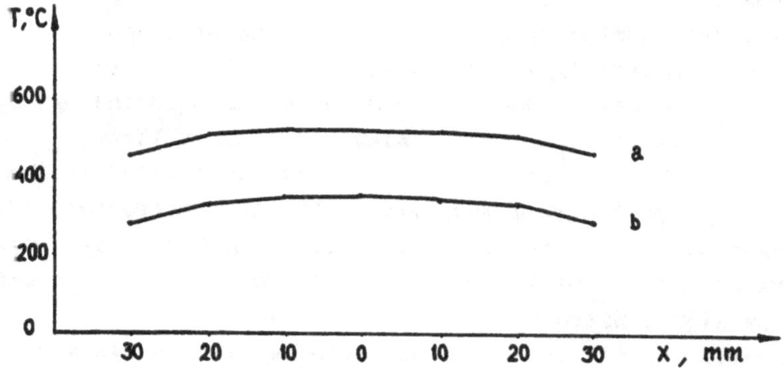

Fig.3. Temperature distribution along the susceptor in the installation: without the reactor
b) with the reactor.

per part of the chamber looks as if the additional quarantce of the atmosphere purity during the process.

The absence of the quartz reactor in the desing of the apparatus, the eployment of the above mentioned spacing of the lamps and the system of the gas supply made it possible to significantly increase the heating speed and the cooling of samples to improve the uniformity and reproducibility of the annealing results of GaAs.

R e f e r e n c e

(I) Yurchenko, V., V.Pivovar, S.Slobodich, I.Karpova: Priborostroenie Minsk 9 (I987) I44-I45.

GaAs-BASED SEMICONDUCTOR COMPOUND SYNTHESIS BY DUAL ION IMPLANTATION

Yurchenko V.A., Zaitsev A.M., Khytko V.I., Pivovar V.S., Ryzhnitsyn A.B.[I)]

The authors (I) have shown that the application of the dual ion implantation technique for maintaining and intentional changing the semiconductors compound stoichiometric ratio enables to increase considerably the electrical activity of the p- and n-type dopants.

The work studies the possibility of using dual ion implantation technique for synthesizing a new $GaAs_{I-x}P_x$ compound semiconductor on GaAs basis.

For comparison several specimens of $GaAs_{I-x}P_x$ were fabricated by the conventional ion implantation technique.

The semi-insulated GaAs substrates were used as the initial structures. Phosphorus ions were used as the main dopant and gallium ions as the additional one. Phosphorus ions were omplanted at 80 keV and gallium ions at I80 keV. During processes the sampless were heated to 300 °C. The dose per each dopant type was $I,2 \cdot IO^{I6}$ cm^{-2}. The ion-implanted structures were annealed at 900 °C during 60 seconds in the hydrogen atmosphere by the flashes of the incondescent halogen lamp.

The quality of the fabricated structures was controlled by the cathodoluminescence spectra with the help of the IR spectrophotometer and liquid-nitrogen-cooled photoelectric multiplier. Electron excitation energy was varied from 5 to I5 keV. During the measurement the samples were cooled by liquid nitrogen

The measurement of all the specimens resulted in the registration of the short-wave radiation of $GaAs_{I-x}P_x$ semiconductor compound with the wavelength $\lambda = 0.73\,\mu m$.

The quality of the fabricated structures was determined on the basis of the relation of radiant intensities (I) at $\lambda = 0.73\,\mu m$ to GaAs band-to-band radiation at $\lambda_b = 0.83\,\mu m$. The radiation bandwidths ($\Delta \lambda$) were compared also. The results of the measurements for the electron excitation energy E_I=5 keV and E_2=I8 keV are shown in the Table.

The Table shows that the relation of radiant intensities at λ and λ_b is better for the sample implanted with gallium and phosphorus ions than for the sample implanted with the phosphorus ions only.

The results obtained show the potential adventages of the dual ion implantation technique over the conventional ion implantation method in the synthesis of the semiconductor compounds.

I) Scientific Research Institute of Radiomaterials, SU-220600 Minsk, USSR

Table

The results of the measurement of radiant intensities and radiation bandwidths

dopants	E_I = 5 keV				E_2 = I5 keV			
	=0.83 m		=0.73 m		= 0.83 m		= 0.73 m	
	I	meV	I	meV	I	meV	I	meV
Ga^+, P^+	24	66	3	30	I60	44	27	56
P^+	50	60	I	-	280	44	20	50

R e f e r e n c e s

(I) Heckingbottom, R., T.Ambrige: Radiation Effects I7 (I973) 3I-36.

INVESTIGATION OF SPECIFIC FEATURES OF GALLIUM ARSENIDE IMPLANTATION WITH HIGH ION DOSES

Yurchenko V.A., Khytko V.I., Korotov V.P., Pivovar V.S., Ryzhnitsyn A.B., Gancharik T.S. [I)]

The formation of the heavily doped n-type regions is one of the most complicated and actual problems in GaAs device and IC technology. When being implanted the violation of the semiconductor compaund stoichiometry results in the generation of the centers compensating the doping effect. The dual ion implantation is shown to contribute to the stoichiometry support and dopant electrical activity rize. Ideally, the dual implants should be made simultaneously or at least in small alternating steps (quasisimultaneously). But for practical considerations, at present the dual ion implantation is carried uout only sequentially, i.e. the step-by-step first dopant full dose set up is followed by the second one (I).

The work described the first attempt to fabricate the heavily doped n-type regions in the semi-imsulating gallium arsenide by the quasisimultaneous dual ion implantation. With the view of comparison the ion-implanted structures were also fabricates by the dual sequential and conventional single ion implantation methods. Silicon ions were used as the main dopant and phosphorus ions as the additional one. Silicon ions were implanted at IOO keV and phosphorus ions at IIO keV to obtain the same penetration depth. Full dose per each dopant was $2,6.IO^{I4}$ cm^{-2}. Elementary dose when performing the dual quasisimultaneous ion implantation was $I,25.IO^{I3}$ cm^{-2}. In this case the full dose per each dopant was picked up by the elementary doses in the alternating manner (firstly phosphorus ions and then silicon ones were implanted, all in all 20 times by each dopant). While carrying out the dual sequential ion implantation the structures were at first implanted by the full dose of the phosphorus ions and then by the full dose of silicon ions. No special measures were undertaken for sample heatimg and cooling. 0.3 m SiO_2 layer was deposited on the samples at 340 °C after the ion implantation. The structures were annealed in the hydrogen atmosphere by the flashes of halogen lamps. The time of the heat treatment varied from several seconds to a few tens of minutes. The quality of the fabricated ion-implanted structures was controlled by the rapid measurements of the carrier concentration profile by C-V method using themercury probe.

Fig.I and Fig.2 illustrate the profiles of the electron distribution in the semi-insulating gallium arsenide after the dual sequential and

---

I) Scientific Research Institute of Radiomaterials, SU-220600 Minsk USSR conventional ion implantation (Fig.I), and also after the dual quasi-simultaneous, dual sequential and conventional ion implantation (Fig.2).

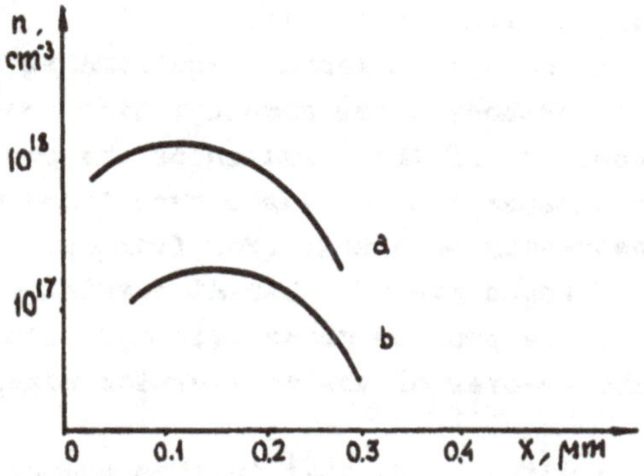

Fig.I. Profiles of the electron distribution across the depth in the semi-insulating gallium arsenide after the dual sequential (a) and conventional (b) ion implantation:
a) P^+ - E=II0 keV, D=2,5.10^{14} cm^{-2}; Si^+ - E=I00 keV, D=2,5.10^{14} cm^{-2};
b) Si^+ - E=I00 keV, D=2,5.10^{14} cm^{-2}.
Annealing: 820 °C, H_2, I min, SiO_2 0.3 μm.

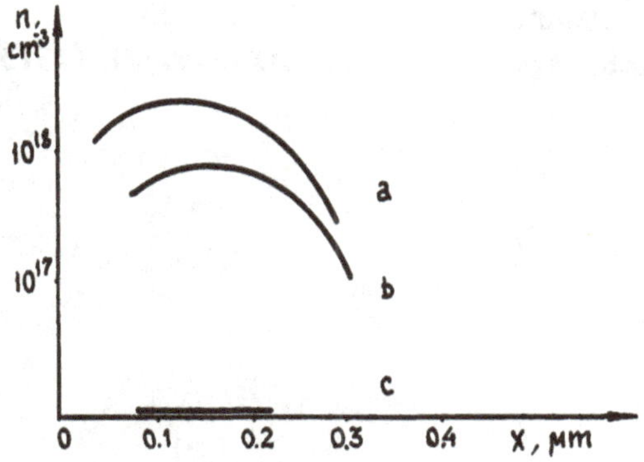

Fig.2. Profiles of the electron distribution across the depth in the semi-inselating gallium arsenide after the dual quasisimultaneous (a), dual sequential (b) and conventional (c) ion implantation:
a) P^+ - E=II0 keV, d=I,25.10^{13} cm^{-2}, N=20; (D=d·N).
Si^+ - E=I00 keV, d=I,25.10^{13} cm^{-2}, N=20; (D=d·N).
b) P^+ - E=II0 keV, D=2,5.10^{14} cm^{-2};
Si^+ - E=I00 keV, D=2,5.10^{14} cm^{-2};
c) P^+ - E=I00 keV, D=2,5.10^{14} cm^{-2}.
Annealing: 750 °C, H_2, I20 min, SiO_2 0.3 μm.

The functions presented reveal that the electrical activity of silicon in gallium arsenide is higher after the dual ion implantation than after the conventional one. And what's more, the layer fabricated by the dual quasisimultaneous ion implantation are characterized by the higher conductance than the layers formed by the dual sequential ion implantation, this factor being the evidence of the higher structure quality and higher dopant utilization coefficient.

Meanwhile it is necessary to note that labour requirements and duration of the ion implantation methods under consideration are also arranged in the given sequence. So, if the duration of the dual sequential ion implantation method exceeds that of the conventional method by a value of time that is necessary to change over from one dopant, to another and to implant the second dopant, then at quasisimultaneous ion implanting the duration of the process grows even more considerably because of the repeated change-over of the accelerator working conditions.

We share the opinion of the authors (2) that furthes and more complete realization of ion implantation method advantages will be in a substantial manner determined by the level of the technological equipment development, i.e. the design of the special-purpose installation for dual ion implantation.

R e f e r e n c e s

(I) Donnelly, J.: Inst.Phys.Conf. 33b-4 (I977) I66-I90.

(2) Heckingbottom, R., T.Ambridge: Radiation Effects, I7 (I973) 3I-36.

MODIFICATION OF THE InP(100) SURFACE BY ION BEAM

J. Zemek and P. Jiříček[1)]

Sputtering of solid surfaces with rare gas ions in combination with surface sensitive techniques is extensively used to produce clean surfaces as well as to obtain depth composition information with high spacial resolution normal to the surface. However, sputtering frequently damages surface structure and alters surface composition (2). AES and XPS measurements of the ion sputtered InP(100) surface revealed high concentrations of indium in the altered surface region (1,5,7,11) within the information depth of both methods (~2-6 nm). However, no detailed information about a depth distribution of phosphorus and indium or about the thickness of the altered region is available apart from the application of the ion-sputtered depth profiling alone (1,3, 12).

In this contribution ,we report on the effect of ion sputtering on the composition of the InP(100) surface and near surface region. The data are obtained in a non-destructive manner by angular resolved photoelectron spectroscopy, taking the In3d and P2p spectra at different take off angles. Thus it is possible to change the information depth of the method and hence to judge the compositional homogeneity of the near surface region. Concentration depth profiles was calculated from angular dependencies.

A mirror-like InP(100) surface was obtained by mechanical etch polishing followed by etching in 0.4% bromine methanol solution. Final cleaninng procedure and measurements were performed in a VG Scientific ADES 400 angle resolved spectrometer. Ion sputtering was accomplished with a differentially pumped ion gun using xenon working gas. The ion energy was set at 3 keV, ion current density at 10^{-5}A, angle of incidence 60° with respect to the surface normal. The target temperature was held at 150 K because the morphology degradation of the InP(100) surface can be significantly reduced by lowering the target temperature during sputtering (3).

Photoelectron spectra were excited by MgK_{α} radiation (1254 eV) at an incident angle of 70° relatively to the surface normal. The energy analyser was operated in the constant energy mode at 20, 50 or 100 eV pass energy. Photoelectron peak areas were determined from the recorded spectra using a computer program which corrected for inelastically scattered backround.

1) Institute of Physics, Czechoslovak Academy of Sciences, 180 40 Praque 8, Na Slovance 2, Czechoslovakia

Fig.1 shows two charecteristic shapes of angular dependecies of the P2p to In3d intensity ratio obtained after sputtering. The first one represented by curves a, b corresponds to a moderate sputtering with the ion dose of $10^{15}cm^{-2}$. The ion dose of 10^{17} cm^{-2} manifests itself by a flat dependence, almost indepedent of θ (curve c). It indicates a homogeneous, indium-rich near surface region. On the other hand, a substantial increase in the intensity ratio (curve a,b) for higher values of take off angle θ reflects a non-homogeneous distribution of phosphorus and indium within the altered layer: a phosphorus depleted sub-surface region with an increase in phosphorus concentration at the surface We note that the technique used is of greater surface sensitivity for higher values of θ. The inelastic mean free path of P2p photoelectrons λ_{P2p} is 2.4 nm and the corresponding probe depths $3\lambda_{P2p}\cos\theta$ achieve 7.6 and 4.6 nm at $\theta = 0^{o}$ and 50°, respectively. The curve a is almost a parallel shift to the higher ratio values due to annealing at 300 K. In comparison, the topmost surface of the annealed sample appears to be more rich in phosphorus. However, the mean P/In concentration ratio within the information depth is still below 1.

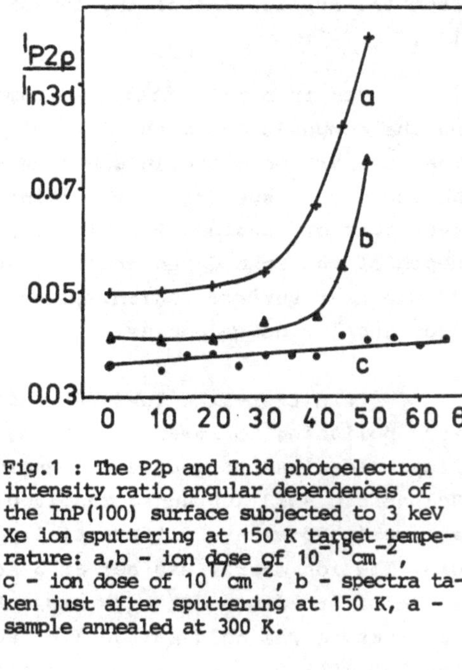

Fig.1 : The P2p and In3d photoelectron intensity ratio angular dependences of the InP(100) surface subjected to 3 keV Xe ion sputtering at 150 K target temperature: a,b - ion dose of 10^{15} cm^{-2}, c - ion dose of 10^{17} cm^{-2}, b - spectra taken just after sputtering at 150 K, a - sample annealed at 300 K.

The dependence of the signal intensity on the depth of analysis was systematically treated for the first time in (4). The determination of the depth concentration profiles $n_P(z)$ and $n_{In}(z)$ is reduced to a solution of the following integral equation (8,9)

$$\frac{I_{P2p}(\theta)}{I_{In3d}(\theta)} = \frac{\sigma_{P2p} \int_0^\infty n_P(z) \exp(-z/\lambda_{P2p}\cos\theta)\,dz}{\sigma_{In3d} \int_0^\infty n_{In}(z) \exp(-z/\lambda_{In3d}\cos\theta)\,dz}, \quad (1)$$

where the left-hand side of (1) represents the measured intensity ratio, σ is the photoionisation cross-section and z is the distance measured from the surface. It was was found in (9) that numerical solution of eq. (1) was higly sensitive to expected experimental errors. This difficulty could be surmounted by application of a priori information on the function n(z) : the element concentration in each layer must be positive and the profile must be monotoneuos. Nefedov and Baschenko (8) described numerical method for profile restoration with the less rigid physical limitations, assuming the function to have no more than one extremum. The method has been tested with the measured angular dependences of Ga3d and As3d intensities of a sputtered GaAs(100) surface revealing arsenic depletion in the sub-surface region with an arsenic-rich topmost surface.

The evaluation of our experimental data was carried out using a FORTRAN program written by Nefedov's group. Results are shown in Fig.2 The phosphorus concentration is constant up to the depth of $2\lambda_{P2p}$. In larger distance from the surface phosphorus concentration gradually decreases. Whithin the information depth the phosphorus is depleted and the thickness of the altered layer seems to be higher than 7 nm. It is approsimately equal to 3 R_p, R_p being the mean projected range for Xe ions of 3 keV kinetic energy. The same conclusion is valid for the depth profile evaluation of curves a and b (Fig.1) except the high phosphorus enrichment in the surface.

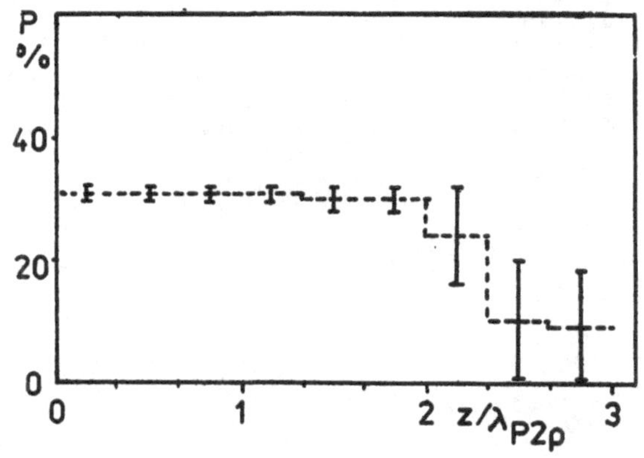

Fig.2: The restored depth concentration profile of the angular dependence c from Fig.1. Restored values are provided with dispersion limits: (——) most probable profile.

Compositional changes induced by ion sputtering may be caused by any of the following processes : ion implantation, Gibbsian adsorption, preferential sputtering, displacement mixing, radiation-enhanced diffusion, and radiation-induced segregation. Because of a large number of competitive processes it is difficult to seperate particular contributions. The phosphorus deficiency in the modified sub-surface region and the phosphorus-rich topmost surface indicate that the preferential sputtering plays a substantial role together with diffusion and segregation. The last two processes are also effective at low target temperature in the presence of high density of structural imperfections. This can produce compositional changes at depths greater than the R_p (6).

In agreement with present results we expect the phosphorus to be deficient within the whole altered layer. The actual phosphorus concentration on the topmost surface will be governed by a dynamic equilibrium between sputtering and radiation-stimulated diffusion/segregation processes.

References :

(1) Bayliss C.R. and D.L.Kirk : J. Phys. D: Appl.Phys.9 (1976) 233-244
(2) Bertz G. and G.K.Wehler in : Sputtering by Particle Bombardment II, Topics in Applied Physics, v.52, 1983, R.Berish (ed.)
(3) Bouadma N., P.Devoldere, B.Jusserand and P. Ossart : Appl. Phys. Lett. 48(1986) 1285-1287
(4) Fadley C.S., R.Baird, W.Sichaus, T.Novakov and S.A.L.Bergstrom : J. Electron Spectrosc.Relat. Phenom. 4(1974) 93
(5) Gruzza B., B.Achard and C.Pariset : Surf.Sci. 162(1985) 202-208
(6) Kirschner J. : Nucl.Instr. and Methods in Physics Res.B7/8 (1985) 742-749
(7) Lau W.M., R.N.S.Sodhi, B.J.Flinn, K.H.Tan and G.M.Bancroft : Appl.Phys.Lett. 51(1987) 177-179
(8) Nefedov V.I. and O.A.Baschenko : J.Electron Spectr. Relat. Phenom. 47(1988) 1-25
(9) Pijolat M. and G.Hollinger : Surf.Sci. 105(1981) 114
(10) Tanuma S., C.J.Powell and D.R.Penn : Surf. Sci. 192(1987) L849-L857
(11) Tu C.W. and A.R. Schlier : Appl.Surf. Sci.11/12 (1982) 355-361
(12) Wada O. : J. Phys. D. : Appl. Phys. 17(1984) 2429-2437

EFFECT OF PROTON BOMBARDMENT INDUCED DEFECTS ON THE OPTICAL ABSORPTION OF GaP AND InP

C. Ascheron, W. Hörig, H. Neumann, V. F. Orlenko
Karl-Marx-Universität Leipzig, Sektion Physik, Linnéstr. 5, Leipzig, 7010, GDR

INTRODUCTION

The intentional use of light ion implantation to create high-resistivity regions in semiconductors has found wide-spread application in various semiconductor technologies including the production of optoelectronic devices. In studying the optical properties of particle bombarded GaP /1/, GaAs /2/ and InP /3/ it has been established that irradiation induced defects give rise to long exponential absorption tails below the fundamental edge. We want to give some information on radiation defect induced absorption of InP and GaP and on the correlation between optical absorption and defects.

EXPERIMENT

The samples were implanted with 0.4 MeV protons and 1.2 MeV He^+ ions. Optical transmittance and reflectance spectra were measured at room temperature in the photon energy range $h\nu = 0.9 - 1.3$ eV for InP and $h\nu = 1.9 - 2.4$ eV for GaP. Damage density was determined by use of the Rutherford backscattering/channeling technique.

RESULTS AND DISCUSSION

Figure 1 shows the fluence dependent variation of the absorption exponent $\alpha_i d_i$ (with α_i as average absorption exponent of the implanted layer and d_i as its thickness) for one selected photon energy (2.1 eV for GaP and 1.1 eV for InP) and the mean damage density within the implanted layers. Figure 1 a further contains data of our experiments with higher energetic H^+ and He^+ ions as well as data for P^+ /4/ and electron irradiation /1/.

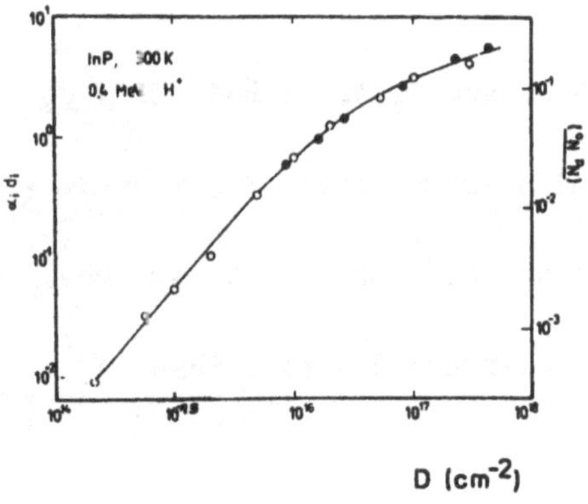

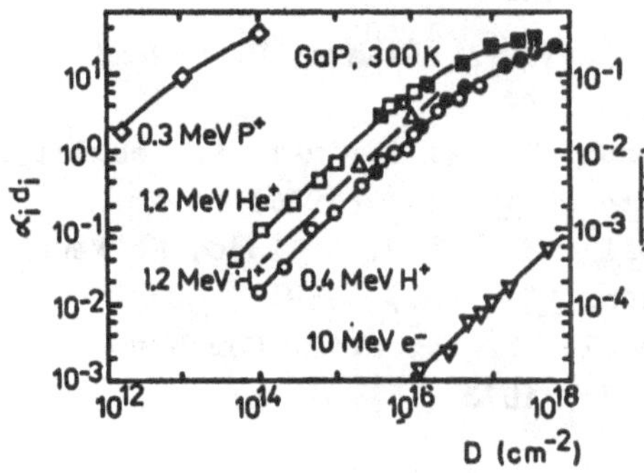

Fig. 1: Fluence dependence of the absorption exponent $\alpha_i d_i$ (open symbols) and of the mean damage density N_d/N_o within the implanted layer (filled symbols) for GaP (a) and InP (b), bombarded with 0.4 MeV protons and other charged particles.

The investigations indicate:

1. In InP and GaP the optical absorption coefficients and the damage density exhibit qualitatively similar sublinear fluence dependences with a tendency to saturating behaviour. This qualitative accordance points to a strong correlation between defect production and optical absorption.
2. In GaP the more strongly damaging heavier particles produce stronger absorbing regions. The fluence dependences for the different particles can be scaled with constant factors. These factors, however, are up to half an order of magnitude smaller than the corresponding ratios of nuclear energy deposition. Thus it can be concluded to a higher optically absorbing efficiency of isolated point-like defects than of point defects involved in defect clusters.
3. It can be supposed that the observed changes in the optical absorption are mainly due to relatively simple defects, such as antisites and small defect complexes involving vacancies, since changes in the optical absorption do also occur in the case of electron irradiation and outside the strongly damaged layer. In Fig. 1 a this is obvious from the stronger absorption in the case of bombardment with higher energetic protons.
4. From the results presented absorption coefficients per point defect are evaluated to be in the order of 10^{-17} to 10^{-16} cm^2. These values are close to those of ion implanted GaAs /2/.

REFERENCES

/1/ Endo, T., A. Sadaki, S. Sasaki, K. Sawa and T. Wada: Rad. Eff. 84 (1985) 89

/2/ Wesch, W., K. Gärtner, E. Wendler and G. Götz: Nucl. Instr. and Meth. B 15, 431 (1986)

/3/ Hörig, W., V. F. Orlenko, H. Neumann and C. Ascheron: Cryst. Res. Tech. 23 (1988) 359

/4/ Wemple, S. H., J. C. North and J. M. Dishman: J. Appl. Phys. 45 (1974) 1578

CHARACTERISTIC PROPERTIES OF HEATING MONOCRYSTALLINE SEMICONDUCTORS BY NANOSECOND LASER PULSES

G.M. Gusakov[1)]

One of the most important results of the thermal model of nanosecond pulse laser annealing (NPLA) is the liquid phase overheating on the semiconductor surface which during laser pulse action significantly exceeds its melting temperature.

Semiconductors such as Si, Ge, GaAs in a liquid state are metals and their reflectivity decreases with the increase of melt temperature. Therefore, studies of the characteristic properties of probing radiation reflection under NPLA enables an estimation of semiconductor melt surface temperature and its evolution as well as comparison with theoretical models.

The outcomes of scarce studies devoted to comparison on monocrystal samples have proved to be contradictory.

We have studied the dynamics of reflectivity (R) and the ellipsometric parameters of Si and Ge during of two-pulse laser irradiation by the first and second harmonics of Nd-YAG lasers /1,2/.

In the experiments the first pulse ($\lambda_1 = 0.53\,\mu m$ and duration (FWHM) $\tau_1 = 50$ ns) initiated a liquid phase on the semiconductor surface, while the more powerful second one ($\lambda_2 = 1.06\,\mu m$ and $\tau_2 = 70$ ns) acted upon the melt surface with a delay τ /2/. It was found that if the pulse energy density lies within $E_m = E = E_v$ (where E_m and E_v are the calculated energy densities as which the surface temperature reaches the semiconductor melting and evaporation temperatures respectively), R and the complex refractive indices of these materials are constant within an experimental error. The measured values coincide with the well-known equilibrium values of the optical constants of liquid Si and Ge near their melting points.

According to the theoretical estimates experimentally corroborated in the equilibrium conditions /2,3/, the temperature coefficient $\alpha = dR/dT$ for the liquid phase is about $(1\div2)\cdot10^{-4}K^{-1}$. That is why the thermal model predicts significant (10 to 20%) reduction of R in the points of maximum melt surface temperature (Fig. 1a).

At the same time, the experimental durations of semiconductor melt coincide with the calculated ones /2/.

Studies of the optical properties of materials near their damage

----

1) SU - 103498, USSR, Moscow, Moscow Institute of Electronic Technology

thresholds also have provided proffs of inadequacy of the thermal model of semiconductor heating by laser pulse.

For example, during irradiation of Ge by a combination of two pulses with total energy $E' = 3.2\ J/cm^2$ a reduction of R is observed (ΔR = 6 through 8%) (Fig. 1b), while when $E'' = 4.2\ J/cm^2$ is reached, the character of reflection changes significantly: after melting of the surface layer, dramatic drop of R is observed, reflection dynamics becoming irregular from pulse to pulse (Fig. 1c). According to the microscopy data, the sample's surface in this mode is damaged. A similar picture is observed for GaAs at $E'' = 1.1\ J/cm^2$, but the passage to damage is much sharper. No peculiarities of dynamics of R have bean found for Si over all the range of E ($E \leq 6\ J/cm^2$).

The experiments, thus, demonstrate that the actual energy densities, under which dynamics of R manifests specific features and the material damage threshold is reached, exceed the theoretical values by the factor of 1.5 - 2 depending on the material.

Here a question aises about the mechanisms underlying the above specific features of reflection which result in a lower temperature of the liquid phase surface kayer with the same life time.

An important, in our opinion, result is as follows. The temporal relations of R (Fig. 1a, b, c) were obtained under single (without irradiation zone overlapping) pulse action upon the semiconductor. We have discovered that if irradiation is carried out with zone "overlapping", the character of reflection changes radically. For the distance between the centers of irradiation zones $r \leq r_m$ (1) (with r_m for the radius of melt boundary) (Fig. 1d) reconstruction of the reflection picture (without overlapping) requires greater pulse energy. For example, the damage threshold for Ge was not reached even at $E'' = 4.5\ J/cm^2$, and melt life time exceeded 4 msecs. Notably, the condition (1) corresponds to the case where the maximum of Gaussian beam is "overlaps" the ripples formed at the boundary of the domain irradiated by the preceeding pulse.

Absence of this effect for PbTe has also been established.

These results suggest that the characteristic properties of heating Si, Ge and GaAs are, possibly, due to generation of surface plasmons /4/ and their relation to the capillary waves on the liquid phase surface because for PbTe the conditions of ripples formation are not satisfied. The results obtained indicate of the importance of nonlinear optical and of hydrodynamic phenomena defining the heating dynamics.

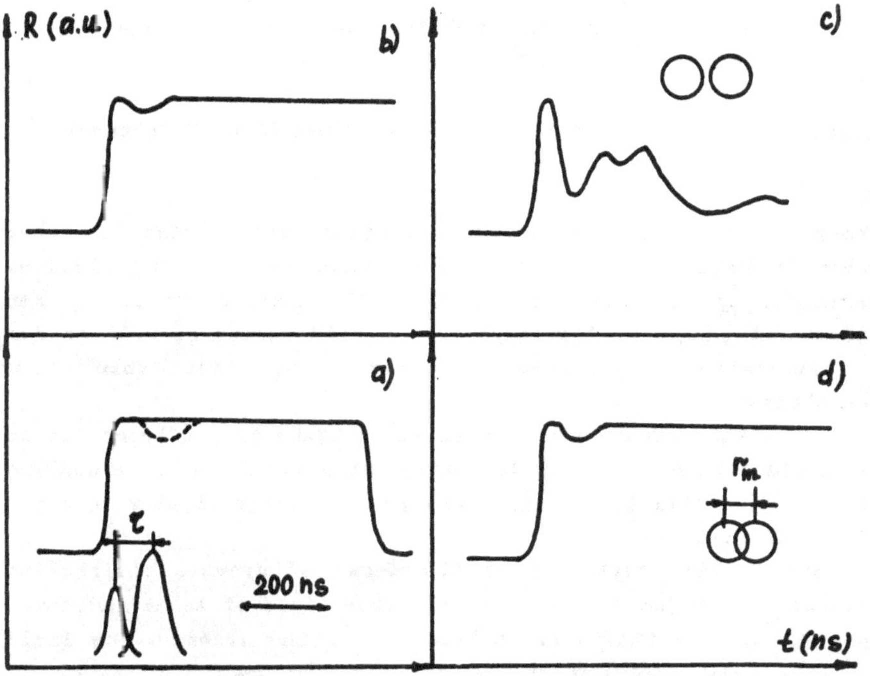

Fig. 1 a, b, c, d .

Time oscillograms of the p-th component of probing ($\lambda = 0.63\ \mu m$) radiation reflectivity for Ge surface for various infrared pulse energy densities ($\lambda_2 = 1.06\ \mu m$), ($\lambda_1 = 0.53\ \mu m$) is fixed and equal to $E_1 = 1.2\ J/cm^2$
a) $E_2 = 0.5\ J/cm^2$, b) $E_2 = 2\ J/cm^2$, c,d) $E_2 = 3\ J/cm^2$
a,b,c) without irradiation zones overlapping, d) with overlapping of irradiation zones (- experiment, --- - theory).

R e f e r e n s e s

(1) Gusakov G.M., Komarnitskii A.A. Letters to JTP 13 (1987) 166
(2) Gusakov G.M., Komarnitskii A.A., Em A.S. Phys. Stat. Sol. (a) 107 (1988) 261
(3) Lampert M.O., Koebel J.M., Siffert P. J. Appl. Phys. 42 (1981)4975
(4) Agranovich V., Mills K. Surface plasmons M 1985

INVESTIGATION OF THE DYNAMICS OF PULSED LASER AN NEALING OF ION-IMPLANTED SILICON BY THE PICOSECOND TRANSIENT GRATING TECHNIQUE

R.Baltramiejūnas, R.Gaška, E.Kuokštis, V.Netikšis, M.Petrauskas*

Investigation of the properties of semiconductors under different treatment using modern technologies such as ion implantation, thermal or optical annealing is of great interest due to the new fundamental phenomena taking place in solid state, as well as due to application importance, in particular, for thin film technology.

The aim of the present paper is to investigate the influence of As ion implantation and the following nanosecond laser annealing (NLA) on the relaxation characteristics of nonequilibrium carriers (NC) in Si crystals.

The time probing technique of the decay of dynamic diffraction gratings (DG) induced by the interference field of laser radiation was used to investigate relaxation of NC. The interference field of light wave exciting pulses induces the periodic change of complex dielectric constant on the surface of semiconductor. This change is caused by the superposition of a number of mechanisms, such as generation of NC, recharge of shallow impurities, free NC absorption, the change of lattice temperature, etc. The intensity of the 1st order diffracted beam for plane wave approximation can be expressed as [1]:

$$I(t) = (1-R)I^*(t) \exp(-\alpha d) \exp(-2A) \, |i \, J_1(F+iA)|^2 \quad , \qquad (1)$$

where $I^*(t)$ is the intensity of probe beam, J_1 is Bessel function of the 1st order, F and A is a modulation depth of phase and amplitude, respectively. The are equal : $F = \pi \Delta n Z^*/\lambda_z$ and $A = \pi \Delta k Z^*/\lambda_z$, where Δn, Δk is the real and the imaginary part of complex refractive index, respectively, d is a sample thickness, and Z^* is an active thickness of DG.

Two identical picosecond pulses of the 2nd harmonic of $YAG:Nd^{3+}$ passively mode locked laser (τ_i = 35 ns, λ_e = 0.53 μm) were directed to the surface of Si plate at an angle of $2\theta \approx 3°$ to each other and were used for excitation. The time probing of the decay of DG with period $L = 0.5\lambda/\sin(\theta) \approx 9$ μm was carried out by the 3rd pulse of low intensity with wavelength λ_z = 1.06 μm. The time delay between exciting and probing pulses could be varied within interval Δt = -100-800 ps. The crystals of Si with the orientation [111] and carrier density at room temperature $3x10^{15}$ cm^{-3} (p-type), as well as As^+-implanted Si crystals (dose $\Phi = 10^{15}-10^{16} cm^{-2}$, ion energy E = 60 keV) were under investigation. The diffraction efficiency (DE) was determined as the ratio of intensities of the 1st order diffracted beam and that of the incident one.

It was established from the measurements of DG decay in Si under different levels of laser excitation that for excitation energy corresponding to the NC density $\Delta N < 10^{19}$ cm^{-3} the DE drop can be described by the dependence $\exp(-\Delta t/\tau_{eff})$ with τ_{eff} = (600±100) ps. The NC life time was evaluated according to (in case $\Delta t \gg \tau_i$):

$$\eta(\Delta t) = \exp(-2\Gamma\Delta t) \qquad (2)$$

where $\Gamma = \tau^{-1}_D + \tau^{-1}_R$; $\tau^{-1}_D = 4\pi^2 D/L^2$. Here Γ^{-1} is DG life time, τ^{-1}_D is a DG decay rate due to diffusion of NC. The life time of NC in Si τ_p = $(2-6)x10^{-9}$s, when a well known value of ambipolar diffusion coefficient D = 10-15 cm^2/s [2] is used.
The decrease of DE up to 50 times was established in Si after As ion implantation with doses $\Phi > 10^{15}$ cm^{-2} that lead to surface amorphization. The DE time dependence was the same within the wide interval of excitation energy. At first, the increase of DE up to the delay of Δt = (0-20) ps was observed. Later, the value of DE stabilizes and remains the same within the time delay under investigation.
In Fig.1 the dependencies of normalized DE (curve 1), coefficients of reflectivity (2) and extinction (3) at wavelength 0.63 μm and ϑ (4) as functions of coordinate within the region of Si<As> (Φ = $9x10^{15}$ cm^{-2}) after NLA (τ_i = 20 ns, λ = 1.06 μm) with energy 1 J/cm^2 are shown. The point x = 1.8 mm corresponds to the center of Gaussian annealing pulse. The normalized time dependencies of DE before and after NLA are shown in Fig.2. The surface points of measuring for curves 1-4 are marked as 1-4 in Fig.1a, respectively. The DE was measured under laser energy density of $7.5x10^{-4}$ J/cm^2.
It was established from the analysis of the obtained data that an increase of the DE after NLA up to 40-50 times in Si<As> with implantation dose $\Phi = 10^{15}-10^{16}$ cm^{-2} is caused by the increase of NC contribution to the change of n due to recrystallization of ion implanted layer and increase of the active DG layer ($\eta \sim Z^{*2}$). The experimental results obtained by DG technique are compared with

ellipsometric studies as well as with measurements of NLA depth. The increase of DE in the region marked by the arrows in Fig.1a are explained by the increase of recrystallized layer of Si during NLA and redistribution of implanted As ions beyond the depth of implantation.

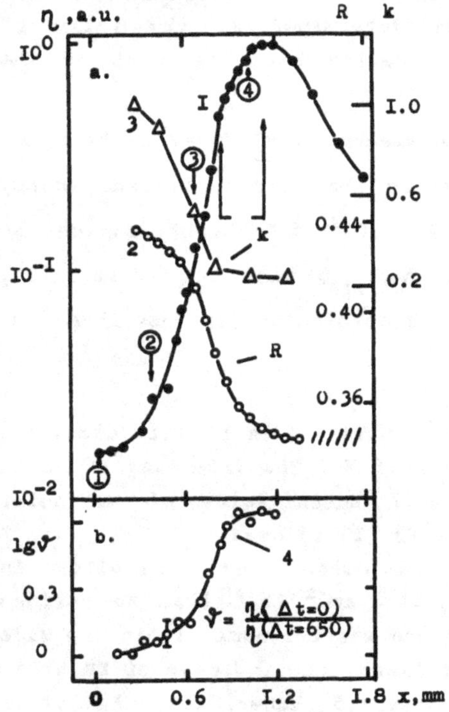

Fig.1

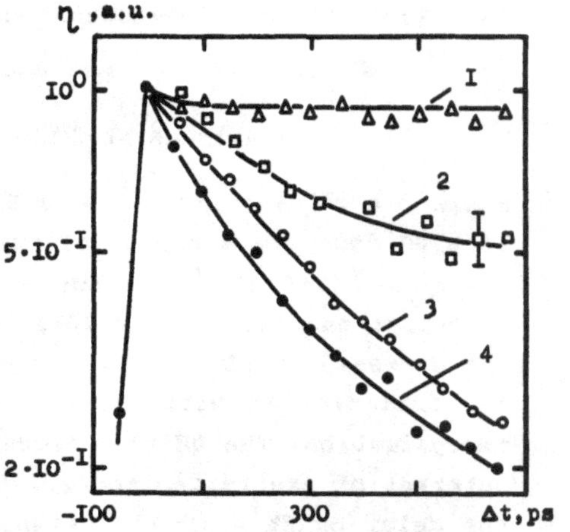

Fig.2

(1) Marine, W., P.Mathiez: IEEE J.Q.El. QE-22 (1986) 1404.

(2) Vaitkus, J., at el.: IEEE J.Q.El. QE-22 (1986) 1298.

*Vilnius, Universiteto 3, Lithuania, USSR

LASER INDUCED PHASE TRANSFORMATION IN SEMICONDUCTORS

Yu. A. Bumay, D. S. Domanewskii, N. E. Domorod, A.G. Ulyashin

Short pulse laser annealing provides a new approach to semiconductor material processing and offers some advantages over conventional furnace and rapid annealing for certain aspects of device fabrication. For understanding of processes involved in laser annealing of semiconductors it is important to predict the melting depths and durations. The intense optical excitation and short duration of laser annealing create the condition far from thermodynamics equilibrium. Therefore, computer simulation of the effect of electron-hole plasma and nucleation on phase transition and liquid-solid interface motion are of special interest for laser annealing theory.

Comprehensive one-dimensional mathematical model for laser heating and melting including the possible plasma and nucleation effects is represented by coupled differential equations for laser intensity $I(x,t)$, penetrating the semiconductor, for carrier concentration $n(x,t)$ and their temperature $T_{eh}(x,t)$ as well as for lattice temperature $T(x,t)$:

$$dI/dx = -\alpha I \tag{1}$$

$$\frac{\partial n}{\partial t} = \alpha_0 I - R(n, T_{eh}, T) + \frac{\partial}{\partial x}\left(D_{eh}\frac{\partial n}{\partial x}\right), \quad x > s \tag{2}$$

$$c_{eh}\rho_{eh}\frac{\partial T_{eh}}{\partial t} = F_{eh}(n, T_{eh}, T) + \frac{\partial}{\partial x}\left(K_{eh}\frac{\partial T_{eh}}{\partial x}\right), \quad x > s \tag{3}$$

$$[c + L\delta(T - T_m)]\rho\frac{\partial T}{\partial t} = F(n, T_{eh}, T) + \frac{\partial}{\partial x}\left(K\frac{\partial T}{\partial x}\right) \tag{4}$$

where $$R(n, T_{eh}, T) = \frac{n - n_i}{\tau_{SR}} + B(n^2 - n_i^2) + Cn(n^2 - n_i^2) \tag{5}$$

$$F_{eh}(n, T_{eh}, T) = \left[h\nu - \left(E_g + 2\frac{3}{2}kT_{eh}\right)\right]\alpha_0 I + h\nu\alpha_{FC} I + \left(E_g + 2\frac{3}{2}kT_{eh}\right)Cn(n^2 - n_i^2) - 2\frac{3}{2}k\frac{T_{eh} - T}{\tau_e}n \tag{6}$$

$$F = \begin{cases} h\nu\alpha_m I, & x \leq s \\ \left(E_g + 2\frac{3}{2}kT_{eh}\right)\frac{n - n_i}{\tau_{SR}} + 2\frac{3}{2}k\frac{T_{eh} - T}{\tau_e}n, & x > s \end{cases} \tag{7}$$

$$K = \begin{cases} K_c, & x > s \\ K_m, & x \leq s \end{cases} \qquad \alpha = \begin{cases} \alpha_0 + \alpha_{FC}, & x > s \\ \alpha_m, & x \leq s \end{cases} \tag{8}$$

Subscripts c and m designate solid and melt states of semiconductor.

---

[1] Belorussian Polytechnical Institute, leninsky Prospect, 65 Minsk, 220027 USSR

Here S - melt front position; α_0, α_{FC} - band-to-band and free carrier absorbtion coefficients; $C_{eh}, C, \rho_{eh}, \rho$ - specific heats and densities of carriers and semiconductor; τ_{SR}^{-1}, B, C - Shockley-Read, radiative and Auger coefficients; D_{eh} - ambipolar diffusivity; n_i - intrinsic carrier concentration; E_g - energy gap; $h\nu$ - photon energy; K_{eh}, K carrier and lattice thermal conductidities; τ_e - energy relaxation time; T_m - melting temperature; α_m, K_m, - absorbtion coefficient and thermal conductivity of melt, L - latent heat of melting; $\delta(T-T_m)$-Dirac delta-function.

Energy and carrier losses from both surfaces of semiconductor can be neglected , so the boundary conditions are: $\partial n/\partial x=0, \partial T_{eh}/\partial x=0, \partial T/\partial x=0$. But the boundary condition for GaAs can be appropriately modified to consider evaporation of As atoms. The boundary conditions for liquid-solid interface (s) for equations (2), (3) can be written as

$$\partial n(S,t)/\partial x = 0, \quad T_{eh}(S,t) = T_m$$

Heat conductivity and diffusion equation are solved numerically by using absolutely implicit sheme of finite difference approximation. When liquid and solid phases are present we solved single for the crystal heat conductivity equation (4) by smoothing method without explicit identification of phase interface. The melt front position is found from isotherm $T=T_m$. Equations (2), (3) are solved only for nonmolten region of semiconductor. The space smoothing interval near the melt front, introduced for approximation of delta-function in equation (4), simulate the nucleation influence and is chosen greater than the size of critical nucleus. Because the equations are nonlinear and coupled iteration is required at each time step for correct solution.

We applied our model to experimental results on time resolved reflectivity during laser annealing of Si and GaAs [1,2]. Temperature and carrier concentration dependences of parameters of solid and molten Si and GaAs, used in the model, are given in the referances [1-6].

The most important parameters in our model is the energy relaxation time which defines the transfer of laser irradiation energy from electron- hole plasma to lattice by means of carrier-phonon interactions. The Fig.1 illustrates the melt front position versus time in Si for cases of finite ($\tau_e = 10^{-12}$ s, 10^{-11} s) and instantenious ($\tau_e = 0$ s) transfer of laser irradiation energy to the lattice. One can see that the time of appearence of melting increases with increasing of τ_e. The melting rate is greater and recrystallization one is lesser for greater τ_e, but τ_e has no strong effect on melting depth. Such behavior of melt front is connected with energy transfer by electron-hole plasma, which decreases the temperature gradients (Fig.2). The lattice temperature, calculated for $\tau_e = 10^{-11}$ s, reaches lower value in comparison with $\tau_e = 0$ s, greater region of semiconductor being heating. High melt temperature reached during laser pulse due to strong light absorption of melt and its low heat coductivity results in the melt depth increasing after laser pulse is over. Note, that after appearing of melt through which the laser irradiation does not penetrate, electron-hole plasma has no effect on heating processes. The time of melt existense, calculated for $\tau_e = 10^{-12}$ s, is in good agreement with the time received from resolved optical reflectivity measurements (50ns) [1]. Analogous behavior of melt front versus time is observed for GaAs (Fig.3). Energy relaxation time for GaAs and Si are close, but GaAs possesses large value of optical absorption and low value of thermal conductivity in comparison with Si, therefore electron-hole plasma effects on heating processes in GaAs more essentially. Fig.3 shows that melting kinetics for GaAs is also affected by evaporation of As atoms.

On the basis of our calculations we can conclude,that kinetics of phase transition and melt front propagation rates during laser annealing may be affected by optically generated hot electron-hole plasma.

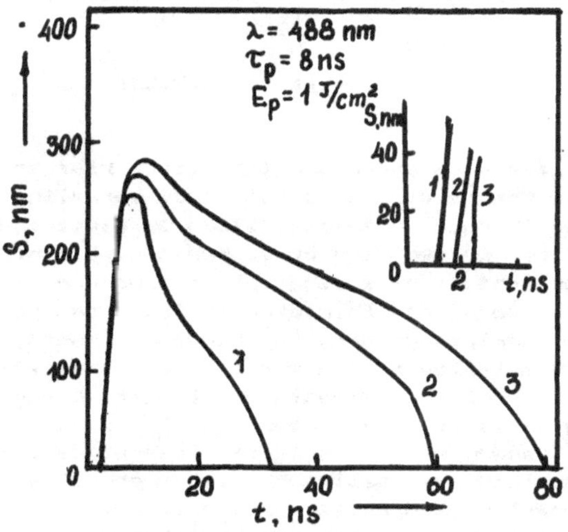

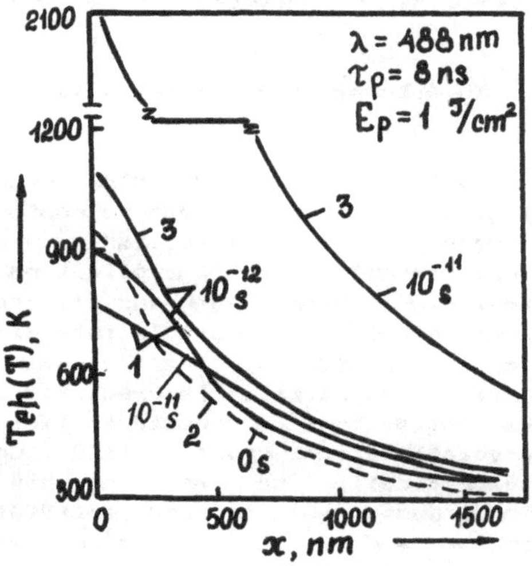

Fig.1. Melt front position versus time for laser annealing of Si, calculated for τ_e =0 s -1, 10^{-12} s-2, 10^{-11} s -3

Fig.2. Profiles of lattice temperature (1,2) and carrier temperature (3) in first nanosecond of laser annealing of Si for τ_e =0 s , 10^{-12} s , 10^{-11} s

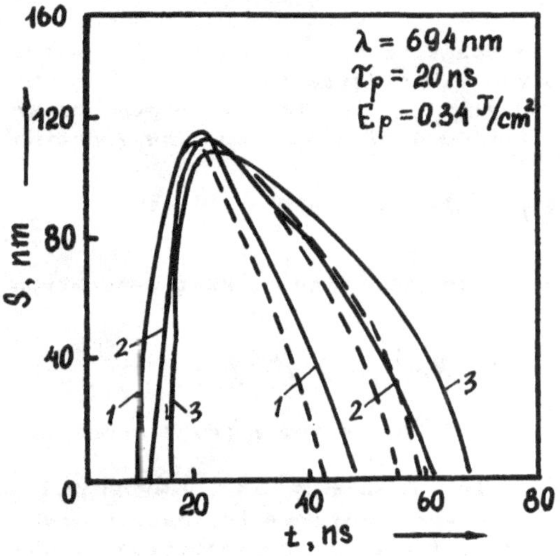

Fig.3. Melt front position versus time for laser annealing of GaAs, calculated for τ_e =0 s -1, 10^{-12} s -2, 10^{-11} s -3, when evaporation of As-atoms from the surface is included (solid lines) or ignored (dashed lines). Experimental time of melt existence is 80 ns [2].

[1] Lo,H.W.,and A.Compaan: Phys.Rev.Lett.44 (1980) 1604 -1607
[2] Pospieszczyk, A., M.A.Harith and B.Stritsker: J.Appl.Phys. 54 (1983) 3176-3182
[3] Meyer,J.R., M.R.Kruer and F.J.Bartoli: J.Appl.Phys. 51 (1981) 513-5522
[4] Blakemore,J.S.: J.Appl.Phys. 53 (1982) R123-R181
[5] Wood,R.F. and G.E.Gales: Phys.Rev. B23 (1981) 2932-2947
[6] Jordan,A.S.: J.Cryst.Growth 70 (1985) 551-558

HEATING OF SEMICONDUCTORS BY ION BEAMS AND ION-BEAM INDUCED PROCESSES

Yu.A.Bumay, D.S.Domanevskii, N.E.Domorod, A.G.Ulyashin, N.V.Shlopak

Implantation of semiconductor materials with energetic ions gives rise to lattice damage. In order to restore the crystal structure some annealing treatment has to be applied such as furnace annealing, laser or electron beam annealing. The application of pulsed ion beams has a number of advantages over laser and electron annealing: a large plate area can be treated with reproducible intensity; metal and insulator surfaces can be easily treated as in the case of semiconductors; the condensed energy profile can be reliable predicted in multilayer structures, etc. It should be noted that as opposed to laser annealing the features of melting and crystallization under pulsed ion effects have been poorly studied theoretically so far. In this paper we investigate theoretically ion-induced melting and solidification of silicon crystal on pulsed proton and B^+, Al^+, P^+, As^+ ions annealing and pulsed proton annealing of GaAs. Pulsed proton annealing of semi-insulating has been carried out and a comparison of our calculations with experimental results is made.
The model describing the pulsed ion annealing is based on the heat-flow equation with the temperature dependent conductivity and for a one-dimensional case has the form:

$$\rho[c(T)+\lambda\delta(T-T^*)]\frac{\partial T}{\partial t}=\frac{\partial}{\partial x}\left(K(T)\frac{\partial T}{\partial x}\right)+Z(E,t,x) \quad (1)$$

where ρ is the density, $c(T)$ is the specific heat capacity, λ is the latent heat of melting, $\delta(T-T^*)$ is the Dirac delta function, T^* is the melting temperature, $Z(E,t,x)=f(t)\cdot P(E,x)$ is the ion-beam power density, $f(t)$ is the function for a time form of the pulse. The function $f(t)$ can be represented in the form:

$$f(t)=\begin{cases}1/\tau\,[1-\cos(2\pi t/\tau)], & 0<t<\tau\\ 0, & t\geqslant\tau\end{cases} \quad (2)$$

where τ the pulse duration. $P(E,x)$ is the function of heat generation which can be written as

$$P(E,x)=Q/\sqrt{2\pi}\,\Delta R_p\,\exp[-(x-R_p)^2/2\Delta R_p^2] \quad (3)$$

where Q is the ion-beam energy density, R_p is the average projected ion path, ΔR_p is the struggling.
The initial temperature distribution in a sample is homogenious. T(x,0)=300 K. Energy losses for radiation and environmental heat removal are negligible for short pulses. Therefore, the boundary condition at the front sample surface has the form: $\partial T(0,t)/\partial x=0$. The opposite surface temperature is constant and equal to 300 K: T(d,t)=300 K.

The approach suggested in [1] has been used for the numerical calculations. It consists of solving unified heat-conduction equation (1) for the melt-solid system without explicit identification of the interface. The heat-conduction equation is solved according to an absolutely implicit scheme of the melting threshold energy and melting front dynamics under the effect of powerful ion-beams. For comparison with laser annealing the results of laser induced melting calculations are presented.

---

1) Belorussian Polytechnical Institute, Leninsky Prospect, 65 Minsk, 220027 USSR

Pulsed proton annealing (E=500 keV, τ =60 ns, Q=0.23;0.60;0.93;1.23 J/cm^2) of semi-insulating Si-doped GaAs with doses 10^{15} cm^{-2} and energies of 100 keV was carried out. Hall effect and capacitance-voltage measurements performed for investigation of the carrier concentration. We observed that energy released by the proton beam was insufficient for melting and recrystallization of the ion-implanted layer (~0.1μm) of Si-doped GaAs. Comparison of our calculation (Fig.1) with the experimental results provides quite a good agreement.

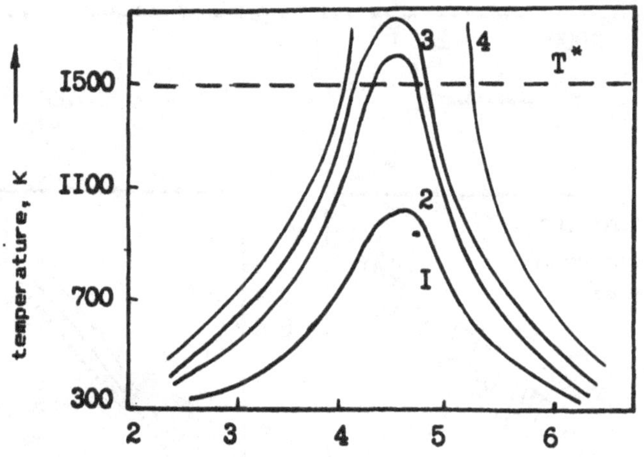

Fig.1. Temperature profiles in GaAs under pulsed proton effect (E=500 keV, τ =60 ns,) with energy densities of 0.23(1); 0.60(2); 0.93(3); 1.23(4) J/cm^2.

The results in Fig.2 are for melt depth in silicon versus maximum temperature of silicon surface affected by a pulsed ion ore laser (λ =0.69μm) beams with different energies and with different pulse duration. As can be seen the model calculation allow to correlate the maximum melt depth and maximum temperature of silicon surface from the beam parameters. Some differences and advantages of pulsed ion annealing may be seen. In particular for pulsed proton beams the melted regions can exist inside the crystal at the depths corresponding to a maximum of energy release without melting of surface. The comparison of our calculatins with the experimental results [2] provides quite a good agreement. The differences in melt depths for B^+,Al^+,P^+,As^+ ions beams with different energies are insignificant.

Thus, the suggested model permits calculations of the heating dynamics, threshold melting energies and melting depths of semiconductors by monoenergetic ion beams.

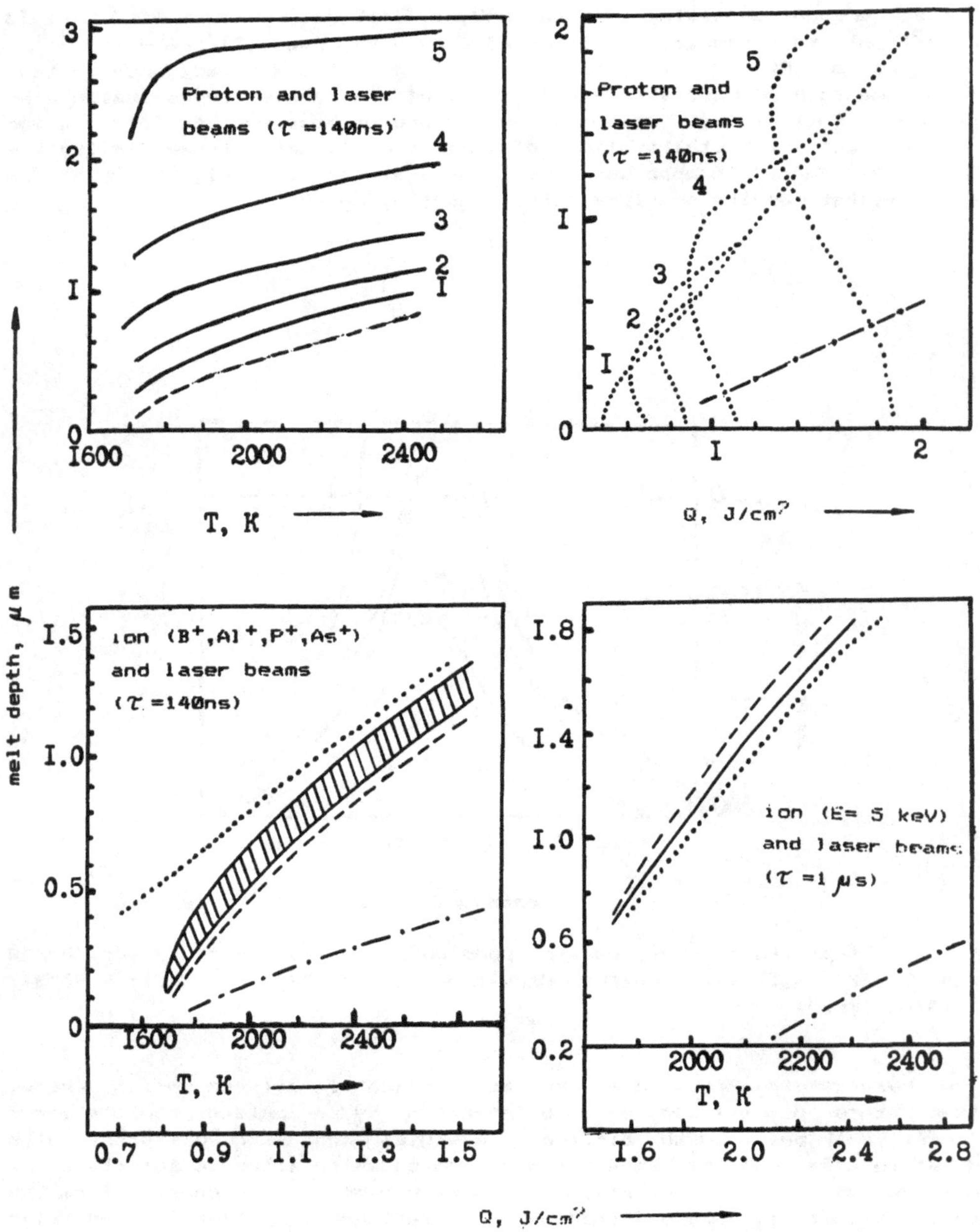

Fig.2. Melt depth versus maximum temperature of silicon surface affected by pulsed ion (———) ore laser (— — —) beams and versus energy density of pulsed ion (.........) ore laser (—.—.—) beams: 1,...,5 - proton pulsed beam with 60, 100, 150, 200, 300 keV, respectively.

[1] Budak B.M., E.N. Solovyeva and A.B. Uspenskyi: Zh.Vych.Mat. i Mat.Fiz. 5 (1965) 828.
[2] Chu W.K., et. al.: Nucl. Instr. and Meth. 194 (1982) 443.

THE FORMATION OF δ-DOPED SILICON LAYERS BY PULSED HEATING RECRYSTALLIZATION

A.V.Dvurechenskii[1)], F.Grötzschel[2)], B.Z.Kanter[1)], Yu.A.Manzhosov[1)], A.I.Nikiforov[1)], P.L.Novikov[1)], and S.I.Romanov[1)]

Recently a considerable interest has appeared to the investigation of semiconductors with δ-doped layers /3,4,1/. But in the formation of these layers with MBE the broadening of impurity depth distribution takes place due to segregation. The purpose of the present work was to obtain the δ-doped Sb layer in silicon by pulsed annealing.

The original structure was prepared in ultrahigh vacuum by the following way. On the atomic-pure surface of the (111)-silicon substrate at the temperature of 700 °C an buffer silicon layer was deposited, then the temperature was lowered to the room one and antimony layer with density 10^{15} cm^{-2} and a 800 Å thick amorphous silicon (a-Si) layer were deposited. Recrystallization was carried out under light action of flash-lamp with 1 ms pulse duration and halogen lamp for a period to 20 s. The samples were studied by He^+ Rutherford backscattering (RBS), RHEED and the Hall effect measurements.

The analysis has shown that there was not full crystallization of amorphous layer at flash-lamp treatment with intensity less than 30 J/cm^2 (Figs 1,2). In the other side irradiation with intensity more than that value result in broadening of impurity profile. As opposed to this treatment by halogen lamp irradiation at 1070 °C ranging from 15 to 20 s result in a-Si film recrystallization without noticable broadening of the RBS measured depth distribution Sb profile (Fig. 3). Monocrystalline film with (111) orientation contains some crystallographic defects, that are typical for solid-phase regrowth. More than half of the total impurities are substitutional, and this value of concentration as was estimated is more than $5x10^{20}$ cm^{-2} exceeding concentration of substitutional Sb in Si layers prepared without pulse heating /1/. The Hall effect measurements of the carrier concentration agree with RBS data. The effective electron mobility in the structure is 40 $cm^2/V.s$.

For comparison segregation effects at liquid-phase and solid-phase crystallization depth redistribution of Sb profiles as measured by RBS in ruby laser annealed (1.5 J/cm^2, 30 ns) Sb^+ implanted Si /2/ and in rapid thermal annealed (1070 °C, 20 s) a-Si-Sb δ-doped layer-(111) Si structure are presented in Fig. 4. One can see a small diffusion and segregation effects at solid-phase crystallization.

Thus, it is shown possibility of obtaining δ-doped layers in silicon with concentration more than $5x10^{20}$ cm^{-2} without noticable impurity redistribution by recrystallization of amorphous layers by light pulse heating.

1) Institute of Semiconductor Physics, 630090 Novosibirsk, USSR

2) The Central Institute of Nuclear Studies, DDR

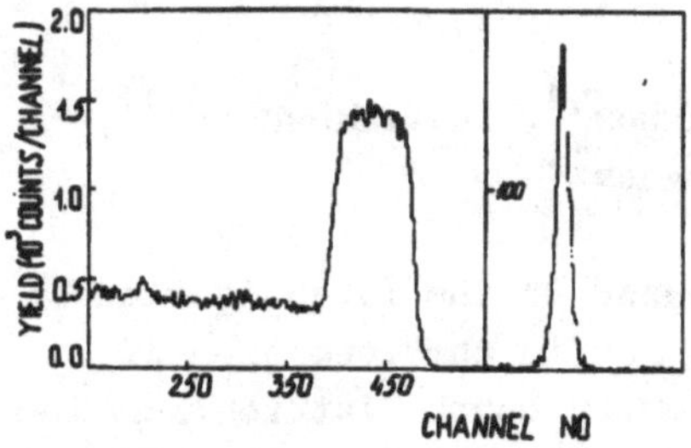

Fig.1. (111)-aligned backscattering spectra for original a-Si-Sb$^+$ δ-doped layer-(111) Si structure.

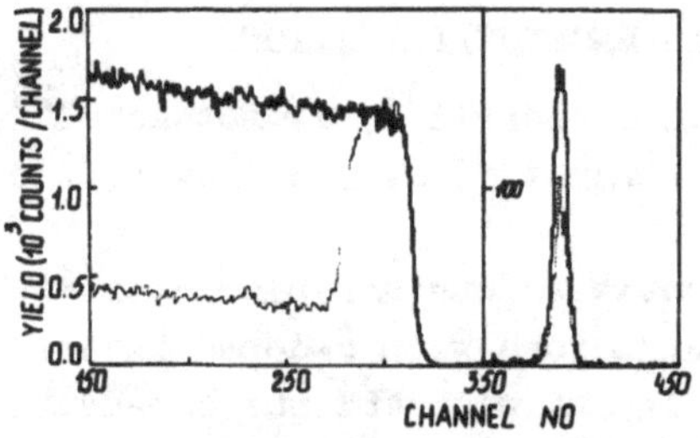

Fig.2. Random (solid curve) and (111)-aligned (dotted curve) backscattering spectra for sample annealed by flash-lamp irradiation with 10 J/cm .

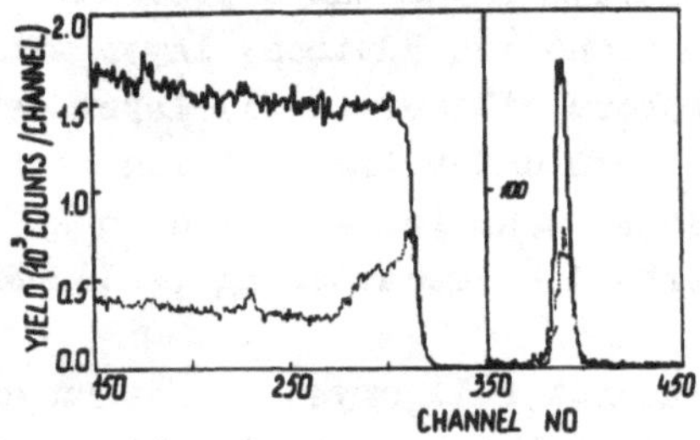

Fig.3. Random (solid curve) and (111)-aligned (dotted curve) backscattering spectra for sample annealed with halogen lamp irradiation at 1070 °C for 20 s.

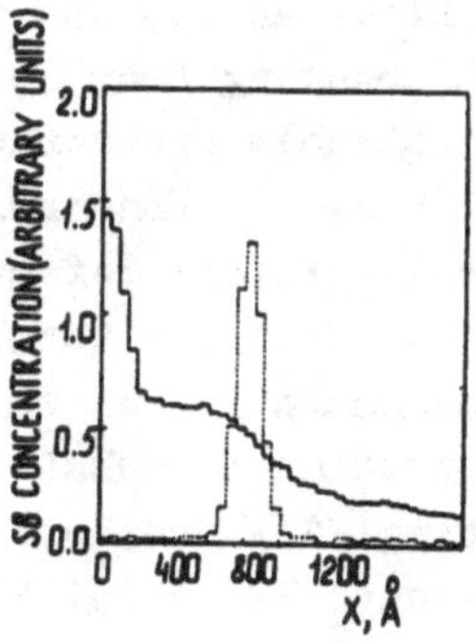

Fig. 4. Depth redistribution of Sb RBS profiles for liquid-phase laser annealing Sb   implanted Si layer (solid curve) and solid-phase thermal annealing of a-Si-Sb$^+$ δ-doped layer-(111) Si structure (dotted curve).

REFERENCES

(1) M.W.Denhoff, T.E.Jackman, J.P.McCaffrey. Appl.Phys.Lett. 54(1989) 1332-1334.

(2) V.A.Gaisler, F.Grötzschel and A.V.Dvurechenskii. Proc. IBMM Conf. Abstr., Kaunas, 1989, 8.

(3) K.Ploog, M.Hauser and A.Fischer. Appl.Phys. A 45(1988) 233-244.

(4) Wenchao Cheng, A.Zenner, Qui-Yi Ye. Semicond.Sci.Technol. 4(1989) 16-19.

INTERMEDIATE CRYSTALLIZATION OF ION-IMPLANTED SILICON DURING NANOSECOND LASER ANNEALING

G.D. Ivlev[1)]

Phase transitions that occur in semiconductors under the action of pulsed laser radiation are often investigated by the time-resolved reflectivity measurements. As follows from many papers, melting of silicon during nanosecond laser annealing is accompanied by approximately two-fold increase in its reflectivity R, the melt life-time being normally less than 1 μs. In the experiments with silicon amorphized by ion implantation (a-Si), nonmonotonicity of the reflection changes at λ =0.63 μm is observed at the stage where increased quasistationary value of R is established [1,4]. A short-term decrease in R at this stage has been explained by some authors [1,4] as resulting from changes in the melt reflectivity caused by significant heating.

It has been shown [5,6] that such a decrease in R is connected with intermediate crystallization due to the metastable liquid phase arising. The metastable melt is formed owing to lower melting temperature T_{ma} of silicon preliminarily amorphized by P^+-ion implantation (70 keV, a-Si layer is about 0.1 μm thick). Studies of the time dependence of the reflectivity at λ =1.06 μm and the pyrometry of pulsed thermal radiation in the green spectral region performed in the experiment with silicon heated by ruby laser monopulse radiation of 70 ns FWHM enabled us to find out [9] that the value of T_{ma} is less than the melting temperature of monocrystal silicon (c-Si) by $\Delta T \simeq 200$ K. The correctness of the pyrometric measurements is confirmed by the results of the computational modelling of the dynamics of laser heating of silicon [9]. The comparison with the experimental data (Fig. 1) shows that, in particular, c-Si melts at a temperature virtually equal to the equilibrium point of the "crystal-melt" phase transition.

The fact that ΔT does not exceed about 200 K is important for considering the nature of the intermediate crystallization which brings about the formation, for a short period of time, of a fine-grained polycrystal (FG p-Si) with the average grain size of 5 to 10 nm. Since such p-Si is formed in about 10^{-8}s, the formation of polycrystal grains must occur at a rate not less than $10^{26}cm^{-3}s^{-1}$. Calculations [8] demonstrate that the rate of homogeneous nucleation approaches this value only for the melt supercooling of about 600 K (Fig. 2). In view of the fact that the metastable melt temperature does not fall below T_{ma} because such melt continues to be affected by the laser action, the prehistory of formation of the FG p-Si can not be explained on the basis of the concept of homogeneous nucleation.

As a result, we came to a conclusion about a heterogeneous nucleation of

1) Institute of Electronics, BSSR Academy of Sciences, 22 Logoiski Trakt, Minsk-90, 220841, USSR

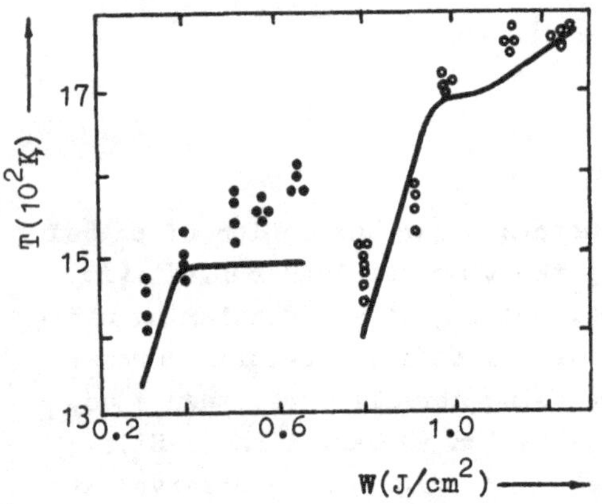

Fig. 1. Experimental points and calculated dependences of the maximum temperature of the silicon surface on the laser pulse energy density W: a-Si(o) and c-Si(o)

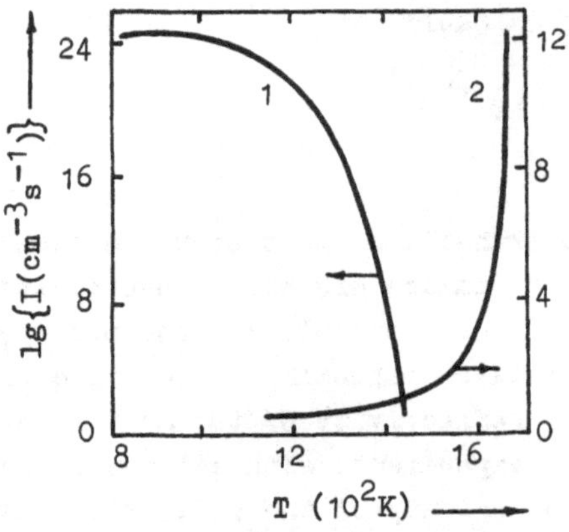

Fig. 2. Calculated temperature dependences of the rate of homogeneous nucleation (1) and the critical radius of the crystalline phase nucleus (2) in liquid silicon

the FG p-Si, namely, about the existence of crystallization centres in the a-Si which do not disappear at T_{ma} owing to their size exceeding the critical size. According to the estimate in [3] , if the heating rate of the metastable melt is higher than about 10^{10} K/s, the intermediate crystallization does not develop because as the supercooling value decreases rapidly the nuclea disintegrate owing to their size changing from the supercritical (Fig. 2) to the subcritical region.

The results of the computational modelling [2,12] of the phase transitions taking place during pulsed heating of a-Si suggest that intermediate crystallization is possible. From the calculations [12] based on the concept of already available crystal nuclea we infer that melting of the FG p-Si after the intermediate crystallization, i.e. the second melting of Si during the laser annealing process, begins from the buried liquid layer forming at a certain stage of heating rather than from the sample surface, which is supported by the experiments [10]. In this case, situation may arise where the surface FG p-Si layer melts at first followed by melting of the layer of the same polycrystal which separates the liquid layer from the monocrystalline substrate. If the second melt front does not reach the substrate, i.e. no conditions are established for subsequent liquid-phase epitaxy, the melt crystallizes into a coarse-grained p-Si with the grain-size of up to 50-100 nm.

We also observed intermediate crystallization in the experiments with silicon implanted with indium ions (80 keV) [7]. It should be noted that, in contrast to phosphorus, indium is characterized by low equilibrium solubility limit in the solid-phase silicon. As follows from the oscillograms of the reflected probe radiation (Fig. 3), the nonmonotonicity of rising R, which is indicative of the intermediate crystallization, is observed for

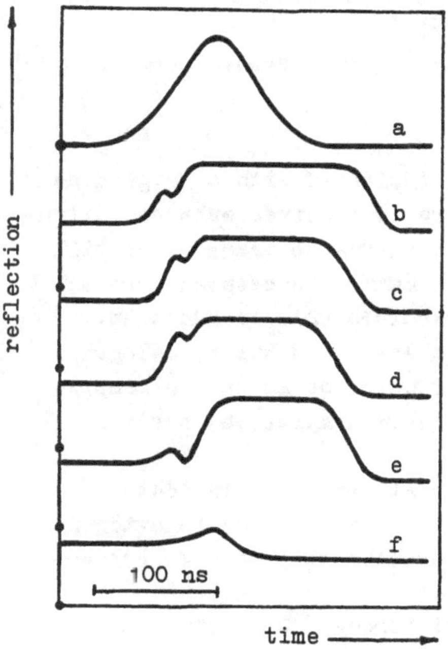

Fig. 3. Oscillograms of the ruby laser pulse (a) and probe radiation (λ = 1.06 μm) reflected from implanted silicon for W=1.3(b-e) and 0.6 J/cm^2(f)

the implantation doses studied: $2 \cdot 10^{16}$(b), $4 \cdot 10^{15}$(c), $6 \cdot 10^{14}$(d), $6 \cdot 10^{13}$ cm^{-2}(e,f). This nonmonotonicity of the reflectivity changes insignificantly as the dose changes from $2 \cdot 10^{16}$ to $6 \cdot 10^{14}$ cm^{-2}, i.e. as long as there is a continuous amorphized layer. During the intermediate crystallization and, in particular, for the highest indium concentration, the displacement of the majority of In atoms towards the p-Si grain boundaries can be expected. Passing from $6 \cdot 10^{14}$ to $6 \cdot 10^{13}$ cm^{-2}, the nonmonotonicity undergoes a marked change since the dose $6 \cdot 10^{13}$ cm^{-2} is sufficient only for formation of a thin buried region of a-Si. The oscillograms in Fig. 3 illustrate that in this case the structural state of silicon resulting from laser irradiation with the energy density of about 0.6 J/cm^2 is the state intermediate in time for a higher value of W. Data of [7] show that the buried a-Si layer (0.6 J/cm^2) undergoes epitaxial crystallization. However, numerous crystal lattice defects are formed. Thus, the intermediate crystallization can be both polycentre and epitaxial in nature. It is observed, also, in the experiments with implanted germanium [11].

(1) Auston, D.H., J.A. Golovchenko, A.L. Simons, C.M. Surko and T.N.C. Venkatesan: Appl. Phys. Lett. 34 (1979) 777-779
(2) Balandin, V.Yu., A.V. Dvurechenskii and L.N. Aleksandrov: Phys. Stat. Sol. (a) 110 (1988) 133-139
(3) Bayazitov, R.M., G.D. Ivlev, I.B. Khaibullin, V.L. Malevich and N.A. Sainov: Fiz. Tekhn. Polupr. 22 (1988) 79-83
(4) Bhattacharyya, A., B.G. Streetman and K. Hess: J. Appl. Phys. 52 (1981) 3611-3617
(5) Ivlev, G.D.: Fiz. Tekhn. Polupr. 19 (1985) 2188-2190
(6) Ivlev, G.D.: Zhurn. Tekhn. Fiz., Pisma 8 (1982) 468-47?
(7) Ivlev, G.D., R.M. Bayazitov, P.I. Gaiduk. et al.: to be published in Poverkhnost. Fiz. Khim. Mekh. (1990)
(8) Ivlev, G.D. and V.L. Malevich: Phys. Stat. Sol. (a) 103 (1987) K87-K91
(9) Ivlev, G.D., V.L. Malevich and V.V. Zhidkov: Phys. Stat. Sol. (a) 106 (1988) K123-K127
(10) Peercy, P.S., J.Y. Tsao, S.R. Stiffler and M.O. Thompson: Appl. Phys. Lett. 52 (1988) 203-205
(11) Zakirov, G.G., G.D. Ivlev and I.B. Khaibullin: Fiz. Tekhn. Polupr. 22 (1988) 947-949
(12) Zhvavyi, S.P. and O.L. Sadovskaya: Phys. Stat. Sol. (a) 112 (1989) K19-K21

ARSENIC SUBLIMATION AND DIFFUSION AT RAPID ELECTRON BEAM ANNEALING OF IMPLANTED SILICON

R. Grötzschel[1], V.A. Kagadey[2], P. Knothe[1], D.I. Proskurovsky[2], N.I. Lebedeva[2]

At rapid thermal annealing (RTA) of Si implanted with a large dose of As^+ ($\sim 5\times10^{15}$ cm^{-2}) by halogen lamps there is observed enhanced diffusion of the impurity into silicon (1-3). The diffusion takes place till the concentration of As at the depth profile maximum decreases down to its equilibrium value at the annealing temperature (2). In addition, many authors note that no losses of As are observed at RTA by halogen lamps. On the other hand, no significant diffusion of As into a sample is observed at electron beam heating of Si<As> samples and certain portion of it sublimates (4).

The present paper describes the further attempts to elucidate the reasons for the aforesaid peculiarities in the As atoms behavior in Si at RTA by electron beam, as well as to investigate probable connections between sublimation and diffusion of As.

As was implanted (1×10^{16} cm^{-2}, 100 keV) through 30 nm SiO_2 in (100) Si wafers. The annealing of samples was made with electron beam of 15 s duration. The maximum temperature of samples observed during the irradiation was 1400 to 1490 K. The dwell time at maximum temperature was ~5 s. Two pairs of samples were irradiated simultaneously. The surfaces of one of the pairs were cleaned of the SiO_2 cap layer by etching in HF, while the other pair of samples was not cleaned of it in order to capsulate the surface at annealing. One sample of each pair was irradiated from the implanted layer side (front to beam), while another one from the wafer side (back to beam). The samples were investigated by Rutherford backscattering spectroscopy (RBS) plus channeling and electrical measurements.

The analysis of RBS spectra yielded the values of χ_{min} ~0.04 at T_{max} = 1400 K, and ~0.034 at T_{max} = 1490 K. These values of χ_{min} show high quality of crystal structure in recrystallized layer. Besides, the quality of the crystal structure does not depend on the direction of irradiation as well as on the presence of SiO_2 cap layer, being only determined by the annealing temperature.

The RBS spectra characterizing the depth distribution of As atoms in Si wafer are depicted in Figs.1 and 2. The results obtained from the analysis of these spectra are presented in the Table. It is seen from these data that, as it was anticipated, the intensity of As sublimation at the annealing is higher for uncupsulated samples compared to that for samples capsulated with SiO_2. In addition it is interesting to note that the amount of sublimated As atoms is smaller for samples irradiated front to beam. This effect is more pronounced for the case

1)Central Institute for Nuclear Research, Rossendorf, DDR
2)Institute of High Current Electronics, Tomsk, USSR

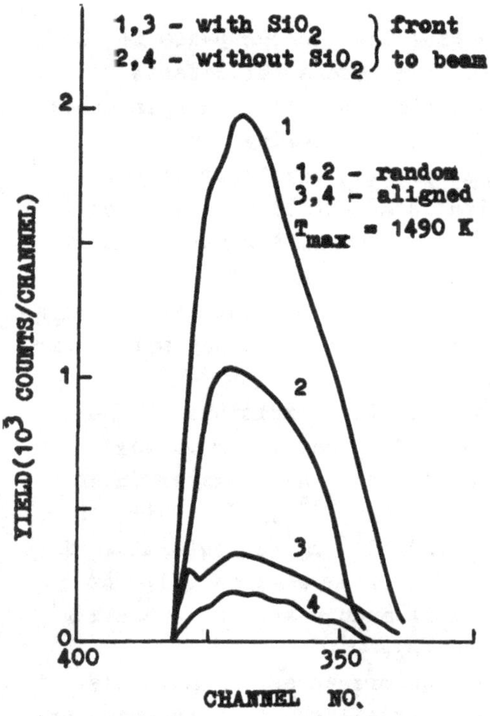

Fig. 1.

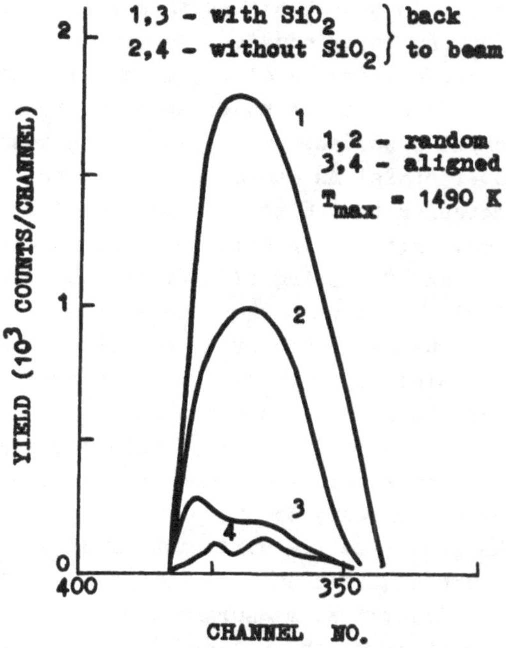

Fig. 2.

Table

Irradiation conditions		N_r 10^{15}, cm^{-2}	N_a 10^{15}, cm^{-2}	f	N_f 10^{15}, cm^{-2}	$N_{subl.}$ 10^{15}, cm^{-2}	N_s 10^{15}, cm^{-2}
with SiO_2	front to beam	9.6	1.9	0.89	8.5	0.4	8.4
	back to beam	8.3	1.0	0.91	7.6	1.7	9.1
without SiO_2	front to beam	4.3	0.8	0.89	3.8	5.7	4.5
	back to beam	4.1	0.5	0.92	3.8	5.9	4.9

N_r and N_a are the random and aligned area density of As atoms, respectively; $f = (1-\chi_{As})/(1-\chi_{min})$ is the fraction of As atoms on lattice sites; $N_f = N_r \cdot f$ is the density of As atoms on lattice sites; $N_{subl.}$ is the sublimated density of As atoms; N_s is the sheet carrier concentration.

of annealing of SiO_2 capsulated samples. This can be explained by the formation on the electron beam irradiated sample's surface of additional film of polimerized oil vapor, which always presents in the residual atmosphere. A mechanism of this film formation was described in (5) for low temperatures (20 to 200 $^\circ$C). In our investigations this effect is observed at higher temperatures also, when according to

(5) the rate of adsorption of molecules from the gaseous phase is too low. It is probable that this polimerized film covers the defects in SiO_2 cap layer appeared due to As implantation resulting thus in about fourfold decrease of the As sublimation at the annealing.

It should be noted that the presence of SiO_2 cap layer on the sample's surface at the annealing results in a twofold increase of interstitial As atoms concentration. Their concentration additionally increases by 1.6 to 1.9 times at irradiation from the implanted layer side. This agrees with our earlier data (4) and shows that the deactivation and following it diffusion of the As atoms is connected with their displacement into the interstices.

The diffusion of As is also dependent on the conditions at the implanted layer surface. At the presence of SiO_2 cap layer on the sample's surface an enhanced diffusion of As into the Si wafer takes place, the diffusion coefficient being $D = 3.6 \times 10^{-12}$ cm^2/s. This D value is within the range of $D = (2.4$ to $12) \times 10^{12}$ cm^2/s determined in the experiments on the photon annealing of similar samples (1). At the same time the diffusion of As into Si is much weaker at annealing of uncapsulated samples ($D \cong 1.3 \times 10^{-12}$ cm^2/s).

Electrical measurement results are also presented in the Table. Comparison of the values N_s and N_f shows that all the As ions which are in the substituting positions are electrically active. The electrical activity in samples annealed under the SiO_2 cap layers, is about 85 to 90 % while that for uncapsulated samples decreases to 45 to 50 % due to the As sublimation.

Thus the comparison of the results obtained at the electron beam annealing of capsulated samples and photon annealing of noncapsulated ones demonstrates similar behaviours of the As ions within the implanted layer in both cases. Note that at photon annealing the samples were not capsulated by SiO_2 previously, but ~10 nm cap layers were formed during the annealing on air, what hampered the As sublimation. There occurs enhanced diffusion of the As atoms into the Si wafers, with the diffusion coefficient being several times greater than its equilibrium value at these temperatures. The As behavior essentially changes at the electron beam annealing of uncapsulated samples. The diffusion of As atoms towards the layer surface in this case is in preference to the diffusion into the depth of the wafer due to sublimation of As atoms. As the result the diffusion into the depth slows down.

References

(1) Narayan J. and O.W.Holland: J. Appl. Phys. 56(10) (1984) 2913-2921
(2) Nylandsted A. et al: Appl. Phys. A33 (1984) 51-58
(3) Kogler R. et al: Nucl. Instr. and Meth.: B19/20 (1987) 516-520
(4) Budishevsky V.S. et al: Physical Research 8 (1988) 262-264
(5) Tkachyuk B.V., V.M.Kolotyrkin: Creation of thin polimer films from the gaseous phase, Khimia, Moscow 1977

INFLUENCE OF IMPLANTED SILICON LAYERS RECRYSTALLIZATION TYPE ON THE MECHANISM OF As DEACTIVATION AT ELECTRON BEAM ANNEALING

R. Grötzschel[1)], V.A. Kagadey[2)], P. Knothe[1)], N.I. Lebedeva[2)], G.E. Ozur[2)], D.I. Proskurovsky[2)]

In many works (see, e.g., (1,2)) it was demonstrated that short pulsed annealing (duration of $\sim 10^{-8} - 10^{-2}$ s) of Si layers implanted with As yield the formation of supersaturated substitutional solution of the impurity. Such annealing is capable of completely activating implanted As concentration in Si at levels up to $(1 \div 2) \times 10^{21}$ cm^{-3}. However further heat treatment, like traditionally in the furnace or by rapid thermal annealing (RTA), results in the decomposition of the substitutional solution and in the deactivation of the impurity. The electrically active concentration of As relaxes to equilibrium value at a given annealing temperature ($\sim 3 \times 10^{20}$ cm^{-3} at 1370 K). Later it was found (3,4) that even after RTA of the implanted layer Si<As> there can occur nonequilibrium concentration of electrically active As provided that certain RTA conditions are realized. This concentration then decreases with the increase of RTA temperature and duration.

The mechanism of As deactivation is widely discussed (1-5). This paper deals with the elucidation of As deactivation peculiarities at thermal treatment after implanted layer recrystallization by submicrosecond or second electron beam annealing.

As^+ ions (6.25×10^{15} cm^{-2}, 60 keV or 5×10^{15} cm^{-2}, 100 keV) were implanted into p-type (100) Si waters. Two modes of annealing were investigated. In the first one the samples were first irradiated by a submicrosecond beam (0.35 μs, $1.25 \div 1.4$ J/cm^2) what resulted in liquid phase epitaxial recrystallization of implanted layer. Then the samples were subjected to additional RTA by a second duration electron beam (15 s, $20 \div 24$ W/cm^2, $T_{max} = 1350 \div 1400$ K). In the second mode the initial samples were subjected to the RTA (5 or 15 s) at once. In this case the solid phase recrystallization of implanted layer took place during the first second, then thermal treatment of the recrystallized layer by the same beam continued.

The samples were investigated using Van der Pauw technique and Rutherford backscattering spectroscopy (RBS) plus channeling.

The results obtained with the samples of Si<As>(6.25×10^{15} cm^{-2}, 60 keV) annealed by a submicrosecond beam are following. The RBS spectra (Fig. 1) demonstrate high quality of the recrystallized layer crystal structure. The same is confirmed by low value of the parameter $\chi_{min} \cong 0.03$. The increase of the beam's energy density from 1.25 to 1.4 J/cm^2 does not practically influence the χ_{min} value. As is seen from the As concentration profile (Fig. 2) ~98% of implanted As dose

1) Central Institute for Nuclear Research, Rossendorf, DDR
2) Institute of High Current Electronics, Tomsk, USSR

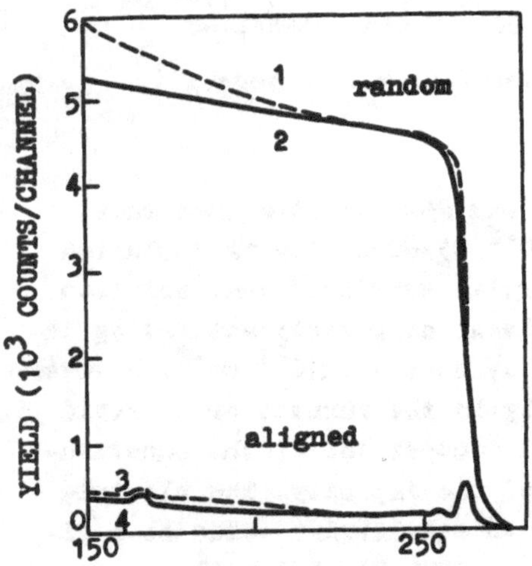

Fig. 1. RBS and channeling spectra of As implanted Si after submicrosecond electron beam annealing. 1,3 - 1.25 J/cm^2; 2,4 - 1.4 J/cm^2.

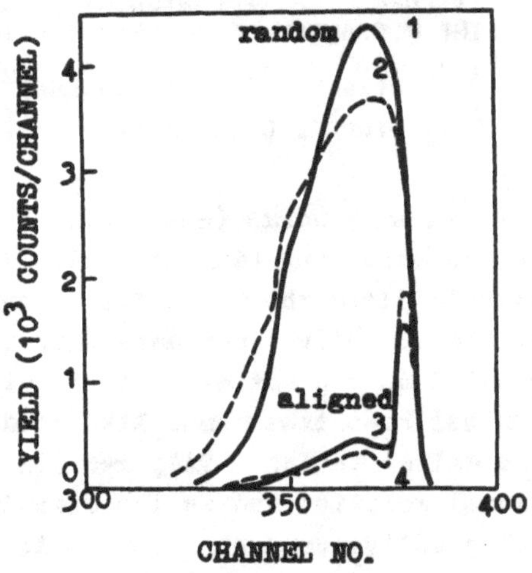

Fig. 2. Arsenic profiles obtained from RBS and channeling spectra after annealing:
1,3 - submicrosecond electron beam (1.4 J/cm^2)
2,4 - the same sample after additional RTA (T_{max} = 1400 K).

is within the sample, that means that no sublimation of As occurs during the submicrosecond annealing. Electrophysical measurements showed almost complete As activation at the annealing in this regime (~95%).

RTA of this samples by an electron beam of 15 seconds duration did not change the quality of crystal structure. According to RBS data the implanted layer contained ~96% of As dose, that means that no significant sublimation of As occurred in this case too. However the electrical activity dropped by ~30%.

If a sample of Si <As> ($5x10^{15}$ cm^{-2}, 100 keV) was only subjected to RTA (in the regime analogous to that in which the sample annealed by submicrosecond pulse was treated) then ~20% of implanted As dose sublimated from the layer as the RBS analysis shows. The electrical activity was ~50% of the implanted As dose. At 5 seconds RTA the electrical activity reached ~80%. Thus, one can see that the impurity deactivation increases at the increase of RTA duration (5).

So it is evident that in both aforesaid modes of thermal treatment deactivation of As^+ ions occurs, but after solid phase recrystallization of the implanted layer As sublimation takes place, while no sublimation is observed in the case of treatment after liquid phase recrystallization. This difference reveals the fact, that two different mechanisms of As deactivation work after the above said types of recrystallization. As to the first mode of irradiation, the results can be explained by the deactivation due to formation of coherent precipitates (2), since no impurity atoms are observed in the interstices. As a result all the As atoms in the implanted layer are "bounded" what hinders their sublimation.

In the second irradiation mode we suggest the following deactivation mechanism (4). The active As^+ ions in excess of the equilibrium active concentration displace from substitutional positions into interstices and through the interstices they abnormally rapidly diffuse to the surface and sublimate.

This difference in deactivation mechanisms can be explained by the fact that significant difference in defectivity of layers formed after liquid and solid phase recrystallizations occurs. In the first case the layer formed after recrystallization is almost entirely free of defects, while in the second case there are more defects in it (6).

References

(1) Lietoila,A., J.F.Gibbons and T.W.Sigmon: Appl. Phys. Lett. 36(9) (1980) 765-768

(2) Popov,V.P., A.V.Dvurechenskii, B.P.Kashnikov and A.I.Popov: Phys. stat. sol. (a) 94 (1986) 569-572

(3) Nylandsted,A., O.Larsen and V.E.Borisenko: Appl. Phys. A33 (1984) 51-58

(4) Budishevsky,V.S., R.Grötzschel, V.A.Kagadey et al.: Physical Research 8 (1988) 262-264

(5) Buzulevitsch,Yu.Yu., R.Grötzschel, V.A.Kagadey et al.: Fiz. Tekhn. Poluprov. 22(12) (1988) P-4827/88

(6) Kagadey,V.A., N.I.Lebedeva, D.I.Proskurovskii and R.Grötzschel: Physical Research (EPM 89), to be published

ANISOTROPIC LOCAL MELTING OF IMPLANTED SILICON BY 0.05 - I5 s DURATION INCOHERENT LIGHT PULSES

Ya.V.Fattachov, I.B.Khaibullin[I)]

There is currently great interest in investigation of important physical effect - the anisotropic local melting of semiconductors, subjected to irradiation by powerful light pulses (see i.g. I,2). This interest is due to the following. Firstly, studing the mechanism and features of local melting allows to obtain valuable physical information about semiconductor properties and processes, occuring in the sample during and after strong light pulse irradiation. Secondly, these studies are directly connected with an important applied problem - regime optimization of pulse light annealing (PLA) of implanted layers (IL), as well as crystallization of semiconductor-on-insulator (SOI). That is why in this work, for the first time, systematic experiments investigating the influence of semiconductor substrate type, power density E_o and light pulse duration τ_p , as well as dose and energy of implantation and the sort of an ion on density, size and shape of recrystallized local molten regions (LMR) on the silicon surface were carried out.

E X P E R I M E N T

(IOO) monocrystalline silicon wafers with a diameter of 76 or IOO mm, of n- and p-type conductivity, 4.5 to IO Ohm·cm specific resistivity were utilized. The ions of III, IV and V group elements B^+, In^+, Si^+, C^+, Ge^+, P^+ with energies from 20 to 80 keV and doses to $3.I \times IO^{I6} cm^{-2}$ were implanted. Light pulse irradiation in regime of local melting was performed, varing τ_p from 0.03 to 20 s and E_o from 40 to 2000 $W \cdot cm^{-2}$ (2,3). The residual defects in annealed IL were investigated by TEM. Surface orientation of silicon wafers and crystallographic axes were determined by X-rays diffractometry.

R E S U L T S and D I S C U S S I O N

The investigation of the dependence of density, size and shape of LMR on τ_p demonstrated that an increase in τ_p (with decreasing of E_o (2,3)) led to a decrease in density (quantity per cm^2) by 2 orders, their sizes increasing not monotonic. When τ_p rises in value the LMR shape is changing from dendritic to near regular geometric, corresponding to the virgin crystal of silicon (fig.I). The aim of our work was the determination of the mechanism of the features observed.

To evaluate the degree of IL recrystallization to the moment of local melting start time-dependences of sample temperature during and after light pulse irradiation with different set of E_o and τ_p were calculated, and compared with the results on pulse-annealed IL structure

I) Kazan Physical-Technical Institute of the Academy of Sciences of the USSR, Sibirskii Trakt, SU - 420029 Kazan, USSR.

investigated by TEM-method. The PLA temperature for all τ_p was chosen to be close to the temperature of local melting T_{LM} as much as possible. It is established by TEM that the increase of τ_p in the interval from IOO ms to IO s (and corresponding decrease of E_0) results in a threefold decrease of dislocation density. For $\tau_p \lesssim$ I s the duration of back front of thermal pulse, determined by the mechanism of cooling - heat radiation, is essentially longer than τ_p . For such τ_p the main contribution in recrystallization process will be due to the back front of thermal pulse. As a result the structural differences detected by TEM are more pronounced at the end of a light pulse, i.e. at the end of the fore-front of a thermal pulse. Thus, LMR density reduction for τ_p varing from O.I to IO s is caused by a decrease of IL defects when the temperature T_{LM} is achieved at longer durations τ_p and lower E_0.

Studing the effect of implantation regimes we observed an increase of LMR density with ion dose (fig.2 a). We attribute this to the quantity increase of the residual defects in IL, when annealing occurs at the temperatures up to T_{LM}. The dependences of LMR density on the tetrahedral covalent radius of implanted elements are analysed (fig. 2b). The increase of LMR density in the interval of r from 0.77 to I.I7 A^o is explained, in particular, by the increase of ion mass and, hence, by the degree of disorder and, finally, of the residual defects. Sharp diminishing of N for Ge and In is likely connected either with the formation of less defective layer after solid phase epitaxial crystallization of IL amorphized by heavy ions, either with the formation in these layers of the specific residual defects less effective in forming liguid nuclei.

For the first time we found out that the dendritic LMR on implanted (IOO) silicon surface is observed only after doping by the ions with the covalent radius essentially smaller than that of silicon : B^+, C^+ (fig. Ic,d; 2c). We explain our results by thermoplastic effects during pulse irradiation i.e. by the mechanisms of primary and conjugate slip (4). There are two possible ways to create relative excess of light induced mechanic stress over elastic threshold. The first one implies stress rising generated by light pulses (at the expense of τ_p reduction and corresponding E_0 increase) when the mechanical properties of the sample remain unchanged. The second way is to lower by some means the elastic threshold of the material. We accomplished both methods. In particular, implantation of B^+ and C^+ creates tensile stresses in the lattice, lowering the elastic threshold and leading to realization of mechanism of conjugate slip at the same stresses, induced by light, that finally will be resulted in dendritic LMR shape. When another ions are doped either tensile stresses lower than the threshold or, on the contrary, compressive stresses appear. In this case thermoplastic effects described do not take place.

Thus the observed feature of local melting of implanted silicon irradiated by light pulses in duration interval 10^{-1} - IO s is explained by the appearence (and interaction) of recrystallization processes and thermoplastic effects (induced by light or stipulated by implantation).

Whe would like to thank T.N.Vasiljeva, E.M.Misyuryov, M.L.Blatt, V.A.Shustov for there essential help.

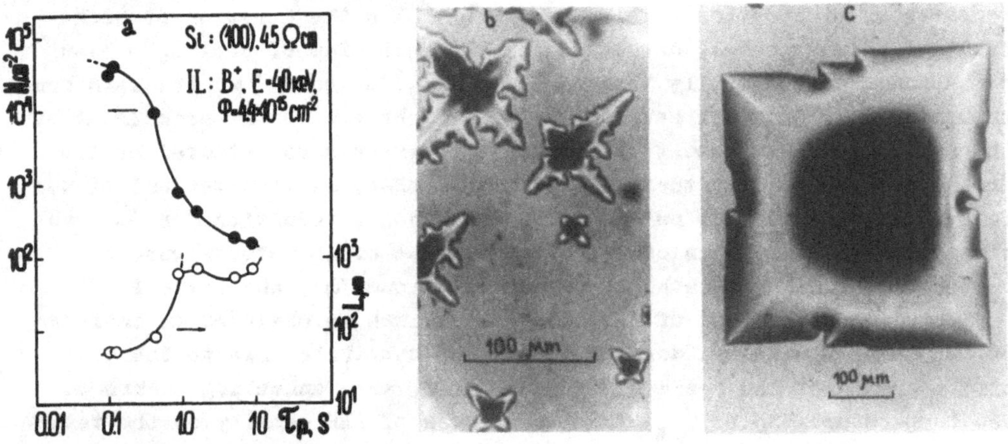

Fig.I. The dependence of density N and maximal size L of LMR on τ_p at different E_o (a) and the shape of LMR for τ_p = 0.42s, E_o = 400 W·cm^{-2} (b), τ_p = 5.01s, E_o = 50 W·cm^{-2} (c).

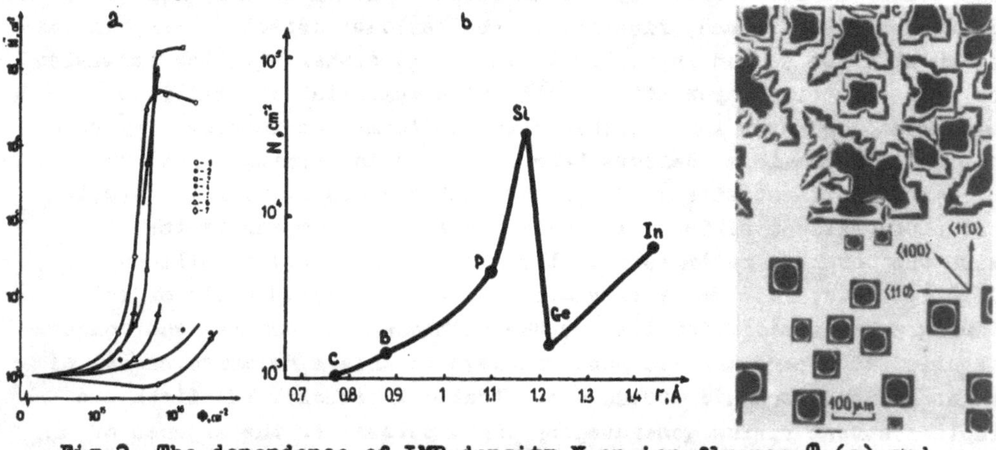

Fig.2. The dependence of LMR density N on ion fluence Φ (a) and tetrahedral covalent radius r (b) of implanted elements; micrograph of the silicon surface : top part implanted by 1.3x10^{16} B$^+$/cm^2 bottom - virgin, τ_p = 0.60 s (c). Ions: 1 - Si$^+$, 2 - P^{++}, 3 - P$^+$, 4 - B$^+$, 5 - In$^+$, 6 - Ge$^+$, 7 - C$^+$ (a). Energy: 1,3,4,6,7 - 40 keV, 2-80keV, 5-30keV(a)

R E F E R E N C E S

(1) Heinig K.-H. - Conf. on Energy Pulse Modif. of Semicond. and Related Materials, Dresden, GDR (EPM'84), 1984, p.265-286.
(2) Fattachov Ya.V., I.B.Khaibullin, R.M.Bayazitov, E.M.Misyuryov. Phys. Stat.Sol. (a). 102 (1989), p699-702.
(3) Fattachov Ya.V., R.M.Bayazitov, I.B.Khaibullin, E.M.Misyuryov, M.A.Mchitarov. Physical Research, v.8. EPM'87. Berlin : Akademie-Verlag, 1988, p. 259-261.
(4) Bentini G., L.Correra, C.Donolato. J.Appl.Phys., 56, (1984), p.2922-2929.

ANNEALING OF THIN ARSENIC IMPLANTED SILICON LAYERS WITH FLASH GALOGEN LAMP AND RF GAS DISCHARGE

A.N.Nazarov[1)],V.S.Lysenko[1)],S.A.Valiev[1)],M.M.Lokshin[1)]
A.S.Tkachenko[1)],I.E.Kunitskii[1)]

The influence of flash lamp annealing (FLA), RF plasma annealing (RFA), traditional thermal annealing (TA) and combination of these treatments on the arsenic implanted SiO_2-Si structures was investigated. The density and the energy spectra of interface electron states (IES), the density and the energy spectra of deep levels in subsurface silicon layer and the free carrier profiles in silicon were studied.

Samples and experimental technique

N-type silicon samples (N_P=$2x10^{15}cm^{-3}$) with 50-nm thermal oxide layer were used. Arsenic was implanted through the oxide layer with the energy of 100 KeV and dose of $1x10^{12}$ions/cm^2. During the implantation process a part of each sample was covered with a screen for control measurments.

Implanted and control st[illegible]ures were treated with galogen flash lamp with the treatment duration ranging from 0,5 to 2,5s. The sample temperature was measured by photodiode; the sample temperature corresponed to 1300°C when a treatment time was 2,5s. The some part of samples was annealed thermally at the temperature range from 450 to 1000°C in the nitrogen atmosphere for 15 minutes. After that Al-SiO_2-Si structures were fabricated.

Al-SiO_2-Si structures, that were thermally and flash lamp annealed, were treated in RF gas discharge with power $2,0x10^3$W for 15 minutes in nitrogen. During this process the additional sample heating to 200°C was used. Other part of structures was annealed additionally at 450°C for 15 minutes in nitrogen (standard postmetallization annealing (PMA)). Besides that part of nontreated Al-SiO_2-Si structuures was annealed in RF gas discharge. RFA power was varied from $6,25x10^2$ to $2,5x10^3$W.

The electrophysical properties of such structures were investigated by high frequency C-V method, C-V free carrier profiling method and thermally stimulated charge release (TSCR) method.

Experimental results and discussion

FLA of the implanted SiO_2-Si systems results in the noticeable decrease of TSCR current peaks. It brings in the evidence of the large decrease of deep level density generated by the ion implantation in subsurface silicon layer. However after such treatments the TSCR

[1)]Institute of Semiconductors, Academy of Scienses of the Ukrainian SSR, Kiev, Prospect nauki 45, USSR spectra are spreading. The control experiments for dividing the IES on the SiO_2-Si interface and the deep levels in subsurface silicon layer show that the spectra spreading was connected with the creation of large IES density after FLA.

Really the measuring of TSCR spectra on the nonimplanted structures shows the initial decrease and then the greate increase of IES density in the upper half of the band gap. The integral IES density from E_c-0,15 to E_c-0,50eV energy gap, calculated from the area under the TSCR spectra, is demonstrated in figure 1. The differential field cleaning of the TSCR spectra [1] allowed to determine the energy density distribution and the energy IES cross-section distribution in the silicon band gap after FLA. It was found that the density of IES had a maximum in the energy range from E_c-0,30 to E_c-0,40eV, and IES capture cross section depended from the energy of IES (E_t) as $S_n(E_t)=3x10^{-18}exp[21(E_c-E_t)]$ [cm^2]. Such energy spectra of IES density were also obtained after the thermal quench and the UV irradiation [3] of Al-SiO_2-Si structures. It was connected with the creation of Si- and SiO- dangling bonds. After TA the noticeable IES density was remained however the density maximum in the studied range of the silicon band gap was not observed.

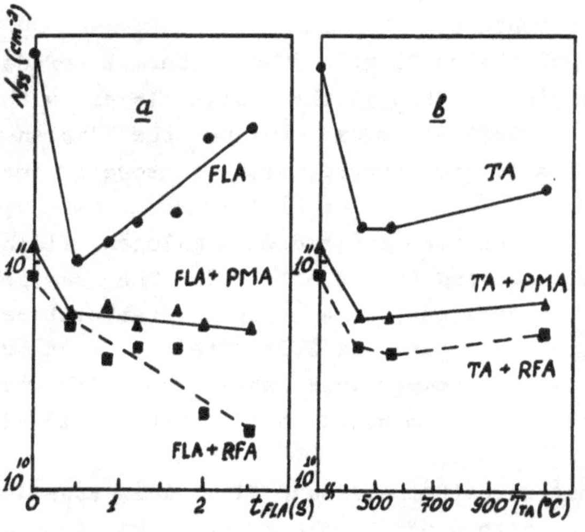

Fig.1

After the standard PMA density of IES, created by FLA or TA was decreased almost to one order of magnitude. But RFA has the more annealing effectiveness of IES in comparison with PMA. We must note that the minimal IES density (about $1,6x10^{10}cm^{-2}$) was achieved after the combination of FLA and RFA (fig.1).

Similar results of deep level annealing were obtained on the implanted structures. Fig.2 shows the total density of IES and deep levels in subsurface silicon layer of the implanted structures in the upper side of forbidden gap in dependence of annealing conditions. The RFA alone did not result in the complete annihilation of studied deep levels and activation of implanted arsenic. Residual total deep level density was $3x10^{11}cm^{-2}$ and the surface free carrier concentration, determinated from C-V profiling method - about $2x10^{11}cm^{-2}$, i.e. 20% of the total implanted impurity concentration. The result is much worse in comparison with the RFA of the phosphorus implanted structures [2]. In that case the residual deep level concentration was about $2x10^{10}cm^{-2}$ and the free carrier concentration reached 60% of the implanted impurity amount. This result is probably connected with the large defect concentration and more complex defect structure in the subsurface silicon layer after arsenic implantation in comparison with phosphorus ion implantation.

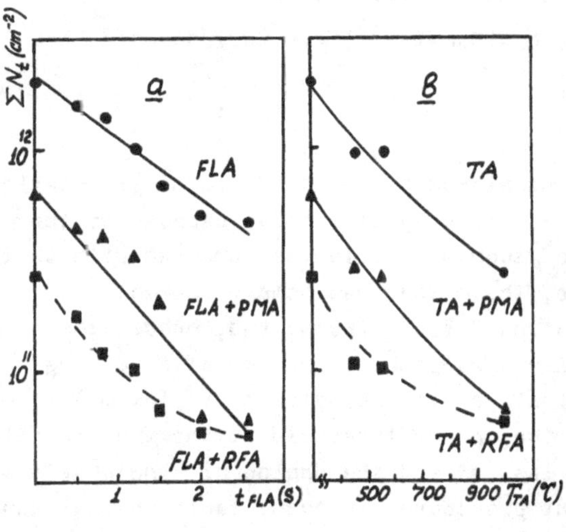

Fig.2

The maximal implanted impurity activation without noticeable spreading of the free carrier concentration profiles was obtained using such treatments as FLA and RFA or TA and RFA (fig.3).

Thus it is shown that the using of combined treatments with RFA as final stage results in minimal concentration of IES, minimal residual concentration of deep levels induced by ion implantation and in maximal activation of implanted impurity. This effect may be due to preliminary thermal or pulsed photon annealing of radiation defects, compensating electrically active impurity, which can not be annealed by low-temperature RFA. These preliminary annealings can also cause the diffusion of implanted impurity from the subsurface layer to SiO_2-Si interface, where the vacancy defect concentration is maximal. During the final RFA residual deep level concentration is being greatly reduced and interstitial impurity atoms are being built in the sites of Si crystal lattice more effectively in comparison with PMA [2].

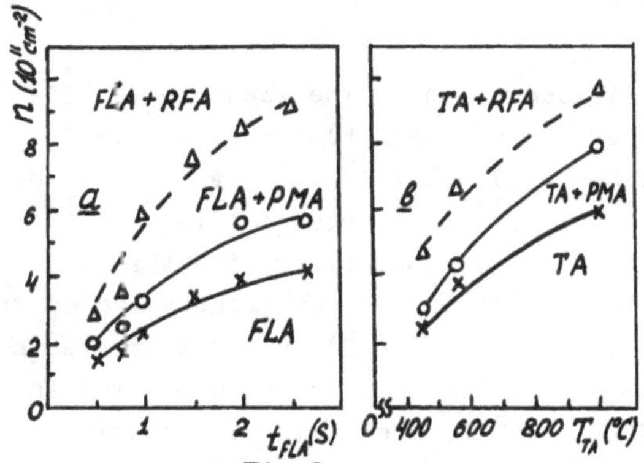

Fig.3

REFERENCES

(1)Lushnikov N.A.,A.G.Zhdan and A.A.Alexandrov:FTP.12(1978)801-803

(2)Lysenko V.S.,A.N.Nazarov,S.A.Valiev and et.al.:Phys.Stat.Sol.(a). 113(1989)807-815

(3)Sah C.T.,J.Y.-C.Sun and J.J.-T.Tzou:J.Appl.Phys.54(1983)2547-2554

RAPID THERMAL ANNEALING OF PROTON TRASMUTATION DOPED SILICON

F.P.Korshunov[1)], A.A.Pavienko[2)], N.A.Sobolev[1)], A.V.Voevodova[1)]

1. INTRODUCTION

The doping of silicon with aluminium by means of proton irradiation in the energy range 20 - 40 MeV /3/ is very attractive because of the much higher doping homogeneity of such a material as compared with that dope during the growth. However, the method has some disadvantages, especially (i) creation of radiation defects (RD's) and, hence, the necessity of their annealing, (ii) simultaneous transmutation doping with Mg donors, and (iii) relatively small doping depth (a few mm) limited by the proton range and the energy threshold of the transmutation reactions (see /3/). Due to the last reason the doping of single wafers appears to be the most convenient procedure. In turn, rapid thermal annealing (RTA) is favourable for the annealing of RD's in the wafers. It is noteworthy that RTA has proved to be a very effective tool for the annealing of neutron transmutation doped silicon wafers /5/.

2. EXPERIMENTAL

In the present work n-type silicon grown by the float zone (FZ) technique with the concentration of phosphorus N(P) = $1.5.10^{13}$ cm^{-3}, oxygen N(O) = 4.10^{16} cm^{-3} and carbon N(C) = 2.10^{16} cm^{-3}, as well as by Czochralski (CZ) method with N(P) = 1.10^{14} cm^{-3}, N(O) = 1.10^{18} cm^{-3}, N(C) = 1.10^{16} cm^{-3} was used. Wafers of these materials with the thickness of 0.5-1 mm were subjected to 25 - 35 MeV proton irradiation up to the doses $1.6.10^{16}$ cm^{-2} (FZ) and $1.1.10^{16}$ cm^{-2} (CZ). RTA was carried out by means of 10 sec long radiation pulses of tungsten halogen lamps in the induced temperature range 500 - 1200 ^oC. Conventional furnace annealing (FA) for 15 min in the temperature range 50 - 1200 ^oC was also made. The photoluminescence (PL) technique at 4.2 K in the near band edge spectral region (1.17 - 1.0 eV) and Hall effect measurements were used for the control of RD's and doping impurities.

3. RESULTS AND DISCUSSION

At various FA and RTA stages of proton irradiated FZ Si a great number.

---

1) Institute of Solid State and Semiconductor Physics, 220726 Minsk, ul. P. Brovki, 17, USSR

2) Institute for Nuclear Research, 252028 Kiev, pr. Nauki, USSR

of centers appear in the PL spectra typical of neutron irradiated FZ Si (see / 6, 7/). The emergence of such PL lines as W, X / 8/ , M36 / 6, 7/ e. a. proves the cascade character of the Si lattice gamage by protons. The same lines are observed in the PL spectra taken in the course of both FA and RTA, though the RTA stages are shifted by 100 - 150°C to higher temperatures as compared to FA. The defect related PL aneals out in the course of FA at $T_a \gtrsim 700°C$ and of RTA at $T_a \gtrsim 850°C$. At higher anneal temperatures only the lines of free excitons as well as those of excitons and multiexciton complexes bound to substitutional P and Al atoms are observed. No line has been found in the PL spectra which could be assigned to the Mg donors. The dependence of the carrier concentration versus FA temperature reveals a peak at $T_a = 600°C$ caused by the generation of so called "600°C acceptors" / 4 / . At $T_a \gtrsim 750°C$ the hole concentration approaches the value of $p = 4.10^{16}$ cm^{-3} and then does not change up to T_a = 1200°C.

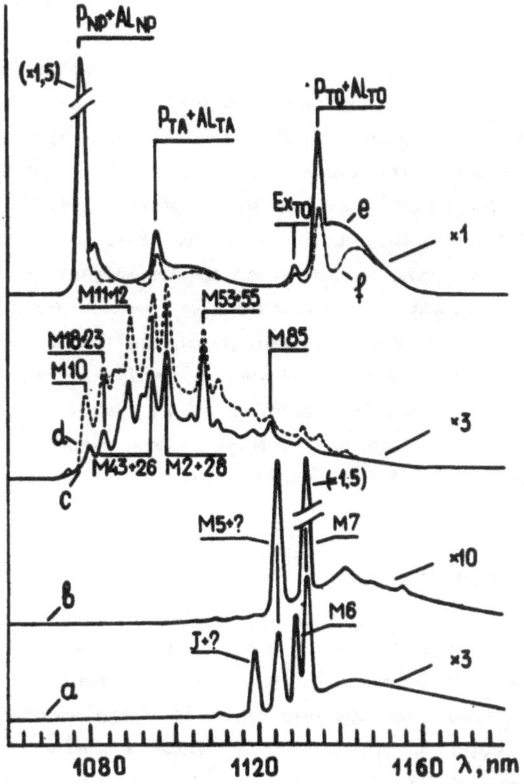

Fig. 1. Survey PL spectra (spectral resolution 1.3 nm) of proton irradiated CZ Si taken at various annealing stages. Annealing temperatures, °C: 400 (a), 450 (b), 600 (c), 720 (d), 800 (e), 900 (f). Annealing modes: FA (a - c, e, f) and RTA (d). Ex - free excitons P and Al - excitons bound to substitutional P and Al atoms. Indexes NP, TA and TO denotes no-phonon transitions and their TA- and TO-phonon replica.

The behaviour of the carrier concentration in the course of RTA of FZ Si differs from that in the course of FA in two aspects: the suppression of the 600°C-acceptors generation (cf. / 5 /) and a shift of the annealing stages by 100 - 150°C to higher temperatures. At $T_a \gtrsim 850°C$ the hole concentration equals to ca. 4.10^{16} cm^{-3} as well. With regard for the relation N(Mg)/N(Al) = 0.4 obtained in / 3 / and for the value N(P) = $1.5.10^{13}$ cm^{-3} we get the aluminium doping efficiency $N(Al)/\Phi = 4.10^{-2} cm^{-1}$ in accordance with our luminescence data and with the previous results / 3 / .

In the PL spectra of CZ Si at various anneal stages many lines observed in FZ Si as well as new ones related to the oxygen containing defect centers M23, 53, 55, 61 72, 85 (see / 7 /) are found. The PL spectra taken at various anneal stages are shown in Fig. 1. Several lines are also observed which have not been identified in the previous studies of neutron irradiated FZ

Si and of electron irradiated FZ and CZ Si / 6, 7 / . In Fig. 1 they are indicated by "?" (they are separated from the known lines at a higher spectral resolution of the PL apparatus). It can be assumed that these lines are caused by complex oxygen related defects of the cascade type. As can be seen from Fig. 1 the PL spectra for FA and RTA are very similar, but the annealing stages are shifted in the case of RTA by 100 – 150 $^{\circ}$C just as for FZ Si. At temperatures of FA $\gtrsim$ 800 $^{\circ}$C and at those of RTA $\gtrsim$ 900 $^{\circ}$C the narrow PL lines caused by the RD's anneal out and only the lines of free excitons and of excitons and multi-exciton complexes bound to the substitutional Al and P atoms are observed. The Hall effect data show that an intense generation of oxygen donors occurs in the course of FA of proton irradiated CZ Si in the range 700 $^{\circ}$C < T_a < 900 $^{\circ}$C. These donors have been observed earlier in neutron irradiated CZ Si / 1, 2 / . They are not fully annealed up to $T_a \not\gtrsim$ 1000 $^{\circ}$C. The creation of donors is also observed in the course of RTA in the temperature range $T_a \cong$ 1000 – 1150 $^{\circ}$C. RTA at T_a = 1200 $^{\circ}$C removes most of the donors.

4. CONCLUSIONS

It is shown by means of PL and Hall effect measurements that RTA for several seconds of the proton transmutation doped silicon allows to remove the radiation defects and to achieve the electrical activation of the transmuted Al acceptors. The necessary RTA temperatures are higher in the case of CZ Si than of FZ Si because of the generation of oxygen-related donors in the former material.

REFERENCES

(1) Babitsky,Yu.M, P.M.Grinshtein, M.A.Il'in: Fiz. tekhn. poluprov. 19 (1985) 2070–2072

(2) Cleland,J.W., P.H.Fleming, R.D.Westbrook e.a.: Neutron Transmutation Doping in Semicond. (Plenum, New York and London, 1979) 261-279

(3) Gaidar,G.P., N.N.Dmitrenko, L.V.Dubar e.a.: Fiz. tekhn. poluprov. 20 (1986) 960–962

(4) Glairon,P.J., T.M.Meese: Neutron Transmutation doping in Semicond. (Plenum, New York and London, 1979) 291–305

(5) Korshunov,F.P., N.A.Sobolev, V.A.Sheraukhov e.a.: Fiz. tekhn. poluprov. 21(1987) 1968–1973

(6) Korshunov,F.P., N.A.Sobolev, V.A.Sheraukhov: Zh. prikl. spectroscopii (Sov. J. Appl. Spectroscopy) 51 (1989) 248–256

(7) Korshunov,F.P., N.A.Sobolev, V.A.Sheraukhov: Proc. Int. Conf. Sci. and Technol. Defect Control in Semicond. (Yokohama, 1989) Ed. K.Sumino, Elsevier, Amsterdam (in the press)

(8) Tkachev,V.D., A.V.Mudryi: Inst. Phys. Conf. Ser. No 31 (Bristol and London, 1977) 231–243

DEFECT PRODUCTION AND ANNEALING IN 11,5 MeV ELECTRON- AND ^{60}Co-γ-IRRADIATED NTD-SILICON

E. Koch[1)], H. Lemke[2)], A. Müller[3)], L. Prager[3)]

INTRODUCTION

Radiation-induced defects in silicon represent efficient recombination centers which can be generated within a large range of concentrations. In order to modify the minority carrier lifetime energy levels of these centers should be located 0,25 - 0,40 eV away from the band edges to minimize the temperature dependence of the reverse current and the injection dependence of the recombination efficiency. The centers should be as temperature-stable as possible.
In this paper we investigated defects generated by 11,5 MeV-electrons or ^{60}Co-γ-quantas. P-n-structures of FZ-NTD-Si treated by high temperature processes under O_2-atmosphere were irradiated using a

- LINAC of type "Electronika" at ZfI Leipzig
- ^{60}Co-Panorama-source at ZfI Leipzig.

The minority carrier lifetime of these devices were measured by LEDERHANDLER-method. Radiation-generated deep levels were observed using the deep level transient spectroscopy technique (DLTS).

RESULTS

A: Defect spectra: The recombination was dominated by

- A-centers consisting of vacancy and oxygen in ^{60}Co-irradiated devices
- divacancies VV in addition to A-centers in 11,5 MeV-electron-irradiated devices:

		recombination efficiency/s^{-1}	N/cm^3
VV	E_c-0,44 eV	2,1 x 10(5)	7 x 10(10)
A	E_c-0,20 eV	1,5 x 10(4)	3 x 10(11)
E_{p1}	E_v+0,35 eV	2,0 x 10(3)	6 x 10(11)

Ratio of concentrations N_{VV}/N_A lightly affected by the dose rate.

1) VEB Mikroelektronik "K. Liebknecht" Stahnsdorf
Ruhlsdorfer Weg, Stahnsdorf, DDR-1533, GDR
2) AdW der DDR, Zentralinstitut für Elektronenphysik
Hausvogteiplatz 5/7, Berlin, DDR-1086, GDR
3) AdW der DDR, Zentralinstitut für Isotopen- und Strahlenforschung,
Permoserstr. 15, Leipzig, DDR-7050, GDR

B: The defect efficiency C in $1/\tau_{irr} - 1/\tau_o$ = CxD which also containing the defect generation rate is weakly dependent on the dose rate:

e - dose rate [Gy/ns]	10^2 C x kGy x µs
1,28	1,25
0,04	1,21

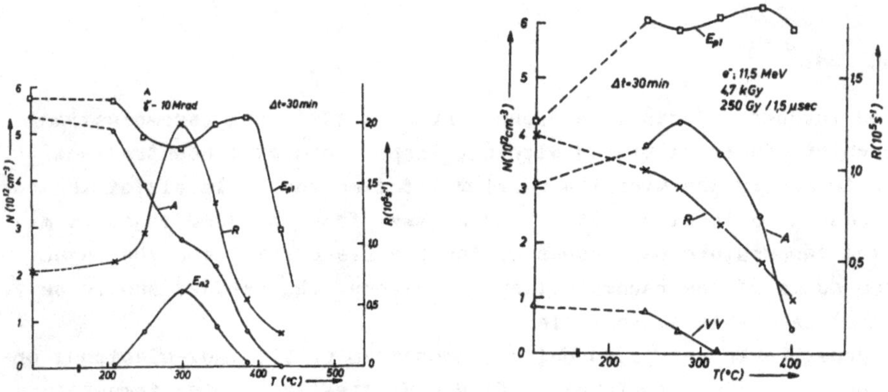

a) b)

fig. 1: Dependence of the defect concentration N and the recombination efficiency R on the isochronal annealing
(a): low dose rate, ^{60}Co-irradiated; E_{n2}: (E_c - 0,37 eV),
(b): high dose rate, electron-irradiated.

C: The annealing behaviour (fig. 1) of the deep levels is very different Time constants for isothermal annealing at 395 °C amount 10 s for divacancies and 10 - 15 min for A-centers, respectively.

The main annealing step depends on the electron dose. A-centers formed by 11,5 MeV electron irradiation (dose rate 3 x 10^8/cm^2 x 20 ms) and by ^{60}Co-γ-quantas are transformed to a new complex at (E_c-0,37 eV) at 250 - 280 °C (fig. 1a). The recombination efficiency of this level is extremely large. A second annealing step of the A-center is found at 330 - 380 °C depending on the dose rate. The complex at (E_c-0,37 eV) is not observed in case of annealed silicon irradiated with electron pulse dose rate of 3 x 10^{10}/cm^2 x 6 ns - 10^{12}/cm^2 x 1,5 µs (fig. 1b).

DISCUSSION

A: Vacancy-interstitial-pairs (FRENKEL-defects) in silicon can be generated by irradiation with electrons of higher energy than 250 keV. At low electron energy (for example 1 MeV generated as COMPTON-electrons at ^{60}Co-irradiation) isolated defects will be induced at first. High-energy electron irradiation (here 11,5 MeV) can lead to multiple collisions producing divacancies.

B: Defect production rate and the ratio of concentrations N_{VV}/N_A will be increased due to growing electron energy. On the one hand, electron energy is affecting single excitation and on the other hand at high dose rate interactions of excitations should be assumed. Overlapping is determined by a following interaction in the diffusion sphere of primary defect. According to the results of KERKOW et. al. /1/ and BOLOTOV et. al. /2/ the diffusion volume of the vacancy is $2{,}4 \times 10^{-15}$ cm^3 and $1{,}4 \times 10^{-11}$ cm^3, respectively. Therefore under the given conditions ($N_A \approx 10^{12}$ cm^{-3}) the A-centers dominantly are produced without remarkable correlation of excitations.

Effects of energy deposited in the silicon lattice on the following excitation could not be detected

C: Variation of the electron pulse dose rate can lead to the disappearence of the level at (E_c-0,37 eV). At high dose rate the A-center annealed at temperatures at which the (E_c-0,37 eV) - level is lacking. It is proposed that the production of defects and their annealing properties are dependent on the charge state of the metastable vacancy-interstitial-pairs and on the A-center. Electron irradiation produces non-equilibrium charge carriers $\Delta n = \Delta p$ in our devices with a lifetime of 40 µs exceeding the pulse duration. These excess carriers shift the quasi-FERMI-levels E_{fn} and E_{fp} according to

$$E_{fn} - E_{fp} = kT \ln \frac{\Delta n \, \Delta p}{n_i^2} \qquad (1)$$

We obtained the generation rate g at which E_{fn} attained the energy level F of the FRENKEL-pairs at (E_c-0,07 eV) and the A-center at (E_c-0,20 eV):

	$E_{fn} - E_{fp}$/eV	$\Delta n = \Delta p/cm^{-3}$	g/cm^3s
F	0,95	2,3 x 10(18)	5,7 x 10(22)
A	0,69	1,5 x 10(16)	3,8 x 10(20)

As an example, the generation rate at pulse excitation of $1{,}8 \times 10^{11}$ electrons/cm^2 x 15 ns will result in $g = 3{,}7 \times 10^{23}$ $cm^{-3}s^{-1}$. Therefore in this case the charge state of primary defects is negative. Otherwise the shift of E_{fn} is low at electron irradiation of low dose rate (for example also at ^{60}Co-γ-irradiation). Therefore we concluded that dose rate effects are caused by excess carrier dynamics.

REFERENCES

/1/ Holldack, K. and H. Kerkow: phys. stat. sol. (a) 98 (1986) 527 - 534
/2/ Bolotov, V. V., A. V. Karpov and V. A. Stuchinskij: Fiz. Tekh. Poluprovodn. 22 (1) (1988) 49 - 55

INVESTIGATION OF THE NEW PHASE FORMATION IN NiSi-Si SYSTEM

Gorbachenko S.G., Zenkevich A.V., Khabelashvili I.D., Kulikauskas V.S., Nevolin V.N.*

An attempt to investigate the initial stage of the solid-phase reaction of silicide formation was undertaken in this work. For this purpose a model, which allows one to connect the nucleation process with the concentration profiles has been developed.

Let us consider solid-phase reaction A + B ⟶ AB, where A is a film deposited on a substrate B. The heating of the initial structure initiates the formation of an AB phase on the potential nucleation centres at the film-substrate interface. The nuclei appear at different moments of time, which results in their nonuniform distribution in size.

The probability of nucleation, W, and the linear growth rate, $v_0 = \frac{dr}{dt}$, are defined as:

$$W = \frac{D_0}{d^2} \exp\left[-\frac{1}{kT}\left(Q + A\right)\right] \quad (1)$$

$$v_0 = \alpha_s d_0 \nu_0 \exp\left(-\frac{U}{kT}\right)\left(1 - \exp\left(-\frac{\Delta\mu}{kT}\right)\right) \quad (2)$$

where D_0 is the diffusion constant for the preferential diffusant of the reaction, d_0 is the lattice constant of the product of the reaction, A is the nucleation work, ν_0 is the atomic frequency in the crystal, U- is the activation energy of the phase transition, $\Delta\mu$ is the change in the thermodynamical potential in the transition /1/. The nuclei growth is followed by the decrease of the A-B interface and the absorption of the potential nucleation centres. At a certain moment of time the nuclei begin to overlap. When the whole A-B interface is covered with the AB nuclei, the A-AB (B-AB) interface will be rough due to the nonuniform distribution of the nuclei in size. The degree of roughness is determined by the relationship between the number of potential centres, the probability of nucleation and the linear nucleus growth rate. Using the exponential law of nucleation and considering the nucleus form to be spherical, one can calculate the distribution of nuclei in size and the shape of the boundary. The rate of reaction at the initial stage, taking into consideration that the nuclei overlap, is:

$$W_r = -\frac{d\alpha}{dt}, \qquad \alpha = 1 - \frac{m}{m_0} \quad (3)$$

---

* Moscow Engineering Physics Institute, 115409, Moscow, USSR.

$$\ln(1-\alpha) = \frac{8\pi\rho N_0 v_0}{W^3 m_0}\left[e^{-wt} + wt - 1 - \frac{(wt)^2}{2} + \frac{(wt)^3}{6}\right] \qquad (4)$$

where ρ is the density of the product, N_0 is the number of potential nucleation centres, m_0 is the initial mass of the film /2/.

At the next stage the growth of the formed AB layer is limited by diffusion of the components of reaction through the grains of the new phase. As the growth rate depends on the diffusion path, the later the grains are nucleated, the faster they will grow. This results in the gradual flattening of the initial boundary of the layer.

For the experimental verification of this model the NiSi-Si system was used. The choice of the system is explained by extremely low difference in the heats of NiSi and $NiSi_2$ formation ($\Delta H = 0{,}03 \frac{\text{kkal}}{\text{at.uet}}$), which results in a rather low probability of $NiSi_2$ nucleation on the NiSi-Si interface, 300 um-thick Si(100) wafers 1 cm^2 in area were used in all the experiments. A 2000 Å-thick NiSi layer was on the surface. The probability of nucleation, W, and the growth rate, v_0 , as a function of T are plotted in Fig.1. It is clear that the most nonuniform distribution of grains in size one can expect at reaction temperatures above T' ≈ 950°C. Using the distribution of grains in size one can calculate the concentration profiles for Ni and Si and compare with the measured RBS ones. In the experiments the samples were subjected to rapid thermal annealing (RTA). Therefore, in all the calculations the heating rate was taken into consideration.

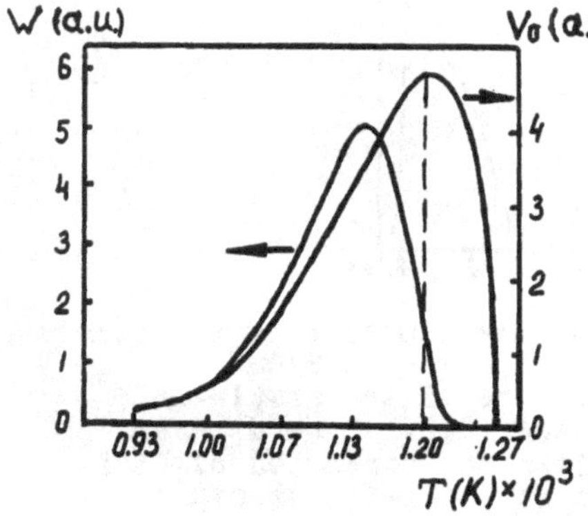

Fig.1. The nucleation probability W and the nuclei growth rate v_0 as a function of temperature (K)

The calculated time dependences of the sample temperature and the reaction rate for different irradiation power densities are plotted in Fig.2.1. The RBS spectra for the samples irradiated with the same power densities are presented in Fig.2.3.

During the electron-beam irradiation of the prepared structures at a power density $P = 20 \frac{W}{cm^2}$ (Fig.2.1a), the temperature does not achieve the characteristic value T' (T' ≈ 950°C for the system under consideration). Hence, the $NiSi_2$-Si interface will be flat to the termination of the reaction (Fig.2.3a). While irradiating the samples with $P = 26 \frac{W}{cm^2}$ (Fig.2.1b), the formation of the $NiSi_2$ layer on the interface will be completed at temperatures higher than T', which does result in the nonuniform distribution of the grains in size. This, in turn, will lead to the lack of the smooth interface between $NiSi_2$ and the Si substrate after the termination of the reaction. The RBS spectra in Fig. 2.3b confirm this assumption. One of the samples was irradiated at the maximum possible power density (excluding melting) $P = 29\ W/cm^2$. When T was about T_{melt} the sample was removed from the heating zone. The most nonuniform distribution in size and the time not enough for reaction to be completed were expected. The strongly "eroded" interface and the presence of two phases, observed in the RBS spectrum (Fig.2.3c) confirm these hypotheses.

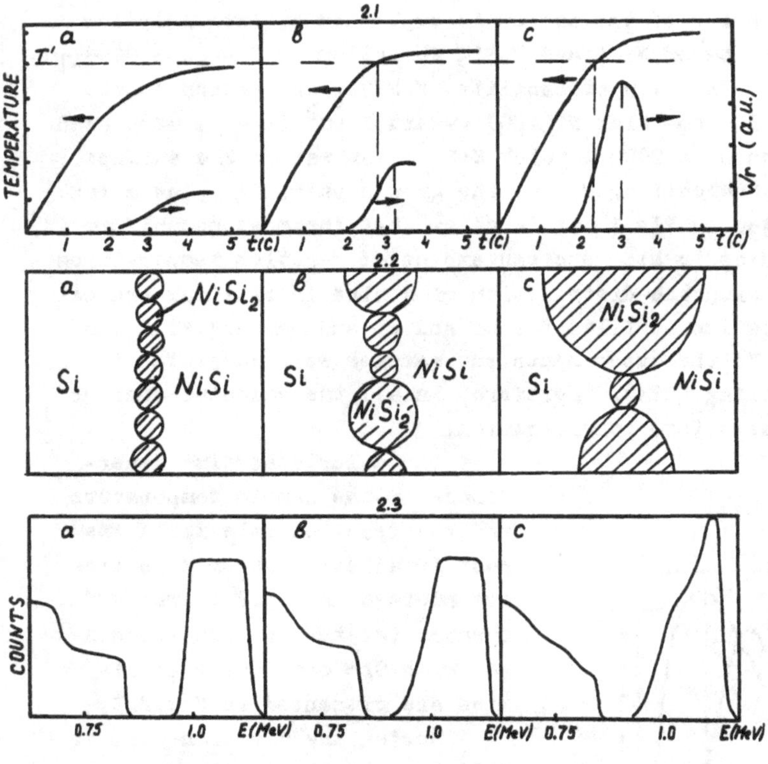

Fig.2. The dependence of the heating rate on the kinetics of rection in NiSi-Si system: (a - $P = 20\ W/cm^2$, b - $P = 26\ W/cm^2$, c - $P = 29\ W/cm^2$); 2.1. - the temperature of the sample and the reaction rate as a function of time; 2.2. - the diagram of $NiSi_2$ grains' size distribution after the formation of $NiSi_2$ layer; 2.3. - backscattering spectra for NiSi-Si structure after heat treatment

Thus, the qualitative agreement between the calculations and the experimental results has been obtained, indicating the validity of the proposed model.

I. Глотенко А.Л., неопубл.

2. Стромберг А.Г., Семченко Д.П."Физическая химия", М., Высшая школа, 1988.

PHASE FORMATION IN LASER DEPOSITED Ta_xSi_{1-x} FILMS

Khabelashvili I.D., Kulikauskas V.S., Nevolin V.N., Prokopenko V.B., Zenkevich A.V.*

The silicides of refractory metals are widely studied because of their potential usefulness in VLSI technology /1-3/. Among them Ta-Si system is of special interest from the physical point of view due to variety of Ta silicides' phases on the phase diagram and a great difference in their melting temperatures.

In this work the reactions of silicide formation in co-deposited Ta_xSi_{1-x} films were studied. In order to exclude the influence of the substrate, 0,19 μm thick Si_3N_4 layer was deposited on mono-Si(100). The Ta_xSi_{1-x} films were prepared by laser sputtering of Ta/Si compound target in 5.10^{-7} torr vacuum at a rate of about 1 Å/c. The concentration ratio Ta/Si varied from 0,25 to 4. The thickness of the films and their element composition were measured by RBS.

The thicknesses of all samples were about 1000 Å. The 4-probe resistometry was used to measure resistivity. The structure of the Ta_xSi_{1-x} films after deposition was studied by TEM.

The resistivity ρ of the Ta_xSi_{1-x} films with different compositions after deposition is presented in Fig.1. In the figure sharply nonmonotonic character of resistivity versus Ta/Si ratio can be seen. This behaviour of ρ can be explained by the nucleation of low resistivity silicide phases at the stage of film deposition. The existence of fine crystalline structure in Ta_xSi_{1-x} films is confirmed by TEM investigation.

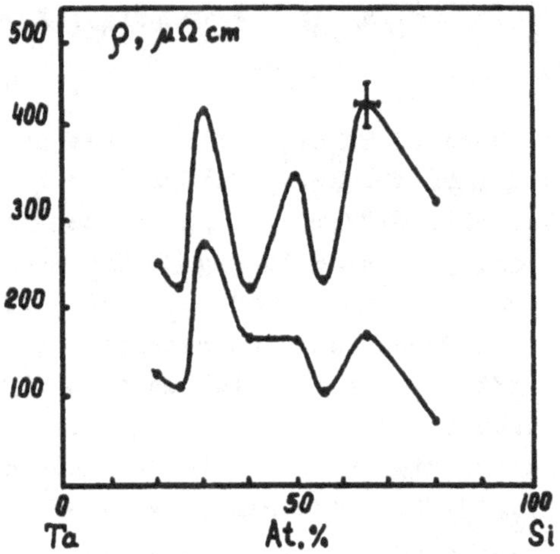

Fig.1. The resistivity of Ta_xSi_{1-x} as a function of their composition:
a - as-deposited; b - after 5-min annealing at 1100°C

* Moscow Engineering Physics Institute, 115409. Moscow, USSR.

The thin-film structures produced were isochronally annealed for 5 min in the temperature range from 500°C to 1100°C in 5.10^{-7} torr vacuum by $\bar{e}$ beam. The resistivity versus the annealing temperature and Ta/Si composition has been studied (Fig.2). The data obtained indicate the correlation of "threshold" temperatures, beginning from which ρ starts to decrease, with the phase diagram data /4/ about the melting points of alloys with corresponding composition Ta/Si. For example, for $Ta_{20}Si_{80}$ film the resistivity fall begins at T = 500°C, while for $Ta_{70}Si_{30}$ - at T = 800°C. The corresponding melting points on the phase diagram are T = 2000°C and T = 2400°C, respectively.

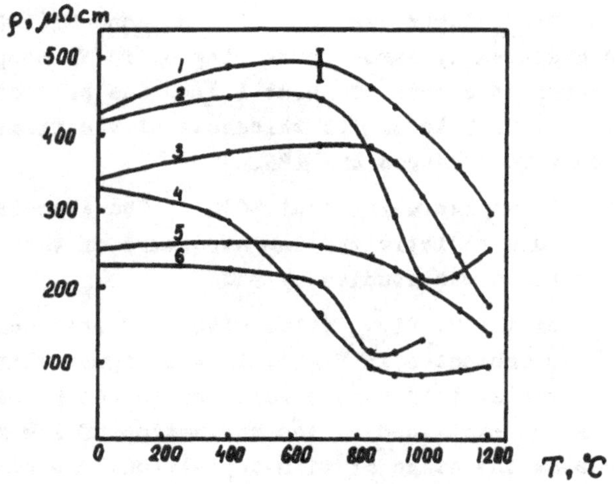

Fig.2. The resistivity ρ of Ta_xSi_{1-x} film as a function of annealing temperature T for different compositions Ta/Si : 1 - $Ta_{70}Si_{30}$; 2 - $Ta_{35}Si_{65}$; 3 - $Ta_{50}Si_{50}$; 4 - $Ta_{20}Si_{80}$; 5 - $Ta_{80}Si_{20}$; 6 - $Ta_{45}Si_{55}$

Phase analysis of annealed Ta_xSi_{1-x} films was performed by X-ray diffractometry and TEM. The X-ray diffraction data on the phase composition of the films with different Ta/Si concentration ratios annealed at T = 1100°C are presented in Fig.3. The quantitative calculation of the phase composition was made according to the intensities of diffraction maxima. These data demonstrate the films to have multiphase composition after annealing with up to 4 different silicide phases. However, when the film composition corresponds to the stoichiometric one, preferentially one silicide phase grows (for example, $Ta_{4,5}Si$, Ta_5Si_3, $TaSi_2$). The resistivity of annealed structures as a function of the element composition also has the minima, corresponding to the films with preferentially monophase composition (Fig.1). The lowest resistivity exhibited $Ta_{75}Si_{25}$ (ρ = 119 $\mu\Omega$.cm), $Ta_{45}Si_{55}$ (ρ = 106 $\mu\Omega$.cm) and $Ta_{20}Si_{80}$ (ρ = 80 $\mu\Omega$.cm).

A relatively high resistivity of these co-deposited films is possibly attributed to the multiphase composition and the small sizes of silicide grains. The resistivity rise at temperatures up to 500°C can be accounted for by the diffusion of dopants (O, C) in the film, because the synthesis of intermetalic compounds for refractory metals wasn't observed.

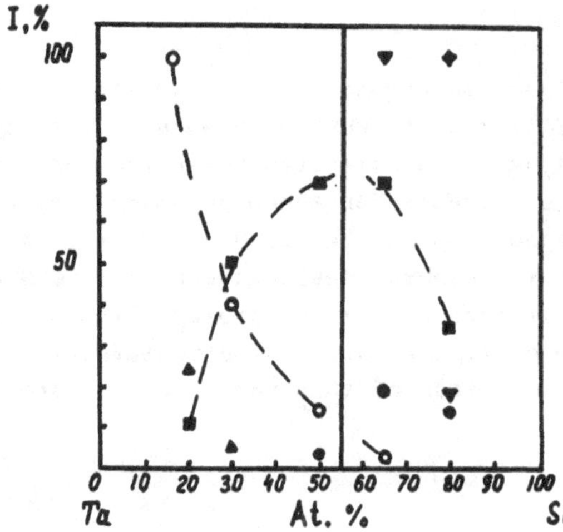

Fig.3. Relative intensities of X-ray diffraction maxima as a function of composition Ta/Si after annealing (1100°C, 5 min) for different silicide phases:

O - $Ta_{4,5}Si$; ▲ - Ta_5Si_3; ■ - Ta_5Si_3 (l.t.);

• - Ta_2Si; ▼ - $TaSi_{0,2}$; ◆ - $TaSi_2$

According to X-ray diffractometry and RBS analysis did not reveal a noticable reaction between Ta/Si film and Si_3N_4 at temperatures up to 1100°C.

References

1. Schulte I., Ferretti R., Hasse W., Danto Y., Ousten Y. Thin Solid Films 140 (1986) 183-188.

2. Gas P., Tardy I., LeGoues F.K., d'Heurle F.M. J.Appl.Phys. 61 (1987) 2203-2211.

3. Gallais P., Hantzpergue I.I., Remy I.C., Roptin D. Thin Solid Films 165 (1988) 227-236.

4. Hansen M., Anderko K. Handbook "Constitution of Binary Alloys", 1958.

LASER DIRECT WRITING OF CONDUCTIVE THIN FILMS FROM GASEOUS PHASES

F. Gänsicke, K. Zimmer, G. Reiße, B. Lämmel [1] and
H. Johansen [2]

1. Introduction

The necessity to develop new techniques for structuring and repairing the interconnection metallization of VLSI - integrated circuits with dimensions smaller than 1 μm has accelerated investigations of laser-induced chemical processes suitable for these purposes. Especially the laser-induced chemical vapour deposition (LCVD) provides some advantages in comparison with conventional methods. The LCVD permits patterning without photolithographic steps, prevents thermal stress of underlaying structures, etc. /1/. We have investigated the laser-induced chemical vapour deposition of thin metal (W, Mo) and silicide films ($TiSi_2$).

2. Experimental

Our "direct writing system" consists of a laser with an optical system, a reaction chamber mounted on a translation stage, a vacuum pump with extracting possibilities for exhaust gases and a gas handling system. The laser is an argon ion laser ILA 120-1 with a maximum power of 3 W.
It can be replaced by a CO_2-laser or an excimer laser. The laser beam is focused onto a substrate surface by three exchangeable objectives generating a spot radius of 15 μm, 4 μm, or 2,6 μm.
A photodetector permits the evaluation of the intensity distribution of the reflected beam and enables us to measure the film thickness in situ. The reaction chamber consists of stainless steel and can be moved in x- and y-direction with scan speeds of 0,02 - 0,2 mm s^{-1}.

---

1) University of Engineering, Mittweida 9250
Platz der DSF 17, GDR

2) Institute of Solid State Physics and Electron
Microscopy of the Academy of Sciences of the GDR
Halle/Saale 4020
Weinbergweg 2, GDR

It contains a substrate holder with heater and entrance windows of glass, quartz glass, or other materials according to the wavelength used.
The oil diffusion vacuum pump guarantees a basic pressure of 10^{-3} Pa and working pressures typical for LCVD applications. The gas handling system is constructed for controlling five different reaction-, carrier-, or buffer gases. It can be modified for special approaches. We have used tungstenhexafluoride (WF_6), tungstenhexacarbonyl ($W(CO)_6$), molybdenumhexacarbonyl ($Mo(CO)_6$), titaniumtetrachloride ($TiCl_4$), and monosilane (SiH_4) for our investigations. The metalcarbonyls are solids at room temperature and were heated to about 50 °C in order to increase the vapour pressure. The metalcarbonyl vapour was transported into the reaction chamber by carrier gases (Ar, H_2). The titaniumtetrachloride was flushed by Ar or H_2.

3. Results and Discussion

The argon ion laser-induced pyrolytical tungsten deposition from tungstenhexafluoride occurs via substitution reaction between tungsten and the silicon substrate according equation (1) /2/

$$2\,WF_6 + 3\,Si \longrightarrow 3\,SiF_4 + 2\,W. \qquad (1)$$

All experiments were done at a WF_6-pressure of 530 Pa, a substrate temperature of 300 K and a laser spot radius of 15 µm. Laser power and scan speed were varied. An increase in film thickness was observed at fixed scan speeds until reaching a maximum value. This self-limiting effect is caused by the inhibited diffusion of WF_6 molecules through the growing tungsten film. Decreasing the scan speed from 0,2 mm s^{-1} to 0,02 mm s^{-1} led to an increase in maximum film thickness as a result of longer illumination times.

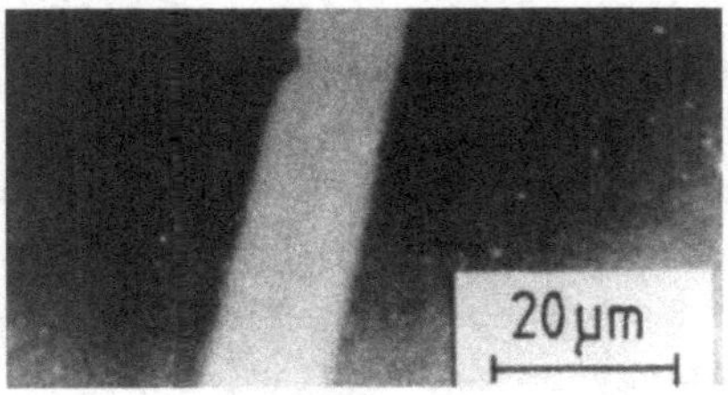

fig. 1: SEM micrograph of a W-line on Si
v_s = 0,1 mm s^{-1},
p_{tot} = 530 Pa, P_L = 3 W,
r_s = 15 µm, T_{sub} = 300 K

We have got 80 nm thick tungsten lines at a scan speed of 0,02 mm s^{-1} and a laser power of 3,5 W.
The minimum film resistivity was 170 µΩcm deposited at a scan speed of 0,2 mm s^{-1} and a laser power of 3,5 W. The resistivity increased to values of 1 mΩcm at slower scan speeds. We suppose this effect occurs by partial silicidation and

film contamination /3/. Tungstenhexacarbonyl was used in a second experiment. The decomposition reaction was proposed as follows /1/:

$$W(CO)_6 \longrightarrow W + 6(CO) \qquad (2)$$

The laser power required for this reaction was 1,2 W at a spot radius of 15 µm and a scan speed of 0,02 mm s^{-1}. The minimum film resistivity showed a value of 12 µΩcm at these deposition parameters. The films exhibited a lower contamination level than those deposited from WF_6. This could be caused by the lower reactor total pressure (300 Pa) and the lower deposition rate of 0,3 µm s^{-1}.
The deposition of molybdenum from molybdenumhexacarbonyl occurs via a step by step dissociation similar to equation (2) /1/. The film thickness and line width increased with increasing laser power.
The maximum deposition rate was 1,1 µm s^{-1}. A typical film resistivity of 2 mΩcm points out the high level of film contamination at these deposition conditions.
A last series of experiments was directed to the deposition of titanium silicide from $TiCl_4$ and SiH_4. The argon ion laser-induced pyrolytic deposition of $TiSi_2$ was carried out at different laser intensities and reaction gas mixtures. We used H_2 and Ar as carrier gases for the titaniumtetrachloride. The film composition depends on the SiH_4 to $TiCl_4 \cdot H_2$ flow ratio. We have got a film stoichiometry of $TiSi_2$ at a ratio of $TiCl_4 \cdot H_2/SiH_4 \approx 0{,}5$. The lowest film resistivity was 30 mΩ cm without thermal annealing. Etching occured with increasing $TiCl_4$ content probably caused by HCl and chlorosilanes. The films were investigated by EBIC and SEM. The results of EBIC analysis show inhomogenities in the electron beam induced conductivity resulting from an inhomogenous film composition and contamination distribution. Further investigations are necessary to explain these effects.

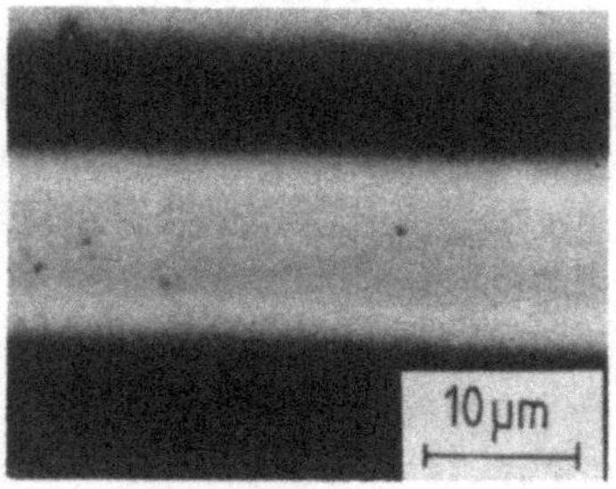

fig. 2: Optical micrograph of a Mo-line on SiO_2/Si from $Mo(CO)_6$
v_s = 0,2 mm s^{-1},
P_L = 2,4 W,
$p_{Mo(CO)_6}$ = 215 Pa,
r_g = 15 µm, T_{sub} = 340 K

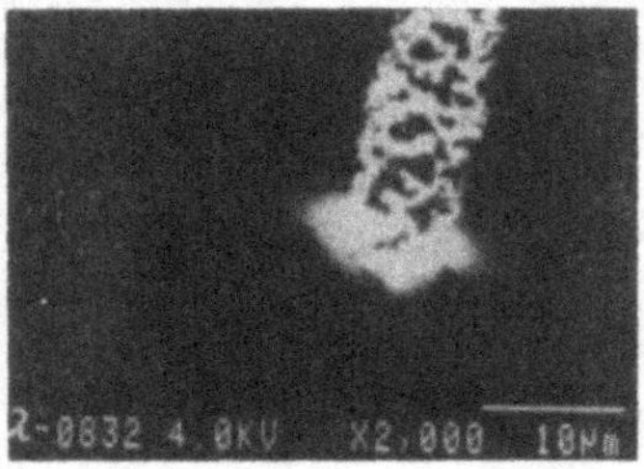

fig. 3: EBIC micrograph of a $TiSi_x$-line on SiO_2/Si
v_s = 0,05 mm s^{-1},
p_{tot} = 2,6 Pa, T_{sub} = 350 K,
P_L = 2,25 W, r_g = 15 µm

/1/ Bäuerle, D.: Chemical Processing with Lasers, Springer Verlag Bln., Heidelbg. 1986.
/2/ Liu, H.C., C.P. Yakymyshin, H.R. Philipp et al., Journ. Vac. Sci. Techn. B 3 (1984)5, 1441-44.
/3/ Zimmer, K.: Dissertation A, Ingenieurhochschule Mittweida 1989.

FORMATION OF A SILICIDE/NITRIDE LAYER SYSTEM BY N^+ IMPLANTATION

B. Weber[1)], J. Danilovich[2)], K. Gärtner[1)], and A. Novikov[2)]

1. Introduction

Silicide on insulator structures will be of interest for the fabrication of isolated conductive structures in three-dimensional integrated circuits. Recently [Dan 89, Gär 89], the formation of a buried nitride layer below epitaxial $NiSi_2$ layers by high dose N^+ implantation was investigated. It was found that an optimum implantation temperature exists to preserve the crystallinity of the silicide layer. The formation of the nitride layer at temperatures below 1075 K which is the limit of thermal stability of $NiSi_2$ will be possible using a high current density regime for N^+ implantation. In this paper the investigations are extended to high dose N^+ implantation into Ni/<111>Si layer systems to form the silicide as well as the nitride layer by only one implantation step.

2. Experimental

Ni layers (15 and 30 nm) were sputter deposited ($P < 10^{-4}$ Pa) onto chemically cleaned <111>Si substrate (1-5 Ωcm). N^+ implantations were performed using high current densities (30 μAcm^{-2}, 120 μAcm^{-2}). The samples were thermally isolated against the sample holder and heated by the ion beam up to 835 and 975 K, respectively. The target temperature during implantation was measured using a thermocouple. The implantation dose was varied between $4*10^{15}$ and $1.4*10^{18}$ cm^{-2}. The investigations of the composition and the crystal quality of the layer systems were established using Rutherford backscattering spectrometry (1.4 MeV He^+, $\Theta = 170°$).

3. Results and discussion

First, a 15 nm Ni/<111>Si layer system was implanted with 100 keV N^+ ions without wobbling of the ion beam (Gaussian like profile) in order to get high current density. The N^+ current was chosen in this way that the ion beam heating provided a target temperature of about 975 K. Because of the inhomogeneous current density the backscattering measurements were always performed at the position of the maximum current density of about 120 μAcm^{-2}. The backscattering spectra measured for different implantation doses D_i (times) show that the silicide layer and the nitride layer are formed during two different intervals ($4*10^{15}$cm^{-2} $\leq D_i \leq 2.4*10^{17}$cm^{-2} and $D_i \geq 1.4*10^{18}$cm^{-2}). For $D_i < 4*10^{15}$cm^{-2} no silicide formation was observed. As can be seen from the random spectrum in fig.1 the formation of

---

1) Friedrich-Schiller-Universität Jena, Sektion Physik
Max-Wien-Platz 1, Jena, DDR - 6900
2) Lenin University Minsk, Department of Physics,
Lenin-Prospect 4, Minsk, USSR 220 080

the $NiSi_2$ layer is already completed at $D_i=2*10^{16}cm^{-2}$, but the crystal quality which is characterized by the Ni parts of the aligned spectra changes non-monotonously up to the dose of $2.4*10^{17}cm^{-2}$. For the lowest dose the silicon near the range of N^+ is heavily damaged. With increasing ion dose the crystal quality of Si improves continuously. With respect to the crystal quality of the epitaxial silicide layer the best result characterized by a backscattering minimum yield χ_{min} of 14% is obtained for a dose of $2.4*10^{17}cm^{-2}$. For the discussion of these results the following facts must be taken into account. First, the target temperature starts from room temperature and reaches the stationary value of 975 K after an implantation dose of $8*10^{17}cm^{-2}$ (corresponding to about 100 s). Second, the densities of nuclear deposited energy in the silicide and in the silicon differ not very much ([Dan 89]), and third, the silicide is more radiation resistent than the silicon [Tsa 82]. Already at the dose of $2*10^{16}cm^{-2}$ the target temperature is high enough to complete the silicide formation. Similar to the processes observed for high current density implantations of Ar^+ into silicon [Kom 86] the crystal quality is influenced by the formation and by the annealing of defects due to the ion beam. For low target temperatures the formation of defects dominates and with increasing temperature the annealing and transformation of defects become more and more important. The maximum defect density is reached in the vicinity of $2*10^{16}N^+cm^{-2}$ and $4.8*10^{16}N^+cm^{-2}$ for Si near the N^+ range and for the silicide, respectively (see fig.1). This difference can be explained by the different radiation resistance.
For $D_i \gtrsim 8*10^{17}N^+cm^{-2}$ stoichiometric concentrations of N in the Si substrate are obtained. The best results with respect to the nitride layer are obtained for $D_i \approx 1.4*10^{18}cm^{-2}$. The distributiones of N, Ni, and Si calculated directly from the backscattering spectrum are given in fig.2. A relatively thick layer having approximately the composition of Si_3N_4,

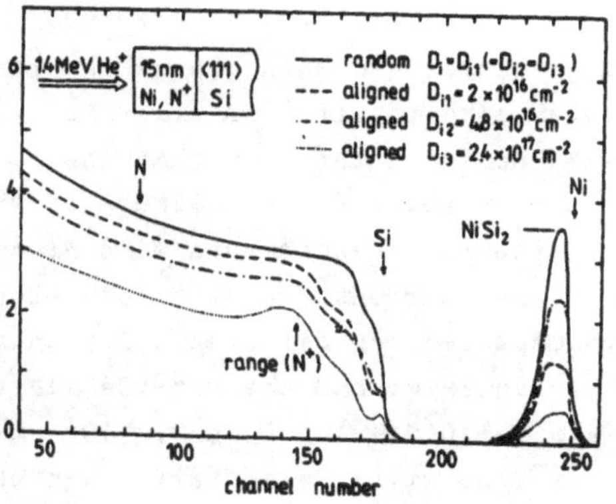

Fig.1. Backscattering spectra for a 15 nm Ni/<111>Si system implanted with 100 keV N^+ ions ($120\mu Acm^{-2}$, 975K) of understoichiometric doses.

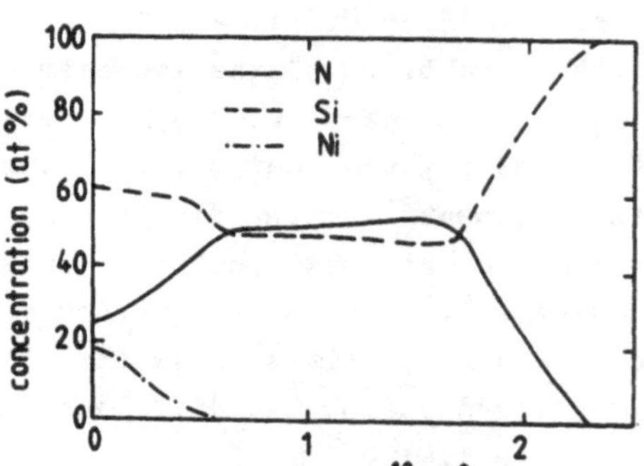

Fig.2. Concentrations of N, Ni, and Si as functions of the areal desity for the same system as in fig.1 implanted with $1.4*10^{18}$ N^+cm^{-2}.

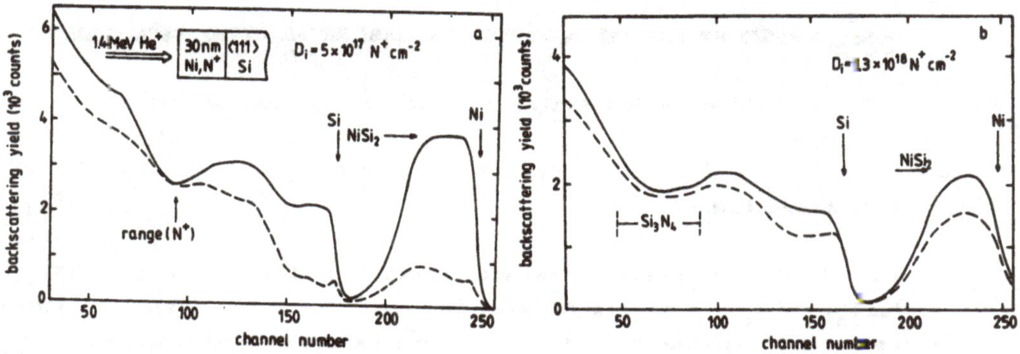

Fig.3. Backscattering spectra for a 30 nm Ni/<111>Si system implanted with 300 keV N^+ ions of under- (a) and overstoichiometric (b) dose.

i.e. the large amount of overstoichiometric nitrogen is completely redistributed by radiation enhanced diffusion. However, the strong nitrogen diffusion to the surface destroys the silicide layer.

To reduce this effect in a second series the ion range was increased relatively to the Ni layer thickness. 300 keV N^+ ions were implanted into a 30 nm Ni/<111>Si layer system with doses between 10^{16} and $1.3*10^{18}cm^{-2}$. The samples were heated by the ion beam up to 835 K. As in the case of the thin Ni layer first the silicide layer was formed. After implantation of $10^{16}\ N^+cm^{-2}$ a NiSi layer was grown without any epitaxy. With increasing dose this NiSi layer transforms into an epitaxial $NiSi_2$ layer. At a dose of $5*10^{17}N^+cm^{-2}$ (understoichiometric) an epitaxial $NiSi_2$ layer is grown (fig.3a) with $\chi_{min} \approx 15\%$. The backscattering spectra show that the N profile lies deep in the silicon substrate (damaged Si layer between $NiSi_2$ and N implanted region). For the overstoichiometric dose of $1.3*10^{18}cm^{-2}$ (fig.3b) a Si_3N_4 layer is formed as in the case of the thin Ni layer but the radiation enhanced diffusion to the surface destroys the epitaxial silicide layer also in this case (χ_{min} deteriorates to about 70%). In addition a decrease of the aligned spectrum in the region of the nitride layer to 90% of the random spectrum is observed. This can be explained with the growth of α-Si_3N_4 crystallites as it was found also for high current implantations into silicon e.g. by [Won 87].

The results obtained show that a complex layer system consisting of an epitaxial $NiSi_2$ layer and an insulating Si_3N_4 layer on silicon substrate can be formed at temperatures below 1075 K by high current high dose nitrogen implantation if the nitrogen diffusion towards the surface can be avoided. This requires further investigations.

References

[Dan 89] phys. stat. sol. (a) 112 (1989) 73
[Gär 89] Proc. REI 5 Conf., to be published in Nucl. Instr.& Meth.
[Kom 86] F.F. Komarov et al. Poverchnost. Fizika Chimija, Mechanika, N6 (1986) 90
[Won 87] S.P. Wong, M.C. Poon, Nucl. Instr. & Meth. B22 (1987) 513

$MoSi_2$ FORMED BY ION IMPLANTATION THROUGH METAL (ITM) TECHNIQUE

E. Vetter, W. Grünewald, J. Tolonics, T. Geßner *)

1. Introduction:

In MOS-VLSI circuits, refractory metal silicides (e.g. $TiSi_2$, $MoSi_2$, WSi_2) have been used to lower the sheet and contact resistance of the source, drain and gate electrodes as well as interconnections. Recently, the self aligned metal-silicide technology (Salicide) has been proven to be a highly desirable process technique for improving the submicrometer device and circuits performance /1/, /2/. The reaction of atoms of a metal layer (e.g. Mo) and silicon will be used in the SALICIDE-technique. It has been reported that uniform Mo-silicide films can be formed in self- alignment with silicon patterns with good reproducibility by using the ion implantation through metal (ITM) technique, which consists of ion beam induced metal-silicon interface mixing and appropriate annealing /3/. In this paper the dependence of the ion beam mixing at the interface Mo/Si on the ion mass will be presented. Further the effects of interface mixing on the formation and growth kinetics of $MoSi_2$ films will be investigated.

2. Experimental:

Molybdenum films with a thickness of about 25nm, 40nm, and 200nm (without ITM-process) were deposited onto (100)-silicon substrates (p-type and n-type, 10-20 Ωcm) by dc magnetron sputtering. The wafers were etched in hydrofluoric acid prior to the deposition to remove the surface oxide (native oxide layer).
As-, B-, BF_2- and P- ions were implanted at doses of 5E15 cm^{-2} and energies varying from 20 keV to 180keV. The ion energies used were chosen to maximize the damage in the region of the Mo-Si interface and to induce amorphization of the Si near this interface. The wafer temperature during implantation was kept

*) Technische Universität Karl-Marx-Stadt, PSF 964, DDR -9010 Karl-Marx-Stadt

below 100°C. Then the samples were annealed at 550°C and 600°C, respectively, with annealing times ranging from 2 minutes to 30 minutes in hydrogen ambient. The mixing of the Mo/Si interface and the implantation damage were analysed by means of RBS and cross-sectional TEM. RBS was combined with channeling technique using 1.2 MeV He ions and a backscattering angle of 120°.
The experimental results were compared with computer simulations of the recoil process by the TRIM-87 code /4/.

3. Results

Fig.1a shows the RBS spectra of four Mo coated Si samples (25nm Mo) implanted with different ions to a dose of 5E15 cm^{-2} each. The solid line in this figure indicates the RBS signal of the sample before implantation. After implantation with B-ions, the Mo and Si edges remain sharp, indicating that there is no appreciable intermixing of Si and Mo during implantation. The intermixing, however, is clearly detected after the implantation with ions of higher masses. The Mo and Si edges are broadened with increasing ion masses. But no indication of compound formation is observed at this room temperature implantation.
Similar results were obtained by the TRIM-87 calculations (Fig. 1b).

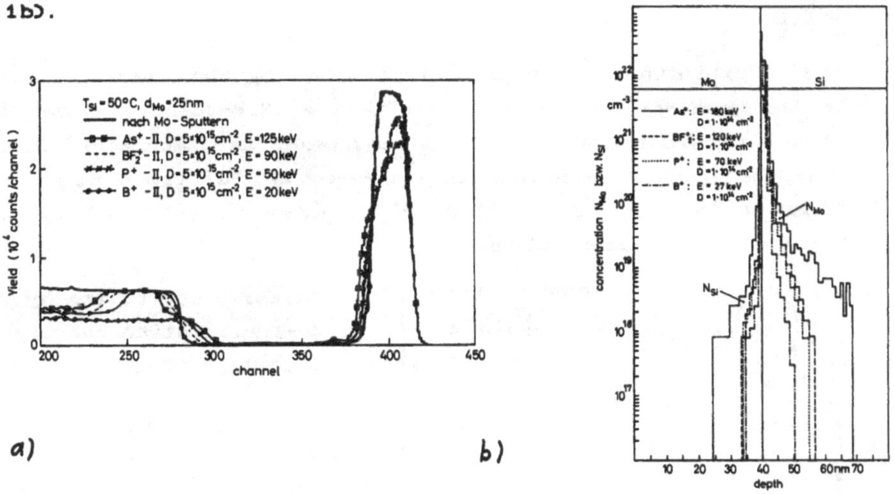

Fig. 1. Intermixing in the Mo/Si double layer at differently ion masses, a) RBS spectra, b) TRIM-87 calculations

The kinetics of the silicide formation was investigated for non-implanted and As- and BF_2-implanted Mo/Si-samples.

Fig. 2 shows an example of the results of the thermal silicidation at 550°C. Obviously the formation of $MoSi_2$ occur most rapidly without implantation. However, the surfaces are markedly rough and a dramatically lateral silicidation over the oxide pattern is observed. The ion implantation prior to annealing retards the silicidation rate, mostly dominating in the case of BF_2 ions. An annealing time of 60min is necessary for a complete reaction of BF_2-implanted Mo-Si layers. Similar results are obtained at an annealing temperature of 600°C.

The advantage of the ITM-process is the formation of silicide layers with sufficent smooth surfaces and reduced lateral overgrowth combined with shallow pn-junctions.

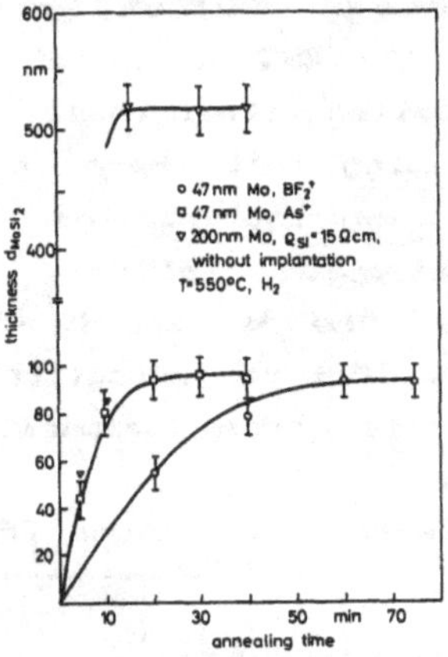

Fig. 2. $MoSi_2$ growth kinetics of As- , BF_2- implanted and non-implanted Mo/Si samples.

Fig. 3 exhibits the damage distribution by the As-implantation in the Mo/Si-system calculated with the TRIM-computer-code. As a first approximation the Si is considered as amorphous (a-Si), when the vacancy concentration exceeds the Si-bulk concentration (5E22 cm^{-3}). The dose of 5E15 cm^{-2} creates an a-Si sublayer of a thickness of about 120nm.

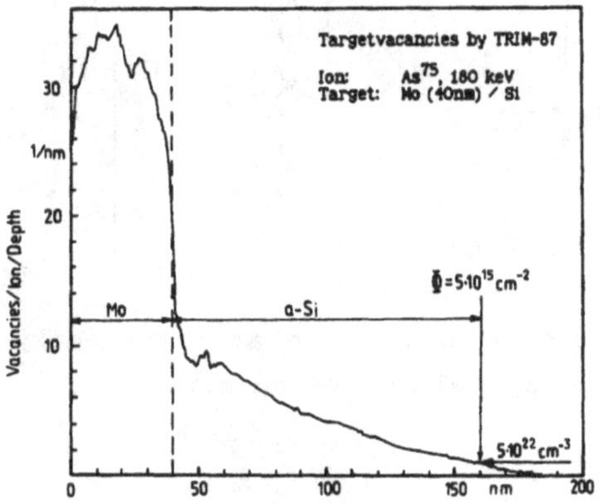

Fig. 3.
Damage distribution by As-implantation calculated with TRIM-87 code

For comparison with experimental results the corresponding cross-sectional TEM bright-field micrograph is shown in Fig. 4a. Under the polycrystalline Mo-layer an intermixed Mo/Si-area and the amorphous Si-layer of a thickness of about 130 nm can be found. An underlying defect zone in the Si-substrate is visible. The corresponding result of the sample after the silicidation annealing for 30min at 600°C is exhibited in Fig. 4b. The $MoSi_2$-formation is completely terminated and the amorphous Si zone is recrystallized, the underlying defects, however, being stable at this temperature.

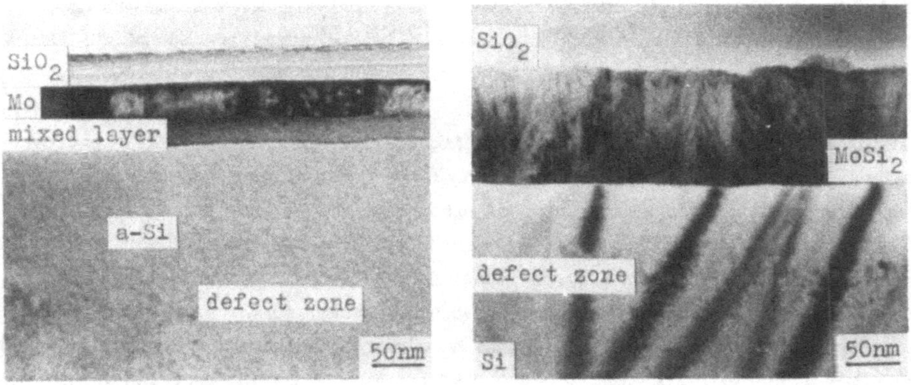

a) b)

Fig. 4. Cross-sectional TEM bright-field micrographs of the as-implanted Mo/Si sample: As, 180keV, 5E15 cm^{-2} (a) and after $MoSi_2$ formation at 600°C for 30min in H_2 ambient (b)

References

/1/ Nagasawa, E. et al.: IEEE Trans. Electron Devices vol. ED-34 No. 3, 1987, 581

/2/ Okabayashi, H. et al.: IEEE Trans. Electron Devices vol ED-31, No. 9, 1984, 1329

/3/ Okabayashi, H. et al.: Layered Structures and Interface Kinetics, Ed.: S. Furukawa, KTK Sc. Publ. Tokyo, 1985, 263

/4/ Ziegler, J.F., Biersack, J.P., Littmark, U.: The Stopping and Range of Ions in Solids, Pergamon Press, New York 1985

QUANTITATIVE RBS ANALYSIS OF SILICIDES AND SILICIDE OXIDES USING RUMP

J.M. Gilpérez[1)], A. Climent[1)], J.M. Martínez-Duart[1)] and J. Perrière[2)]

INTRODUCTION

Rutherford backscattering spectrometry has been widely used in the characterization of silicides and silicide oxides because of the favourable condition they offer for this technique, M.R. Herman (1984), in which a thin film of a heavy element (the metal) is deposited on a lighter element sustrate (Si). In this case, precise information on the composition of the different layers forming the sample under analysis can be easily provided. As the physical phenomena involved in RBS analysis are well understood, computer programmes have been developed which help in the quantification of the RBS spectra. Among them RUMP program, L.R. Doolittle (1985), (1986), is widely used.

One of the problems that arise when fitting the RBS spectra to obtain quantitative information is the error in the determination of the composition or thickness of the films under analysis. Possible sources of error may arise either for uncertainties in the experimental setup such as the energy and total dose of the incident ions, solid angle of the detector, etc., or from the inaccuracy of stopping power values. The latter appears as the most important source of uncertainty in the simulation of RBS spectra. Several reviews about their values have been published, D.C. Santry et al (1980), D.I. Thwaites (1985), J.F. Ziegler (1973), (1977). The values incorporated in our version of RUMP (3.31) are the J.F. Ziegler et al (1982) empirical values.

In this work we analyze RBS spectra of silicides and silicide oxides considering the mentioned uncertainty contributions. Current stopping power of Si, Ti, Ni, Ta and W and Bragg's rule in the silicides and silicide oxides formed are checked.

SAMPLE PRERARATION AND RBS ANALYSIS

Thin films of Ti, Ni, Ta and W and the corresponding silicides or metal-silicon mixtures were deposited on base Si single crystal wafers or Si wafers covered with SiO_2 by sputtering. The oxidation of the silicides was carried out in an oxygen plasma ambient at 2.5×10^{-2} mbar oxygen pressure and temperatures around 700°C.

RBS spectra were obtained using the 2.5 MeV Van der Graff acelerator facility at the Université Paris VII. The energy of the incident $^4He^+$ ions was 2 MeV ± 1% and a scattering angle of 165° was used. The solid angle of the detector, Ω, was measured with a reference sample composed by a thin film of Ta_2O_5 containing 100×10^5 Ta atoms/cm^2 ± 2% on Si. The average value obtained for several samples was 1.45 msr ± 1.1%. The dose, Q, (20 or 30 μC) of incident ions was acurrately known as the scattering

1) Universidad Autónoma de Madrid, Cantoblanco, E-28049-Madrid, Spain.
2) Université Paris VII, 2 Place Jussieu, F-75251-Paris Cedex 05 France.

chamber was electrically insulated acting as a good Faraday cup, G. Amsel et al (1971). The error estimated in the calibration line (KeV/channel and energy of channel zero is 0.5 %. The single independent uncertainty contributions sum up quadratically to 2.5 % relative total uncertainty.

RESULTS AND DISCUSSION

Pure elements:

Simulation spectra of Ti, Ni, Ta and Si as pure elements were made using the experimental ones. As an example, the Ti peak is shown in fig. 1. With the linear scale factors (sc) of the stopping power available from RUMP, fitting of the experimental and simulated normalized yields is reached. The scale factor values found for a best fitting of the spectra are 0.95, 0.90, 0.95 and 1.00 for Si, Ti, Ni and Ta and W respectively. The uncertainty in these scale factors is affected by the uncertainty of the experimental parameters and the statistic noise in the yield, $1/\sqrt{N}$, being N the total amount of counts in a channel.

Since lower values of stopping power were considered, the simulated thickness was increased. For samples with thickness in the order of $10^3 \times 10^{15}$ atoms/cm^2, this increment goes from 2% for WSi_2 to 10% for Ti. The accuracy in the simulated thickness depends on the statistical error. In our experimental conditions we have found that for W and Ta the thickness uncertainty is about 5×10^{15} atoms/cm^2, whereas for Ti and Ni is about 10×10^{15} atoms/cm^2. Energy straggling was also considered, finding a good agreement with Bohr's theory, N. Bohr (1915).

Silicides and silicide oxides

In the simulation of W, Ta, Ni and Ti silicides and silicide oxides we found again the peaks in the simulated spectra smaller than the experimental ones. So, the use of stopping power corrections was necessary. Since for compounds the program uses the stopping powers given by Bragg's rule, corrections in the stopping power of one element modify the heights of all elements proportionally to said heights. In this way, correction in the Si stopping power was used to fit the W and Ta peaks in the WSi_2 and $TaSi_2$ spectra. For $NiSi_2$ and $TiSi_2$, the use of the correction factors for Ni and Ti besides the one for silicon, drives us to a satisfactory fitting of such spectra. The Si and Ti peaks in the $TiSi_2$ spectrum are shown in fig. 2 as an example. The presence of oxygen in silicide oxides does not affect the previous considerations significantly and we obtain the same accuracy in the fitting of the oxides spectra using the corrected stopping powers. Therefore, the use of the same scale factors estimated for pure elements in the fitting of the compounds, give us a good agreement with Bragg's rule.

The composition accuracy depends on the statistical noise of the heavier element of the compound. So, we shall have a higher accuracy for WSi_2 and $TaSi_2$ than for $NiSi_2$ and $TiSi_2$. Concentration ratios for assumed stoichiometric WSi_2 and $TaSi_2$ were found $C_{Si}/C_W = C_{Si}/C_{Ta} = 2 \pm 0.025$ whereas for $NiSi_2$ and $TiSi_2$ are $C_{Si}/C_{Ni} = 2 \pm 0.050$ respectively. The same arguments can be used for silicide oxides.

CONCLUSIONS

The influence of uncertainties in the experimental parameters and the inaccuracy of stopping power values in the quantification of RBS spectra with the help of the RUMP simulation program has been studied. With the estimated uncertainty of the experimental parameters of 2.5%, we have found that in our experimental conditions a correction is needed for the stopping power of pure elements such as Si, Ti and Ni. With these corrections it is possible to obtain satisfactory fittings of silicide thin films of Ti, Ni, Ta and W and their oxides. Besides, we have found a good agreement of our results with Braggs, rule.

ACKNOWLEDGES

The authors thank R. Pérez-Casero for his assistance in RBS measurements.

REFERENCES

(1) Amsel, G., Nadai, J.P., D'Artemare, E., David, P., Girard, E. and Moulin, J.: Nucl. Instr. and Meth. 92 (1971) 481–498.

(2) Bohr, N.: Philos. Mag. 30 (1915) 581.

(3) Doolittle, L.R.: Nuc. Instr. and Meth. Phys. Res.B9 (1985) 334–351. Nucl.Instr. and Meth. Phys. Res. B15 (1986) 227–231.

(4) Herman, M.H.: J. Vac. Sci. Technol. B2 (1984) 748–755.

(5) Santry, D.C., Werner, R.D.: Nucl. Instr. and Meth. Phys. Res. 178 (1980) 523–530.

(6) Thwaites, D.I.: Nucl. Instr. and Meth. Phys. Res. B12 (1985) 84–89.

(7) Ziegler, J.F.: Hellium Stopping Power and Ranges in All Elements, Pergamon N.Y. 1977.

(8) Ziegler, J.F., Brodsky, M.H.: J. Apply. Phys. 44 (1973) 188–196.

(9) Ziegler, J.F., Biersack, J.P., Littmark, U.: Empirical Stopping Power for Ions in Solid, IBM Research Report RC 9250 (1982).

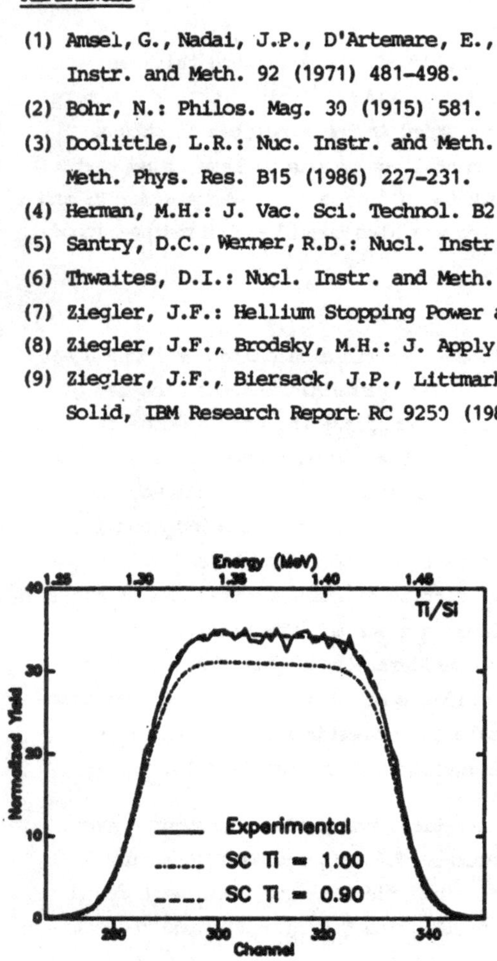

Fig. 1. Stopping power modification in Ti as pure element.

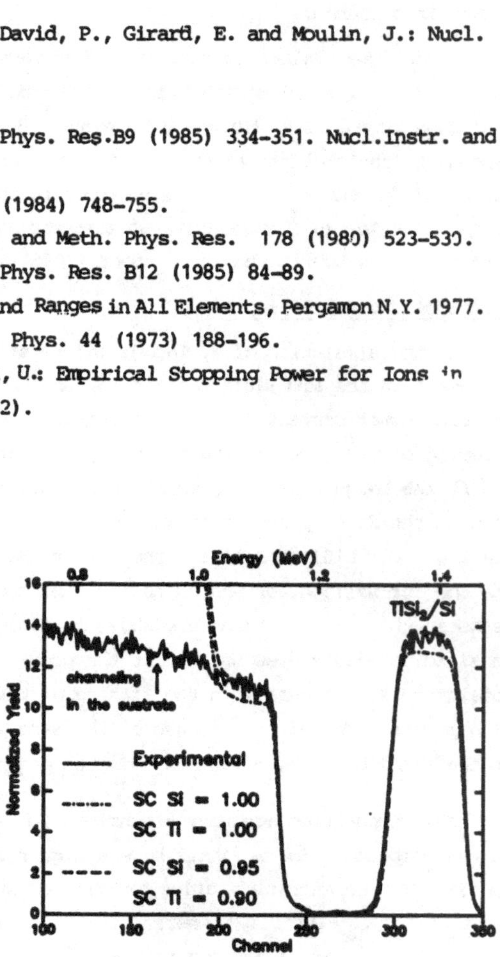

Fig. 2. Stopping power modifications in Ti silicide.

REDISTRIBUTION OF B AND As BY RTP AFTER IMPLANTATION INTO $TiSi_2$ ON Si

D. Panknin[1], E. Wieser[1], L. Drechsler[1], E. Bugiel[2], F. Richter[2], F. Wolf[2], E. Landi[3], and S. Solmi[3]

For shallow junctions in VLSI applications the formation of self aligned silicide contacts on source or drain and the gate electrode by the so called salicide technology is very promising. In this case doping of source and drain may be accomplished either by implantation through the metal before silicide formation or by implantation through the silicide after the selective etching.

In this contribution some anomalous effects are reported observed in the redistribution of As and B as well as in the depth distribution of carriers caused by the second high temperature treatment of the salicide process.

The dopants were implanted through a $TiSi_2$ layer formed by RTP at 750 °C and furnace silicidation at 690 °C, respectively, in N_2 of 70 nm Ti deposited on (100) Si by sputtering. Selective etching was carried out before the implantation. The thickness of the silicide layer amounted to about 120 nm. The implantation energy was choosen in such a way that the peak of the implantation profile was near or at the interface silicide/silicon (150 keV for As, 30 keV for B). The implanted dose was 1×10^{16} cm^{-2} in each case. For the high temperature treatment following the implantation rapid thermal processing (RTP) for 10 s at 1000 °C and 1100 °C, respectively, was used.

Dopant and carrier depth distributions were determined by SIMS and incremental sheet resistance and Hall effect measurements, respectively. In the latter case stepwise sheet removal by anodic oxidation and HF etching was used. The distribution of residual defects was determined by cross section TEM.

Fig. 1 shows the depth distribution of As implanted with 150 keV after annealing at 1100 °C (curve a). In the Si region below the silicide interface additionally two As profiles obtained by computer simulation are also shown. The starting profile for this computer simulation was constructed very similar in its shape to the part of the as implanted profile extending into the silicon beyond the silicide. It was scaled in such a way that the integral As content was equal to the As sheet concentration in the silicon region of the experimental profile measured after annealing.

For calculation of curve b a programme is used for redistribution and electrical activation of As in Si /1/ using a diffusion coefficient according Hoyt and Gibbons /2/. The diffusion coefficient contains tree terms describing the diffusion according to neutral as well as single and twofold negatively charged vacancies.

1) Central Institute of Nuclear Research, Rossendorf, PF 19, DDR - 8051 Dresden

2) Institute of Semiconductor Physics, Walter-Korsing-Straße 2, DDR - 1200 Frankfurt/O.

3) C.N.R. Istituto LAMEL, Via de Castagnoli 1, I - 40126 Bologna

A comparison of this calculated profile with the measured ones shows clearly an anomalous enhanced diffusion observed in this experiment. In order to get agreement between experiment and computer simulation an enhancement of the diffusion coefficient by a factor 5 is necessary. Curve c represents a profile calculated by using this enhancement factor. After annealing at 1000 °C the diffusion enhancement by a factor 15 has been observed.

This anomalous enhanced diffusion compared to the normal diffusion behaviour of implanted As during the annealing treatment can be explained assuming a supersaturation of the silicon region under the silicide with point defects emitted from the interface. The enhancement of the As-diffusion due to large point defect concentration is discussed in /3/.

Fig. 2 shows the depth distribution of carrier concentration and mobility in the silicon for As implanted with 150 keV through the silicide and annealed at 1000 and 1100 °C, respectively. The silicide layer has been removed by HF etching before the profile measurement.

From a comparison of the dopant concentration measured by SIMS with the carrier concentration an average electrical activation of the As contained in the silicon layer of 95 % after 1000 °C and 75 % after 1100 °C annealing has been evaluated.

The carrier profiles show a decreasing carrier concentration in direction to the surface, this means to the original interface silicide/silicon. The mobility increases in this depth region. The decrease of carrier concentration is in contrast to the flat shape of the dopant profile. The decreasing flank becomes broader with increasing annealing temperature. This observation points on the formation of electrically active defect configurations in the region near to the interface. The more pronounced effect after annealing at 1100 °C points to a reaction between silicide and silicon substrat which broadenes the defective interface region.

Fig. 3 shows B depth distributions obtained by implantation at 30 keV into the silicide/silicon system and subsequent annealing at 1000 and 1100 °C, respectively. The shoulder in the depth profile corresponding to the 1100 °C annealing indicates clearly a non mobile part of B in the interface region. The maximum concentration of mobile B, marked by the shoulder position is in good agreement with the maximum carrier concentration of about 9×10^{19} cm^{-3}. The formation of non mobile and electrically inactive B precipitates during annealing if the maximum B concentration exceeds the temperature dependent solubility limit is a well known fact. But, the steady state solubility limit at 1100 °C corresponds to 1.9×10^{20} cm^{-3} /4/. In RTP experiments non equilibrium states corresponding to a higher solubility have been observed (3.3×10^{20} cm^{-3} /5/). Therefore, from the lower solubility observed here it can be concluded that the defect states in the region at or near the interface silicide/silicon promote the boron precipitation.

In Fig. 4a is shown a cross section TEM image of an As-implanted sample (1000 °C, 10 s). Near the interface, where a diminished electrical activation has been observed, no extended defects are visible. This means the compensating defects are point defect complexes. The observed defect band between 80 and 160 nm below the inferface is well above

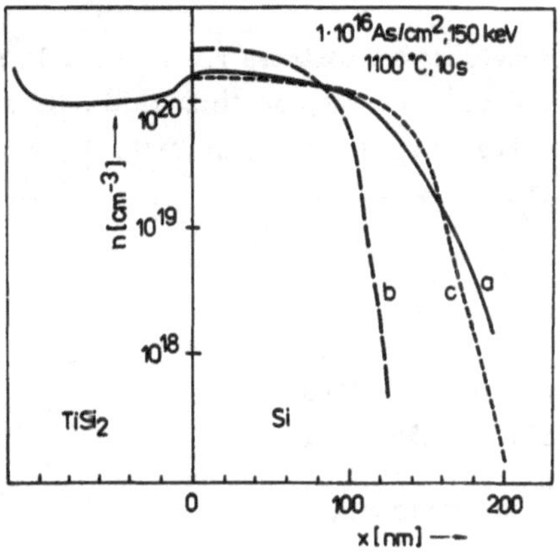

Fig. 1

As depth distribution after annealing at 1100 °C
a) SIMS measurement
b) computer simulation using the diffusivity according to /2/
c) computer simulation for a diffusion enhancement by a factor 5

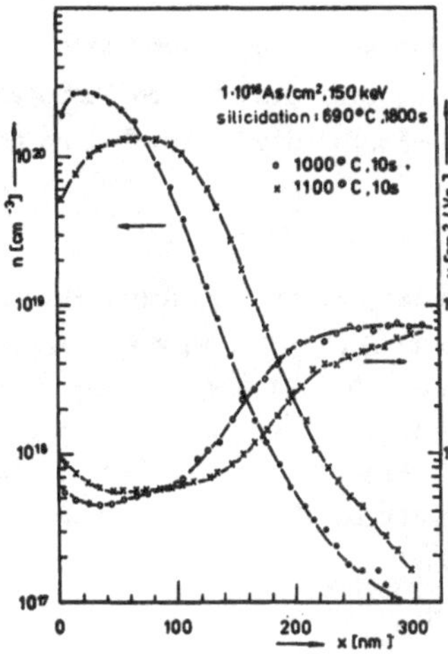

Fig. 2

Depth distribution of carrier concentration and mobility after As-implantation and annealing at 1000 and 1100 °C

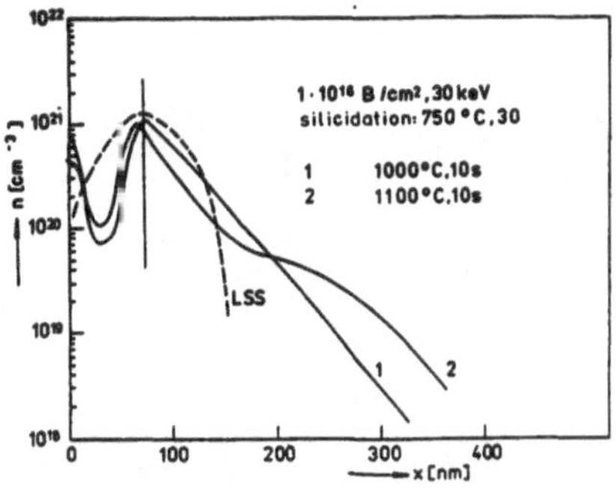

Fig. 3

B depth distribution after annealing at 1000 and 1100 °C

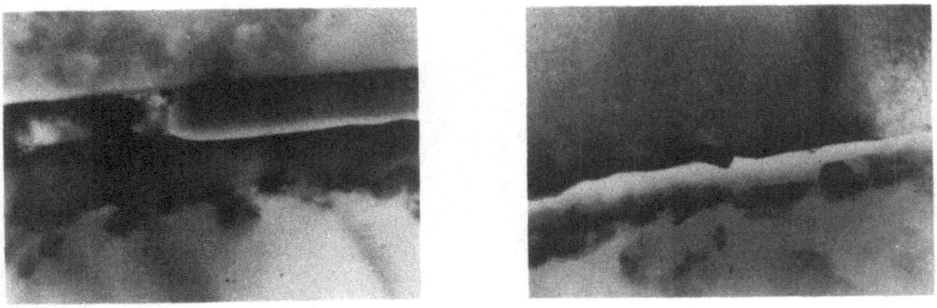

b)

Fig. 4

Cross-section TEM
a) As implantation and annealing at 1000 °C, 10 s
b) B implantation and annealing at 1100 °C, 10 s

the pn-junction.

In the case of B-implantation (Fig. 4b) dislocations and precipitates are detectable near the interface in agreement with the discussed diffusion hindering in this depth. Here also a band of dislocation loops above the pn-junction has been observed after 1000 °C. It transforms into a dislocation network during 1100 °C annealing.

References:

/1/ Kögler, R. et al., Appl. Phys. A (in press)

/2/ Hoyt, J.L., J.F. Gibbons. Mat. Res. Soc. Symp. Proc. 52 (1986) 15

/3/ Fahey, P. and R.W. Dutton, Proc. Semicond. Silicon (1986) 671

/4/ Armigliato, A. et al., in Semiconductor Silicon 1977
Ed. H.R. Huff and E. Sirtl, Electrochem. Soc. Inc. Princeton (p. 638)

/5/ Guimaraes, S. et al., Phys. Stat. Sol. (a) 93 (1986), 549

SURFACE RELIEF DEVELOPMENT AND FATIGUE CRACK INITIATION IN IMPLANTED AND NONIMPLANTED METALS

C. Blochwitz and E. Preuß *

Introduction

To understand the processes leading to an extension of life of cyclically deformed metals by ion implantation, it is nessesary to investigate the development of surface phenomena during fatigue. For this purpose we report the results of surface replications of different fatigue stages. The portion of grains containing slip bands, the surface portion of slip bands in slip active grains and the azimuthal distribution of slip traces of implanted and nonimplanted specimens are compared, using polycrystalline nickel specimens.

Generally, the smaller surface roughness after fatigue of implanted material is considered to be responsible for delayed crack nucleation and longer fatigue life.

The experimentally observed slip activities are connected with the surface roughness and consequently with the visibility of slip bands at the surface. Supposing that the height of a slip step is proportional to the sine of the angle γ between slip direction and its projection into surface, a solution of the problem of visibility of slip traces in implanted surfaces is given. This attempt uses numerical calculation of the distribution of the γ angle in relation to the azimuthal distribution of slip traces in the specimen surface. In implanted material the azimuthal distribution of v i s i b l e slip band traces is characteristically changed, and the portion of observable active slip bands is reduced in comparison with nonimplanted metal. This should be taken into account when different experimental results will be compared!

Experimental

spec.	IP regime	grain size	fatigue regime	cycles to failure
4	5x1[illegible] N^+/cm^2 50 KeV T<3[illegible]3 K	115 µm ± 55 µm	ε_{pa} = const. = 1×10^{-4}	195000
5	non-implanted	100 µm ± 50 µm	ε_{pa} = const. = 1×10^{-4}	68000

Table 1 Experimental data of the reported specimens (Ni polycrystals with 99,9 % purity)

The specimens were deformed in symetric push-pull test with constant amplitude of plastic strain. At different stages of fatigue surface replications were made.

*Pädagogische Hochschule Dresden, 8060 Dresden Wigardstraße 17, Sektion Physik

Experimental Results

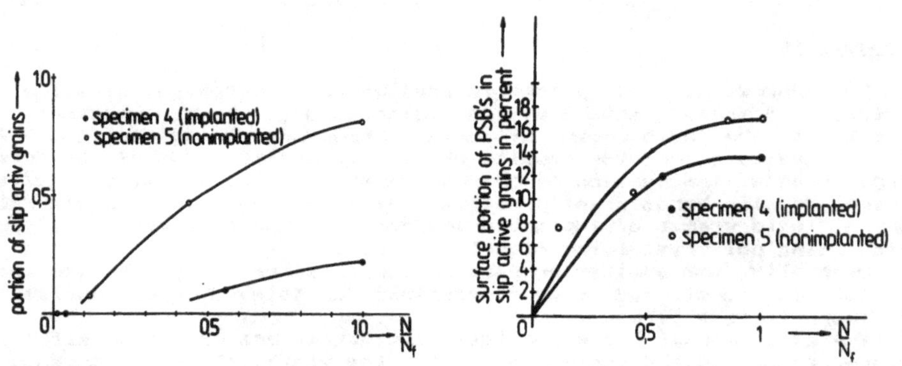

Fig. 1 Development of the portion of grains containing persistent slip bands (left) and development of the surface portion of slip bands in slip active grains (right) of spec. 4 and 5

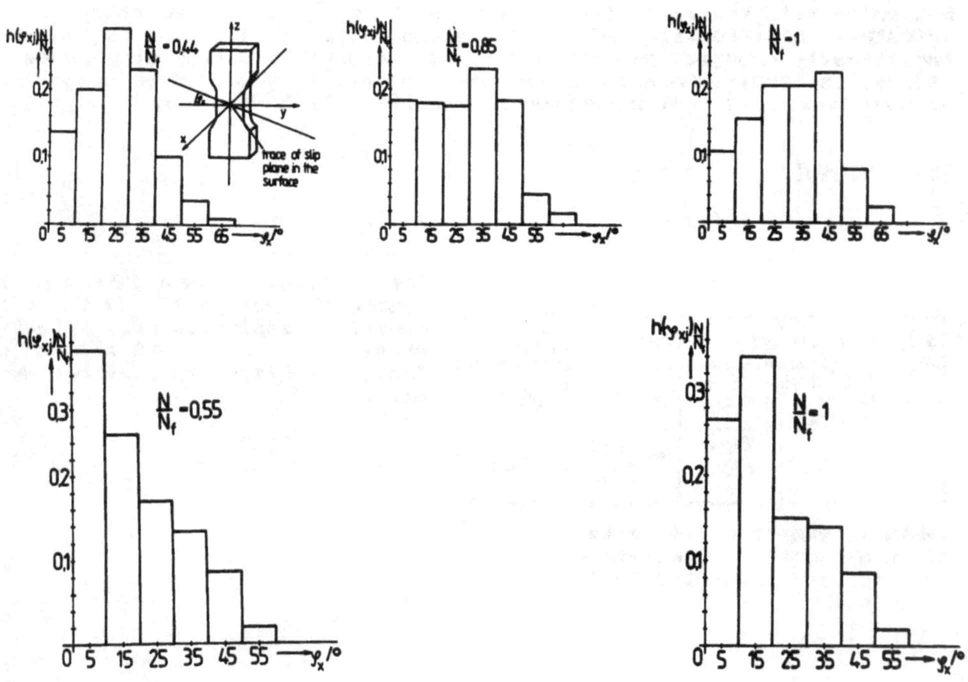

Fig. 2 Development of the azimuthal distribution of slip traces of the nonimplanted specimen 5 (upper series) and the implanted specimen 4

$h(\varphi_{xj})N/N_f$ ...frequency of groups of slip traces with azimuth φ_x in class j at cycle number N (in relation to the number of cycles to failure N_f); $\Delta\varphi_x = 10°$

Fig. 3 Numerical calculation of the azimuthal distribution of slip traces for different slip conditions (minimal orientation factor of the slip systems to become active); $\Delta \varphi_x = 10°$
The best fitting to the experimental results for nonimplanted material gives $\mu_{min} = 0,4$ with respect to the height of the maximum of the distribution and maximum of azimuth

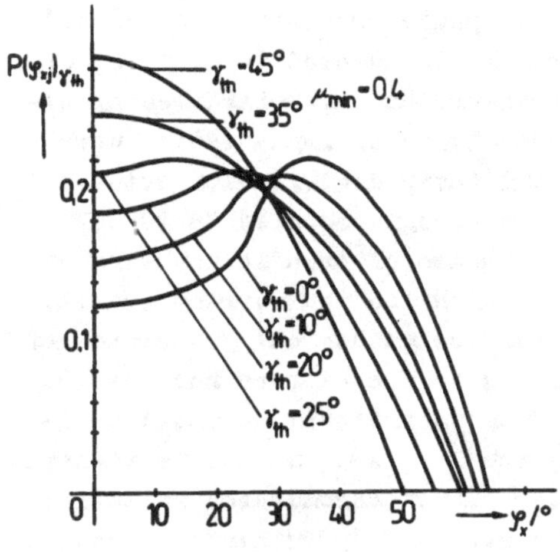

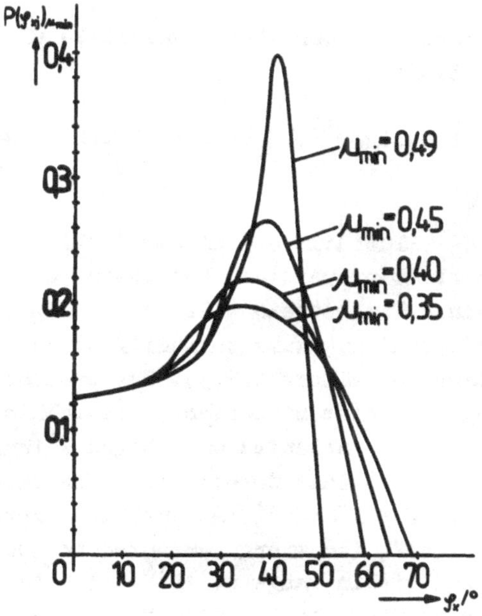

Fig. 4 Correction of the azimuthal distribution of the slip traces for slip condition $\mu_{min} = 0,4$ by reduction of all slip systems giving an angle between slip direction and the surface smaller then γ_{th}

Conclusions

Implanted material shows a characteristically changed azimuthal distribution of the traces compared to nonimplanted material. Numerical calculation of the distribution of the angle between slip direction and surface gives the possibility to correct the calculated azimuthal distribution of the slip traces, being similar to the analytically calculated results in /3/. This explains the measured difference between the azimuthal distributions of implanted and nonimplanted materials as result of the nonappearance of slip traces, produced by slip directions with a small angle with the surface, after implantation.

This is supported by the result that the correction of the azimuthal distribution reduces the absolute portion of cases leading to slip traces in the same way as ion implantation reduces the portion of observed slip active grains what is considered to be the main cause of life extension like reported in/1,2/. Further this agrees with the other result that the PSB-covered portion of the free surface of slip active grains is not decreased so hardly by ion implantation (see fig.1).

The development of the azimuthal distribution during fatigue seems to be a superposition of the decrease of γ_{th} (minimum of γ to produce visible slip bands) and of the decrease of μ_{min} caused by general and local hardening during fatigue.

R E F E R E N C E S

/1/ Hohmuth, K., Richter, E., Rauschenbach, B. and Blochwitz, C.: Mat. Sci. Engng., 69 (1985) 191 - 202

/2/ Vardiman, R.G.: in Hochman, R.F. (Ed.), Ion Plating and Implantation Applications to Materials (1986) 107 - 113

/3/ Preuß, Heinz H.W.: Cryst. Res. Technol., 21 (1987) 241-250

CHANGE OF MECHANICAL PROPERTIES OF Al-Mg ALLOY UNDER THE ION BEAM IRRADIATION

V.P.Goltsev[1)], G.Götz[2)], V.V.Khodasevich[1)], V.V.Uglov[1)], G.H.Choi[1)]

Aluminium and Al-based alloys have the complex of technically important properties, but their relatively low strength and wear resistance limit their wide use in new technologies. Ion implantation is one of promising ways to modify structure and phase composition of Al and Al-based alloys(1,2,3). The present article is devoted to studying changes of element content, crystallic structure and microhardness of Al-Mg alloy, implanted with B^+ and N_2^+ ions. Condition of implantation were chosen so that defect formation rates and their distribution coinside both for B^+ and N_2^+ implantation. Boron ion energy amounted to 40 KeV while N_2^+ ion energy amounted to 100 KeV. Doses of irradiation were varied within range of 10^{16}-$5\cdot10^{17}$ ions/cm^2 with the ion current density no greater than 10 μA/cm^2. Changes of surface mechanical properties of irradiated layers were estimated according to the Vickers hardness(H_μ) measurements. A range of charges on a diamond indenter amounted to 2-150 g with sonded depth interval being equal to 2-8 μm. X-rays investigation were carried out on difractometer in monochromatased radiation of cooper. Element composition analysis was made by Rutherford backscattering(RBS) and electron microprobe methods. Al-Mg alloy is multicomponent system, containing Mn,Fe,Si,Cu besides Al and Mg. Experimental

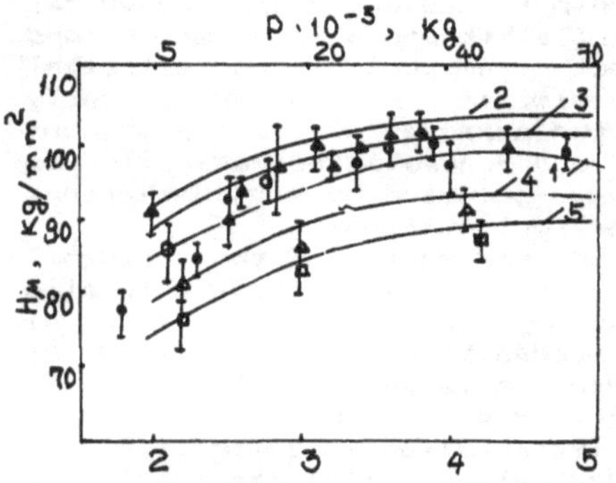

Fig.1. Relative change in hardness against depth for the Al-Mg alloy unimplanted(1) and implanted with various doses of N_2^+ ions: 2-10^{16}, 3- $3\cdot10^{16}$, 4- $6\cdot10^{16}$, 5-10^{17} ions/cm^2.

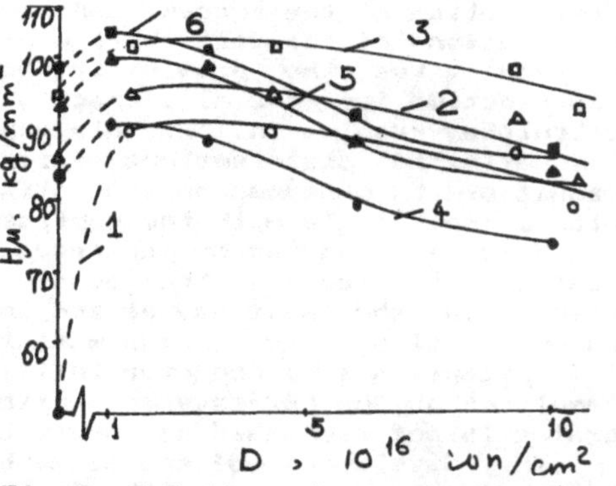

Fig.2. Relative change in hardness against doses implanted with ions B^+(1,2,3),N_2^+(4,5,6) in selected impress depth: 1,4- h=2 μm; 2,5- h=3 μm; 3,6- h=4 μm.

1) Byelorussian State University, Lehinskii Pr. 4, 220080 Minsk, USSR
2) Friedrich Schiller University, Jena, GDR

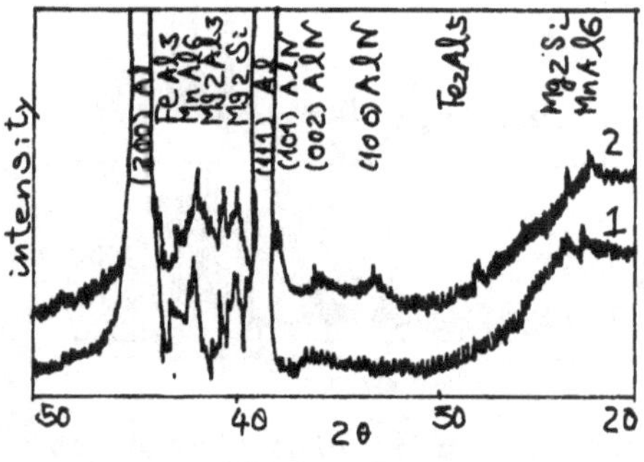

Fig.3. XRD patterns for the Al-Mg alloy unimplanted(1) and implanted (2) with dose of $10^{17} N_2^+/cm^2$.

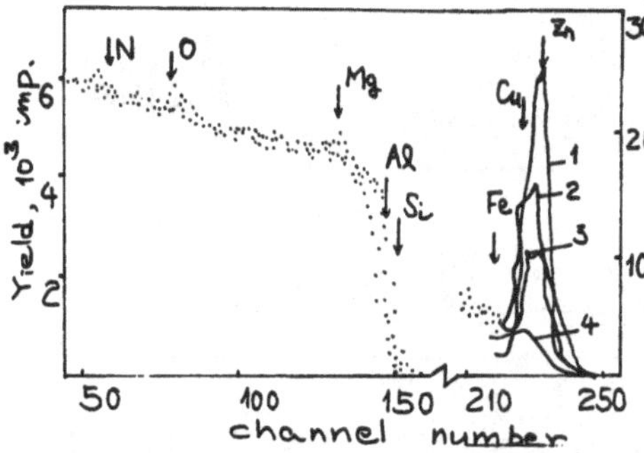

Fig.4. RBS spectrum of 1.4 MeV He^+ ions alloy Al-Mg unimplanted(1) implanted with various doses of N_2^+ ions: 2- 10^{16}, 3- $3 \cdot 10^{16}$, 4- $6 \cdot 10^{16}$ ions/cm².

results have shown a non-linear character of hardness changes as a function of dose and depth. Maximum H_μ grouth amounted to 60% with $4,5 \cdot 10^{16} B^+/cm^2$ at 2 μM. The X-rays phase analysis has revealed the presence of the following compounds in alloy: Al, β-Al_3Mg_2, Mg_2Si, $FeAl_3$, $MnAl_6$. During B^+ and N_2^+ irradiation alongside with redistribution of pattern intensities of these phase goes the formation of phases AlN and AlB_2. In addition, formation of phases Fe_2Al_5 and η-Fe_5Si_3 with nigh content of iron was noted. It should be noted that the formation of AlN is known during implantation of polycrystalline Al(99,99) in N_2^+ ions doses range 10^{16}-$10^{18} N_2^+/cm^2$ at 50,100,150 KeV(1). The paper has shown that as a result of implantation redistribution and segregation of doping impurities occured both in the range of 5 μm(result of RBS for Mg and Zn) and in the range 40 R_p(result of microprobe for Fe,Si,Mg). It was established that radiation-stimuated procceses of segregation and redistrbution of dopped impurities Mg,Fe,Si,Zn in Al-Mg alloy are controlled not only irradiation dose but also depend on implanted ion type and charac-

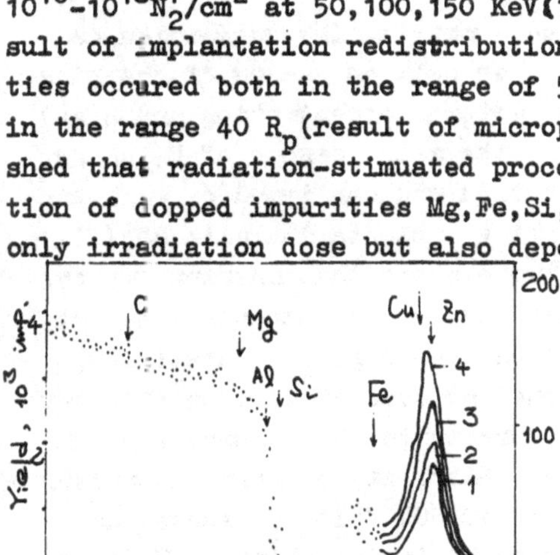

Fig.5. RBS spectrum of 1.4 MeV He^+ alloy Al-Mg implanted with various doses of B^+: 1-$1.5 \cdot 10^{16}$, 2-$4.5 \cdot 10^{16}$, 3-$9.1 \cdot 10^{16}$, 4-$1.5 \cdot 10^{17}$ ions/cm².

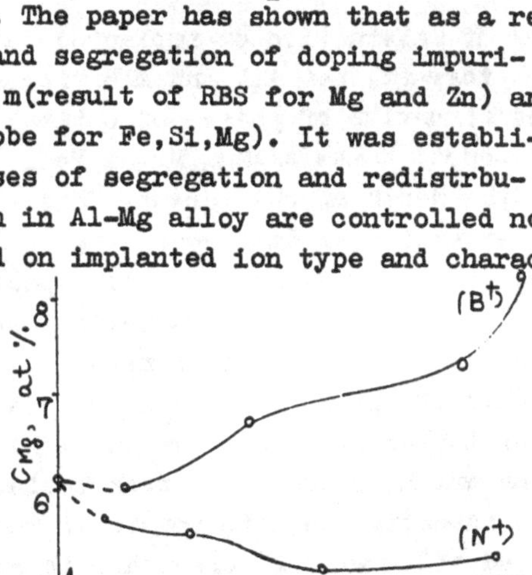

Fig.6. Dosent relative change of concentration alloy Al-Mg against doses N_2^+(1) and B^+(2) ions.

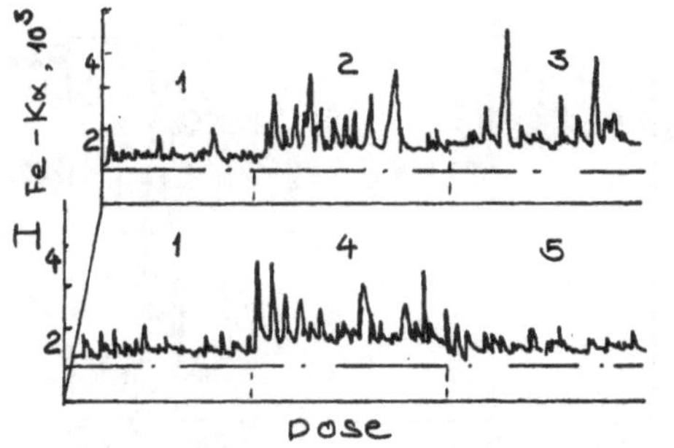

Fig.7. Intensity distribution of characteristic X-rays(Fe-Kα) in scanning line by surface of alloy Al-Mg unimplanted(1) and implanted with $N_2^+$and B^+ions:2-10^{16} N_2^+/cm^2, 3- 6·$10^{16}$$N_2^+$/cm^2,4-4.5·$10^{16}$ B^+/cm^2, 5-1.5·10^{17} B^+/cm^2,6-phone.

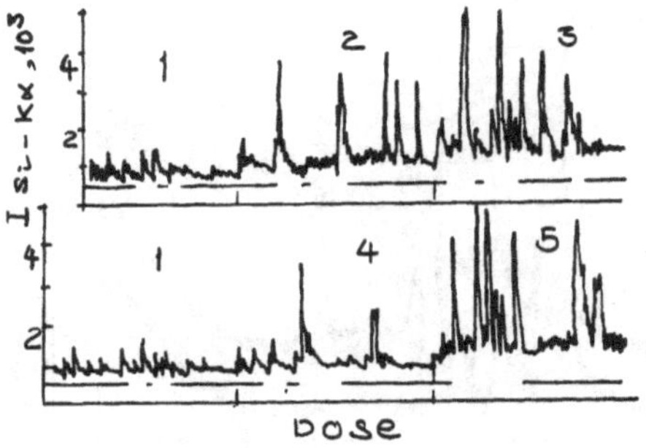

Fig.8. Intensity distribution of characteristic X-rays(Si-Kα) in scanning line by surface of alloy Al-Mg unimplanted(1) and implanted with N_2^+ and B^+ ions:2-$10^{16}$$N_2^+$/cm^2, 3- 6·$10^{16}$$N_2^+$/cm^2,4-4.5·$10^{16}$ B^+/cm^2, 5-1.5·10^{17} B^+/cm^2,6-phone.

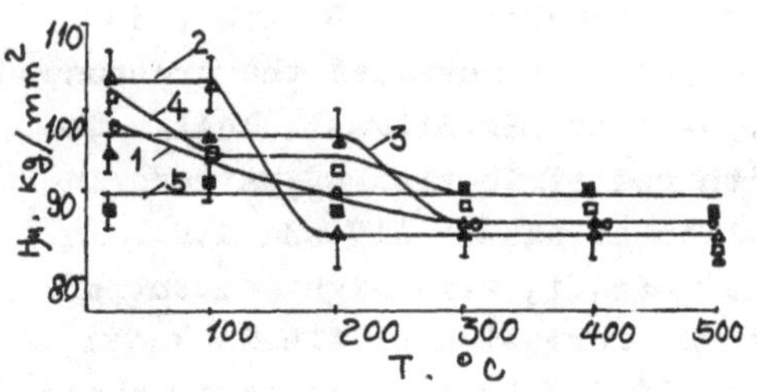

Fig.9. Relative change in hardness against temperatures of isochronous annealing for alloy Al-Mg unimplanted(1) and implanted with N_2^+ and B^+:2-1.5·$10^{16}$$B^+$/cm,3-1.5·$10^{17}$ B^+/cm,4-$10^{16}$$N_2^+$/cm^2,5-$10^{17}$$N_2^+$/cm^2.

ter of interaction of implanted ion and a matrix. The suggestion was put forward that AlN and AlB_2 formation as well as phases of changing and formation of phases of optimal parameters(concentration and size) of second phase(Al_3Mg_2,Mg_2Si,$FeAl_3$) are the main reasons of H_μ change on the depth of the target. This siggestion was confirmed by well-known experimental data according to Al alloy implantation with Sb^+(2) and Cu^+(3) ions. In these papers were shown that maximum improvement of mechanical characteristics(wear resistance, hardness) was at dose 10^{16}ion/cm^2. Author noted that second phase precipitates(AlSb, $FeAl_3$) of optimal size 1,2 μm were formed at dose 10^{16}ion/cm^2(2). Further implantation dose growth leads to precipitation parameter extention which influence on wear resistance. Stability of formed structures was investigated intemperatures range of 20-500°C.It was shown that irradiation with B^+ions leads to recovery beginning shift of H_μ to the range of higher temperatures. The magnitude of the shift is about 100°C.

References

(1) Iwaki M. Ohira S.:Nucl. Instrum. and meth. B119/20(1987)162-166
(2) Madakson P.B.:J. Appl. Phys. V 55(9)(1988) 3308-3314
(3) Madakson P.B., Dearnaley G.: Proc. Int.Ion Eng.Congr.-ISIAT'83, Kyoto, Japan 1983

STRUCTURAL-PHASE CHANGES IN FE FILMS IRRADIATED SUCCESSIVELY BY NITROGEN AND BORON IONS

GOLTSEV V.P., DRAKO V.M., KHODASEVICH V.V., UGLOV V.V., KULESHOV A.K.[1)]

Transmission electron microscopy was employed to study phase changes in thin Fe films subjected to successive implantation of nitrogen and boron ions and then to annealing in vacuum at temperatures ranged from 400°C to 600°C. Thin ion films were deposited on fresh NaCl chips by the electron-beam evaporation technicue. The film thickness did not exceed 100 nm. An ion energy was chosen so that the positions of maxima for the energy released by nitrogen and boron ions coincided. It amounted to 100 kV for molecular nitrogen ions, and 40 kV, for boron ions. Implantation doses varied from 10^{16} to $3\cdot10^{16} N_2^+/cm^2$ and 10^{16} to $6\cdot10^{16} B^+/cm^2$, respectively.

The experiments performed have revealed that nitrogen ion implantation at the chosen irradiation conditions results in formation of the phase ε-$Fe_{2+x}N$ in thin Fe films (fig. 1).

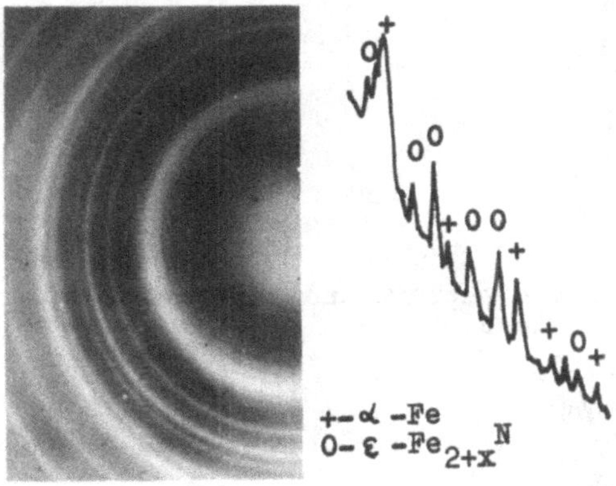

Fig. 1. Transmission electron diffraction pattern and the corresponding photometer scan obtained from 6E16 N_2^+/cm^2 implanted iron.

Boron ion implantation with doses 10^{16} ion/cm² and $3\cdot10^{16}$ ion/cm² leads to the Fe_2B phase formation (fig. 2). Diffrractional patterns from this films implanted with dose $6\cdot10^{16} B^+/cm^2$ nöt differ from initial ones. Amorphous layer of 5 nm was formed on the films surface (fig. 3).

Successively implantation of nitrogen and boron ions generates a phase cmposition with is not a simpe superposition of the phases formed by separate irradiation by nitrogen or boron ions.

The structure cinsisted of α -Fe and FeB is formed when nitrogen and boron doses correlation is $3\cdot10^{16} N_2^+/cm^2$ and $10^{16} B^+/cm^2$. A great

1) Byelorussian State University, Leninskii Pr. 4, 220080 Minsk, USSR

number of FeB precipitations are formed for doses correlation 1E16 N_2^+/ cm^2 and 3E16B^+/cm^2 (fig. 4).

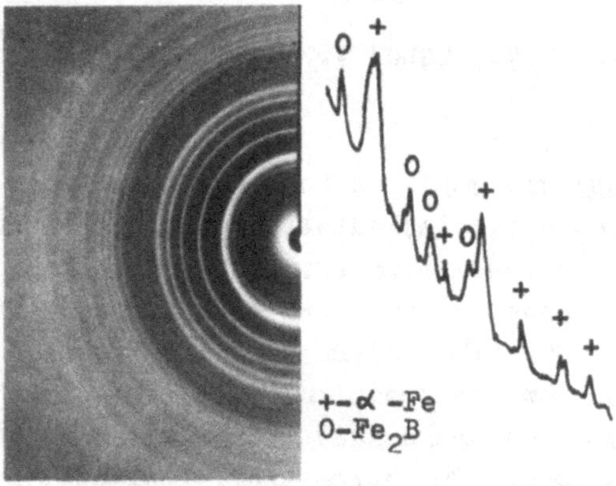

Fig. 2. Transmission electron diffraction pattern and the corresponding photometer scan obtained from 1E16 B^+/cm^2 implanted iron.

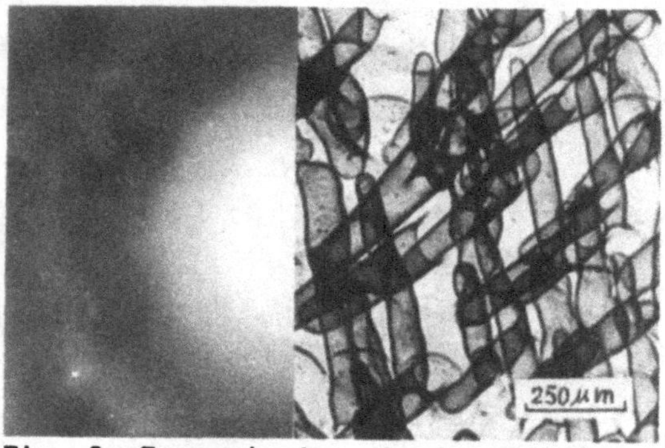

Fig. 3. Transmission electron diffraction pattern and microphotograph of Fe implanted with B^+(6E16 ions/cm^2) ions.

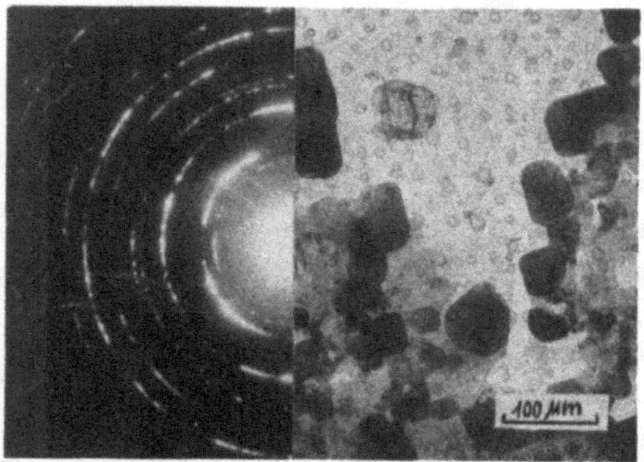

Fig. 4. Transmission electron diffraction pattern and microphotograf of Fe irradiated successively by N_2^+(3E16 ions/cm^2) and B^+(1E16 ions/cm^2) ions

With nitrogen and boron implantation doses 3E16 N_2^+/cm^2 and (1 to 6)E16 B^+/cm^2 respectively, a structure forms to be composed of α-Fe, $-Fe_{2+x}N$ and FeB (fig. 5).

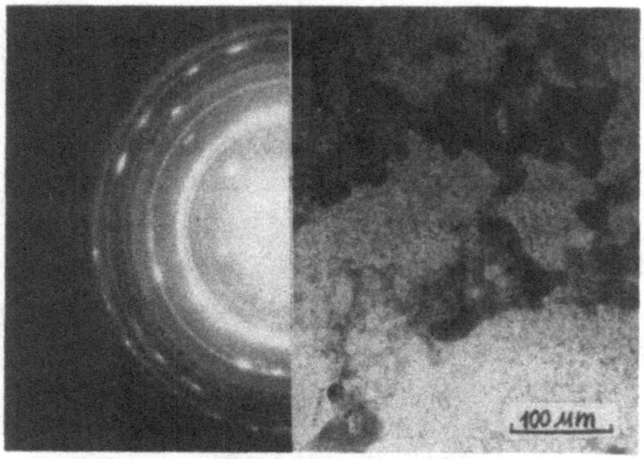

Fig. 5. Transmission electron diffraction pattern and microphotograf of Fe irradiated successively by N_2^+(3E16 ions/cm^2) and B+(6E16 ions/cm^2) ions.

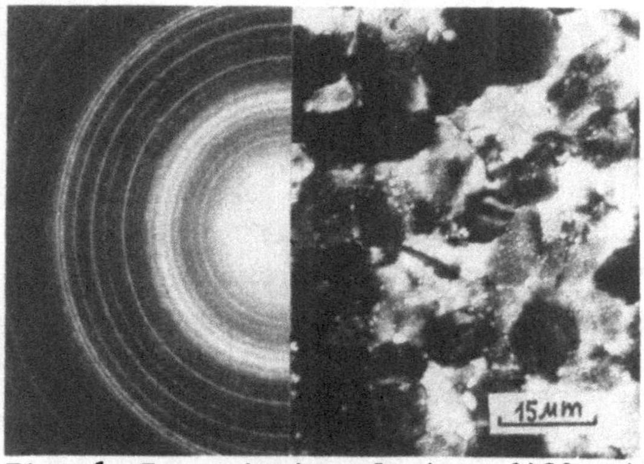

Fig. 6. Transmission electron diffraction pattern and microphotograf of Fe irradiated successively by N_2^+(3E16 ions/cm^2) and B+(6E16 ions/cm^2) ions and post annealing at 600°C.

Annealing of the films with this phase composition preparated by successively ion implantation reveals that the phase ε $-Fe_{2+x}N$ is stable in the total test temperature range. Besides, starting from 500^{0}C the phase γ'-Fe_4N develops. At the same time for all the annealing temperatures FeB signals allows us to suggest decomposition of this phase in the test temperature range (fig. 6).

IMPURITY REDISTRIBUTION DURING HIGH TEMPERATURE CREEP OF Sb^+-IMPLANTED Al-ALLOY

V.P.Goltsev, S.I.Zhukova, I.V.Kasko[1)]

The investigation of creep processes in metals and alloys has a scientific and practical interest. The knowledge of the determing creep mechanisms in the selected regimes of deformation gives a possibility to prognosticate the durability of material under load and change its properties. It is well known that the treatment of the surface leads to the considerable change of the behaviour of materials during the creep, so far as at the surface the plastic flows form. For example, the creep resistance increases after thin film deposition [1]. In this sense the use of the ion implantation as a way to modify the surface is promising.

The effect of the ion irradiation on creep of metals has practicaly not been studied [2], most of the papers study the possibility using of ion treatment to increase wear resistance, stability to corrosion and tribological properties [3].

The authors of [4,5] studied the influence of Sb ion implantation on parameters of the steady-state creep of D16 alloy. The doping of 10^{16} Sb^+/cm^2 was shown to cause the strength increasing. The following dose increasing returns the alloy properties to the initial values.

For the chosen range of temperature and stresses the steady-state creep rate $\dot{\varepsilon}$ can appropriately be described by the Weertman equation:

$$\dot{\varepsilon} = A\tau^n exp(-U/kT),$$

were U is the activation energy of the steady-state creep, n is the creep rate reponse to the stress τ . The values of the steady-state creep parameters are presented in the Table [4]:

Dose, Sb^+/cm^2	U, eV	n
0	1.32	3.0
10^{14}	1.29	3.0
10^{15}	1.33	2.7
10^{16}	1.32	2.0
$2 \cdot 10^{16}$	1.29	2.8
$4 \cdot 10^{16}$	1.34	3.3

The change of n shows the increase of the creep resistance up to dose 10^{16} Sb^+/cm^2 and the following loss of strength of alloy. The activation energy in the range of its error determination does not depend on implantation dose, it suggests the invariable mechanism wich controls the creep at the experiment. The dependence of microhardness on the dose of implantation correlates good with the

1) Byelorussian State University, Leninskii Pr.4, 220080 Minsk, USSR

results of the creep investigation (Fig.1). It is to be noted that the values of U, n and H depend on initial condition of the alloy. In [5] the investigated samples were annealed before the irradiation, whereas in [4] they were irradiated in the delivery condition.

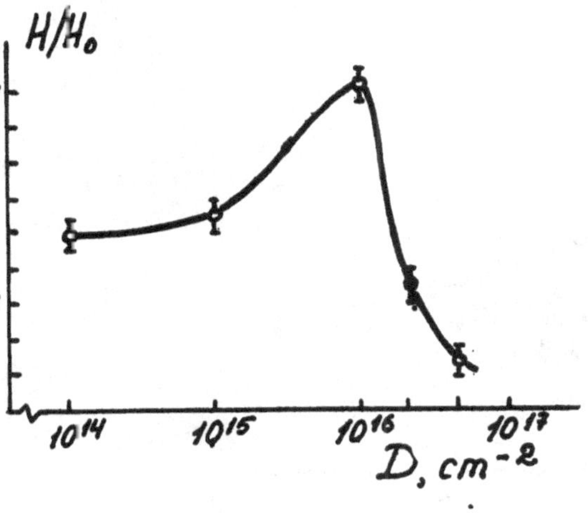

Fig.1 Dependence of microhardness on dose of the Sb^+ implanted D16 alloy

The initial alloy D16 is related to the ageing alloys and contains cooper as the main dopant and in addition -- magnesium, manganese and ferrum. The strength properties of the alloy are caused by the formation of surface compounds wich contain cooper - $CuAl_2$ and $CuMgAl_2$, so it is important to trace the cooper distribution in the implanted layers to have a possibility to interprete the change observed in the mechanical properties of the alloy.

In the present work the results of the investigation of the element composition change in the surface layers during high temperature deformation of Sb^+ implanted Al-alloy are presented. The implantation energy was 35 keV and the dose range - $10^{14}... 4 \cdot 10^{16}$ Sb^+/cm^2. The deformation was performed by using four-point bending of the samples at temperatures higher $0.5T_m$ and stresses up to 20 MPa.

The Rutherford backscattering of 1.4 MeV helium ions was used. The spectra from the initial and implanted with 10^{16} Sb^+/cm^2 samples before and after bending deformation in the state of creep at T = 600K and τ = 20 MPa were analised. The element depth distributions were calculated from the basic element signals and then they were made presise by fitting of the simulated and experimental spectra taking into account all elements.

The distributions of cooper and implanted stibium are shown in Fig.2. These results indicate that during the high temperature deformation the cooper diffuses to the surface of the sample and its concentration increases by a factor 1.3 in comparison with the average bulk value (curves 2, 2'). After the doping of 10^{16} Sb^+/cm^2 the relative Sb content at the distribution's maximum reaches 4%, but the perseptible redistribution of the cooper in the implanted layer (curves 1, 3) is not observed.

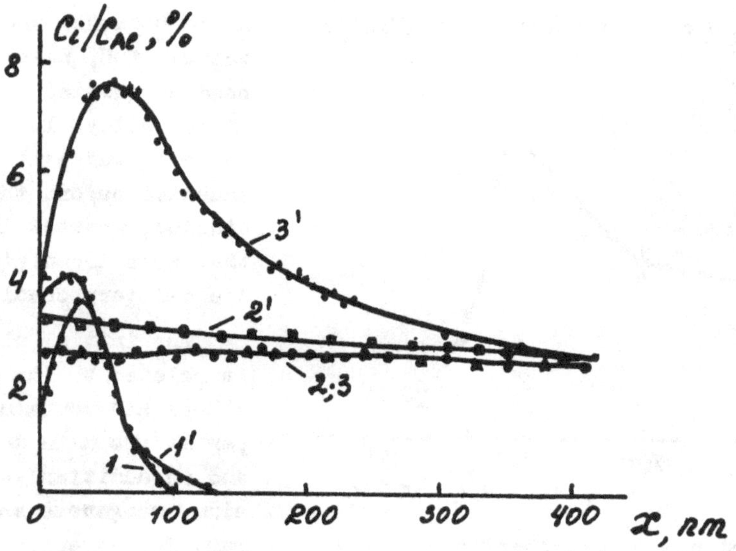

Fig.2 Depth distributions of stibium (1) and cooper (2,3) in the initial (2,2') and implanted alloy (1,1',3,3') before (1,2,3) and after creep (1',2',3')

But in the implanted samples after deformation the cooper segregation in the surface layer with the thinckness of $\geq 3\bar{R}_p$ is observed. The relative cooper content at the distribution's maximum at the depth of 56 nm is 7.8% and only at the depth of 400 nm it reaches the volume concentration. The cooper segregation in the surface layers of deformed samples caused by raising diffusion in the strength field of bended sample, so implanted Sb greatly enhances this process. Polygonization processes which control creep in the selected regimes of deformation are more intensive, because dislocation network taking place then in polygonization and promoting enhanced diffusion along dislocation tubes to the sample surface is formed during implantation. Increased Cu content causes alloy hardening as a result of dislocation sources blockage and stopping of mobile dislocations.

REFERENCES

(1) Yaroshevich P.Yu., Belomytsev Yu.S., Bulgak M.V. and Muchin I.P.: Metalicheskie monokristaly. Polychenie i issledovanie svoistv, Nauka, Moscow(1976) 104-107

(2) Goltsev V.P., Zhukova S.I. and Anishchik V.M.: Phys.Stat.Sol.(a), 96(1986) 135-139

(3) Hirvonen J.: Ion Implantation, Metallurgiya, Moscow 1985 (in Russian)

(4) Zhukova S.I. and Seichas E.: Thesises 1 Vsesojusnoi conferencii"Modificacija svoistv construkcionnich materialov puchkami zarjagennich chastits",ch.1, Tomsk(1988) 183-184

(5) Goltsev V.P. and Zhukova S.I.: Phys.Stat.Sol.(a),108,K1(1988)K1-K4

THE BORON IMPLANTATION EFFECT ON STRUCTURAL-PHASE CHANGES IN THE SYSTEM Zr-O/Ni PRODUCED BY ATOMIC MIXING

V.P.Goltsev[1), G.Götz[2), V.V.Khodasevich[1), I.I.Prikhodko[1), V.V.Uglov[1)

The ion beam mixing in solids is stipulated by three possible mechanisms. First, the mobile defects generated by the irradiation may give rise to an enhancement of diffusion over simple thermal diffusion. Second, the mixing may by induced by many succesive displacements of atoms in the solids by the collision cascades. Finally, if the cascades are gense cnough and sufficently long-lived then virtually all of the atoms in the cascade volumes may be instantaneously in motion what is known as "energy spikes". This also give rise to substantial mixing. All the processes and the phase transformation as well can result in the modification of properties of the implanted materials [1,2] .

The bynary system Zr-O/Ni was mixed by argon ions with the energy of 150 keV and dose of 10^{16} ion/cm^2. The next bombardment by boron ions with an energy of 60 keV was carried out in a dose range of (4 to 10)$\cdot 10^{16}$ ion/cm^2. An ion energy was chosen to have a maximum of the energy releasedin elastic collisions near the bynary system interphase. An elementary composition was studied by the method of Rutherford back scattering (RBS). A structure and a phase composition were examined by transmission electron microscopy (TEM). A sputtering rate for the interaction with argon plasma was determined experimentally from the secondary ion mass-spectroscopy (SIMS) data. Isochronous annealing of samples was performed in vacuum in a temperature range of 300 to 1500 K.

The elementary analysys has revealed that under the argon ions bombardement conditions the region of nickel and zirconium intermixing, developswith its dimension being of $2\cdot 10^{17}$ cm^{-2}in units of Nt(Fig.1).

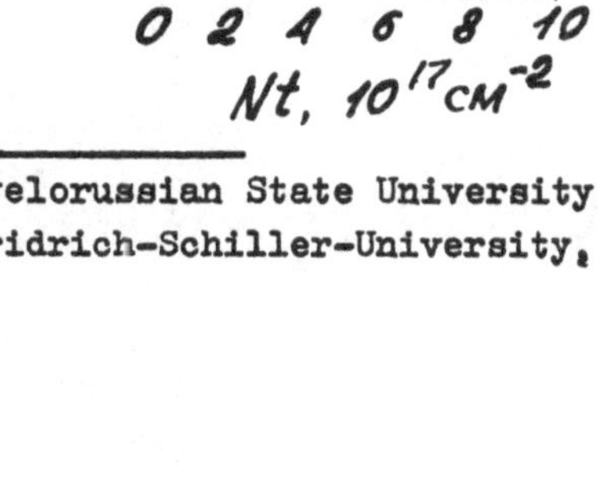

Fig.1 Nickel, zirconium and oxigen concentration profiles in films mixed by argon ions

1) Byelorussian State University, Leninskii Pr. 4, 220080, Minsk, USSR
2) Fridrich-Schiller-University, Max-Wien-Platz 1, 6900, Jena, GDR

Oxigen in the atomic-mixing samples concentrates in the upper zirconium film and its concentration at the surface attains 40 at.% to result in zirconium oxidation. The phase analysis has showed the obtained structure to be a mixture of two phases, i. e. Ni and ZrO_2 (Fig.2).

With the subsequent implantation of B^+ ions an interstitial boron profile forms an increase of the dose of implanted ions is accompanied by boron displacement from the surface layers of ZrO_2 into substrate Ni (Fig.3). TEM reveals that boron implantation leads to boride phase formation (Fig.4).

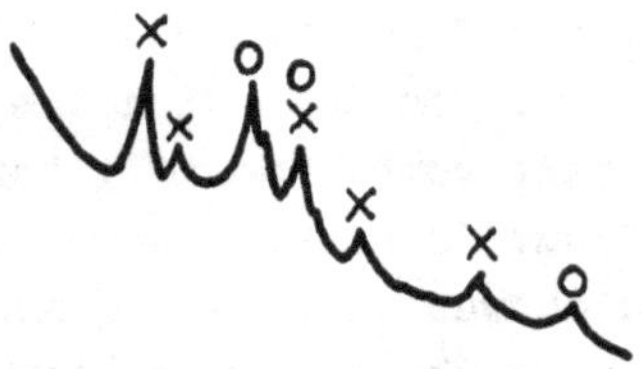

Fig.2 : Electron difraction pattern of atom-mixing films: × - ZrO_2; o - Ni

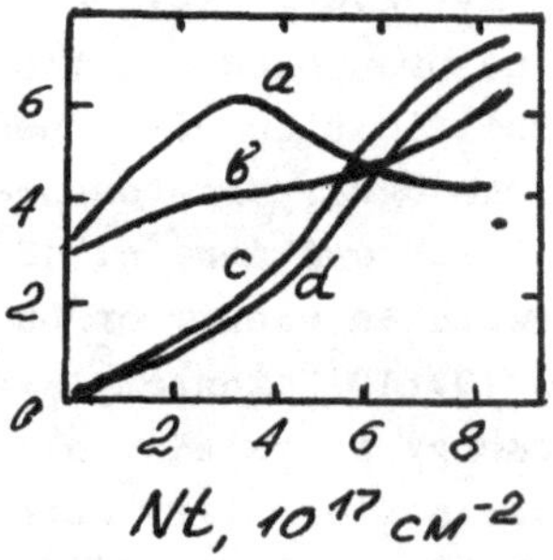

Fig.3: Concentration profiles of boron implanted in the atom-mixed samples with doses: a- $4\cdot10^{16}$; b- $5\cdot10^{16}$; c- $7\cdot10^{16}$; d- 10^{17} ion/cm^2

At the same time an abrupt decrease (by a factor of 4) in a sputtering rate of the ZrO_2 surface layer by low-energy (5 keV) argon plasma is observed (Fig.5). As far as boron is displaced from the surface layers, the sputtering rate increases. Simultaneously, microhardness of samples enhances, it increases two times as compared to the required samples.

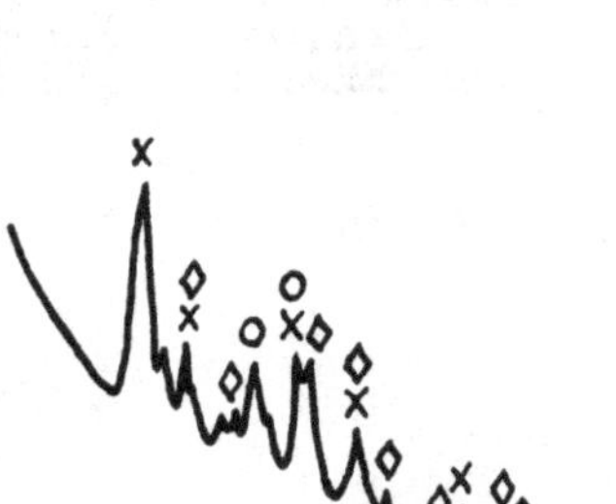

Fig.4 : Electron difraction pattern of atom-mixing films next boron implantation: × - ZrO_2; o - Ni; ◊ - Ni_4B_3

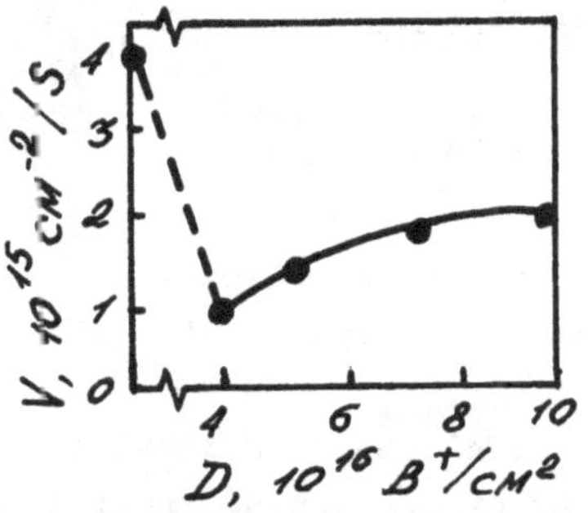

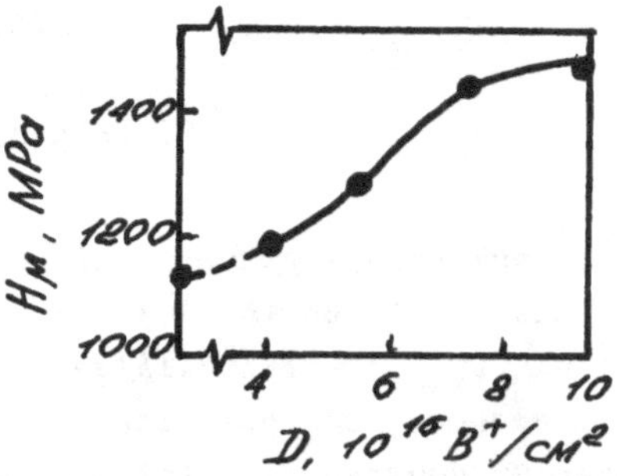

Fig.5: The sputtering rate of ZrO_2-layer under interaction by argon plasma versus boron implantation dose

Fig.6: The microhardness of the atom-mixed samples versus the boron dose at the subsequent implantation

Annealing has demonstrated that the structure and the phase composition do not change at temperatures ranged from 300 to 1100 K. With a temperature above 1100 K Ni_4B_3 decomposition and formation of a new intermetallic compound $NiZr_2$ are observed. At the same time the mixing region enlarges. The implanted boron retards its growth (Fig.7).

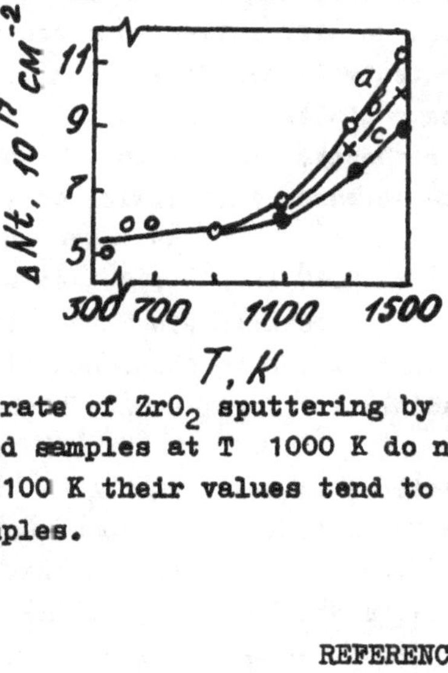

Fig.7: Altaration of the mixed region size versus the temperature in the atom-mixed samples (a); in the samplesaadditionally up to dose $7 \cdot 10^{16}$ (b) and 10^{17} (c) B^+/cm^2

A rate of ZrO_2 sputtering by argon plasma and microhardness of annealed samples at T 1000 K do not change. With a further increase of T 1100 K their values tend to quantities typical of non-irradicated samples.

REFERENCES

(1) V.V.Khodasevich, I.I.Prikhodko, V.V.Uglov: XVII All.Union Conf. Interaction of Charged Particles with Crystals, MGU, Moscow(1988)150-152

(2) V.P.Goltsev, V.V.Khodasevich, G.Götz, I.I.Prikhodko, V.M.Drako, V.V.Uglov: Phys.Stat.Sol.(a), 111(1989)K13-K16

IMPLANTATION INDUCED TEXTURE

K. Helming and B. Rauschenbach[1)]

The anisotropy of physical properties of polycrystalline materials depends mainly on the texture, i.e. on the orientation distribution function (ODF) of crystallites.

In general, the pole figures of bulk material can be determined by X-ray or neutron diffraction. For the study of texture of small areas (some μm) or thin layers ($< 1 \mu m$ thickness), the electron diffraction shall have to used, nevertheless the measurement of the electron intensity and its interpretion are very difficult.

Polycrystalline layers of titanium were implanted with 50 keV nitrogen ions at room temperature. The ion current densities were maintained below 5 $\mu A/cm^2$ and the ion fluence was $1*10^{17}$, $5*10^{17}$ and $1*10^{18}$ ions/cm^2.

After implantation, the layer structure was investigated by the transmission electron microscopy in combination with an in-side entry goniometer. Texture analysis using electron diffraction is geometrically similar to the X-ray film technique (details see Refs. /1,2/). If a sample with texture tilted through an angle then, in general, the measured intensity of the Debye ring is changed. From this, the hkl-pole figure is obtained by plotting the measured intensities on a Wulff net. A series of h_i-specific diffraction diagrams may be taken while tilting the sample by small angular steps (2.5^o or 5°) about an axis perpendicular to the primary beam direction (details see Refs. /1,2/).

The high-fluence nitrogen ion implantation in hexagonal titanium forms f.c.c. titanium nitride precipitates for fluences higher than $1...5*10^{16}$ ions/cm^2 /3/.

As example, a series of diffraction patterns is shown in Fig. 1, which must be taken while tilting the sample by small angular steps about an axis perpendicular to the primary beam direction. We found a radial-symmetrical behaviour (fibre texture, fibre axis parallel to the implantation direction) for the whole range of the measurement, i.e. the pole figures were independed on the azimuth angle.

For determination of the ODF, the strongest diffraction rings of α-TiN were chosen. In order of increasing interplanar spacing, the (111), (200), (220), and (311) reflections were used. The comparison of the intensities as function of the tilt angle was carried out on the base

----

[1)] Zentralinstitut für Kernforschung Rossendorf
DDR-8051 Dresden, PSF 19

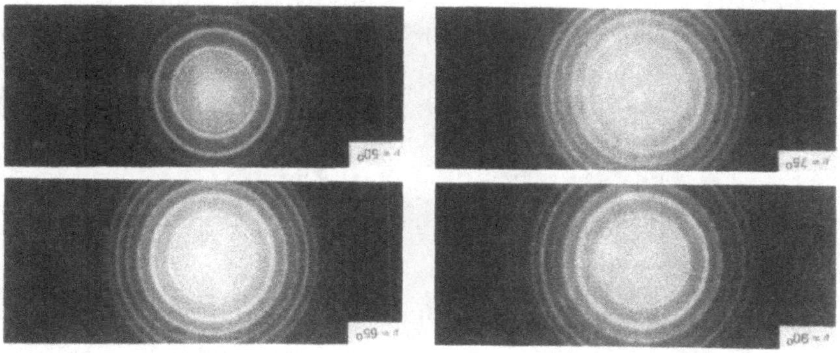

Fig. 1: Series of electron diffraction patterns (SAD) of nitrotitanium after tilting (ϑ is the tilting angle, $\vartheta = 90^{\circ}$ is parallel to the implantation direction)

of relative intensities, i.e. the diffracted intensities were normalized. In Fig. 2, the normalized intensities for the four reflections (1010)-Ti, (1011)-Ti, (111)-TiN, and (200)-TiN are illustrated as function of the

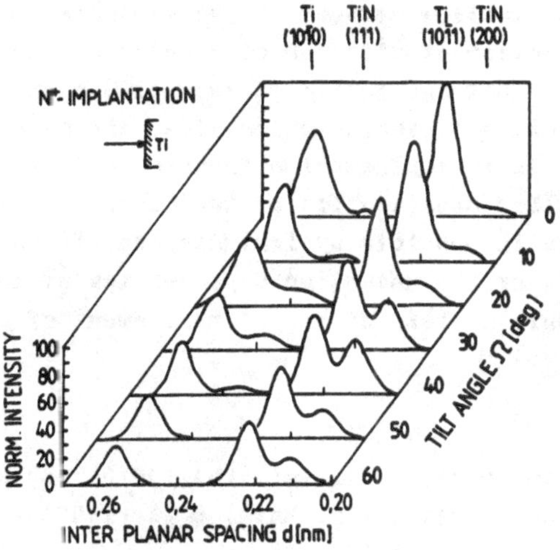

Fig. 2:
Diffracted and normalized intensities of selected Ti- and TiN-reflections versus the tilt angle for Ti after N-ion implantation at room temperature. The implantation direction was perpendicular to the sample surface.

tilt angle in 10°-steps after perpendicular implantation. It is clearly demonstrated, that the intensities are strongly depended on the tilt angle.

Then, an ODF was determined on basis of the (200), (220), and (111) pole figures. The associated ODF is exhibited in Fig. 3. The main component (111) is mark distinctly in contrast to the random distribution (about 12 times).

The microstructure of the titanium precipitates is characterized a preferred orientation (fibre texture), where the most intensive texture component, the (111) plane, is perpendicular to the direction of implan-

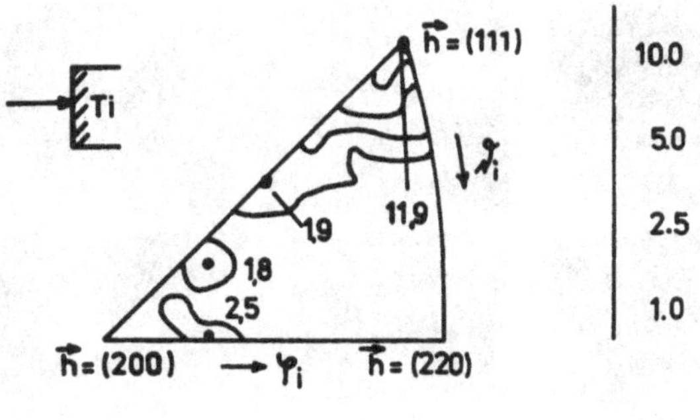

Fig. 3: Orientation distribution of normals of crystal planes h_i parallel to the surface normal of TiN-sample after implantation perpenticular to the surface

tation parallel to the surface of the sample. The aligment effect was explained by the so-called seed selection process through ion channeling i.e. the implanted ions should be destroyed a major fraction of grains and preserve only grains of of the implantation-induced TiN of crystallographic orientation where the lattice damages can be minimized by ion channeling (e.g. the |111| direction of titanium nitride precipitates).

The measurements show that the normal of the (111) plane is strongly coupled with the direction of implantation, i.e. the high-fluence ion implantation leads to an implantation induced texture. The results indicate that a controlled modification of the texture of implantation induced precipitates is possible by implantation. It follows the possibility of modification of the anisotropic properties of polycrystalline material, for example in respect to an improvement of wear or corrosion resistance.

References

/1/ K.Helming and B.Rauschenbach: phys. stat. sol. (b) 138 (1986) K1
/2/ B.Rauschenbach and K.Helming: Nucl. Instr. Meth. B 41 (1989)
/3/ B.Rauschenbach: J. Mater. Sci.: 21 (1986) 395

SYNTHESIS OF COMPOUNDS IN MOLYBDENUM DURING ION IMPLANTATION COMBINED WITH THERMAL AND PULSED ELECTRON ANNEALING

Kh.R.Kazdaev, M.T.Akchulakov, E.M.Bayadilov [1)]

Majority of experimental works oriented to solving the task of establishment of phase formation mechanism in ion implanted layers are performed on thin films whose thickness can be compared with projected range of bombarded ions. At the same time the data on phase transformation proceeding for ion implantation in subsurface layers of massive sample can differ from similar results on films and arouse greatest practical interest. Moreover, it is necessary to perform versatile investigation relating not only to problem of revealing the peculiarities of phase transformation depending on implantation conditions but also search of postimplantation methods directed to change of formed structure state or obtaining new ones.

The paper is dedicated to studying the possibilities that can be presented by ion implantation method in combination with thermal and electron impulse annealing for chemical compound synthesis in subsurface layers of transition metals and to investigation of character of phase emerge and decay depending on implanted impurity type for different conditions of thermal treatment.

For investigation were used massive samples of single crystal molybdenum implanted by chemical active elements (see Table). Ion energy values were chosen for obtaining identical distribution profiles. Thermal annealing was carried out with 100° step in 200-1200°C temperature range. Impulse annealing was made by 140 keV energy electron beam at 15 $A \cdot cm^{-2}$ current density. Impulse length varied in experimentally defined range (6-16 μks) with 2 μks step. Structure investigation of implanted layers was carried out by means of X-ray diffraction analysis by using glancing beam. Sample surface state control after impulse annealing was performed by scanning electron microscopy (SEM).

The results of analysis after implantation and subsequent thermal annealing allows one to distinguish some peculiarities. a) Depending on the character of phase formation elements used may be divided into three groups. Nitrogen and carbon atoms forming corresponding compounds with Mo during implantation are related to the first group. The atoms of S, P and Si that form molybdenum compounds as a result of subsequent thermal annealing at temperatures exceeding definite threshold values are related to the second group. Boron is the transition element between these two groups whose compounds with Mo arise both at implantation and as a result of subsequent thermal annealing. b) The phases

[1)] Institute of Nuclear Physics, Academy of Sciences of Kazakh SSR Alma-Ata 480082, USSR

TABLE

Phase transformation in ion implanted molybdenum with variant regimes of thermal treatment.

Ion	E, keV (dose 5×10^{17} cm^{-2})	Thermal isochronous annealing				Impulse annealing	$\frac{r_x}{r_{Mo}}$ [1]
		Observed phases	Structure	Temperature, °C of formation	Temperature, °C of decay	Observed phases	
N	45	δ-MoN	hexagonal	after im-plan-tation	700-800	decay	0.50
		γ-Mo_2N	fcc		800-900	γ-Mo_2N	
C	40	γ'-MoC	hexagonal		800-900	α-MoC_{1-x}	0.55
		Mo_2C	hexagonal		1200	Mo_2C	
B	35	MoB_2	hexagonal		700	decay	0.58
		α-MoB	bct	700	1200	decay	
		Mo_2B	bct	1100	1200	Mo_2B	
S	105	Mo_2S_3	monoclinic	700	1000	Mo_2S_3	0.75
P	100	Mo_3P	bct	900	1200	Mo_3P	0.79
		Mo_4P_3	o'rhombic			Mo_4P_3	
Si	90	Mo_3Si	cubic	900	1200	Mo_3Si	0.84

appearing during implantation process have more close-packed lattice as compared to the substrate materials ones; in case of phases formed at additional thermal annealing the structures may have various lattice types. c) Implantation elements of the second and transition groups cause amorphization of the implanted layers while in the case of the first group such clearly expressed effect is not observed.

The regularities found allow one to make few suppositions relative to phase formation mechanisms in ion implanted molybdenum. The comparison of implanted particle atomic size with matrix one (Table) shows that element division into groups depending on phase formation character is in good correlation with Hagg rule for transition metals according to which the structure is determinated by radius relation $r = r_x/r_{Me}$, where r_x and r_{Me} are impurity and metal radii, respectively. If r is less than 0.59, the simple structures are formed, in the rest cases it forms complex ones. It gives rise to the supposition that under ion implantation the character of phase formation and their structure in transition metals will be determined by radius relationship. If r is less than 0.59, the phases with more close-packed lattice will arise in ion implantation process. In the rest cases the compounds will be formed after additional annealing only and their structure depending on implanted impurity concentration is determinated by corresponding diagrams of phase states. In the region, where r is near 0.59 both the first and second phase formation character are expected. The reason of the regularity observed is apparently in the fact that the stresses arising in implanted atomic arrangement can be removed by means of polymorphous transformation of more close-packed structures [2] while r does not exceed 0.59 value. For further increase of the implanted atomic size the disordered displacement of matrix atoms takes place. Therefore, the formation of compounds will occur at additional annealing and be characterized by phase transition temperature of solid disordered solution to the chemical compound. The structure of the phases is to be formed according to the state phase diagrams and determinated by implanted impurity concentration, which is steadily formed in implanted layers for reaching phase transition temperature.

The experimental data analysis on impulse annealing was performed on the basis of the experimental dependence of heating temperature on impulse length obtained in supposition of melting the surface when eutectic temperature of compounds is reached. The results show that in molybdenum samples implanted with the elements of the first group (N,C) the increase of temperatures to eutectic values causes the transformation of the structure compound state from texture to polycrystal; for the recrystallization from melt the block structure appears. For the second element group (S, P, Si) and boron the synthesis of the corresponding compounds takes place for temperature exceeding threshold values found in the experiments on thermal annealing. And the stoichiometry and state of the arising structures appear to be mainly similar to those formed at usual annealing. Further, increase of heating temperature causes such structure state changes of compounds as in the first element group. The comparison of synthesis compound results by ion implantation in combination with thermal and impulse annealing shows that using impulse annealing is more preferable since it helps to avoid significant redistribution of impurity concentration in the process of heating up to transition temperature. Moreover, impulse annealing gives possibility to vary the state of the compound structures avoiding the decay due to great heating and cooling rates of ion implanted layers.

REFERENCES:

(1) Samsonov V.G.: Tugoplavkiye soyedineniya, Metallurgizdat, Moscow, 1969.
(2) Umanskii Ya.S.: Karbidy tverdykh splavov, Metallurgizdat, Moscow, 1947.

PRACTICAL APPLICATIONS OF ION IMPLANTATION FOR WEAR PROTECTION

A. Kolitsch, E. Richter, U. Scholz

The possibility of applying the ion implantation for a selected modification of some surface sensitific properties of materials has been well known and investigated in the last time. Properties such as hardness, wear and corrosion resistance, friction or fatigue can be improved without affecting the bulk properties of the materials. The discovery of the process is generally acknowledged to be in 1973 (1) and the first industrial use of ion implantation on the production level was in 1985 for the treatment of titanium-based hip joints (2). The current state of the ion beam technology and present prototype processing equipments in hand enable the treatment of precision products and expensive tools. However, a broad industrial application is still very limited.

The ion implantation process has many advantages against other surface technologies such as physical vapor deposition (PVD), electron beam hardening or laser assisted deposition :

- surface properties can be modified selectivly and independently from the materials bulk properties,
- the concentration of implanted ions can be controlled easily,
- in contrast to conventional coating processes ion implantation change only the near-surface composition without production of discrete layers on the materials surface,
- There is not a sharp interface in the implanted surface layer, no problems of bonding failure or delamination of coatings will be,
- Alloys coated by ion implantation are not limited by classical thermodynamic parameters,
- Ion implantation is a low temperature process, capable of being applied to finished tools without changing its dimensions.

The important disadvantages of the implantations process are the very high costs of the equipment, the small thicknes of the implanted layer (1um) and the intricated target handling systems for complicate formed tools; the implantation process is a line-of-sight-process.

Academy of Sciences of the GDR, Central Institute of Nuclear Research Rossendorf, P.O. Box 19, Dresden DDR-8051

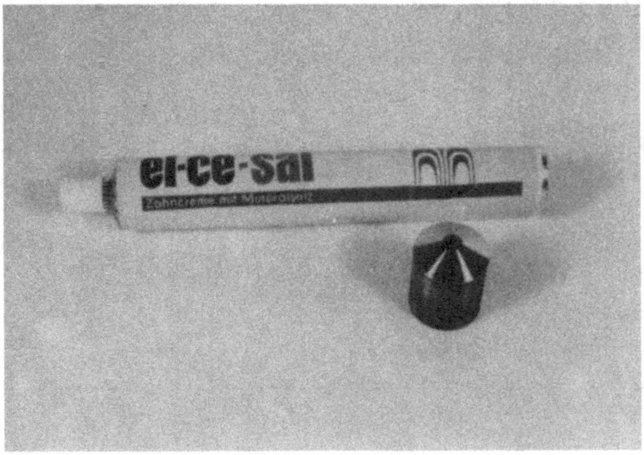

Fig. 1
punching tool for aluminium tubes
(lifetime x 4)

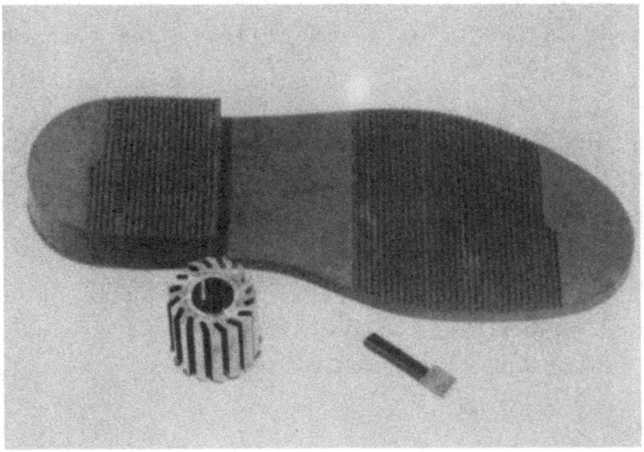

Fig. 2
Cutting tool for plastics
(lifetime x 2-4)

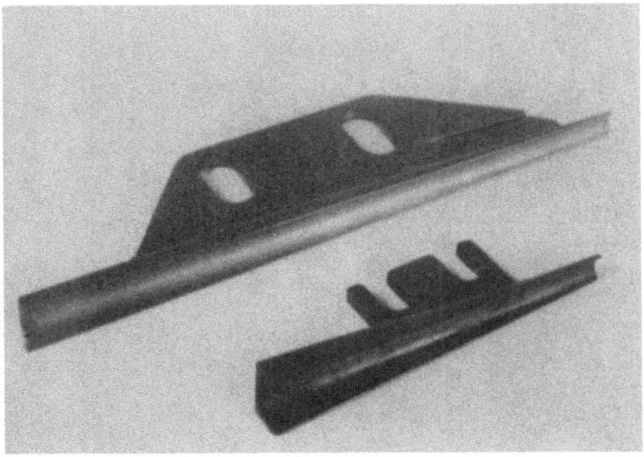

Fig. 3
Forming tool for cigarettes
(lifetime x 2-5)

The simplest commercial implantation process is the implantation of nitrogen. The possibilities of the nitrogen ion implantation for modifying tribotechnical properties of steels we have demonstrated not only in fundamental experiments or laboratory tests but also for different tools in industrial trials. In table 1 a review is given on the investigated tool types.

Table 1
Examples of investigated industrial tools
- implanted dose: $8x10^{17}Ncm^{-2}$, energy : 50 keV -

tool	materials	lifetime factor
drills	W-Co	2 - 4
puncher for tubes	steel 210Cr46	3 - 4
forming tool for cigaretts	steel 210Cr46	2 - 5
cutter for plastics	steel 90MnV8	4 - 10

Figure 1 - 3 show some industrial tools after implantation of nitrogen ions (dose : $8x10^{17}Ncm^{-2}$, energy : 50keV) for practical application. The improvement of the lifetime is given in the legende.

Summary

The examples show that nitrogen ion implantation improve the wear resistance of some industrial tools proofed under true production conditions by factors between 2 and 10.

References

(1) Hartley, N.E.W. et al., J. Mater. Sci 8 (1973) 900

(2) Sioshansi, P., Nucl. Instr. & Meth. B 19/20 (1987) 204

RESIDUAL STRAIN ANALYSIS OF Ar^+ and N_2^+ IMPLANTED 304 SS.

D. C. Kothari*, L. Guzman* C. Cestari*, L. Lutterotti** and P. Scardi**

It is reported by various groups that the martensitic transformation takes place during nitrogen implantation on 304 SS . However there are few attempts to measure strain in such a system. The present work concentrates on the elaboration of X-ray diffraction (XRD) spectra of argon and nitrogen Implanted 304 SS to obtain the homogeneous and local strain (i. e. rms-microstrain) values, texture and grain size. The possible physical state is deduced from the above elaborations and also the mechanisms of such physical changes have been proposed. Ar implantation is performed in order to differentiate between the physical and chemical effects present during nitrogen implantation on 304 SS.

The specimens used for implantations were electropolished 0.5 mm thick foils of 304 SS. The implantation parameters are listed in the table shown later. The details of XRD analysis are described in reference [1]. The peak separation was performed using a least square method. The line profile analysis, to obtain rms-microstrain and crystallite size, was performed using the Single Peak (SP) [1, 2] and/or Warren-Averbach (WA) [3] methods. The preferred orientation parameter f was determined with respect to the (111) peak, using the method desribed in reference [1]. In this formalism f takes values between -1 to +1 and $f = 0$ indicates no preferred orientation whereas $f = -1$ or $+1$ indicate complete preferred orientation in either ways. The homogeneous strain is calculated for the (111) peak using the formula, strain = (change in d)/d.

The XRD spectrum of unimplanted specimen (No. : 1) exhibited very narrow peaks due to austenite indicating no strain and large grains size. The XRD spectrum of nitrogen implanted specimen (No. : 2) showed broad peaks due to austenite and additional peaks due to

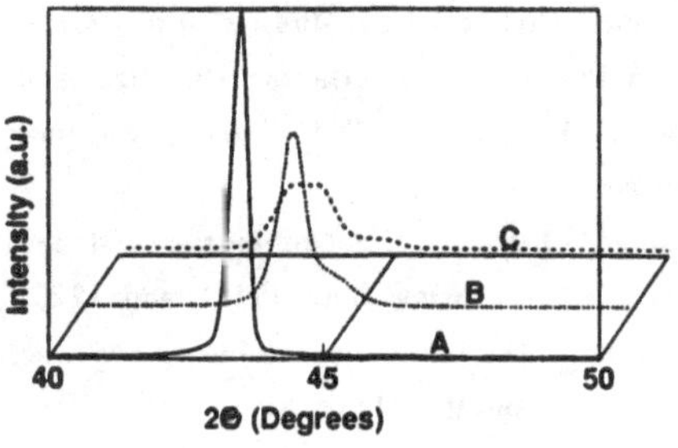

Fig. 1 XRD spectra exhibiting (111) austenite and (110) martensite peaks. : A) unimplanted, B) nitrogen implanted C) argon implanted.

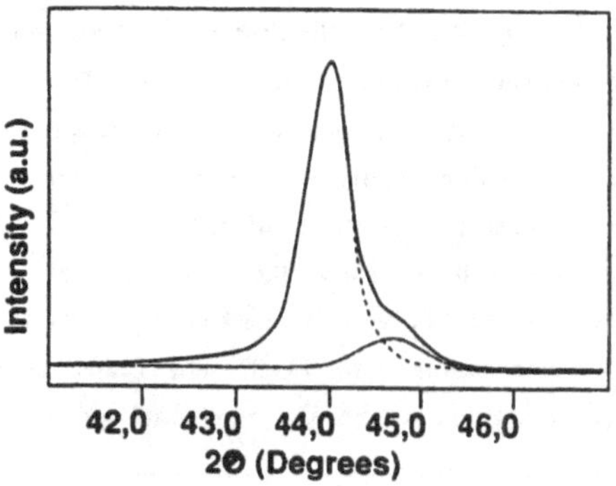

Fig. 2 Austenite (111) and martensite (110) peaks separated for sp. 2.

* : I. R. S. T , I-38050, Povo, Trento, (Italy). ** : Deptartimento di Ingegneria, Universita di Trento, I-38050, Mesiano di Povo, I-38050, Trento, (Italy).

martensite. It was deduced that the nitrogen implantation reduces grain size, introduces local strain and produces martensite. The argon implanted specimen (No. : 3) also had similar effects. In addition to these effects, in this specimen, there were additional peaks near all the peaks due to austenite. These peaks are interpreted to be due to an additional austenite component having larger d values. This also indicates that, the presence of Ar causes homogeneous strain, in addition to local strain, in the specimen. In Fig. 1 all the above effects can be seen prominently. After both N_2^+ and Ar^+ implantations, the (200) peak disappeared from the XRD spectrum. Thus the preferrred orientation parameter f = 1, for this peak. This effect is present upto the depth of X-ray penetration, which is two orders of magnitude larger than the range of implanted species. This texture effect is ascribed to recrystallization during implantation. It is possible that during recrystallization some planes are preferrred because of the presence of implanted species. For (220) peak a moderate preferrred orientation is observed (for sp. 2, f = + 0.16 and for sp. 3, f = - 0.13).

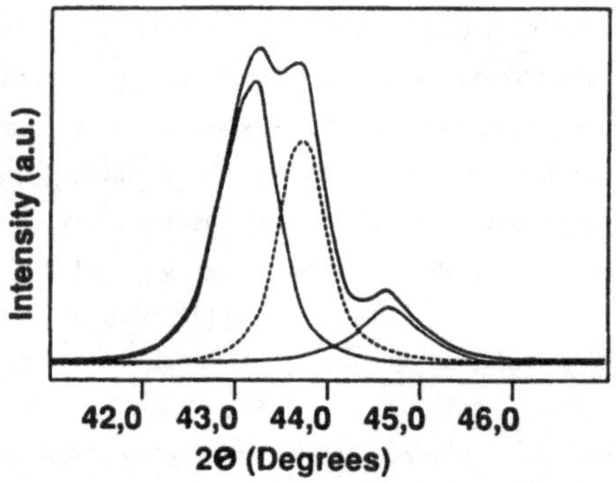

Fig. 3 Two austenite (111) and martensite (110) peaks separated for sp. 3.

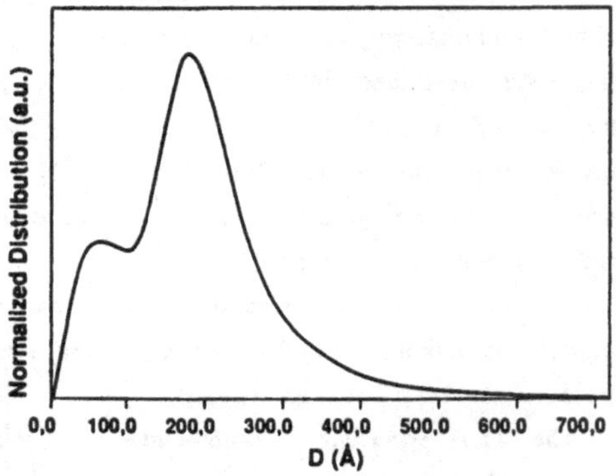

Fig. 4 Normalized rms-microstrain destrbution computed using the WA method for sp. 2.

For sp. no. 2, the crystallite size distribution obtained by the WA method is bimodal indicating the presence of crystallites with two sizes probably situated at two different depth regions (Fig. 4). The distrbution of rms micro-strain was observed to have an intermediate trend between the constant one and $(D)^{-1/2}$ dependent one. This can be due to the mixed contribution from the presence of dislocations and stress fields due to interfaces. The size and strain distributions could not be obtained for the specimen no. 3, since the WA analysis was not possible because of overlap of three peaks in the XRD spectra.

Figs 5 to 8, show the strain and size values computed for austenite, martensite and the second austenite (which is marked as Aust+Ar in the figs.) phases, using both (111) and (220) peaks of austenite and (110) and (220) peaks of martensite. The following table gives the details of specimens and the results of profile analyses carried out on the (111) XRD peak :

Spec No.	Ion	En. keV	Dose atoms/cm^2	profile analysis method	which peak	Cryst size Å	Local strain x10^3	Homogeneous strain
2	N_2^+	30	$1x10^{18}$	SP, WA	Aust.	190	3	
3	Ar+	30	$1x10^{18}$	SP	Aust.	175	2.5	-
					2nd Aust.	140	4.5	0.013

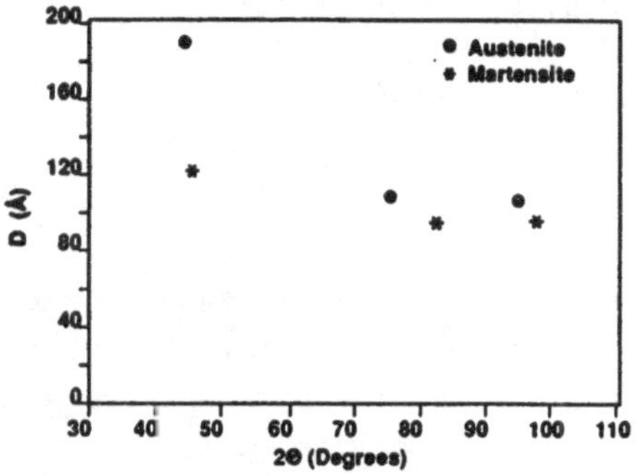

Fig. 5 Grain size of nitrogen implanted specimen computed using various XRD peaks.

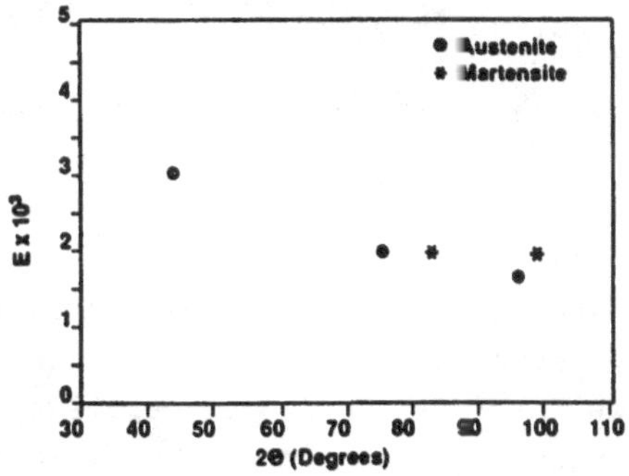

Fig. 6 RMS microstrain of nitrogen implanted specimen computed using various XRD peaks.

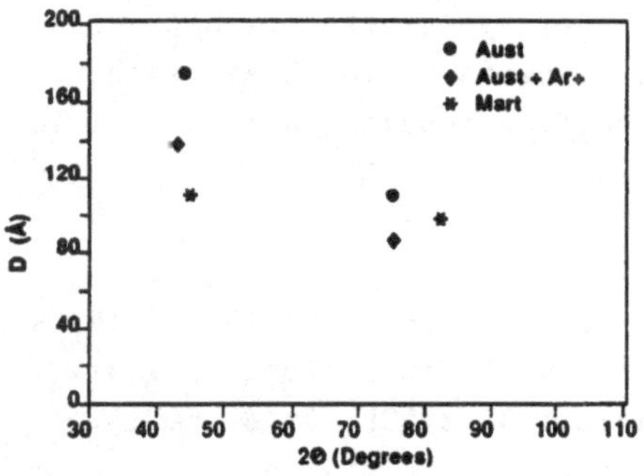

Fig. 7 Grain size of argon implanted specimen computed using various XRD peaks.

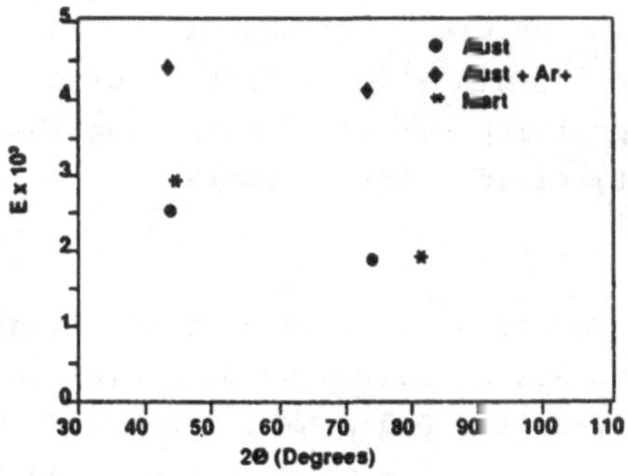

Fig. 8 RMS microstrain of argon implanted specimen computed using various XRD peaks.

It can be seen that, both the implantation species refine the microstructre (i. e. reduce the grain size) perhaps by creating defects. The strain and size values are almost isotropic, except for specimen 2, both the values are higher for the (111) austenite peak. The martensite formed in argon implanted specimen (no. 3) can be stress induced, since moderately high local strain and very high homogeneous strain are observed. The second austenite component has a higher local strain and lower crystallite size than the first one. It is possible that Ar bubbles are present in the specimen. In such a case it is highly probable that the second austenite component is in the vicinity of bubbles. In nitrogen implanted specimen the formation of martensite need not be stress induced, since no homogeneous strain is present and the local stress is not very high. Possible Cr-depletion, either because of formation of Cr-nitrdes or radiation induced segregation of Cr, could cause chemical induced ferritic transformation (it is not possible to destinguish between the ferrite and martensite using the present XRD data).

REFERENCES

[1] D. C. Kothari, P. Scardi, S. Gialanella and L. Guzman, Phil. Mag., To be published.
[2] R. K. Nandi, H. K. Kuo, W. Schlosberg, G. Wisler, J. B. Cohen and B. Crist Jr., J. Appl. Cryst., **17** (1984) 22.
[3] B. E. Warren and B. J. Averback, J. Appl. Phys., **23** (1952) 1059.

WEAR RESISTANT COATINGS PRODUCED BY C^+ IMPLANTATION

C. Neelmeijer[1], P. Knothe[1], M. Posselt[1] and E. Richter[1]

1. INTRODUCTION

Ion beam techniques are modern methods to improve metal surface properties such as hardness, resistance to wear and corrosion. In this connection the implantation of reactive ions into a surface layer to introduce chemical compound formation as well as atomic mixing at the surface layer - substrate interface is a promising technique /1,2/. To create a customary hard coating we performed C^+ implantation into a titanium layer which was vapour deposited onto steel. Pin-on-disk measurements show a surprisingly high wear resistance exceeding the expected value be more than two orders of magnitude. The complementary application of both RBS and $^{12}C(d,p_o)^{13}C$ nuclear reaction analysis discerns C^+ ion beam assisted deposition (IBAD) of carbon on the sample surface together with a C-Ti ion beam mixing (IBM) process.

2. EXPERIMENTAL

By means of electron beam evaporation a titanium layer of 105 nm thickness was deposited on mechanically polished 210Cr46 tool steel (2.1 at%C; 0.32 at%Si; 0.3 at%Mn, 0,02 at%P; 0,02 at%S; 11,31 at%Cr; 0.16 at%Ni). After that C^+ ions were implanted in two steps E_1 = 50 keV and E_2 = 20 keV. Low beam currents I_1 = 12 μA and I_2 = 5.5 μA, hence long irradiation times t_1 = 810 min and t_2 = 760 min, had been used to implant the C^+ ion fluencies $\phi_1 = 8\cdot10^{17}$ at/cm^2 and $\phi_2 = 4\cdot10^{17}$ at/cm^2, respectively, into the 4.3 cm^2 target area. Oil diffusion pumps coupled with lN_2 baffles generated a 10^{-3} Pa vacuum inside the specimen chamber.

The wear behaviour of the nonimplanted and C^+ implanted Ti-deposited steel samples was examined by means of a pin-on-disk arrangement. The volume abrasion was determined versus the sliding distance of a fixed hardened (HR_C = 60) bearing ball (3 mm diameter, 0.8 N load) on the rotating steel disk (1 m/s velocity) in air without lubrication.

Fig. 1 compares the high energy segments of three RBS spectra. With regard to spectrum NI (nonimplanted) after C^+ implantation three significant changes are obvious from spectrum I (implanted):

1) Central Institute for Nuclear Research Rossendorf, PF 19, Dresden, DDR - 8051, GDR

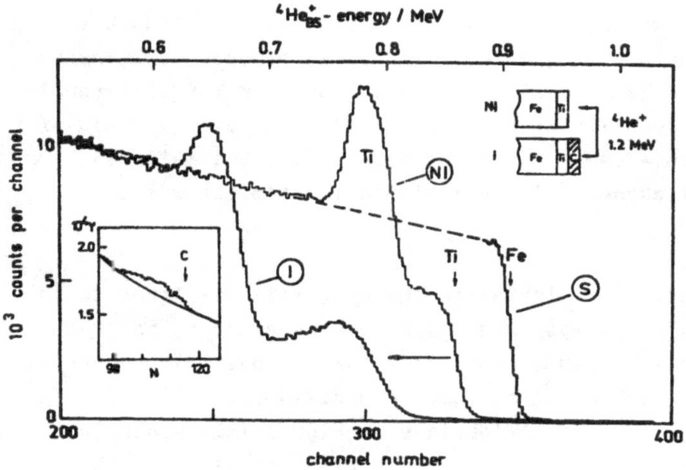

Fig. 1

RBS spectra observed at 1.2 MeV ^4He random incidence
$\vartheta = 170°$,
$\Omega/4\pi = 1.1$ msr
Q = 50 μC

S - steel substrate
NI - Ti on steel, nonimplanted
I - Ti on steel, C^+ implanted

Ti edge shift due to carbon covering ⟵

- the Ti surface signal is shifted towards lower energies,
- the high energy edge is broadened and
- a carbon signal appears in the low energy part of the spectrum (insert of Fig. 1).

For explanation carbon growth was assumed to take place during the implantation process changing the intended surface layer structure.

As shown in Fig. 2 oblique deuteron incidence and glancing proton detection ($\alpha = 75.5°$, $\vartheta = 170°$) enabled the $^{12}C(d,p_0)^{13}C$ reaction to observe a well resolved carbon depth profile within the 200 nm near sur-

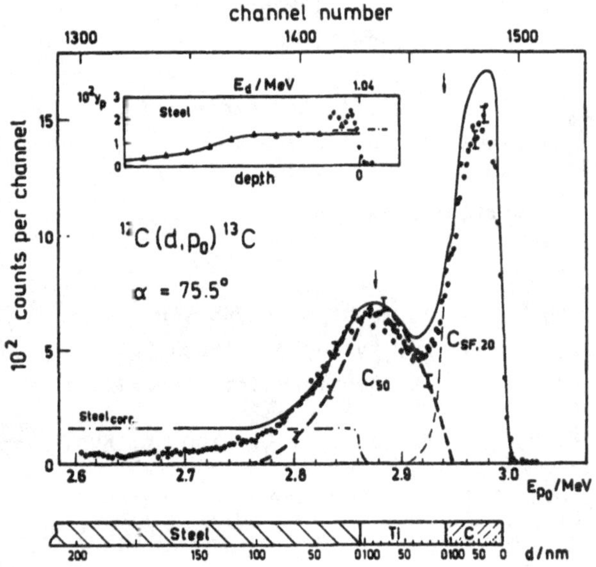

Fig. 2
$^{12}C(d,p_0)^{13}C$ reaction

p_0-spectrum: 105 nm Ti on steel after C^+ double implantation (50 keV, 20 keV)

• as measured, —— multiplied by $1/\sigma(E_d)$, —·— substrate contribution, ▬ ▬ ▬ 50 keV implantation profile C_{50}, - - - C surface layer +20 keV implantation profile $C_{SF,20}$, insert: C background due to the C content within the steel matrix face region of interest. In accordance with RBS (Fig. 1) a carbon covering of 106 nm thickness (C as stopping material /3/) was deduced from the energy width of the surface layer signal C_{SF}. Starting from the real sample structure, steel substrate - Ti(105 nm) - C(106 nm), the energy scale of Fig. 2 was converted into a depth scale for each individual material. After corrections /4/ the real shape of the C implantation profile was deduced.

3. RESULTS

After Ti deposition and C^+ implantation the wear rate was found to be lowered by a factor of 1000. A wear reduction of such a degree had been discussed only in connection with diamond-like carbon, a-C:H layers or Ti-carbide coatings prepared by reactive vapour deposition /5/.

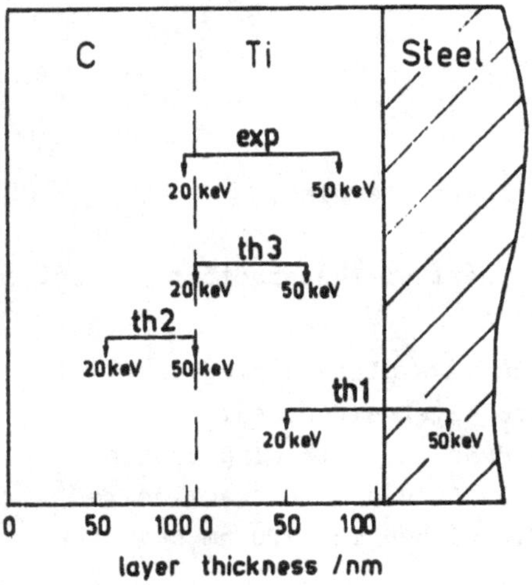

Fig. 3

C^+ profile maxima

exp - experimental positions
th1 - theoretical, no C covering
th2 - theoretical, 106 nm carbon
th3 - theoretical, dyn. C growth

As obvious in Fig. 3 TRIM Monte Carlo calculations allow to describe the observed positiones of the implantation profile maxima if dynamic C growth is taken into account /4/. Moreover, the shape of the double implantation profile as measured and calculated (th3) is found to be very similar. Consequently, the desired sample structure (th1) was drastically changed (exp) due to systematic carbon deposition during the implantation. Originally IBM via 50 keV C^+ implantation was intended. However, really C^+ IBAD of carbon takes place accompanied by IBM at the C-Ti interface forming Ti-carbide. Both processes are attributed to create the large increase of wear resistance observed for C^+ implantation using low ion beam currents and vacua of sufficiently high C partial pressure.

REFERENCES

/1/ Smidt, F.A., B.D. Sartwell, Nucl. Instr. Meth. B6 (1985) 70

/2/ Rauschenbach, B., R. Küchler, M. Posselt, Vakuum 38 (1988) 987

/3/ Ziegler, J.F.: Handbook of stopping cross sections for energetic ions in all elements: Pergamon Press New York, Vol. 5 1980

/4/ Neelmeijer, C., E. Hensel, P. Knothe, M. Posselt, E. Richter, Nucl. Instr. Meth. B42 (1989) 369

/5/ Weissmantel, C., G. Reisse, H.J. Erler, K. Bewilogua, H. Ebersbach, Le Vide Suppl. 196 (1979) 487

STRUCTURE-PHASE TRANSPORTATION IN BIMETALLIC SYSTEMS UNDER POWER PULSE INFLUENCE

*Plotnikov S.V.,*Rusin Y.G.,*Stuckert Y.A.,**Remnjev G.E.,
**Pogrebnjak A.D.

*Ust-Kamenogorsk Institute of Civil and Highway Engineering.
**Scientific Research Institute of Nuclear Physics.Tomsk.

Introduction

One important aspect of the power ion beam influence (PIB) on the surface layer of the sample coated by thin film of the other matter is mixing of matrix.Under certain conditions the influence of PIB is the result of the surface microcratering.The character of beam influence on the matter is largerly determined by the convectionally heat and mass-exchange processes in the crater zone.

Experimental

There had been studied the multilayer coated samples on Bcc-Fe base,the samples P6M5 and Y9 steel and as well the samples of Bcc-Fe,Cu and Al after the influence of PIB.

The sample irradiation was performed by "TEM" accelerator worked in technical regime with carbon ion beam parameters: energy E=(0,2-0,4) MeV,pulse current density j=(20-250) A/cm^2,pulse life r=(60-100) ns,pulse number is of I to 5.Beam consisted of 70 % of C and 30 % of H$^+$.

Transmission electron microscopy (TEM), raster electron microscopy (REM),Rutherford backscattering analysis (RBS) and non-standard methods on determining wear and corrosive resistance of irradiated samples were used for this studies.

The inverstigations showed that microcraters of I0 to 50 mkm in size were formed on the sample surface under the influence of PIB at the interval of 50-I20 A/cm^2.Crater density and its appearance depend on the Thermophysical characteristics of material,the deficiency of structure,the beam current density and on the distance from accelerator focus (fig. I a).

The crater formation is connected with energy releasing on defects of structure.The investigation of new ground samples showed that in the process of radiation under the surface layer appear structures resembling the crater section (fig. I b).

The size of this local sections and the depth of their deposition corresponds to the crater size and to the depth of their possible nucleation.The local energy release in such region leads to the origin of vapor-liquid phase in a small volums in which convectional mass-transport intensively takes place. The peculiarity of mass-transport in the process of PIB radiation was studied on Bcc-Fe samples with multilayer coatings (~100) nm of dissolved and insoluble in it components. The RBS data showed that in the process of systems irradiation intermixing of the components with the matrix occur. In the Fe-Pb system were found isolations of 10 to 60 nm in size. The studies of electrone microdifraction pattens of selected area showed, that the availability of isolations is connected with electronogram reflexes and correspond to the interfacial distances d_{III}=0,26 nm and d_{220}=0,10 nm. This proves the existance of a new FePb phase having cubical superstructure and the lattice period of which is about 0,45 nm (fig. 2).

After PIB irradiation in practically insoluble Fe/Ag system layer mixing occur. The surface studies by TEM method showed, that any compounds and structural regulations were not found by this. Ag is arranged in the form of separate isolations similar to spherical form (fig. 3). After irradiation in partly dissolved in a solid state Fe-Cu system a component mixing also occur. As a result of PIB influence the magnitude of Cu dissolvability in Fe greatly exceeded the dissolvability limit and became equal to 40-50 at.% Cu. The rest part of Cu disposed in the form of separate isolations (fig. 3).

Conclusions

Powerful ion beam irradiation gives the possibilities to modify the surface layer of materials, to synthetize new "banned" by phase diagram compounds, and to form nonequilibrium amophous stfuctures. Powerful ion beam irradiation gives rize to a considerable (substantial) extention of solubility limit of second components.

The peculiarity of formed structures gives the possibilities to raize many technological characteristics of the surfaces of steel and alloys. Synthesis of compounds in geterogeneity systems is possible under realization of two interchangeable conditions: firstly to mix the components and secondly to fix the new compound. Of all the possible mechanisms of mixing (thermodiffusion, barodiffusion ... ets.), the most probable is the convectional mass-transport in microcrater zones.

The analysis demonstrates, that mixing occures during the time of the order of 10^{-7}-10^{-8}s, the coefficient of diffusion is about $10^{-3}cm^2$, the formation of nonequilibrium amorphous structures occur under pressure about (10^3-10^5) bar and under the temperature considerably exceeding the temperatures of substance melting. The formation of a given structures is fixed owing to high velocity of cooling (more 10^6K/s).

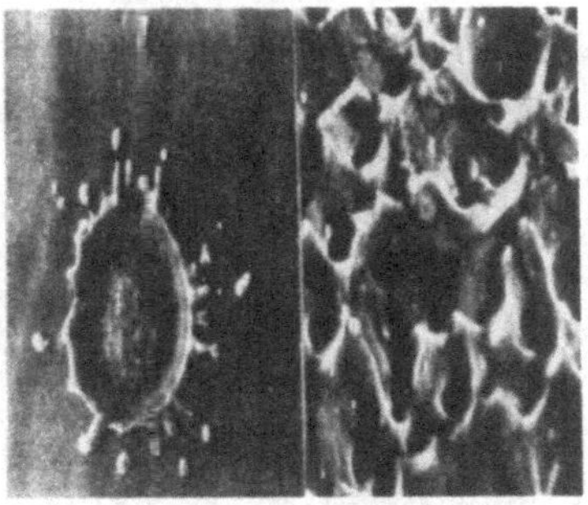

Fig. 1a. Surface of Al after ion irradiation by j= 80 A/cm^2 and j= 120 A/cm^2 (SEM), x 300

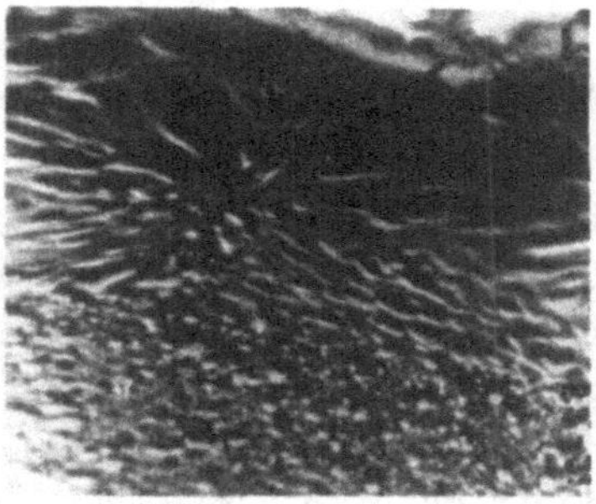

Fig. 1b. Ground sample of steel R6M5 after ion irradiation (crater local section) (SEM), x 500

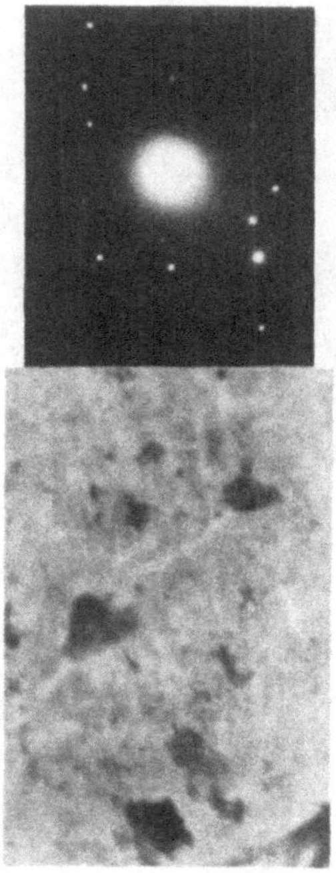

Fig. 2. Fe/Pb system after ion irradiation, x 2000 (TEM)

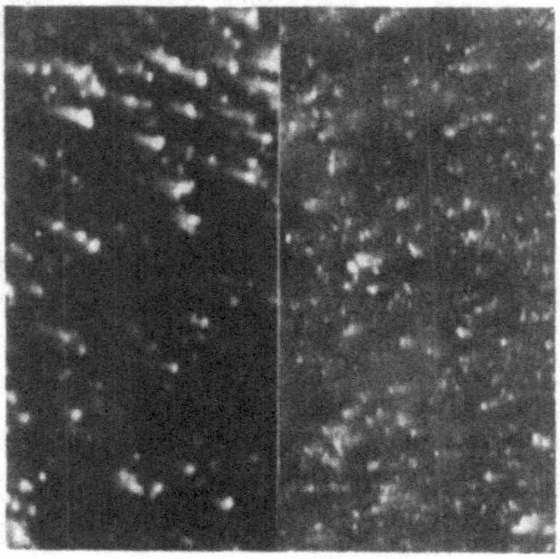

Fig. 3. Fe/Ag and Fe/Cu systems dark field imagines (TEM)

STRUCTURAL AND PHASE CHANGES IN NEAR SURFACE LAYERS OF HIGH POWER ION BEAM IRRADIATED HARD ALLOYS

A.D.Pogrebnjak*, O.Radjabov, Sh.M.Ruzimov, A.Emiraliev, V.P.Yanovski*

The paper deals with the investigation of hard alloys TN-20 (80% TiC, 16% Ni, 4% Mo) and VK8 (92% WC, 8% Co) prepared in the form of tetrahedral rods which have been irradiated by high power carbon ion beams on the TONUS accelerator [1] with the beam parameters : ion energy of 0.2 to 0.5 MeV, current density of $j = 30$ to 250 A/cm^2, pulse duration of 60 to 80 ns . To measure microhardness, we employed the MICROMET device (USA). The element analysis was performed using the apparatus for secondary ion mass-spectrometry (SIMS), with 10^{-5} % element sensitivity and about 30 Å layer by layer resolution. To investigate a near surface structure of alloys, the X-ray analysis and

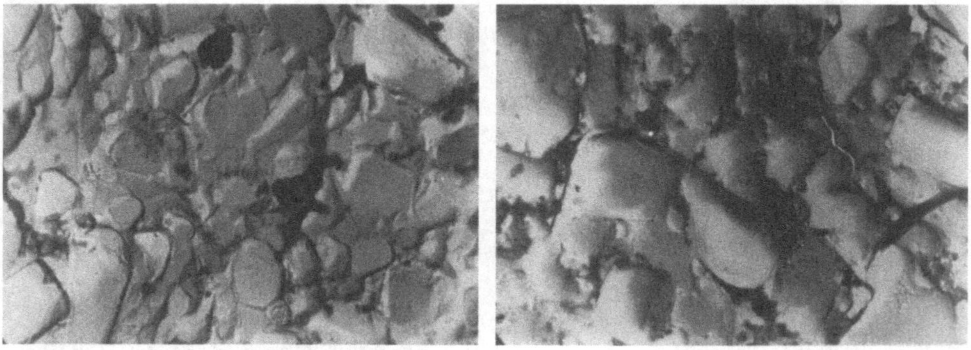

a b

Fig. 1 .
Electron microscopy photos for initial (a) and irradiated (b) VK8 samples

raster electron microscopy were employed. The lattice parameter of a carbide phase practically remained unchanged in irradiated TN-20, and the lattice parameter of a metal phase increased significantly . The diffraction analysis and electron microscopy photos allowed us to conclude about significant amorphization of a metal phase in TN-20 surface and about high temperature etching in it. As a result, deeper

Research Institute of Applied Physics, Tashkent, USSR
*Nuclear Physics Institute, 634050 Tomsk, P.O.Box 25, USSR

layers enriched by titanium and with higher hardness characteristics appeared in the surface The latter is supported also by microhardness measurements performed in irradiated TN-20 which show 1.3 to 1.8-fold increase in microhardness depending on the current density. The incfease in the lattice parameter of a metal phase may be stipulated by titanium enrichment from the carbide phase as a result of irradiation, i.e. the irradiation stimulates the substance deposition within the alloy. This is supported by the SIMS analysis which shows changes in the element composition in TN-20 surface after irradia - tion. The electron micro scopy photos of VK8 alloy (Fig.1a) show distinct carbide grains of various forms and dimensions (0.5 - 12 μm) Irradiation (Fig.1b) results in fusing the grain boundaries, high temperature etching and as a result, intergrain boundaries get outcropp- ed. Figure 2 shows X- -ray patterns for initi- al and irradiated sampl- es VK8. One can see a cubic tungsten carbide (a= 0.422 nm) having been formed as a result of irradiation which is stable at high tempera- tures and fixed at RT by quenching with a very high speed. Diffraction lines of this phase (Fig. 2 c) are fairly broad which evidences a shallow dispersivity of formations of this phase. Microhardness investigated in a modified layer in which a for- mation of a metastable phase (WC_x) was observed showed that HPIB ir- radiation resulted in increased microhardness by a factor of 1.2 to 1.5. The questicn about formation of a metastable cubic carbide phase and its role in improving the mechanical characteristics after HPIB irradiation are of the most interest, since appearance of the third phase in alloys of VK-series (free carbon, η-phase) usually results in deterioration of their mechanical characteristics [2] . Probably, a cubic tungsten

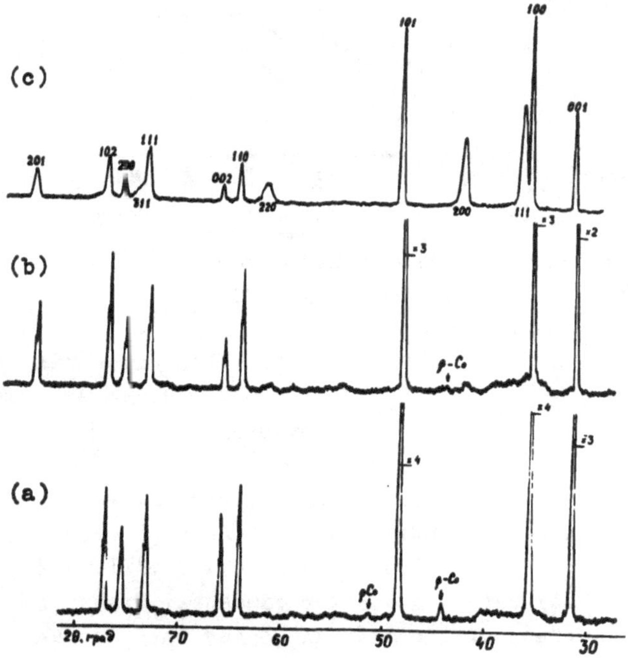

Fig.2 .
X-ray patterns for VK8 alloy : a non-irradiated sample (a), a non-irradiated sample (b), an irradiated sample (c), faces of HPIB irradiated sample . Superscripts for a hexagonal carbide and subscripts for cubic carbide .

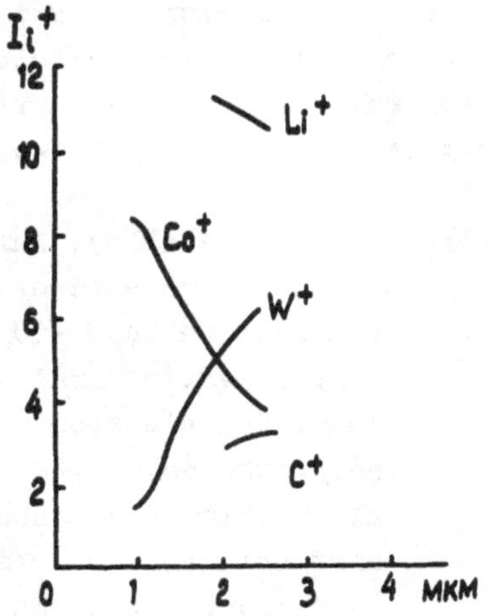

Fig. 3a .

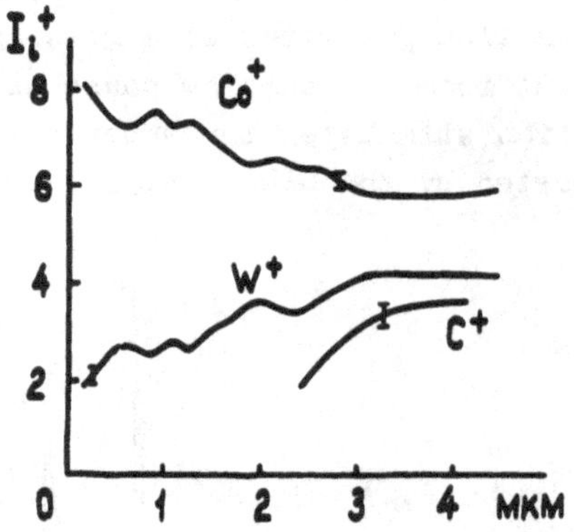

Fig. 3b .

Secondary ion current vs the etched layer thickness in VK8 alloy obtained using SIMS analysis

a) the initial state :
1. O - Li^+ ;
2. • - Co^+ ;
3. + - W^+ ;
4. × - C^+ ;

b) after HPIB exposure: $j=120$ A/cm^2, $n=3$ pulses : 1. Co^+ ; 2. W^+ ; 3. C^+ ;

carbide is produced from the hexagonal one after ion bombardment though we do not eliminate its formation in the intergrain regions as a result of a decay of a solid solution Co(WC). Figure 3 displays profiles of an element composition of VK8 alloy before and after exposure. High Co amounts are observed near the surface of the initial Co and they decrease in the depth. After irradiation cobalt and tungsten are redistributed, cobalt is redistributed more uniformly near the surface and at 3 μm depth shows its saturation. As for tungsten, which was low near the surface after irradiation, it is also stabilized in the depth. One should note that free C_{12}^+ is very low in the surface both in initial state and after irradiation. This carbon is found in the tungsten carbide formations mainly .

REFERENCES

(1) Logachev E.I., Remnev G.E., Usov Yu.P. : Prybory i Technika Experimenta 1 (1983) 22 - 24 .

(2) Tretiakov V.I. : Osnovy Metallovedenia i technologii Proizvodstva Spechennykh Tverdykh Splavov (USSR) , Metallurgia, Moscow 1976.

EFFECT OF DEFECT STRUCTURE ON INCREASED MECHANICAL AND FRICTION PROPERTIES IN HIGH POWER ION BEAM IRRADIATED α- Fe

A.D.Pogrebnjak, Yu.P.Sharkeev*, D.V.Lychagin*, M.A.Makhmudov,
G.V.Pushkareva*, I.G.Romanov, V.N.Kolodii

The polycrystal α- Fe (with 400μm mean grain dimensions) has been investigated in this paper. The carbon concentration did not exceed 0.001 wt.% . The dislocation density in an initial state after recrystallization annealing (950°C per 2 h) was equal to 0.5×10^9 cm^{-2} . Samples (square plates) of 1 cm^2 area and 0.025 to 0.30 mm thickness have been irradiated on the TONUS and TEMP accelerators 1 . The beam current density, j , varied as 60, 100, 200 A/cm^2 as well as a number of pulses from 1 to 11 . For the electron microscope analysis of a microstructure which has been formed in a near surface target layer after exposure, foils have been prepared using the near surface part of the target. The distance from the analyzed layer to the irradiated surface is within 0.5 to 1.0 μm . The electron microscope TESLA-540 was employed to study the films. Scalar dislocation density was plotted vs number of pulses and ion current density. The MICROMET device was used to measure microhardness. To investigate vacancy defects, the positron annihilation technique was used and curves of annihilation photon angular distribution (APAD) were measured, those procedures had been described in 2 . To interprete and identify defects, we applied isochronous annealing of HPIB irradiated samples .

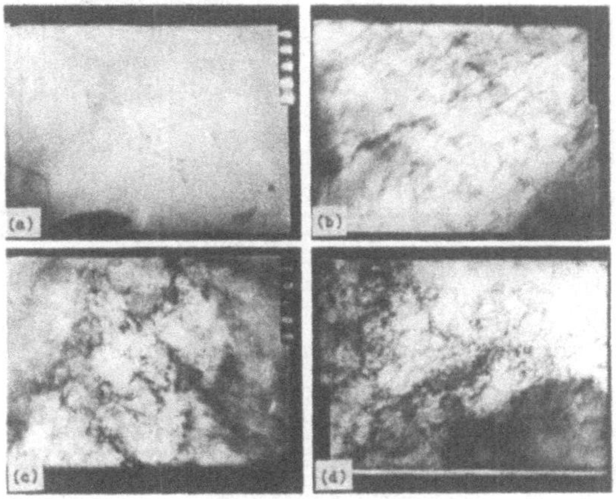

Fig.1 .
Dislocation structure of α- Fe .
a) the initial state; b,c) after HPIB exposure. b) j = 60 A/cm^2, n=1pulse; c) j = 200 A/cm^2, n=5pulses; d) in the deformed sample, ε = 8.5 % .

Friction coefficients have been also measured vs the applied loads,

Nuclear Physics Institute, 634050 Tomsk, P.O.Box 25, USSR
*Institute of Civil Engineering, Tomsk, USSR.

see 3 for details. The initial α - Fe shows low amount of uniformly distrubuted dislocations. After pulsed irradiation, a developed dislocation structure was formed in a near surface layer. The dislocation density increased by one order of a magnitude and higher in comparison with an initial state. Figure 1 shows typical images of dislocation structures which have been formed in α - Fe after pulsed ion bombardment and plastic deformation in one-axis compression. Dislocation structures are uniform in cross-sections perpendicular and parallel to irradiated target surface and change from complex dislocation weaves to uniform dislocation distribution. Increase in ion current density to 200 A/cm^2 and number of pulses to 5 is accompanied by essential reconstruction in a

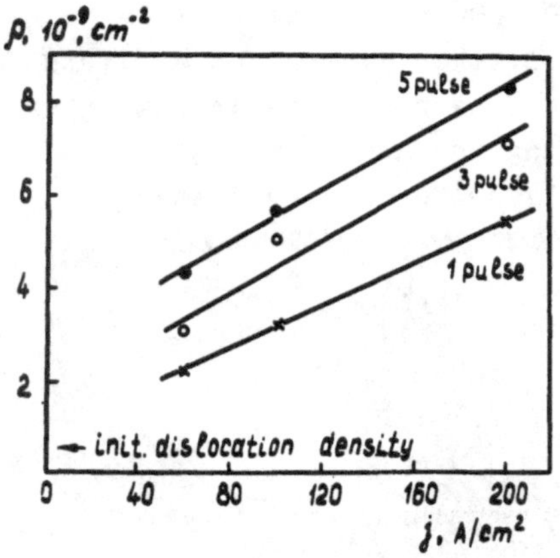

Fig.2

The scalar dislocation density in the near surface layer α- Fe vs the density of the HPIB current ; number of pulses : curve 1 for n=1; curve 2 for n= 3; curve 3 for n= 5;

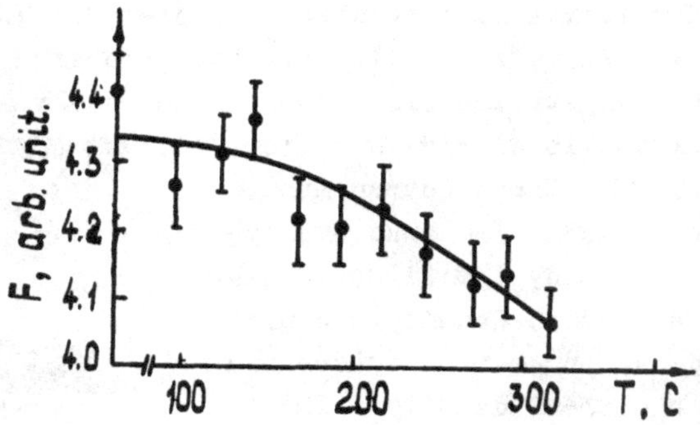

Fig.3

The f-parameter of APAD curves vs the temperature of isochronous annealing in α- Fe samples irradiated by HPIB of j = 100 A/cm^2 and n= 3pulses.

dislocation structure. One can observe an initial stage of dislocation cell formation in images of Fig.1 b . Figure 2 dislpays the dependence of a scalar dislocation density on ion current density for various pulse numbers. It is seen that the dislocation density increased linearly with ion current density. Increased number of pulses is also accompanied by the increased dislocation density. The analysis of dislocation structures as well as their comparison with structures in the deformed metals (in particular, for α- Fe, see Fig. 1 d) shows that their nature is a result of plastic deformation, initiated by a pulsed ion irradiation. One of the most probable reasons of this is a stressed state in a near surface region, which is a result of a shock wave produced by pulsed bombardment and its propagation in a target. The investigations performed using the positron annihilation show that the current density essentially affects formation of vacancy clusters and loops rather than number of pulses (the irradiation dose). The HPIB irradiation results in essential decrease in FWHM of the APAD curves. Isochronous annealing of the HPIB irradiated α- Fe samples showed that the annealing stage which we interprete as coagulation of clusters into vacancy loops appeared above 100 - 300°C , Fig. 3 . Figure 4 a displays the microhardness for the HPIB irradiated α- Fe vs the current density. The obtained dependence has a linear character and correlates well with changes in dislication density in a near surface layer of irradiated samples. One should notice 1.5-fold increase in the surface microhardness. Figure 4 b shows tha obtained experimental dependence of microhardness on the irradiation dose (number of pulses) for α- Fe . The analysis shows that two segments can be defined in this case. The first linear segment of the curve shows growth in microhardness (a sloping part) and the second one shows saturation (at above 3.5×10^{14} cm^{-2}) .

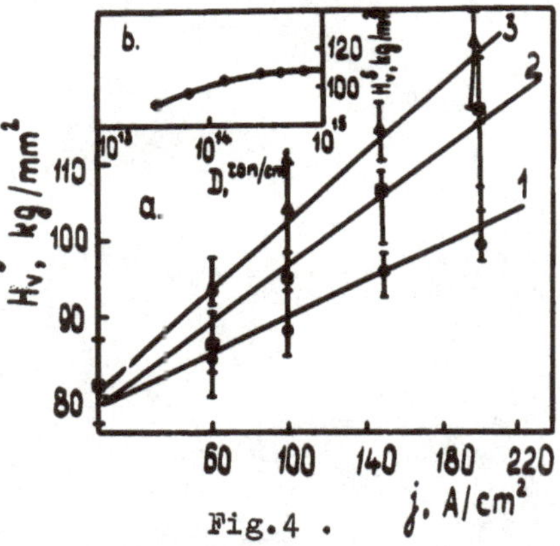

Fig.4 .

Vickers microhardness in HPIB irradiated α- Fe surface at 5 g load: a) vs the current density (E= 0.2-5 MeV, τ =80 ns); b) vs the irradiation dose (number of pulses).

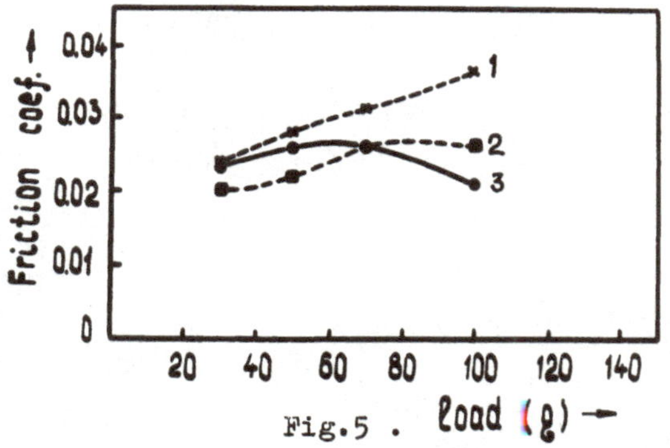

Fig.5 .

The friction coefficient on the load applied to the sample α- Fe : * initial state;
● j = 150 A/cm^2, n= 3 pulses ;
■ j = 150 A/cm^2, n= 7 pulses.

Thus we conclude that to modify mechanical properties (microhardness), it is required 5 pulses which correlates well with change in frictional properties (measurements of the friction coefficient) . Figure 5 shows that the friction coefficient changes insignificantly at low loads (at the HPIB dose of 10^{14} cm^{-3}) . Its value decreased with the load (curve 2) and reached 0.18 with 100 g which is two times lower than in the initial non-irradiated sample . The increase in the number of pulses results in a low decrease of a friction coefficient at low loads (curve 3), but its value reached 0.25 even at 100 g load

An additional analysis of elements in the α- Fe surface modified layer did not show any significant increase in oxygen and carbon (SIMS and Auger), i.e. oxidic or carbide layers had not been formed in - Fe sample surface at higher current densities. Therefore all the changes in mechanical properties : microhardness increase, a decrease in friction coefficient, we believe, depend strongly upon the disloca - tion density and concentration of shallow vacancy loops.

REFERENCES

(1) Pogrebnjak A.D., Remnev G.E., Chistiakov S.A., Ligachev A.E. : Physica (USSR) (1987) 52; Akerman D.R., Isakov I.F.,Kolodii V.N., Opekunov M.S., Remnev G.E. : First All-Union Conf.on Modification of Structural Mater. Properties by Charged Part. Beams, 1 (1988), Tomsk, Polytechnical Institute 1988 .

(2) Pogrebnjak A.D. : Phys.Stat.Sol. A 85 (2) (1984) 191 .

(3) Kutuzov V.L., Ovsiannikov M.Yu., Romanov I.G., Pogrebnjak A.D., Remnev G.E. : Phys.Stat.Sol. 112 (1989) 361 .

AMORPHIZATION MECHANISMS IN ION-BOMBARDED METALLIC ALLOYS: EXPERIMENTAL CONFIRMATION OF MODEL PREDICTIONS

L. Thomé[*], A. Benyagoub[+] and F. Pons[o]

[*]C.S.N.S.M., Bat. 108, 91405 Orsay, France.
[+]Hahn-Meitner-Institut, P.O. Box 390128, 1000 Berlin 39, F.R.G.
[o]G.A.N.I.L., B.P. 5027, 14021 Caen, France.

I - INTRODUCTION

The amorphization mechanisms of crystalline metallic alloys by ion bombardment have been extensively studied in the last years (3-9). Recently, a model, based on the assumption that the elementary mechanism is the formation of amorphous clusters as soon as a favorable short-range order (threshold ion concentration) and a given topological disorder (threshold defect concentration) are locally reached, was developed (1). It is shown that the fraction of amorphous volume (amorphous fraction) α present at a given depth of an ion-bombarded crystal can be written as a product of two terms (1):

$$\alpha = f(d,d_c,T)\ g(c,c_c,T) \qquad (1)$$

where $f(d,d_c,T)$ and $g(c,c_c,T)$ are functions depending respectively on the damage production (d,d_c) and on the implanted species (c,c_c), as well as on the bombardment temperature (T).

At temperatures sufficiently low to avoid implanted ions and radiation defects to migrate into the bombarded layer, it is possible to express functions f and g from statistical considerations concerning the spatial distributions of ions and defects (1):

$$g(c,c_c,T<T_i) = \sum_{N_i=N_{ic}}^{\infty} \frac{(\bar{N}_i)^{N_i}}{N_i!} \exp(-\bar{N}_i) \qquad (2)$$

where $\bar{N}_i$ is the mean number of ions (corresponding to a mean concentration c) implanted into an elementary volume v of the target, N_{ic} is the critical number of ions (corresponding to a concentration c_c) above which v becomes amorphous and T_i is the temperature at which implanted ions become mobile.

$$f(d,d_c,T<T_d) = \sum_{N_d=N_{dc}}^{\infty} \sum_{n_i=0}^{\infty} \frac{(\bar{n}_i)^{n_i}}{n_i!} \exp(-\bar{n}_i) \frac{(n_i\bar{n}_d)^{N_d}}{N_d!} \exp(-n_i\bar{n}_d) \qquad (3)$$

where $\bar{n}_i$ is the mean number of ion impacts into an elementary volume v, $\bar{n}_d$ is the mean number of defects created by each ion impact in v (so that $\bar{n}_i\ \bar{n}_d$ corresponds to a mean defect concentration d), N_{dc} is the critical number of defects (corresponding to a concentration d_c) above which v becomes amorphous and T_d is the temperature at which radiation defects become mobile. Equation (3) can be simplified in the two extreme cases where the mean number of defects created by each ion impact in a volume corresponding to the mean size of an amorphous cluster is either very small or large compared to unity.

$$f_P(d,d_c,T<T_d) = \sum_{N_d=N_{dc}}^{\infty} \frac{(\bar{n}_i\bar{n}_d)^{N_d}}{N_d!} \exp(-\bar{n}_i\bar{n}_d) \qquad \text{if } \bar{n}_d \ll 1 \qquad (4)$$

$$f_C(d,d_c,T<T_d) = \sum_{N_d=N_{dc}}^{\infty} \frac{(\bar{n}_i)^{n_i}}{n_i!} \exp(-\bar{n}_i)\ \delta_{N_d,n_i\bar{n}_d} \qquad \text{if } \bar{n}_d > 1 \qquad (5)$$

where δ is the Kronecker index.

channeling experiments performed on the Ni-B system (and analyzed with the procedure described in Ref. 2) in order to check the validity of the model predictions in the three cases where the amorphization process is dominated by $g(c,c_c,T)$ (implantation experiments), $f(d,d_c,T)$ (irradiation experiments) or a combination of both expressions.

II - IMPLANTATION EXPERIMENTS [$g(c,c_c,T)$]

In implantation experiments, a species B is implanted into a pure target A up to a concentration x in order to form a system of composition $A_{1-x}B_x$. At the ion fluences generally required (higher than 10^{16} atoms cm^{-2}) to modify noticeably the short-range order of the implanted alloy, the topological disorder is already saturated ($d>d_c$), so that $f(d,d_c,T) = 1$ in equation (1) and $g(c,c_c,T)$ governs the amorphization process.

Figure 1 presents RBS results recorded on a Ni single crystal implanted at low temperature with increasing fluences of B ions. It is clear that the amorphization kinetics, which displays a sigmoidal shape with the presence of a marked ion concentration threshold, is well reproduced with equation (2). The fit gives the following values for the critical B ion concentration c_c and critical volume v_c of the amorphous clusters formed: $c_c = 0.093$; $v_c = 1.5 \times 10^{-21}$ cm^3, comparable to those obtained in other metal-metalloid systems (1).

III - IRRADIATION EXPERIMENTS [$f(d,d_c,T)$]

Irradiation experiments consist in bombarding a target with ions which do not modify the chemical nature of the bombarded system. This is generally achieved by implanting low fluences (less than 10^{15} atoms cm^{-2}) of chemically inert ions or by using high energy ions which do not come to rest in the amorphized layer. In this case, since the alloy composition does not change and is favorable for amorphization ($c>c_c$), $g(c,c_c,T) = 1$ in equation (1) and $f(d,d_c,T)$ governs the amorphization process.

Figures 2 and 3 present RBS results recorded on a Ni_3B single crystal irradiated at low temperature with increasing fluences of D ($\bar{n}_d \ll 1$) and B ($\bar{n}_d > 1$) ions, respectively. The shape of the amorphization kinetics displayed on these figures obviously depends on the nature of the irradiating ion and can be reproduced with either equation (4) (D irradiation) or equation (5) (B irradiation), in agreement with the model predictions. The critical defect concentration d_c and critical volume v_c of the amorphous clusters formed following D irradiation, deduced from the fit, are: $d_c = 0.11$; $v_c = 4.3 \times 10^{-22}$ cm^3. On the other hand, the fit of B irradiation results leads to the following value of the parameters: critical number of ion impacts $n_{ic} = 2$; critical defect concentration $d_c = 0.026$; amorphization cross-section $a = 1.5 \times 10^{-14}$ cm^2.

IV - COMBINED EXPERIMENTS [$f(d,d_c,T)\ g(c,c_c,T)$]

The concomitant influence of the two terms of equation (1) in the amorphization process of ion-bombarded Ni-B can be checked in the following experiment. A partially amorphous alloy is formed by low temperature B implantation of a Ni single crystal (amorphous fraction $\alpha_1 \sim 0.5$), annealed above the crystallization temperature (amorphous fraction reduced to α_0 close to zero) and post-irradiated at low temperature with increasing fluences of B ions. Figure 4 presents the variation of $\Delta\alpha = \alpha-\alpha_0$ as a function of the B ion fluence during the last post-irradiation sequence. It is worth noting that this variation is first governed by the function $f(d,d_c,T)$ and then (as the value of α_1 is reached) mainly determined by the function $g(c,c_c,T)$. Experimental data are well reproduced using a combination of equations (5) and (2) with the following values of the fitting parameters: critical concentration $c_c = 0.095$ and critical volume $v_c = 1.4 \times 10^{-21}$ cm^3 of the amorphous clusters formed; number of ion impacts $n_{ic} = 1$, amorphization cross-section $a = 3.75 \times 10^{-14}$ cm^2. It is remarkable that these values agree well with those found in the simple implantation or irradiation experiments reported above.

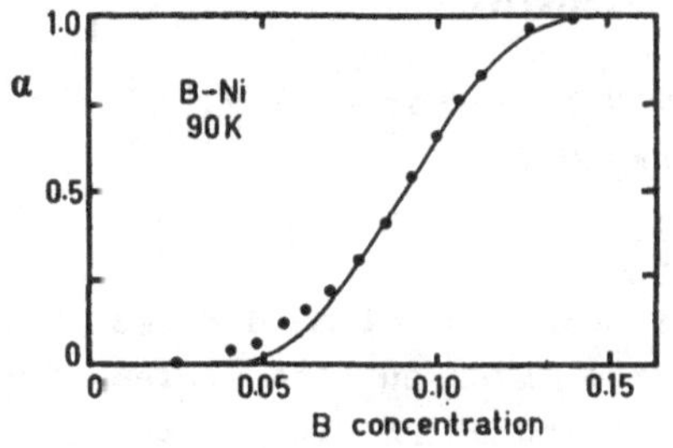

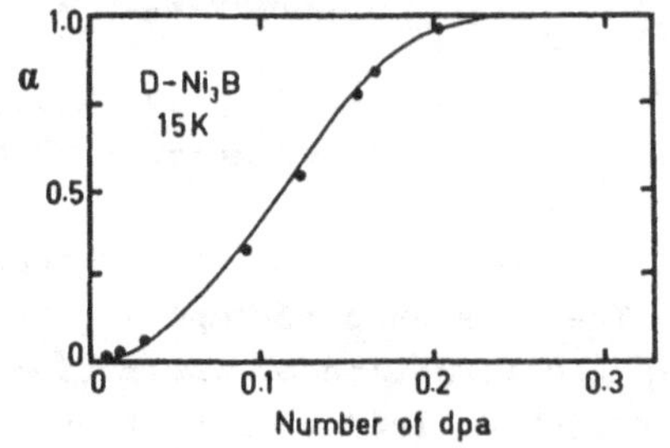

Fig. 1: B *fluence dependence of the amorphous fraction* α *for 90 K implanted Ni-B alloy. The fluence scale has been converted into a B concentration scale according to RBS profiles.*

Fig. 2: D *fluence dependence of the amorphous fraction* α *for* Ni_3B *alloy irradiated at 15 K with* D *ions. The irradiation fluence has been converted into a number of dpa with* TRIM.

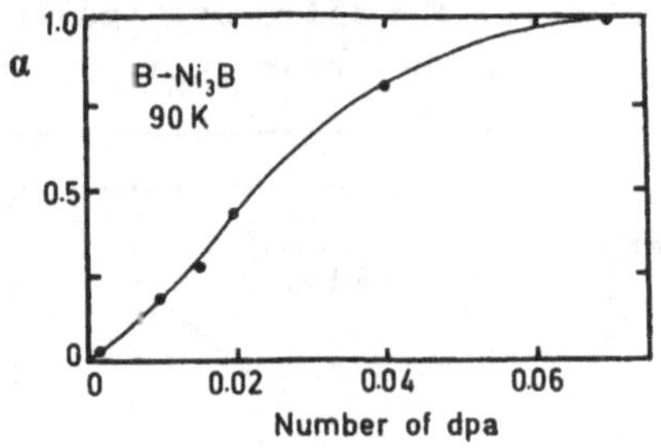

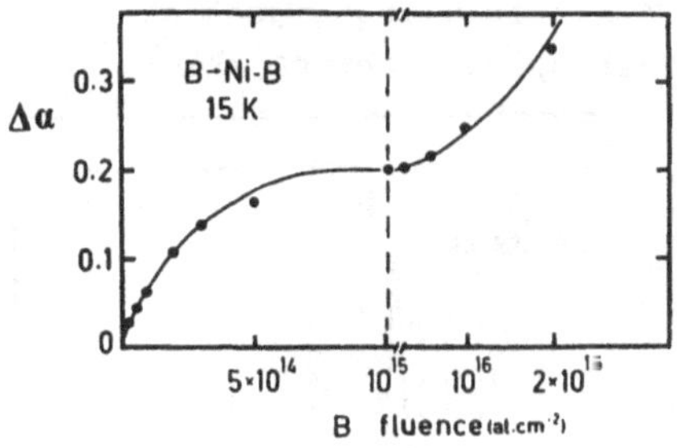

Fig. 3: B *fluence dependence of the amorphous fraction* α *for* Ni_3B *alloy irradiated at 90 K with* B *ions. The irradiation fluence has been converted into a number of dpa with* TRIM.

Fig. 4: B *fluence dependence of the amorphous fraction* Δα *for* Ni-B *alloy formed by 15 K-B implantation, annealed at 400°C and post-irradiated at 15 K with B ions.*

REFERENCES

(1) Benyagoub, A., and L. Thomé: Phys. Rev. B 38 (1988) 10205.
(2) Benyagoub, A., and L. Thomé: Radiat. Effects 105 (1987) 9.
(3) Cohen, C., A. Benyagoub, H. Bernas, J. Chaumont, L. Thomé, M. Berti and A.V. Drigo: Phys. Rev. B 31 (1985) 5.
(4) Heera, V., and B. Rauschenbach: Radiat. Effects 91 (1985) 71.
(5) Linker, G.: Nucl. Instrum. Meth. B 19/20 (1987) 526.
(6) Ossi, P.M.: Mater. Sci. Engin. 90 (1987) 55.
(7) Pedraza, D.F.: Mater. Sci. Engin. 90 (1987) 69.
(8) Simonen, E.P.: Nucl. Instrum. Meth. B 16 (1986) 198.
(9) Thomé, L.: Nuclear Physics Applications on Materials Sciences, Vol. 144 of NATO Advanced Study Institute, ed. E. Recknagel and J.C. Soares, Klumer, Norwell, MA (1988) 183.

ETCHING AND PATTERNING OF POLYCRYSTALLINE DIAMOND FILMS BY PULSED UV AND IR LASER RADIATION.

Ageev V.P.,Chapliev N.I.,Konov V.I.,Kuzmichov A.V., Pimenov S.M.,Ral'chenko V.G.

Introduction.

The existence of impurity absorption in CVD diamond films (DF) as well as the formation of absorptive graphite layer on the surface during laser radiation lead to effective laser etching of DF in spite of the photon energy hn is less then diamond band gap E = 5.4 eV. In this paper we describe the results of investigation on surface morphology changes of polycrystalline DF due to UV and IR laser radiation.

Experimental

Free-standing and grown on W and Si substrates from mixture of methane and hydrogen at pressure < 1 atm (15 - 60 μm thick) DF were used in this experiments [1]. Raman scattering studies of the films reveal characteristic diamond peak at 1330 cm^{-1} (Fig.1). The films are opaque in the visible, with abrupt absorption edge taking place near 4 μm.

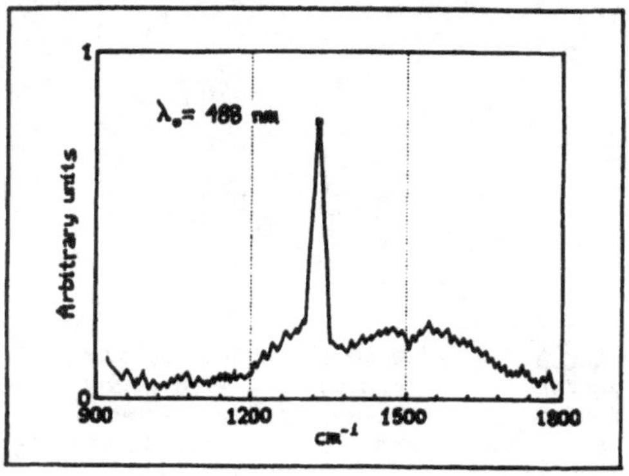

Fig.1 Raman spectrum of DF.

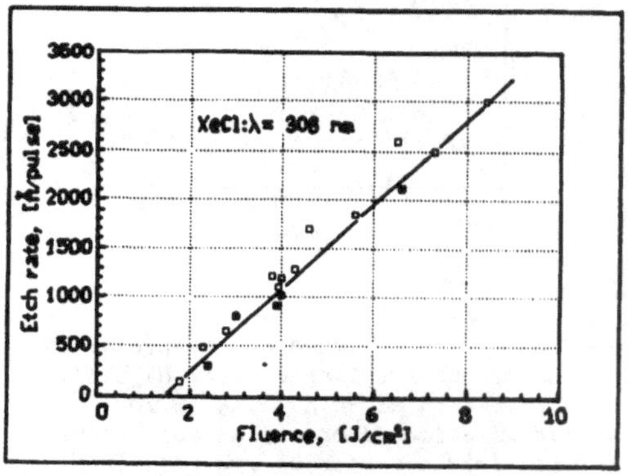

Fig.2 Etching rate vs fluence
□ air; ■ Xe.

Pulsed repetition rate TEA CO_2, XeCl and KrF excimer lasers (λ = 10.6 μm, τ = 150 ns, λ = 308 nm, τ = 20 ns, λ = 248 nm, τ = 20 ns respectively, repetition rate - 1- 10 Hz) were used in experiments. The intensity distribution of UV radiation on the irradiated surface was near uniform, with energy density having been reached 10 J/cm^2. The energy density in the center of linearly polarized single mode CO_2-laser beam was up to 200 J/cm^2. Irradiation of DF with UV radiation was performed in air and in Xe atmosphere (200 Torr),and IR radiation in vacuum (10^{-6} Torr) by the method stated elsewhere [2].

Institute of General Physics,USSR Academy of Sciences.
Vavilov str.,38, Moscow 117942, USSR.

UV laser irradiation of diamond films.

To determine the etch rate dependence on incident energy density **E** a number of laser pulses **N** needed to form a hole through the free-standing **DF** 15 μm thick was mesured. The etch rates $v = h/N$ (h - thickness of the film) in atmosphere of air and **Xe** proved to be nearly the same and depended linearly on **E** (Fig.2). The presence of oxygen in gaseous environment does not influence the etching rates, but hinders a redeposition of graphite around the laser spot due to the chemical reactions with carbon [4]. A sublimation of material of the film becomes noticeable at E_{sub} = 1 J/cm^2 and the etch rate is equal to v = 3000 Å/pulse at **E** = 8 J/cm^2 (Fig.2). Obtained etch rates exceeds by an order of magnitude those one of natural diamond irradiated by laser pulses at λ = 193 nm [3], under fluences been equal.

In our opinion, the reason of this difference is a lower thermal conductivity of **DF** (k = 410 W/mK [1]) in comparison with that of diamond monocrystal (k = 2000 W/mK) leading to a greater pulsed heating of the film surface, as well as lower optical transparency [3]. Transmittance of **DF** which was mesured during etching (i.e. decreasing film thickness), does not change up to the formation of the hole in the film and corresponds to effective absorptivity α = 2000 cm^{-1}. This indicates on the presence of the absorptive graphite layer in accordance with two-stage etching process [3].

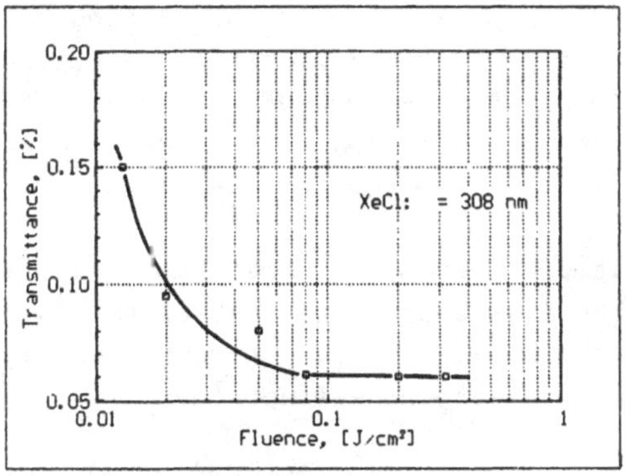

Fig.3 Transmittance **DF** vs fluence of UV radiation.

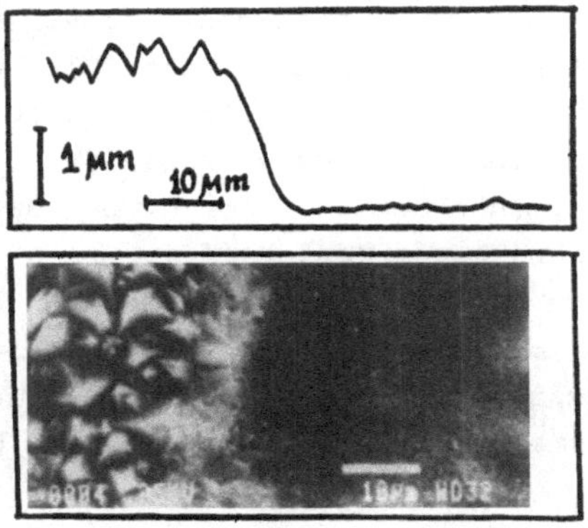

Fig.4 Surface morphology after UV irradiation.

The surface layer of film is initially graphitized and then evaporated with simultaneous moving the graphite "piston" into the bulk of film. The evaluations showed that the surface temperature of graphite (but not of diamond)reach the sublimation temperature of graphite **T** = 3700 °K at the threshold energy **E** ≈ 1 J/cm^2. The graphitization threshold was determined by measuring of the optical transmittance of laser radiation under low fluences **E** = 0.02 - 1 J/cm^2 (Fig. 3). Transparency of the film begins to decrease at **E** ≈ 0.02 J/cm^2. Since this value is small for essential heating of diamond surface, it should be supported that graphitization starts on some absorptive centers situated, for instance, at the intercrystallite boundaries.

The surface morphology studies using the profilometry and scanning electron microscopy showed that the initial roughness of **DF** surface reduced from the value of 1 μm to that of 200 Å (Fig. 4). So the UV laser beam could be used for effective polishing of **DF**.

It also should be noted that the laser induced graphitization would be used for writing a conductive roads on the **DF** surface.

<u>CO_2- laser irradiation of DF.</u>

The peculiarities of CO_2-laser action on **DF** are due to a volume absorption with effective absorptivity α = 200 cm^{-1} . It was found that a character of laser induced graphitization and laser damage of **DF** strongly depended on whether the film irradiated is a free-standing state or on the substrate.

Interaction mechanism of CO_2-laser radiation with a free-standing film is the same as the two-stage etching process described above. The optical transmittance at laser wavelength λ = 10.6 μm versus laser energy density has the same behavior as for UV laser irradiation [2]. But graphitization and sublimation thresholds, being equal to E_{gr} = 8 J/cm^2 and E_{sub} = 53 J/cm^2 (E - energy density at the center of laser spot) are more higher than those for UV. The multiple irradiation of **DF** surface in this case (that is at $E \geqslant E_{sub}$) leads to the etching and patterning of diamond film [2]. Using of polarized radiation is a reason of formation of linear periodical structures on **DF** surface. The formation of such kind of light-induced gratings is well-known phenomenon and it could be used in the fabrication techniques of diamond-based devices [5].

In the case of irradiation of diamond coating on the substrate a part of laser energy passing through the film heated the substrate. As a consequence the graphitization of diamond was accompanied by thermal cracking of the coating [6] and the etching was not observed.

<u>References</u>.

(1) Builov L.L., Aleksenko A.E., Botev A.A., Spitsyn B.V.: Dokl.Akad. Nauk,287(1986),p.888,in Russian
(2) Konov V.I., Chapliev N.I., Pimenov S.M.: Izv. Akad. Nauk, seria phys.,53(1989), 423 - 430,in Russian
(3) Rothschild M., Arnone C., Ehrlich D.J.: J.Vac.Sci.Technol. B4(1986),p.314
(4) Ageev V.P.,Builov L.L.,Chapliev N.I.,Konov V.I.,Kuzmichov A.V., Pimenov S.M., Prokhorov A.M., Ral'chenko V.G., Spitsyn B.V.: Dokl. Akad. Nauk 303(1988), 598-601, in Russian
(5) Geis M.W., Efremov M.N., Rathman D.D.: J.Vac.Sci.Technol., A3(1988), p.1953
(6) Builov L.L., Chapliev N.I.,Konov V.I., Pimenov S.M., Spitsyn B.V.: Poverkhoct', to be published, in Russian.

THE ALLOYING OF Fe AND V BASE WITH W, Nb AND Al BY ION-BEAM MIXING

Choupyatova L.P., Komarovsky I.A., Peksheva T.E., Murzin I.G. [1]

The aim of this work is to evaluate experimentally the mechanisms providing ion mixing (IM) for bilayer systems with different combination of such parameters as the heat of elements mixing (ΔHmix), the radiation enhanced diffusion (RED) domination temperature (T_c) and the mass ratio of ion and mixing atoms.

Experimental

Ion-beam mixing (IM) was performed at 300 K in bilayer systems consisting of Fe and V base coated with W, Nb, Cu and Al films of the thickness 10-30 nm. Samples were bombarded by Ar^+ and Xe^+ ions for doses ranged from 6×10^{14} to 1×10^{17} cm^{-2} at flux density of 0.05-10 $\mu A/cm^2$ in the energy range 40-350 keV. Auger probe JAMP -10S with 1 keV Ar^+ milling was used to get mixing element concentration depth profies.

Results and discussion

TRIM program /1/ was used to obtain the distribution of the deposited energy Fd which determines the quantity of the radiation defects. The Fd depth profiles in W/Fe Nb/Fe and Al/Fe system after Ar^+ and Xe^+ bombardment are shown in Fig. 1. The profiles for V base systems are similar qualitatively. It is seen from Fig.1 that near to interface the Fd (Nb)$\approx$Fd (Fe) Fd (W)>Fd (Fe) and Fd(Al)<Fd(Fe).Fig.2 represents AES distribution

----

1) Dept. Metallography, Inst. of Steel & Alloys
Leninsky pr. 4, 117936 Moscow, USSR.

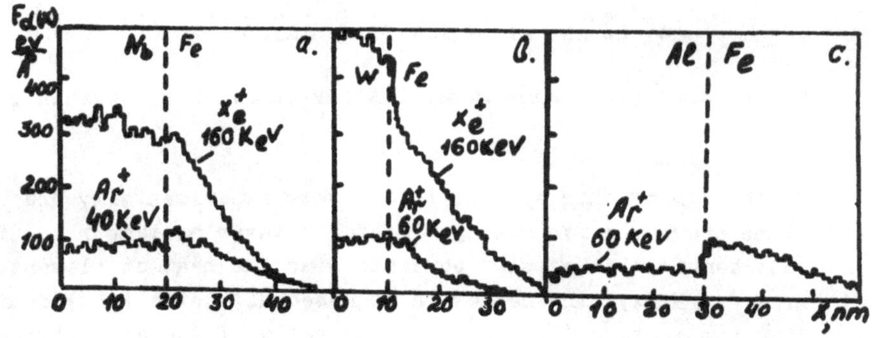

Fig.1 The depth profiles of Fd calculated by TRIM program /1/
a) Nb/Fe system b) W/Fe system c) Al/Fe system

profiles of mixing elements for the system investigated before and after IM. One can see that IM of Fe and Nb (ΔHmix=-25 kJ/mol /2/, $T_c > 300$ K /3/) Fe and W (ΔHmix= 0 kJ/mol, $T_c > 300$ K) takes place within radiation damage region. The Ar^+ bombardment results in the lower thickness of mixing the region than Xe^+ bombardment in both systems. The element spreading of Nb/Fe system from the interface is practically the same (Fig. 2a). For W/Fe system the Fe atoms diffuse through the W layer and segregate to the surface and accumulate here with dose increasing (Fig. 2b).

On the contrary for Al/Fe system (ΔHmix =-32 kJ/mol, $T_c <$ 300 K) the preferential spreading of Al in Fe is observed. The Al penetrating depth is much greater than the defect formation depth (~5 times higher for $\Phi = 1 \times 10^{17}$ Ar^+/cm^2). At Al/Fe concentration profiles the constant composition step may be seen which corresponds to the stoichiometry of Al_3Fe (Fig. 2c). In the case of Cu/Fe system (ΔHmix= 22 kJ/mol, $T \gtrsim 300$ K) there is no any mixing what correlates with high positive ΔHmix. The mixing in V base system demonstrates qualitatively similar behaviour.

The dependence $\sigma \sim \Phi^{1/2}$ is received for all systems It enables us to consider the IM as the random walk process and to describe it by means of the cascade mixing mechanisms and the RED mechanisms. In the system W/Fe and Nb/Fe where the spreading of elements is limited by cascade region and value Fd exceeds the local melting energy (Em) IM apparently is reali -

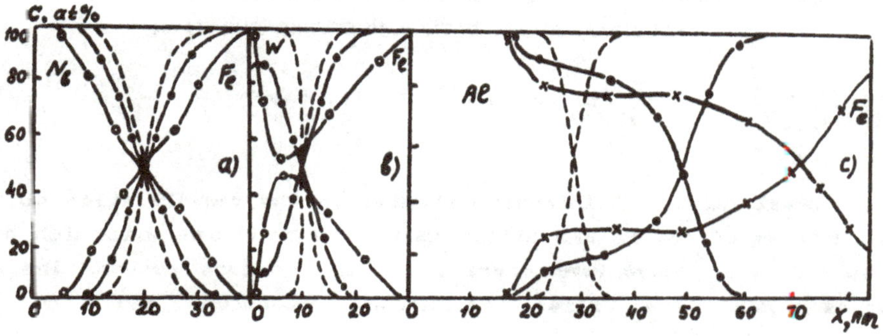

Fig. 2 AES depth profiles before (---) and after $1x10^{16}Xe^+/cm^2$(-•-), $1x10^{16}$ Ar^+/cm^2 (-••-), $5x10^{16}Ar^+/cm^2$ (-x-) irradiation

zed by thermal spike mechanism. This is confirmed by σ decrease with ion flux density increasing above 0,2 $\mu A/cm^{-2}$ /3/. In the systems Al/Fe and Cu/Fe where the vacancy mobility is high enough ($T_c \lesssim 300$) the mechanism of RED dominates. It is seen from the fact that spacing of elements exceeds the radiation damage range (Al/Fe system) or IM is absent at all (Cu/Fe system).

The Fig. 1 and Fig. 2 comparison shows that mixing atoms migrate towards the region with the higher value of Fd. Apparently the direction of the migration is determined by defect flow in the field of temperature gradient near to interface. There wasn't revealed any influence ΔHmix on the mixing efficiency in the systems with the high T_c (W/Fe and Nb/Fe) what disagrees with the conclusions of /4/. The value of ΔHmix plays the main role in the systems with low T_c where RED is dominant. In those systems ΔHmix assists (as in Al/Fe) or prevents (as in Cu/Fe) the element mixing.

REFERENCES

(1) Ziegler J.F., Biersack J.P., Littmark V.: The Stopping and Range of Ion in Solids. V.1, ed. by J.F. Ziegler, Pergamon Press, New York, 1985, p.202.
(2) Miedema A. Philips Tech. Rev. 36 (1976) 217-231.
(3) Choupyatova L.A. et al. Surface, to be published.
(4) Workman T.W., Cheng Y.T., Johnson W.L., and Nicolet M.A. Appl. Phys. Lett. 50 (1987) 1485-1487.

SURFACE MODIFICATION OF STEELS BY IRRADIATION OF HIGH CURRENT ELECTRON PULSE WITH NANOSECOND DURATION

E. Dancheva[1], I. Trifonov[1]

A great number of intensive studies on the possibilities to modify the surfaces of metals and alloys using energy fluxes with high density of 10^5-10^{10}W/cm^2 have been carried out in the last decade. The energy sources used are laser radiation, electron and ion beams [1],[2],[4],[5],[7].

The aim of this paper is to investigate the changes in the surface layers of steels with various carbon content (0.14÷1%) as a result of high current electron beam irradiation of nanosecond duration. The experimental work was performed on the high current accelerator "Perun-2" [3] with parameters: electron energy 250-300 keV, current density 8-12kA/cm^2 and pulse duration 15x10^{-9}s. The metods of metalographical analysis, scaning electron microscopy and X-ray difraction were used. An analysis of the heat processes during the pulse action and after it was done.

After a single pulse irradiation of the samples with a power density (2-3)x10^{9}W/cm^2, the surface was quickly melted the rate of temperature rise being about 10^{10}-10^{11}K/s. Rapid solidification and quenching of the melt with a superhigh rate follows. The metalographic study of the cross sections from the irradiated samples shows that the melt depth is about 20-30μm. The structure of the solidified melted zone is formed of martensite and residual austenite, which is confirmed also by the X-ray analysis of the samples. Next to this zone is a zone with more dispersive nature of the grains than that of the main matrix. The authors of [5] have observed a sharp differentiation of five zones due to the long times of action (300μs, q_{max}=2x10^{5}W/cm^2). Such a differentiation is not observed at our conditions (τ=10-15ns) due to the peculiarities in the development of the termal processes in depth. The averaged values of the surface steel microhardness increase with increasing of the carbon content. The maximum microhardness of about 2000kg/mm^2 observed by the authors [5] for St 45 could not be observed in our studies. In fact in the surface modified layer the microhardness averaged for about 80% of the points is approximately 920kg/mm^2 but at some of these points the microhardness reaches 1600kg/mm^2. The remaining 20% show considerably lower values than those cited above and they correspond to the austenite

1 Sofia University,Faculty of Physics,A.Ivanov str.5,Sofia 1126, Bulgaria

microhardness. The higher values of the microhardness are due to the small size martensite grains, produced during the superhigh rate quenching from the liquid phase, that have 3-4 times higher microhardness than those produced during the quenching from the solid phase [6]. The variation of the St 45 microhardness in depth is shown in fig 1. It is seen that the microhardness attains a maximum value at a depth of about 5μm, subsequently decreasing monotonously.

An additional theoretical analysis of the heat transfer processes, caused by nanosecond pulsed irradiation, the beam power density being high enough to ensure melting of thin surface layer, is made. It is shown that the layers, situated much deeper than the electron penetration depth, are impossible to be heated above the temperature of the perlite-austenite phase transition. The oscilating character of the microhardness, observed by other authors, should be explained by phase structural and mechanical modifications due to powerfull accoustic and detonation waves caused during the adiabatical heating of the thin surface layers. The temperature variation in depth of the sample for different periods of time after the pulse end, under the conditions of our study is given in fig.2. It is seen that for a period of time of $t \leq 100\mu s$ after the end of the pulse, surface temperature decreases to $0.5xT_{melt}$. The temperature gradient on the boundary of the heated layer strongly changes from 10^6K/cm for $t=2\mu s$ to 10^4K/cm for $t=200\mu s$. After a period of time of about 600μs the temperature becomes uniform all over the sample. These results are graphically shown in the fig.3. It is seen that the rapid heating and cooling of layers situated deeper than the depth of the melted zone, are possible only in a very narrow region with a depth of about the depth of the melted zone. Phase structural changes and due to them changes of the physico-mechanical and chemical properties in our experiments [3] are mainly observed in the frames of the melt zone and in the adjacent layers.The rate of the temperature changes in the material is of the greatest importance for the processes. After the pulse end when the melted zone is solidified begins the cooling of steel layers in depth. In the period of time from 2μs to 50μs the solidified zone cools with a rate more than 10^7K/s.

The investigation made in this work shows that after irradiation of steels, using high current electron pulse of nanosecond duration and beam power density of about 10^9W/cm^2, the main physico-mechanical changes are possible principally in the frames of the melted and quickly cooled zone. Phase structural changes of the layers, situated in depth of about several times larger than the depth of electron penetration, is impossible.

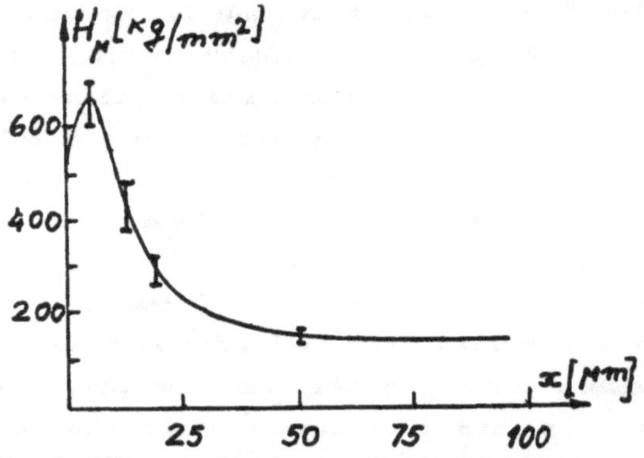

Fig.1. The variations of microhardness as a function of depth for St 45.

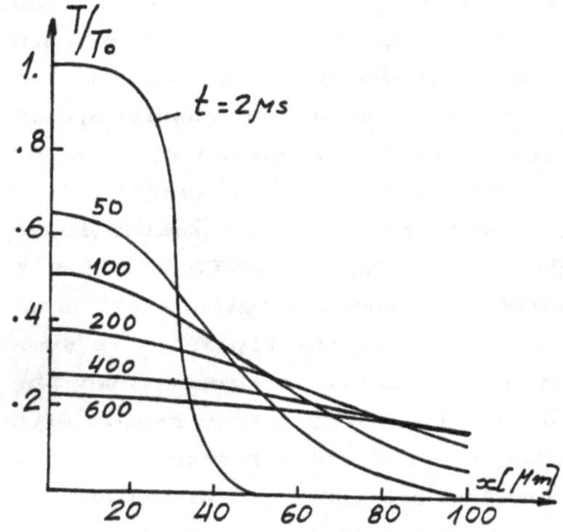

Fig.2. Temperature distribution in depth. Parameter: time after the pulse end.

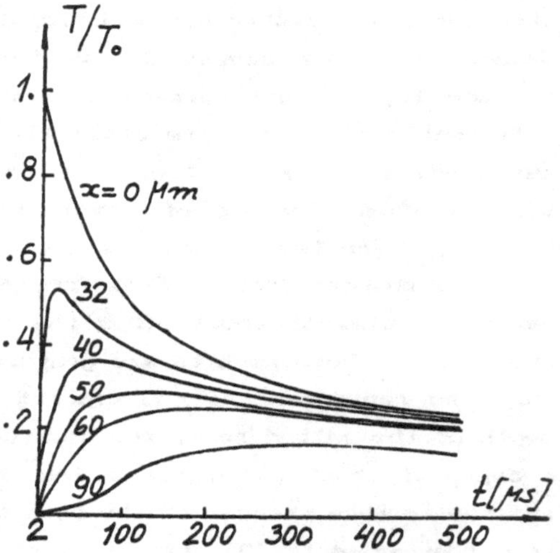

Fig.3. Temperature distribution as a function of time after the pulse end. Parameter: depth of layer in the sample.

References:

1. Ashby, Easterling: Acta metall. 32, 11 (1984),1935; 34, 5 (1986),1333.
2. Bikovsky,V.F. et al.: Transactions of Second International Conference on Electron Beam Technologies 1988, Varna, Bulgaria.
3. Dancheva,E.,Trifonov,I.: Bulg. J. Phys. 16, 3 (1989),294.
4. Didenko,A.N. et al.: Effects of charged particle beams on the metal and alloy surfaces, Energoatomizdat, Moscow 1987.
5. Ityn,V.I. et al.: Izvestia VUZ.Physica 8 (1985) 39.
6. Cahn,E.W., Haasen,P., eds.: Physical Metallurgy, North Holland Physics Publishing 1983.
7. Mesyatz,G.A.,ed.: High-current pulsed electron beams in technology, Nauka, Moscow 1983.

THE PHYSICAL PROPERTIES OF LAYERS IN GLASSES FORMED BY ION MIXING AND IMPLANTATION

A. A. DESHKOVSKAYA[1)], V. P. KOMAR[2)], I. V. SKORNYAKOV[2)]

Ion bomradment of glass surface in implantation and mixing regimes is an effective technique for modification of optical and mechanical glass properties. The changes in the refraction index, reflectivity, mechanical strength and the coating-to-glass surface adhesion clearly testify to that. The complicated physical and chemical processes that are induced by high energy beams within the near-surface layers underlie the modifying action of ion bombardment.

This study is one of the series. Three glass compositions (quartz SiO_2) low-alkaline aluminoborosilicate (Pyrex) and high-alkaline aluminosilicate) were the object of the study. Different species of ions (Gr. I-IX of the Periodical System) were used as implanted impurities. Single-layer film coatings of metals, nonmetals and their compounds served as impurities in mixing.

Ion bombardment at the dose $D=10^{14}$ to $6\cdot10^{17}$ cm^{-2}, ion energy E= 30-300 keV, $j<5$ $\mu A/cm^2$ was carried out.

The influence of ion bombardment on the optical properties of glasses were investigated analysing the IR reflection spectra recorded by a Perkin-Elmer-180 spectrophotometer over the range of 1500 - 400 cm^{-1}.

The spectral differences of doped and nondoped glass samples were manifested by the shifts of the band with a maximum at 1100 cm^{-1} (stretching vibration of Si-O-Si bond). Ion bombardment (E=30-300 keV) leads, as a rule, to a 2 to 18 cm^{-1} shift of this band in implantation regime and to a 60-155 cm^{-1} shift in mixing. The experiments have shown that the magnitude of the frequency shift depends not only on the parameters of ion beam (energy and dose) but also on the bombarding ion species and the chemical composition of target glass (fig.1, a,b,c).

The observed relations can be explained qualitatively, if the frequency shift $\Delta\nu_{1100}$ is taken as a measure of chemical interaction between the implanted impurities and the atoms of target glass.

As the mass and the serial number of the dopant implanted at definite energy and dose increase, the frequency shift diminishes. This

1) Minsk Radioengineering Institute, P.Brovki 6, 220069 Minsk, USSR
2) Institute of Physics, Byelorussian Academy of Sciences, Minsk, USSR

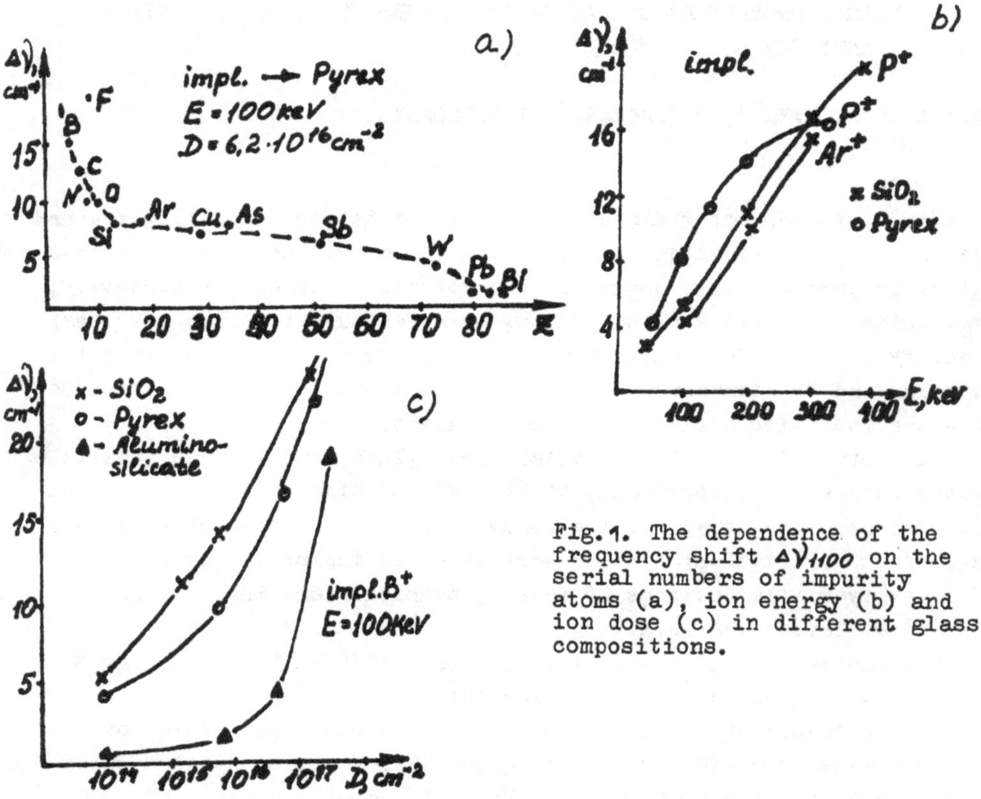

Fig. 1. The dependence of the frequency shift $\Delta\nu_{1100}$ on the serial numbers of impurity atoms (a), ion energy (b) and ion dose (c) in different glass compositions.

may be explained, first, by decreasing the implantation depth, second, by the fact that for heavier particles the nuclear stopping that leads to atomic displacements is a dominant mechanism of atomic collisions at 100 keV, whereas the contribution of electronic stopping, responsible for ionization processes, is small. Therefore, a probability of chemical interactions between the heavy particles and surrounding atoms is low. The unusually large frequency shift observed in fluorine implanted glass can be ascribed to a very sensitive chemical activity of the element.

The comparison of IR spectra for implanted glasses that have equal projective path lengths, i.e. equal effective thicknesses of modified layers, shows that larger frequency shifts are observed for the systems with heavier implanted impurities. It should be remembered, however, that the energy of heavy ions, required for the implantation depth to be equal to that in case of light ions, is very high. As the energy rises, the contribution of inelastic processes, that result in a higher probability of chemical reactions due to more intensive ionization, becomes larger.

Incorporation of impurity atoms (in contrast to removal of atoms out of it) should result in a displacement of the band under study towards the low wave numbers irrespective of the structural function done by the incorporated impurity vitrifier or modifier in glass, in other words, irrespective of either the jointing of structural nit involves the impurity or the impurity "heals" (neutralizes) one of the ends of the Si-O-Si bond.

The analysis of IR reflection spectra shows that multicomponent high-alkaline glasses with a weaker structure (as compared with that of quartz glass) are less sensitive to implantation ion bombardment, whereas in mixing the efficiency of modifying the glass properties is higher which may be explained by a high activity of Na atoms in atomic mixing at the glass-coating interface. At high doses of bombarding atoms, the mixing efficiency is so high, that Na^+ go out of the coating layer, and the release of Na and its compounds can be visualized as a white deposit on the metal film surface.

Fig. 2 shows the plots of the behavior of refraction index at 400-1200 cm^{-1} for quartz glass implanted with P^+ at various energies. Though the shape of dispersion curve remains generally the same, both increase and decrease of refraction index and a marked shift of all maxima on the curve towards the lower wave number are observed on some of its sections. Ion mixing causes essential increase of refraction index (fig. 2b).

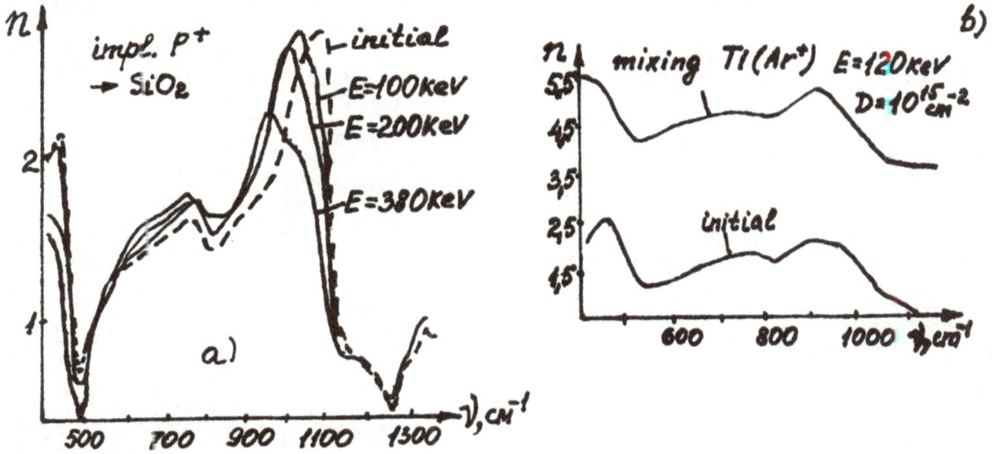

Fig. 2. Plot of the refraction index vs. the wave number in quartz (a) and alkaline aluminosilicate (b) glasses under ion bombardment.

The complicated behavior of the refraction index in implanted glasses cannot be ascribed to radiation-induced packing or ion bombardment-induced loss of sodium atoms. It may be suggested that the changes are caused by forming new chemical bonds involving the implanted impurities.

Ion bombardment in implantation regime produces, as a rule, the anti-reflection coating on all glasses studied, i.e. $R_L/R<1$ where R_L and R are the reflectances of the dopes and untreated glasses.

In case of ion mixing we only got anti-reflection coatings in quartz glass, whereas in other glasses both anti-reflection and reflecting layers were produced (R_L/R varied from 0.2 to 2.26).

STRESS WAVES AND STRUCTURAL MODIFICATIONS IN METAL AND ALLOYS AT HIGH CURRENT PULSED ELECTRON BEAM IRRADIATION

V.I.Itin[1], S.V.Lykov[1], G.A.Mesyats[2], P.B.Popov[1], D.I.Proskurovsky[2], V.P.Rotshtein[2]

Action of an intense pulsed electron beam on metals yields the formation of a stress wave or a shock wave propagating from the irradiated surface into a sample bulk [1]. The wave amplitude can be quite sufficient to produce the plastic deformation and marked strengthening of the target material [2].

This paper discusses the investigations into the mechanism of excitation as well as into the basic regularities and parameters of the interaction between a crystal lattice and stress waves occurring in the metals due to their irradiation by a pulsed electron beam with the power density up to 10^8 to 10^9 W/cm^2. Particular attention was paid in this study to the evolution of the wave profile and to the structural modifications of the target material within the region where the wave propagates.

The maximum energy of electrons in a pulse reached 240 keV, pulse duration being 60 ns. The targets used in the study were the foils made of pure iron and the alloy 36НХТЮ. Thickness of the foils varied from 30 to 200 μm in a 10 μm increment. The wave was recorded with a piezoelectric transducer placed on the rear side of a foil. The beam power density of $\sim 2\cdot 10^8$ W/cm^2 initiates the surface melting of the sample. In this case the thermoelastic mechanism of the stress wave excitation takes place due to rapid expansion of the heated surface layer. At the power density about 10^9 W/cm^2 there occurs evaporation of the target material and formation of craters on the irradiated surface. In this case the recoil momentum of the flying apart vapor also contributes to the wave formation along with the thermoelastic effect.

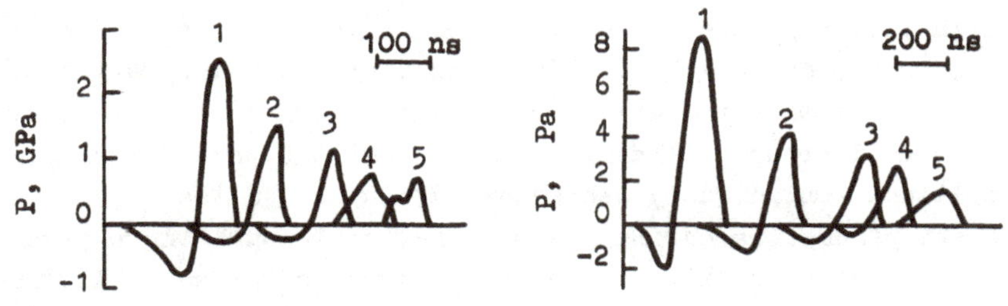

Fig.1. Changes of the stress wave profile in the case of iron foils of different thickness: 1 - 30 μm; 2 - 50 μm; 3 - 80 μm; 4 - 150 μm; 5 - 190 μm; a) the regime of the surface melting; b) the regime of intense melting and evaporation.

The irradiation in the regime resulting in the surface melting

1) Siberian Physico-Technical Institute, Tomsk, USSR
2) Institute of High-Current Electronics, Tomsk, USSR

yields the bipolar profile of the stress wave passed through the foil which involves the compression and dilatation pulses (see Fig.1a). The dilatation pulse appears due to the relief of the irradiated surface. Incidently, it is important to note that the amplitude of the compression pulse sharply decreases at the depth of 30 to 40 μm. The amplitude of the dilatation pulse vanishes by the depth about 150 μm. At the depth about 500 μm the wave becomes elastic and can be represented by almost perfect sinusoid.

At simultaneously acting factors of intense melting and evaporation of the surface layer, the contribution of the recoil momentum of flying apart vapor results in the increase of the compression pulse amplitude and duration (see Fig.1b). But nevertherless the view and all peculiarities in the parameters evolution are the same as in the case of purely thermoelastic initiation of the wave.

The irradiation of 36HXTЮ alloy foils in the regime of surface melting produces the wave with characteristically bipolar view after passing the foils (see Fig.2a). An increase of these foils thickness is accompanied by the effects analogous to those observed in the case of iron targets. The difference is that the sharp decrease of the compression pulse amplitude occurs at the depth 120 to 140 μm.

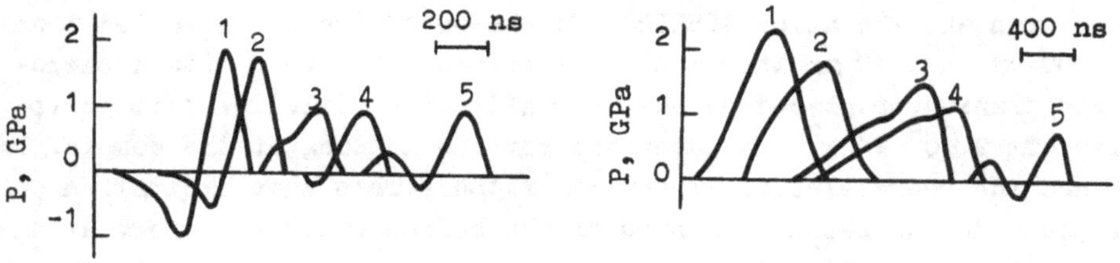

Fig.2. Changes of the stress wave profile in the case of foils of 36HXTЮ alloy of different thickness: 1 - 80 μm; 2 - 110 μm; 3 - 140 μm; 4 - 170 μm; 5 - 190 μm; a) the regime of the surface melting; b) the évaporation régime.

At the irradiation in the evaporation regime the wave is in fact the pulse of compression (Fig.2b). Under conditions when the target surface is loaded due to evaporation process no dilatation pulse occurs. The foil thickness increase yields the increase of the wave duration, while its amplitude continuously decreases. Evolution of the stress wave profile and parameters during its propagation through the crystal is connected with the dissipation of energy stored in the wave. Attenuation of small intensity nonlinear waves in metals is determined by elastico-plastic behavior of the medium. Particular mechanisms of the wave energy dissipation are determined by irreversible structured changes due to the creation of different crystal lattice defects and their displacements and interactions. The structural effects accompanying the stress wave propagation have been studied in [3] for the case of massive samples of iron and 36HXTЮ alloy using the methods of electron - positron annihilation and transmission electron microscopy. It was found that there exist a direct connection between the regularities of the wave parameters changes and peculiarities in the crystal lattice microstructure.

The phenomenon of sharp decrease of the compression pulse amplitude is related to the wave energy dissipation due to plastic deformations. It is observed in the iron samples that the life time of positrons τ is maximum at the depth of 30 to 40 μm from the irradiated surface (Fig.3) (the value of τ is proportional to concentration of the vacancy type defects). This is caused by deformation processes occurring due to the intense dissipation of the wave energy in the region. The increase of the defects concentration near the surface is connected with the hardening by high temperatures.

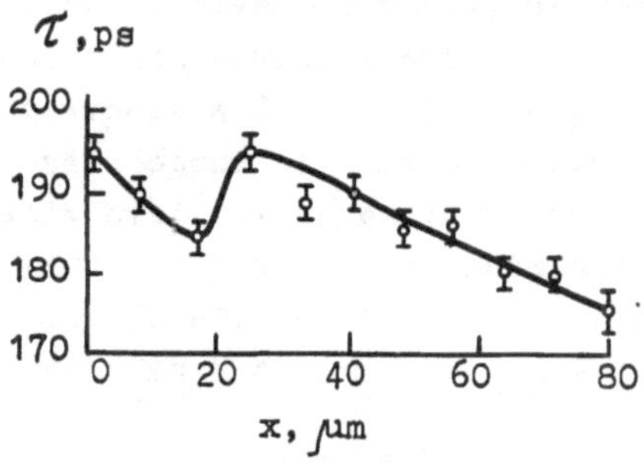

Fig.3. Life time of positrons as a function of the distance to the irradiated surface for iron samples.

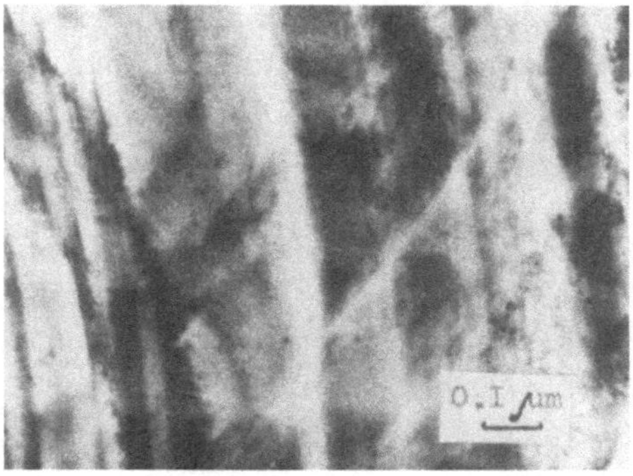

Fig.4. TEM photograph of 36HXTIO alloy at the depth of ~ 140 μm.

In the case of samples of 36HXTIO alloy irradiated in the regime of surface melting (Fig.2a) an enhanced concentration of dislocations was revealed. In these samples there was observed the dominating quantity of plane clusters that is characteristic of metals with the FCC lattice after a rapid deformation. However, the maximum plastic deformation is observed at the depth 120 to 150 μm in the region of sharp decrease of the stress wave amplitude. Here were revealed multiple bands of glides, significant fragmentation and bands of the type of the adiabatic shear bands (Fig.4). The developed localization of the plastic deformation in the region of intense dissipation of the wave energy well agrees with the "wave delay" theory [4].

REFERENCES

(1) Akkerman, A.F. et al.: Zh Eksp. Teor. Fiz. 3 (1986) 852-859

(2) Itin, V.I. et al.: Izv. Vyssh. Uchebn. Zaved. Fiz. 6 (1985) 38-43

(3) Lykov, S.V. et al.: Dokl. Akad. Nauk SSSR (1989)(to be published)

(4) Olson, G.B. et al.: Shock Waves and High-Strain-Rate Phenomena in Metals, ed. by M.A. Meyers and L.E. Murr, Plenum Press, New York and London 1981.

MELTING AND MIXING OF FINE METAL LAYERS AFTER EXPOSURE TO LOW - ENERGY HIGH CURRENT ELECTRON BEAM

Yu.Yu.Krjuchkov, V.M.Milutin, N.N.Nikitenkov, A.D.Pogrebnjak,
A.A.Garapadskii, G.E.Ozhur*, D.I.Proskurovskii*

In this paper we report original results on electron beam melting and mixing in double layer and three layer metal structures. We investigated Ag-Fe, Cu-Fe, Al-Fe, Fe-Al, Cu-Mo, Au-Cu double systems and Au-Cu-Mo three layer system. α-Fe samples of about 200 μm thickness were used as substrates and a layer of about 50 to 200 nm was deposited on the surface by the electron beam. To investigate mixing process, we employed RBS, AES, Mössbauer spectroscopy, electron raster and transmission microscopies. Simultaneously microhardness (Micromet device, USA) and the friction coefficient vs the applied load were measured for those samples. An electron gun described in paper [1] was used. The electron beam formed in the double layer between the cathode and the anode plasmas and was transported to the target in an external magnetic field with H = 1 to 1.5 kOe. The beam current fairly exceeded the limiting vacuum current due to compensation of space charge in the anode plasma filling the drift gap. Prefilling of the accelerating gap (d = 1 to 2 cm) by

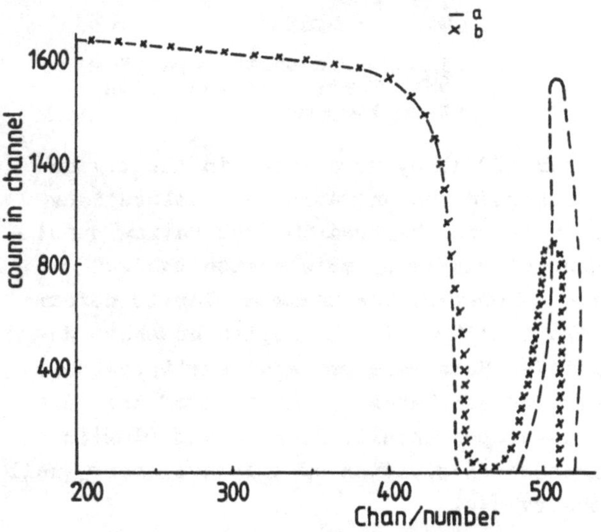

Fig.1. RBS energy spectra for Au-Cu system in initial (a) and irradiated (b) with W = 5.2 J/cm^2 states.

Nuclear Physics Institute, 634050 Tomsk, P.O.Box 25, USSR
* Tomsk Institute of High Current Electronics

the anode plasma allowed us to produce the electron beam with a fair good uniformity from the cold cathode at characteristic values of the accelerating strength amplitude $U \approx 30$ kV . The calorimeter was used to measure the energy density distribution over the beam cross-section. The obtained results were compared with the results of graphic integration of the beam current oscillograms and accelerating voltage oscillograms . Both methods showed fair good coincidence within the limits of the measurement inaccuracy (about 10%). The beam energy density varied with in 0.5 to 5 J/cm^2 with 0.3 to 1 μs [1] beam current duration. Single pulses were used in our experiments. Figure 1 displays RBS energy spectra for Au-Cu system for the initial state (a) and for high current electron beam irradiated samples (b) with the energy flow 5.2 J/cm^2 . A thin Au film of 20 nm was present in an initial sample, and the $Au_{10}Cu_{90}$ layer of about 150 nm with an average stoichiometry was formed after the irradiation, i.e. gold was deposited in the depth exceeding 100 nm. One should pay attention to good mixing ability.

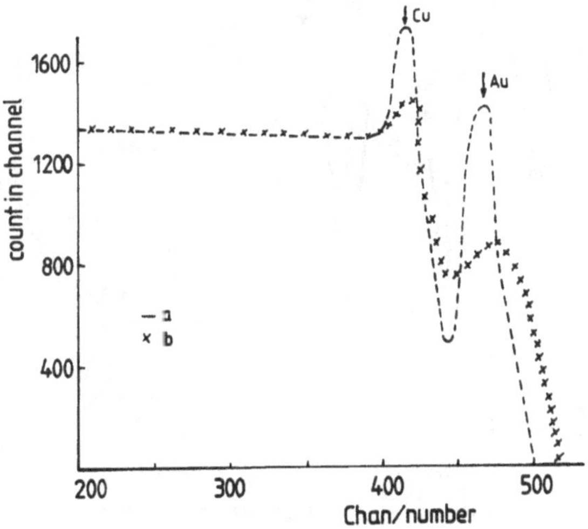

Fig.2. RBS energy spectra for Cu-Au-Mo system in the initial (a) and irradiated (b) with W = 5.2 J/cm^2 states.

Figure 2 displays the energy spectra RBS for three-layer Cu-Au-Mo system in the initial state and after the irradiation. The deposited Cu layer thickness was about 240 nm and the gold film thickness of 35 nm . The spectra show pronounced Cu peaks in Mo (410 - 434 channel) and the characteristic Cu peak (493 - 503 channel). Mixing of those films occurs after high current electron beam exposure (W = 5.2 J/cm^2) and stoichiometry changes. Au_{20}-Cu_{15}-Mo_{65} compound has been obtained.

Figure 3 displays the Mössbauer spectra for Ag-Fe double system in the absorption geometry before (a) and after (b) high current electron beam exposure with 4.5 J/cm^2 . An isomeric shift and sixtet broadening were observed in the spectrum after irradiation which evidenced formation of a chemical compound composed of Ag and Fe. Thus the mentioned examples show that electron-beam mixing may be successfully employed to form new compounds and even with essentially different thermophysical and thermodynamical properties. Also we should like to note that the results obtained using the electron beam pulsed irradiation in definite regimes (optimum energy flows) correlate well with the results on ion mixing and melting by high power ion beams (HPIB) [3,4] .

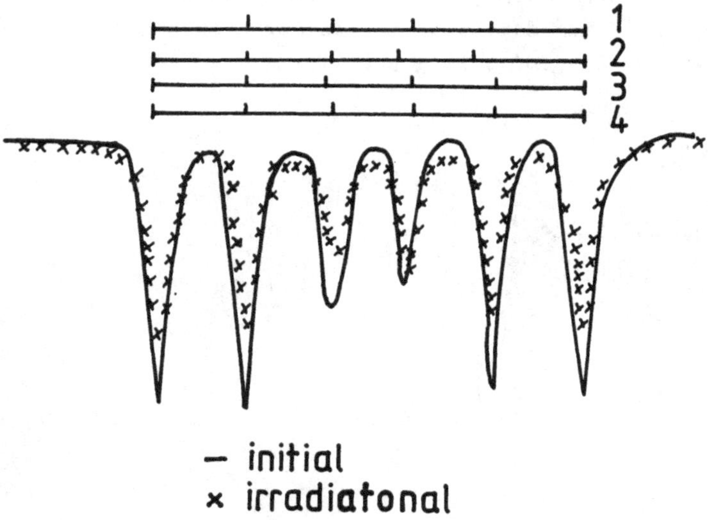

Fig.3. The Mössbauer spectra for Ag-Fe system : initial state (1,2) and after high current electron beam exposure (3,4), with $W = 4.5\ J/cm^2$.

1. G.E.Ozhur, D.I.Proskurovskii: Pisma v ISTF, 14 (1988) 5, 413-416.
2. G.E.Ozhur, D.I.Proskurovskii, E.B.Jankelevitch: Proceedings of the 2-nd Intern.Conf.on Electr.Beam Technol-88, Varna,Bolgaria (1988) 216-- 221.
3. A.D.Pogrebnjak, G.E.Remnev, I.B.Kurakin, A.E.Ligachev: Nucl.Instrum. and Meth, B36 (1989) 286-305.
4. A.D.Pogrebnjak, G.E.Remnev, S.V.Plotnikov: Mater.Sci. and Eng.(1989) 114, in the press.

EFFECTS OF ION FLOWS ON STRUCTURE AND PROPERTIES OF CERTAIN INORGANIC DIELECTRICS

Pichugin V.F., Feodorov A.N., Shmyrin A.J.
Tomsk Polytechnical Institute, 634004,
Tomsk, USSR

1. Introduction

A number of papers studying ion implantation effect on electric conductivity of some organic and inorganic dielectrics appeared recently (1,2). A sufficient growth in dielectric conductivity after ion irradiation and a noticeable structure change of implanted layers were noted in these and other papers.

This paper presents investigation results of the implantation effect of different ion types on surface conductivity of UF ceramics (Al_2O_3, SiO_2 and other oxides) and crystal quartz.

2. Experimental

Samples of UF ceramics and crystal quartz are plates of about 10x10 mm^2. All the sample surfaces were mirror-polished. Implantation was performed by ions of H^+, He^+, C^+, O^+, Al^+, Ar^+, Cu^+, Cd^+, Pb^+ using both stationary and pulse beams with doses from $1 \cdot 10^{14}$ ions/cm^2 to $5 \cdot 10^{16}$ ions/cm^2. The bombarding ion energy ranged from 20 to 150 KeV. The beam current density was about 10 $\mu A/cm^2$. The target temperature in the ion implantation process was controlled by a tungsten-rhenium thermocouple and didn't exceed 60°C.

The surface conductivity of the ion modified layers was measured by a two point electrode method. The electrodes were concentric rings. Ohm's law is observed in the range of electric field strength values used. All the measurements were carried out in the temperature range from 300 to 600 K the vacuum was 10^{-5} torr.

3. Results

Quartz and UF ceramic surface conductivity before irradiation is known to have ion character. The ion implantation leads to colour changes in the samples which obtain metallic lustre. The phenomenon may be due to free electron concentration increase, that is, the electron conductivity prevailed in the ion implanted layers of the materials studied that coincides with the conclusions made in the paper (2) for sapphire.

The measurements showed that the surface conductivity values for both UF ceramics and quartz increased about 10^7-10^{12} times under Ar ion irradiation with a dose of 10^{16} ions/cm^2. It is worth mentioning that a

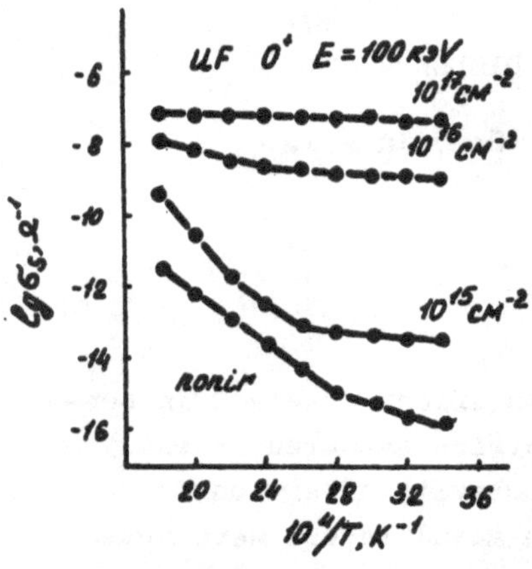

Fig. 1. Conductivity of UF ceramics implanted with oxygen ions depending on 1/T.

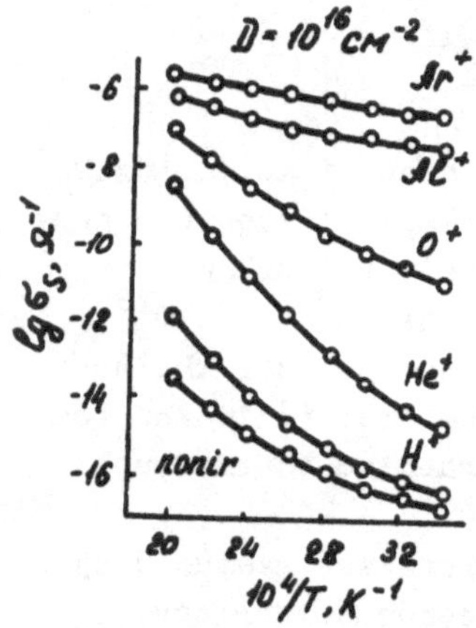

Fig. 2. Quartz conductivity when irradiated with different ions depending on 1/T.

considerable jump in the conductivity took place in the dose range from 10^{14} ions/cm^2 to 10^{16} ions/cm^2. A tendency to conductivity saturation is observed when doses exceed 10^{16} ions/cm^2.

Fig. 1 gives surface conductivity temperature (σ_S) of the UF ceramic samples irradiated with different doses of oxygen ions (O^+). The lg σ_S is seen to be a linear function of 1/T. In addition one or several linear sections in the dependence of lg σ_S on 1/T can be observed depending on the dose. The conductivity activation energy is observed to decrease with the dose increase.

The ion implanted layer conductivity of the materials studied depends not only on the dose but on the implanted element atomic mass as well. This is shown by fig. 2 which gives the surface conductivity temperature dependence for quartz irradiated with different ions (the dose was 10^{16} ions/cm^2). Both the conductivity values and the dependence slopes of lg σ_S on 1/T are determined by mass number M of the implanted ions. The character of the conductivity change is the same both in the case of inert gas implantation and in case of metal ion implantation (curves 5,6 fig. 2). The activation energy of the surface conductivity changes from 1 eV for the non-irradiated quartz to 0.02 eV for the Ar ion irradiated quartz.

The surface conductivity dependence for the quartz on the mass number of the implanted ions are given in fig. 3. The measurement was carried out at room temperature. The dose was 10^{16} ions/cm^2, the ion energy was 100 KeV. The surface conductivity increases with the growth of the mass number of the implanted ions in the range of M values from 1 atomic mass unit (a.m.u.) to 40 a.m.u. The further increase of the bombarding ion mass leads to the decrease in the conductivity of the ion implanted quartz layer. The optimal value of the mass number M of the bombarding ions causing the maximum change in the surface conductivity of the implanted layer is within the range of 35-40 a.m.u.

The conductivity studies of the quartz and UF ceramic modified layers when the energy of the bombarding ions changed from 50 KeV to 150 KeV showed that both the conductivity value and the character of its temperature dependence is not determined by the implanted ion energy.

4. Discussion

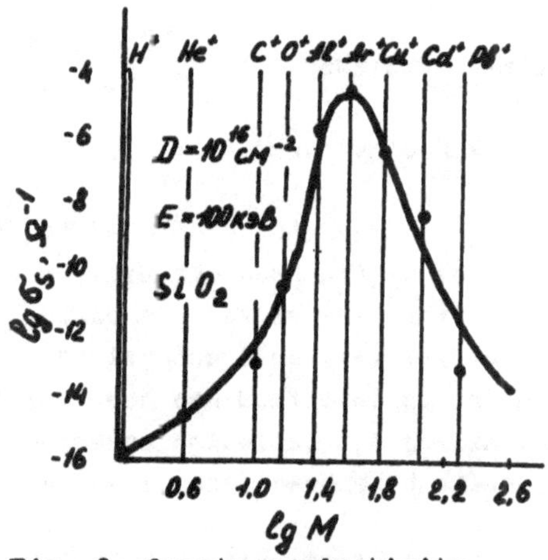

Fig. 3. Quartz conductivity depending on the mass number of the implanted ions.

In accordance with the results of the implanted ion type effect on the surface conductivity its increase is caused rather by a rise in the concentration of radiation damage induced by the ion implantation than by a doping impurity.

We think that to understand the results given it is possible to use the conductivity model formulated in the papers (5,6) which is correct to amorphous semiconductors. In accordance with this model the dependence of state density on energy is represented in fig. 4. Boundary energies E_c and E_v separate localized states from non-localized ones. The density of the localized states in a mobility slot is defined by radiation damage concentration. This model assumes two conductivity mechanisms: the conductivity connected with carriers induced in the non-localized states and the conductivity connected with carriers making hops between the localized states. Depending upon the dominance of this or that process either one or two sections on the conductivity temperature dependence of the ion doped layer can be observed.

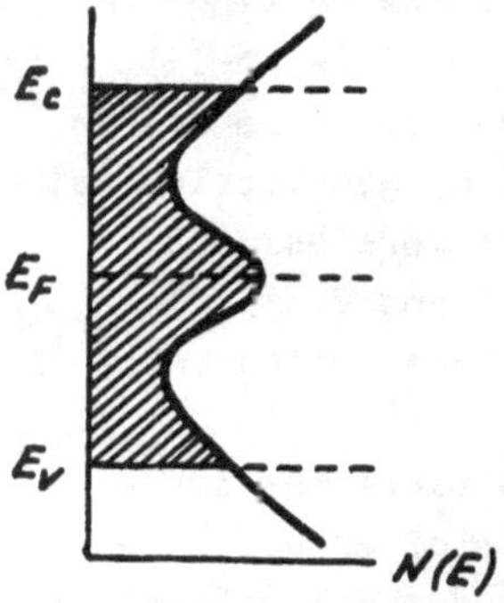

Fig. 4. The model of state density in an ion modified dielectric layer.

We suppose the basic mechanism of charge transfer in the studied temperature range to be carrier hops between the localized states. When ion doses are 10^{16} ions/ cm^2 and higher the concentration of the lattice radiation damage made under ion deceleration becomes very high and tends to saturation. In accordance with this the further implantation leads to a weak change in the density of the localized states and, consequently, to the surface conductivity saturation depending on the dose.

REFERENCES

1. J.Davenas, C.Dupuy et al. Radiation Effects, v. 74, 1983, p. 209-217.
2. S.Sato, M.Iwaki. Nucl. Instr. and Meth. in Phys. Res., 1333(1988), p. 145-149.
3. H.Jain. Surface Science 186(1987), p. 256-266.
4. N.S.Kostyukov, V.V.Maslov et al. Dielectric Radiation Stability, Tashkent (1981).
5. N.F.Mott,E.A.Davis.Electron processes in non-cryst.mater.,Oxf.,1979.

ION BEAM MIXING OF Al/Fe BINARY SYSTEMS

B.Rauschenbach[1)], R.Küchler[1)], M.Posselt[1)], P.Knothe[1)] and R.Dietsch[2)]

The atomic mixing of thin layered metal systems by ion beam irradiation has become a notable field of research during the last years. The basic mechanisms of ion beam induced mixing are not yet very well understood. In general, the ion-beam induced broadening of an interface has been interpreted on the basis of ballistic atomic mixing and radiation enhanced diffusion. In addition, chemical driving forces have been recognized to contribute to the ion mixing phenomena.

In this paper we report on mixing experiments with Al/Fe bilayers. Thin bilayers of Al and Fe were deposited in high-vacuum by evaporation. Two different configurations of the bilayers were prepared having Al or Fe on the top. The first set of samples consisted of 28 nm Al, 63 nm Al, 93 nm Al or 113 nm Al layers on iron and the second set of 13 nm Fe, 34 nm Fe, 38 nm Fe or 49 nm Fe on aluminium. The ion mixing was carried out with 200 keV Xe-ions at room temperature. Ion currents were below 0.5 $\mu A/cm^2$ and the ion fluences were ranged between $5*10^{15}$ and $7.5*10^{16}$ $ions/cm^2$. The samples were analyzed by Rutherford backscattering technique (RBS) and Auger electron spectroscopy (AES).

A problem is the determination of the sputtered depth since the sputtering rate is generally matrix-dependent. On the base of accurate measurements of the individual thickness of the evaporated metal layers, it is possible to define a sputtered depth scale /1/. The determinaton of the composition before and after mixing as function of depth near the metal/metal interfaces is difficult to achieve with sputter depth profiling techniques. Here, it is necessary to separate the interface broadening which results as a consequence of the Ar-ion sputter profiling by AES analysis from the interface broadening which results as a consequence of the high-energetic Xe-ion beam induced interface broadening during mixing. Recently, we have demonstrated a technique for determination of composition under such conditions /1,2/.

Ion beam induced mixing of Al/Fe bilayers is investigated as a function of the thickness of the metal top-layers and in dependence on the ion fluence of mixing. For example, Fig. 1 shows concentration profiles of Al after mixing with $1*10^{16}$ $Xe-ions/cm^2$. For depths higher 10...20 nm,

----

[1)]Zentralinstitut für Kernforschung Rossendorf
DDR-8051 Dresden, PSF 19
[2)]VEB Hochvakuum Dresden
DDR-8017 Dresden, Niedersedlitzer Str. 63

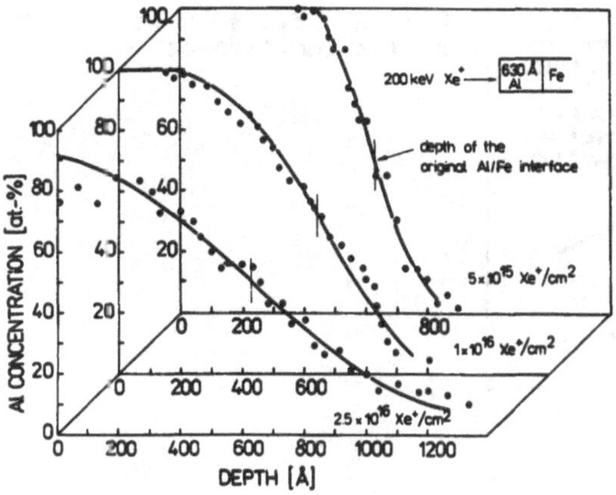

Fig. 1:
AES depth profiles of Al top layers with different thickness on Fe before and after mixing (solid points) with Xe-ions obtained by 3 keV Ar-ion sputtering. The experimentally determined concentration distribution were smoothed by cubic spline functions (before mixing - dashed lines, after mixing - solid lines).

we have a nearly pure binary Al/Fe system, because the oxygen component should be negligible and the concentration profiles of these both components should be mirror-images. The dashed lines in Fig. 1 characterize the concentration distribution of Al before ion mixing. The shape of these profiles is caused by Ar-ion sputtering (see Refs. /1,2/). Profiles before and after mixing with $1*10^{16}$ Xe-ions/cm² are similar in deeper regions. With a decreased distance from the surface, the differences between the profiles before and after mixing is increased /2/.
Series of AES measurements of bilayers have been performed at various fluences, starting at $5*10^{15}$ ions/cm² up to $7.5*10^{16}$ ions/cm². A change of the profile shapes are observed as a function of the Xe-ion fluence. The tendency towards smear out of the original concentration distribution is increased with increasing ion fluence.
We evaluate the diffusion coefficients as functions of concentration from the in-depth profiles by Boltzmann-Matano method /3/. It is evident that the diffusion coefficients are far from being constant /2/. The results show that the diffusion coefficients of Al are higher than the diffusion coefficient of Fe. For low fluences, the maximum of the diffusion coefficients of Al in Fe and also Fe in Al are found near the Al/Fe interface /2/. The differences of the diffusivities at both sides of the interface are reduced with increased fluences. Probably, the radiation induced point defect concentration and also the radiation enhanced diffusion are more pronounced in Fe than in Al (see discussion in Refs. /1,2/).
For determination of the mixing efficiency, the nuclear energy deposition at the interface is obtained from the Monte-Carlo simulation program TRIM and the diffusion coefficient is estimated from the results of the Boltzmann-Matano analysis. The results are illustrated in Fig. 2. The notable feature of this figure are the followings. In dependence on ion fluence, the mixing efficiencies for Al in Fe range between 75 and 30 $Å^5$/eV and for Fe in Al between 35 and 15 $Å^5$/eV. Fig. 2 shows an ap-

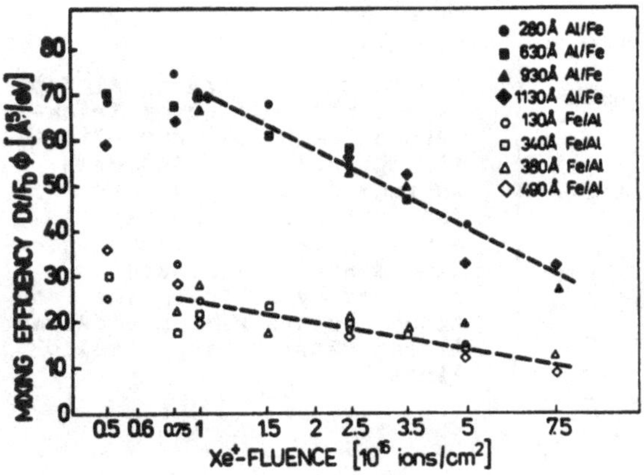

Fig. 2:

Mixing efficiency as function of the ion fluence

proximately logarithmic decrease of the mixing efficiency with increasing ion fluence. Gaboriaud et al. /4/ have observed a similar blocking effect of mixing for fluences $1{*}10^{16}$ Xe-ions/cm². Up to now, the effect of decreasing of the mixing efficiency as function of fluence is not completely understood. Ballistic, structural and thermochemical effects may be considered /2,4,5/.

In summary, the main results are:

(1) The interface broadening which is a consequence of the Ar-ion sputter profiling by AES analysis is clearly separated from the interface broadening which is caused by high-energetic Xe-ion beam mixing.

(2) The experimentally determined concentration depth curves exhibit a mixing effect for all fluences and a tendency towards saturation of the mixing effect at high fluences.

(3) The mixing induced profiles in the interface region deviate significantly from the Gauss error function. Therefore, the diffusion coefficients are evaluated from the concentration-depth profiles as a function of concentration by the Boltzmann-Matano method. The average diffusion coefficients vary between 5 and 50 Å²/s for Al on Fe and between 1 and 20 Å²/s for Fe on Al.

(4) For low fluences, the depth profiles of the diffusion coefficient corresponds to the depth distribution of the nuclear energy deposition.

(5) A variation of the mixing efficiency with the ion fluence and with the depth of the interface are found. Ballistic structural, thermal and thermochemical effects influence the mixing behaviour and its dependence on fluence and on thickness of the top layer.

References

/1/ B.Rauschenbach, R.Küchler, M.Posselt and R.Dietsch: Vacuum 38 (1988) 967

/2/ B.Rauschenbach, M.Posselt and R.Küchler: Appl. Phys. A 48 (1989) 347

/3/ P.G.Shewmon: Diffusion in Solids, McGraw-Hill, New York 1963

/4/ R.J.Gaboriaud, J.J.Grob and F.Abel: Nucl. Instr. Meth. B 19/20 (1987) 648

/5/ R.S.Averback: Nucl. Instr. Meth. B 15 (1986) 675

LASER BEAM MODIFICATION OF OPTICAL FILMS

D. Schäfer[1], B. Brauns[1], P. Höfner[1], R. Wolf[2], B. Steiger[2], J. Bärmann[2]

High refractory oxides are still of current interest for the manufacture of high power coatings in the visible and near infrared region. In order to enhance their performance we treated SiO_x-, ZrO_2- and Sc_2O_3-films with laser pulses to modify the optical properties. The films were produced in a conventional vacuum evaporation plant equiped with a turbomolecular pump. The SiO was evaporated at a oxygen partial pressure of about $1.3 \cdot 10^{-2}$ Pa from a molybdenum boat by resistance heating (RE). For the deposition of ZrO_2 and Sc_2O_3 we applied an electron beam evaporation technique (EB). Experimental details of the reactive thin film deposition process are given in Table 1.

After deposition the layers were irradiated in air with uv-laser pulses (25 ns half width) delivered from an excimer laser operating with KrF at 248 nm.

The unfocused laser beam impinges directly on the film surface with a fluence of approximately 30-60 mJ/cm^2. We investigated optical transmission, absorption and laser damage.

The damage threshold was determined with an acousto-optical Q-switched Nd:YAG laser (λ = 1.06 μm) with a pulse duration of 85 ns (half width) /3/. The arithmetical mean of the maximum fluence at which no damage occurs for all irradiated sites and the fluence at the onset of damage is defined as the damage fluence D_o, which corresponds to a damage probability of 0 %. Analogous to this, the damage threshold D_1 is defined as the fluence at which the damage probability is 100 %.

Absorption was measured by means of photothermal displacement spectroscopy /1/ at the same wavelength λ = 1.06 μm.

---

[1]AdW der DDR, Zentralinstitut für Optik und Spektroskopie, Rudower Chaussee 6, DDR-1199 Berlin

[2]Ingenieurhochschule Mittweida, Platz der DSF, DDR-9250 Mittweida

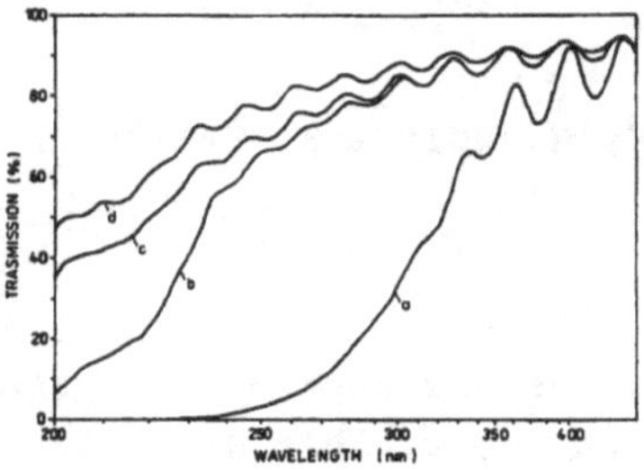

Fig. 1 Transmission of SiO_x-films v.s. wavelength after laser irradiation
a - nonirradiated
b - 1000 pulses, 40 Hz
c - 20000 pulses, 100 Hz
d - 20000 pulses, 40 Hz

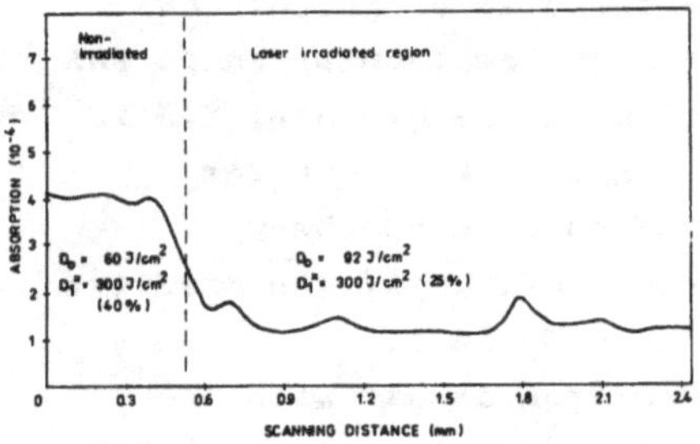

Fig. 2 Absorption scan across the SiO_x-film (the damage thresholds in the irradiated and nonirradiated region are given; D_1^* damage probability in per cent at the fluence of 300 J/cm²

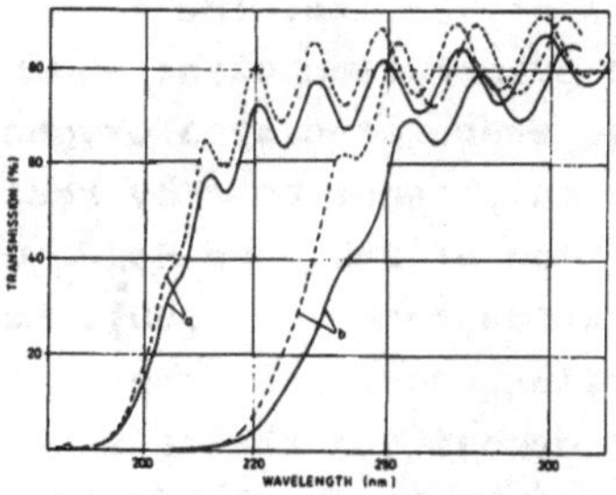

Fig. 3 Transmission of Sc_2O_3(a) films and ZrO_2(b)-films (———) and after laser irradiation (-----)
Sc_2O_3 :60 mJ/cm², 2000 pulses, 2Hz
ZrO_2 :45 mJ/cm², 2000 pulses, 2Hz

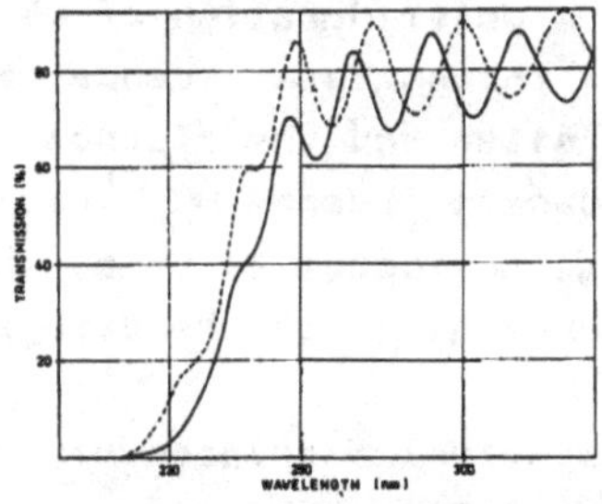

Fig. 4 Transmission of a ZrO_2 film (———) and after annealing 3 h at 400 °C in air (-----)

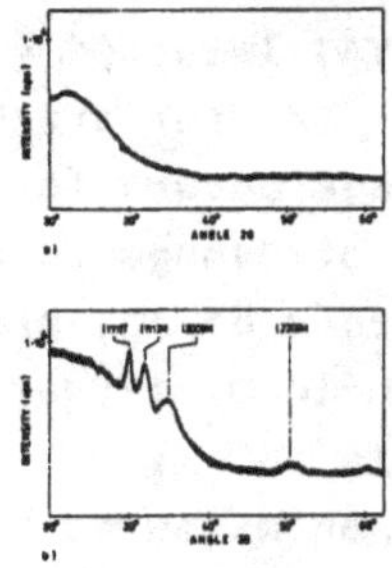

Fig. 5 X-ray diffraction of ZrO_2 -films
a - after laser irradiation
b - after annealing 3 h, 400 °C in air (T - tetragonal, M - monocline)

In Fig. 1 the change of the transmission in dependence of the wavelength after irradiation with 1000 and 20000 pulses for SiO_x-films is shown. Fig. 2 shows the absorption scan across the film surface whereby a decrease of absorption at the irradiated region of the sample is clearly perceivable. Results of laser damage measurements are also contained in Fig. 2 and prove an increase in damage resistance at the laser treated region.
The refractive index of the SiO_x-films was lower after uv-irradiation. The reason is the further oxidation of the films /2/.
Fig. 3 presents some results of the change of uv-transmission of ZrO_2- and Sc_2O_3-films. The decrease of the absorption was smaller as for SiO_x-films. By annealing the ZrO_2-films we found similar effects of transmission change (Fig. 4). X-ray investigations revealed the amorphous structure of the laser irradiated sample but a crystalline phase after annealing (Fig. 5).
Summarizing we can say that an enhanced oxygen invorporation in oxide films can be reached by uv laser irradiation and this immediatly influences the deposition of stochiometric films with increased laser damage thresholds.

References

/1/ Beierlein, Th., B. Steiger, R. Wolf, G. Zscherpe und D. Schäfer: Feinwerktechnik & Meßtechnik 97 (1989) 23-27

/2/ Blum, S.E., K.H. Brown and R. Srinivasan: Appl. Phys. Lett. 43 (1983) 1026-1027

/3/ Wolf, R., G. Zscherpe, E. Welsch, V. Göpner and D. Schäfer: Optica Acta 33 (1986) 919-924

Table 1

Film	Deposition method	Deposition rate (mm/s)	Oxygen partial pressure (Pa)	Substrate tempera- ture (K)	Film thickness (nm)
SiO_x	RE	0,4	$1,3\cdot10^{-2}$	293	700
ZrO_2	EB	0,9	$9\cdot10^{-3}$	293	560
Sc_2O_3	EB	1,4	$9\cdot10^{-3}$	423	600

MIXING OF Cu - Ni SYSTEM BY GAS - METAL ION BEAM

G.V.Potemkin, A.E.Ligachev, G.M.Novikova, O.S.Kuzmin, V.P.Yanovskii

Nuclear Physics Institute, 634050 Tomsk, P.O.Box 25, USSR

The process of dynamic mixing by a stationary gas-metal beam of the system of two metals having the same atomic radii has been studied. The current density and ion dose were investigated to affect the efficiency of the component mixing in the film deposited on the matrix.

The separator-free equipment was used for irradiating the samples by charged particles with the energy to 50 keV, the current was up to 100 mA. Ion beam involves the ions of working gas of a sputtered substance and the neutral component. The equipment was provided with ion source caused by sputtering a solid target by gaseous ions of glow discharge in ExB fields. A predominant component of ion beam is N_2^+ , Ti ions constitute about 30 to 40 %. The neutral component deposited on a sample as Ti - ions with the energy less than 100 eV , is comparable in order of the magnitude with the integral dose.

The samples, in the form of mickel substrate with the 75 nm-thick Cu film deposited by the electrolytic method, were bombarded with the i ion beam of 40 keV energy and current density of j = 50; 150; 250 ion/cm^{-2} , the integral doses ranging from about 10^{17} to 10^{18} ion/cm^2 . The dose and current density on samples were measured using the Faraday cup and temperature - by tungsten-rhenium thermocouple.

The samples were irradiated on a potential electrode of equipment, and they were not exposed to a forced cooling. Radiation heating of the samples resulted in temperature increase in samples during bombardment up to thermo-dynamical equilibrium state. The equilibrium temperature depending on the current density ranged from about 100 to 500°C and the time of equilibrium onset - from a few minutes to tens of minutes . The structure of a near-surface layer was analyzed by means of the X-ray diffractometer DRON-3 with Cu-K radiation and glancing beam geometry. The concentration profiles were measured using SIMS.

Figure 1 (a-d) shows the results of measuring the profiles of the initial and irradiated samples; each plot shows the dose - N_0 , the ion current density - j, irradiation time - t , and equilibrium temperature- - T .

At the doses of about 10^{17} ion cm^{-2}, appreciable mixing of deposited layer with matrix was observed. At these doses the difference between the concentration profiles of samples irradiated with j_1 = 0.15 mAcm^{-2} and j_2 = 0.25 nAcm^{-2} is of interest. Equilibrium temperature and efficiency of mixing are essentially higher for high density; this is confirmed by Ni and Ti profiles at the depth up to 20 nm. The higher temperature (T_2 500°C) at j_2 , as compared with (T_2 210°C) at j_1 , and the ion current density cause higher concentration of defects in the near surface layer favouring their higher mobility. This leads to more

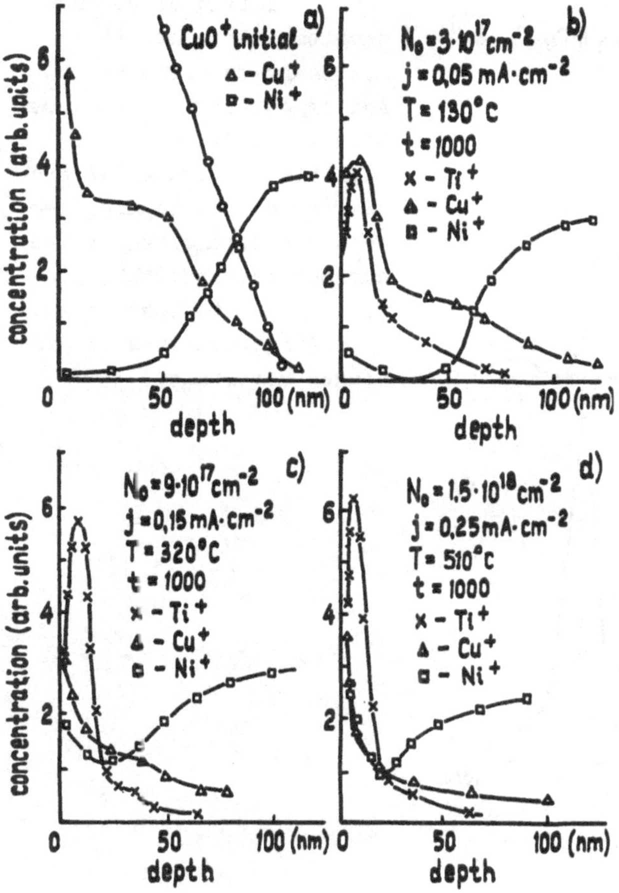

effective mixing and destruction of phase interfaces. At the same j , the mixing intensity depends on integral dose, N_0 , within the studied interval of 10^{17} N_0 10^{18} ion/cm^{-2} and increases with N_0 increasing .

Fig.1 (a-d). SIMS - - profiles for samples.

For the samples in Fig. 1 (a-d), the shape of concentration Ti pro - file is typical in contrast to the Ni - profile. For the depths compared with the value of a projective path of accelerated ions, Ni - concentration increases while reaching a boundary. As compared to Ni the more effective accelerated ion beam sputtering of a copper film leads to substitution of its atoms for both deposited and accelerated Ti atoms and to Ni atom enrichment of the layer too/1/ As the dose increases, formation of a solid solution of substitution. Fig.2 (b), with further formation of intermetallic compound takes place, Fig.2 (c) . Higher current density of processing promotes at the same time the annealing process, i.e. ordering of a mixed layer leading to generation of a continuous series of solid solution of substitution, to disappearance of the film-matrix interfaces, and to a complete film dissolution in matrix.

From cur investigation the dependence of mixing depth on the current density is ambiguous though a high density must promote deeper mixing. The rival processes of ordering seems to mask this phenomenon at high temperatures. For an industrial application of ionic mixing it is apparently advisable to put into operation an ion beam processing of samples with minimum N_0 and maximum j . The values of N_0 and j_{max} are limited by properties of initial and irradiated materials. When

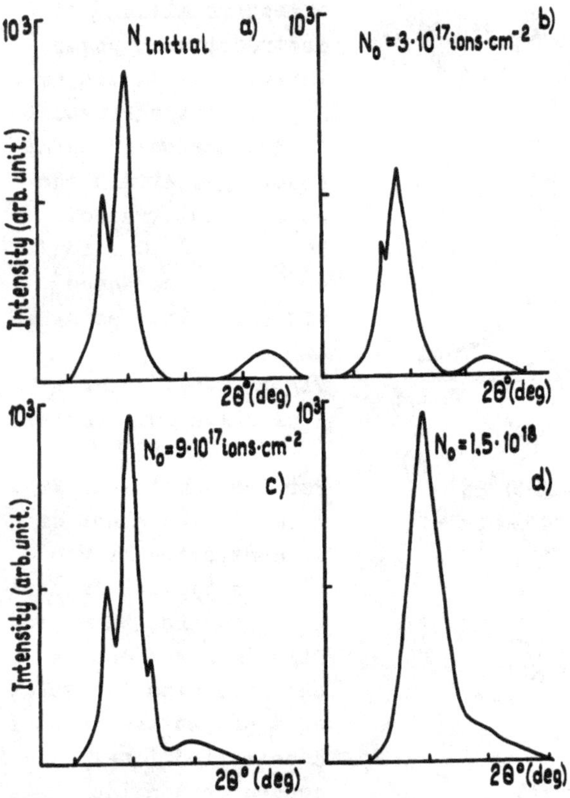

Fig.2 (a-d). X-ray pattern for samples.

forced cooling of processed samples is absent, it is necessary to optimize N_0 and especially j because both the efficiency and possible defectness of processing depend on high temperature determined by the current density; this sometimes leads to loss of some specific properties of processed materials.

REFERENCES

1. Sputtering by Particle Bombardment I, ed. R.Behrisch, p. 232-234.

STRUCTURE TRANSFER AND PHASE CHANGE BY GENERATION OF THIN FILMS BY MEANS OF LASER-INDUCED DEPOSITION METHODS

H.-J. Dietze[1], S. Becker[1], G. Keßler[2], W. Pompe[2]

For the application of laser-induced deposition methods in generating thin films (e.g. hard amorphous boron nitride, diamond-like carbon or high-temperature superconducting films) the knowledge of possible structure transfers from the target to the substrate and of phase changes by the interaction of the laser beam with the target material is of importance. Plasmadiagnostic investigations of the cluster formation in laser plasmas by means of laser ionization mass spectrometry /1/ showed for several original structures of the target material (e.g. for carbon /2/ or boron nitride /3/) that by the laser induced plasma deposition (by the deposition of atomic, molecular or ionic species of the laser plasma on a substrate) no structure transfer occurred.
Stoichiometric and structure transfer from the target to the substrate in the sputter regime /4/ are observed. The chemical bonds in the sputtered particles (diameter 100 nm - μm) being largely conserved.

LASER-INDUCED DEPOSITION METHODS

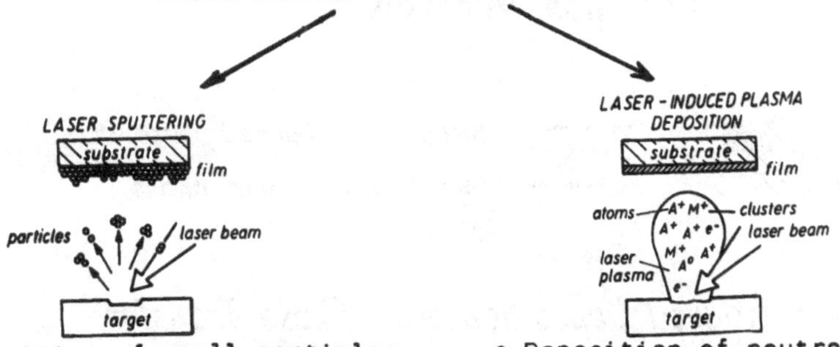

- Deposition of small particles with the original structure of the target material in the film
- No dissociation of the chemical bonds of particles
- Transfer of stoichiometry of the target material

- Deposition of neutrals, atomic and cluster ions via a plasma state in the thin film
- The structure is destroyed
- Dissociation of chemical bonds
- Change of stoichiometry

1) Central Institute of Isotope and Radiation Research, Academy of Sciences of the G.D.R., Leipzig (G.D.R.)

2) Central Institute of Solid State Physics and Material Science, Academy of Sciences of the G.D.R., Dresden (G.D.R.)

Phase Transfer on a Target Surface
Laser Sputtering and Laser-Induced Plasma Deposition

I. Photon - Solid State Interaction

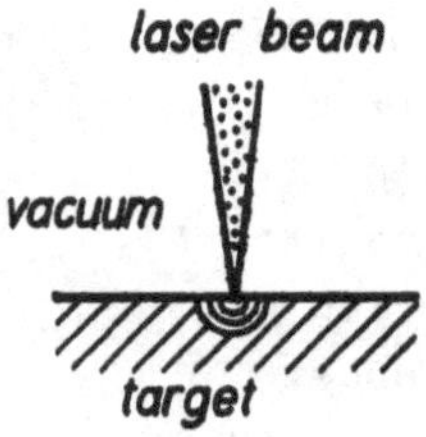

→ heat up
→ pressure increase

} in laser focus

II. Plasma - Solid State Interaction

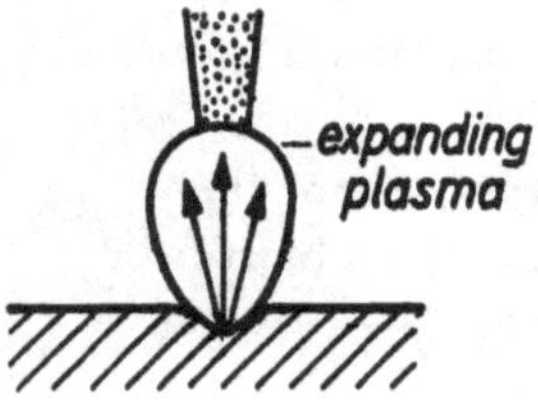

→ shock wave produces a high pressure

→ solid - solid phase change (high temperature)

III. Results of Interaction Processes

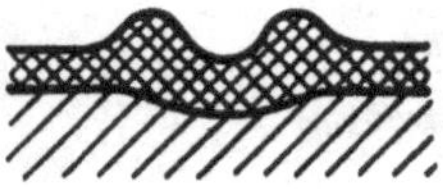

— target with changed phase /5/

— target with original phase

IV. Laser - Induced Deposition after Phase Transfer

photon - solid state interaction
plasma - solid state interaction

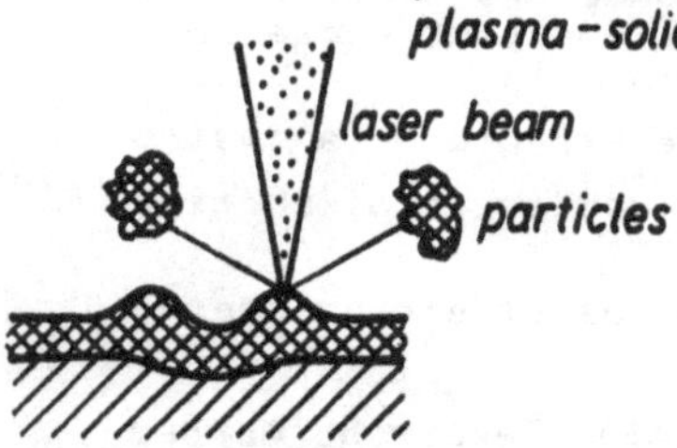

laser sputtering

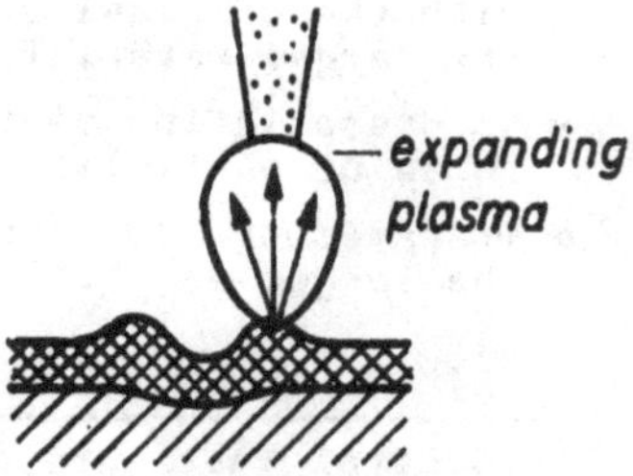

laser - induced plasma deposition

- transmission of target structure in the film

- structure of the film is independent of the target

REFERENCES

/1/ H.-J. Dietze, S. Becker, Beiträge zur Clusterforschung, ZfI-Mitt. 134 (1988) 1-174

/2/ S. Becker, H.-J. Dietze, W. Pompe, Z. Chem. 26 (1986) 453

/3/ S. Becker, G. Keßler, H.-J. Dietze, H.-D. Bauer, W. Pompe, Thin Solid Films (in press)

/4/ S. Komuro, Y. Aoyagi, T. Morikawa, S. Namba, Jap. J. Appl. Phys. 27 (1988) L34

EMISSION OF EXCITED MOLECULES DURING ION BOMBARDMENT OF METALS

POP S.S. and BANDURIN Yu.A.[I]

At present time only one of the components of ion-photon emission (IPE) of metals is considered to be investigated sufficiently enough. This component is the radiation of spectral lines from excited secondary atoms and ions, which is produced when these particles escape from the surface. The component of IPE which is conditioned by the radiation of the surface itself, i.e. ionoluminescence, has not been practically studied yet. One more component of radiation is observed in IPE spectra. This is continuous radiation (CR), which consists of molecular bands or wide continuum. This type of radiation is caused by sputtered excited particles which form the extent shining region in front of the bombardment target.

Many experiments of CR investigation have been carried out in Uzhgorod State University during several last years. The results of these experiments are reviewed in these report. Much more details of CR investigations are discussed in book[3]. CR has been studied also by many other researchers (see, for example, the reviews[2-5]). But so far there is no the general opinion even about nature of CR emitter, that is very impotent for the determination of their formation mechanisms.Therefore the first aim of the experiments carried out in our laboratory was to identify the different emitters of CR.

We varied both the interacting systems (ion-target), and the conditions of interaction (bombardment energy, angle of incidence and the coverage of metal surface by particles of active gases). We used beams of K^+,He^+,Ne^+,Ar^+,O_2^+,N_2^+,NO^+ ions with energy I - 20 keV. The bombardment of different metallic targets (Ag,Cr,V,Ti,Fe,Ni,Zr,Mo,Nb,Hf,Ta,W, Gd,Tb,Ho,Yb,Tm) was carried out both in the residual vacuum conditions (when sufficient surface roughness was achieved) and at the controllable leak-in of O_2,N_2,CCL_4,SF_6,CF_4,NF_3 gases. Fig.I showes,for example the spectrograms of radiation resulting from K^+ bombardment of Ho target in residual gas pressure $5\cdot IO^{-8}$ Torr (the lower curve) and in atmosphere of O_2,SF_6 and CCL_4 at gas pressure $I\cdot IO^{-5}$ Torr [I].The analysis

[I] Uzhgorod State University, 46,Gorky Street, 294000 Uzhgorod, USSR of such data for different systems ion-metal as well as identification of observed spectral peculiarities with known spectroscopic information gave the possibility to determine the nature only narrow bands in IPE spectra. These bands are emitted by simple excited molecules such as MO,MF,MCLand MN, where M is the atom of metallic target. Broad bands and continuum radiation which were observed in the IPE spectra for many d- and f-metals in presence of chemically active gases were not identified. Evidently this radiation is emitted by more complex molecules M_nX_m type, where M is target atom and X is an atom of adsorbed gas.The indexes n and m point the number of atoms in molecules.

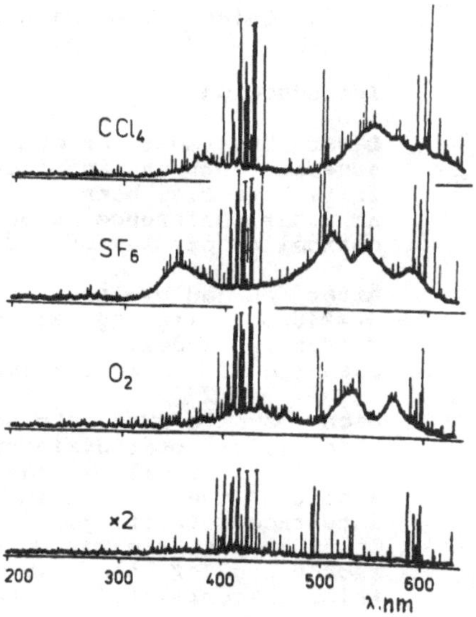

Fig.I. Ho-target radiation spectrograms under K^+ ion bombardment of I2 keV energy.

It was found that there was no band emission of mentioned molecules when the target surface was not covered by adsorbed particles.In this cases it was observed only weak band radiation of Mo_2 and Nb_2 dimers[3]. The angle of incidence, enrgy and mass of bombarding ions didn't sufficiently change the form of band spectra. The magnitude of absolute photon yield from excited molecules is rather large, especially for rare-earth metals, when the target surfaces are covered by a monolayer of adsorbed particles. For example, the photon yield is $2 \cdot 10^{-4}$ photon per incident ion in the spectral region 496 - 602 nm for case $K^+ \rightarrow Ho+O_2$ interaction. Ion energy was I5 keV and dynamic balance coverage of Ho surface by oxygen was approximately 0,6. At the same time the photon yield for spectral line HoI493,3 nm emitted by sputtered excited Ho atoms is two order of magnitude less. We considered that the excited molecules are sputtered from the surface layer at the last phase of the linear collisions cascade.

(I) Braslavets,V. and etc.: Lett.Zh.techn.Fiz.I2(I986)543-547.
(2) Gritsyna,V.: Radiotechn. and Electron.26(I98I)I969-I983.
(3) Pop,S. and etc.: Ion-Photon Emission of Metals,Fan,Tashkent I989.
(4) Thomas,E.: Progr. in Surf. Sci.I0(I980)384-490.
(5) Veksler,V.: Zh.techn.Fiz. 52(I982)I273-I279.

CO_2 LASER TREATMENT OF ZnO THIN FILM OPTICAL WAVEGUIDES

M.Bertolotti[1], A.Ferrari[2], M.Laschena[1], M.Rossi[1], L.S. Quian[3], F.Quaranta[4], and A.Valentini[4]

Introduction

Laser treatment of optical thin films has been considered by several authors and found to affect optical properties. Silica films have f.e. been studied by S.E. Blumm [1] and more recently at this conference D.Schafer et al. [2] have investigated the optical properties of SiO_x, ZrO_2 and Sc_2O_3 films after pulsed UV-laser irradiation from a KrF excimer laser at λ=248nm, finding effects on both optical absorption and refractive index.
Scanning CO_2 laser treatment of a variety of oxide thin films has been shown to produce a decrease of optical losses in a number of thin films over different substrates [3]. In the case of ZnO, Bertolotti et al. [4] have confirmed these results. S. Dutta [5] ascribed the reduction of optical losses to a reduction of interface scattering.
In the present paper we show that in the case of ZnO a partial reordering of the film is produced under the laser treatment which is correlated to the optical losses reduction.

Experiment

ZnO films prepared using a r.f. diode sputtering system were deposited on both Corning glass 7059 or c-axis oriented sapphire. The ZnO/Corning film had in general a polycristalline columnar structure with a preferred orientation of the crystallite with the c-axis perpendicular to the plane of the substrate [6].
The samples on sapphire gave a nearly uniform ring structure without any appreciable preferential orientation and the deposition of layers of thickness larger than about 1.5 µm was more difficult than the corresponding Corning glass case and less reproducible.
Laser annealing was performed with a cw CO_2 laser (λ=10.6 µm) of about 4.4 W focussed on the ZnO film with a germanium lens over a spot of about 200 µm diameter.
The laser beam was scanned in front of the sample at variable velocities between 3 and 6 cm/s. At the used power density, higher scanning velocities gave no appreciable changes in the optical losses, and velocities lower than 5.5 cm/s broke the ZnO/Corning glass samples, while ZnO/sapphire samples were broken at velocities lower than 3 cm/s.

1) Dipart. di Energetica, Universita' Roma I, Via A.Scarpa 16 Roma, Italy and GNEQP/INFM
2) Dipart. di Elettronica, Universita' Roma I, Italy
3) Changchun Inst. of Optics and Fine Mech., Academia Sinica, P.R. China
4) Dipart. di Fisica, Universita Bari and GNSM/CISM, Bari, Italy

The samples during the laser treatment were taken at a temperature of about 140 ^{o}C to prevent any damage to the substrate due to thermal stresses.
In both cases of Corning glass and sapphire substrates a decrease of optical losses was found for all the propagated modes in the film as shown in table I.

TABLE I

MATERIAL losses (dB/cm)	ZnO Corning glass before	after annealing	ZnO sapphire before	after annealing
TE_0	3.51 (23)	1.85 (0.32)	7.70	2.35
TE_1	5.29 (31)	3.44 (17)	10.59	4.28
TE_2	11.10 (41)	5.89 (26)	--	--
TM_0	8.27 (26)	3.56 (2)	11.10	4.49
TM_1	11.97 (36)	5.88 (27)	17.01	8.45
TM_2	20.41 (--)	11.42 (--)	--	--

The data obtained after laser annealing refer to a scanning velocity of 5.5 cm/s for the ZnO/Corning glass samples and of 4.5 cm/s for the ZnO/sapphire samples.
Values in parenthesis for ZnO/Corning glass samples refer to previous measurements by Bertolotti et al. [4] .
Some observations can be immediately done. The first one is that reduction of losses is relatively larger for higher order modes than for lower order modes. The second point is that in previous investigation [4] larger losses reduction was obtained but starting from films with initially higher losses than the films used in the present experiment. Finally it can be seen that the loss reduction is relatively larger for ZnO/sapphire films than for ZnO/Corning glass.
Observation with RHEED showed that in both cases of Corning glass or sapphire a marked reordering was produced under laser irradiation. This can be seen in Fig. 1 where RHEED patterns for

a ZnO/sapphire sample are shown for as grown (a) and irradiated (b) specimens.
In the irradiated specimen it clearly appears the reduction of length of the arcs visible in correspondence of the polycristalline rings which is evidence of an increase of order.
This narrowing of the arcs was observed in all cases after the laser irradiation, so that the increase of order can be correlated with the optical losses reduction.

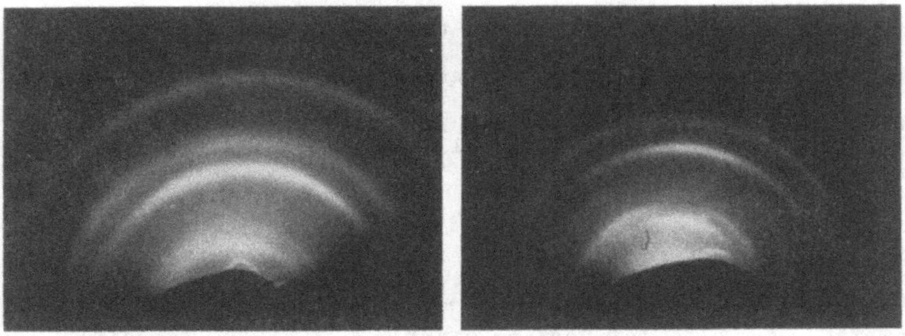

(a) (b)

Fig.1 RHEED patterns for an as grown ZnO/Corning film (a),and the same specimen after the laser irradiation (b)

III Conclusions

The presented results show that the measured decrease of optical losses is accompanied by an increase of ordering upon laser annealing of the film structure. This reordering depends on the laser irradiation conditions and has been found to optimize at some special value of the scanning velocity.
Comparison with previous results obtained with "worse" samples seem to indicate that laser annealing can be effective to produce a larger degree of order in "bad" films which have initially poor orientation.
In "good" films that have already a strong preferred orientation no improvement or a modest one should be expected by the laser treatment.

Acknowledgement

This work has been partially supported by the National Research Council of Italy in the project MADESS.
The authors wish to thank prof.L.Vasanelli for helping providing the ZnO samples.

Bibliography

1 S.E.Blum, K.H.Brown, R.Srisivesan, Appl.Phys.Lett. 43, 1026 (1983)
2 D.Schafer, B Brauns, P.Hofner, R.Wolf, B.Steiger, J.Barman, EPM 1989
3 S.Dutta, N.E.Jackson, I.T.Boyd, F.S.Nickernell, R.L.Davis, Appl.Phys. Lett. 39,206, (1981)
4 M.Bertolotti, A.Ferrari, A.Jaskow, A.Palma, E.Verona,J. Appl.Phys. 56,2943 (1984)
5 S.Dutta,IEEE J Quantum Electr. QE/18 1982 800

RESIDUAL GAS ADSORPTION AT THE FILM-SUBSTRATE INTERFACE DURING ION-BEAM ASSISTED DEPOSITION

J.Budinavičius,L.Pranevičius,S.Tamulevičius I)

Introduction

Ion irradiation during thin film growth takes place in many technological processes which are connected with plasma-assisted deposition. Energy transfer to the substrate and condensed atoms essentiably changes all stages of film formation and allows us to form thin solid films with unique properties (I,2,3).The elementary processes (such as adsorption of film and residual gas atoms) influence on the nucleation processes and consequently determines the size and orientation of the crystallites.The present work concerns the adsorption of the oxygen atoms from the residual gas during the deposition of metallic film on semiconductor substrate under high energy ion beam irradiation.The using of high energy ions allows us to pick out the influence of ion energy losses on the processes mentioned above.

Experimental

The silver films were deposited onto the crystal silicon substrate by condensation of silver atoms ,obtained by ion sputtering of a pure silver target.The silver atom flux density was equal to $4xIO^{I4}$ atoms $cm^{-2}xs^{-I}$.The growing films were irradiated by 50-I75keV He^+,Ne^+,Ar^+ ions.Ion flux density was equal to $5xIO^{II}$-$5xIO^{I2}$ ion $cm^{-2}xs^{-I}$.The temperature of the substrate was varied from 300K to 800K.The experiment was carried out in an ion implanter chamber with a residual gas pressure of IO^{-4}Pa, which corresponds to oxygen atoms flux density to the substrate of $IxIO^{I3}$ atoms $cm^{-2}xs^{-I}$.The formed structures were studied by Auger electron spectroscopy sputter profiling.

Results,discussion,conclusions

We observed that in all cases the adsorbed oxygen layer at the film-

I) Kaunas Polytechnic Institute ,V.Juro-50,Kaunas,Lithuania,(USSR) substrate interface exists. Some of the adsorbed oxygen atoms form a chemical bond with the substrate atoms , forming non-stoichiometric silicon oxide (4). From the experimental results (Fig. I, 2) we can see that the amount of adsorbed oxygen atoms depends on the ion and atom flux density, ion mass and substrate temperature . The existence of maximum in Fig. I shows that two mechanisms determine the amount of adsorbed oxygen: oxygen atoms come to rest on the surface as a result of adsorption and leave the surface owing to the ion beam activated removal. In the low flux region (Fig. I) increasing oxygen content shows that oxygen adsorption is an ion-activated process. The comparison of results presented in the Fig. I and Fig. 2 allows us to conclude that electron energy losses as well as nuclear energy losses are of great importance in the oxygen adsorption process. Electron losses are more pronounced in the low ion flux density region.

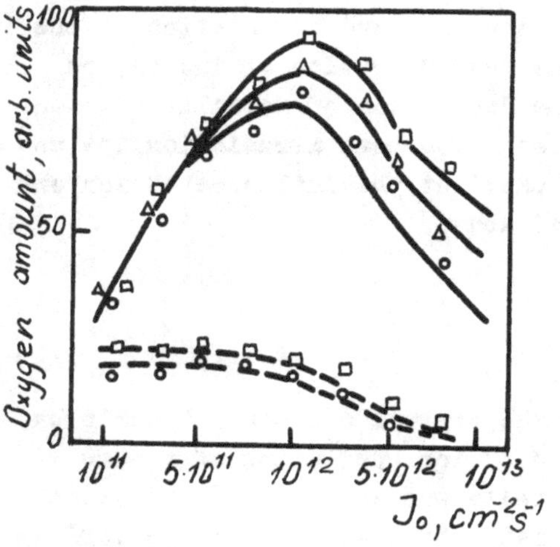

Fig. I. Amount of adsorbed oxygen atoms versus the Ar^+ ion flux density Io. Ion energy: □ -50 keV, ○ -100keV, △ -175keV. — total amount of oxygen, - - - amount of oxygen in chemical bond.

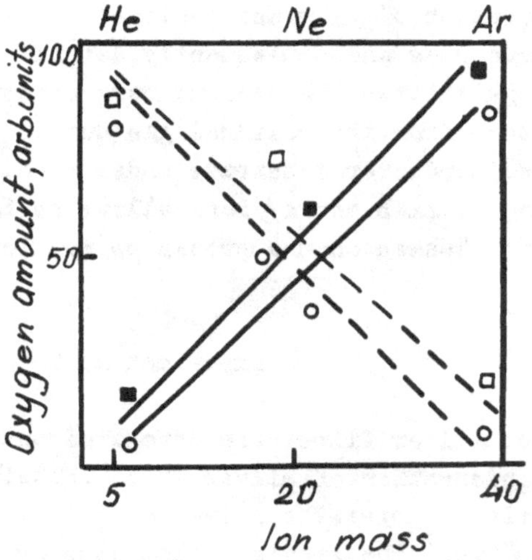

Fig. 2. Amount of adsorbed oxygen atoms versus the ion mass. $I_0=10^{12}$ $cm^{-2}s^{-1}$ Ion energy: □,■ -50keV, ○,● - 175keV. - - - chemically adsorbed oxygen, — physically adsorbed oxygen.

We know that in ion energy interval of 50-175 keV the electron stopping power increases (from 0,2 to 0,31 arb.units) and nuclear stopping power decreases (from 0,4 to 0,29 in arb.units) with increasing Ar^+ ion energy in Ag and Si substrates. A monotonic reduction of adsorbed oxygen atoms amount with increasing ion energy enables us to assume that radiation defects act as additional adsorption centres. The surface defects concentration decreases with increasing ion energy. The agreement between the curves with decreasing ion flux density and increasing amount of chemically bonded oxygen allow us to conclude that the ion-activated atoms are either atoms which get part of ion energy in ellastic collision events and atoms with excited extra-nuclear electrons.This statement is evidenced by results presented in Fig.2.The greater amount of oxygen atoms which form the chemical bond was noticed in the case when the irradiation with light ions take place.

As we said ,the existence of maximum in Fig.I evidences that two processes determine the amount of adsorbed oxygen .One can see that the position of curve maximum drifts to the lower ion flux density region with the increasing ion energy.It means that the removal of oxygen atoms from the surface begins in lower ion flux density for greater ion energy.This fact enables us to assume that ion activated desorbtion is the main mechanism of removal of oxygen atoms from the surface.

The facts given above evidence that the adsorption of atoms from the residual gases is determined either by electron and nuclear stopping power.The surface defects as well as surface atoms with excited electron shells determ the amount of adsorbed oxygen.The desorbtion is the main mechanism of oxygen removal from the surface during ion irradiation.

References

(I)Budinavičius ,J.,L.Pranevičius and S.Tamulevičius:Mater.Sci.Eng.B2 (I989) 203-206

(2)Harper,J.M.E.,J.J.Cuomo and R.J.Gambino:Nucl.Instrum. and Meth.Phys. Res. B7/8 (I985) 886-892

(3)Takagi,T.:J.Vac.Sci.Technol. A2 (I984) 382-388

(4)Weissmantel,C.:J.Vac.Sci.Technol.I8 (I98I) I79-I85

ANNEALING OF SURFACE AND INTERFACE LAYERS CAUSED BY THE PLASMA/SUBSTRATE INTERACTION DURING THE DEPOSITION OF PECVD a-Si:H FILMS

G. Suchaneck, R. Deltschew, K. Schade, M. Albert [1)]

1. Introduction

Radio frequency (r.f.) glow discharges are used to deposit thin films of insulators, semiconductors and conductors in the fabrication of microelectronic integrated circuits, semiconductor-electronic, optoelectronic and optical devices by Plasma Enhanced Chemical Vapor Deposition (PECVD). High energy electrons generated in the discharge dissociate and ionize gaseous molecules, thereby forming chemically active radicals and ions. Because of the difference in mobility between electrons and positive ions, all surfaces in contact with the discharge assume a negative potential with respect to the plasma. The potential drop from the plasma potential to the electrode or wall potential in the sheath between the glow and the surfaces (typical time-averaged values 10 -500V for grounded surfaces) accelerates the positive ions that diffuse to the sheath edge to all surfaces in contact with the plasma. Secondary electrons generated by the incident ions on the surfaces are accelerated into the plasma forming an electron beam like component generating ions near the sheath edge. Plasma electrons whose energies are larger than the potential drop in the sheath reach the surfaces with reduced energy. Since the momentum associated with these electrons and also with the discharge-generated photons is much less than that of positive ions, their influence on surface phenomena is negligible. Therefore the PECVD process can be used to study the influence of low energy ion beams on material properties if the substrate bombardment by other excited heavy particles is small. This is the case in the used low pressure argon-silane glow discharge.
Because of the small ion-induced damage and minimal morphological changes under typical PECVD conditions other film properties must be taken into consideration. We take advantage of the very sensitive to ion bombardment photoconductivity of amorphous hydrogenated silicon films /1/.
Due to the plasma/substrate interaction, amorphous hydrogenated silicon (a-Si:H) films deposited by r.f. glow discharge are anisotropic in the growth direction. This anisotropy may be described in a first approximation by a three-layer model: an interface layer (IL), the bulk of the a-Si:H film and a surface layer (SL) /4/. Different properties of the IL and SL cause different values of the photoconductivity measured by illuminating different sides of the film /3/. Investigating the spectral dependence of the photoconductivity, information on IL and SL defect densities and thicknesses may be determined /7/. This paper deals with the annealing of ion-induced defects in the IL and SL during the deposition in dependence on the substrate temperature.

---

1) Technische Universität Dresden, Sektion Informationstechnik
Mommsenstraße 13, Dresden 8027, GDR

2. Experimental details

Photoconductivity measurements on r.f. (f = 1.6 MHz) glow discharge a-Si:H films deposited using a gas mixture of 11.4% silane in argon in a parallel plate radial flow reactor on glass and fused silica substrates were performed on coplanar aluminium structures with a length of 4 - 10 mm, a width of 1 mm and an electrode gap of 0.4 - 1.25 mm. The applied voltage was 200 V. Film deposition and deposition equipment were described in detail elsewhere /5/. The illumination wavelength was varied by means of interference filters between a high pressure Hg lamp with an IR cut and the sample. Measurements were carried out on samples, which were stored under atmospheric and daylight conditions for few weeks after the deposition. Previous to the measurements the dependence of photoconductivity on the light intensity, the time of illumination and the applied voltage was determined /2/. Measurement conditions were selected in a manner that metastable effects were negligible. The mean light penetration depth for a given wavelength was determined as the reciprocal of the optical absorption coefficient α. Details of optical measurements were described in /6/.

3. Results and discussion

The dependence of the photoconductivity ratio σ_i/σ_s, where σ_i and σ_s are the photoconductivities measured by illuminating from the substrate side and the surface side, respectively, on the substrate temperature was investigated as a measure of the IL and SL defect density ratio.

At a low light penetration depth $\alpha^{-1} < 50$ nm, σ_i/σ_s is constant at low substrate temperatures ($t_s \leq 240°$ C), increases to 270° C and decreases at higher temperatures (fig. 1). At a large light penetration depth $\alpha^{-1} \geqslant 200$ nm ($\lambda \geqslant 577$ nm), σ_i/σ_s is independent of the substrate temperature. For a given substrate temperature σ_i/σ_s increases at lower wavelengths. This is consistent with a higher defect level in the SL in comparison with the IL. The shown dependence is caused by a steeper rise of σ_i than that of σ_s in the temperature range of 250 - 270° C (fig. 2).

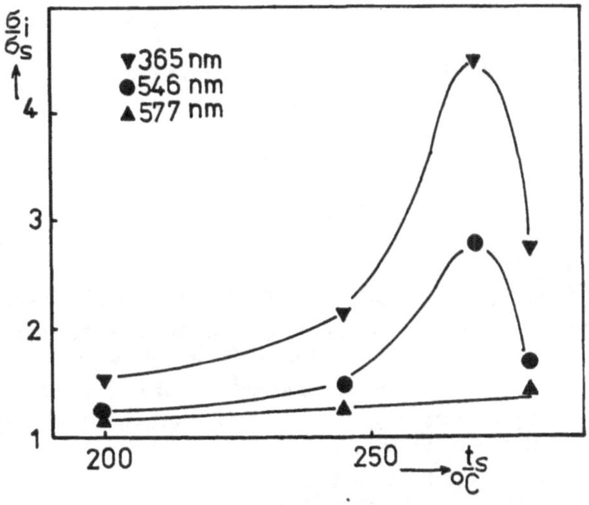

Fig.1: Substrate temperature dependence of the photoconductivity ratio for various wavelengths of the incident light

Because of the isochronical deposition processes the obtained results may be interpreted in the following manner: Substrate ion bombardment during the deposition causes a more defective SL. By further deposition the SL is transferred to the bulk and its properties may be changed. At low temperatures the defects generated in the IL during the initial growth stage are stable and a constant photoconductivity ratio is obtained. With rising substrate temperature these defects are annealed out and the photoconductivity ratio is increased. In the high temperature range the defects in the SL are annealed out, too. This causes a decrease of the photoconductivity ratio probably until a constant value σ_i/σ_s is reached.

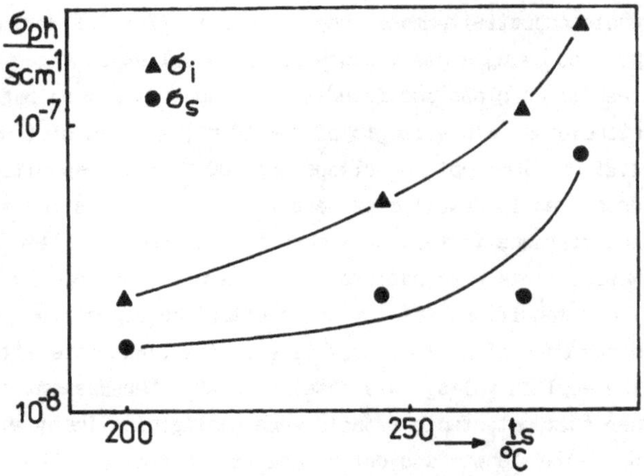

Fig.2 : Substrate temperature dependence of the photoconductivities at a wavelength of 365 nm

/1/ Ando, K., M. Aozasa and R. S. Pyon : Appl. Phys. Lett. 44(1984) 413-415

/2/ Deltschew, R., G. Suchaneck and M. Albert: Schichtsysteme für zukünftige Bauelemente der Mikro-, Opto-, Bioelektronik und Optik, TU Karl-Marx-Stadt 1989

/3/ Hasegawa, S., S. Shimizu and Y. Kurata : Phil. Mag. B 49(1984) 511

/4/ Mailhiot, C., J. F. Currie, S. Sapiena, M. R. Wertheimer and A. Yelon : J. Noncryst. Solids 35/36(1980) 207-212

/5/ Paul, W., M. Albert, J. Kuske, G. Suchaneck and K. Schade : Wiss. Z. TU Dresden 38(1989) 192-197

/6/ Suchaneck, G., K. Schade. H. Stötzel, M. Albert and J. Kuske : Wiss. Z. TU Dresden 38(1989) 198-199

/7/ Xu, X., A. Morimoto, M. Kumeda and T. Shimizu : Jpn. J. Appl. Phys. 26(1987) L 1818

SCANNING CO_2 LASER INDUCED OXIDATION REACTION IN Ti THIN FILM

J.Dudonis and D.Furmonavičius

INTRODUCTION

Laser microtechnology is widely used for localized modification of solids, for induction of chemical reactions in surfaces of solids (ref 1, 2). Though perspectives of microelectronic applications of lasers are very wide, chemical reactions induced by lasers are not well enough investigated. In this work the process of oxidation of Ti thin film by scanning lighting of CW CO_2 laser is studied.

EXPERIMENTAL

Ti thin films (0,2 ... 0,5 μm thick) were deposited by planar magnetron on highly doped (ρ =0,3 Ω·cm) n-type (~300 μm thick) Si wafers. The sputtering was carried out under the conditions listed here: pressure of Ar $p_{Ar} = 3 \cdot 10^{-2}$ Pa, current I = 2 A, voltage U = 350 V. Before the deposition substrates were cleared by ion radiation, bias voltage was -100 V. Chemical reactions in Ti thin films were induced by focused beam of CW CO_2 laser (power up to 40 W). System for investigation consists of a microspectrophotometer (MPV - 2 type, Leitz) with scanning table and a CO_2 laser as shown in Fig. 1. The system allows to measure the local change of reflection from a surface, the smalest diameter of which is 10 μm. The diameter of laser beam was 500 μm and the velocity of scanning table was 2,5 mm/s. The power of a laser beam was being changed during the experiment. During the lighting of samples by laser, localized race of a temperature was measured by optical spectrum of thermal radiation. The thickness and the composition, the change of optical properties of the formed thin films were investigated after the laser lighting.

RESULTS AND DISCUSSION

In the lighting of samples by laser, localized short heats of Ti thin film and Si substrate are seen because of imbibement of light by Si free electrons. The temperature of heats, evaluated by spectrum of thermal radiation is 700 ... 1000 °C and depends on a power of CO_2 laser beam and a velocity of the scanning table. Optical properties of the films where laser beam have fallen were strongly changed in comparison with the

Physics Department of Kaunas Polytechnical Institute, Lithuania
Donelaičio 73, Kaunas 233006

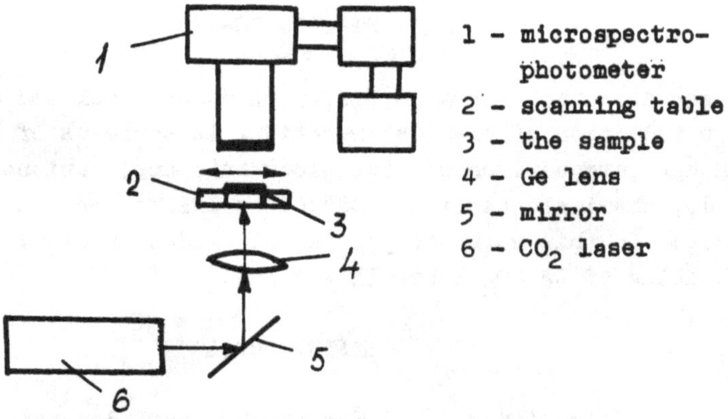

Fig. 1 System for investigation of the process of Ti thin film oxidation by scanning lighting of CO_2 laser.

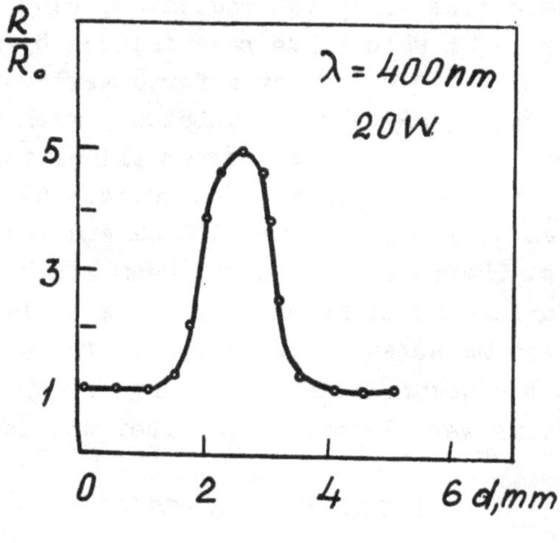

Fig. 2 The change of reflection factor of thin film across the surface of sample.

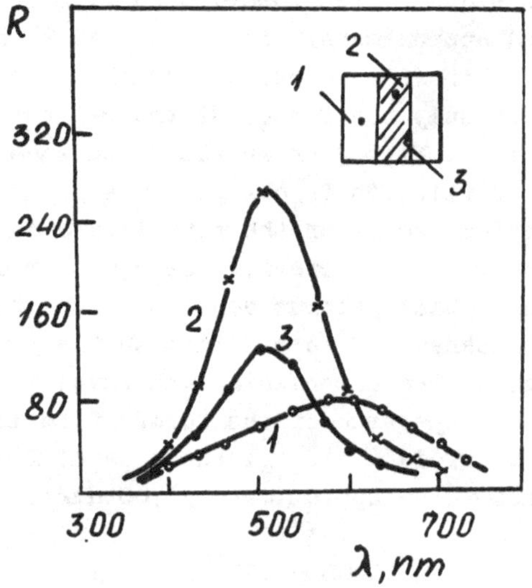

Fig. 3 The dependence of reflection in the different spots of film (1 - Ti film, 2,3 - TiO_x film)

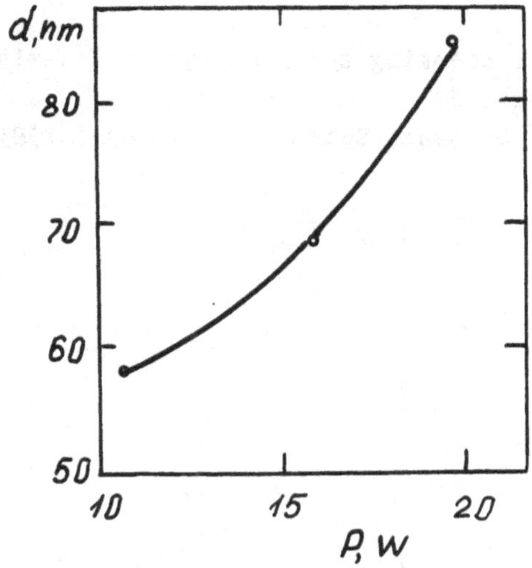

Fig. 4 Dependence of thickness of formed film on the power of scanning beam.

not lighted zones. Due to the large difference of reflection factor of the formed film and Ti thin film, formed film is clearly seen in the fone of Ti film. The width of the formed thin film depends on the power of the laser beam and approximately is 1 mm. Fig. 2 shows the change of reflection factor of thin film across the sample. Due to the power of the laser beam changed not only the factor of the reflection but also the color of the formed film. Fig. 3 shows the dependence of reflection in the different spots of film. While changing the power of the beam, the dependence of refraction factor on the wave length changes. This shows that a clear film forms and an interference spectrum of reflected and falling beams is seen. This spectrum depends on the thickness of thin film.Dependence of thickness of formed film on the power of scanning beam is shown in Fig. 4. The composition was investigated by Auger electron spectroscopy (spectrum of the formed film is shown in Fig. 5) and it was found to be oxide - $TiO_{1.9}$. The rate of TiO_x formation in Ti matrix is estimated to be approximately 300 nm/s.

CONCLUSIONS

Localiced oxidation of Ti film on Si substrate is induced by scanning CW CO_2 laser beam in air. This allows to form TiO_x islands of disirable configurations in Ti films matrix.

LITERATURE

1. W W Duley : Laser Processing and Analysis of Materials. Plenum Press. New York and London 1983
2. R M Osgood and T F Deutsch, Science, 227, 709 (1985).

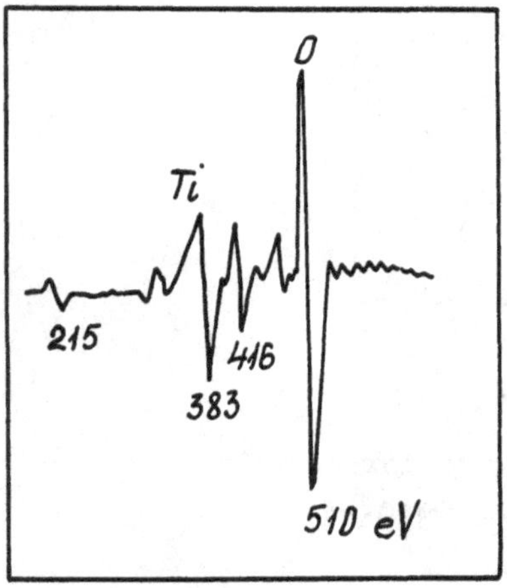

Fig. 5 AES spectrum of formed TiO_x film.

ION BEAM ASSISTED DEPOSITION OF Al ON Fe.

D. C. Kothari*, L. Guzman*, A. Tomasi*, A. Miotello**, A. Gupta° and G. Principi°.

It has been reported that the amorphous-AlFe (a-AlFe) alloy can be formed by Ion Beam Mixing (IBM) and transformed into the icosahedral-AlFe (i-AlFe) phase by post annealing[1]. The transformation from the a-AlFe to i-AlFe is the subject of the present work. Instead of IBM, in the present work a more versatile ion beam technique called, the Ion Beam Assisted Deposition (IBAD) [2] is used, in which the deposition and implantation are carried out "in situ" to obtain thicker ion beam mixed layers without external contamination. The Differential Scanning Calorimetry (DSC) technique is used to thermally characterize the AlFe alloy prepared by IBAD. The Conversion Electron Mössbauer Spectroscopy (CEMS) and X-ray Diffraction (XRD) techniques are used to identify the phase transformations seen on the DSC curve.

The specimens were prepared using a six-step sequential process of deposition of Al and bombardment of Ar^+ ions, on the Fe substrate. In each step, the thickness of deposited Al was 100 A and the fluence of 30 keV Ar^+ ions was $7.2x10^{16}$ ions/cm^2. Because of sputtering the thickness of final mixed layer is estimated to be less than 500 Å. Since the ion fluence was sufficiently high and the projected range and straggling for 30 keV Ar^+ ions are 314 and 119 Å respectively (greater than the thickness of Al layer in each sequential step), the mixed layer is

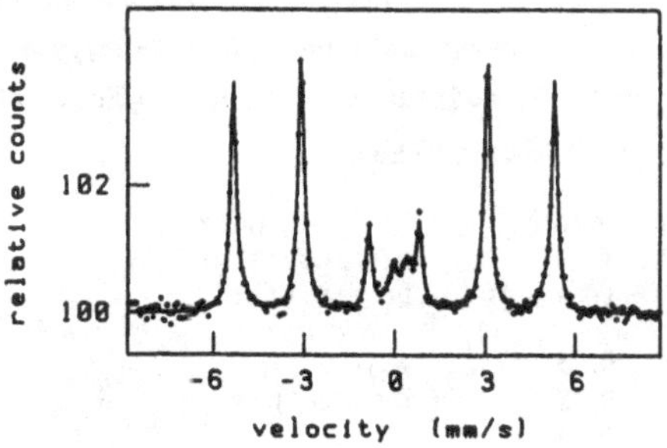

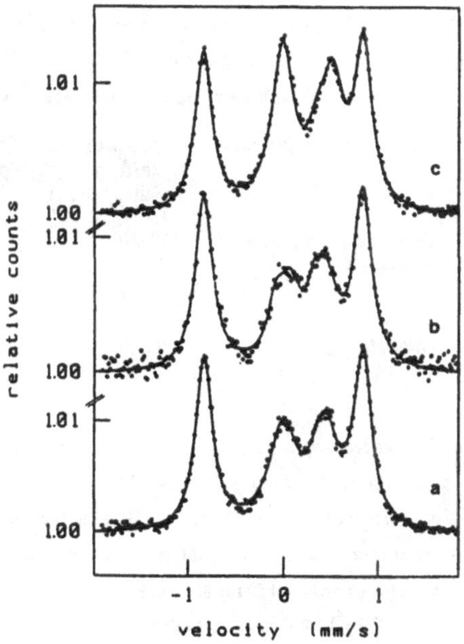

Fig. 1 : CEM spectrum before annealing.

Fig. 2 : CEM spectra : a) unannealed; b) annealed at 300 °C; c) annealed at 480 °C. The solid lines represent the least squares fits.

* : I. R. S. T., I-38050, Povo, Trento, (Italy). ** : Dept. of Physics, Univ. of Trento, Povo, Trento, (Italy). ° : Dept of Mech. Engg., Univ. of Padova, Via Marzolo 9, I-35131, Padova, (Italy).

expected to be uniform. The Al/Fe ratio is estimated to be between 5 to 7 in the mixed layer, except near the interface where Fe rich AlFe alloy is expected. The DSC measurements were carried out using the Perkin Elmer DSC-VII calorimeter. The heating in the calorimeter was carried out at a rate of 40° C/min., in Ar gas atmosphere at 760 torr. Five specimens were used at a time for one calorimetric measurement. The XRD spectra were recorded in the Seeman-Bohlin geometry suitable for thin films using Cu K_{α} radiation. The room temperature CEM spectra were measured using a conventional spectrometer with a flowing gas (95% helium, 5% CH_4) proportional counter and a source of ^{57}Co in a rhodium matrix. A standard least squares minimization routine was used to fit the spectral profiles as a superposition of Lorentzian lines. The XRD and CEMS spectra were also recorded after annealing the specimens for half an hour in vacuum (10^{-6} torr.) at 300 °C and 480 °C.

Fig.1 shows the CEM spectrum of the unannealed specimen, which displays a doublet in the central region, in addition to the six line pattern due to Fe substrate and to a very weak magnetic component due to the expected Fe rich AlFe alloy. The parameters of the doublet (D_u) are similar to those due to the a-AlFe phase reported by Schurer et al [3]. Fig 2 shows the CEM spectra before and after annealing recorded with an expanded velocity scale. The annealing of the specimen at 300 °C made the doublet (average parameters, D_{av}) slightly assymetric with the isomer shift δ = 0.20 mm/sec. and the quadrupole splitting Δ=0.40 mm/sec. These parameters are close to those reported by Dunlap et al [4] for the i-AlFe phase. The slight assymetry could be possibly due to two doublets (D_1 and D_2). The fitting with two doublets improved the quality of the fit. This implies that there are two sites for Fe in the i-AlFe phase. This result agrees well with the more recently reported parametrs by Dunlap et al [5]. The CEMS spectrum of the specimen annealed at 480° C, shows an increased assymetry and could be fitted with the parameters (D_3 and D_4) of Al_5Fe_2 [6] and the crystalline-$Al_{86}Fe_{14}$[4] compounds. The Mössbauer parameters are listed in the following Table :

No.	Specimen	Doublet	Isomer shift δ (mm/sec.)	Quadrupole splitting Δ (mm/sec.)	Width Γ (mm/sec.)	Possible interpretation [references]
1	Unannealed	D_u	0.20	0.44	0.35	a-AlFe [3]
2	Annealed 300 ° C	D_{av}	0.20	0.40	0.38	i-AlFe[4]
		D_1	0.12	0.42	0.19	,,[5]
		D_2	0.26	0.39	0.30	,,
3	Annealed 480 ° C	D_3	0.26	0.54	0.29	Al_5Fe_2[6]
		D_4	0.18	0.36	0.26	crystalline-$Al_{86}Fe_{14}$[4]

The XRD spectrum (Fig. 3) of the unannealed specimen showed the peaks due to α-Fe related to the substrate, no peak due to Al and unusually high background counts (higher than normally observed in the Seeman-Bohlin geometry) specially at low 2θ values, possibly related to the presence of an amorphous structure on the surface. After annealing at 300 °C, the XRD spectrum shows an additional very broad peak at 2θ=37.7° corresponding to d=2.38 Å and a shoulder at 2θ=43.8° corresponding to d=2.11 Å. The first peak could be attributed to the (111) plane of strained Al phase having very small crystallite size. The shoulder could be due to the diffraction peak from a reciprocal lattice vector (100 000) of the i-AlFe phase. These results are in agreement with the earlier results [1,7] and the XRD results of Dunlap et al [7]. Thus by

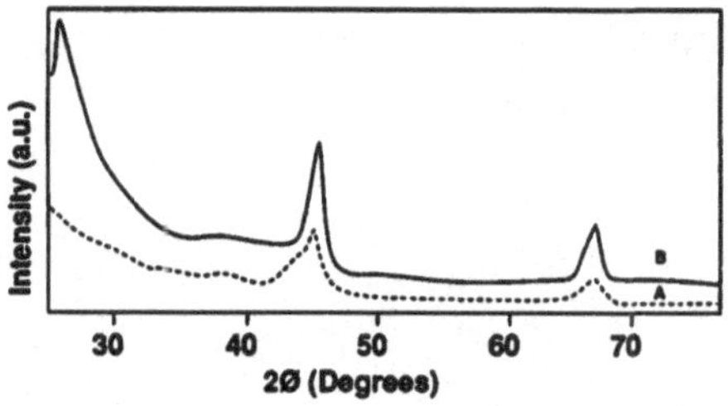

Fig. 3 : XRD spectrum : A) annealed at 300 °C; B) unannealed.

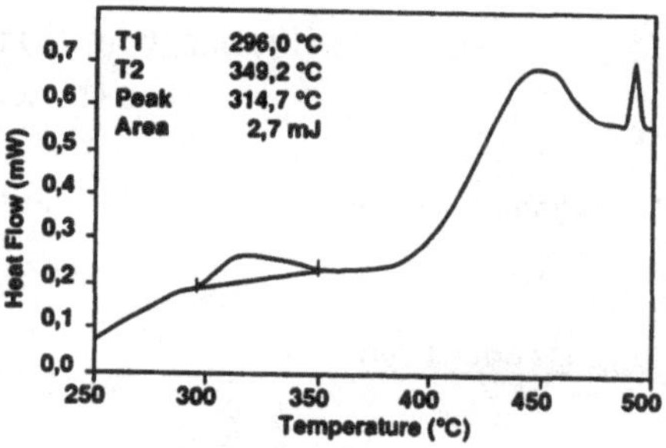

Fig. 4 : DSC scan.

annealing possibly the a-AlFe has transformed into the phase mixture of small grains of Al and i-AlFe. This is in agreement with literature reports that the fcc Al is usually present along with the i-phase in rapidly cooled AlFe alloy.

In the DSC scan (Fig. 4) one can see three exothermic peaks. The first two peaks are broad and the height of the first peak is low. From the above XRD and CEMS results it is safe to conclude that the first peak observed in the DSC scan is due to the transformation from a-AlFe to i-AlFe and the second broad peak can be attributed to the icosahedral to crystalline transformation, which is in agreement with the results of Chen et al [8] who observed similar transformation in the same temperature ranges for AlMn and AlFe alloys. The third sharp peak in the DSC scan remains unidentified at present, but a similar peak observed in the temperature range of 500 to 550 ° C by Dunlap et al [8] was attributed to the transformation $Al_6Fe \rightarrow Al_3Fe+Al$. Considering the area of the first peak and by estimating the weight of a-AlFe phase on the substrate, the total enthalpy of transformation was evaluated to be not more than 2000 J/mole which is in agreement with the results of Chen et al.[8] who reported the total enthalpy of transformation for AlMn alloy to be equal to 1050 J/mole. Theoretical analysis of the grain growth process yielded the activation energy to be equal to about 2 eV/at., similar to the value 2.5 eV/at. obtained by Chen et al [8]. The differences between our results and those of Chen et al [7] regarding the total enthalpy of formation, could be due to the lower initial radii of micro- quasicrystallites or different geometrical configuration in our case since the alloy and the technique of preparation are different. Further work is in progress in this direction.

REFERENCES

[1] K. Hohmuth, V. Heera and B. Rauschenbach, Nucl. Instr. and Meth., **B39** (1989) 136.
[2] D. C. Kothari, P. Scardi, S. Gialanella and L. Guzman, Phil. Mag., To be published.
[3] F. J. Schurer, B. koopman, F. van der Woude and P. Bronsveld, Solid state Comm, **59** (1986) 619.
[4] R. A. Dunlap, K. Dini, G. Strink, G. S. Collins and S. Jha, Hyp. Int., **28** (1986) 963.
[5] R. A. Dunlap, D. W. Lawther and D. J. Lloyd, Phys. Rev, **B38** (1988) 3649.
[6] M. A. Z. Vasconcellos, S. R. Teixeira, P. H. Dionisio, W. H. Schreiner and I. J. R. Baumvol, Presented at the International Conference on the Industrial Applications of Mossbauer effect, Parma (Italy), september, 1988.
[7] R. A. Dunlap and K. Dini, Can. J. Phy., **63** (1985) 1267.
[8] L. C. Chen and F. Spaepen, Nature, **336** (1986) 366.

LASER LIGHT INDUCED OXIDATION OF METALS AND SEMICONDUCTORS IN EXTERNAL ELECTRIC FIELD

R. Vajtai[1], L. Nánai[1] and I. Hevesi[1]

I.Introduction

The investigation of oxidation of metals and semiconductors is a well-known area of laser-material interaction. There are a lot of publication in topic of experimental laser induced oxidation. The extra effects of laser beam has been shown; the oxidation rate is considerable larger, the oxidation process and the heating of the sample may be localized, and new morphology of oxide can be formed.

The heterogenous reactions are passed off in several steps (adsorption of oxigen, transport of reagents, and the chemical reaction itself). The reaction rate is given by the rate of slowest step of reaction. In case of thick oxide layer the slowest process is the transport of reagents through the oxide. There are a few theoretical models of reaction kynetics. (Cabrera-Mott, Wagner theoty, etc.).

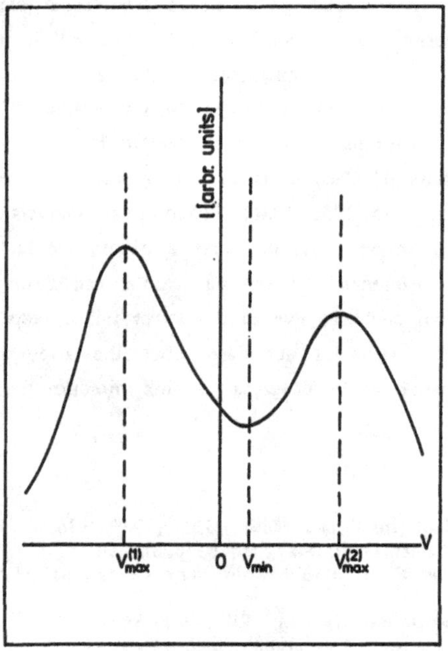

Fig.1. The oxide thickness vs. strenght of external electric field (t=3min.)

The external electric field can change the reaction rate if the patricipiants in the transport are charged. In this case the formed oxide thickness can be calculated as the function of time if the concentration of cemisorbed oxygen at the oxide-gas interface doesnot depend on strenght of electric field. In many other cases the dependence of oxidation rate on electric field possible to represent in another way: we will study the oxide thickness as funtion of electric field at constant illumination time. This time in our cases was much more longer than the activation time of reaction. This result has been shown in Fig.1.

[1] Dept. Exp. Phys. of Jate University, H 6720 Szeged, Domt. 9, Hungary

II.Experimental

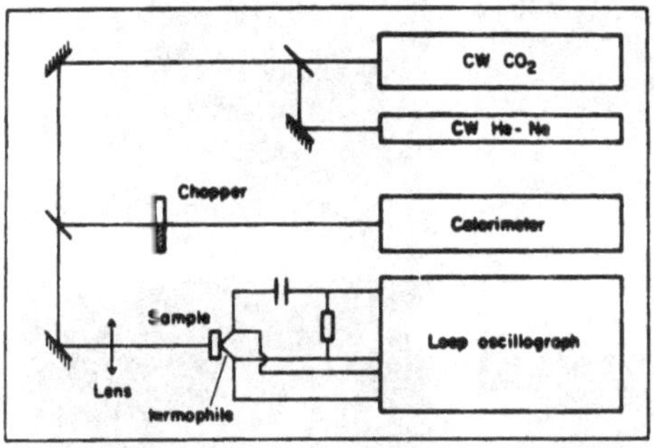

Fig.2. The experimental set-up

The scheme of the experimental set-up is presented in Fig. 2. The light of a CO_2 or a Nd:YAG laser was directed through a NaCl or a glass prism, respectively, onto the surface of the sample plate. The vanadium plates were spherical with size of 0.7 cm^2 surface area and 0.3 mm thickness. The silicon plates were square with size of 7x7 mm^2 surface area and 0.5 mm thickness. Lens was not applied in this way the sample was practically covered by the spot of laser beam and the spot size was much more larger than the thickness of the sample. The sample was put onto the ceramic holder and with it into a cylindrical capatitor. The power of laser beam and value of high voltage were registered.
A Chromel-Alumel thermocouple was embedded into the back side of the vanadium samples for recording the temperature change during illumination.

III.Results

The oxidation rate of vanadium was measured by weighing of the sample before and after irradiation. The theoretical dependence of mass growth (Fig.1) is in good agreement with the experimental results. A strong influence of the field on the temperature was observed, namely the stationary temperature of the sample suddenly decreased when the electric field was switched on. The reasons for this variation could be connected with the convection flow in the liquid oxide and the change of oxidation heat.

The optical microscopical photographes of the preilluminated silicon samples are shown in Fig. 3. Three min. illumination time and laser power of 50 W was applied. The silicon melted on the central part of the sample and hillocks grew on the both side of it. Developing of these hillocks could be bound up with oxidation in two way: the heat of oxidation can change the distribution of temperature; the SiO_2 can dissolve in melted silicon and can change its surface tension. The Fig.3.f. shows crater-like structure of the hillock.

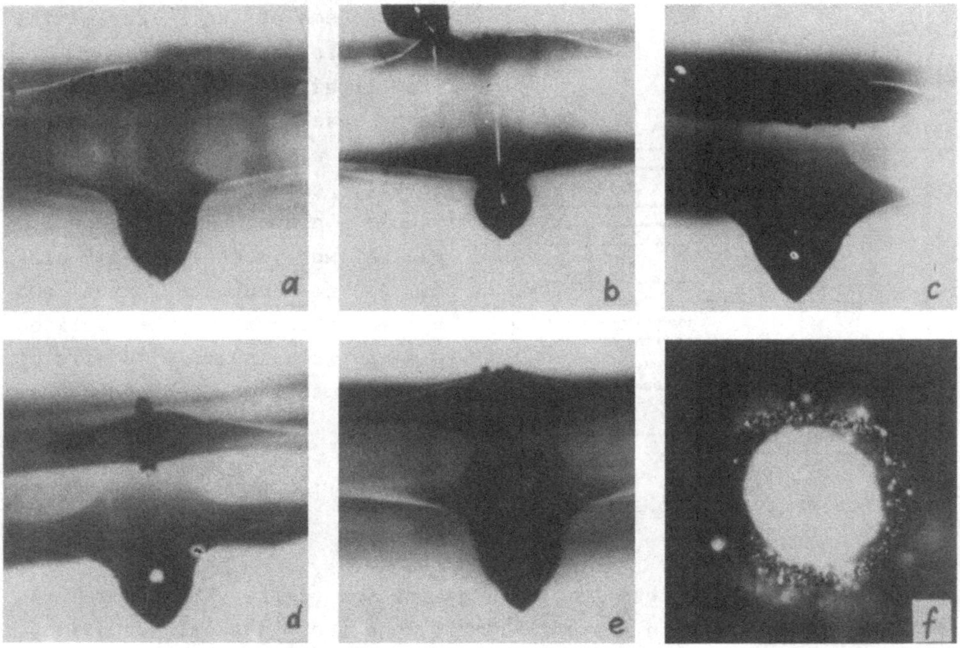

Fig. 3. Microscopical pictures of silicon samples after the illumination
(λ=1.06 μm, P=50 W, U=0 h_1=30μm, h_2=420μm (a),
U=1.0 kV, h_1=360μm, h_2=400μm (b), U=2.0 kV, h_1=43μm, h_2=425μm (c),
U=3.0 kV, h_1=140μm, h_2=435μm (d)
U=5.0 kV, h_1=82μm, h_2=460μm (e and f))

Acknowledgement - we would like to acknowledge the OMFB for financial support to the experiments of silicon oxidation.

References

(1) Bauerle, D.: Chemical Processing with Lasers, Springer-Verlag (1986)
(2) Boyd, I.W.: Laser Processing of Thin Films and Microstructures Springer-Verlag (1988)
(3) Bunkin, N.F.: Thesys
(4) Jorgensen, P.J.: Jour. Chem. Phys.37(4), (1962) 874-877
(5) Nánai,L. et al.: Appl. Phys.A ,(1989)
(6) Nánai,L. et al.: Appl. Phys.A ,(1989)
(7) Young, E.M.: Appl. Phys.A47,(1988) 259-269

METHODS INVESTIGATION OF METAL-POLYMER COMPOSITION TEMPERATURE STABILITY INCREASING

V.A.Petrov,[1] G.H.Koishibaeva,[1] A.O.Savchenko,[1] V.I.Kopchenov[1]

Within the problem of metal protection a great attention is paid to polymer coatings. One of the requirements to such metal-polymer compositions is their thermal stability together with high cohesion strength of the composition components.

The general drawback of the existing polymer coatings on metals is low temperature stability caused by low temperatures of depolymerization and low kinetic properties of polymer material (2) which restrict the sphere of their usage.

The paper presents a study of ion implantation influence on properties of a metal-polymer composition in the condition of heating under vacuum and in the air. Polyethylenglycol, polypropylene, polycaproamide, fluoroplastic and other polymer materials were taken out. The polymers were dissolved in some corresponding dissolvents and applied on a titanium foils. Polyethylenglycol is well-dissoluable in the water and is quite appropriate not only for the investigations but also for the production of thin films for a survey by the electron microscope EMV-100B (USSR). To check the presence of the amorphous state in the polymer, the picture of optical diffraction on helium-neon laser were observed (λ = 632,8 nm).

The compositions obtained like that were implanted by Cu, Al, Ti, La, Ni and Fe ions having the doses from 10^{15} to 10^{17} sm^{-2}. To study the thermal-composition stability, they were annealed in the air and under the vacuum within the temperature limit to 700°C. Protection coating stability was tested during the metal-polymer composition creeping probation within the temperature limit from 350 to 600°C.

Fig.1 presents a microdiffraction of the initial polyethylenglycol which evidences in favour of the amorphous structure of the polymer. According to the picture of optical diffraction, the spherolites dimensions were found using the formula:

$$R = \frac{\lambda}{\mathrm{Sin}(\Theta\ \mathrm{max}/2)} ,$$

where R is the radius of the spherolite, Θ - the angle defining the maximum diffusion position. The spherolite dimensions of the initial

1)The Ust-Kamenogorsk Teacher's Training Institute,492036,the USSR state polyethylenglycol were 20-50 mu m (fig.2).

Ion implantation of polyethylenglycol by Ti ions didn't cause any changes in the thin polymer structure, as for the diffraction picture (fig.3), it revealed considerable changes, testifying to the decrease of the spherolites dimensions. This phenomenon is connected with the polymer heating during ion implantation.

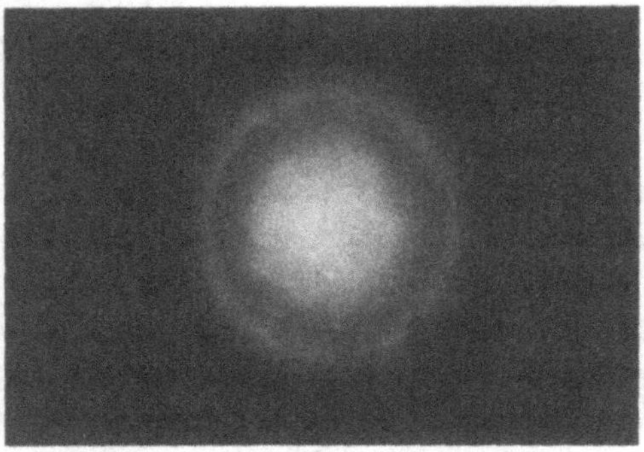

Fig.1. Microdiffraction pattern of the initial polyethylenglycol.

The study of the metal-polymer compositions thermal stability in the initial state revealed that in the first case with the temperature higer than 160°C the decomposition of the polymer occured with further evaporation from the surface, as for the samples implanted, they could stand the temperature to 650°C without any polymer-destruction effects.

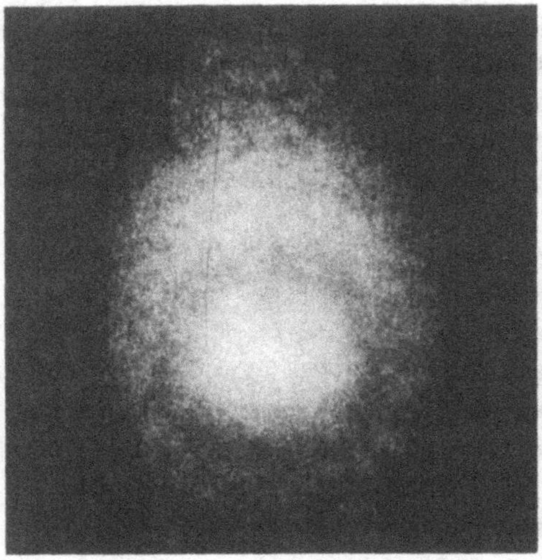

Fig.2. Optical diffraction picture of the initial polyethylenglycol.

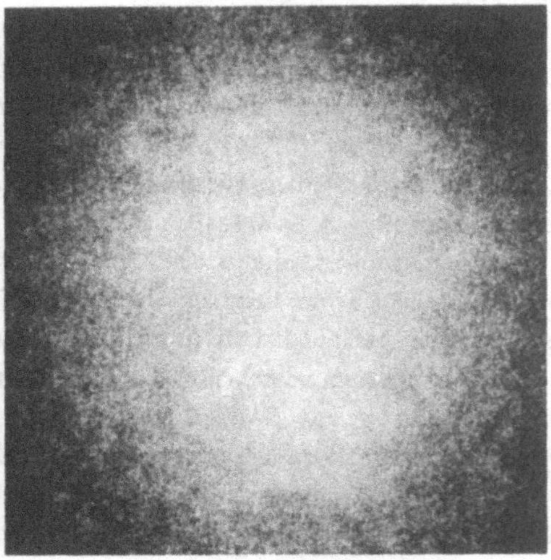

Fig.3. Optical diffraction picture of polyethylenglycol implanted by Ti^+.

Keeping in mind a strong hemosorbtion capability of titanium to hydrogen, one can suppose the following mechanism of transformations of polyethylenglycol implanted by titanium after its heating:

$$(-CH_2-CH_2-O-)_n \xrightarrow{t^\circ} (-CH_2{=}CH-O-)_n + 2H^{\bullet}\uparrow$$

$$2H^{\bullet} + Ti \text{ ---- } TiH_2 .$$

The high thermal stability of the polymer can be connected with the formation of new compositions, such as, in particular, the formation of the titanium hydride and the appearance of conjugated bonds, according to the chemical bond character the titanium hydride formed is similar to the metals and provides thermal stability of the composition. The revealed effect was observed for the implantation of other ions, having a high hemosorbtion capability to hydrogen.

As it has been above mentioned, the other indirect method for thermal stability analysis was a creeping test.

It was chosen due to the fact, that a thin titanium foil is used as a material for the discharge window of electron accelerators and is exposed to strong oxidation from the atmosphere in the result of heating by the electron beam. The oxidation processes cause considerable increase of creeping velocity and the window destruction. The pneumatic method described in ref.(1) was used as a creeping test.

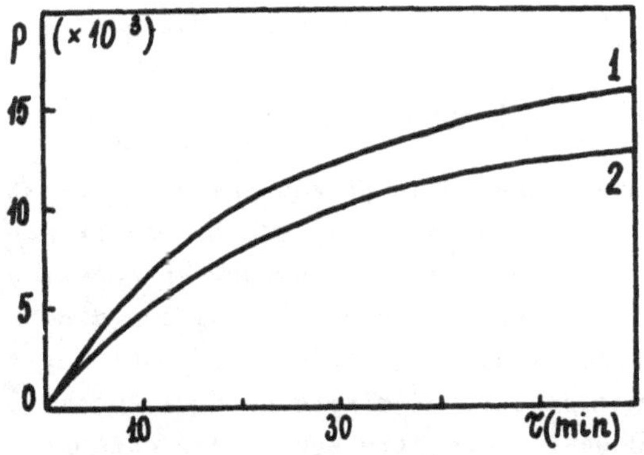

Fig.4. The creeping curves for Ti, covered with polyethylenglycol (curve 1) and after further implantation of Ti^+ (curve 2), the test temperature being 400°C.

The creeping curves (fig.4) for the titanium with polymer coating (curve 1) and after further implantation (curve 2) evidence in favour of considerable decrease of the velosity in the stage of fixed creeping, which allows to use protection polymer layers after implantation as an effective means for the protection of the discharging window of the accelerator.

REFERENCES

(1) Bugayev,S., Ju.Kolobov, A.Savchenko, V.Petrov, V.Chervinsky, S.Chuhlantsev: In the book Interaction of atom particles with the surface of solid substance, Tashkent (1989) 187
(2) Madorsky S.: Thermal dissociation of polymers, Mir, Moscow (1967)

CATHODIC ARC TECHNIQUE - PRESENT STATE AND DEVELOPMENTS

B. Rother, J. Vetter and J. Siegel [a)]

Introduction

The cathodic arc technique (CAT) is an effective fast rate evaporation and deposition technique the principle of which is also suitable to be applied in getter pumps and in ion sources. It serves a particular meaning as ion enhanced deposition technique of hard and protective coatings. Features of the technique are growth rates in the range of 1 to 20 nms^{-1}, film growth from highly ionized vapours and gases as well as low specific evaporation power together with an effective energy transfer from the electric power supply of evaporation to the generated free particles. In comparison with other ion supported deposition techniques these data provide considerable advantages of CAT /1/.

Characteristics and recent developments of the technique

Cathodic arc evaporation sources are operated at dc and pulsed modes at currents of about 100 A and 10 A and appropriate voltages of approximately 20 V and 300 V respectively. At both modes high vacuum or reaktive environment up to 1 Pa may be applied. The considered vacuum arc discharges burn within the highly ionized vapour of the cathode material which is generated by the high power density of minute cathode spots carrying the whole arc current and moving randomly across the cathode surface. Figure 1 is a schematic illustration of cathodic arc discharges with indicated characteristic parameters. The eroded cathode material is composed of ions, neutral vapour and macroscopic droplets. The ion component representing roughly 10 % of the arc current is characterized by considerable amounts of multiple charged ions and a velocity directed away from the cathode. The therefrom resulting suitability to ion enhanced deposition techniques is limited, however, by the occurance of macroscopic droplets of the cathode material being incorporated into growing films. Recent developments of CAT are therefore related to that problem.

a) Technische Universität Karl-Marx-Stadt, Sektion Physik/Elektronische Bauelemente, PF 964, Karl-Marx-Stadt, 9010, DDR

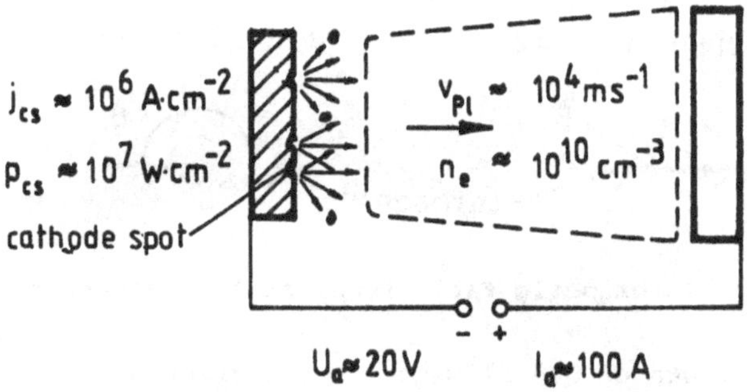

Figure 1: Principle arrangement and operation of cathodic arc discharges
j_{cs}, p_{cs} current density and power density of cathode spots
v_{pl}, n_e velocity and density of arc plasmas
U_a, I_a voltage and current of the discharge

Decreasing droplet emissions are repoted to be obtainable by decreased arc current, increased reactive gas pressure, accelerated spot motion and by the application of hot, thermoemitting cathodes /1/. In addition, proper separation procedures within the transport space are further means to depress the impingment of the droplets to the coating. External magnetic fields employed to modify the cathode spot motion meet several aspects of droplet reduction.

The application of external magnetic fields

The spot motion as well as the expanding spot plasma are strongly dependend upon the action of external magnetic fields. In particular, the following points are of importance to CAT:

- confinement of the cathode spots to the erosion face,
- leveling of the erosion profile of the cathode,
- stimulation and acceleration of the cathode spot motion,
- minimization of the arc current,
- deflection of the expanding plasma beam.

Principle magnetic field forms at the cathode erosion face as illustrated in figure 2 are (i) essentially straight fields with varying components in relation to the face normal and (ii) tunnel fields. In both cases the field component perpendicular to the face normal causes stimulated spot motion across the cathode.

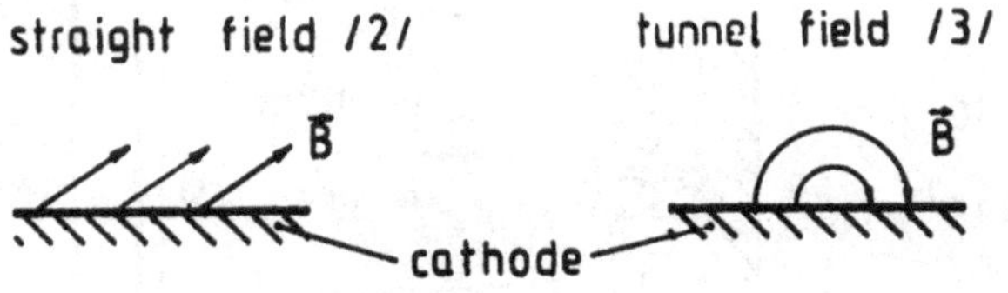

Figure 2: Principle magnetic field forms at the cathode erosion face

Straight field lines of 0.5 mT magnetic induction with the main component perpendicular to the face normal are suitable to stimulate the spot motion of 100 A dc discharges at metallic cathodes such as titanium and aluminium. Simultaneously, modifications of the erosion profile, of the substrate current density as well as decreased droplet diameters were verified /2/. At graphite cathodes similar effects were observed at magnetic inductions of more than 5 mT.

In addition to the spot drive tunnel fields increase the density of the expanding spot plasma in the environment of the spot thus stimulating the formation of new spot areas. As a result the minimum arc current of dc cathodic arc discharges may be considerably lowered resulting in an effective reduction of macroparticle erosion /3/.

Summarizing, the application of external magnetic fields to cathodic arc evaporation acts together with appropriate designs of evaporation sources and operation regime on the droplet ejection from cathode spots decreasing them to minimum values.

(1) Rother, B.: Surface Engineering 4(1988) 335 - 341

(2) Rother, B., J. Vetter and J. Siegel: 7th Int. Conf. on Ion and Plasma Assisted Techniques, Geneva 1989

(3) Gurevitsh, L.S., D.A. Karpov, S.L. Potekhin and G.L. Saksaganskij: On the motion of cathode spots in external magnetic fields (in Russian), NIIEFA, Leningrad 1982

MODIFICATION OF ADHESION PROPERTIES OF SOLID FILMS BY ION IMPLANTATION

E.M. SHPILEVSKY[1), A.A. DESHKOVSKAYA[2), I.A. DILKO[1)

The effect of argon, phosphorus, boron ion implantation on the adhesion of copper films to glass substrates was studied. The films were produced by vacuum deposition at the residual air pressure P 10^{-4} Pa. The film thickness was 30 to 80 nm, because of which they become commensurable with the average projective ion paths. Ions with 30,100,120, 150 keV energies were implanted at doses of $1.0 \cdot 10^{18}$, $6.2 \cdot 10^{19}$, $2.8 \cdot 10^{20}$, $3.8 \cdot 10^{21}$ ion/m^2. Adhesion was measured by the method of physico-chemical action of the active medium /1/. Water solution of disodium salt of ethylenediamineacetic acid served as active medium. The time, during which a definite portion (50%) of selected patches over the film area was separated served as a measure of adhesion (see fig. 1)

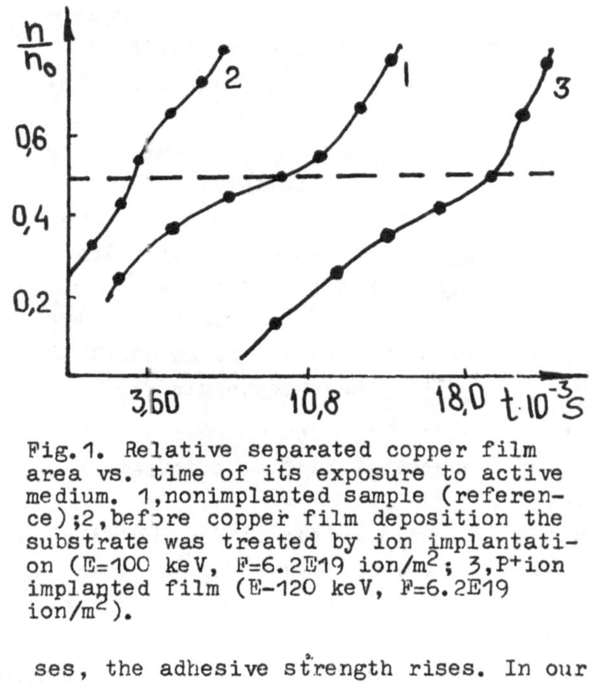

Fig. 1. Relative separated copper film area vs. time of its exposure to active medium. 1, nonimplanted sample (reference); 2, before copper film deposition the substrate was treated by ion implantation (E=100 keV, F=6.2E19 ion/m^2; 3, P^+ ion implanted film (E-120 keV, F=6.2E19 ion/m^2).

Implantation of different ion species (B^+, P^+, Ar^+) essentially increased the film to-substrate adhesion. Ions implanted (P^+) into the glass substrate before the film deposition reduced the adhesion, whereas UV irradiation increased it (Table). Ion implantation (P^+) into the samples with UV treated and untreated substrates led to the same results.

Thus, ion implantation into the substrate irradiation change the substrate surface state that affects the adhesion.

It is known /2/ that as the surface impurity decreases, the adhesive strength rises. In our case it may be suggested that UV irradiation cleans up adsorbed water and gas molecules from the substrate, while ion implantation "contaminates" the substrate surface.

The adhesion of ion implanted copper films to the substrate changes with the implanted ion dose. In fig. 2 phase curves of separation time

1) Byelorussian State University, Leninskii Pr. 4, 220080, Minsk, USSR
2) Minsk Radioengineering Institute, P. Brovki 6, 220069, Minsk, USSR

Table

The ratio of the copper film separation time under the action of active liquid to the separation time of initial (nonimplanted) substrate (E=100 keV, F= 2.8E20 ion/m^2).

Action	Implantation into substrate	Implantation into film		
Substrate	P^+	B^+	P^+	Ar^+
Glass	0.20	6.1	6.4	9.0
Quartz glass	-	6.6	-	9.8
Glass+UV	-	6.0	6.4	8.9

for P^+ and Ar^+ ion implanted (E=100 keV) copper films, 50 nm thick, are given.

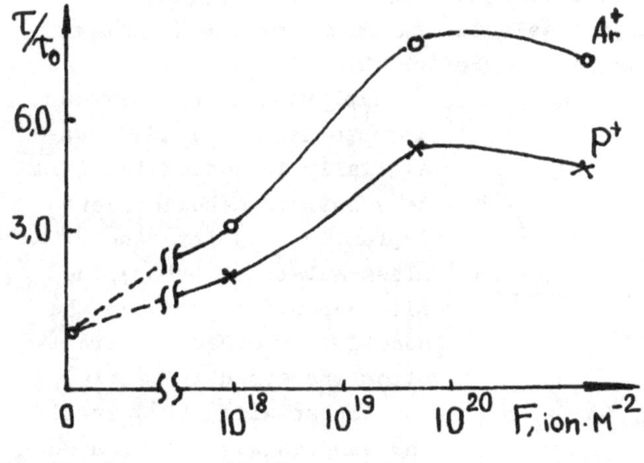

Fig.2. Changes of adhesion (the relative separation time in the active liquid medium) vs. implanted ion dose (d=50 nm, E=100 keV).

Metallography and ion probing of the substrate surface have shown that high doses (10^{20}- 10^{21} ion/m^2) make the implanted films separate at the film - ion mixed layer boundary but not at the initial film-substrate interface.

The adhesion of nonimplanted copper films slightly increases with their thickness. For the thickness range studied the changes did not exceed 15 %. In ion implanted films the adhesion strengths changed several times, increasing or diminishing with increasing the film thicknesses depending on ion species and energy (fig.3).

Analysis of the relation between the adhesion strength and the film thickness shows that the relation may be ascribed to arising compression or extension stresses, or to mixing. In any case, the depth, at which the concentration of implanted ions is highest, is of great importance.

Following /3/ in the calculation of average projective ion paths, the results obtained for the adhesion changes can be expressed as the function of projective paths (R_p), measured in the units of film thickness(d) (fig.4).

The relations obtained here are more than surprising as, in spite of the neglected effect, the ratio R_p/d appeared cruicial for the modifi-

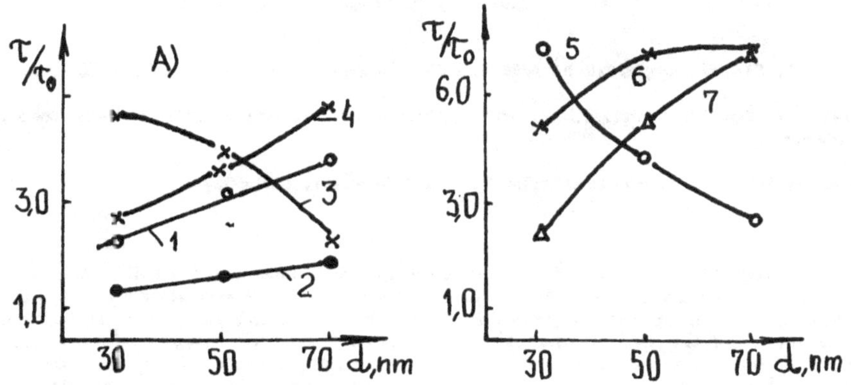

Fig. 3. Changes of the adhesion of ion implanted films vs. thickness (F=6.2E20 ion/m^2). 1, B^+, E=30 keV; 2, B^+, E=100 keV; 3, P^+, E-30 keV; 4, P^+, E=100 keV; 5, Ar^+, E=30 keV; 6, Ar^+, E= 100 keV; 7, Ar^+,E=150 keV.

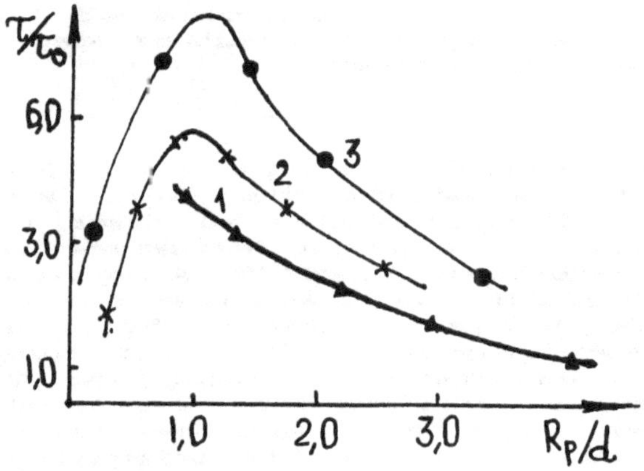

Fig. 4. Changes of the adhesion of implanted films (F=6.2E20 ion/m^2. 1,B^+; 2, P^+; 3, Ar^+.

cation of adhesive properties of the films.

Increasing the adhesion of UV irradiated films from light (boron) to heavier (argon) ions may be attributed to the changes of mixing efficiency.

REFERENCES

1. Author Certificate no.993108,Cl.St01 no.19/04,Bull.Izobretenii no.4, 30.01.1984.
2. Zimon A.D. Adhesion of Films and Coatings. Khimiya,1977, 327 pp.
3. Spatial Distributions of Energy Reserved in a Cascade of Atomic Collisions in Solids. Energoizdat, Moscow, 1985, 248 pp.

FILM DEPOSITION BY LASER INDUCED VACUUM ARC EVAPORATION

P. Siemroth[(1)], H.-J. Scheibe[(2)], H.-D. Bauer[(2)], W. Pompe[(2)], D. Schulze[(3)], R. Wilberg[(3)]

(1) Zentralinst. f. Elektronenphysik d. AdW d. DDR, Hausvogteipl. 5-7, Berlin, DDR-1086

(2) Zentralinst. f. Festkoerperphysik u. Werkstofforschung d. AdW d. DDR, Helmholtzstr. 20, Dresden, DDR-8027

(3) VEB HOCHVAKUUM DRESDEN, Niedersedlitzer Str. 63, Dresden, DDR-8017

1. Introduction

At present, high attention is given to the plasma-assisted deposition in all fields of thin film deposition because of the special film properties which are attained by the influence of plasma, i. e. by the influence of high-energy ionized particles during the deposition process. Vacuum arc deposition (VAD) as well as the laser pulse vapour deposition (LPVD) hold an extraordinary place since high deposition rates can be realized with these methods. While vacuum arc evaporation generally uses a continuous target/cathode evaporation, the LPVD process is performed by pulse evaporation. With VAD, an uncontrolled motion of the arc over the cathode might by detrimental because the energy input is unevenly distributed over the cathode area.
This paper deals with a new method, the laser-assisted vacuum arc deposition (laser-arc). The favourable features of LPVD are combined with those of the approved technique of vacuum arc evaporation. The pulsed deposition of LPVD is to be combined with the efficiency of the vacuum arc. Besides the test arrangement, first results are presented to characterize the deposition process and the deposited layers.

2. Experimental details

The experimental arrangement is similar to that used in / 4 /. In a vacuum chamber ($p < 10^{-4}$ Pa) a negatively biased target (cathode) is irradiated by a highly focussed laser pulse. In the experiments a Nd-YAG-laser with a pulse energy of 6 mJ and a duration of 90 ns was used. A power density of about 10^{8} Wcm^{-2} were obtained. Tests were made with single pulses as well as with repetition rates between 5 Hz and 1 kHz. A ring-shaped electrode/anode is arranged parallel to the cathode. Cathode and anode are connected with a pulsed current source. The applied voltage was varied between 0 and 500 V. The laser induced plasma initiates a vacuum arc discharge as soon as it reaches the anode. This arrangement was used to measure the parameters of the laser induced plasma, to study the transition range between laser plasma and the developed arc and to carry out the first film deposition tests. The cathode-anode-distance was varied between 5 and 30 mm. The distance between target and the substrat was 50 mm. Depositions were carried out with arcs of 700 A in maximum and a duration of about 10 µs. Graphite, tungsten and titanium were used as target/cathode materials. Deposition was performed on Si (111) wafers and NaCl substrates which had been precleaned in usual manner. The film deposition was started after a substrate cleanig by glow discharge in a N_2-atmosphere of 5 Pa. Investigations on the film morphology and structure were made by mean of the transmission electron microscope (TEM) and electron diffraction method.

3. Results and discussion

3.1 Arc ignition

The anode current was measured to study the plasma parameters as well as the arc development initiated by the laser pulse. With a potential below the burning voltage of a vacuum arc no initiation may occur, and the anode collect an electron current from the plasma. The current signals (plasma and arc current) are presented in fig. 1. It can be seen that the starting phase of the current signal is independent on whether initiation occurs or not. After reaching a certain level of plasma current, transitions into the vacuum arc starts. This critical plasma current for all studied materials is in the range between 40 and 100 mA. In dependence on the value of the pre-arc plasma current a different behavior can be observed:

- If the plasma current reaches this critical value only at its maximum, arc current starts there (e. g. carbon).
- In the case that the plasma current exceeds the critical value remarkably (as with titanium) the transition starts already during the rising stage of the plasma current.
- No arc initiation occurs at all if the plasma current does not reach the initiation value (e. g. tungsten).

From this behavior it is possible to give an analysis of the processes underlying the arc initiation. The decisive factor for the arc ignition is obviously the current density from the laser plasma to the cathode surface. Therefore it can be concluded that the arc initiation is a result of a specific interactions between the laser plasma and the cathode rather than by the direct heating of the target material. Arc ignition due to plasma-surface interaction -the so-called unipolar arc- was studied in detail and it was found /1/,/2/ that there are minimum values of net ion current density to the cathode surface to trigger a unipolar arc discharge. Moreover, if the current exceeds this critical value, it must flow for a definite time before the arc struck. The following critical values were established:

$$j_{init} \simeq 10^4 \ldots 10^5 \ \mathrm{Am^{-2}}$$

$$q_{init} \simeq 3 \cdot 10^{-2} \ \mathrm{Amm^{-2}}$$

The critical current density and integral in our test were calculated. The following values were obtained:

$$j_{init} = 3.7 \cdot 10^5 \ \mathrm{A/m^2} \quad \text{corresp. to 40 mA}$$

$$q_{init} = \begin{cases} 5.5 \cdot 10^{-2} \ \mathrm{As/m^2} & \text{for Ti} \\ 4.6 \cdot 10^{-2} \ \mathrm{As/m^2} & \text{for C} \end{cases}$$

It can be seen that the critical values found in our experiments are near to the minimum conditions of unipolar arc ignition. The initiation is followed by a short transition phase of strongly fluctuating but fast rising current (dashed line in fig. 1). The unstable current rise is followed by a phase of exponentialy rising current. The time constant is independent of the electrode separation. After a certain time which depends on the electrode separation the current rise becomes steeper and it is now determined only by the parameters of the supply circuit.

3.2 Film deposition

Films of C and Ti in the thickness range of up to 500 nm were deposited. By light microscopic observation of the whole substrate surface, only isolated droplets could be observed in the µm range. This allows to suggest that deposition with both materials is homogeneous. For investigation of the film morphology and structure, thin films up to a thickness of 100 nm were removed from the substrates and studied in an electron microscope (Type JEM 200 cx). TEM-images of C show that the films are smooth and dense. The electron diffraction pattern show two diffuse rings of $d_1 = 0.208$ nm and $d_2 = 0.123$ nm are characteristic for hard amorphous (diamond-like) C-films as obtained with LPVD /3/. By means of ellipsometric investigations a refractive index of 2.06 at $\lambda = 0.6328$ µm was measured.

The Ti-films (removed from NaCl) show structures within the nm-range and particles of equal size. Slight impurities of Cu, Zn (anode components) can be detected. The diffraction pattern showes the film to be microcrystalline in the form of nm-sized crystallites. The diffraction reveals a tendency towards TiC/TiO rather than Ti. This is a result from the influence of residual gas. Other investigations are under preparation.

4. Conclusions

A) A laser produced plasma can be used for the ignition of a stable vacuum arc. At moderate laser power densities of $5 \cdot 10^7 - 10^8$ cm^{-2} arc ignition is observed for carbon and titanium.

B) A necessery condition for arc ignition is that the plasma current to the target/cathode exceed the threshold level for unipolar arc initiation. Under our experimental conditions this level is equivalent to a plasma current of 0.04 A.

C) Because the arc ignition is stable beyond this threshold level it is possible to use the principe of laser sustained arc initiation for a repeated ignition of vacuum arcs.

D) It was shown that films of high quality may be deposited by means of rapeated ignition of pulsed high current arc discharges. Especially the formation of larger droplets can be diminished by using the presented technique.

Literature

/1/ V. M. Antonov, L. B. Gevorkyan, A. N. Ponomarenko, ZhTF(pisma) 4(1978)995.

/2/ R. Hancox, Brit. Journ. Appl. 11(1960)468.

/3/ W. Pompe, H.-J. Scheibe, A. Richter, H.-J. Bauer, K. Bewilogua and Ch. Weissmantel, Thin Solid Films, 144(1986)77.

/4/ W.Pompe,H.-J.Scheibe,P.SiemrotH,R.Wilberg,D.Schulze and B.Buecken, 16th International Conference on Metallic Coatings, April 17-21 1989, San Diego, USA.

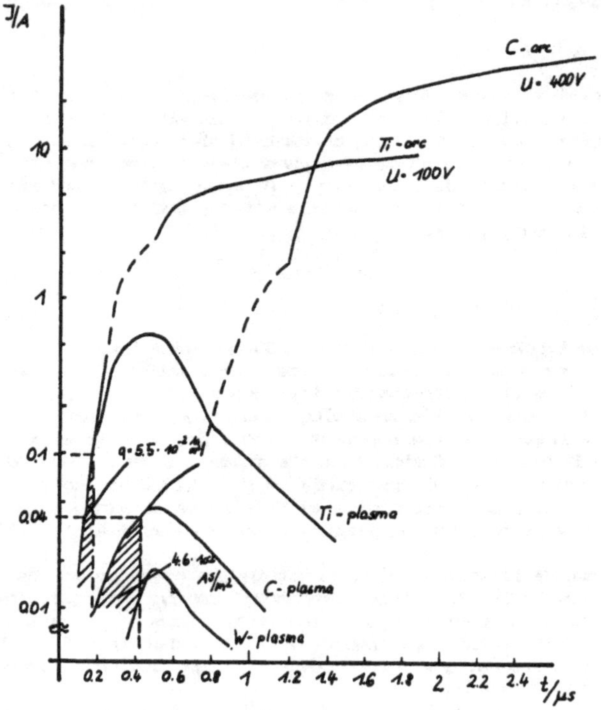

Fig. 1 Time resolved laser plasma/arc current signals of different cathode materials.

HETEROGENEOUS ION SYNTHESIS OF INSULATING LAYERS IN SILICON

A.B. Danilin, V.N. Mordkovich, M.Y. Barabanenkov

Institute of Microelectronics Technology and High Purity Materials, USSR Academy of Sciences, 142432 Chernogolovka, Moscow District, USSR

The method is based on the effective use of the heterogeneous processes of nucleation and growth of new phase inclusions [1]. Nucleation of inclusions in a very thin layer allows minimizing the role of coalescence and, hence, reducing consumption of energy in layer synthesis. The main parameters of the later are the dose (the dose rate) and synthesis temperature. The yield of the reaction of heterogeneous phase formation is nonmonotonic function of these parameters [2].

Heterogeneous ion synthesis can be simulated. Let spatial distribution of the reagent $C^{o}(X)$ described by Pearson IV-type distribution be induced in the target material by ion implantation. When using substoichiometric implantation doses the maximum value of $C^{o}(X)$ is always less than impurity atom concentration $C_{n.ph.}$ characteristic of the new phase. The chemical reaction between the matrix atoms and impurity giving rise to new phase precipitates proceeds only at $C^{o}(X) > C_{cr}$ where C_{cr} is the critical impurity concentration essential to nucleation. The transition of impurity to the new phase composition is described by $C_b(X,0) = \alpha \cdot K(X) \cdot C^{o}(X)$ (Fig.1), where reaction constant K takes the form

$$K(X) = \frac{C^{o}(X) - C_{cr}}{C_{n.ph.} - C_{cr}}$$

α - being the numerical parameter, $K=0$ at $C^{o}(X) = C_{cr}$ and $K < 0$ at $C^{o}(X) < C_{cr}$ which corresponds to the assumption about annealing dissolution of nuclei occuring in regions with $C^{o}(X) < C_{cr}$. The model assumes that new phase nuclei are stationary, shaped as spheres of radius ρ_o and do not interact immediately.

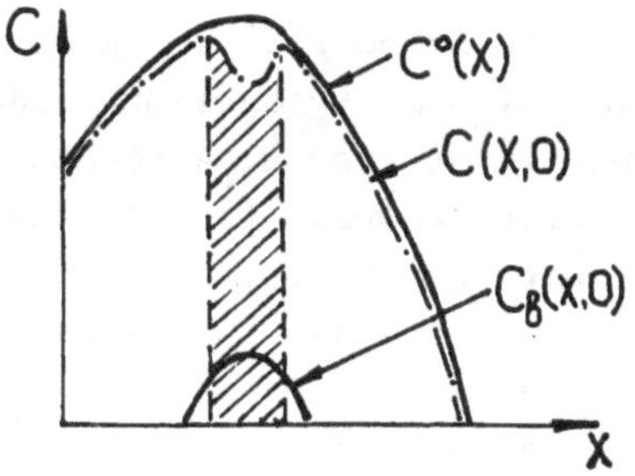

Fig.1 Transition of free impurity to the bound state

The nucleus growth is described by the stationary diffusion equation. Mass balance occurs on the new phase centre surfase (Fig.2), i.e. the amount of the material penetrating into the nuclei in a definite time is equal to that of the material brought from the initial phase to its surface.

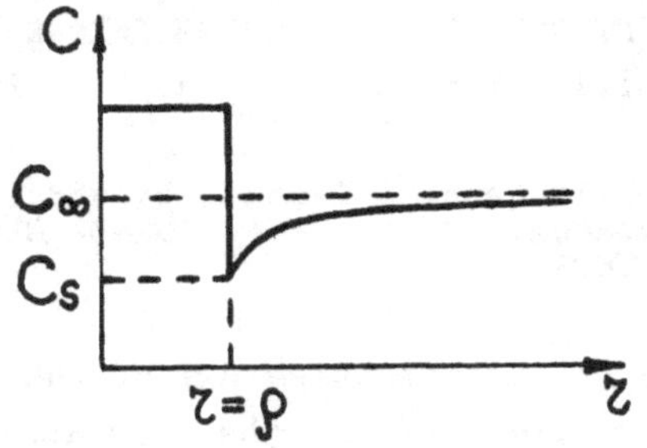

Fig.2 Mass balance on the new phase inclusion surface

The basic equation of the model takes the form:

$$\frac{\partial C(X,t)}{\partial t} = D \cdot \frac{\partial^2 C(X,t)}{\partial X^2} + \omega\left[C(X,t)\right] \qquad (1)$$

$$\omega = -4\pi D \cdot N(X)\left[C(X,t) - C_s\right] \cdot \rho(X,t) \qquad (2)$$

$$\rho^2(X,t) = \rho_0^2 + \frac{2D}{C_{nph} - C_s} \cdot \int_0^t \left[C(X,t') - C_s\right] dt' \qquad C(X,0) = C^{\circ}(X) - C_b(X,0)$$

where C(X,0) is free impurity distribution; ω - distributed impurity drain power, i.e. the ability of the impurity drain to absorb a definite amount of the reagent; ρ - the nucleus radius; D - the free impurity diffusion coefficient; C_s - the matrix solubility of the reagent; N(X) - the distribution of new phase inclusion density defined as

$$N(X) = \frac{C_b(X)}{C_{nph} \cdot 4/3 \pi \rho_0^3}$$

By distributed drain we mean the set of new phase pricipitates. Expression (2) for drain power differs from the analogous expression [3] by the inhomogeneous matrix depth distribution of new phase centre density and a high impurity oversaturation. Nucleus growth is thought to occur owing to chemical segregation of reactive impurity on the nucleus surface. Solution therewith is found for two cases. In the first case the growth rate is limited by the segregation process, and in the second one by reagent diffusion. Fig.3 (a,b) shows comparison of the calculation data with the experimental results [4] of determination of concentration profiles of bound nitrogen atoms. The calculations were made using the following values of constants: $D = 4.8 \cdot 10^{-12}$ cm^{-2}/s [5] ; $C_s = 5 \cdot 10^{15}$cm^{-3} [6] ; $C_{cr} = 7 \cdot 10^{20}$cm^{-3} [7] ; $C_{n.ph.} = 5.3 \cdot 10^{22}$cm^{-3} [8] ; $\rho_0 = 5$ nm [9] . At the initial state of solid solution dissociation the process is well described by chemical impurity segregation on the nuclei (Fig.3 a), whilst when increasing duration of annealing, account should be taken of impurity diffusion at distances exceeding appreciably the size of new phase inclusions to achieve good agreement between calculation data and experimental results (Fig.3b).

The process can be based on repeated implantation-annealing cycles in order to attain continuity of synthesized layers [10]. Yet, on account of the strong dependence of local phase formation on presence of radiation defects as well as relative easy crystallization, fabrication of continous buried Si_3N_4 layers is complicated and requires that the temperature and annealing modes of implantation cycles should be optimized.

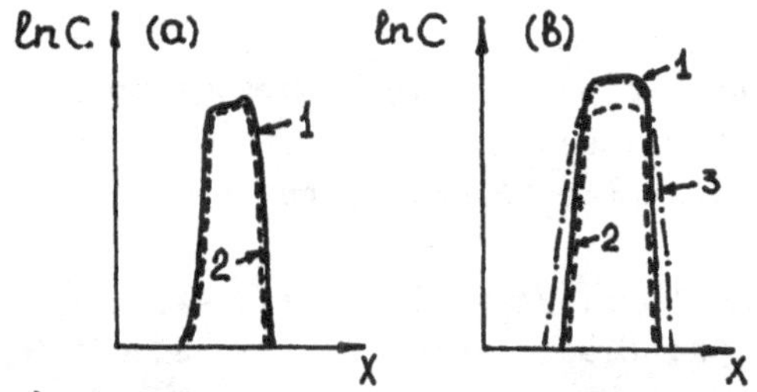

Fig.3(a,b) Comparison of the calculated and experimental concentration profiles of nitrogent in silicon: 1 - experimental data from 4, E = 200 keV, F = $7.5 \cdot 10^{17} cm^{-2}$, T_a = 1200°C; 2 - calculation neglecting macrodiffusion; 3 - calculation with regard to macrodiffusion; a - annealing time 20 min, b - annealing time 2 h.

The use of the effect of mutual attraction of oxygen and nitrogen atoms implanted in silicon simultaneously and sequenteally [11] and the effect of oxygen retardation of Si_3N_4 phase crystallization made possible synthesising thin (< 0.1 mkm) insulating layers of silicon oxynitride even at an implantation dose of $2.4 \cdot 10^{17} cm^{-2}$ [13]. Thicker layers (0.2 - 0.25 mkm) were obtained with doses (4-6)$\cdot 10^{17} cm^{-2}$ (Fig.4).

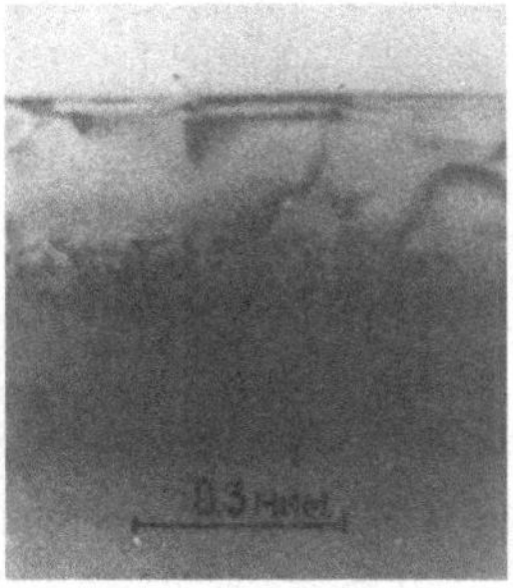

Fig.4 Microphotograph of cross-section of a silicon sample with a buried dielectric layer obtained by sequential implantation of oxygen and nitrogen atoms. E_{O^+,N^+} = 150 keV; F_{O^+} = $1.0 \cdot 10^{17} cm^{-2}$; F_{N^+} = $5.0 \cdot 10^{17} cm^{-2}$; T_{ann} = 1200°C; t=2 h.

The best structural properties were exhibited by buried layers with a nitrogen-oxygen concentration ratio C_N/C_O = 2.

REFERENCES

1. Danilin A.B., V.N. Mordkovich: Int. Conf. on Ion Impl. in Semicond. and Other Mater., Lublin, 1988, p.51.
2. Danilin A.B., Surface Physics, chemistry, mechanics. 1989. N 5. 132-136
3. Lyubov B.Ya: Diffusion processes in nonhomogeneous solids, Nauka, Moscow, 1981, (in Russian).
4. Kilner J.A. and R.J. Chater Nucl. Instr. and Methods B 15 (1986), 214-217
5. Pavlov P.V. et al. Phys. Stat. Sol. (a) v.35, (1976) 11-16
6. Yatsuruki Y. et al. J. Electrochem. Soc. v.120, (1973) 975-978
7. Tsujide T. et al. J. Appl. Phys. v.51, (1980) 1605-1610
8. Properties of Silicon: IEEE, N.Y. 1988
9. Belz J.: Dissertation Universitat Dortmund, 1986
10. Belogorohov A.I. et al.: All-Union Conference. Ion-beam modification of Materials Kaunas (1989) 177 (in Russian)
11. Borun A.F. et al. Rad. Effects v.107, (1988), 9-13
12. Borun A.F. et al. Electron. Techn. Ser.3 Microelectronics v.1 (130), (1989) 39-41 (in Russian)

A TWO-DIMENSIONAL MODEL OF SOI STRUCTURES CRYSTALLIZATION BY PULSE NANOSECOND HEATING

L.N.Aleksandrov, V.Yu.Balandin, A.V.Dvurechenskii, and O.A.Kulyasova[1)]

The processes of silicon layer melting and crystallization in the $Si-SiO_2-Si$ structure by nanosecond pulse heating have been simulated by the solution of a two-dimensional heat-conductance equation and logic function of phase transitions. The lateral movement of the crystallization front has been observed along the SiO_2 surface from the seeding window. By simulation data, the crystallization front is inclimed to the SiO_2 surface (~30°), the velocity during crystallization varies (1-5 m/s) and depends on the initial substrate temperature.

The onset of heterogeneous nucleation on the SiO_2 surface after melt formation and non-monotonous variation of crystallization rate have b een found.

Introduction

One way to obtain a SOI structure is by recrystallization of the poly-crystalline (poly-Si) film lying on an insulator with windows. By pulse heating, the initial poly-Si layer melts, and oriented crystallization begins from seeding windows in SiO_2 propagating along the dielectric layer /5/.

The simulation of temperature changes and of phase transitions in the SOI structure within a one-dimensional model of thermal pulse heating has allowed us to find conditions that are favourable for the lateral growth of the crystal film /1/. We have found optimal original thicknesses of poly-Si and SiO_2 layers for which the lengthes of crystallites is maximum /2/ and determined the values of pulse durations and energy densities of a laser beam that do not lead to layer melting at entire melting of poly-Si and uniform melting of the seed /1/.

However, within the one-dimensional approximation, it is impossible to describe the dynamic of the movement of the lateral crystallization front, estimate the rate of lateral and normal growth of crystallites, calculate heat flows near the seeding window, detect the character of the effect of seeding windows size on the kinetics of crystallization.

Calculation Results and Discussion

In the present work we perform simulation of processes by the solution of a two-dimensional equation of heat conductance and logic function of phase transitions:

$$c\rho \frac{\partial T(x,y,t)}{\partial t} = \frac{\partial}{\partial x}\left(k\frac{\partial T(x,y,t)}{\partial x}\right) + \frac{\partial}{\partial y}\left(k\frac{\partial T(x,y,t)}{\partial y}\right) + P(x,y,t) + Q(S)$$

$$S = S(T,\Phi)$$

1) Institute of Semiconductor Physics, 630090 Novosibirsk, USSR

Here Q(S) is the function of latent heat phase transitions; C, ρ, K are heat capacity, density and heat conductance respectively; P(X,Y,t) is the function of heat sources produced by the laser in the nanosecond range of durations. The coordinate X is counted off the poly-Si surface, and Y off the middle of the window.

Boundary conditions are:

$$\left.\frac{\partial T}{\partial x}\right|_{x=0}=0; \quad \left.\frac{\partial T}{\partial y}\right|_{y=-d_2}=0; \quad \left.\frac{\partial T}{\partial y}\right|_{y=d_3}=0; \quad T\Big|_{d_1}=T_s,$$

where d_1 is the thicknes s of the SOI structure; d_2 is the distance between the middle of the window and the edge of dieletric layer; d_3 is the distance between this edge and the middle of the dielectric layer; T_s is the ambient temperature.

Initial conditions are $T(X,Y,0)=T_s$.

The problem was solved by the finite-difference technique using the three-point implicit circuit.

The results of model calculations allow to observe the position of the lateral crystallization front in different time points after the onset of pulse action (Figs 1a,b,c,d) It is seen that the lateral crystallization front is nonperpendicular to the SiO_2 surface, the angle of inclination is ~ 30°. The lateral front movement is blocked by heterogeneous nucleation on the melt-SiO_2 interface (Fig. 1b) leading to the growth of the polycrystal (Fig. 1c,d). As in /2/, when calculating the nucleation rate, the wetting angle was taken equal to 90°, according to /3/.

The size of grown crystallites from the seed above SiO_2 is determined by the lateral crystallization front velocity during the melt lifetime on the SiO_2 surface. The growth rate has been found to vary non-monotonously with time at different substrate temperatures. The initial points of the curves of growth rate dependence on time correspond to the moment when the lateral front comes out to the SiO_2 surface after the pulse action begins. For example, at T_s=300 K the crystallization front movement along the SiO_2 surface begins 150 ns after the start of the pulse (Fig.2, curve 1). The increased velocity of the crystallization front at the initial stage is caused by increased undercooling in the front due to heat outflow to the seed region. However heterogeneous nucleation and polycrystals growth is accompanied by heat release which leads to a retarded lateral growth and subsequently to the convergence of the two fronts (Fig. 1d). Thus, for T_s=300 K (Fig.2, curve 1) the convergence of opposite fronts is observed already 270 ns after the onset of pulse action, and the size of crystallites grown from the monocrystalline seed does not exceed 0.5 μm. With substrate temperature increase to 1500 K the melt duration on the SiO_2 surface before the first nucleus appearance increases, and the length of the grown crystallites is enhanced.

Conclusion

Therefore, the dynamics of silicon melting and crystallization in the seeding windows region of the SOI structures have been analyzed by thermal pulse heating simulation. It has been found that the dependence of the lateral crystallization growth rate on time is shaped as a curve with th maximum. The lateral growth rate of crystal film varies from 5 to 1 m/s at T_s=300-1400 K. With the substrate temperature increase T_s from 300 to 1500 K, the length of the grown crystallites increases. This type of dependence on T_c is in agreement with the experimental data of /4/, for example.

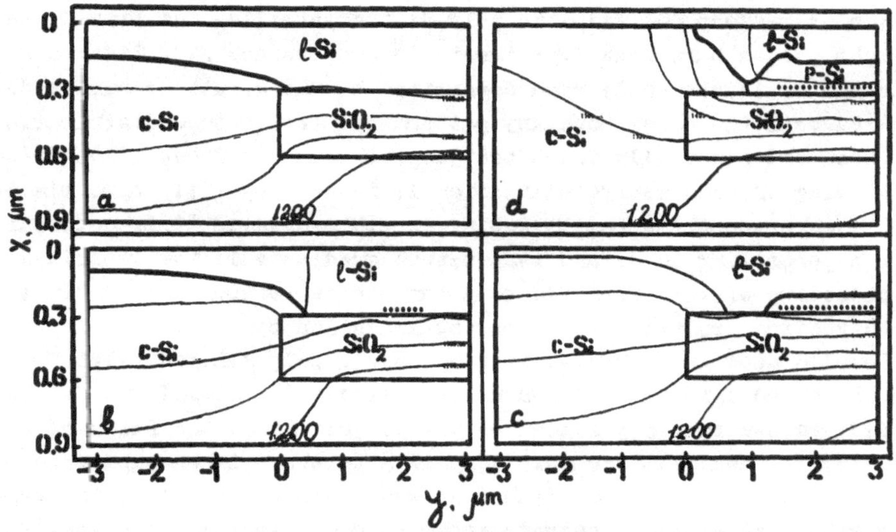

Fig.1. Dynamics of lateral crystallization front movement at different time points after the beginning of pulse action (30 ns, 1.5 J/cm): 180 (a), 210(b), 240(c), 300 ns (d). Points indicate the places where nuclei appear. T_s=300 K. c-Si-monocrystalline silicon; l-Si-silicon melt; p-Si-polycrystalline silicon.

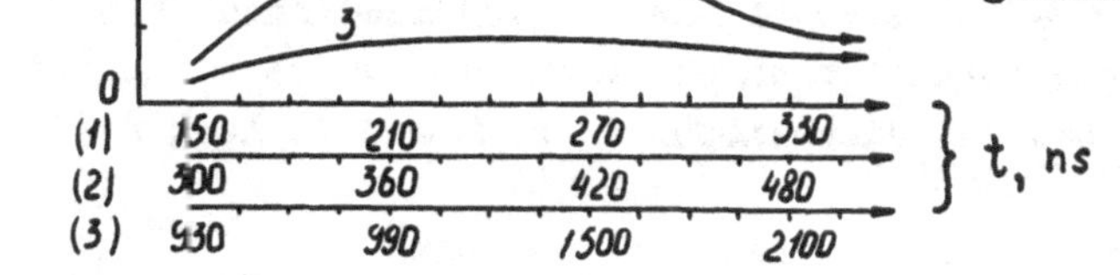

Fig.2. The time dependence of the lateral growth of crystallites at T_s, K: 300(1), 1000(2), 1400(3). The initial points of curves denote the times of the lateral front escape to the SiO_2 surface after the beginning of pulse action.

REFERENCES

(1) L.N.Aleksandrov, V.Yu.Balandin, A.V.Dvurechensky. O.A.Kulyasova. In: Eur.SOI Workshop from Met. to Dev. 1988, Neylon (France)

(2) L.N.Aleksandrov, V.Yu.Balandin, A.V.Dvurechenskii, O.A.Kulyasova. Thin Solid Films 171(1989)235.

(3) L.R.Hamilton, D.L.Barret, H.Whrli, A.I.Bennet. J.Cryst.Growth 7(1970)296.

(4) Yu.A.Manshosov, G.A.Kachurin, A.B.Plotnikov. Poverkhnost' Fiz., Khi., Mekh. 2(1989)138.

(5) L.O.Wilson, G.K.Celler. J.Electrochem.Soc.:Sol.St.Sc.132(1985)2748

TEMPERATURE DISTRIBUTION IN MELT AND IN GROWING CRYSTAL DURING SILICON-ON-INSULATOR FORMATION BY MILLISECOND HEATING

S.N.Kolyadenko, A.V.Dvurechenskii[1], D.Stock[2] and V.A.Zinovjev[1]

The method for silicon-on-insulator (SOI) film production through poly-silicon recrystallization by pulsed heating enables to create structures for three-dimensional integrated circuits. The conditions for the formation of perfect SOI films in poly-Si recrystallization by millisecond light pulses have been experimentally ascertained /1/. Silicon film formation is obviuusly much determined by the growth mechanism dependent on both the phase boundary pattern and temperature distribution. It has been experimentally found that crystallization front (CF) during pulse heating induced recrystallization is faceted by (111) type planes. But to be sure about the really acting growth mechanism it is necessary to know supercooling on CF and temperature gradients in its vicinity.

The objective of this work was to study the temperature fields in the melt and growing crystal during recrystallization by pulsed heating.

The samples used in recrystallization have a poly-silicon layer 1.0 μm thick on the oxidized silicon wafer surface. The SiO_2 layer 0.5 μm thick had windows regulary spaced at 40-150 μm. Since the films of the growing crystal and melt are thin comparing to the silicon substrate (400 μm), the temperature in films, as was shown by the analysis, varies largely with that of the substrate-separating dielectric interface. Thus $t \simeq 10^{-6}$ s later after introduced temperature perturbation $dT/dt = dT_o/dt$ where T_o is the substrate-SiO_2 interface temperature and T is the temperature of the Si film (T_s) or of the melt (T_L). Now if in the time interval (Δt) T_o changes little, dT_o/dt can be substituted to $\Delta T_{oj}/\Delta t_j = A_j$. In our calculations T_o has been found from the solution of heat conductivity equation for the $Si/SiO_2/Si$ substrate structure, and the value of Δt was $(1-2)\times10^{-5}$ s. From the solution of the same problem it has been also found that T_s and T_L variation in the direction perpendicular to the substrate surface was negligible which permitted us to focus on T_s and T_L distribution only along crystallization direction. The estimations show that for 10^{-5} s the size of the crystallized band changes negligibly comparing to the total lehgth which enables us to treat the growing film and melt as two one-dimensional rods, l and b long respectively.

The heat conductivity equation for them has the form

$$\frac{d^2T_i}{dx^2} - \frac{H}{K_i d_{Si}}(T_i - T_o) + \frac{A_o}{K_i d_{Si}} - \frac{\varepsilon_i \sigma}{K_i d_{Si}}(T_i^4 - T_{amb}^4) = \frac{1}{\varkappa_i}\frac{dT_i}{dt} \qquad (1)$$

Here the term $H(T-T_o)$ describes the heat loss to the substrate across the SiO_2 film boundary. A_o and $\varepsilon\sigma(T^4-T^4_{amb})$ respectively are the ener-

[1] Institute of Semiconductor Physics, 630090 Novosibirsk, USSR
[2] Friedrich-Schiller-Universität Jena DDR

gy absorbed by the film from the heat source and reirradiated from its surface to the ambient with T_{amb}. Since $dT/dt = dT_0/dt = A_j$ and at T_{amb} not much different from T_i (which takes place under experimental condition), (1) can be reduced to

$$\frac{d^2T_i}{dx^2} - B_i(T_i) = -\Delta_i \qquad (2)$$

where

$$B_i = \frac{1}{K_i d_{Si}}(H + 4\varepsilon_i \sigma T_{amb}^3); \quad \Delta_i = \frac{1}{K_i d_{Si}}\left[(1-R_i)P_j + 4\varepsilon_i \sigma T_{amb}^4\right] + \frac{1}{æ_i} A_j$$

index $i = L$ refers to the melt and $i = s$ to the crystal, index j stands for the number of the time intervals Δt into which the process is broken; R_i is the reflectivity coefficient; P_j refers to the lower of the source heating the structure during Δt; ε is the black-body radiation power; σ is the Stefan-Boltzmann constant.

At the crystallization front the conditions

$$T_L = T_S = T_{CF} \qquad X = 0 \qquad (3)$$

$$Q_{(CF)} = K_S\, dT_S/dx - K_L\, dT_L/dx \quad X = 0 \qquad (4)$$

should be satisfied, and on other rod ends the boundary conditions

$K_L \cdot dT_L/dx = 0$, $x = b$ (because of symmetry) for the melt,

$K_S \cdot dT_S/dx = -C$, $x = -\ell$ where C is the flow from crystal to seed determined from the equation similar to (2) should be satisfied. The solution of equation (2) for the melt has the form

$$T_L = (T_{CF} - \Delta_L/B_L)\frac{ch\,\mu_L \cdot (b-x)}{ch\,\mu_L \cdot b} + \Delta_L/B_L \qquad (5)$$

where $\mu^2 = B$ and for the crystal

$$T_S = (T_{CF} - \Delta_S/B_S)\frac{ch\,\mu_S \cdot (\ell+x)}{ch\,\mu_S \cdot \ell} + \frac{C \cdot sh\,\mu_S x}{ch\,\mu_S \cdot \ell} + \Delta_S/B_S \qquad (6)$$

where K_s and K_l are heat conductivities for the crystal and melt; d_{Si} is the thickness of Si films and melt. T_{CF} was found from the combined solution of (4), (5), (6). The function of the heat source $Q(_{CF})$ has been selected according to the growth mechanism considered:

$$Q_{(CF)} = K_1 \cdot \Delta T^2, \quad \Delta T = T_m - T_{CF}$$

for dislocational growth mechanism and $Q(T_{CF}) = K_2 \exp(-K_3/T)$ for layered growth. The coefficients K_1, K_2, K_3 were taken from /2/. Now, knowing T_{CF}, the crystallization rate $V_j = Q_j/æ \cdot \rho$ was obtained, where $æ, \rho$ is the latent crystallization heat and silicon density. The change of the crystallized layer size has been found as $\Delta \ell_j = V_j \Delta t$ in the Δt interval and during t time, as $\ell = \sum V_j \Delta t$. Such an approach can be justified by that, according to our estimations, the change in ΔT within 10^{-5} s is less than 0.1 which corresponds to rate change by 5-10%.

The performed analysis of temperature distribution in the growing crystal and melt during pulsed heating recrystallization has shown:
a) the necessary conditions (temperature distribution and undercooling at CF) for the layer-by-layer growth are always satisfied during recrystallization. That is, $K_S \frac{dT_S}{dx}\big|_{x=0} + K_L \frac{dT_L}{dx}\big|_{x=0}$ is always greater

than $-\frac{\Theta \cdot \mathcal{L} \cdot V}{C}$, where Θ is anisotropy of the kinetic coefficient and $\Theta > 1$ for the growth of the singular facet, and C is the heat capacity of the melt /3/. So we can ascertain that at least at the initial stage the crystallization process at the pulsed heating follows the layer mechanism of growth. The faceting of CF observed in the experiment at the concluding stage of the growth is perhaps due to the transition to the normal growth mechanism conditioned by the multiple nucleation with high values of ΔT; b) the heat release across the separating dielectric becomes dominant already when the size of the film grown is over 5 μm; c) the growth rates in recrystallization by pulsed heating depend on the position of the crystallization front, distance between the neighboring seed regions and attaine a value of 1 to 5 cm/s.

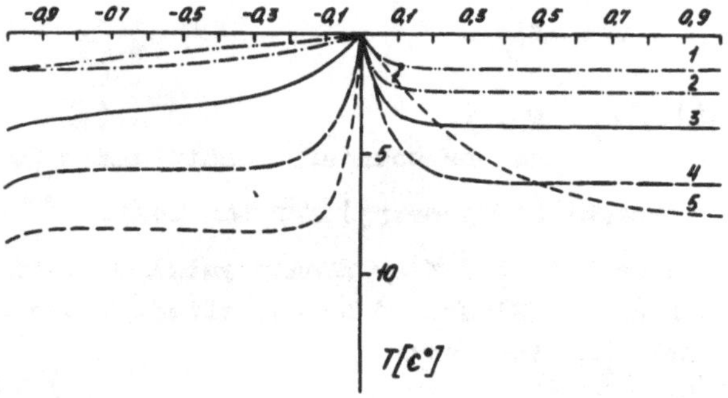

Fig.1. Temperature distribution in the growing crystal (from 0 to 1) and in the melt (from 0 to 1) at CF positions with relation to the window edge in SiO_2, μ: 2.5(1), 5(2) 10(3), 25(4). The temperature point of reference is at CF. From (-1) to 1 is 50 μm, the duration of the heating pulse is 10 ms.

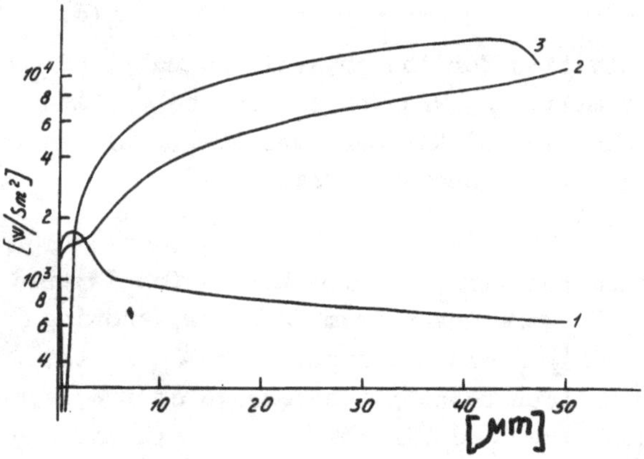

Fig.2. The dependence of heat flows from CF to the window in SiO_2(1), to the crystal(2), to the melt (3) vs the CF position. Zero is the edge of the seed window.

REFERENCES

(1) S.N.Kolyadenko, A.V.Dvurechenskii, A.L.Vasilyev. Elektronnaya prom. N4, 3(1989).

(2) V.V.Voronkov. Kristallographia, 17(1972)909.

(3) A.A.Chernov. In: Growth of Crystals, v.II. Izd. of Yerevan Univ. Yerevan, 1975.

SOI STRUCTURES FORMATION BY PULSED HEATING

S.N.Kolyadenko, A.V.Dvurechensky, and A.L.Vasilyev +

1. Introduction

The application of silicon-on-insulator structures (SOI) provides the new perspectives in the creation of three-dimensional, high-speed and radiation hardening integrated circuits. The zone-melting-recrystallization (Geis et al) and recrystallization techniques by means of scanning laser (De Veirman et al) or electron beams (Davis et al) are used extensively for SOI manufacturing from poly-Si layers. In the first method a substrate must be held at high temperature during recrystallization, in the other methods a substrate temperature is low, but the application of high-energy focused beams produces the high-temperature gradients and associated thermal stresses. That leads to the defects formation in a seed and under SiO_2-layer. This paper presents results of structural investigations of SOI obtained by pulsed heating recrystallization technique.

2. Experimental Procedures

In the experiments (001) Si wafers with thermal oxide layer 0.3 - 0.5 μm thick were used. In the SiO_2-layer the seeds 5 - 50 μm wide were opened photolithographically, their sides were oriented along [1$\bar{1}$0] substrate directions. On the substrate prepared in this manner a poly-Si layer 0.6 - 1.0 μm thick was deposited with help of low-pressure CVD technique.

The heating and melting of poly-Si layer on the entire surface of the wafer were simultaneously carried out by incoherent light pulse

Fig.1 Defectless area width d as function of pulse duration τ (the poly-Si thickness: 1-0.6 μm, 2- 1.0 μm)

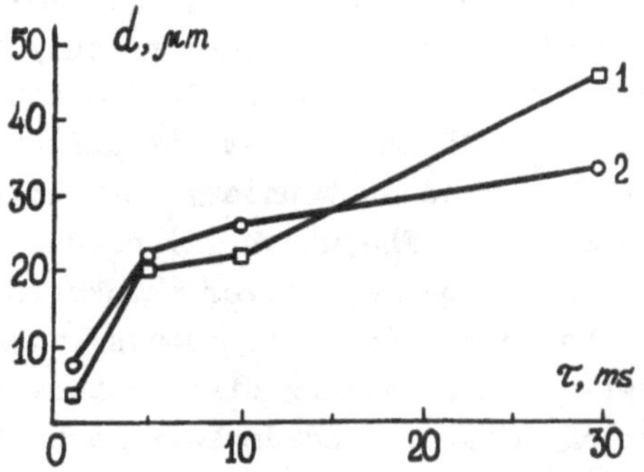

(+) Institute of Semiconductor Physics, Novosibirsk 630090, USSR.

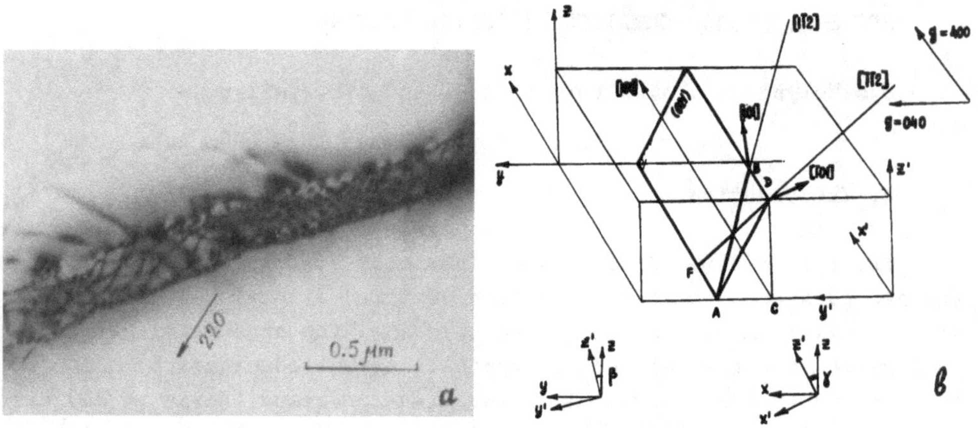

Fig.2 a) TEM micrograph of the subgrain boundary SB-2 in the SOI layer,
b) Schematic presentation of the SB-2 structure ($\beta = 19'$, $\gamma = 34'$, arrows indicated the Burgers vector of dislocations)

from 1 to 30 ms duration. The wafer was heated up to 1100°C directly before irradiation (Kolyadenko et al). The recrystallized layers structure had been investigated by the optical and electron microscopes, HREM of cross-section was used for structural investigation of Si/SiO_2 interfaces on atomic level.

3. Results of Structure Investigations and Discussion

Crystallization of melted Si layer takes place via lateral moving of a crystal-liquid front from seeds in the [$\bar{1}$10] direction. A tilt dislocation boundary SB-1 directed along [110] forms in the meeting place of the crystallization fronts moving from neighbouring seeds. According to crystallization conditions there is a critical distance d_s between the seeds at which (and less) a single crystal Si layer with (001) surface orientation is growing on the top of amorphous SiO_2 layer. The growing layer contains the only one defect - SB-1. Misorientation of SOI areas separated by SB-1 doesn't exceed 0.5° in this case.

If $d_s > 150 \mu m$ then at the distance d from the seed appear the low-angle mixed subboundary SB-2 which directed along [010] and terminated at SB-1. The width (d) of defectless area dependent on irradiation conditions and layer thickness is shown in Fig.1.

The formed SB-2 represents either one-dislocation boundary or two system of intersecting dislocations (see Fig.2a). Analysis of the diffraction contrast and dark-field weak-beam images of the SB-2 reveal that SB-2 lies in the planes near the (021) plane. SB-2 consists of 30°- dislocations directed along [$\bar{1}$12] and [1$\bar{1}$2] with the Burgers vector b= a/2[101] and b= a/2 [$\bar{1}$01] in accordance with Fig.2b.

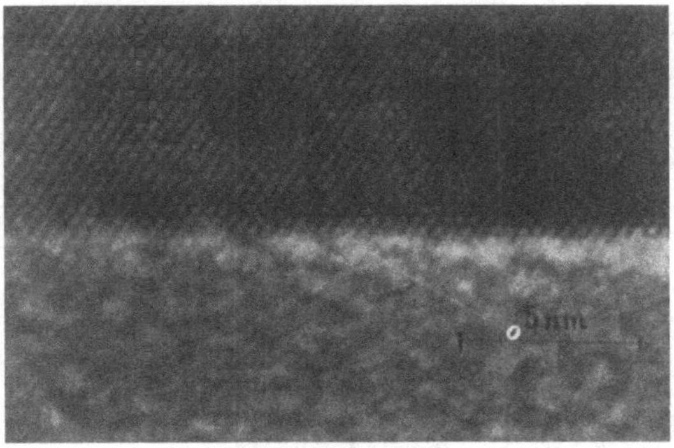

Fig.3 Cross-section HREM micrograph of the recrystallized Si/SiO_2 interface.

Misorientation of the neighbouring subgrains separated by SB-2 has been defined by the Kikuchi-line shift method and doesn't exceed 1.5°. From SB-2 the 60°-dislocations with the Burgers vector b=a/2[011] go away along [110] directions. They glide in the {111} planes, and their glide distance increases approximately from 1 to 5 μm as the subgrain misorientation increases. From SB-1 such dislocations glide at 20-25 μm.

Traditional and high-resolution cross-section electron microscopy denote that after recrystallization the extented defects (dislocations, stacking faults) don't take place in the area of seeds and under SiO_2 layer. The recrystallized Si/SiO_2 and SiO_2/Si-substrate interfaces are planar, the interface roughness doesn't exceed 0.5 - 0.8 nm (see Fig.3).

Several standard technological operations were held on SOI structures - thermal oxidation, As^+ and B^+ implantations with further annealing, phosphorus diffusion. According to the results SOI undergo the same structural changes as bulk Si.

The SOI surface morphology analysis near the SB-1 proves that low angle boundary SB-2 appear at the macroscopic growth step on the crystallization front, which faceted by {111} planes. It is supposed that interior corner of this step is the place of the stress concentration in the growing Si layer. The crystallization front faceting is observed at long distances between seeds. But it doesn't lead to appearence of dislocations conditioned by contamination clusters in the traces of the interior corner of the facets. Thus, contamination influence on formation SB-2 is secondary. Non-linear temperature distribution in the growing layer is the main reason for stresses causing the SB-2 formation. It is estimated that stress magnitudes are sufficient for the appearence of the dislocations.

Authors are happy to acknoledge the gratitude to A.L.Vasiliev (Institute of Crystallography, Moscow) for HREM investigations.

Davis J.R.,R.A.McMahon and H.Ahmed: J.Electrochem.Soc. 132 (1985) 1919
De Veirman A.,D.J.Wouters et al: Inst.Phys.Conf.Ser.No.93, 2(1988) 117
Geis M.W.,H.I.Smith, and C.K.Chen: J.Appl.Phys. 60 (1986) 1152-1160
Kolyadenko S.N.;A.L.Vasilyev.A.V.Dvurechensky,to be published in the Thin Solid Films

FORMATION MECHANIZMS AND STRUCTURES OF BURIED Si_3N_4 LAYERS, PRODUCED BY HIGH-INTENSITY IMPLANTATION AND RAPID THERMAL ANNEALING

R.V.Gribkovskii, F.F.Komarov, A.P.Novikov [1)]

INTRODUCTION

The formation of SiO_2 and Si_3N_4 buried layers in Si by high current ion implantation of oxygen or nitrogen is one of the promising techniques to fabricate silicon-on-insulator (SOI) structures. However, there are some processes in silicon crystals under high-intensity implantation conditions which are not well understood /1-3/. The present work taken up with the aim of studying the dependences of the implanted nitrogen distribution and buried layers structure on the nitrogen ion dose and beam current density.

EXPERIMENTAL

The silicon single crystals, 1x1 cm^2, were implanted with a stationary beam of N_2^+ ions at anergy of 140 keV to the doses ranging from 10^{17} to 10^{18} at/cm^2. The temperature due to ion beam heating were estimated to be about 600, 800 and 1000°C for the ion curren density of 50, 100 and 250 $\mu A/cm^2$. Postimplantation annealing was performed by the halogen lamp system /4/ at 1300°C for 20s with ramp-rate of 130°Cs^{-1}. Before and after annealing samples were examened by RBS, IR-transmission, TEM and THEED.

RESULTS AND DISCUSSION

The distribution of nitrogen atoms in silicon implanted with the lowest and the highest doses of N_2^+ ions is shown in Fig.1. At the initial stage of irradiation an increase in ion current density is accompanied by assimetry disappearence and diffusive blurring of the profiles. These effects are due to temperature growth with ion current density. High implantation doses couse the profiles shape transformation from a Gaussian like to trapecoidal one after the implantation temperature reached 800°C. The constant concentration region is formed from the maximum area exclusively towards the surface. Simultaneously a smoll general profile shift is observed, which is also directed towards the surface. With the implantation temperature 1000°C nitrogen saturation takes plase at a somewhat smaller concentration while the profile shift is slightly

----------------1) Inst.Appl.Phys.Problems,Minsk-64,Kurchatov str.7,USSR

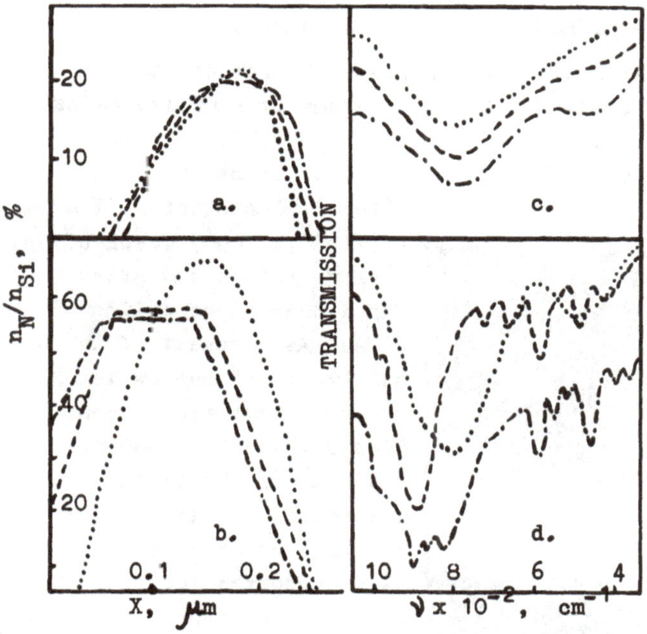

Fig.1. The depth distribution of N_2^+ implanted in Si with doses a)-1.5x10^{17}, b)-9.5x10^{17} and correcponding IR spectra.

greater. The distribution shift towards the surface may be due to a more effective sputtering of silicon as the beam power increase. The transformation of distribution is caused by simultaneous action of high temperature and effective generation of nonequilibrium point defects. The direction of understoichiometric atoms of nitrogen redistribution is determined by an increased concentration of mobile defects in the area of the maximum elastic energy deposition.

The same figure shows the IR-transmission spectra of this samples. The spectra obtained for the samples irradiated with the lowest doses of nitrogen have a wide absorption band with the maximum of 800 cm^{-1}, which corresponds to the oscillation energy of Si-N bonds. As the ion current density increases, the absorption on the N-Si-N bonds also increases. Therefore one can say that there exist a more effective generation of stoichiometric inclusions Si_3N_4 .The spectra obtained for samples irradiated with the highest doses at implantation temperature of 800 and 1000^oC have a lineage structure, characteristic for ß-Si_3N_4 /5/.Thus the formation of buried layer of stoichiometric composition Si_3N_4 is accompanied by its crystallization into ß phase.

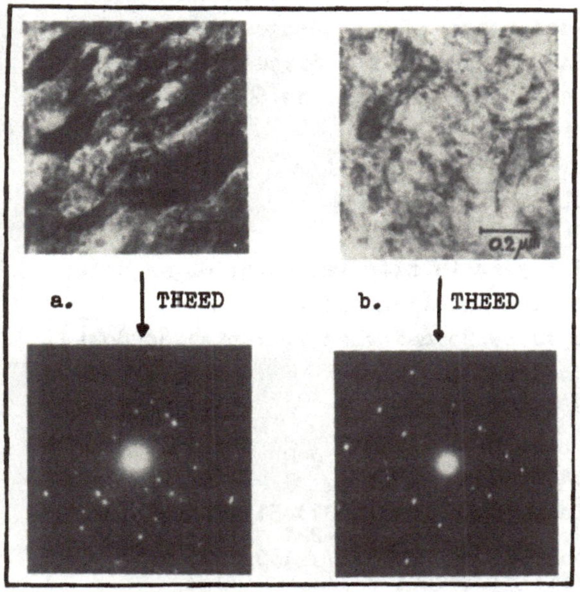

Fig.2 Evolution of nitride structure

Fig.2 shows the evolution of the nitride layer structure formation. Near the back interface (Fig 3a.) the buried layer has a clearly defined like-block structure, disappearing sa the layer growth. Near the front boundary the ß-Si_3N_4 layer is epitaxial. The annealing of these samples does not cause a significant redestribution of implanted nitrogen atoms or a modification changes in buried ß-Si_3N_4 layers.

Fig.3a shows the tipical distribution of nitrogen atoms in silicon after implantation medium (~$7x10^{17}$ at/cm^2) doses of N_2^+ ions before and after rapid thermal annealing (RTA). As a result of thermal treatment narrow layer of stoichiometric composition Si_3N_4 are formed in the area of the implant distribution maximum. One can also observe an increased consentration near the surface, associated with nitrogen tending to migrate along the extended defects. It follows from fig.3b,c that the layers being formed under such conditions are crystallized into ⍺-modification. The tipical structure of these layers is also represented on fig. 3c. Thas, an effective redestribution of implanted atoms accompanied by a change in byried layer modifications is observed after RTA of the samples, irradiated near the stoichiometry threshhold with implantation temperatures 600...1000°C.

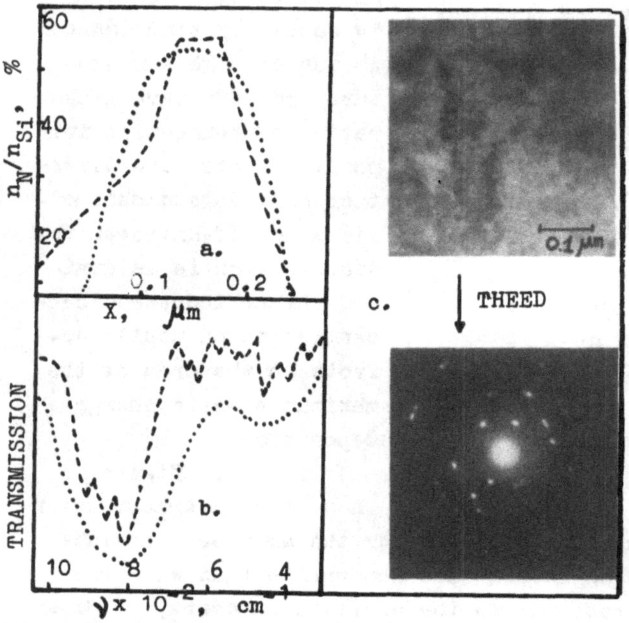

Fig.3 Formation of Si_3N_4 layers by RTA

CONCLUSIONS

1. The formation of ß-Si_3N_4 buried layers in sity implantation is determined by an ion current density and implantation dose.
2. RTA enables thin ⍺-Si_3N_4 layers to be formed when implantation dose is about the stoichiometric threshhold.

REFERENCES

/1/. S.P.Wong and M.C.Poon: Nucl.Instr.Meth. B22(1987) 513-519
/2/. P.L.F.Hemment: Mat.Soc.Symp.Proc. 53(1986) 207-221
/3/. D.H.Douglas-Hamilton: Nucl.Instr.Meth. B21(1987) 158 162
/4/. V.A.Labunov and V.V.Gribkovskii: El.Techn. 2/3(1984) 73-77
/5/. Yu.N.Volgin and Yu.I.Ukhanov: Optics and Spectroscopy 38(1975)727-31

THE ROLE OF ADDITIONAL NITROGEN TRAPS IN ION SYNTHESIS OF SILICON NITRIDE BURIED LAYERS.

G.A.Kachurin, I.E.Tyschenko, V.P.Popov, A.E.Plotnikov, S.A.Tija

1. At present to create the SOI structures with the buried dielectric layers the doses of N^+ or O^+ ions are raised up to $\sim(1-2)\cdot10^{18}$ cm^{-2}. This causes excessively long implantation times and requires special high-current implanters. Long bombardment even at usual target temperatures $T_i \simeq 500-600°C$ strongly damages the upper silicon layer. To obtain the layers SiO_2 or Si_3N_4 of practically acceptable thiknesses of 10-30 nm the above mentioned doses are not necessary. Their application results from the necessity to get in the maximum of the impurity distribution the critical concentration, which provides the formation of a continuous dielectric layer during annealing.

2. The present paper deals with two problems. The first one is to study the possibility of dielectric ion synthesis at the elevated (up to 1000-1100°C) T_i, which should suppress the accumulation of defects in the top Si layer. The second one is to stimulate the precipitation of dielectric phase by creating the additional sinks for implanted atoms.

3. N^+ ions with E=135 keV were implanted into Si at the current density $j \simeq 2-5$ $\mu A/cm^2$. The samples were heated independently up to T_i=600-1100°C. The implanted structure properties were controlled by the Rutherford backscattering (RBS) of He^+ ions with E=1.5 MeV, layer-by-layer Auger electron spectroscopy and cross-sectional electron microscopy.

4. Fig.1 shows that temperature rise from T_i=900°C to T_i=1000°C stops nitrogen to accumulate in the buried layer and it sinks to the surface /2,3/. It was found previously, that $T_i \simeq 900°C$ is the limiting one for introduction into Si the residual damage by light ions /1/. Basing on these data, an assumption was made, that the defects serve as sinks for nitrogen, and their creation would stimulate the formation of nitride phase. According to /1/, havier ions can create damage in Si at T_i 900°C, which have to provide phase formation at T_i 900°C. A test experiment was performed for N_2^+ ions at T_i=1000°C. It was supposed, that a partial overlap of two spikes from fragments of N_2^+ molecule would provide the local density of displacements, sufficient for the formation of stable damage. Though, fig.1 shows that this failed to happen.

5. Nitrogen sinks stable to high T_i were obtained by preliminary implantation 10^{16} cm^{-2} ions Ar^+ (Fig.2). The layer-by-layer Auger electron spectroscopy shows that in this case nitrogen accumulates in

---

Institute of Semiconductor Physics, Ac. Sci. USSR, Siberian Branch.
630090 Novosibirsk, USSR.

the depth at T_i=1000°C and partially even at T_i=1100°C. Maximum distribution of nitrogen corresponds to the mean range of N^+ ($R_p \simeq 320$ nm), but not to the maximum of elastic losses of Ar^+ ions ($R_d \simeq 100$ nm). At the depth 300–400 nm the elastic losses of Ar^+ are less than 10^{-2} of their maximal value, but it is necessary also to take into account the defect formation in depths by the dechannelling Ar^+ions. Channelling for hot implantation is rather real up to great doses ($R_{max} \simeq 800$ nm).

6. The other way, based on results /1-3/, is to form the nitrogen traps by initial implantation of small dose N^+ at $T_i < 900$°C. In this case light ions can introduce gross structural defects, so that nitrogen atoms will have additional sinks during further high temperature ($T_i > 900$°C) implantation of the main dose. The results of such an experiment are presented in figs. 3 and 4. They confirm the formation of buried nitride layer and high perfection of top Si structure. In the aligned RBS spectrum the scattering from the top Si layer is close to the scattering from the virgin Si, and electron microscopy fails to find dislocations in the cut-off region.

7. The limits were studied, in which one can vary the dose and T_i of the preliminary N^+ implantation. If the dose is left at the level $1 \cdot 10^{17}$ cm^{-2}, and T_i is elevated up to 850°C, the additional traps are introduced and the possibility preserves to form buried nitride layer by implantation of the main dose at T_i=1000°C (Fig.5). But after preliminary N^+ implantation at T_i=900°C nitrogen capture was found to be not sufficient, so a part of successively implanted at T_i=1000°C nitrogen sinked to the surface. If T_i=600°C, and preliminary dose is decreased to 10^{16} cm^{-2}, then the efficiency of nitrogen capture decreases even more (Fig.5). It is clear, that for the certain concentration of nitride nuclei to form not only traps (lattice imperfections) are necessary, but also the sufficient quantity of nitrogen atoms. Otherwise the nuclei do not gain the critical size and dissociate under high temperature treatment.

8. Conclusions. Rise of the temperature of nitrogen implantation up to T_i=900°C allows one to diminish essentially accumulation of defects in the cut-off silicon layer. At $T_i > 900$°C implanted nitrogen does not precipitate in the buried layer and sinks to the surface. Lattice imperfections, caused by ion bombardment, serve as nitrogen traps and as nitride phase nuclei. Stable up to T_i=1000°C traps can be introduced either by heavy ion bombardment or by preliminary implantation of small dose (10^{16}–10^{17} cm^{-2}) of N^+ ions at $T_i < 900$°C. The studies of other possible sinks (as δ-layers or interfaces) relatively to their ability to accumulate nitrogen atoms at high temperatures are of obvious interest. These studies being successful, one would have an opportunity to synthesize very thin layers of buried dielectric by implantation of small ion doses.

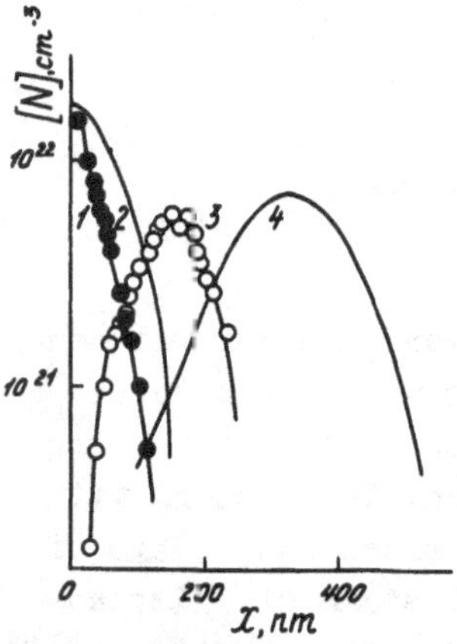

Fig.1. Nitrogen depth profiles after implantation N_2^+(1,3-Auger and N^+(2,4-RBS data) at the target temperatures T_i=1000°C (1,2) and 900°C(3,4). U=135 kV, j≃2-5uA/cm².

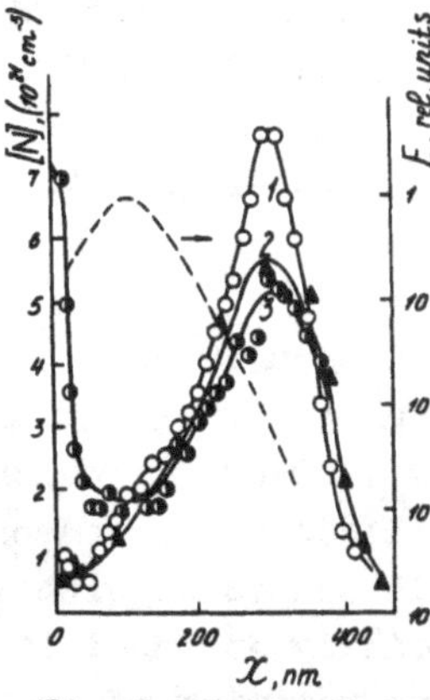

Fig.2. Depth distributions of nitrogen (Auger data) implanted at different T_i after preimplantation of Ar^+ ions, 145 keV, 10^{16}cm^{-2}, T_i=800°C. Nitrogen implantation temperatures,T_i,°C: 1-900, 2-1000, 3-1100 Dashed line-elastic losses of Ar^+ ions.

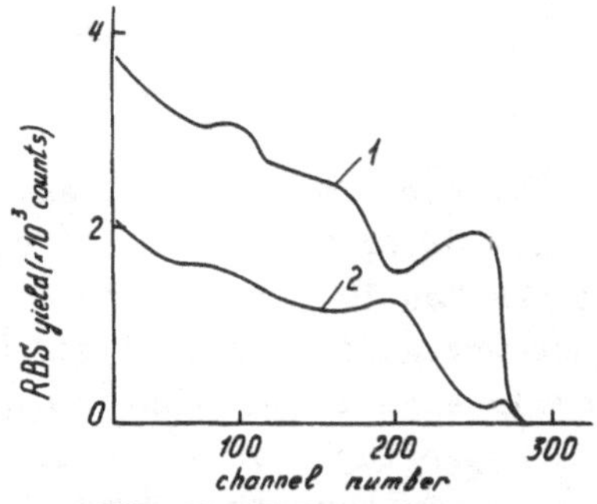

Fig.3. RBS spectra (He^+, 1.5 MeV) of the sample implanted with nitrogen 1·10^{17}cm^{-2}, E=135 keV at T_i=600°C followed by 7·10^{17}cm^{-2} at T_i=1000°C. 1-random, 2-aligned.

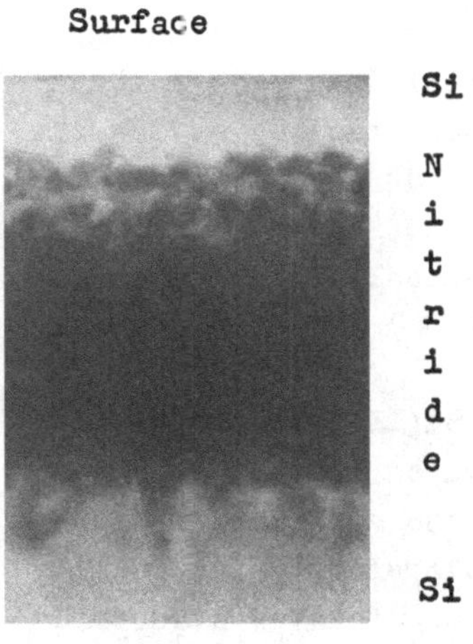

Fig.4. TEM cross section of the sample implanted as indicated Fig.3.

100nm

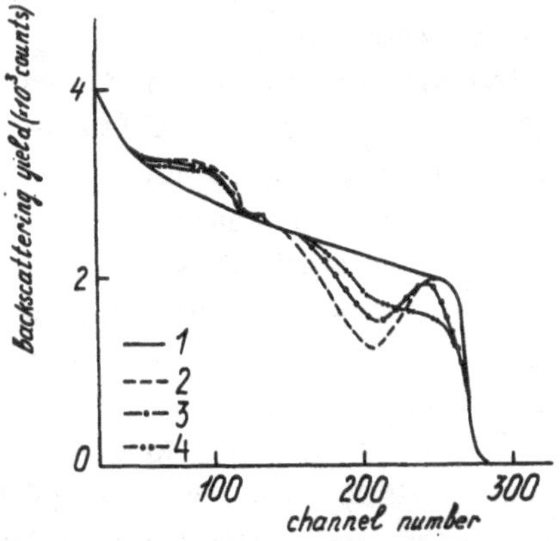

Fig.5.RBS random spectra (He^+, 1.5 MeV) of nonimplanted silicon (1) and from the samples implanted with ions (E=135 keV) in two steps. First: 2-10^{17}cm^{-2}, T_i=850°C; 3-10^{17}cm^{-2}, T_i=900°C; 4-10^{16}cm^{-2}, T_i=600°C. Then ~8·10^{17}cm^{-2} at T_i=1000°C in all three cases.

(1) Kachurin G.A., Tyschenko I.E., and Fedina L.I.: Poverkhnost. 7(1989)
(2) Kachurin G.A., Tyschenko I.E., Popov V.P., and Tijs S.A.: Phys.Stat. Sol.(a). 113(1989)
(3) Kachurin G.A., Tyschenko I.E., Tijs S.A., Plotnikov A.E., and Popov V.P.: Fiz.Tekhn.Poluprov. 23(1989)434–438.

SYNTHESIS OF Si_3N_4 BURIED LAYERS AT LOW DENSITIES OF ION CURRENT AND INDEPENDENT HEATING OF SILICON TARGETS.

G.A.Kachurin, I.E.Tyschenko, A.E.Plotnikov, V.P.Popov.

1. For the ion synthesis of dielectric buried layers the current densities O^+ or N^+ of some tens $\mu A/cm^2$ are traditionally used. This leads to the heating of targets up to T_i=400-600°C, which is necessary for the monocrystalline upper layer of silicon to be conserved. The heating with the intensive ion beam though is connected with an essential damage of the upper layer of silicon, especially at the initial stage of implantation. To the other hand, the productivity of an implanter is determined by total current of ions, but not by the current density.

2. For the SOI structures to be finally formed, the high-temperature anneal is required: about 1200°C after implantation N^+ and about 1400°C after O^+. The synthesized layers of SiO_2 are always amorphous, and the layers of Si_3N_4 crystallize as a result of anneal. The formation of polycrystalline structure makes the electrophysical characteristics of the Si_3N_4 layers worse, though as to many other properties the nitride buried layers are preferable to the oxide ones. In both cases, after the anneal in the top Si layer the high density of dislocations is preserved, i.e. up to $\sim 10^9$ cm^{-2}.

3. The present paper considers the formation of buried nitride layers at the N^+ ion current densities $\sim$3-5 $\mu A/cm^2$ and at independent initial heating of samples up to 1000°C. It was supposed that the use of low ion current densities and an external heating of the substrates during implantation make it possible to improve the structure of the upper Si layer and to diminish the temperature of additional anneal. The studies were carried with transmission electron microscopy (TEM) and Rutherford backscattering (RBS).

4. N^+ ions with E=135 keV were implanted into Si(III) at temperatures T_i= 600, 700, 750, 850, and 900°C. It was shown earlier, that at higher T_i nitrogen sinks from the buried layer to the surface /1,2/. According to the RBS data, the increase of T_i in temperature interval 600-900°C causes an essential improvement of the top Si layer structure (Fig.1). The structure of the nitride buried layer essentially changes, too (Fig.2). After T_i=600°C a homogeneous finegrained structure can be seen, which causes the difraction corresponding to the amorphous state. With the increase of T_i dendritic excrescens are seen, whose density increases with the dose. After dose $5\cdot 10^{16}$ cm^{-2} (T_i=900°C) the nuclei of nitride phase are easily observed. Microdifraction showed that these are cristallites of α-Si_3N_4. The cross-sectional microscopy for T_i=850°C

Institute of Semiconductor Physics, Ac. Sci. USSR, Siberian Branch, 630090 Novosibirsk, USSR.

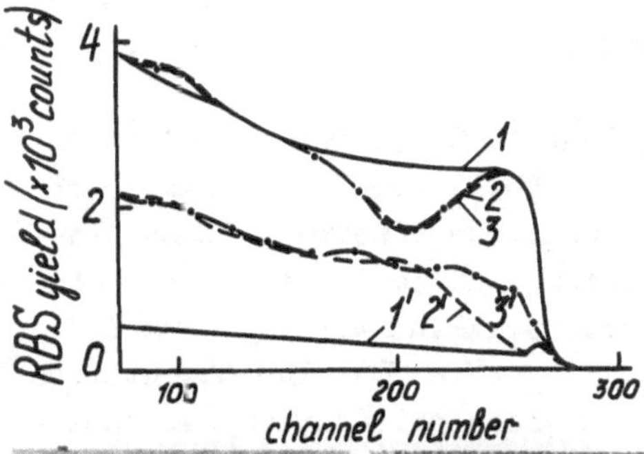

Fig.1. RBS spectra (He^+, 1.5 MeV) of silicon samples implanted at 135-keV and substrate temperatures T_i=600°C(2,2') and T_i=850°C(3,3'). (1,1')-virgin Si. 1,2,3-random, 1', 2', 3'-aligned.

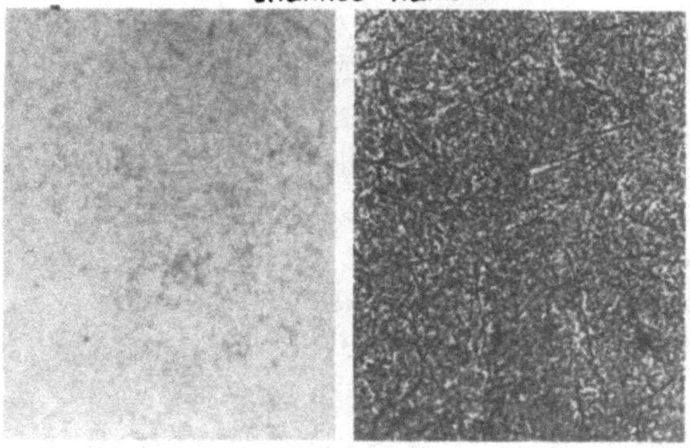

Fig.2. TEM micrographs of silicon samples implanted with nitrogen dose $\sim 8 \cdot 10^{17} cm^{-2}$ at 135 keV and substrate temperatures T_i=600°C(a) and 900°C(b).

1 μm

a b

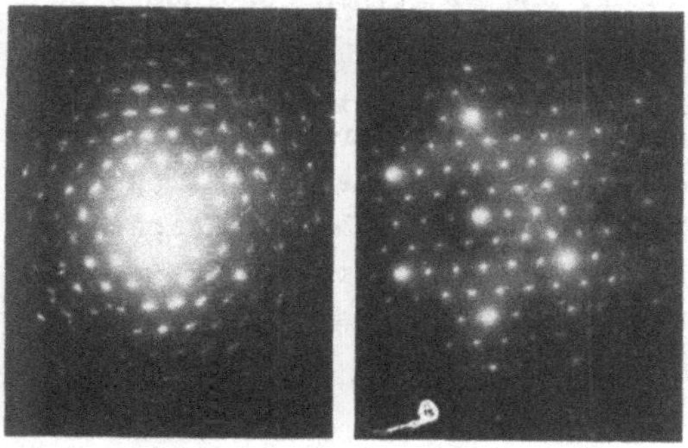

Fig.3. Crystallization of buried Si_3N_4 layers resulting from annealing 1100°C (a-3 hours, b-5 hours) of the samples implanted at T_i=600 °C.

a b

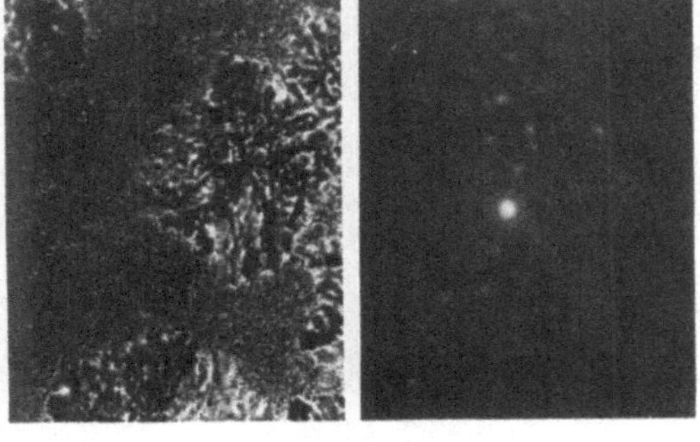

Fig.4. Crystallization of buried Si_3N_4 layers resulting from annealing of the layers implanted at T_i=850°C (a) and 900°C(b). Annealing regime: a-3 h at 1100°C, b-1 h at 1200°C.

1 μm

a b

was made. It is found, that the upper layer of Si contains defects, though according to RBS (Fig.1) and to reflection diffraction, the Si surface is sufficiently perfect.

5. The implanted samples were annealed at 1100-1200°C for 1-3 hours in nitrogen. If T_i=600°C, then after anneal 1100°C for 1 hour large nitride grains appear, and after 3 hours a continuous monocrystalline layer α-Si_3N_4 forms (Fig.3a). The transmission diffraction after 6-hour annealing gives clear reflections of monocrystalline top Si layer and of continuous monocrystalline layer α-Si_3N_4 (Fig.3b). The orientation of α-Si_3N_4 layer is correlated with the silicon orientation. An analogous result was obtained after implantation at T_i=700-750°C with additional annealing 1100°C x 3 hours. But implantations at T_i=850°C and 900°C retard the formation of continuous monocrystalline α-Si_3N_4 layer. In this case annealing 1100°C x 3 hours resulted the formation of crystalline spherolitic grains α-Si_3N_4, surrounded with the amorphous matrix (Fig.4a). It is only annealing 1200°C x 1 hour that causes the formation of the continuous layer of monocrystalline α-Si_3N_4 in such samples (Fig.4b). At the same time, the annealing improves the quality of the top Si layer.

6. Conclusions. The use of low densities of ion current and of independent heating of targets influences the formation of SOI structures in two ways. First, it prevents the accumulation of the defects in silicon. The improvent of silicon matrix allows one to obtain continuous buried monocrystalline matrix oriented layers of α-Si_3N_4 at rather moderate additional anneal (1100°C x 3 hours). Second, at $T_i \geqslant$800°C the ion bombardment already on initial stages stimulates the formation of numerous α-Si_3N_4 crystalline nuclei, which serve as sinks for nitrogen atoms and as centers of crystalline grain growth. For the total monocrystallization of a buried layer in this case the annealing 1200°C is required. The optimum T_i is in the interval 600-800°C.

---

(1) Kachurin G.A., Tyschenko I.E., Popov V.P., and Tijs S.A.: Phys. Stat. Sol.(a) 113(1989)

(2) Kachurin G.A., Tyschenko I.E., Popov V.P., Tijs S.A., and Plotnikov A.E.: Fiz. Tekhn. Poluprov. 23(1989)434-438.

OPTIMIZED SOI-PROCESSING BY ION BEAM SYNTHESIS OF BURIED SILICON OXYNITRIDE IN SILICON

W. Skorupa[1] M. Voelskow[1] R. Grötzschel[1] J. Albrecht[1]

Jan Vanhellemont[2] Ann De Veirman[3]

INTENTION

At present, ion beam synthesis is accepted to be a leading candidate for the formation of SOI(silicon-on-insulator)-structures. Trying to combine the advantegeous properties of silicon nitride (diffusion inhibition, gettering) and silicon dioxide (better electrical properties), we investigated the formation and properties of buried silicon oxynitride layers to perform a SOI-oriented "Insulator Engineering". Generally, this includes the formation of a silicon layer free of crystal defects and unwanted impurities supported by a insulating layer which is neighboured by interfaces with microstructure and electrical properties adequate to a selected application.

EXPERIMENTAL

$^{14}N^+$-and $^{16}O^+$-ions were implanted with an energy of 330 keV into the central (2x2)cm^2 part of 2 inch silicon wafers (n-type, 3-5 Ohm cm, <100>) coated with a thermally grown silicon dioxide layer of 108 nm thickness. The compound layers were formed by a totally implanted dose of $1.2 \times 10^{18} cm^{-2}$. To form silicon oxynitride this dose was implanted with two different N:O-ratios: 1:1 and 3:1. For both ratios, the implantation sequence of the two elements was also changed joining a recent study of silicon oxynitride formation by ion beam synthesis at 60 keV /1/. External target heating during the implantation and post implantation annealing (PIA) by halogen lamps were performed at 500°C and at 1380°C/15 min/air, respectively. For the purpose of surface protection the wafers were covered before PIA with a silicon dioxide layer of 500 nm thickness by chemical vapour deposition at 400°C.

Rutherford backscattering and channeling techniques with 1.7 MeV $^4He^+$ ions were used to assess the atomic profile and the radiation damage of the silicon top layer, respectively. Auger electron spectroscopy (AES) and spreading resistance profiling (SRP) were used to get chemical information and the electrical carrier profile, respectively. To study the microstructure, Transmission electron microscopy at cross sectional samples (XTEM), High resolution electron microscopy (HREM) and Spectroscopic ellipsometry (SE) were applied. The simulation of the SE spectra based on experience already got at SIMOX samples by a 5-layer model /2/. In a companion paper the results of a detailed Infrared spectroscopy study of the samples described above are reported /3/.

PRELIMINARY RESULTS

An overview on all RBS-and channeling spectra is given in Fig.1. In each spectrum the random and aligned spectra of unimplanted silicon are also indicated. The dip of the random spectra in the channel region (CHR) 135-235 points to the silicon depletion because of the implanted impurity distribution.

The as-implanted state

Gaussian profile shape for all cases (Fig.1,a,c,e,g);low damage near to the surface; broad defect band of dislocation loops and polycrystalline silicon joining the buried layer (see also Fig.2); (113)-defects typical for Si-interstitials on the bottom side of the buried layer for all four layer types (HREM-study); buried layers generally amorphous; contrast

[1]Central Inst. for Nuclear Research, PF 19,DDR-8051 Dresden, GDR
[2]IMEC Leuven,Kapeldreef 75, B-3030 Leuven, Belgium
[3]RUCA, Groenenborgerlaan 171, B-2020 Antwerpen, Belgium

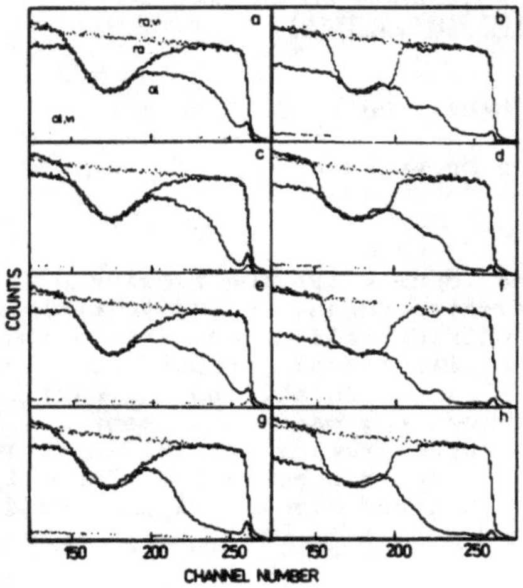

Fig.1: RBS-spectra after implantation (a,c,e,g) and annealing (b,d,f,h) for N:O=3:1(a,b),1:1(e,f)-and O:N=1:3(c,d), 1:1(g,h)-implantation. The total dose is $1.2\times10^{18}cm^{-2}$.

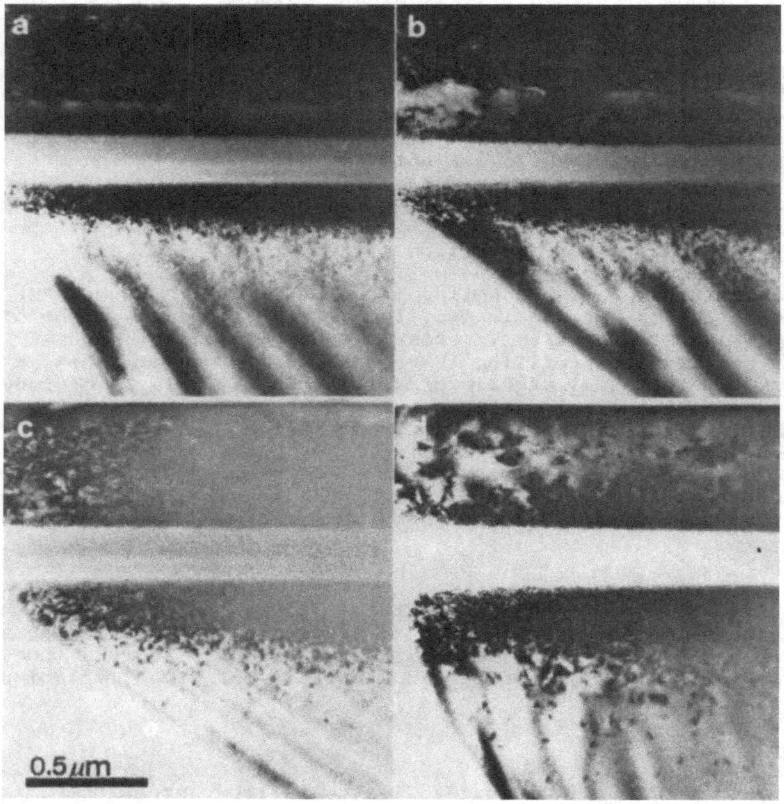

Fig.2: XTEM-micrographs after implantation for N:O=3:1(a), 1:1(b)- and O:N=1:3(c), 1:1(d)-implantation. Total dose: $1.2\times10^{18}cm^{-2}$

---

change due to bubbles within these layers for the N:O=3:1-implanted layers (Fig.2a,c);

The annealed state

Flat topped profile shape with steeper tails; no complete redistribution as in the case of nitrogen implantation /4/; for N-O implantation (Fig.1,b,f) small dips (CHR 220-230) due to impurities (O/N) in the silicon top layer (CHR 200-265) are correlated with defect bands (Fig.3,a,b); for O-N implantation such regions rich in impurities join immediately the buried layer (CHR 200-220) (Fig.3,c,d); distinct defect bands visible as indicated in the channeling spectra (Fig.1b,d,f,h) and the XTEM-micrographs (Fig.3); the lowest residual damage due to XTEM for the N-O=1:1-implantation; generally precipitates of silicon nitride below the buried layer for N-O-implantation (only visible in Fig.3a), for the case of N-O=3:1-implantation a narrow defect band of stacking fault tetrahedra immediately below the buried layer (Fig.3a);

A TEM/HREM study concerning special defect configurations of the samples reported here will be published elsewhere /5/.

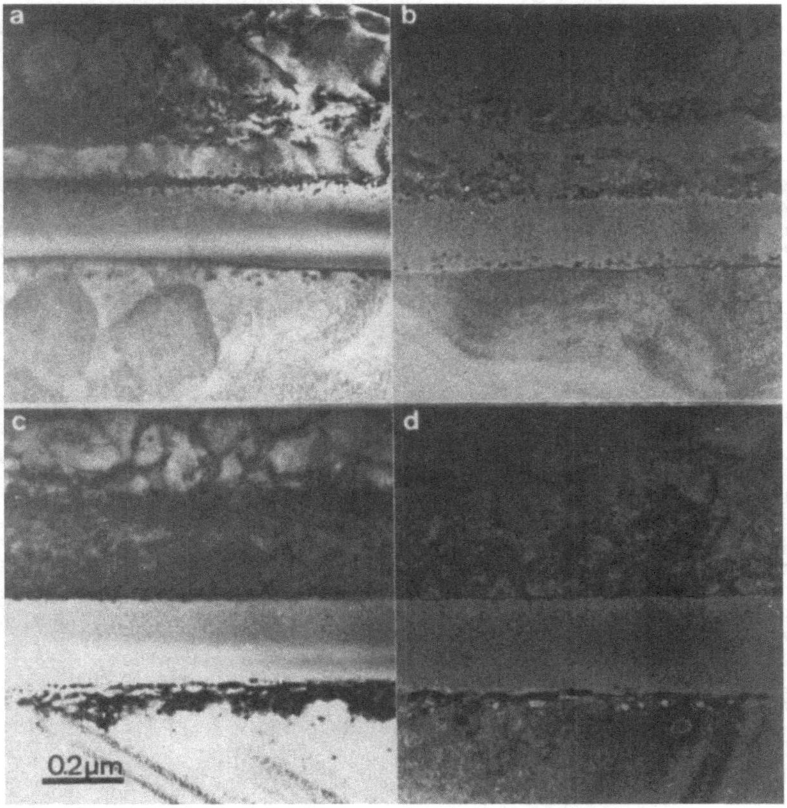

Fig.3: The same as in Fig.2, but after annealing.

Table 1: The thickness values (nm) of the silicon top layer and the buried compound layer evaluated by XTEM, SE and SRP after annealing:

Implantation parameters	Silicon top layer XTEM	SE	SRP	Buried compound layer XTEM	SE	SRP
N:O=3:1	440	457	450	180-210	234	200
N:O=1:1	480	478	460	170	265	260
O:N=1:3	485	455	460	220	243	280
O:N=1:1	460	462	460	230	301	250

Especially, the SE results for the silicon layer correspond very well to that of the other methods. The buried layer thicknesses differ in some cases although a three layer model was used to simulate the buried compound layer /6/. It was remarkable that the best simulation for both the 3:1-implanted layer types was got without assuming any silicon dioxide content and a higher silicon nitride density.

Discussion and conclusions

1. The best microstructure was reached for the N:O=3:1-implantation. This can be explained by the higher stopping power of silicon rich in silicon nitride due to the higher density of this material (2.9-3.18 g/cm^3) as compared to that of silicon (2.33 g/cm^3) and silicon dioxide (2.25 g/cm^3). This leads to a reduced straggling of the secondly implanted oxygen and explains that by Infrared spectroscopy the highest content of a silicon oxynitride compound was found /3/. Likewise, this explains the formation of silicon nitride precipitates below the buried layer only for the N:O-implantation sequence.

2. The annealing energy used here is not sufficient to fully dissolve all precipitates from the surrounding silicon into the buried compound layer. This was of importance for a better study of the process, but must be further optimised for application purposes. On the other hand, this underlines the fact that no infomation exists concerning the properties of silicon oxynitride precipitates in silicon.

3. The necessity for assuming a high silicon nitride density to simulate the SE spectra supports recent findings that only the use of such higher values delivered quite good results in calculating the nitrogen profile by a theoretical model /7/. Likewise, the results of step height measurements by Talystep of volume swelling after nitrogen implantation could only be explained by the formation of silicon nitride with a higher density /8/.

REFERENCES

/1/ W.Skorupa, K.Wollschläger, R.Grötzschel, J.Schöneich, E.Hentschel, R.Kotte, F.Stary, H.Bartsch, G.Götz: Nucl.Instr.Meth.B32(1989)440

/2/ J.Vanhellemont, H.E.Maes, A.De Veirman: J.Appl.Phys. 65(1989)4454

/3/ K.Wollschläger, W.Skorupa, U.Kreissig: these proceedings

/4/ W.Skorupa, K.Wollschläger, U.Kreissig, R.Grötzschel, H.Bartsch: Nucl.Instr.Meth.B19/20(1987)285

/5/ A.De Veirman, J.Van Landuyt, Jan Vanhellemont, W.Skorupa: to be published

/6/ A more detailed SE study will be published by J.Vanhellemont, H.E.Maes, A.De Veirman, J.Van Landuyt, W.Skorupa, M.Voelskow at the Spring Mtg. of the Electrochem.Soc.,Montreal, May 1990

/7/ E.Sobeslavsky, W.Skorupa: phys.stat.sol.(a) in press

/8/ E.Hensel and W.Skorupa: Thesis, AdW der DDR, 1985

STRUCTURAL CHARACTERIZATION OF THICK ZONE-MELTED SOI-LAYERS

K. Hoeppner, B. Tillack, R. Banisch, H.H. Richter[1]

In the last years substantial efforts have been directed toward development of the zone melting recrystallization as a technology which enables to produce large-area silicon-on-insulator (SOI) layers with defined orientation and good crystallographic quality.
Beside the generation, characterization, and application of thin SOI-films (thickness up to 4 um) thick SOI-films are of special interest because of the possibility to get an alternative to the conventional dielectric isolation technology for high voltage and power devices.
The generation of thick SOI-layers zone melted by a halogen lamp strip heater is investigated with the aim to get informations about the crystallization mechanism.
The results of the crystallographic characterization by means of etch pit technique and defect etching demonstrate that the seeding recrystallization (Fig. 1) of thick poly-Si-films results in single-crystalline Si-films containing an area free of subgrain boundaries (SGB`s) near the seed.

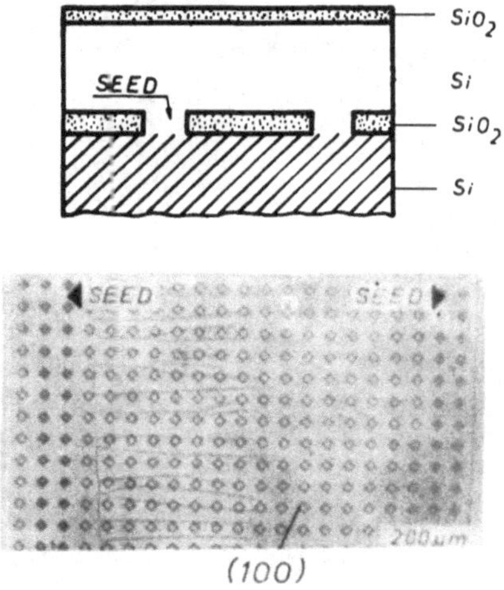

Fig. 1

Schematic illustration of sample structure for seeding recrystallization.
In the optical micrograph the quadratic etch-pites and their uniform arrangement indicate a single crystalline film with (100)-orientation.

[1] Institut fuer Halbleiterphysik der Akademie der Wissenschaften der DDR, Frankfurt (O), Postfach 409, DDR-1200 Frankfurt (O), German Democratic Republic

The extent of this area increases with increasing film thickness /1/ and in a certain distance to the seed the subgrain boundaries set in.

By detailled TEM-investigations of thick films with (100)-orientation and <011>-scan direction it was established that the area free of SGB`s contains dislocation arrangements with characteristic properties. In the region 1 near the seed very long edge-dislocations lie parallel to the surface (length up to 25 um, b = <110>). With increasing of the distance to the seed the kind of dislocations in the regions 2 changes and their density increases. Inclined dislocations are identified as screw dislocations. The subgrain boundaries in region 3 origin from this dislocation arrangement. Similar results for regions 1 and 2 have been founded for the recrystallization of thin SOI-layers without seed /2/. It was found that all subgrain boundaries rotation vectors 0 are mainly parallel to the <011>-scan direction and that they exhibit "pure tilt boundaries" or "partial twist boundaries" with a high portion of tilting. During the first 10 um the tilt angles measured with the displacement of the Kikuchi-poles in the diffraction patterns beside the SGB are smaller then 0,2° and increase later up to 2°.
From the characteristic properties of the arrangement along the scan direction in the region 1 - 3 a direct connection to the decreasing of the thermal gradient with increasing distance to the seed can be concluded /1/. Furthermore, other investigations planed in future will indicate the influence of the substrate orientation ((100), (111)), scan direction (<010>, <011>, <112>) and layer thickness on the defect structure of SOI-layers caused by the thermal gradient at the liquid/solid interface.

References

/1/ B. Tillack, R. Banisch, H. H. Richter, and K. Hoeppner, Applied Surface Science 36, (1989) 579-587

/2/ L. Pfeiffer, A. E. Gelman, K. A. Jackson, and K. W. West, Mat. Res. Soc. Symp. Vol. 74

THE INVESTIGATION OF ION BEAM SYNTHESISED SILICON OXYNITRIDES BY IR-SPECTROSCOPY

K.Wollschläger,W.Skorupa,U.Kreißig [1])

The formation of buried oxynitride layers with improved physical properties is possible by changing the chemical composition. In addition to traditional analytical methods such as RBS, ELMI, AES (4) the infrared spectroscopy (IR) is able to give hints concerning the compound synthesis.It is a non-destructive method.

<u>Experimental</u>

O^+, N^+ ion implantation energy		330 keV
Implantation temperature		500°C
Orientation of the substrates		<100> Si
Dose (10^{17} cm^{-2})	6 N^+;6 O^+ respectively 9 N^+; 3 O^+	

<u>Superposition of idealized IR-bands</u>

In the IR-spectra of buried insulator layers appear interference maxima and minima. Firstly a mathematical elimination of this effect is necessary. In dependence on the implantation sequence more or less broad oxynitride bands are observed. For a semi quantitative description of differences the spectrum may be viewed as a sum of the components SiO_x, $Si_3O_3N_2$, Si_2ON_2 and SiN_y. The corresponding band positions of stoichiometric oxynitrides are situated between those of the pure components SiO_2 and Si_3N_4 (3) - see Fig. 1a,b. (SiO_x is the sum of SiO, Si_2O_3, and SiO_2). The dottet line in Fig. 1 is the accounted sum of the components. In the vicinity of the maximum of the measured spectrum the sum curve does not exceed it. In this way other phases - for example crystalline nitride - can be seen, which disturb the concept of compound superposition.

<u>Evaluation of the concentration profiles</u>

The model for a rough estimation is simplified compared to other one, which describe the oxygen respectively nitrogen implantation alone (2,5). The diffusion broadening is neglected, because the contrary process - the redistribution by segregation dominates at T >500 °C.

Constants used in fig. 2 for	oxygen and	nitrogen
projected range	620 nm (6)	710 nm (1)
straggling	100 nm	100 nm
average volume (10^{-23} cm^3)	1.22	0.33
sputter yield	0.15	0.2

<u>Discussion</u>

The thickness of the crystalline silicon overlayer (as impl.) is 550nm corresponding to the value derived from electron microscopy. It can be calculated from the distance of the interference maxima in the IR-spectra using a refractive index of 3.4 .

---

[1]) Central Institute for Nuclear Research Rossendorf, Academy of Sciences of GDR; PF 19, DDR-8051 Dresden ; GDR

Implantation sequence N^+; O^+:

In the estimated concentration profile (Fig. 2) the nitrogen atoms are not as deep in the substrate as in the case of the other implantation sequence. This is caused by sputtering during the nitrogen and oxygen implantation. The implanted oxygen leads to volume swelling and broadens somewhat the nitrogen profile.

Already after the implantation the IR-spectrum shows a remarkable portion of a nitrogen rich oxynitride (see Fig. 3a). Pure components SiO_x and SiN_y originate in the wings of the concentration profile. In this region is one Si-O bond or one Si-N bond surrounded with excess silicon.

The probability for the formation of stoichiometric Si_3N_4 arises with a greater nitrogen dose. The high nitride density diminishes the projected range of subsequent implanted oxygen (1) and prevents the diffusion of it. Post-implantation annealing leads to partial crystallization of the nitride.

Implantation sequence O^+;N^+:

With analog arguments one can show, that the distance between the oxygen and nitrogen distribution is somewhat greater. The amount of formed oxynitride is smaller (Fig. 1b).- In order to inhibit the nitride crystallisation a pre-implantation of $3*10^{17}$ cm^{-2} oxygen at 500°C is sufficient because of the diffusion broadening of substoichiometric oxygen.

High temperature annealing:

Annealing at 1380°C brings the system near to the thermodynamic equilibrium. The oxynitride Si_2ON_2 seems to be a stable modification (Fig. 3). The apparent exception in Fig. 3c is due to a disturbation of the components concept by crystalline silicon nitride (it has one band at 950 cm^{-1}).

References

(1) Burenkov,A.F., F.F.Komarov, M.A.Kumakhov and M.M.Temkin : Prostranstvennye Raspredeleniya Energii Vydelenoi v Kaskade Atom. Stolknovennii v Tverd. Tel., Energoatomizdat, Moskva 1985

(2) Jäger,H.U., E.Hensel, U.Kreißig, W.Skorupa and E.Sobeslavsky Thin Solid Films 123 (1985) 159-169

(3) Rand,M.J., Roberts,J.F., J.Electrochem.Soc. 120 (1973) 446-453

(4) Skorupa,W., M.Voelskow, R.Grötzschel, J.Albrecht, J.Vanhellemont, A.De Veirman elsewhere in these proceedings

(5) Sobeslavsky,E., H.U.Jäger, U.Kreißig, W.Skorupa and K.Wollschläger, phys.stat.sol. (a) 105 (1988) 387-396

(6) Ziegler,J.F., J.P.Biersack and U.Littmark: The Stopping and Ranges of Ions in Solids, Pergamon Press, New York 1985

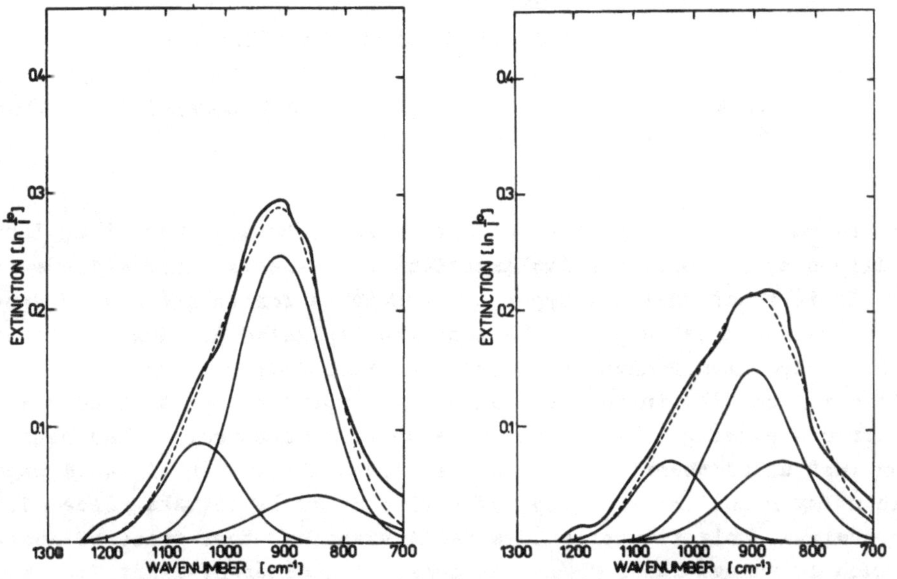

Fig.1 Superposition of idealized bands of IR-spectra (see Fig.3a,b)
a) N^+ ; O^+ b) O^+ ; N^+ (both after annealing)

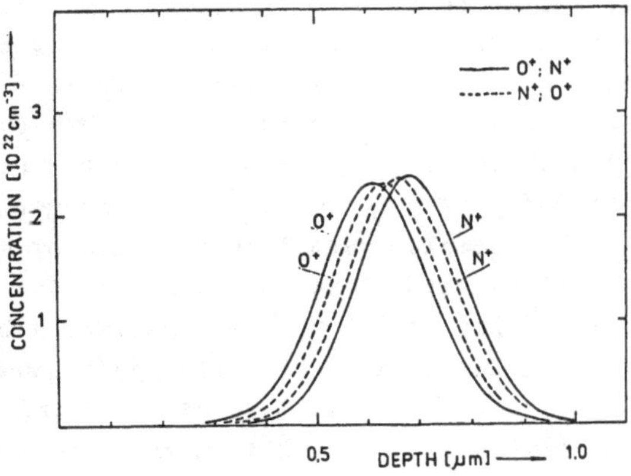

Fig. 2
Estimated concentration profiles for oxygen and nitrogen atoms
---- N^+;O^+
____ O^+;N^+
(as implanted)

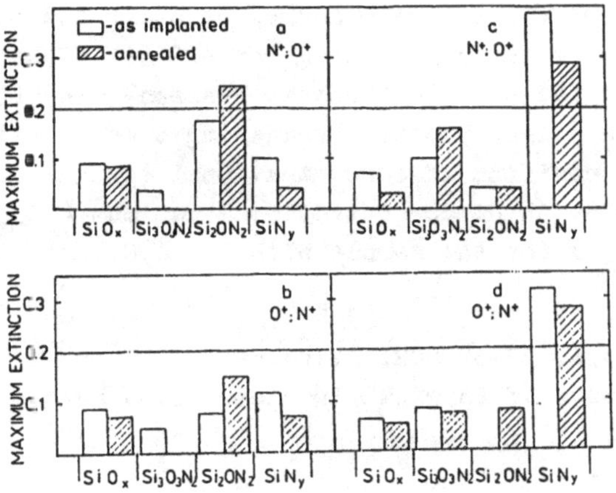

Fig. 3
Histogramm of maximum values of the IR-components

HIGH DOSE IMPLANTATION OF NITROGEN INTO SILICON: RBS AND NUCLEAR REACTIONS MEASUREMENTS

V.Y. Arbuzov, I.Y. Grande, V.P. Korobeinikov, T.M. Pyatkova, V.N. Baghaev[1)] and T.I. Starikova[2)]

In recent years the problem of synthesis of buried insulating layers in silicon by nitrogen ion implantation has attracted increasing attention. It is found that the types of structures formed are highly dependent on the sequence in which the ions are implanted and samples annealed. Numerous experiments have been performed with nitrogen implantation into silicon followed by furnace annealing at temperature up to 1200°C and for periods ranging from some minutes up to some hours. It has been shown that under these conditions the recrystallization of top damaged silicon layer and the formation of buried Si_3N_4 layer take place /1,2/. Nevertheless, only few works have dealt with the other types of annealling such as pulsed-laser /3/, heat pulse by incoherent light /4/, etc. In this connection it is of great interest to investigate scanning electron beam annealing which has the advantages of pulse treatment and, on the other hand, decreases the influence of beam inhomogenuity /5/.

The present communication deals with the results of the study of nitrogen profiles and implantation induced lattice damage in silicon as-implanted and after annealing by scanning electron beam. P-type (100) silicon wafers were implanted with 350 keV ^{15}N ions and at a current density of 0.01 mA/cm^2 to the doses of $0.6-4.0x10^{17}$ cm^{-2}. Target temperature did not exceed 200°C. The samples were then annealed by scanning electron beam (15 keV, 3 mA) and simultaneous heating of the back side of the plates by incoherent light to 750°C. The irradiation by electrons resulted in the cycle influence with specific power density of $1.0x10^3$ W/cm^2 and 1 ms exposition, the total time of annealing being some seconds. As a result, the plates were heated as a whole to about 900°C and thus temperature gradient which caused defects was excluded.

Nitrogen concentrational profiles were measured by using the resonant nuclear reaction $^{15}N(p,\alpha\gamma)^{12}C$ at a proton energy of 429 keV. It appears that in samples as-implanted the profile shape is Pearson-like (Fig.1). The dose increase leads to a slight profile broadening towards silicon implanted surface while the position of concentrational maximum remains constant. The values of the depth of concentrational maximum (R_{max}) and the profile width (ΔR_{max}) for the sample with D = $4.0x10^{17}$ cm^{-2} are shown in table 1.

[1)]Ural Polytechnical Institute, SU-620002, Sverdlovsk, USSR
[2)]Scientific Research Institute of Physical Problems, SU-103460 Moscow, USSR

With the help of nonresonance nuclear reaction $^{15}N(p,\alpha_0)^{12}C$ at a proton energy of 480 keV the measurements of the profiles were also carried out. The standard Si_3N_4 film of 200 nm thickness on silicon with natural content of ^{15}N was used as a reference. By comparison of α-particles spectrum for the sample implanted to $D = 4.0 \times 10^{17}$ cm^{-2} with the spectrum for standard film the values of R_{max}, ΔR_{max} and the maximum concentration (N_{max}) were obtained (see table 1). The advantage of using $^{15}N(p,\alpha_0)^{12}C$ as compared with $^{15}N(p,\alpha\gamma)^{12}C$ reaction consists in higher efficiency of α-particle registration by a detector. This makes it possible to analyze the lower concentration of ^{15}N in a natural mixture. However, to derive the real profile the special technique of deconvolution is needed /6/.

Channeling backscattering proton spectra for as-implanted samples show that silicon top layer remains monocrystalline but highly defective and reaches almost full amorphization at a depth of ~500 nm (Fig.2). This high damage of silicon is consistent with the low implantation temperature and dose range. The width of damaged region considerably exceeds the width of nitrogen concentrational profile. To get the quantitative information the simulation of RBS random spectra was performed. If the values R_{max} = 780 nm, ΔR_{max} = 280 nm and N_{max} = 1.9×10^{22} cm^{-3} (see table 1) are used in simulation, the best agreement of experimental and estimated spectra is observed (for $D = 4.0 \times 10^{17}$ cm^{-2}).

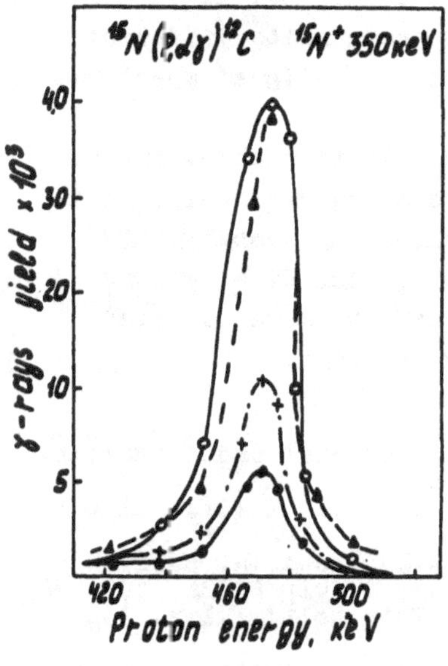

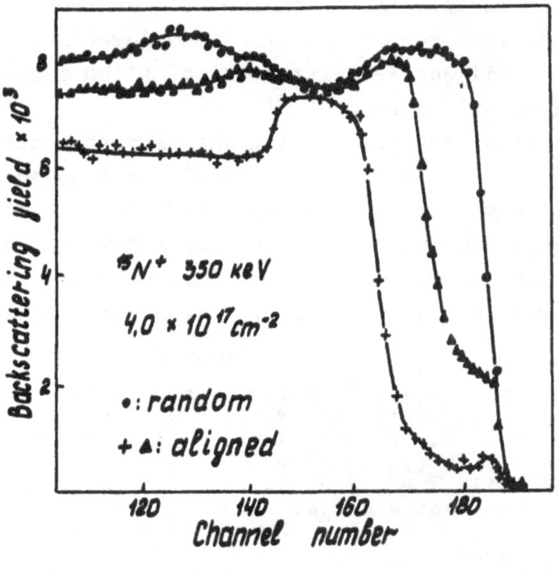

Fig.1. Nitrogen concentrational profiles obtained from silicon samples implanted at 350 keV $^{15}N^+$ to the doses of 0.6×10^{17} cm^{-2} (•), 1.0×10^{17} cm^{-2} (+) and 4.0×10^{17} cm^{-2} (○), as-implanted and after annealing by scanning electron beam ($D = 4.0 \times 10^{17}$ cm^{-2}) (▲).

Fig.2. RBS spectra (0.6 MeV, H^+) of silicon samples implanted at 350keV to the dose of 4.0×10^{17} cm^{-2}, as-implanted (•,▲) and after annealing by scanning electron beam (+).

Table 1

The parameters (R_{max}, ΔR_{max}, N_{max}) for silicon sample(350 keV,$4x10^{17} cm^{-2}$) obtained by nuclear reactions and RBS methods and the data of /7/.

Method	R_{max},nm	ΔR_{max},nm	N_{max},cm^{-3}
$^{15}N(p,\alpha\gamma)C^{12}$	800	330	-
$^{15}N(p,\alpha_0)C^{12}$	900	330	$1.6x10^{22}$
RBS	780	280	$1.9x10^{22}$
data /7/	780	130	-

As it is seen from the table, the data obtained by three methods mentioned above are in good agreement. It allows to apply this or that method depending on the experimental conditions and the sample type. It should be noted that the significant excess of experimental profile width over tabulated ΔR_{max} /7/ arises, perhaps, from the profile diffusion expension due to a great number of radiation defects. At the same time the Pearson-like experimental profile shape shows that under given parameters of implantation before annealing, the nitrogen redistribution caused by nucleation of Si_3N_4 phase is negligeable.

Annealing by scanning electron beam results in a slight narrowing of nitrogen profile while the lower border remains fixed (Fig.1). This aspect is quite significant because nitrogen redistribution against the concentrational gradient takes place. Usually the electron annealing leads to the profile broadning if chemical reactions and compound formation do not occur /5/. Channeling backscattering spectra for annealed samples show the reduction of silicon top layer to the crystalline state which is close to the initial one (Fig.2). In this case the recrystallization initiates from the silicon implanted surface. It is also seen that buried nitrogen-rich layer has a structure quite different from silicon substrate one, of which random value of aligned spectrum gives evidence.

Thus, the present studies show that scanning electron annealing results in recrystallization of top damaged silicon layer by solid-state epitaxy from silicon implanted surface. This annealing probably also stimulates the nucleation and subsequent growth of the Si_3N_4 grains which are accompanied by nitrogen redistribution. The more detailed investigations of the structures formed are in progress.

REFERENCES

(1) Skorupa W., K. Woiischager, U. Krissig, R. Grotzchel and H. Barsch: Nucl. Instr. and Meth. B19/20 (1987) 285-289

(2) Belz J., E.H. Te Kaat, G. Zimmer and H. Vogt: Nucl. Instr. and Meth. B19/20 (1987) 279-284

(3) Smith T.P., P.J. Stiles, W.M. Augustyiak, W.L. Brown, R. Hull, D.C. Jacobson and R.A. Kant: J. Appl. Phys. 58(1), 1 July (1985) 193-196

(4) Liental Z., R.W. Carpenter and J.C. Kelly: Thin Solid Films. 138 (1986) 141-150

(5) Dvurechensky A.V., G.A. Kachurin, E.V. Nidaev, L.S. Smirnov: Pulse Annealing of Semiconductor Materials. ''Nauka'' Moscow 1982(in Russ.)

(6) Ziegler J.F., G.W. Cole and J.E.E. Baglin: J.Appl. Phys. vol. 43 Nom. 9 (1972)

(7) Komarov F.F., A.F. Burenkov, M.A. Kumakhov, M.M. Temkin: Tables of Parameters of Distributions Ion-Implanted impurities. BGU (1980) (in Russia)

HYDROGEN INCORPORATION IN PROTON-IMPLANTED ELEMENTAL AND COMPOUND SEMICONDUCTORS

C. Ascheron, D. Lehmann, C. Neelmeijer[+], A. Schindler[++]

INTRODUCTION

For the use of the method of hydrogen implantation in device technology the knowledge of the incorporation behaviour and of the distribution of the implanted hydrogen is necessary in order to optimize the implantation process.

EXPERIMENT

$A_{III}B_V$ compound and elemental semiconductors were implanted with 0.3 MeV protons. The depth profiles of the hydrogen distributions were revealed by bevelling with 1 keV Ar^+ ions (slope 3 μm/10 mm). The hydrogen concentrations were determined with the techniques of elastic recoil detection (ERD) and of particle induced gamma spectroscopy (PIGE). The analyzing beams were 1.7 MeV $^4He^+$ and 17.1 MeV $^{19}F^+$ (methods described in /1,2/).

RESULTS

The concentration profiles detected on as implanted samples are summerized in Fig. 1. It should be mentioned that both methods deliver coinciding results. For the purpose of comparison Figure 1 contains the theoretical hydrogen profiles computed with the model of Biersack /3/. The figure shows that the hydrogen distributions exhibit pronounced maxima in a "buried" layer of about 0.5 μm thickness in depths of 2.2 to 2.8 μm, which roughly correspond to the theoretical projected ranges. The differences in half widths of the hydrogenated layers are due to different range straggling. It can be seen that the experimentally detected profiles are satisfactory described by the model calculations.

Karl-Marx-Universität Leipzig, Linnéstr. 5, Leipzig, 7010, GDR
[+]Zentralinstitut für Kernforschung Rossendorf, PF 19, Dresden, 8051, GDR
[++]Zentralinstitut für Isotopen-und Strahlenforschung, Permoserstr. 15, Leipzig, 7050, GDR

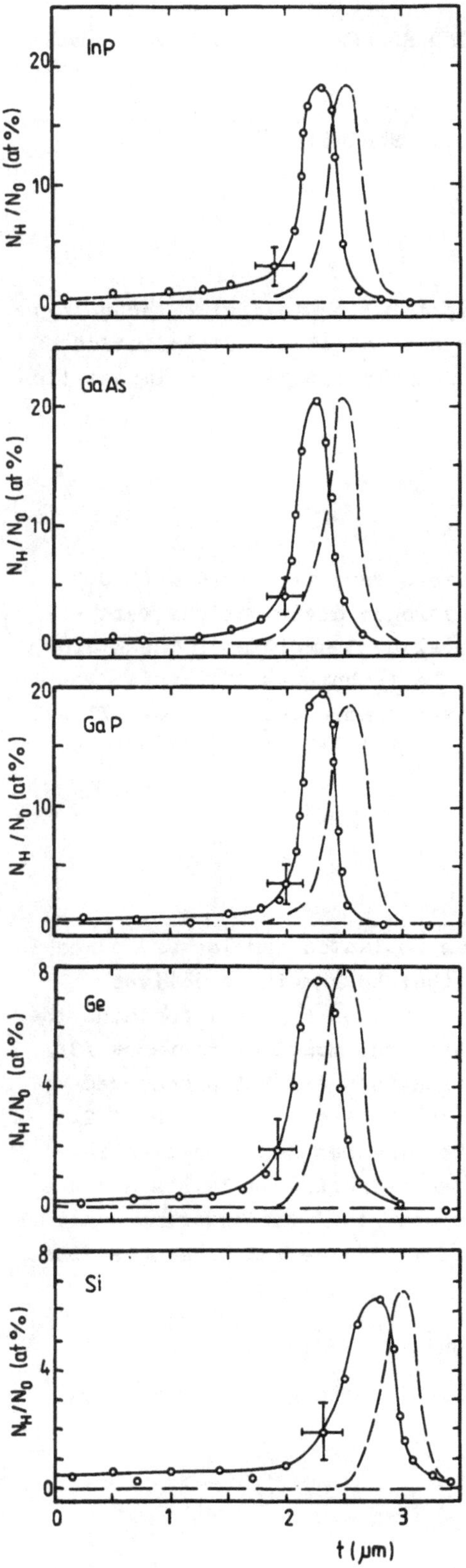

Figure 1:
Experimentally detected (o) and computed (--) depth profiles of hydrogen implanted with fluences $D = 3 \cdot 10^{17}$ cm^{-2} in InP, GaAs, GaP and with fluences $D = 1.8 \cdot 10^{17}$ cm^{-2} in Si and Ge.

The integration over the profile shows that nearly the whole implanted hydrogen is retained within the samples up to relatively high fluences. This means that all these materials have a relatively high hydrogen storaging capacity, nearly as high as in hydrogenated silicon /4/. The fact, that less than 10% of the implanted hydrogen escaped, though the solubility of hydrogen in this materials is drastically exceeded and though only a small fraction of the implanted hydrogen is chemically bonded /4,5/ points to an effective trapping of gaseous hydrogen within voids. This has been indicated by TEM studies on GaAs /6/, GaP /7/ and Si /8/, which show the formation of hydrogen filled platelets near the projected range /see Fig. 2/. At higher fluences and at higher temperatures blistering and flaking of the implanted regions occurs (see Fig. 3). This means: The maximum hydrogen incorporation is determined by the mechanical strength of the material. The lowest "resitivity" against flaking is observed in InP and the highest "resitivity" in GaP and Si.

CONCLUSIONS

1. The hydrogen storaging capacity of the considered materials is very high. It is restricted by their mechanical strength.
2. The mobility of hydrogen in $A_{III}B_{V}$ and elemental semiconductors is very low at room temperature conditions.

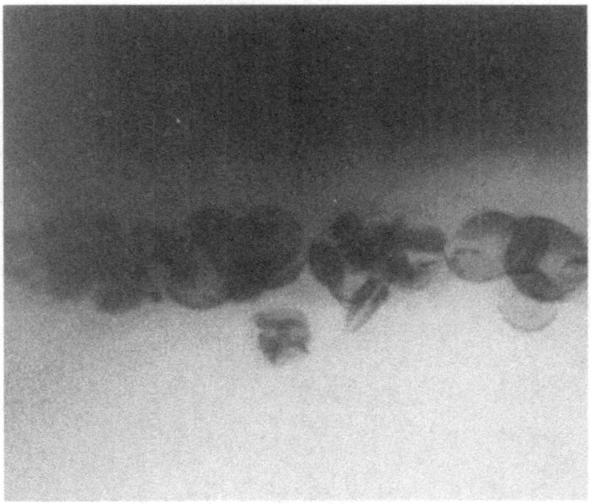

Fig.2

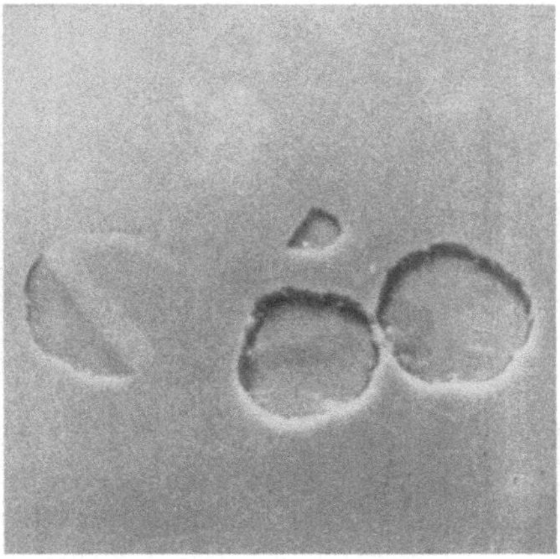

Fig. 3

Figure 2:
Cs - TEM pattern of high-fluence implanted GaP showing hydrogen filled platelets in the region of the projected range of the protons /7/.

Figure 3:
SEM patterns /7/ of high-fluence implanted beam spots on GaP exhibiting blisters (a) and flaking (b).

REFERENCES

/1/ D. Lehmann, C. Ascheron, C. Neelmeijer:Proc. 4th Meeting Nucl.Anal. Meth., Dresden 1987, p. 110

/2/ Rudolph, W., C. Bauer, R. Gippner, D.Grambole, C. Heiser, H.-J. Thomas: Proc. 3d Meeting Nucl. Anal. Meth., Dresden 1983

/3/ Biersack, J., L. Haggmark: Nucl. Instr. and Meth. 174 (1980) 257

/4/ Imura, T., A. Hiraki: Jap. Rev. Electr., Comp., Telecomm., Semicond. Tech. 8 (1983) 1955
/5/ Ascheron, C., C. Bauer, H. Sobotta, V. Riede: phys. stat. sol. (a) 89 (1985), 549
/6/ Neethling, J. H. et al.: J. Appl. Phys. 6 (1988) 7o4
/7/ Ascheron, C., H. Bartsch, A. Setzer, A. Schindler, P. Paufler: Nucl. Instr. and Meth. B28 (1987) 350
/8/ Chu, W. K., H. Kastl, R.F. Lever, S. Mader, B. J. Masters: Phys. Rev. B16 (1977) 3851

FOCUSED-ION-BEAM MODIFIED STRUCTURES OF METAL ON GLASS

L. Bischoff[1], T. Chudoba[2] and F.K. Naehring[1]

Basic steps of the focused ion beam (FIB) repair of photolithographic masks are investigated.
After a short review of the microelectronic mask repair demands (1) the FIB mask repair cycle (2) is outlined. From the sputtering of chromium-on-glass layers (3) a condition to get sharp edges (4) is derived.

1. Defects in photolithographic masks

In photolithography the fabrication of fully defect-free masks is just impossible for very large scale integrated circuits. On average, a density of 0.1 defects/cm^2 is obtained in mask production. To get a reliable output of good chips the photolithographic masks have to be nearly defect-free. Defects are iteratively repaired in cycles of detection and correction. To realize only one repair cycle a high-reliable defect detection and a stable repair process are desired. As the structure details go beyond the 1 um level the LASER repair has to be replaced by FIB repair. It is expected that masks for 4 and 16 Mbit dRAM's (Table 1) and similar structures will be repaired by a mixmatching of LASER and FIB repair (Table 2).

Table 1: Tolerable defect sizes of dRAM's - the challenge of repair technology

dRAM	technol. level	minimal feature size		defect size isolated		defect size on edges		position accuracy	
mask scale	:	1:1	5:1	1:1	5:1	1:1	5:1	1:1	5:1
256k	4	1.5	7.5	1.8	2.3	0.9	1.8	0.2	0.3
1M	5	1.0	5.0	<1.0	1.7	0.6	1.2	0.15	0.2
4M	6	0.8	4.0	?	1.2	?	0.8	?	0.15
16M	7	0.6	3.0	?	0.8	?	0.5	?	0.1

(1:1 technology is used only for uncritical masks with structure details that exceed the minimal feature size.)

Table 2: Data of modern LASER and FIB repair stations

repair of	pinspots on edges		pinholes on edges	
LASER ¦ FIB	Nd : YAG	Ga+ FIB	Ar-ion	Ga+ FIB + hydrocarbons
min. reparable defect size / um	0.5	0.1	1.0	0.1
position accuracy / um	0.1	0.05	0.1	0.05

2. FIB repair of chromium defects

The repair of chromium-on-glass masks is the first application of the focused ion beam technique. Some special FIB repair stations are on the market (2,3,5,6).
The FIB repair technology is outlined in Fig. 1.

- The repair area is selected and treated by computer control and electrostatic deflection of the beam.
- The defect location data are transferred from a mask inspection system to the FIB.
- Prior to repair the defect area has to be imaged.
- The transmissivity of the sputtered glass surface has not to be changed for more than 5 % to accomplish the printing process. Best results have been optained by chemical assisted sputter etching (4).

1) Zentralinstitut fuer Kernforschung Rossendorf, PF 19, Dresden, 8051
2) Zentrum fuer Mikroelektronik Dresden, Grenzstr. 28, PF 34, Dresden, DDR 8080

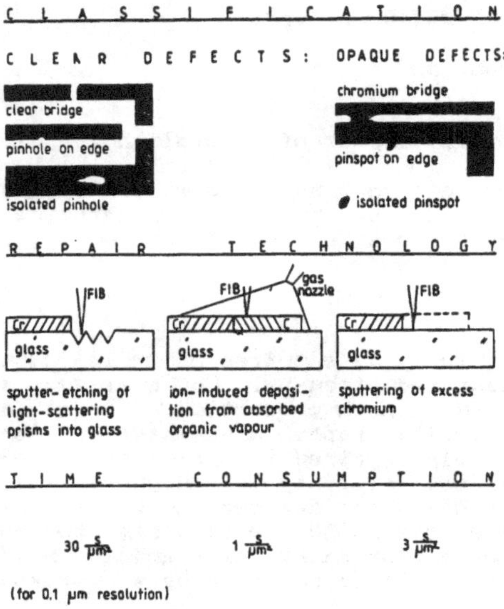

Fig. 1: FIB repair of clear and opaque defects on photolithographic chromium-on-glass masks.

3. Results

With a Ga+ FIB of 8 nA, 40 keV and 1 um diameter lines and rectangular openings (1) have been sputtered into chromium-on-glass (Fig. 2).

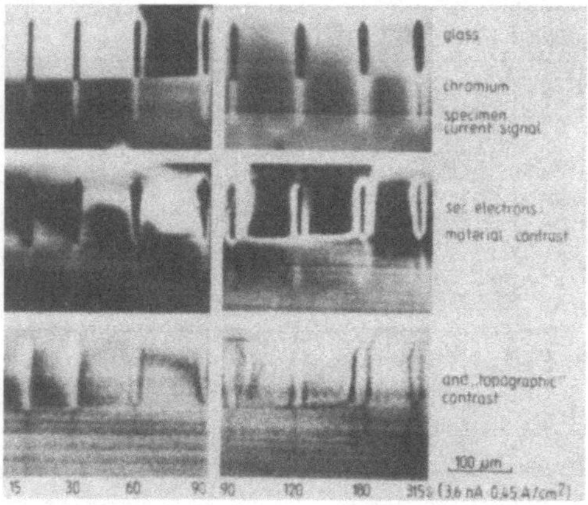

Fig. 2: Ion induced SE images taken after sputter erosion of lines across the chromium edge.

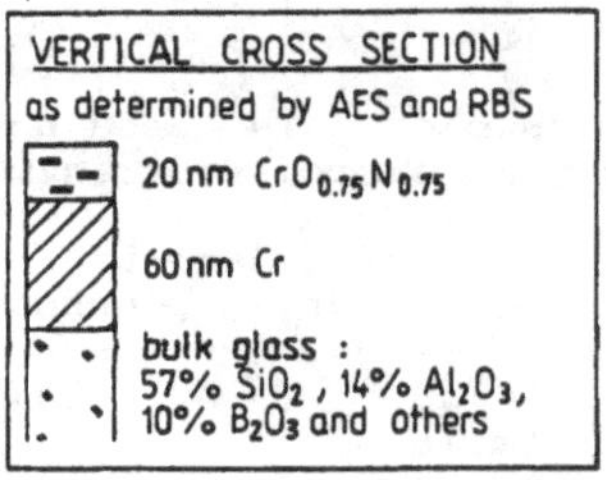

Fig. 3: Cross section of the chromium-on-glass masks.

Before and after "repair" the "defect" area has been imaged. The target current as well as the ion induced secondary electron (SE) signal, collected and amplified by a segmented Si semiconductor detector, are used as information channels in a raster scan mode (1). The following features have to be mentioned:

- The layer of chromium and chromium oxinitride (Fig.3) is sputtered within 30 s by a 3.6 nA beam. This corresponds to a dose of 3.6 * 10^17 Ga+/cm^2.
- The sputter yields were found to be 2.3 for chromium and 1.4 for chromium oxinitride.
- The sputter yield of glass is 0.78 atoms/ion.
- A loss in glass transmissivity of approximately 20 % has been observed due to gallium bombardment.
- Prolonged large-area sputtering (>100 um * 100 um) causes chromium layer break and rupture caused by charging.
- Even gallium-implanted doses as low as 10^13 Ga/cm^2 are visualized by the SE and target current detection system (1).
- The mentioned limitations by charging are thought to be overcome by simultaneously flooding the mask surface with an electron gun.

4. Discussion on getting sharp edges

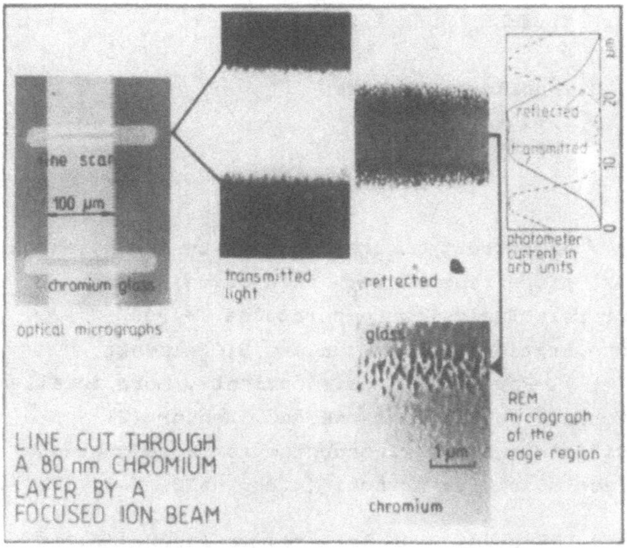

Fig. 4: Reflection and transmission profiles of FIB line cuts in chromium

As can be seen from Fig. 4 the edges of the chromium openings appear not as a straight line. This is due to the ion beam profile and the sputter conditions. The FIB profile effects the frame of the chromium windows. It causes a um-wide chromium glass transition region. Throughout thinning the grain structure of the layer becomes evident. The crystal orientation dependence of the erosion speed leads to different thinning-up of the individual chromium-grains. The situation is summarized in Fig. 5:

- A prolonged sputtering is needed to remove the slowly-etching grains too.
- The process time should be selected so that the grains with fast and slow erosion will meet to steep part of the ion beam profile.

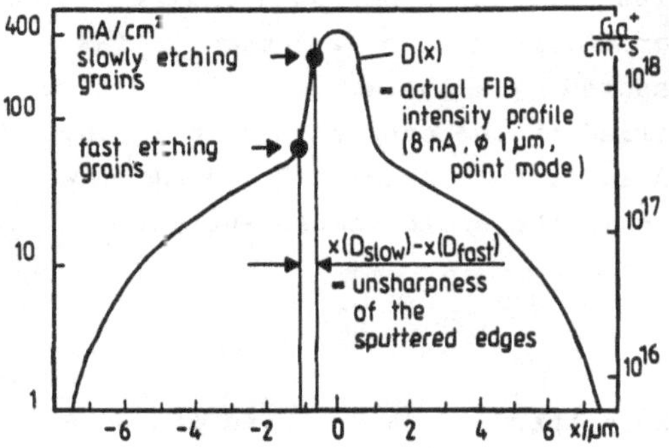

Fig.5: The unsharpness of a sputtered edge results from the inteference of the grain structure of the sputtered layer with the ion beam profile.

5. References

(1) Bischoff, L. and Naehring, F.K., Annual Report 1988, ZfK-667(1989) No. 4.24 and 4.35

(2) Kaufmann, H.C., Thompson, W.B. and Dunn, G.L., Proc. SPIE Vol. 632, Electron-Beam, X-Ray and Ion-Beam Lithographies V (1986) 60

(3) KLA short report, Electronics Jan. 6 (1986) 69

(4) Nakagawa, Z., et. al., J. Vac. Sci. Technol. B6 (1988) 1030

(5) Quantronix Corp. data sheet, Darmstadt (FRG) 1987

(6) Robinson, W.P. and Parker, N.W., Proc. SPIE Vol 773, Electron-Beam, x-ray and Ion-Beam Lithographies VI (1987) 216

IN SITU XPS AND LEED STUDY OF Si/SiO_2 INTERFACE CHANGES INDUCED BY ArF EXCIMER LASER IRRADIATION

V. Cháb, P. Jiříček, J. Kubátová, I. Lukeš [1)]

Introduction

Modern VLSI technology provoked the high interest in the research of physical and technological properties of very thin oxide layers ($<$ 20 nm) produced on Si at relatively low temperatures ($<900^{o}$ C). New methods were used for preparation of high quality SiO_2 layers (laser induced oxidation (1)). For further developmment , more detailed knowledge about Si/SiO_2 interface formation is needed. Zehner (2) demonstrated that laser modification of semiconductor surfaces could be effectively studied by surface sensitive methods (LEED, AES, XPS, UPS).

In this paper LEED and XPS studies of ArF excimer laser-induced modification of Si(100) surface and Si/SiO_2 interface are presented.

Experimental

Laser irradiation and subsequent XPS and LEED measurements were performed in the analysing chamber of a VG ADES 400 angle-resolved photoelectron spectrometer with the basic pressure lower than 1.10^{-10} Torr. ArF excimer laser pulses (LAMBDA PHYSIK EMG 100, pulse width 10 ns at FWHM) were transmitted into the spectrometer chamber through a UV grade fused silica window. The samples were cut from the standart Si(100) 0.35 mm thick wafers (Czochralski grown, p-type, 5-12 ohm.cm). The energy density of the laser beam measured by a thermopile calorimetric joulemeter (PHOTON CONTROL) was varied by the optical system in the range of 0.3 - 1.4 J/cm^2. A focused laser spot area ($\sim$2 - 4 mm^2) was scanned manually by the manipulator over the whole sample surface (5x5 mm^2).

Results

On as received Si(100) no LEED patterns were observed and only common contamination was detected. Peak A at 99 eV (Fig.1a) corresponds to Si2p emission from the Si substrate and peak B at 103 eV characterizes Si in the native oxide layer. Depending on laser pulse energy density, following processes were observed. After irradiation by laser pulse with energy density in range of 0.8-1.2 J/cm^2, i.e. above the melting threshold 0.4 J/cm^2 (3) and below the evaporation threshold 1.4 J/cm^2 (3)

[1)] Institute of Physics, Czechoslovak Academy of Sciences, Na Slovance 2, 180 40 Prague 8, Czechoslovakia

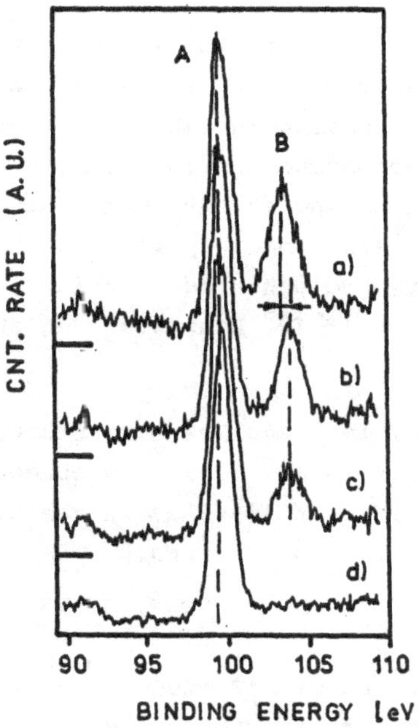

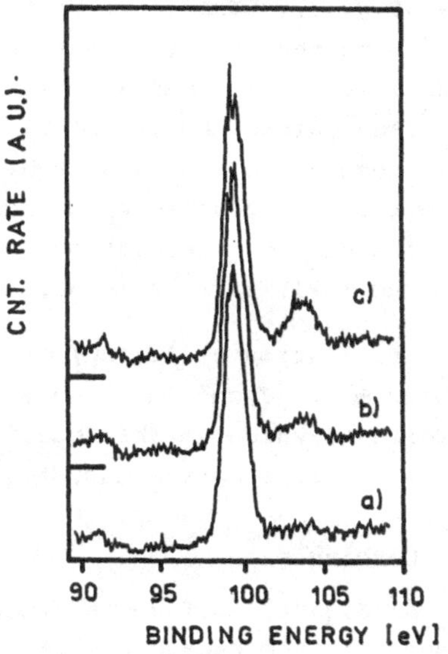

Fig.1. XPS spectra of Si2p electron states: a) as inserted, b) after 1 laser pulse with E = 1.2 J/cm², c) after 3 pulses (1.2 J/cm²), d) after 1 pulse (1.4 J/cm²).

Fig.2. XPS spectra of Si2p electron states of the laser cleaned surface: a) exposed to oxygen, b) irradiated in O_2 (2 pulses E = 0.8 J/cm²), c) irradiated in O_2 by four pulses (E = 0.8 J/cm²).

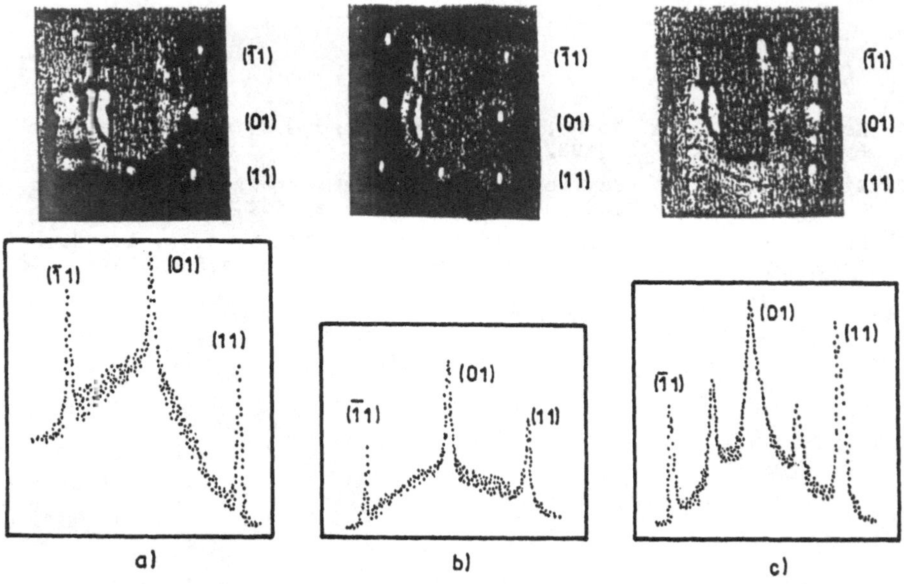

Fig.3. LEED patterns (49 eV) and spots intensity profiles of laser prepared Si(100) surface: a) 1 pulse (E = 1.2 J/cm²) - recrystallizated native oxide layer, b) 1 pulse (1.4 J/cm²) - clean surface, c) after heating clean surface at 400° C.

peak B is shifted to higher binding energy and decreased in the intensity (Fig.1b). LEED (1x1) patterns are observed (Fig.3a). By the laser pulse with the energy density near the evaporation threshold the native oxide layer is removed and Si(100)(1x1) reconstruction is observed on the clean surface (Fig.1d,3b). By subsequent heating at 400° C (30 min.), (1x1) reconstruction is switched (Fig.3c) to (2x1) one, typical for the clean surface prepared by thermal procedure. No enhanced sticking coefficient for molecular oxygen is observed ($\lesssim 10^5$ Langmuir) on such prepared (1x1) surface (Fig.2a).

Irradiation of Si(100)(1x1) surface by laser pulses with energy density $\lesssim$ 1.0 J/cm² in the oxygen atmosphere ($\sim 10^{-5}$ Torr) produces oxide layer which is characterized by the peak at 103.5 eV in the Si2p photoelectron spectrum (Fig.2b-c) and by (1x1) LEED patterns.

Conclusions

a) Si(100) surface is effectively cleaned by ArF excimer laser pulses with the energy density near the evaporation threshold. The (1x1) reconstruction was prepared for the first time.

b) With this type of laser epitaxial layers of SiO_2 were formed either by recrystallization of the native oxide layer or by laser induced oxidation in a low pressure oxygen atmosphere.

References

(1) Boyer, P.K., G.A. Roche, W.H. Ritchie and G.J. Collins: Appl.Phys.Lett. 40 (1982) 716-719

(2) Zehner, D.M., in: Semiconductors and Semimetals , vol. 23, eds. C.W. White and R.T. Young, Academic Press 1984, p.405

(3) Lukeš, I., R. Šášik and J. Kubátová in: SPIE Proc., Vol.1033, "3rd Intern. Conf. Trends in Quantum Electronics,Bucharest 1988

EMISSION PROPERTIES OF ALLOY GOLD-BERYLLIUM FIELD ION SOURCE

Novak Drandarov, Borislav Nikolov*

A stripe geometry liquid metal field ion source, which combines the advantages of both the filament and the reservoir sources, have been constructed. The emission properties of these sources have been investigated, using a gold-beryllium alloy, the latter being of some interest in microelectronics as well. This alloy has been obtained according to its component binary phase diagram. The alloy consists of 98.4 wt.% Au and 1.6 wt.% Be.

Figure 1 shows the current-voltage curves of these sources. Three regions can be distinguished. The first region includes a sharp increase of the emission current at the threshold voltage and a more smoothly changing part of the curve. The second region is transitional, here the ion current is changing unevenly with increasing the applied voltage. The third region includes a rapid increase of the emission current with increasing the voltage. Another peculiarity of this curve is the presence of hysteresis. Furthermore experiment shows that with increasing the temperature the difference between the three regions gradually disappears. In view of this, we suppose that these peculiarities of the curve are connected with the density and homogeneity of the metal alloy stream from the reservoir to the emitter tip.

Having used the "freezing" method, we have shown that these trends of the current-voltage curve are also connected with the change of the structure of the emitter (figures 2,3,4). It is obvious from these photos that the rapid increase of the ion current is connected with the appearance of a small cone on the emitter tip. Several three dimensional structures appear in the transitional region and their number and magnitude increase in the third region (figure 4).

The mass-spectrum of the emitted ions has been determined using Wien filter. It is obvious (figure 5) that the following ions are

---

*Institute of Electronics of the Bulgarian Academy of Sciences, 72, Blvd.Lenin, Sofia, Bulgaria

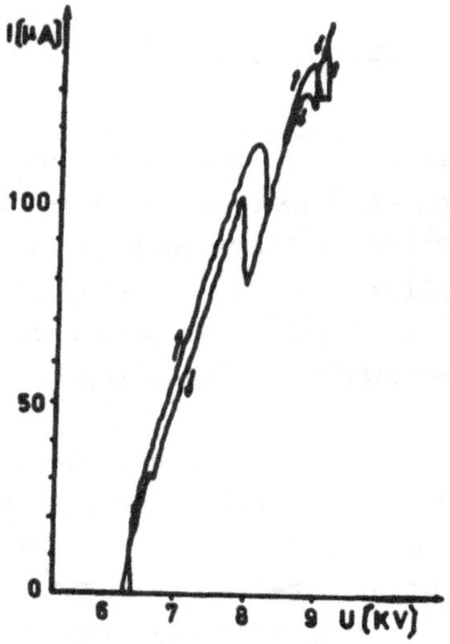

Figure 1. Current-voltage curve of a AuBe stripe geometry field ion source, working temperature 1200°C.

Figure 2. "Frozen" emitter in the first region of the current-voltage curve.

Figure 3. "Frozen" emitter in the transitional region of the current-voltage curve.

Figure 4. "Frozen" in the third region of the current-voltage curve.

emitted: Au_2^+, Au^+, Au^{2+}, Be^+, Be^{2+}. By means of the same filter the temperatures for optimum emission of the different ions have been determined. It has been experimentally proved that for Gold ions this temperature is 1300°C and for the Beryllium ions it is 1250°C.

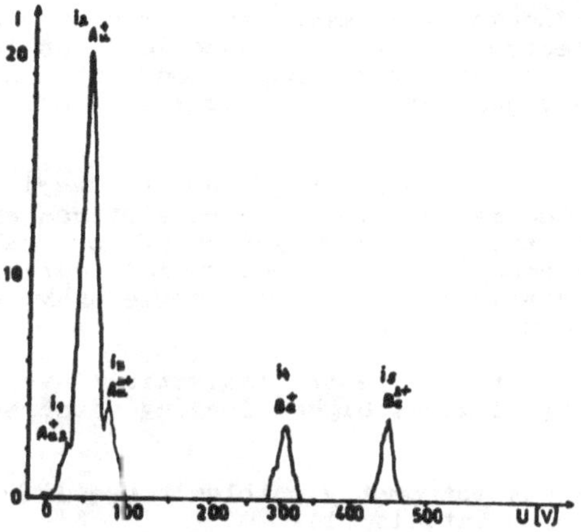

Figure 5. Mass-spectrum of the emitted ions by the AuBe field ion source.

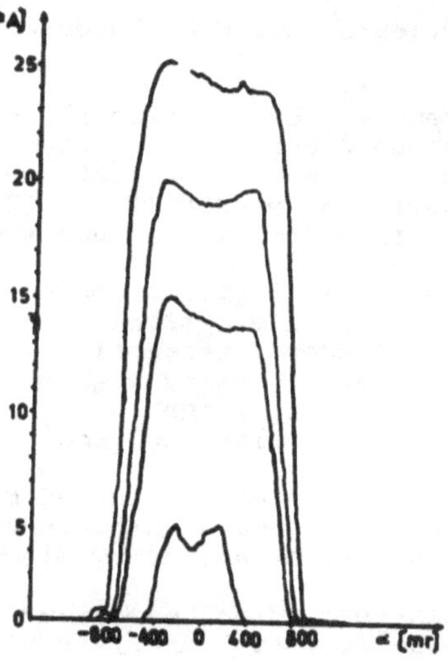

Figure 6. Angular distribution of the emitted ions for different values of the ion current.

By means of a device, constructed especially for this purpose, the angular distribution of these ions has been measured. Figure 6 shows that the angular distribution is strongly dependent upon the magnitude of the ion current. When the ion current rises from 5 μA to 25 μA, the semiangle of the emission increases from 400 mrad to 800 mrad.

These experimental results can be explained neither by the Swanson[1] model, nor by the Sikharulidze[2] model. They are still not satisfactory enough to enable us to suggest a new model of working of the alloy field ion sources.

References.

1. Kingham D., Swanson W., Vacuum, v.34, No.10/11 (1984), 941-945
2. Sikharulidze G., Burmiy Zh., Visokochistiye veshtestva, No.2, (1987), 171-179.

This work is sponsored by the Science Committee of P.R.B., Research Contract No.108.

A LOW ENERGY ION MICROSCOPE FOR DIAGNOSTICS AND PREPARATION OF MICROELECTRONIC STRUCTURES WITHIN AN MULTIANALYSIS SYSTEM

W. Driesel[1] and F.K. Naehring[2]

Presently the beneficial uses of focused ion beams in scanning ion microscope mode are explored for electronic device diagnostics. The ion beam interaction effects result in a variety of image contrasts and preparation possibilities (2). They add some new properties to multi-analysis ultra high vacuum systems.

- The five contrast modes installed now open the view to various features of electronic circuits. One may use ion induced electron and ion images, respectively, that are displayed by fast TV or slow computer processed high S/N pulse counting, and specimen ion current images. The image contains informations on specimen topography as well as on elemental and voltage distribution.
- In the case of insulating material, for instance passivation layers, the ion induced secondary ion signal gives higher imaging contrast than that of secondary electrons.
- High-resolution secondary ion mass spectrometry enables three dimensional material analysis of submicron details (1).
- The surface may be sputter-cleaned by the scanned ion beam during analysis.
- The passivation layer of solid state circuits can be locally opened prior to analysis of elemental composition, voltage measurements (5) and electron beam voltage contrast imaging, that are helpfull in failure analysis.
- Different wiring layers may be locally connected by focused ion beam sputter deposition (3,4).
- Unwanted connections are sputtered through for repair, trim or circuit programming (3,4).
- All these steps of micro surgery are controlable by full parallel ion beam analysis.

A low energy ion microscope has been built. The energy of 1 to 15 keV implies a penetration depth of 2 to 15 nm, low intermixing during sputter erosion, shallow doping, and analysis of surface defects. The ion optical column consists of a Ga liquid metal ion source with triode optics, an asymmetrical electrostatic condenser lens, an octupole for astigmatism correction and beam centering, an electrostatic Einzel objective lens, a second octupole for scanning deflection of the ion probe, and a channeltron detector. In ion optical calculations the probe diameter is minimized with respect to the emitter-condenser distance and to the working distance required between the specimen and the ion optical column. The 10 keV beam is focused to 0.3 um diameter and current density of 0.7 A/cm^2. Liquid metal ion sources offer high erosion speed.
Investigations on the wiring layers of integrated circuits are presented as a first performance test of the instrument. The system is equipped with a scanning electron microscope, a scanning Auger microscope, and a photoelectron emission microscope.

1) Institut fuer Festkoerperphysik und Elektronenmikroskopie, Weinberg 2, Halle, DDR 4020
2) Zentralinstitut fuer Kernforschung Rossendorf, PF 19, Dresden, DDR 8051

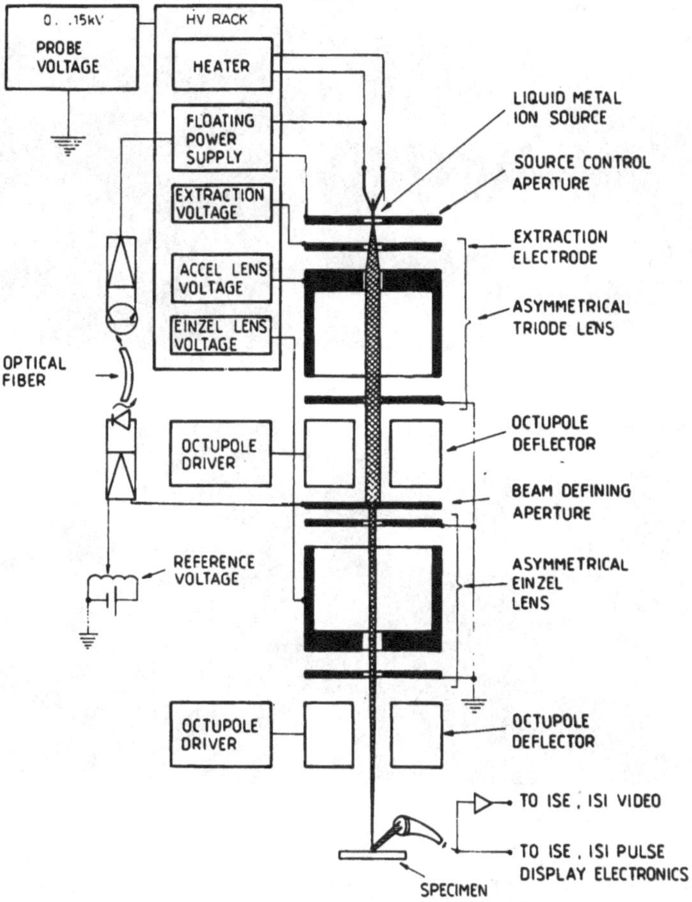

Fig. 1: Scheme of the ion gun

Fig. 2: Ion gun module

Fig. 3: Performance test: Resolution measured by using a 5/10 um grid. The horizontal amplitude is 42 um/8.3 V; SE-image in TV-mode.

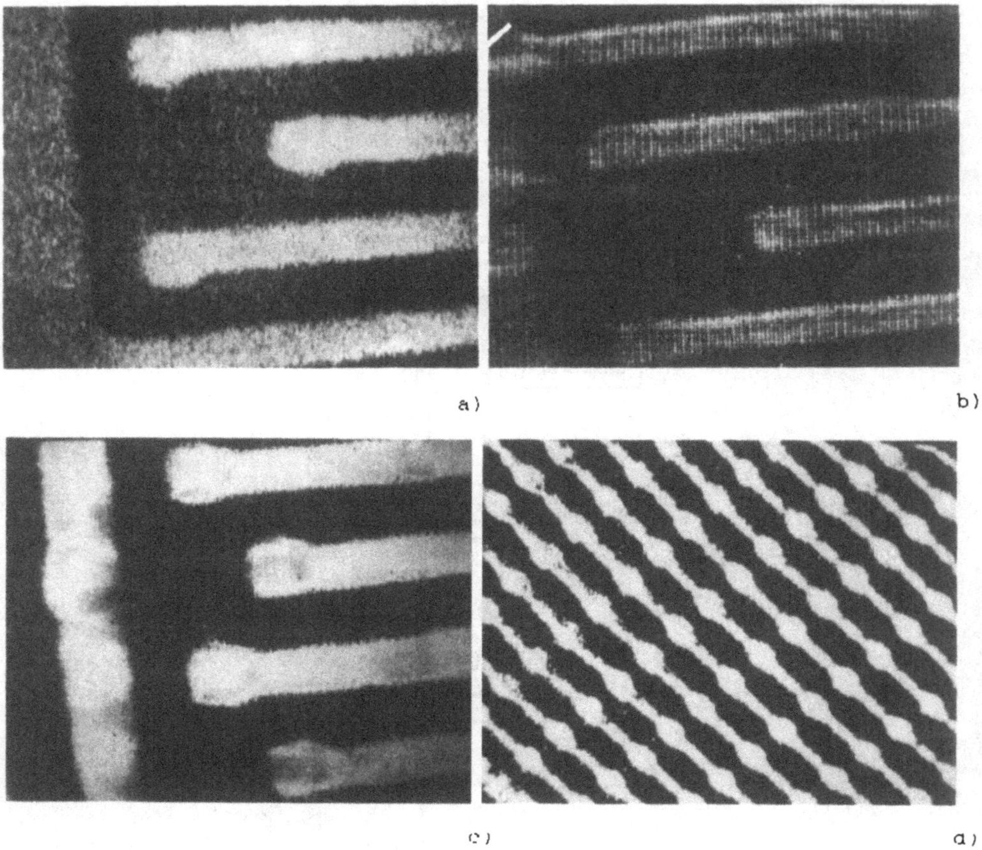

Fig. 4: Section of an integrated circuit: Al-conducting pathways (bright) on SiO2 (dark).
a) ISE; TV,
b) ISE; pulse counting,
c) ISI; pulse counting,
d) ISI; specimen current, AS converted and imaged with pulse counting

References

(1) Levi-Setti, R., Crow, G. and Wang, Y.L., Scanning Electron Microscopy 1985/II, 535
(2) Melngailis, J., J. Vac. Sci. Technol. B5 (1987) 469
(3) Melngailis, J., et. al., J. Vac. Sci. Technol. B4 (1986) 176
(4) Sudraud, P., Benassayag, G., Bon, M., Microel. Eng. 6 (1987) 583
(5) Swanson, L.W., Tuggle, D. and Jai-Zheng, Li, Thin Solid Films 106 (1983) 241

DRIFT TUBES RESONANT ACCELERATORS WITH ACCELERATING FIELD FOCUSING FOR THE ION IMPLANTATION

A.N.Antropov, V.V.Belousov, B.Yu.Bogdanovich,
N.M.Gavrilov, A.A.Kostjunin, S.A.Minaev, A.V.Shalnov*

Up to date electrostatic accelerators with energies up to 0.5 MeV /5/ are mainly used in industry ion-beam installations. They have high efficiency and allow to accelerate any types of ions with the precision parameters of beam. However high voltages results in different kinds of exploring difficulties such as large dimensions and necessity of radiation shielding. Taking into account up to date interest to the accelerated ions beams with the energy up to several MeV's it seems reasonable to appeal to the principle of high-frequency accelerating. One of the main question needed to solve on creating high-frequency ions accelerator at small energies is the focusing problem. Next question is possibility of regulating and transferring from one type of ions to the other.

In the up-to-date high-frequency ions accelerators at the energy to several MeV's either signchangable focusing by the accelerating field or electrostatic quadrupoles are used. In this case both the construction and the parameters of accelerating system are defined by the chosen type of focusing. All known methods of focusing by the accelerating field are devided into 2 groups. In the first one focusing fields have the quadrupole symmetry in the second one they are axial-symmetrical. So called space-similar or radio-frequency quadrupole focusing (RFQ) /3/ applies to the first group.

The second group of focusing methods uses the axial-symmetrical focusing fields and usually is realized in the systems with drift tubes and accelerating gaps. Such systems are simple in construction and have some advantages in technology and small sizes. The different varieties of the alternating phase focusing (APF) are related to this group /4/. On the basis of the phase variable focusing methods one may create as the accelerators at fixed energy in the form of the united multislit resonator so the module systems with deep energy relugating and retuning on the ions type, consisting of the consecutive situated independently phased resonators. One of the example of the alternating phase focusing accelerator on the basis of multislit resonator may serve the accelerator of protons at the energy of 1.2 MeV stanted in 1981 in MEPhI /1/. Its dimensions are the following L = 75 cm, diameter D = 20 cm, the injection energy W_1 = 90 KeV, the impulse power of HF generator is about

*Moscow Engineering Physics Institute, Moscow, USSR

80 kW. The impulse accelerating current is $I_i = 0.5$ mA, the slittness of the work regime is $S = 10^4$. The used generator allows the work regime $S = 10^2$, $I_a = 10$ mkA. In order the radio-frequency alternating phase focusing accelerator may be used for applying purposes it is necessary to improve the focusing method and to transfer it into the uninterrupted or quasi-uninterrupted regime of working. The parameters of the drift tube accelerating systems for protons and one-charged B-ions, which are the result of calculating and modeling are represented in the Table 1:

Table 1

Ion	W_i KeV	W_o MeV	λ m	P grad	L m	Ra mm	Ec MV·m	N
$_1H^+$	30	1.5	2	240	1.3	6÷8	5	29
$_{11}B^+$	25	1.2	6	220	1.2	6÷8	5	22

here W_i and W_o are initial and output ion energies; λ is the radio-frequency wave length; P is the phase acceptance; L is the accelerator length; Ra is the aperture radius of the drift tubes; Ec is the electric field intensity in the centres of accelerating gaps; N is the number of gaps. These results are got in the system with the original construction of focusing periods where alongside with the flight phase of the bunches at the centres of accelerating gaps, the gaps flight angle is changed periodically and the maximum value of the flight angle may approximate to 360°. Rather simple and effective two-wired resonator is chosen as the accelerating system for the noninterrupted radio-frequancy accelerator. It consists of two horizontal wires placed into the cylindrical metal screen. To these wires the drift tubes are connected by turns. The created resonator may generate in the range of frequency 50÷150 MHz that allows to accelerate not only protons but the heavier ions as well. The cooling system can provide the structure temperature within 10°C limit at the dissapated power 100 KW. The ion accelerator capacity on the basis of multislit resonator is limited by impossibility of fluent regulating of energy and accelerating the different types of ions. That's why the construction of the accelerator consisting from the consecutively located independently placed accelerating moduls every of which supplied by separate generator of comparativly small power (at 10 KW) is rather perspective. Such scheme in combining with electrostatic focusing lenses is used in the commercial accelerator for the ion implantation of the firm EATON /2/. Using one-typed double-slits resonator sections with the same dimensions of the so called "accelerating part" (i.e. gaps and tubes) at the frequencies 13.6 MHz and 27.2 MHz it is possible to accelerate ions in the range of the given masses (1÷130) up to the energies at some MeV's.

The specific character of the ions dynamics in the given above systems consists in the fact that the flight phases of the accelerating gaps centres can not be the same. In increasing of the ions speed the flight angles of the drift tubes are reduced and the phase jump from gap to gap strives to π . Hence already in the mode of constructing of the accelerator there are the elements of the alternating phase focusing. The values show that combining these elements with the periodical phase flight change at the tube centres of the separate modules at the expense of their proper phasing it may be possible pass over outer focussing arrangements. In this case not less than 30% injected ions will be captured into the accelerating regime. So resonance accelerators with the accelerating field focusing may be created as on the basis of multigap resonators so in the form of flexible module systems. The module systems though yield to multigap resonators in energetic efficiency they are more convertible at the expense of possibility of quick retuning.

References

(1) Abramenko N. et al.: Jornal Technicheskoi Fiziki, 53(5)(1983)858-862.
(2) Glavish H.: Nuc. Instr. and Meth. in Phys. Res. B 24/25 (1987) 771-776.
(3) Kapchinsky I.: Theory of lin. res. acc. Energoizdat, Moscow 1982.
(4) Murin B.: Linear ion accelerators, Atomizdat, Moscow 1978.
(5) Simonov V.: Ion implantation equipment, Radio & svjaz, Moscow 1988.

QSIMS DEPTH PROFILING IN THERMALLY OXIDIZED POLYSILICON LAYERS

U. Gunst and B. Müller [1)]

1. Introduction

During the oxidation process of highly doped silicon wafers or polysilicon films a redistribution of impurities across the Si/SiO_2-interface takes place. The annealing of implanted or insitu LPCVD-doped silicon for activation may also lead to the segregation and redistribution of dopands. In case of polysilicon gates this effect can cause the degradation of the gate oxide /1/.

Most investigations were carry out by means of Auger sputter profiling /2, 3/. In addition, SIMS measurements were performed recently /4, 5/. Quantification at the interface, however, is difficult due to the variations of the secondary ion yield and the sputter rate. The bombardment conditions should be chosen to minimize the matrix effect. To continue our preliminary measurements on steam oxidized polysilicon layers /6/ we have improved some details of our QSIMS equipment and then tested it by the use of a new series of samples having dry oxidized phosphorus doped polysilicon layers. The oxidation temperature was variable. With the help of replica TEM- and SEM-micrographs the grain size of the polysilicon surface was evaluated after etching the oxide films.

2. Experimental

SIMS depth profiles were measured with our small quadrupole-based SIMS equipment using 3 keV Ar^+ primary ion (PI) beam, 60 ° angle of incidence, rastered over 1 x 2 mm^2 with positive secondary ion (SI) collection from the central area of 300 x 600 μm^2 /7/. All measurements were performed with the O_2-jet directed on the sample surface to transform the Si surface into SiO_2, thereby avoiding ionization changes in the interface and in the polysilicon (saturation). Fig. 1 shows the ^{28}Si-intensity versus the sputter time for a 140 nm thick oxide layer /Si-structure. We can see only small matrix intensity changes in the surface and interface region. In /6/ we have discussed the influence of the charge built up on insulator films on the measured intensity due to the shifting of the energy distribution of the SI.

1) Technische Hochschule Ilmenau, Sektion PHYTEB, PSF 327, Ilmenau, DDR 6300

In our case of modified plate capacitor in front of the used analyzer the energy window was $\Delta E \approx 4...6 eV$. In Fig. 1 one can also see the needed target bias to shift the sample surface potential in the optimum of the SI-intensity. Under these conditions we have measured the intensities of $^{31}P^+$ (+ $^{30}SiH^+$), $^{28}Si^+$, $^{16}O^+$ and $^{23}Na^+$, the $^{23}Na^+$-intensity maximum was used to define the interface position. The origin quadrupole spectrometer QMG 311 (Balzers) was controlled by a computer. Using the software program CASIMS and the external silicon standards we can calculate the impurity distribution under the assumption of a linear transition of the sputter rate at the interface. In Tab. 1 the oxidation parameters used and the oxide thicknesses obtained are listed. The SIMS samples were cut out from the middle of the 3" wafers and then chemically cleaned and heated in the SIMS-apparatus to 450 K.

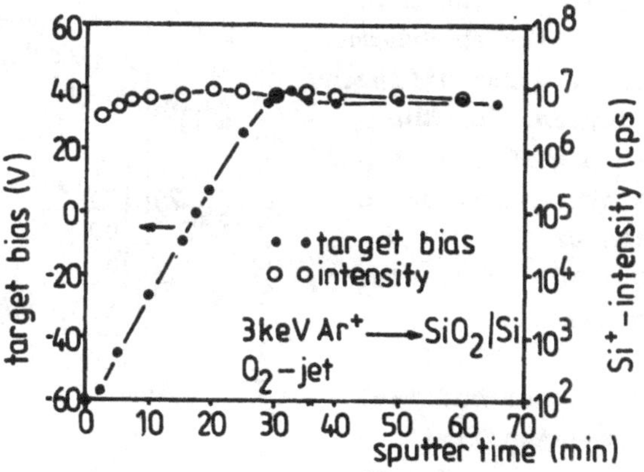

Fig. 1 Target bias and $^{28}Si^+$-intensity versus sputter time, CVD-Ox/Si-structure, d_{Ox} = 140nm

Tab. 1: Oxidation parameters and oxide thickness of the measured poly-SiO_2/poly-Si-structures, thermal oxidation in dry O_2, 3 % HCl ambient.

sample	oxid. temp. T_{ox} (K)	oxid. time t_{ox} (min)	oxide thickness d_{ox} (nm)
POx 2	1173	240	55
POx 6	1223	140	68.6
POx 9	1273	65	71.3
POx 12	1323	30	71.6
POx 15	1373	14	81

3. Results

The phosphorus detection limit is determined by the $^{30}SiH^+$-cluster formation and depends on the gas background in the near of the sample surface. Because the $^{30}SiH^+$ - intensity, however, is constant in every region we have used a simple correction to reach the accurate phosphorus concentration.

In Fig. 2 there is shown the phosphorus profile for different samples. For the sample POx2 we found a concentration in the oxide of $C_P = 10^{19} cm^{-3}$, clearly above the detection limit, for the other one the detection limit is too high. Furthermore we can see the phosphorus pile up in the polysilicon near the interface and the different slopes of the profile. In Tab. 2 there are listed the most important results for the first measured series.

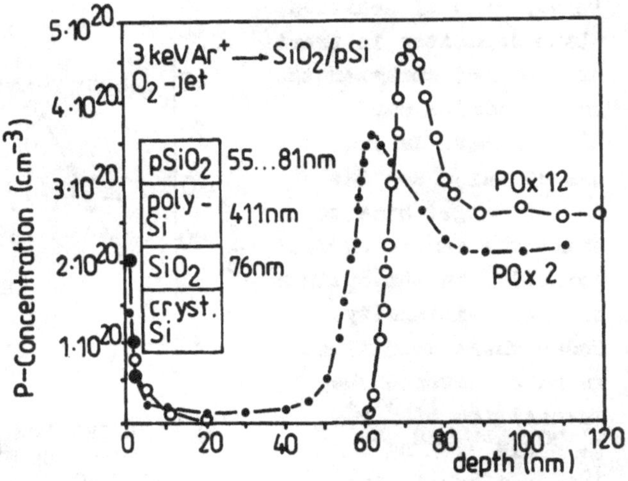

Fig. 2 Phosphorus profile of sample POx 2 (d_{ox}=55nm) and POx 12 (d_{ox}=71,6nm)

Tab. 2: phosphorus concentration in the polyoxide and in the polysilicon (C_{Pmax}, C_{Pbulk}), pile up factor $q_P = C_{Pmax}/C_{Pbulk}$ and detection limit.

sample	P - conc. oide (cm^{-3}) $\cdot 10^{19}$	detect. limit (cm^{-3}) $\cdot 10^{18}$	max. P-conc. (cm^{-3}) $\cdot 10^{20}$	P -"bulk" conc. (cm^{-3}) $\cdot 10^{20}$	pile up factor q_P
POx2	1.1	1.3	3.7	2.08	1.8
POx6	0.35	1.1	4.1	2.50	1.6
POx9	-	1.6	4.4	2.20	2.0
POx12	-	1.0	4.9	2.73	1.8
POx15	-	1.0	5.3	2.88	1.8

With the exception of sample POx9 we can see an increase of C_{Pmax} and C_{Pbulk} with the temperature of the oxidation, but the pile up factor remains nearly constant ($\bar{q}_P \approx 1.8$). In relation to other measurements /2/ there is no influence of the oxide thickness on the C_{Pmax}- value (saturation).

Fig. 3 shows carbon/Pt.-replica transmission micrographs obtained from the polysilicon surface after etching the oxide layer.

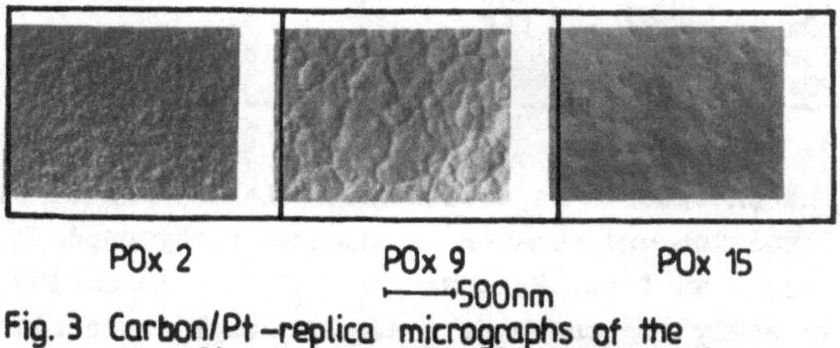

Fig. 3 Carbon/Pt-replica micrographs of the poly-Si surface after oxide layer etching

For samples POx2, 9 and 15 the evaluated grain sizes are 110 to 370 nm. The oxidation temperature and the maximum P-concentration effect the increase of the grain size with this temperature. But the variation of the grain size does not correlate with the surface roughness. The surface from sample POx15 is very flat in relation to the other ones.

4. Discussion

Together with the QSIMS-measurements there were performed electrical investigations (Fowler-Nordheim plots) on the same samples /8/. The field multiplication factor m ($M^{\pm}$) = E_{th}/E_{0x} ± is here a measure of the change of the really used field strength to the theoretical Fowler-Nordheim value at given current density. We found a remarkable correlation between the surface roughness and the distance of the P-peak from the interface (Na^+-peak) as a function of the oxidation temperature on the one hand and between the maximum P-conzentration and m(M^+) on the other hand. Both the roughness of the interface and the P-incorparation in the polyoxide should increase the factor m (M^+) for lower oxidation temperatures.

References

/1/ Taylor, M.A., D.L. Flowers and R.J. Cosway: Electrochem. Soc. 134 (1987) 11, 2935 - 2937.
/2/ Schwarz, S.A., R.W. Barton, C.P. Ho and C.R. Helms: J. Electrochem. Soc. 128 (1981) 5, 1101 - 1106.
/3/ Chang, C.C., T.T. Sheng, and T.A. Shankoff: J. Electrochem. So . 130 (1983) 5, 1168 - 1171.
/4/ Sakamoto, K., K. Nihi, F. Ichikawa and S.J. Ushio: J. Appl. Phys. 61 (1987) 4, 1553 - 1555.
/5/ Morgan, A.E. and Ph. Maillot: SIMSVI Sept. (1987) 709 - 716.
/6/ Marquardt, H.U., U. Gunst and B. Müller: Physical Research, EPM 87 (1987) 485 - 488.
/7/ Müller, B., H.U. Marquardt and Th. Stauden: Proc. 7th Czechoslovak Conf. on Electr. and Vac. Phys. 2 (1985) 377 - 386 and 405 - 413.
/8/ Schulze, H.: private communication.

PERFORMANCE OF A BRAGG IONIZATION CHAMBER FOR DEPTH PROFILING AND SURFACE ANALYSIS

E. Hentschel, R. Kotte, H.G. Ortlepp, F. Stary and D. Wohlfarth

Central Institute for Nuclear Research, Rossendorf
DDR - 8051 Dresden, PF 19, GDR

Nowadays, elastic recoil detection (ERD) is a well established technique for depth profiling. A substancial improvement of the ERD has been obtained by using time of flight (TOF) measurements [1] or a $\triangle E$-E telescope [2] to perform a particle identification. We present in our paper a third method of particle identification which has been developed for the investigation of heavy ion reactions. The main ingredients are a Bragg ionization chamber (BIC) and an electronic equipment for deriving simultaneously the energy E and the Bragg-peak height BP which is approximately proportional to the atomic charge number Z of the detected particle.

In fig.1 a schematic presentation of the experimental arrangement is shown. The recoiled or sattered ions pass a thin entrance window (stretched polypropylene, 30 to 50 $\mu g/cm^2$ thick on a supporting brass grid) of the BIC and produce a locally dependent ionization (Bragg curve) which is indicated by the dashed area in fig.1. The electrons drift in the electric field which is parallel to the tracks to the Frish grid (FG), pass the grid and are collected at the anode. The entrance window is the cathode (typical voltage -3kV) and the FG is grounded. In the drift volume between the cathode and the FG the electron drift velocity is 3.8 cm/μs for pentane gas. As indicated in fig.1 the electrical signal at the anode of the BIC looks like a mirrored picture of the Bragg curve.The area of the signal is proportional to the particle energy and the maximum corresponds to the Bragg-peak height BP which is at least for light and medium mass nuclei nearly proportional to Z.In fig.2 a (E,BP)-plot is shown. At high energies one can find well separated branches for the different Z values. The low-energy limit E_{lim} of this operative energy region is given by

$$R(E_{lim}) = R(E_{max}) + 1.5 \varrho v_D \triangle\tau_{eff} \qquad (1)$$

where R is the range in the BIC gas , E_{max} the energy at the stopping power maximum, ϱ the mass density of the gas, v_D the electron drift velocity and $\triangle\tau_{eff}$ the effective shaping time [3]. The arrows in fig. 2 indicate the E_{lim} values for different Z numbers. The most advanced method of signal processing is to sample the amplified signal by means of a Flash-ADC (FADC) and to derive by a fast digital processor (DP) the two parameters E and BP which are adressed to the memory of a 256x256 colour display. The details are described in refs.[4,5].

Fig. 2 illustrates the performance for depth profiling. The data were taken from the memory of the colour display and the events are plotted vs.the parameters Bragg-peak height BP and energy E. An Al target was bombarded with ^{35}Cl-ions (45 MeV) for about 1h. The Al has natural contents of C, O and Si and was implanted with $10^{17} N^+/cm^2$ (150 keV). The energy resolution $\triangle E/E$ is between 1 and 2% (FWHM)and depends mainly on the BIC window. This resolution corresponds to a depth resolution of about 15 nm. The Z resolution $\triangle Z/Z$ is between 2 and 2.5 % (FWHM). For counting rates up to 1000/s the background caused by pile up is practically negligible. The upper limit of the background in

the operative region corresponds to a concentration about four orders of magnitude below a stoichiometric concentration. An important advantage of the Z-selection is the possibility of combining in one measurement the spectroscopy of recoiled ions and scattered projectiles. This enables a complex material research including depth profiling and surface analysis. In contrast to the widely used semiconductor detectors the BIC with it's continuous gas exchange has no radiation damage. The window thickness can be measured accurately and recombination in the gas is expected to be smaller than in silicon. The energy calibration for heavy ions can be done in a more satisfying way. According to first test runs the measurement of heavy elements on a surface with a layer thickness of about one ng/cm^2 with a statistical accuracy ≈ 10% is possible. The pollution of water and air could be measured even in the presence of a strong background radiation.
After studying target systems well known by their preparation (e.g. buried SiO_2) the BIC has been applied to investigate buried insulating layers in silicon formed by high dose implantation of nitrogen and oxygen at 60 keV [6] and nitrogen implantation profiles in Al-targets.
In order to understand the principial limitations and conditions of a quantitative analysis a closed formula for the measured ERD spectra has been derived . Details are described in [3].
For the practical work a computer program has been developed. The strategy is to put in assumed functions $c_i(d)$ and to compute in a calculation step all energy spectra in a realistic way including all the energy absorptions in the target and the BIC window, the kinematics and the folding with the detector response. By comparing the calculated and the measured spectra changes of the the model functions $c_i(d)$ are introduced. Their choice is guided by experience.
Summarizing our results so far we can state that in the operative region given by eq.(1) the BIC is a powerful tool for analyzing particularly light ions up to Z about 30. The BIC allowes a combination of ERD and RBS in one measurement. The future trend in Bragg-curve spectroscopy is to use a faster sampling and a more sophisticated digital processing to get a more detailed characterization of the detected particle. Improving the window preparation it was already possible to obtain with a BIC a better energy resolution than obtained by a surface barrier detector [7].

Acknowledgements
The authors are indebted to Dr. W. Skorupa and Dr. B. Rauschenbach for valuable discussions on solid state properties and for making available the various targets.

References

(1) Whitlow,H.J.,G.Possnert and C.S.Petersson: Nucl. Instr. and Meth. B27 (1987) 448-457
(2) Behrooz,A.M. R.L.Headrick, L.E.Seiberling and R.W.Zurmühle: Nucl. Instr. and Meth. B28 (1987) 108-112
(3) Hentschel,E., R.Kotte, H.G.Ortlepp, F.Stary and D.Wohlfarth: Nucl. Instr. and Meth. B43 (1989) 82-91
(4) Ortlepp,H.G. and A.Romaguera: Nucl. Instr. and Meth. A276 (1989) 500-508
(5) Fromm, W.D. and M.Kahlenbach: ZfK - 530 (1984)
(6) Skorupa,W., K.Wollschläger, R.Grötzschel, J.Schöneich, E.Hentschel, R.Kotte, F.Stary, H.Bartsch and G.Götz: Nucl. Instr. and Meth. B32 (1988) 440-445
(7) Oed,A., P.Geltenbort, F.Gönnenwein, T.Manning and D.Souque: Nucl. Instr. and Meth. 205 (1983) 440-445

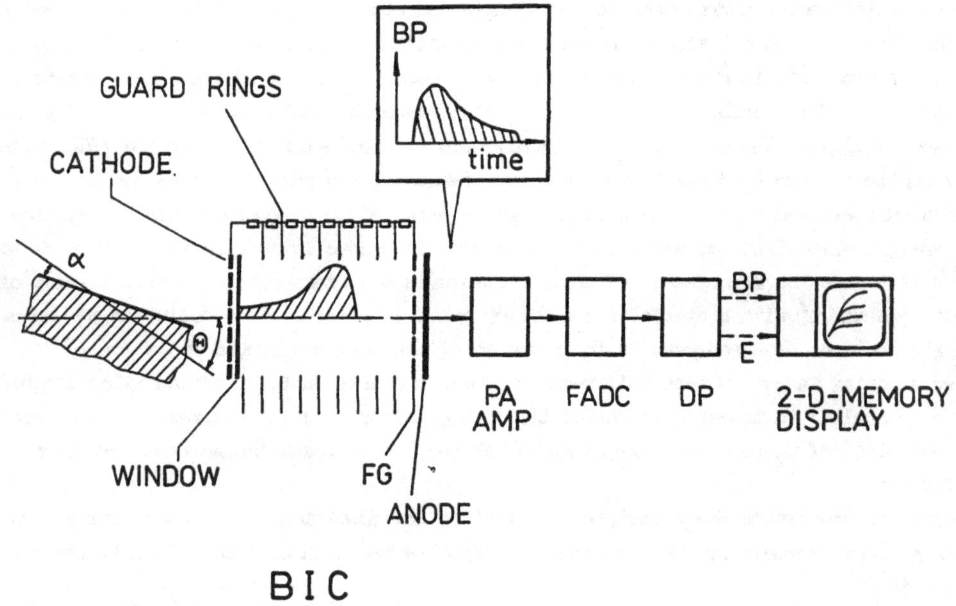

Fig. 1. A schematic plot of the experimental arrangement. The signal from the anode of the Bragg ionization chamber (BIC) is amplified by the preamplifier (PA) and the amplifier (AMP). After sampling by a flash ADC (FADC) the energy E and the Bragg-peak height BP are derived by a fast digital processor (DP).

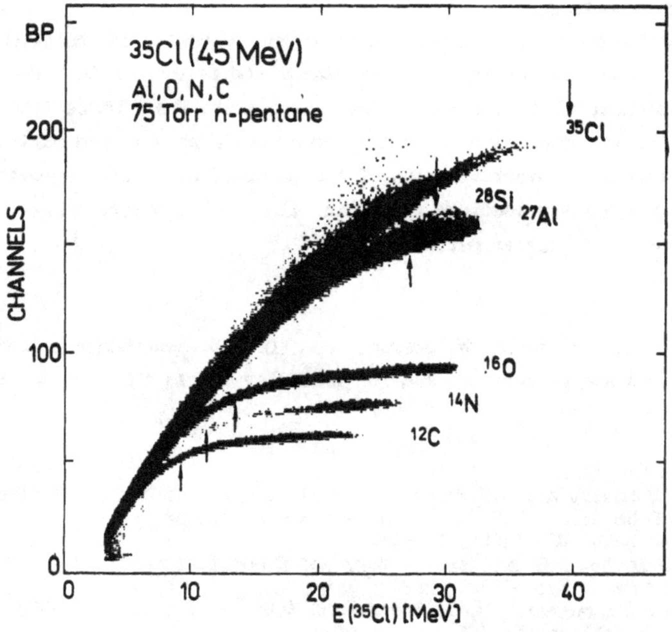

Fig. 2. Plot of the Bragg-peak height BP vs. the energy E. The arrows indicate the low-energy limit E_{lim} calculated by means of eq. (1) for the different species of ions.

A LITHIUM LIQUID METAL ION SOURCE

E. Hesse and F.K. Naehring[1)]

The potential advantage of lithium liquid metal ion sources in focused ion beam lithography

In the lithographic process ion beam exposure of resists has the opportunity of very high resolution structuring because the ions deposit their energy in a very narrow range, producing secondary electrons of very low energy and therefore short range. Thus there is no backscattering and minimal proximity effect as compared to electron irradiation (6). Light ions should produce small and deep damage regions because of the prevailing electronic stopping. For 100 keV Li ions in PMMA a 527 nm deep (Rp = 339 nm; ΔRp = 188 nm) and 84 nm wide (ΔRt = 42 nm) electronic energy deposition region is calculated (5). That is why these light ions should be especially usefull for lithography.
The fact that an ion deposits some 100 times more energy than an electron in a smaller volume is the basis for much of the interest in the development of ion beam lithography as the method needed for structuring at nanometer dimensions (6). Siegel (6) calculated, that writing 50 nm-lines, ion beam lithography would have a limiting rate two magnitudes faster than electron beam lithography (100 MHz versus 1 MHz).
Both gaseous field ionization and liquid metal ion sources provide high-resolution, high-current-density focused ion probes that could be used to write patterns in resists in the same manner as electron beams. Using liquid metal ion sources chromatic aberration limits the size of the focused beam (3). The diameter of the probe increases linearly with the energy broadening ΔE. In the expression of Knauer (2) ΔE is proportional to the cubic root of the ion mass. The first metal in the table of elements, lithium, should have the smallest broadening.

Lithium: abundances of the isotopes: 6Li: 7,52 % and 7Li: 92,48 %
melting point: 186 °C
vapour pressure at melting point: 2 * 10^-8 Pa

Because of this not too high values of melting temperature and vapour pressure it is possible to use the pure element as source material, contrary to other low mass metals like boron and beryllium. A smaller energy broadening of the low mass ions is expected from pure element sources than from source material alloyed with high mass elements as AuSiBe. Moreover, if a certain probe diameter is wanted, i.e. a given ΔE has to be setteled, the elemental ion source will provide a higher probe current compared to an alloy source the additional material of which has to be ionized too. Therefore a liquid metal ion source whith lithium should have a small energy broadening (Mair and Mulvey calculated about 2 eV near the onset current (3)) and a high intensity in a lithium focused ion beam is expected.

The ion source

First experimental results with liquid lithium ion emission have been obtained in a simple source arrangement (Fig.1). To restrict evaporation loss and ionization of thermally evoporated atoms a closed reservoir is necessary. The emitter consists of a capillary reservoir and a movable tungsten needle. This needle of 300 µm diameter was etched in NaOH to form the emitter tip. The intensitive reaction of lithium alkali metal demands preparation in an inert environment.

1) Zentralinstitut fuer Kernforschung Rossendorf
PF 19, Dresden, DDR-8051

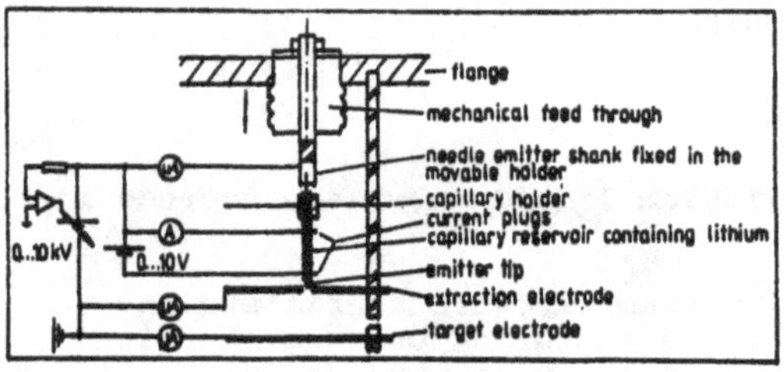

Fig. 1: Source arrangement for liquid lithium ion emission
a) Scheme, b) capillary reservoir with current plugs

Preparation of the capillary reservoir with lithium

At first the corrosion crust is cut off from the lithium block immersed in benzine. Some cubic millimetres of the lithium block are punched into the stainless steel capillary. When the capillary is taken out of the benzine to fix it in the capillary holder and to fasten the current plugs a thin film of benzine remains on its surface and protects lithium from air. The flange with the source arrangement is put onto the vacuum chamber for evacuation.
In the evacuated chamber lithium has been molten by direct heating of the capillary reservoir. In the absence of any film pure alkali metals always will spread over any pure solid metal (1).
After wetting the needle is moved to a position 0.5 mm behind the extraction electrode. The extraction voltage is raised until the emission starts.

Parameters of the lithium ion emission
Table 1

ion	Li^+
extraction voltage	3.8 kV
ion current	20 µA
heater current	5.2 A (15 W)
stability	1 %
(during 15 min., current control mode)	
life time	140 min
full emission angle	90

Results

During the life time the I-V characteristics of the source changed for serveral times. Therefore with constant extraction voltage the ion current is not constant. This current has been electronically stabilized by shifting the voltage. With 20 µA emission current the extraction voltage varied between 3.8 and 4.2 kV. Near the end of the life time the voltage of the stabilization circuit increased fast to 10 kV, the supply limit.

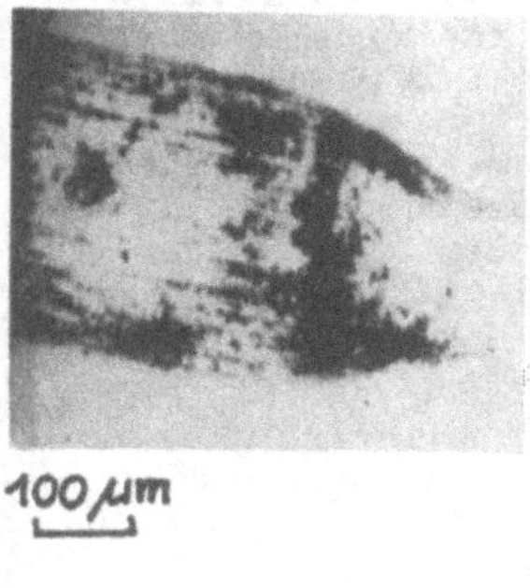

Fig. 2: Scanning electron micrograph of the emitter tip depleted from lithium at the end of emitting time. Material contrast mode, 200 fold magnification. The backscattering coefficient of tungsten is much higher than that of lithium. Therefore bare tungsten is bright in comparison to lithium-wetted regions.

At the end of the emitting time the emitter tip was deplected of lithium (Fig.2). The short life time is a result of hindered liquid flow, because the tip may be simply refreshed: The needle is newly wetted by drawing it back into the reservoir and the emission can be started again after pushing it on to its original position.
The reasons of unsufficient liquid flow may be (1):

- growth of Li3N layers from residual gas molecules or from nitrogen dissolved in the melt,
- formation of hydrocarbon layers from molecules of the residual gas by secondary electrons.

Supposing the life time to be fully determined by adsorbtion of nitrogen from the residual gas an increase from 140 minutes to 100 hours may be reached by a pressure change from 5*10^-4 Pa to 10^-5 Pa. The amount of nitrogen dissolved in lithium could be reduced by using a protecting argon atmosphere during preparation.

Lithium implantation into a silicon target

In an additional step it has been proved that mainly Li+ ions are emitted by the source. This was done by the evaluation of the depth distribution of 7Li in a silicon sample irradiated by the Li source. Two depth profiles of implanted Li were measured by SIMS. One has been registrated by (O2)+ primary ions about two month after implantation. The other has been traced by Ar+ primary ions about three month after implantation (Fig.3). Comparison of the measured profiles with an calculated implantation profile for Li+ ions (4) provides the same ranges. This means that the majority of the implanted ions are Li+ ions.
To sum up, the main ion emission of the lithium source is Li+.

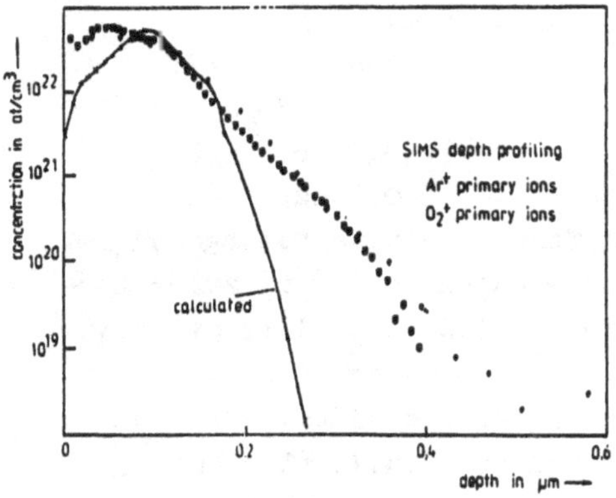

Fig. 3: SIMS depth profiles of a lithium implanted (100) silicon target (dose 7 * 10^17 ions/cm^2)

References

(1) Addision, C.C.: The chemistry of the liquid alcali metals, J. Wiley & Sons, Chichester 1984
(2) Knauer, W.: Optik 49 (1981) 335-354
(3) Mair, G.L.R., Mulvey, T.:Microelectronic Engineering 3 (1985) 133-146
(4) Otto, G.: privat comm.
(5) Posselt, M.: privat comm., The calculation bases on a version of the "TRIM"-scheme, developed by Biersack etal..
(6) Siegel, B.M. in "VLSI Electronics Microstructure Science" (N.G. Einspruch, ed.), Vol. 16, chapter 5, Academic Press, Orlando 1987

The samples were SIMS analysed by Mrs. L. Drechsler. Dr. L. Bischoff imaged the emitter tips by SEM. These contributions are kindly acknowledged.

PULSE EXPLOSION ION BEAM SOURCE WITH ONE PULSE REGIME FOR SURFACE MODIFICATION OF MATERIALS

Korenev S.A.
Joint Institute for Nuclear Research, USSR, 101000, Moscow,
P.O. Box 79.

The pulse ion beam source for surface modification of materials are used today/1/. The diode explosion schemes with two pulse regimes are generally used/2/. The first pulse of negative voltage is necessary for cathode plasma forming, and the second pulse of positive voltage for ion beam elongating.

The investigation of plasma bunches forming methods in vacuum conditions led to an idea of explosion ion source creation with one pulse supply regime (positive pulse voltage for source)/3/. The first experimental researches have shown its working ability/4/ and its simplicity in comparison with the other sources.

The variant of explosion ion beam source of that type for surface modification of materials will be discussed in this report.

Ion source consists of high-voltage Arkadiev-Marx pulse generator and vacuum diode with irradiation chamber.

The pulse generator has got following parameters:

1) pulse amplitude of voltage	100 - 300 kV;
2) electric energy	10 - 125 J;
3) duration of current pulse	300 ns;
4) pulse repetition rate	0,2 Hg.

At the output of pulse voltage generator forms positive voltage pulse.

Vacuum diode consists of stainless steel vacuum chamber, anode support with anode plasma initiator and cathode with fine cloth net. with vacuum chamber of diode and has substrates for plasing of irradiation samples. The collimator is placed before irradiational samples. Residual gas pressure was $5 \cdot 10^{-5}$ Tor in the diode and irradiational chamber. The ion beam current, falling at the sample, was registered by an integrating Rogovski transformer and high voltage - by high-resistance voltage devider with impulses, recording on the broad band double-trace oscillograp.

The anode plasma initiator materials determine the ion sort. To produce carbon ions a carbon-fibrius initiatir/5/ is used, to produce three component stream-ions of niobium, titanium and copper - niobium-titanium superconductive cable with partial pickled copper matrix/6/ is used. As the result we can say that this ion source can be used to produce multicomponental ion streams.

The explosion ion source can be plane or coaxial one, depending on functional purpose. For working ability of ion source it is necessary to realize the conditions for forming an anode plasma, that for plane and coaxial form should have a following form:

$$E_p = \frac{KU}{D} \gtrsim E_{thr} = 10^7 \text{ V/cm} \quad (1)$$

$$E_c = \frac{K\,U}{r\ln(R/r)} \gtrsim E_{thr} = 10^7 \text{V/cm} \quad (2),$$

where: E_p, E_c - average electric intensity in the plane and coaxial diodes;
K - amplification factor of electric intensity on the initiator of anode plasma;
D - distance between anode and cathode;
r, R - radius of anode and cathode;
E_{thr} - threshold electric intensity in the diode, at which anode plasma is formed by explosion effect.

The ion current yield satisfies Child - Langmuir and the ion current density was realised by changing the cathode gap. Dependence of carbon ion current density from the distance between anode and cathode you can see at the fig.1.

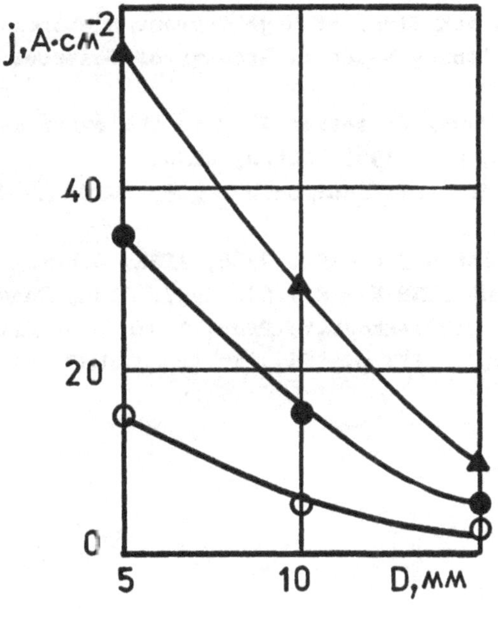

Fig.1. Carbon ion current density dependence from the anod - cathode dis - tance, diode voltage is: o - 100 kV; ● - 200 kV; ▲ - 300 kV.

It is necessary to mark that ion source, considered in this paper can be used for vacuum deposition facings of thick$\sim$ 0,5 - 100 mkm. The pulse charakter of substance's mass transfer (time of mass transfer is smaller than diffusion time of changing of elements stoichoimetric relationship) allowed to get multicomponent films from one pulversed target. The ion source has been used for spraying films of high temperature superconductors Y-Ba-Cu-O /7/. The last experiments on five component high temperature superconducting Bi-Ca-Sr-Cu-O films production gave good results. Stoichoimethric relationship of Y-Ba-Cu, Bi-Ca-Sr-Cu after the material was transferred from the target to the substance has been conserved with a high presision.

In the conclusion we can say:

- the explosion ion source of this construction and principle of work is more simple thah pulse sources of this type;
- the ion source can form a singlecomponent as well as multicomponent streams;
- the ion current density regulation has been realized by changing the diode gap corresponding to Child-Langmuir;
- this ion source can be used for vacuum deposition of singlecomponent and multicomponent facings, so that sprayied target is the anode plasma initiator itself.

References.

1. Didenko A.N., Ligachov A.E., Kurakin I.B. Chargea particles beam action on the surface of metal and alloy. Energoatomizdat, Moscow. 1987.
2. Isakov I.F., e.a. Proc. of the 5th Simp. of High Current Electronics, P.II, p.60, IHCE Syberian Publishing House of Academy of Sciences of USSR, Tomsk, 1984.
3. Korenev S.A. Proc. 2 Seminar Young Scientist JINR in the experimental physic field, NR15-85-862, p.4, 1985, Dubna, JINR.
4. Korenev S.A. Proc. 3 Seminar "Plasma Electronic", p.93, CFTI AS USR, Kharkov, 1988.
5. Korenev S.A., e.a. Preprint JINR N13-88-292, 1988, JINR, Dubna.
6. Vavra I., Korenev S.A. Preprint JINR N13-87-751, 1987, JINR, Dubna.
7. Korenev S.A., Valentovich D., Sikolenko V.V. Proc. 1 Conf. of High Temperature Superconductivity, ,P.III,p. 166., FTILT AS USR, Kharkov, 1988.

ULTRASONIC SURFACE WAVES FOR STUDYING THE PROPERTIES OF THIN FILMS

R. Kuechler *)

Introduction

The mechanical properties of most vacuum deposited thin films are different from those of the bulk material. Evaporated films have in many cases poor adhesion to the substrate and are highly stressed [1]. In order to study the mechanical properties and stress profiles of thin films surface acoustic waves (SAWs) are a convenient tool. SAWs successfully have been used to study stress profiles in glass surface layers [2,3].
SAWs propagate along the free surface of a solid. The penetration depth of these inhomogeneous waves is approximately one wavelength, so the thickness of the surface layer through which the wave propagates changes with the applied frequency ($\Lambda = v/f$). In this report the modification of the spreading velocity of ultrasonic surface waves by thin film deposition is described by an approximation of the nonlinear equation of motion. It is assumed that the residual stress is given as a step function in the range of the penetration depth of the SAW.
The ratio of the layer thickness to the wavelength determines first of all the spreading velocity change of SAWs after surface modification.

Calculations to the problem

The starting point for describing the effect of residual stress in deposited films on the modification of the spreading velocity are the nonlinear equations of motion.
These equations contain the interaction between residual stress and the SAW. It is sufficient for the practical calculation to start with the linear approximation of these equations [4]:

$$\varrho \ddot{U}_i = (C_{ik,jl} + S_{ik,jl,mn} U^0_{m|n}) U_{i|kl} \qquad (1)$$

with $S_{ik,jl,mn} = C_{ik,jl,mn} + C_{kl,mn}\delta_{ij} + C_{kn,jl}\delta_{im} + C_{ik,lm}\delta_{jm}$

ϱ : mass density
U_i : particle displacement vector (wave)
U^0_i : particle displacement vector (residual stress)
$C_{ik,jl}$: second order elastic coefficients

The $C_{ik,jl,mn}$ are the third order coefficients in the expansion of the strain energy. An isotropic material requires three independent third order coefficients (l,m,n).

Geometry and Parameters

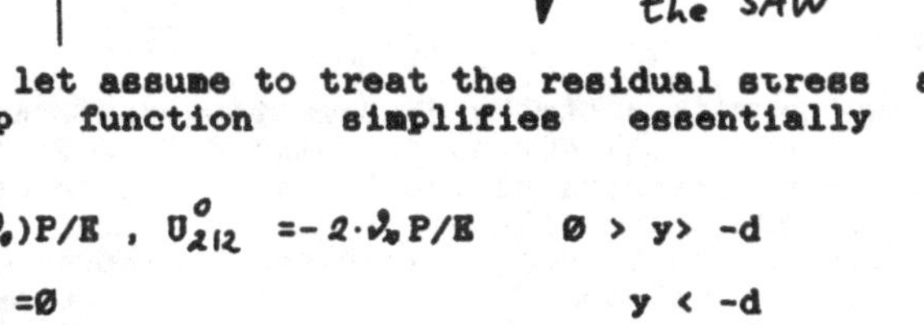

The film on a substrate let assume to treat the residual stress as a step function. The step function simplifies essentially the calculation:

$$U^0_{1|1} = U^0_{3|3} = (1-\vartheta_0)P/E\,,\quad U^0_{2|2} = -2\cdot\vartheta_0 P/E \qquad 0 > y > -d$$

$$U^0_{1|1} = U^0_{2|2} = U^0_{3|3} = 0 \qquad y < -d$$

*) Central Institute for Nuclear Research, Rossendorf

Within Ø > y > -d tne corresponding equations (1) are given by [4]

$$
\begin{aligned}
\rho\ddot{U}_1 &= (\lambda+2\mu+\varepsilon_1)U_{1|11} \quad +(\mu+\varepsilon_3)U_{1|22} \quad +(\lambda+\mu+\varepsilon_4+\varepsilon_5)U_{2|12} \\
\rho\ddot{U}_2 &= (\lambda+2\mu+\varepsilon_2)U_{2|22} \quad +(\mu+\varepsilon_3)U_{2|11} \quad +(\lambda+\mu+\varepsilon_4+\varepsilon_5)U_{1|21}
\end{aligned} \tag{2}
$$

with

$$\varepsilon_1 = 2P/E((1-2\nu)(1+2m)+2(1-\nu)(m+2n)+3/2(\lambda+2\mu)-2\nu(\lambda+\mu))$$

$$\varepsilon_2 = 2P/E((1-2\nu)(1+2m)-4\nu(m+2n)-2(1-\nu)\lambda)$$

$$\varepsilon_3 = 2P/E((1-2\nu)(m+n)-\nu n+1/2(1-\nu)\mu)$$

$$\varepsilon_4 = 2P/E((1-2\nu)(1+m)-\nu m+1/2(1-3\nu)\lambda)$$

$$\varepsilon_5 = 2P/E((1-2\nu)(m+n)-\nu n+1/2(1-3\nu)\mu)$$

λ, μ Lame coefficients, E Young modulus, ν Poisson ratio
For y < -d system (2) is valid with $\rho_0, \lambda_0, \mu_0, \nu_0$ and P=Ø.
The solution of the system (2) can be written in the form

$$
\begin{aligned}
U_1 &= A(y)\cos(kx-\omega t) \\
U_2 &= B(y)\sin(kx-\omega t)
\end{aligned} \tag{3}
$$

Substituting Eq.(3) in the Eq.(2) and taking into cosideration the boundary conditions (the plane y=Ø is stress less and A,B→0 for y→-∞) leads to a transcendental equation which determines the velocity of SAW. An analytical solution of the transcendental equation is only possible for $d/\Lambda \ll 1$:
Velocity of SAW $V_R = V_R^0 + \Delta V$

with $V_R^0 = (.87+1.12\,\nu_0)/(1+\nu_0)V_T^0$, $V_T^0 = \sqrt{\frac{\mu_0}{\rho_0}}$, $V_L^0 = \sqrt{\frac{\lambda_0+2\mu_0}{\rho_0}}$

$$
\begin{aligned}
\Delta V = -4\pi \frac{d}{\Lambda} V_R^0 \left(\frac{V_R^0}{V_T^0}\right)^2 \frac{\mu\, e^{-(2.2+\nu_0)\nu_0}}{\mu_0} \Bigg[& \left(\frac{V_R^0}{V_T}\right)^2 (s_0+q_0) - 4 s_0 \frac{(\lambda+\mu)}{(\lambda+2\mu)} \\
& - \frac{1}{\mu}\left(s_0 \left(\varepsilon_1 + \frac{\lambda^2}{(\lambda+2\mu)^2}\varepsilon_2 - \frac{2\lambda}{(\lambda+2\mu)}\varepsilon_4 \right) - 2 q_0 (\varepsilon_5 - \varepsilon_3) \right) \Bigg]
\end{aligned} \tag{4}
$$

with $q_0 = \sqrt{1-\left(\frac{V_R^0}{V_L^0}\right)^2}$, $s_0 = \sqrt{1-\left(\frac{V_R^0}{V_T^0}\right)^2}$

Figure 1 shows the spreading velocity change ΔV calculated from Eq.(4)($P=Ø \rightarrow \varepsilon_i=0$) in comparison with the accurate solution for four different films overlaying on elastic halfspace as a function of the ratio of the film thickness (d) to the wavelength (Λ). The elastic coefficients of the film were changed in the following manner:

	λ [GPa]	μ [GPa]	ρ [g/ccm]
elastic halfspace	1Ø7.4	8Ø.3	7.7
film 1	125.2	93.1	8.1
film 2	143.9	1Ø6.5	8.45
film 3	164.2	121.7	8.84
film 4	184.9	139.1	9.24

The number with a stroke denotes the exacte solution.
Figure 2 shows ΔV-curves calculated from Eq.(4) as a function of d/Λ with the compressive stress (P) in the film as a parameter. Thus, only P was changed in the film.
Although the ratio d/Λ is very small a large compressive stress results in a measurable modification of the spreading velocity. As third order elastic coefficients the values of steel C45 [5] were used.

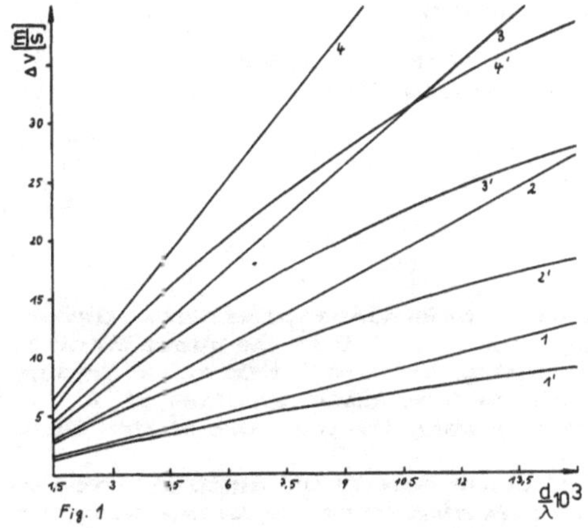

Fig. 1

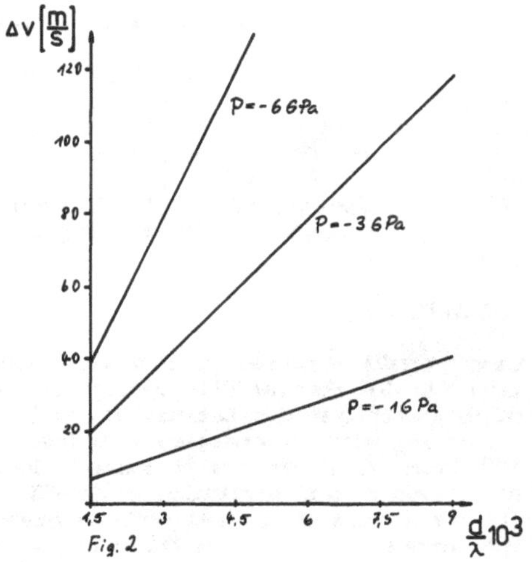

Fig. 2

Result of the measurement

The investigated sample was a steel bar (10cm*1cm*1cm) with a 1 μm thick TiN-film deposited by PVD. The velocity change measurement was performed by the sing-around method [6]. The frequency was 4 MHz. For V the average value 7.9+/-2.1 m/s was measured . Variation of the free parameters in formula (4) yielded P∼-8 GPa for the residual stress in the TiN-film. This is a reasonable value.

Conclusion

1. The stress in thin film plays an important role in surface hardening. The measurement of this stress is possible by using ASWs.
2. The presented calculations are useful for the determination of the mechanical parameters of both coated and implanted metal surfaces.
3. Although the ratio between the thickness and the wavelengh is very small the modification of the spreading velocity can be measured for example by the sing-around method.

References

[1] K.L. Chopra: Thin Film Phenomena" Mc Graw-Hill Book Comp. New York 1969
[2] J.A. Bucaro and C.M. Davis: J. Appl. Phys., 43(1972) 2151
[3] R. Küchler und R. Richter: Bericht ZfK-472, Zentralinstitut für Kernforschung, Rossendorf, März 1982
[4] R.N. Thursten and K. Brugger: Phys. Rev. A133 (1964) 1604
[5] D.H. Tietz: Ultraschallmesstechnik, Berlin, 1974
[6] S. Haussühl: Kristallphysik, Leipzig, 1983

PROBE MEASUREMENTS IN LASER PRODUCED PLASMA

H. Müller[(1)], P.Siemroth[(2)], K. Wetzig [(1)]

(1) Zentralinstitut für Festkörperphysik und Werkstofforschung der AdW der DDR, Helmholtzstr. 20, Dresden, DDR-8027.

(2) Zentralinstitut für Elektronenphysik der AdW der DDR, Hausvogteiplatz 5-7, Berlin, DDR-1086.

INTRODUCTION

Laser produced plasma is used widely for deposition tasks (laser pulse vapor deposition - LPVD). Special film properties can be attained with LPVD if the plasma consists of ions with energies between 10 and 100 eV. In contradiction with theoretical predictions /4/ such ion energies were found even with low laser power densities of 10^7 ... 10^8 W/cm^2 /2/.Therefore it seems to be important to study the processes of the laser stimulated plasma production in detail.
In the present experiment probe measurements of plasma density and electron temperature were carried out simultaneously with in situ observation of the surface structure of the target between single laser shots.

EXPERIMENTAL

The present experiments were carried out in the vacuum chamber of a commercial scanning electron microscope /6/. A Nd-glass laser beam (λ=1.06 µm) was introduced into the microscope and focused to a power density of 10^7 ... 10^8 W/cm^2 on the sample surface. The laser pulse duration was about 50 ns. Between single shots the interaction area on the target can be observed with the SEM.
Langmuir-probe measurements of plasma density and electron temperature were carried out in expanding plasmas produced by irradiating targets of copper, titanium and superconducting material ($YBa_2Cu_3O_x$). The probe was located in a distance of about 3 mm apart from the laser focus. It had an effective area of 0.1 mm^2.

RESULTS AND DISCUSSION

Time resolved measurements of plasma density (for all target materials) and electron temperature (for Ti and superconducting material) were carried out. With a laser power density of about 10^8 W/cm^2 the following maximum plasma densities were found:

for Cu	$3.6\cdot10^{20}$ m^{-3}
for Ti	$6.0\cdot10^{21}$ m^{-3}
for supercond. mat.	$7.6\cdot10^{21}$ m^{-3}.

The time interval between the plasma production due to a short laser pulse and its registration at the probe was treated as the time of flight of ions with a certain energy from the small region of dense collision dominated plasma (laser spot) to the probe. Therefore the plasma density at a definite time corresponds to the amount of ions which leave the primary dense plasma cloud with the corresponding energy. It was found, that the energy distribution of the plasma ions can be fairly well described by a Maxwellian distribution. The "temperatures" corresponding to the fitted distributions are about 17 eV as well for Cu as for Ti unless the densities differ for more than one order of magnitude (see fig. 1). The density signals found with superconducting material are in good agree-

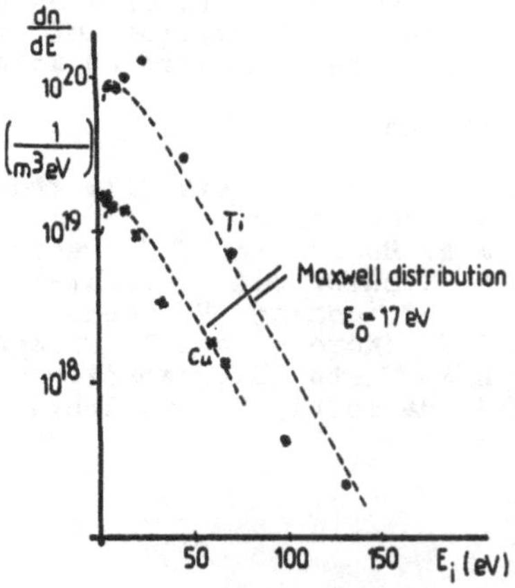

Fig. 1 Energy distribution of laser produced plasma ions for Ti and Cu targets and the fitted Maxwellian distribution for E_0=17 eV.

ment with a superposition of Maxwellian distributed ion energies for each component (Y, Ba, Cu, O) with the same "temperature" of 17 eV.

The ion energies of about 17 eV as measured in the present experiments and similar results from other authors (e. g. /2/) are in contrast to theoretical predictions. At the used wave length of the Nd-glass laser $\lambda = 1.06\ \mu m$ and for a power density of 10^8 W/cm^2 the temperature of the laser produced plasma near the focal point was estimated to be in the range of $T \approx T_e \approx T_i = 5 - 8$ eV /4/. An explanation may be given from the following discussion of the erosion pattern found on copper.

Simultaneous and comparing observations of laser pulse intensity, plasma density and surface morphology show, that the process of plasma production is of highly stochastical nature. Fig. 2 a ... d reproduces a series of laser spots on a untreated copper surface together with the Langmuir-probe. Only a small part of the laser spot is effected during one single shot (e. g. the process from fig. 2 a to fig. 2 b), moreover a lot of shots have no visible impact on the surface and no plasma production at all. The typical erosion structure on copper are small craters of about 10 µm in diameter, which flock together to a more or less homogeneous, larger spot area after a certain number of shots (fig. 2 d). This small craters seem to be a result from the plasma-surface interaction rather than from the direct laser-surface interaction. Calculations gave evidence that due to the high reflectivity of copper the laser light intensity is too low to heat a clean copper surface up to melting temperature but it is quite sufficient to produce plasma from inclusions and impurity layers. This plasma will be heated by the laser light up to temperatures of about 5 eV and interact with the metal surface. It was shown earlier that such a plasma is able to initiate with a certain probability arc discharges, the so called unipolar arcs /5/, /3/. This arcs burn between the plasma cloud and the metallic surface. They were fed by the potential difference between the plasma and the target surface and transfer plasma energy into electrical current which erodes the surface. The unipolar arcs may exist as long as the laser irradiation heats the plasma cloud. The ions with mean energies of about 17 eV originate obviously from the arc spots. Ion energies from the cathode spot of vacuum arcs are in the same range (e. g. /1/).

Therefore we conclude that at least in the power density range studied in our experiments unipolar arcing may increase the plasma-surface interaction remarkably and thereby the laser power input into small surface regions. On the other materials (Ti, supercond.) the absorption coefficient is high enough that during each single laser pulse a

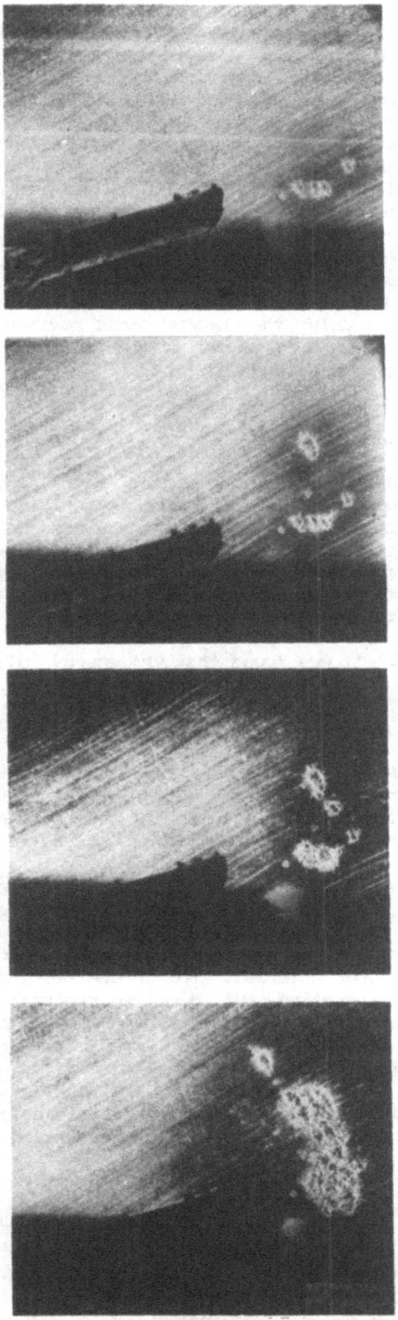

Fig. 2 Erosion pattern on copper target after
- -a- one laser shot
- -b- two laser shots
- -c- three laser shots
- -d- five laser shots

thin surface layer of some µm thickness was molten quite homogeneously within the whole laser spot. Examples are given in fig. 3 (Ti) and fig. 4 (supercond. mat). With this materials also plasma can be observed during each pulse, but the density signals differ remarkable from shot to shot. Therefore it is important, to control the plasma parameters permanently, if such plasmas shall be used for deposition tasks (LPVD). Probe measurements seem to be a simple and effective method for in situ control. The internal structure of the surface seems to be a result of only the hydrodynamic movement of the molten surface due to the repulsing force of the vaporized material.

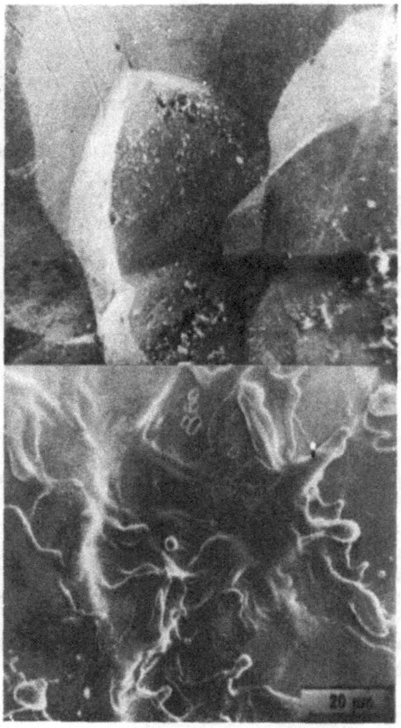

Fig. 3 Surface morphology of the Ti target
-a- before laser treatment
-b- after one shot

CONCLUSIONS

* The kinetic energies of the plasma ions, produced with all target materials, used in the present experiment are remarkable higher than predicted from only the laser-plasma interaction but they are in the same order of magnitude as the ion energies from the cathode spot of vacuum arcs.

* It was found that the energy distribution of the laser plasma ions is Maxwellian-like. The mean energy (temperature) is for all studied materials about 17 eV.

* The probe signals of $YBa_2Cu_3O_x$-material can be described as a superposition of signals of the components, each of them also consisting of ions with Maxwellian distributed energies.

* If we take into account the low absorption coefficient of copper, it can be shown that the laser power density in the present experiment is to low for an evaporation of any copper material. But in contrast, not only evaporation but also a remarcable plasma production was observed. Moreover the erosion pattern, found in our experiments are similar to the cathode spot erosion of vacuum arcs. Therefore we conclude that the laser-target interaction at low power densities is enhanced by a process of unipolar arcing.

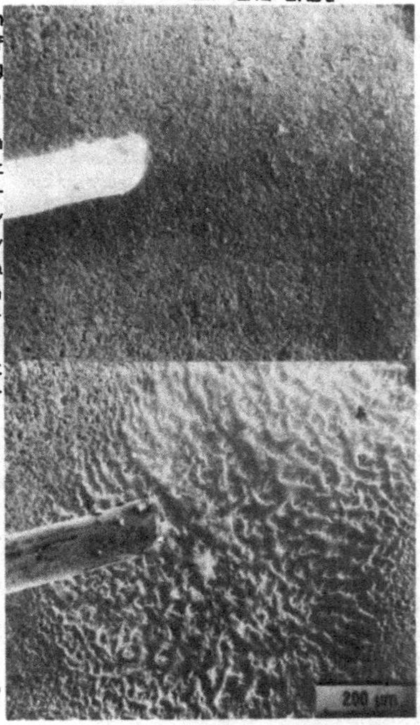

Fig. 4 Surface morphology of the $YBa_2Cu_3O_x$ target
-a- before treatment
-b- after 23 shots

LITERATURE

/1/ J. E. Daalder; J. Phys. D:Appl. Phys. 9(1976)2395.
/2/ D. Lubben et al.; J. Vac. Sci. Technol. B3(1985)968.
/3/ A. V. Nedospasov, V. G. Petrov; Nucl. Fus. 21(1981)413.
/4/ H. Puell; Z. Naturforsch. 25A(1970)1807.
/5/ F. Schwirzke; J. Nucl. Mater. 128&129 (1984)780.
/6/ K. Wetzig, J.Edelmann, W. Fischer, H. Müller; Scanning 9(1987)99.

PLASMATRON ION SOURCE FOR ION IMPLANTATION

H. Reuther+, J. Strümpfel++

For ion implantation in metals high current ion sources are needed. Up to now only sources for gases are available. In the this work an ion source for generating metal ion beams also from high melting elements is presented. Operating of source is based on a magnetron discharge as used for high rate sputtering [2]. However, contrary to the known magnetron for sputter deposition the plasma zone over the source target is bounded by a special shaped shield containing a hole for the beam extraction (Figure 1). The schematic electrical setup is shown in Figure 2.

A working gas is necessary for operating the source. Besides the metal ions also a large fraction of ions of this gas are generated. With argon as working gas the source was used to produce Cu, Al, and Ti ion beams, however, the production of, for instance, tungsten or molybdenium is possible too in the same way. The total ion current amounted to up to 40 mA (Table 1) , the metal ion fraction lay between 15% for Cu and 40% for Al (Table 2).

Table 1: Maximum ion currents (without mass separation)

Cu	35 mA	(power 713 W)
Ti	27 mA	(power 592 W)
Al	42 mA	(power 656 W)

Table 2: Maximum fraction of metal ions

Cu	15 %
Ti	19 %
Al	40 %

Without beam handling the source produces a very broad ion beam. For this reason the distance between source and target should be short for simple applications.

In Figures 3, 4, and 5 total ion current and metal ion fraction vs. discharge power of the source at different pressures in the argon pipe are shown. The investigation of the mass spectra produced by the plasmatron source was performed in a special device for testing ion sources described in [1]. This device enables maximum currents of 10 mA and energies of 20 keV. The distance between source and Faraday cup was about 110 cm, therefore only little currents were detected. The maximum currents were measured in another setup without mass separation where the distance between source and sample was about 40 mm.

References
[1] Hentschel, R. and D. Henke: ZfK Report, in press
[2] Schiller, S., U. Heisig and K. Goedicke: Thin Solid Films 40 (1977) 327-334

+ Zentralinstut für Kernforschung Rossendorf der Akademie der Wissenschaften der DDR, Postfach 19, Dresden, 8051, GDR
++ Forschungsinstitut "Manfred von Ardenne" Zeppelinstr. 7, Dresden, 8051, GDR

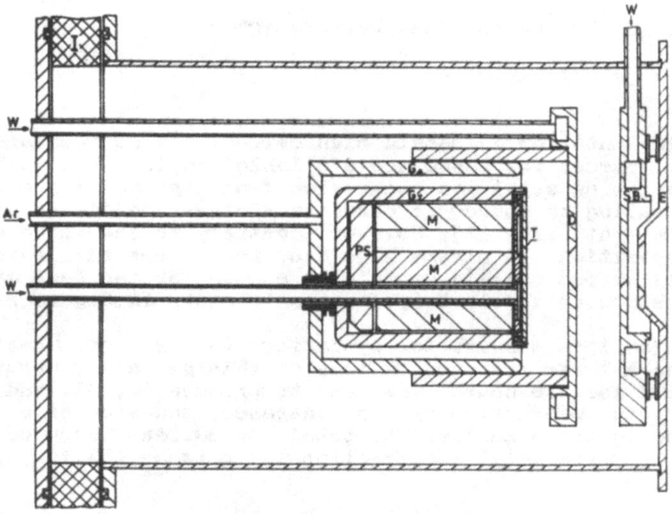

Figure 1: Plasmatron ion source
(W water, M magnet, PS pole piece, T magnetron target, D magnetron seal, E extraction electrode, I insulator, SB secundary electron screening, GI inner magnetron screening, GA outer magnetron screening)

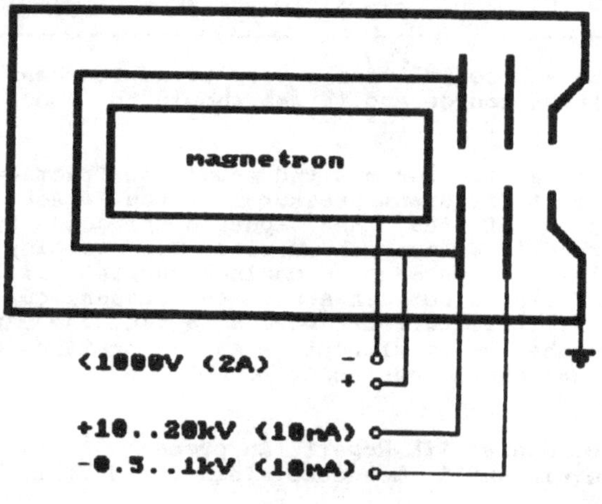

Figure 2: Schematic electrical setup

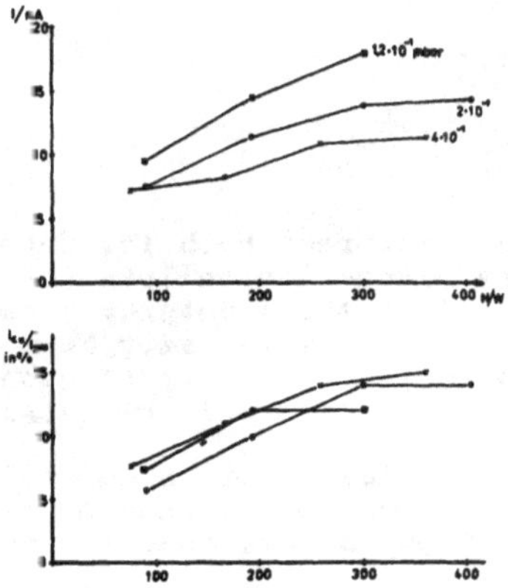

Figure 3:
Metal and relative metal ion current of copper in dependence of the pressure in the argon pipe

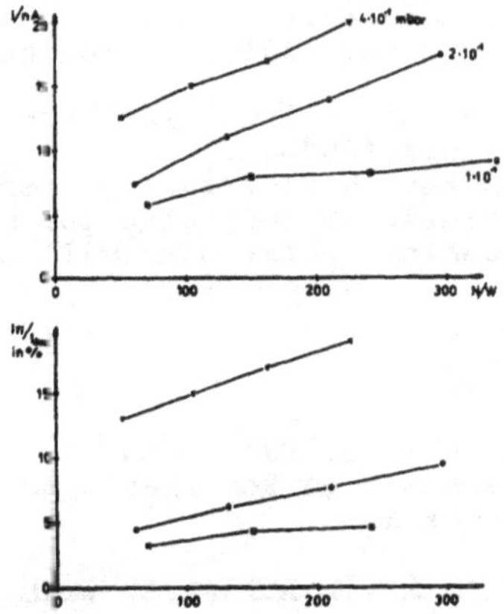

Figure 4:
Metal and relative metal ion current of titanium in dependence of the pressure in the argon pipe

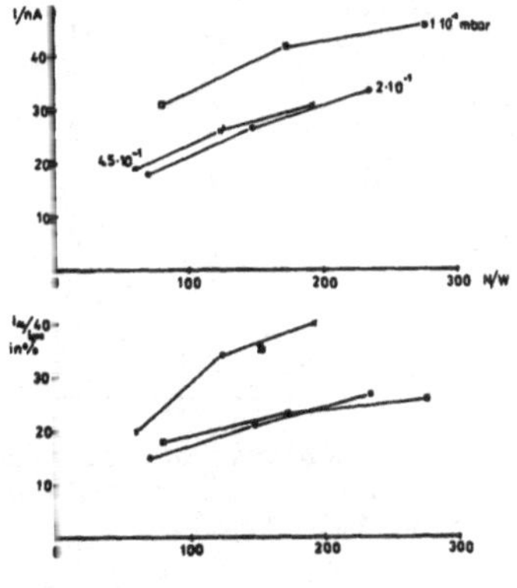

Figure 5:
Metal and relative metal ion current of aluminium in dependence of the pressure in the argon pipe

CEMS STUDY ON ALUMINIUM IMPLANTED IRON

H. Reuther, E. Richter+

1. Introduction

The system iron-aluminium is of some interest both for fundamental research and for technical reasons. The iron-rich alloys show very interesting magnetic properties [1], and in the aluminium-rich region quasicrystalline phases were detected [2]. Moreover, amorphous alloys can be obtained in this system. The corrosion behaviour of such alloys is very favourable, and it seems to be that they will be applied as nuclear reactor materials.

Strong efforts were made to produce thin layers of such alloys on normal materials. Technologies used are ion beam mixing or other ion assisted methods and different kinds of vapour deposition. In some cases multilayers are produced [3-5].

However, there is a further method to obtain thin layers, namely direct ion implantation. According to the desired concentration range either iron has to be implanted in aluminium or aluminium has to be implanted in iron. The layer influenced by ion implantation has some 100 nm in thickness. Because there is no homogenous doping as a result of the implantation process it could be necessary to anneal the sample after the implantation to get a smoothing of the initial concentration profiles.

In the present work aluminium implantation in iron is investigated by conversion electron Mössbauer spectroscopy (CEMS).

Implantation conditions are chosen in that manner that concentration profiles different in depth and maximum aluminium concentration could be studied. The investigation of the annealing behaviour will be the subject of another paper [6].

2. Experimental

Implantation conditions are given in Table 1. Due to the low current densities it could be guarantied that there was no remarkable warming up of the samples during the implantation process.

Table 1: Parameters for aluminium implantation in iron

Sample	Dose (cm^{-2})	Energy (keV)	Current density ($\mu A/cm^{-2}$)
FA1	5×10^{16}	50	2
FA2	7.5×10^{16}	50	3
FA3	1×10^{17}	50	5
FA4	2×10^{17}	50	4
FA5	5×10^{17}	50	4
FA6	5×10^{16}	100	1
FA7	7.5×10^{16}	100	2
FA8	1×10^{17}	100	3
FA9	2×10^{17}	100	3
FA10	5×10^{17}	100	4
FA11	2×10^{17}	200	3

For the Mössbauer measurements the same conditions as described earlier [7] were used. The Mössbauer spectra were fitted with the NORMOS least squares computer program written by R.A. Brand.

+Central Institute for Nuclear Research Rossendorf of the Academy of Sciences of the GDR, Post Box 19, Dresden, 8051, GDR

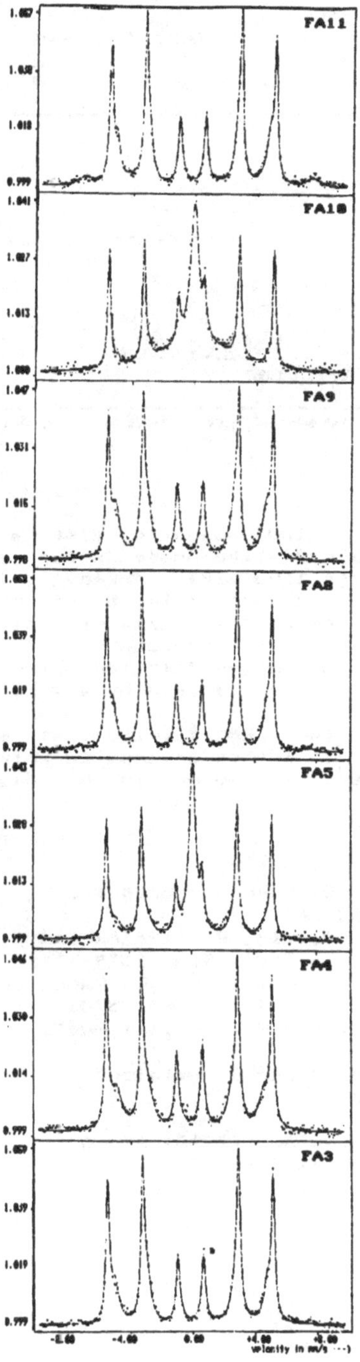

Figure 1: CEM spectra

3. Results

In Figure 1 the fitted CEM spectra of several aluminium implanted iron samples are shown. Only the spectra of samples implanted with doses of 1×10^{17} cm^{-2} and more were selected because there is no difference of principle between the spectra of samples implanted with 1×10^{17} cm^{-2} and the spectra of those samples implanted with lower doses.

In the spectra of samples implanted with doses up to 2×10^{17} cm^{-2} only magnetic components were found, however in the CEM spectra of samples FA5 and FA10 implanted with 5×10^{17} cm^{-2} nonmagnetic fractions were detected too.

In Table 2 the hyperfine parameters of the fitted spectra are given.

4. Discussion

The mean projected range of aluminium in iron is 29, 58, and 117 nm for ion energies of 50, 100 and 200 keV, respectively. For this reason it is to take into account that in the CEM spectra, especially in the case of the 50 keV implantation, a fraction generated by the bulk material behind the implanted region is included. The bulk material is normal alpha iron, in Table 2 represented by magnetic component (1). Magnetic component (2) is due to the aluminium implantation. Its hyperfine parameters let conclude that a Fe_3Al like structure was formed. The isomer shift increases and the magnetic field decreases slightly with increasing implantation dose. This is the consequence of the increasing aluminium content as already reported in [8]. While the line width of the magnetic component (1) is only somewhat larger than that of normal alpha iron the lines of the second component are much broader and increase with increasing implantation dose. This is an evidence of the increasing disorder in the implanted region with a transition into the

Table 2: Hyperfine parameters of the CEM spectra

Sample	Magnetic components												Nonmagnetic components					
	(1)				(2)				(3)				(1)			(2)		
	A	IS	HI	HW	A	IS	HI	HW	A	IS	HI	HW	A	IS	HW	A	IS	HW
FA1	74.2	0	33.1	0.264	25.8	0.029	30.8	0.445										
FA2	63.7	0	33.1	0.263	36.3	0.037	30.8	0.332										
FA3	62.8	0	33.0	0.264	37.2	0.045	30.4	0.566										
FA4	54.2	0	33.0	0.263	45.8	0.047	29.5	0.817										
FA5	46.6	0	33.0	0.250	23.6	0.069	29.8	0.877					29.8	0.226	0.509			
FA6	76.8	0	33.2	0.249	23.2	0.027	31.1	0.442										
FA7	71.7	0	33.2	0.252	28.3	0.026	31.1	0.408										
FA8	62.7	0	33.2	0.253	37.3	0.034	30.9	0.479										
FA9	47.3	0	33.2	0.250	52.7	0.043	30.4	0.654										
FA10	33.0	0	33.2	0.249									21.8	0.189	0.647	45.2	0.280	6.008
FA11	53.4	0	33.2	0.263	34.0	0.034	30.9	0.478	12.6	0.302	46.2	1.059						

A is the relative area in %, IS the isomer shift in mm/s, HI the magnetic hyperfine field in T, and HW the line width in mm/s.

amorphous state.

In samples FA5 and FA10 besides the magnetic also a non-magnetic component was found indicating a FeAl-like phase which should be amorphous too because of its large line width. Moreover, in the spectrum of sample FA10 a further component noticable by a general increase of underground symmetrically to the zero velocity was detected. Similar to iron implanted with the same amount of phosphorus [7] this subspectrum should be ascribed to an amorphous fraction without any preferred short range order resulting in a widely spread hyperfine field distribution.

The magnetic component (3) found in the CEM spectrum of sample FA11 indicates the presence of iron oxid. It is not obvious in which way the oxygen could reach the sample, however, it seems to be happened somehow during the implantation process.

References

[1] Chacham, H., E. Galvao da Silva, D. Guenzburger and D.E. Ellis: Phys. Rev. B35 (1987) 1602-1608

[2] Dunlap, R.A., D.J. Lloyd, I.A. Christie, G. Stroink and Z.M. Stadnik: J. Phys. F: Met. Phys. 18 (1988) 1329-1341

[3] Godbole, V.P., S.M. Chaudhari, S.V. Ghaisas, S.M. Kanetkar, S.B. Ogale and V.G. Bhide: Phys. Rev. B31 (1985) 5703-5711

[4] Iwaki, M., Y. Okabe, K. Takahashi and K. Yoshida: Nucl. Instr. Meth. 209/210 (1983) 941-945

[5] Jaouen, C., J.P. Eymery, E.L. Mathe and J. Delafond: Mat. Sci. Eng. 69 (1985) 483-488

[6] Reuther, H.: to be published

[7] Reuther, H.: Nucl. Instrum. Methods B30 (1988) 61-66

[8] Stearns, M.B.: Phys. Rev. 147 (1966) 439-453

CEMS MEASUREMENTS ON ION IMPLANTED TOOL STEELS

H. Reuther+, E. Richter+, I.S. Szücs++

Introduction

Fe-57 Mössbauer spectroscopy on iron and iron containing samples is very suitable for investigating atomic surroundings of the iron atom. Steel alloys generally show complicated Mössbauer spectra composed by many subspectra with hyperfine parameters very strongly depending on every single component of the alloy.

On the application of the conversion electron Mössbauer spectroscopy (CEMS) for the study of iron implanted with boron, carbon, and phosphorus had been reported in [3]. Using this technique not the 14.4 keV-gamma-rays as in the case of the conventional Mössbauer transmission experiments but the conversion and Auger electrons are detected. These electrons have a maximum energy of 7.3 keV. The Mössbauer signal can be obtained from a surface layer of about 300 nm. But the main fraction of the signal, namely 65 %, comes from the first 50 nm. This range is just interesting for ion implantation.

In the present paper a CEMS study is given relating two tool steels (210Cr46 and 90MnV8) which showed a pronounced improved wear behaviour after ion implantation with carbon and nitrogen, respectively [1].

Experimental

The amounts of the main components of the investigated steels are given in Table 1.

Table 1: Composition of the investigated steels (wt.%)

Steel	C	Cr	Mn	V
210 Cr 46	2.1	11.5		
90 MnV 8	4.0		1.8	0.1

Steel 90MnV8 was implanted with $8x10^{17}$ cm^{-2} nitrogen ions at 50 keV. Implantation was performed without mass separation. The ratio of nitrogen atom to nitrogen molecule ions was 40 to 60.

Steel 210Cr46 was implanted in two steps with $5x10^{17}$ cm^{-2} carbon ions at 50 keV and with $2x10^{17}$ cm^{-2} carbon ions at 20 keV. The ion beam was mass separated. An additional covering of the sample with carbon took place as a result of the vacuum conditions during the implantation process.

The Mössbauer measurements were carried out at room temperature with a conventional spectrometer in constant acceleration regime. The samples were mounted on the backplate of a proportional gas-flow counter. The gas mixture was He with 5 % CH_4. The gas flow rate was about 20.5 cm^3/min, and the counter voltage was 760 V. Samples had a diameter of 32 mm and a thickness of 2 mm. The investigated area had a diameter of 10 mm.

The Mössbauer spectra were fitted with the MM/Sirius least squares fit program.

+ Central Institute of Nuclear Research Rossendorf of the Academy of Sciences of the GDR, Post Box 19, DDR-8051, GDR
++ Central Research Institute for Physics H-1525 Budapest, P.O. Box 49, Hungary

Results

The CEM spectra are shown in Figures 1 and 2, and the hyperfine parameters of the fitted spectra are given in Tables 2 and 3.

Table 2:
Hyperfine parameters of the CEM spectra of the steel 210Cr46

	Untreated				Implanted and covered with carbon			
	A	IS	HI	QS	A	IS	HI	QS
sextet 1	23	-0.011	33.9		8	-0.012	33.8	
sextet 2	16	-0.015	32.9					
sextet 3	28	-0.022	30.7		21	0.017	30.9	
sextet 4	15	-0.011	27.9					
sextet 5					5	0.266	23.6	
sextet 6					45	0.236	16.3	
singulet	3	-0.110			4	-0.094		
doublet	15	0.126		0.550	17	0.159		0.713

Table 3:
Hyperfine parameters of the CEM spectra of the steel 90MnV8

	Untreated				Implanted with nitrogen			
	A	IS	HI	QS	A	IS	HI	QS
sextet 1					25	0.237	47.2	
sextet 2	60	-0.015	33.1		29	-0.012	33.2	
sextet 3	31	-0.015	31.0		7	-0.014	32.2	
singulet	2	-0.103			19	0.300		
doublet	7	0.088		0.537	20	0.416		0.298

A is the relative area in %, IS the isomer shift in mm/s, HI the magnetic hyperfine field in T, and QS the quadrupole splitting in mm/s.

The Mössbauer spectrum of the untreated steel 210Cr46 shows about 82% martensitic (sextets) and 18% austenitic fractions (doublet and singulet). Martensitic structure gives rise to a great deal of different iron surroundings, and sextet 3 is due to the introduction of chromium into the iron lattice which leads to a decrease of the hyperfine field by about 2.2 T [4].

The austenitic fractions and also the fraction due to the chromium are influenced only a little as a result of the carbon implantation. However, the martensitic fractions strongly decrease and two new subspectra (sextet 5 and 6) appear. By reason of their hyperfine field parameters they can be ascribed to the crystalline phase of Fe_2C. This phase should be responsible for the improved wear behaviour of the sample.

The Mössbauer spectrum of the steel 90MnV8 shows in the untreated state 91% martensitic (sextets) and 9% austenitic fractions (doublet and singulet).

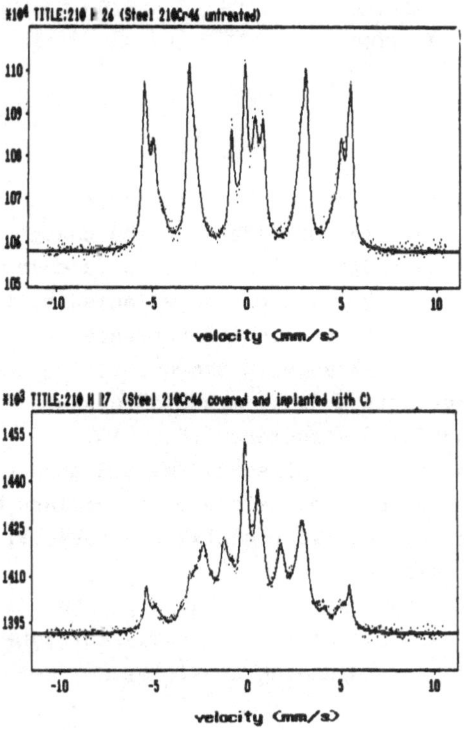

Figure 1:
CEM spectra of steel 210Cr46 untreated and implanted

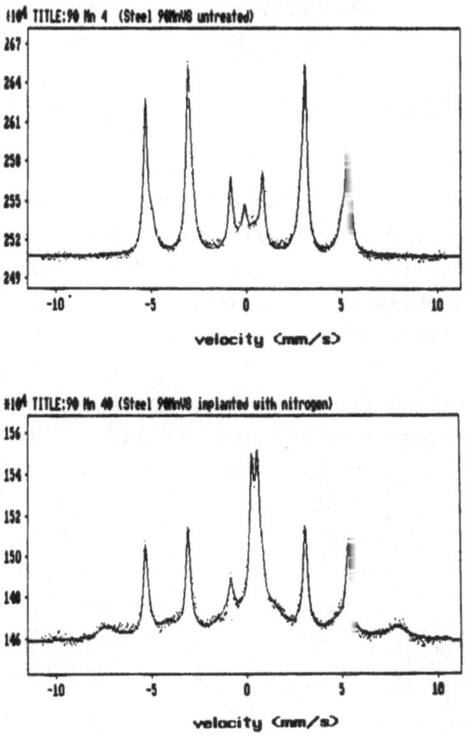

Figure 2:
CEM spectra of steel 90MnV8 untreated and implanted

Due to the implantation with nitrogen ions the martensitic fractions are decreased and a component with a composition around Fe_2N is produced. Moreover, there is an evidence for iron oxide which eventually overlapes still existing austenitic portions or non-stoichiometric iron nitrides. It is not obvious in which way oxygen could reach and oxidize the sample. The phase Fe_2N is mostly found if iron or steel are implanted with high doses of nitrogen and the sample temperature during implantation is kept low [3]. It seems to be responsible for the improved wear behaviour.

References

[1] Kolitsch, A., E. Richter and U. Scholz: this conference, 5.13
[2] Reuther, H.: phys. stat. sol. (a) 93 (1986) 201-205
[3] Reuther, H., E. Richter, M. Fujinami and Y. Ujihira: Proc. EPM'87, Akademie-Verlag, Berlin, 1988, p. 231-233
[4] F. Van der Woude and K.W. Maring: Proc. Int. Conf. Mössbauer Spectroscopy, Bukarest, 1977, Vol. 2, p. 133

LOW-FREQUENCY CAPACITANCE-VOLTAGE CHARACTERISTICS OF A SIS STRUCTURE WITH A THIN SILICON SUBSTRATE

N. Sieber[1], G. Otto[1]

It is well known that electrical properties of the MOS structures fabricated in thin silicon layers on insulator are influenced by the charge coupling between front and back silicon interfaces. Thus, the space charge capacitance characteristic and the threshold voltage can be affected by the silicon thickness and the charge at the interface thin silicon/underlying insulator. The estimation of the charge associated with the underlying interface can be done by using capacitance-voltage (C-V) measurements of the silicon - underlying insulator - silicon (SIS) structures (Fig. 1).

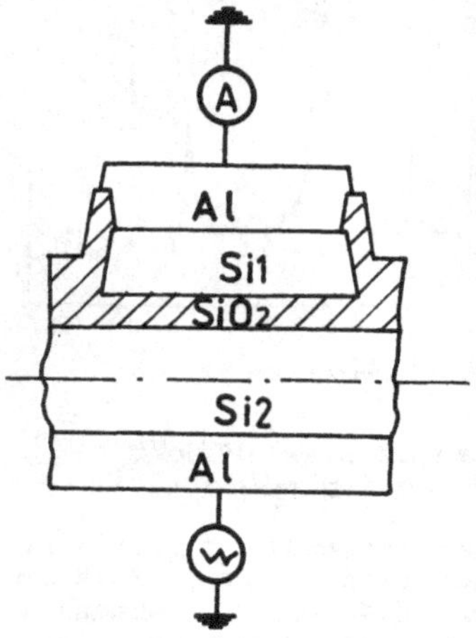

Fig. 1
Semiconductor 1 - insulator - semiconductor 2 (SIS) test structure

For a given voltage applied to the SIS structures the interface charge can be determined by a comparison of the experimental C-V curve with the theoretical curve.
In /1,2/ the theoretical C-V curves of the SIS structures were calculated for different conductivity types and doping densities in semiconductors.
The total capacitance of an ideal SIS structure (without interface states and fixed oxide charge) normalized to the oxide capacitance C_{ox} can be written as

$$C/C_{ox} = 1/[1 + C_{ox}/C_{SC1} + C_{ox}/C_{SC2}] \qquad (1)$$

where C_{SC1} and C_{SC2} are the space charge capacitance in semiconductor 1 and 2.
The estimation of semiconductor space charge capacitance and its dependence on the surface potential U_S requires the solution of Poisson's equation:

$$d^2U(x)/dx^2 = -2qn_i/\varepsilon_{Si}\varepsilon_o(\sinh U_F - \sinh U(x)) \qquad (2)$$

where U and U_F are the normalized electrostatic and bulk Fermi potentials, q is the electronic charge, n_i is the intrinsic carrier concentration, ε_o is the permittivity of free space and ε_{Si} is the dielectric constant of the simiconductor.

1) Central Institute for Nuclear Research Rossendorf, Academy of Sciences of GDR, PF 19, DDR - 8051 Dresden

After integrating Poission's equation the normalized electric field can be written in the form:

$$F = [e^{U_{F1}}(e^{-U} + U - 1) + e^{-U_{F1}}(e^{+U} - U - 1)]^{1/2} \qquad (3)$$

Using charge neutrality in thermodynamic equilibrium ($Q_{SC1} + Q_{SC2} = 0$) the space charge capacitances and the relationsship between the voltage applied to the gate U_G and the surface potentials U_{S1} and U_{S2} can be written as

$$C_{SC1} = C_{SCO} \cdot dF_{S1}/dU_{S1};\ C_{SC2} = C_{SCO} \cdot dF_{S2}/dU_{S2} \qquad (4)$$

$$U_G = (U_{S1} - U_{S2} + \phi_{S21} + (\mathrm{sign}U_{S1})(\varepsilon_{Si}d_{OX}/\varepsilon_{OX}L_{Di})*F) \qquad (5)$$

where C_{SCO} is the Debye capacitance,

ϕ_{S21} is the work function difference between Si1 and Si2,

ε_{OX} is the dielectric constant of the oxide and d_{OX} is the oxide thickness.

In the case of thin Si films where the maximum depletion region width extending from an inverted surface is comparable with the thickness of semiconductor the integration of Poissons's equation can be done under following condition

$$U\,|_{X=d_{Si1}} = 0;\quad dU/dx\,|_{X=d_{Si1}} = E_o$$

In this case the normalized electric field F_1 can be calculated as

$$F_1 = |e^{U_{F1}}(e^{-U_1} + U_1 - 1) + e^{-U_{F1}}(e^{U_1} - U_1 - 1) + E_o^2)|^{1/2} \qquad (6)$$

For a given semiconductor thickness d_{Si1} the normalized electric field E_o for U=0 can be determined by the integral

$$\left| \int_0^{U_{S1}} dU/F(U,U_F,E_o) \right| = d_{Si1}/L_{Di}$$

The calculation of the space charge capacitances and of applied voltage U_G can be done in analogy to the thick semiconductor.

The calculated low frequency capacitance characteristics of a p-i-n type strukture with impurity concentration of $2*10^{15}$ cm^{-3} are shown in Fig. 2. The decreasing base thickness results in a changing shape of the CV-curves and in increasing minimum normalized capacitance C_{MIN}.
In the p-i-p case the CV characteristics exhibit two depletion regions resulting in two minima (Fig. 3).
The decreasing thickness of the semiconductor 1 affects also the width of the low frequency capacitance characteristic. The left part of the CV curves remains almost unchanged. In this case two interfaces can be well separated one from the other. Thus, for the investigation of the interface thin semiconductor/underlying insulator it is advantageous to produce the test SIS structures with semiconductors both of the same type of conductivity.

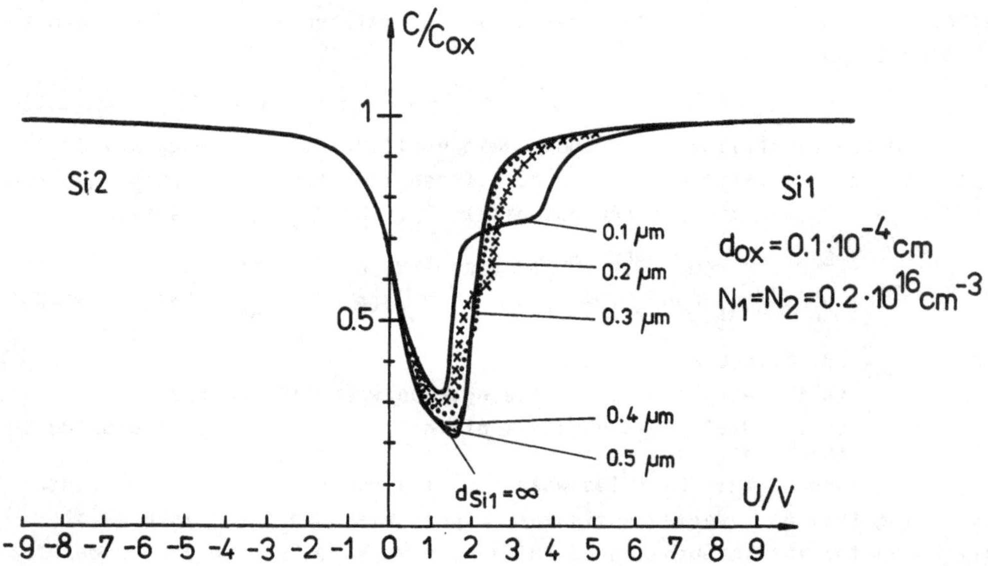

Fig. 2

Calculated low frequency capacitance characteristics of a p-i-n structure for varying thickness of semiconductor 1

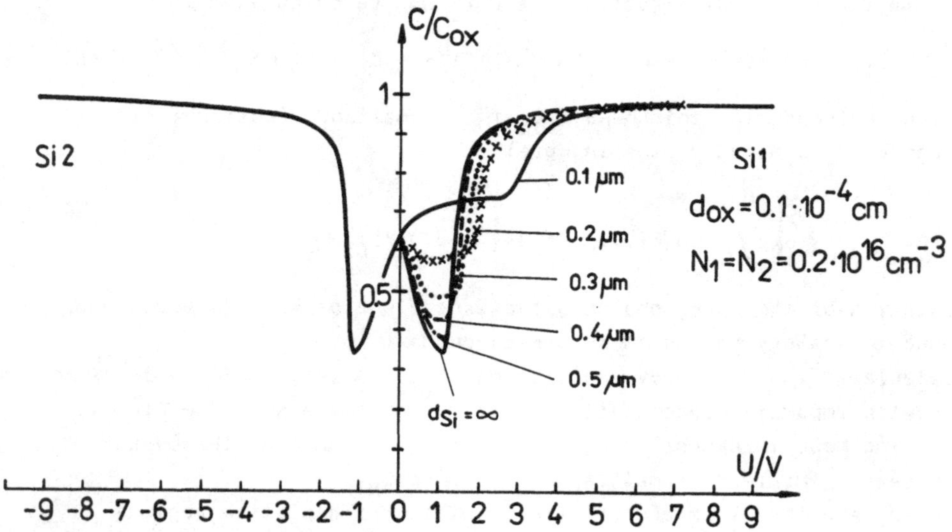

Fig. 3

Calculated low frequency capacitance characteristics of a p-i-p structure for varying thickness of semiconductor 1

REFERENCES

/1/ Nagai Kiyoko et al., Solid-State Electron. Vol. 28, 8 (1985) 789

/2/ Temple Viktor A.K. and Shewchun John, IEEE Trans. Electron. Dev. ED18, 4 (1971) 235

INVESTIGATION OF ARSENIC IMPLANTED SILICON BY OPTICAL REFLECTOMETRY

Vl.Stavrov,R.Varbanov,O.Vasiliev,D.Berova[1)]

1.Introduction.

Optical methods for investigation of ion implanted layers nowadays are widespread.Their precision allows their usage for dose and uniformity estimation (1).However,one has to keep in mind that the processes responsible for changing the optical and (for example) electrical characteristics of the silicon are principally different. Comparison of results of different measurements allows estimation of parasitic process influence on layer's characteristics.Such investigations, including larger number of ion implanted layers' parametres, would help us to increase the quality of layers and repeatability of results.

We examined two groups of parametres-directly and not directly included in defect distribution and found a very high sensitivity of reflectance spectra vs.ion implantation parametres.

2. Experiments and spectra computation method.

p-type,100 mm,(100) oriented,CZ grown silicon wafers were used for all experiments.

Reflectance spectra were measured by automatic spectrometre in the range of 500 - 800 nm.Correction for light source's and detector's spectral characteristics were introduced by computing program.In order to avoid the influence of external factor,such as power supply instabilities and reflectance of virgine silicon,we examined difference of spectra after and before implantation.

Ion implantation was performed by medium current,wafer-by-wafer,without cooling machine.Doses were varied from 1E14 to 7.5E15cm, energies-from 50 to 140 keV and current density- from 0.1 to 5.6 uAcm.

All computations were accomplished assuming linear relation of optical characteristics of silicon vs. defect concentration and Gaussian defect depth distribution.We have used multilayer model for spectra computation.More details about computing procedure are given elsewhere (4).

3. Results and discussions

3.1. Influence of parameters directly included in defect depth distribution(E and D).

In order to avoid heating during implantation in that part of experiment we have used very low current density - 0.1 μAcm^{-2}.

Experimentally measured and computed spectra for dose D=5E14 cm^{-2} and energy from 50 to 140 keV are shown in fig.1 a and b. For best fitting of computed spectra we have choisen parametres a=0.78, b=1.48 and N_{th}=7E18 cm^{-3} for all energies.

We found very good agreement between experimentally measured and computed reflectance spectra if we examine interference minima and maxima,but there was some disagreement between the values of the reflectance index change.We believe that is due to the usage of incorrect data for refractive index $\hat{n}_a$,especially for its imaginary part k_a,characterising absorbance.There is a great difference between measured values for k_a by different autors.It

---

1)Kombinat "Microelectronics",2141 Botevgrad,Bulgaria

can be caused by measurement methods as well as the method for amorphous layer production (CVD, magnetron sputtering, ion implantation).

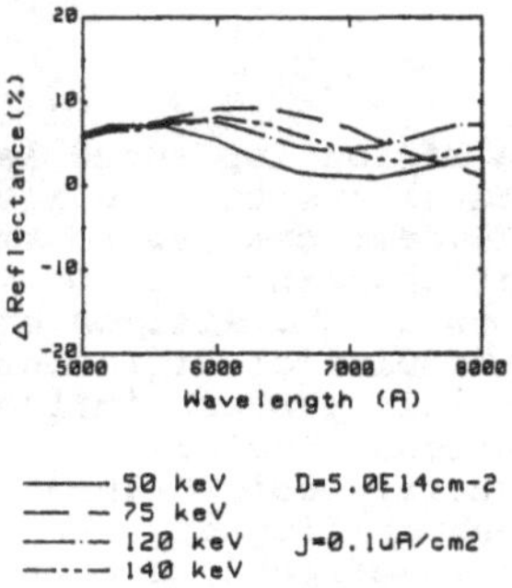

Fig.1a

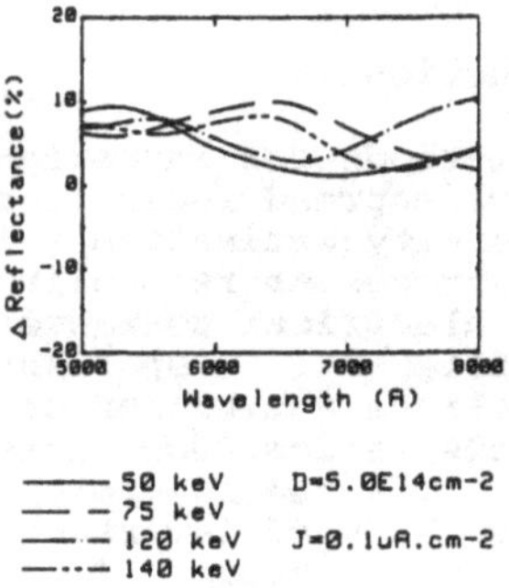

Fig.1b

Experimentally measured spectra for E=120 keV and doses from 1E14 cm^{-2} to 7.5E15 cm^{-2} are shown in fig.2. We found best fitting of computed spectra for the same value of a,b and N_{th} .

3.2. Influence of parameters not directly included in defect depth distribution (T,j).

In order to investigate high current density ion implantation influence upon reflectance spectra we have used different ion current density. We have no possibility to find out clear dose-rate effects because of the absence of wafer cooling system.

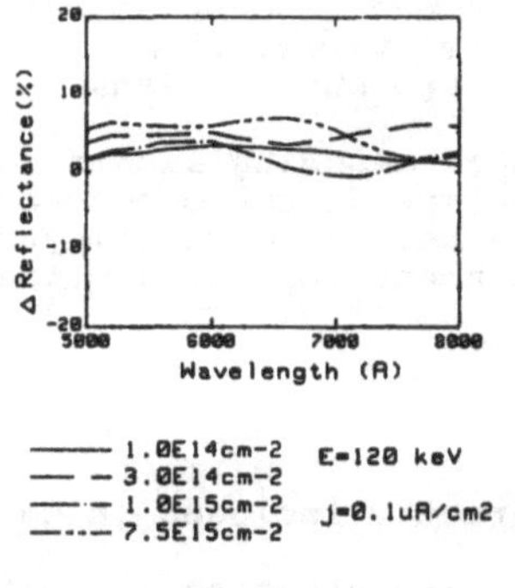

Fig.2

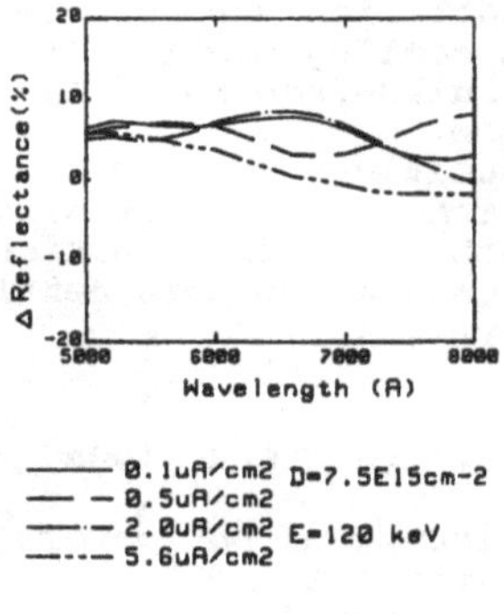

Fig.3

For dose D=7.5E15 cm^{-2} and energy E=120 keV, different dose rate irradiations were performed. Experimentally measured spectra for j from 0.1 to 5.6 μAcm^{-2} are presented in fig.3 and values for N for best fitting of computed spectra are given in Table 1.

For all computations we have used a=0.78 and b=1.48.

One can see that if flux density exeeds 5 uAcm ,amorphous layer does not occur at all and there are not any interference minima and maxima. N increases with the increase of current density and the amor-phisation is strongly affected even for j=0.5 μAcm^{-2}.

Fig.4 and Table 2 show results for dose D=5E14cm^{-2}, energy E=120kev and current density j from 0.1 to 5.6 μAcm^{-2}.

We can point out following features:

- amorphous layer width depends slighly on current density.

- N_{th} decreases with current density increasig.

First feature can be explained by a relatively small heating during implantation with dose D=5E14 cm^{-2} because of the small time for implantation and by the heat exchange between silicon wafer and wafer holder.

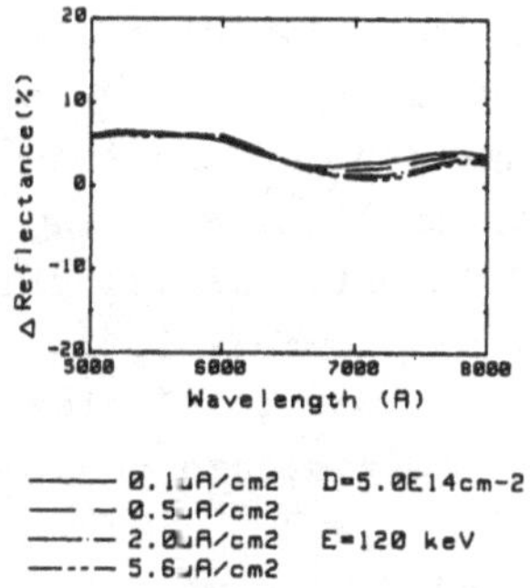

Fig.4

J,uA/cm2	0.1	0.5	2.0	5.6
Nth,cm-3	7E18	8E19	6E20	>1E21

Table 1

J,uA/cm2	0.1	0.5	2.0	5.6
Nth,cm-3	7E18	6E18	5E18	4,5E18

Table 2

The second one cannot be explained by thermal effects but it is in a good agreement with RBS measured data (2) and offered by (3) model for crystal-amorphous interface evolution during implantation.

On the basis of these results,in order to avoid harmful effects of self-annealing during ion implantation we offered double implantation procedure for creation of heavily As doped,fully amorphized layers - first with dose D_1=D-D_2 where D is nesesary for technology dose and second-D_2=5E14 cm^{-2}-nesesary for a complete amorphisation.In that case ion current is not restricted by self-annealing effects,so we have to be careful only about dose uniformity.

4. Conclusion.

On the basis of linear dependence of opticall characteristics vs. defect concentration and Gaussian distribution of ion implantation created defects a model for reflectance index computation is offered and applied to different cases.

We proved the possibility of modelling the influence of ion implantation parameters on reflectance spectra.Information about defect reactions during implantation can be obtained from the threshold defect concentration N_{th},introduced into the model.The method appears to be very sensitive and useful for the production of heavy As implanted layers for VLSI industry

References:

(1) Golin,J.,N.Schell,J.Glaze and R.Ozarski:Nucl.Instr.and Meth.B21(1987) 542-549

(2) Holland,O,J.Narayan and D.Fathy: Nucl.Instr.and Meth. B7/8 (1985) 243-250.

(3) Leiberich,A.,D.Maher,R.Knoell and .Brown: Nucl.Instr.and Meth. B19/20(1987) 457-461

(4) Stavrov,Vl.,R.Varbanov: VEIT'89,Varna (1989)

HIGH DOSE ION IMPLANTATION SYSTEM FOR COMPOUND SEMICONDUCTORS

Yurchenko V.A., Zaitsev A.M., Checha V.P.[1)]

The development of the compound semiconductors ion implantation technology with high ion doses brings about the necessity for working out the equipment providing the sample implantation with the ions of the two dopants to the same depth either simultaneously or in small alternating steps (I). The possibility for the maintaining and intentional changing the semiconductor stoichiometry during the process will enable to increase considerably the electrical activity of the dopant. It'll also allow to use ion implantation more effectively for the compound semiconductors synthesizing.

The system for implantating ions of the two dopant is described. The special features of the system include the electromagnetic mass-separator with two adjustable inlets and one outlet, two simultaneously working ion sources, programmer, device for chanking the ingector working conditions and accelerator. The system can be used for the simultaneous implanting in small alternating steps the ions of the two dopants with different A/Z values. During the quasisimultaneous implantation the ions are implanted by the cycles, consisting of the small alternating doses, each ion type being implanted at a definite energy.

Fig.I shows the layont of the system.

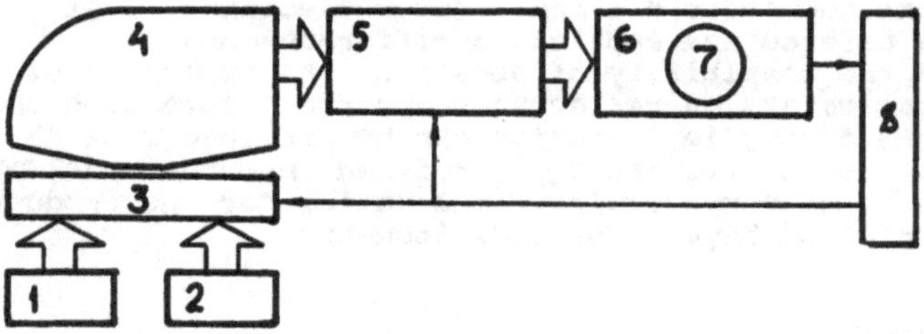

Fig.I. Layont of the system for dual ion implantation:
I,2 - ion sources; 3 - ion beam selector; 4 - mass-separator;
5 - accelerator; 6 - loading chamber; 7 - ion current sensor;
8 - programmer

I) Scientific Research Institute of Radiomaterials, SU-220600 Minsk, USSR

Ion beam trajectories come from the ion sources I,2 through the mass-separator 4 along the axis of the accelerator 5 to the loading chamber 6 for positioning the samples to be processed. Ion beam selector 3 is plased between the ion sources and mass-separator. Ion current is controlled by the sensor 7 installed in the loading chamber and connected to the programmer 8. The programmer outlet is connected to the beam selector and accelerator supply.

When the implantation process is carried out in small alternating steps the ion beam selector procides the passage of the ions of one dopant from two through the mass-separator. This type of ions is conformed by the definite accelerator voltage value. After the sample is implanted to the set dose the ion beam selector terminates the passage through the mass-separator of the ions of another type. In this case the corresponding accelerator voltage value is changed automatically. The second dopant is implanted till the second set dose is reached and here one ion implantation cycle is completed. The number of the cycles and the dose ratio in the cycle are determined on the basis of the full implantation dose by each other dopant.

When the implantation process is carried out continiously the ion beam selector provides the simultaneous passage of the two dopant ions through the mass-separator. The specific feature of the given working conditions is that the two dopant ions are driven by the same accelerating potential in contrast to the quasisimultaneous conditions and the total ion current is registered in the loading chamber. In this case it is advisable to use the ions with close mass/charge values and to stabilize more profoundly the ion sources working conditions.

The system model has been used for the dual ion implantation of Si/P, Si/As, Ga/Se, Ga/P etc.

R e f e r e n c e s

(I) Donnelly, J.: Inst.Phys.Conf. 33b-4 (I977) I66-I90.

AN OXYGEN ION SOURCE OF DUOPLASMATRON TYPE

R. Hentschel[1] and D. Henke[1]

1. INTRODUCTION

Among the SOI-(silicon-on-insulator)technologies the stoichiometric implantation of oxygen ions in silicon (SIMOX = separation by implantation of oxygen) has good chances to become an industrial process in microelectronics. For this oxygen ion sources with a high lifetime on the required beam current and atomic fraction level will be needed. We investigated oxide cathodes on platinum carriers in a Duoplasmatron ion source to get a reliable oxygen ion source of the mA range for a research implanter.

2. APPARATUS

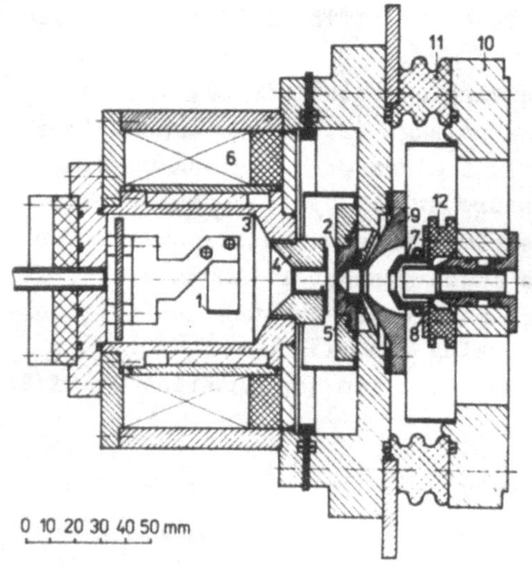

Fig. 1
Duoplasmatron ion source

The ion source is shown in Fig. 1. It is based on a fine beam Duoplasmatron modified by adding an expansion cup 5 of 10 mm diameter and a three-electrode extraction system 7, 8, 9 with Pierce geometry. The accel-decel extraction system guarantees a maximum extraction voltage of 20 kV. The extraction gap has a width of 8 mm. The anode 2 consists of hermet-WCu 300, it has a hole of 2 mm diameter and is directly cooled with deionized water. The oxide cathode 1 consists of a 256 meshs/cm^2 grid made from 0.12 mm platinum wires. The grid is formed to a three-layer belt of 11 mm width and 55 mm length which is clamped with its ends into well-cooled steel holders and coated with a mixture of an oxide or carbonate, binder and solvent according to (1). Activation is not necessary since the cathode will soon be poisoned when it operates in oxygen.

1) Academy of Sciences of the GDR, Central Institute for Nuclear Research Rossendorf, PF 19, Dresden, DDR - 8051, GDR

3. RESULTS

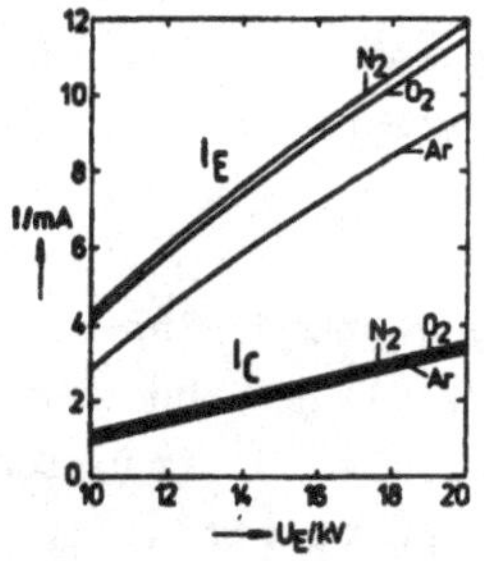

Fig. 2

Dependence of the total beam current I_E and beam current I_C in a faraday cup with the acceptance angle of the implanter Vesuv-8M on the extraction voltage U_E

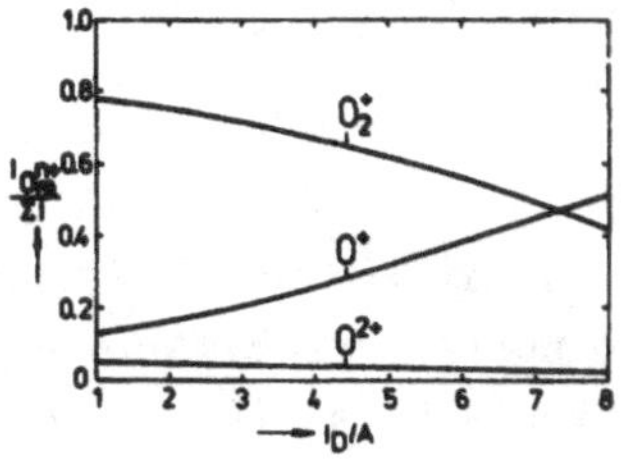

Fig. 3

Fractions of different oxygen ions in the beam in dependence of the discharge current I_D

Fig. 2 shows beam currents and Fig. 3 the composition of the oxygen beam. Both from atomic fraction and beam current considerations (magnetic stray fields in the expansion cup deteriorate the beam quality at high source magnet fields) the discharge current must be high. Routine operation parameters are: cathode current 70 A, discharge current 8 A, source magnet current 0.1 A, source pressure 6 Pa, extraction voltage 20 kV, electron barrier voltage - 1 kV.

The results of the oxide cathode investigations are summarized in fig. 4 which gives mean discharge current densities at the hot cathode in the source in dependence of cathode temperatures for some oxides preferentially in O_2 atmospheres. For lack of equivalent curves in the literature a comparison is made for H_2 operation (2). Strontium oxide and mixed oxide, a great share of which is strontium oxide (1), had by far the lowest emission temperatures in oxygen. These cathodes lasted more than 60 hours at routine conditions, the cause of failures was sputtering. Barium oxide was also emitting sufficiently below the

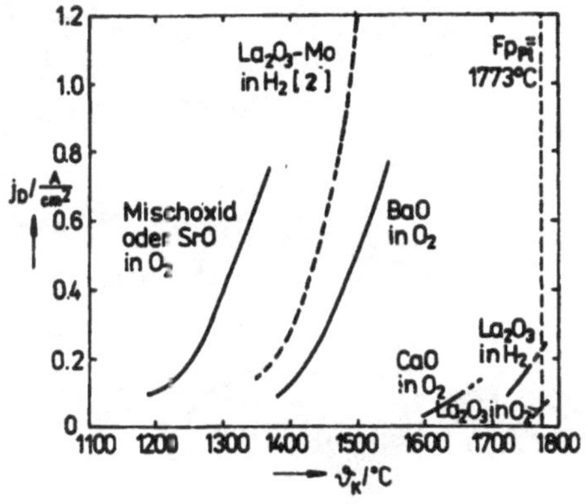

Fig. 4

Dependence of the mean discharge current density j_D at the cathode in the ion source on the cathode temperature platinum melt-point, but the lifetime was limited to about 10 hours by vapourization of the oxide. Other emission substances on platinum carriers are irrelevant to oxygen ion sources. The anode and other parts of the ion source did not fail in 200 hours of oxygen operation.

4. SUMMARY

An oxygen ion source of Duoplasmatron type is presented. The source delivers a beam current of 3 mA into the acceptance angle of the research implanter Vesuv-8M at 20 keV injection energy with an O^+ fraction of more than 50 %. An oxide cathode with an emitting layer of strontium oxide on a platinum carrier is used. This ensures in connection with a well-cooled anode a lifetime of more than 60 hours with pure oxygen gas inlet.

REFERENCES

/1/ Ardenne, M.v.: Tabellen der Elektronenphysik, Ionenphysik und Übermikroskopie, Band 1, VEB Deutscher Verlag der Wissenschaften, Berlin 1956

/2/ Itoh, Y., T. Sugawara: Journ. Nucl. Sci. Technol. 23 (1986) 191 - 196

LASER-INDUCED PLASMA DEPOSITION OF THIN HIGH-TEMPERATURE SUPERCONDUCTING Bi-Sr-Ca-Cu-O FILMS

S. Becker, H.-J. Dietze

The discovery of high-T_c superconductors has activated an enormous research potential in the world. The future of high-T_c superconducting films is connected with their successful application in microelectronics.

Among the variety of applied film deposition techniques the laser-induced deposition methods are excellent techniques for the generation of high-T_c superconducting films /1-3/. Thin high-T_c superconducting films with the best physical and electrical properties were obtained on single-crystal $SrTiO_3$ substrates with perovskite structure. In this paper, we report the growth of thin films of Bi-Sr-Ca-Cu-O systems deposited by pulsed Nd-YAG laser evaporation and ionization of superconducting targets on the technologically important silicon substrate materials.

By reason of the chemical reactivity of silicon with the chemical elements of thin films during the high-temperature annealing process in order to prepare the superconducting phase of the Bi-Sr-Ca-Cu-O system the formation of good superconducting thin films directly on the silicon substrate is difficult. Diffusion effects, contamination of the thin films with the chemical elements of the substrates and interface reactions can be avoided by the preparation of thin buffer layers between the substrates and the thin superconducting films.

By reason of similar structure parameters of $SrTiO_3$ and Bi-Sr-Ca-Cu-O superconducting films it is possible to apply thin $SrTiO_3$ buffer layers with perovskite structure between the substrate and the thin film.

A detailed description of the experimental arrangement for laser-induced plasma deposition of thin films was given previously /4/. The high-temperature superconducting $Bi_2Sr_2Ca_1Cu_2O_x$ was irradiated by a Nd-YAG laser (wave length. 1060 nm, repetition rate: 5 kHz, pulse duration: 100 ns, laser power density: $5 \cdot 10^8$ $W \cdot cm^{-2}$) in a vacuum chamber.

X) Central Institute of Isotope and Radiation Research, Academy of Sciences of the G.D.R., Leipzig (G.D.R.)

The substrate temperature was either held at 450°C or at room temperature. For heated substrate a better adherence of the films to the substrate was observed. But the Bi concentration in the thin films decreases with increasing substrate temperature. By reason of the low melting point of metallic Bi a depletion by 40-50 % in Bi content in the thin films deposited at 450°C compared with the Bi concentration in the initial target material was detected by means of X-ray fluorescence spectrometry. For the preparation of stoichiometric thin Bi-Sr-Ca-Cu-O films on heated substrates the stoichiometry of the target material must be appropriately changed.

In Fig. 1 the scheme for the generation of high-T_c superconducting Bi-Sr-Ca-Cu-O films on silicon is shown. In the first step a SrO/TiO mixture is deposited on a silicon single-crystal substrate (100) with a thin silicon oxide layer (the thicknesses of the SrO/TiO_2 and SiO_2 thin films are about 100 nm and 10 nm, respectively).

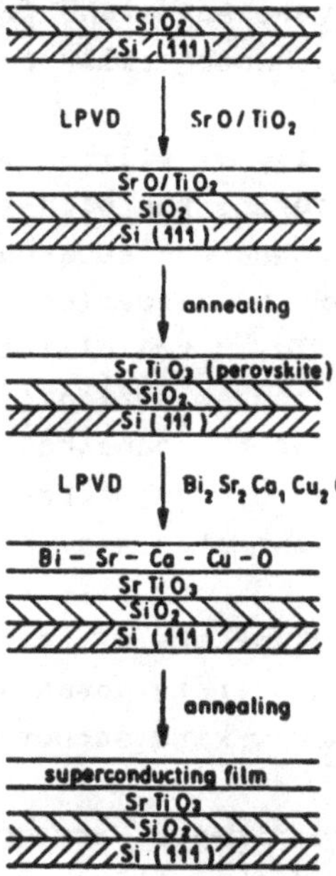

Fig. 1: Schematic representation of a possible way for the preparation of superconducting thin films

By short annealing of the amorphous SrO/TiO_2 film at 800°C and slow cooling a polycrystalline strontium titanate film is formed as demonstrated by X-ray diffraction analysis. In the second step a superconducting target is used to produce thin Bi-Sr-Ca-Cu-O films. The post-annealing of the films is carried out at 875°C in air for several minutes. During the annealing the resistance of the films drops over several orders of magnitude. The final resistance of annealed films at room temperature is about some mΩ.

Fig. 2 shows the typical resistance versus temperature curve of a thin high-temperature superconducting film on silicon with a $SrTiO_3$ buffer layer. We observed a metallic behaviour of our films a decrease of resistance as the temperature was lowered to 100 K and a sharp transition into the superconducting state.

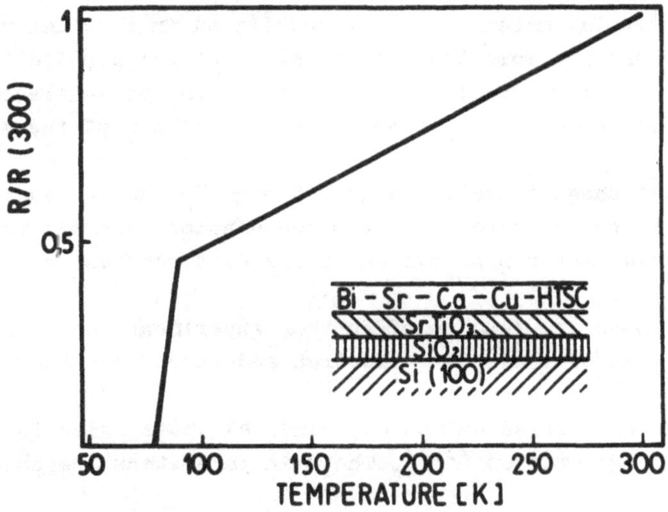

Fig. 2: Temperature dependence of the resistance of Bi-Sr-Ca-Cu-O thin films on Si(100)/SiO_2/$SrTiO_3$

Thin superconducting Bi-Sr-Ca-Cu-O films with thicknesses less than 500 nm can be prepared with a T_c close to that of bulk ceramic target materials by means of the laser-induced plasma deposition method.

REFERENCES

/1/ T. Venkatesan, X.D. Wu, A. Inam, J.B. Wachtman. Appl. Phys. Lett. 52 (1988) 1193

/2/ H.S. Kwok, P. Matrocks, L. Shi, X.W. Wang, S. Witanachchi C.Y. Ying, J.P. Zheng, D.T. Shaw, Appl. Phys. Lett. 52 (1988) 1825

/3/ S. Komuro, Y. Aoyagi, T. Morikawa, S. Nambo, Jap. J. Appl. Phys. 27 (1988) L 34

/4/ H.-J. Dietze, S. Becker, phys. stat. sol. (in press)

EXPERIMENTAL STUDIES OF EFFECT OF HIGH CURRENT PULSE ELECTRON AND ICARBON ION BEAMS ON THE HIGH TEMPERATURE Y-Ba-Cu-O, Bi-Ca-Sr-Cu-O SUPERCONDUCTORS

Korenev S.A., Sikolenko V.V., Chernakova A.
Joint Institute for Nuclear Research, USSR, 101000, Moscow.
P.O. Box 79.

High T_c superconducting materials are investigated in many laboratories of the world. There are some tendentions of practical application of high T_c superconductivity. The effect of high current pulse electron and ion beams irradiation on high T_c superconductors is one of the most actual questions.

Resent studies have shown a stabilization of high T_c superconducting Y-Ba-Cu-O surfaces and their protection from degradation by creating a melting protective layer after high current pulse electron beam processing /1,2/.

In this report we show the results of further experiments on investigation of effect of high current pulse electron and carbon ion beams investigation.

Experiments have been carried out on experimental installtion for the electron and ion beams generation /3,4/. The main parameters are shown in the table 1.

Table 1.

current yield regime	kinetic energy, keV	current pulse duration, ns	current density, A/cm^2	pulse frequency, Hz
electron	100 - 300	300	5-1600	0,2
carbon ions, C^+	100 - 300	300	1 - 60	0,2

Y-Ba-Cu-O and Bi-Ca-Sr-Cu-O samples have dimensions 1x5x20 mm^3. The vacuum in the apparatus was $5 \cdot 10^{-5}$ tor.

Basic characteristics of high T_c superconductors:

a) measurement of temperature crossing;
b) resistance and critical current;
c) structure analysis;
d) element analysis;
e) topography of surface.

The resistivities were measured by four-probe method on frequency f=250 - 500 Hz and inductivity method on frequency f=5 - 10 kHz. Critical current were measured by original pulse mrthod using R-C generator where current was measured by non-contact method (with help current transformer). X-ray difraction analyses were carried out by the use of X-ray difractometer, and element analyses were received by the use of semiconductor X-ray spectrometer on electron microscope YJSM-840. Surface topographic was received with the help of JSM-840.

Fig.1a,b shows the curve of resistance R/R_1 versus temperature for two bulk samples Y-Ba-Cu-O before irradiation (curve 1) and after irradiation by electron pulse beam with current density $j \sim 1000 A/cm^2$, kinetic energy 250 keV, current pulse duration $\sim$ 300 ns (curve 2).

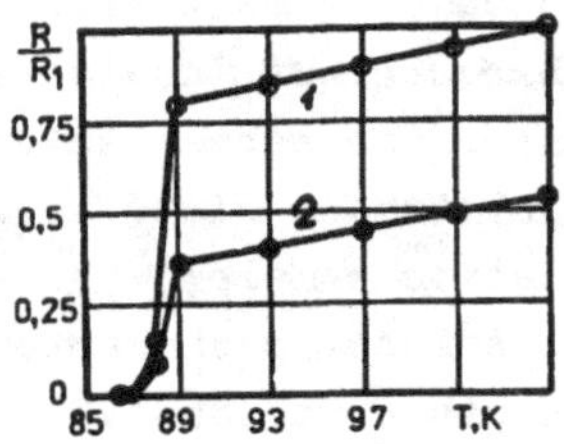

Fig. 1a

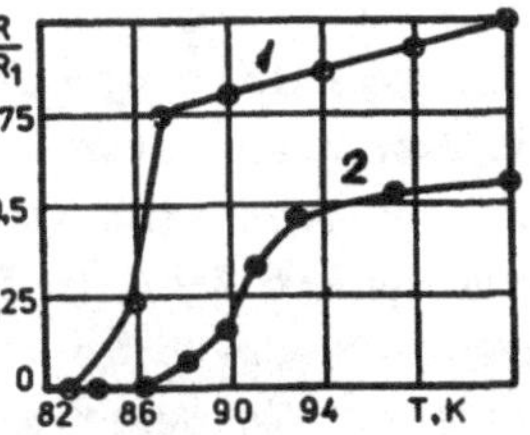

Fig. 1b

Fig. 2a shows the curves of resistance R/R_1 versus temperature for polycrystal samples Bi-Ca-Sr-Cu-O, fig. 2b shows such curves for bulk samples Bi-Ca-Sr-Cu-O.

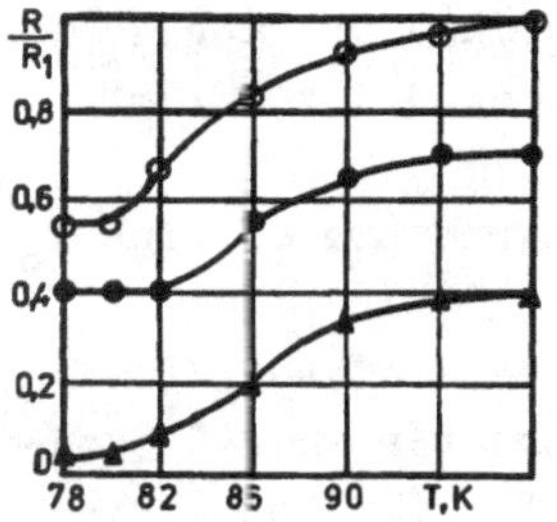

Fig. 2a. Resistivity R/R_1 as a function of temperature T before irradiation (o) and after irradiation by electron beam $j \sim 1000$ A/cm^2, E $\sim$200 keV, $\tau_p \sim$300ns: N = 10 pulses (•), N=20 pulses (▲).

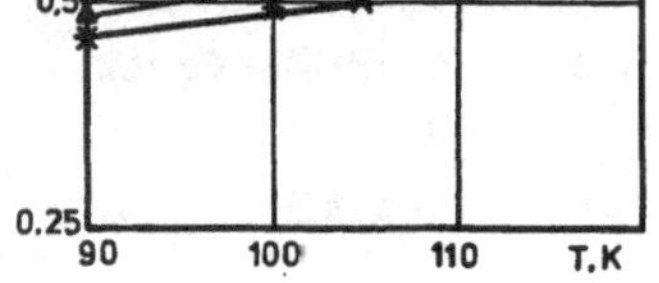

Fig. 2b. Resistivity R/R_1 as a function of temperature T for Bi-Ca-Sr-Cu-O before irradiation (o) and after irradiation by E $\sim$200 keV, τ_p =300 ns electron beam with N=140 pulses, j=150 A/cm^2 (•) and j=7A/cm^2 (▲) and by carbon ion beam with j=35A/cm^2, N=80 pulses (x), N=250 pulses (*), $\tau_p \sim$300 ns.

Results of measurements R=f(T) shows, that effect of high current electron and carbon ion beams irradiation decrease consists of integral resistance of samples.

Surface melting layer on electron penetration depth were observed for electron pulse density $P_o \gtrsim 10^7 W/cm^2$ for Y-Ba-Cu-O samples and for $P_o \gtrsim 4 \cdot 10^6 W/cm^2$ for Bi-Ca-Sr-Cu-O samples. Analysis of melting surface have shown, that the more homogeneous melting had place in Bi-Ca-Sr-Cu-O ceramics. Increasing of critical current from $J_c \sim 80-90A$ to $J_c \sim 200$ A was observed in samples with melting surface layer.

There was no melting layer in the case of carbon ion irradiation, but mixing effects on superconductor's surface were observed.

X-ray difraction analysis shown that amorphyzation in Y-Ba-Cu-O and Bi-Ca-Sr-Cu-O was not observed. Y-Ba-Cu and Bi-Ca-Sr-Cu stoichiometric analysis has shown that irradiation doesn't change the stoichiometry.

The results can be explained by pulse character of interaction of high current electron and ion beams with substance,. The babeam interaction time is less than the process time in the substance. Since τ_p is less than diffusional development time in irradiation samples, the stoichiometry in Y-Ba-Cu and Bi-Ca-Sr-Cu is stable. Adiobatic character of heat processes in charged particle penetration depth picks out all power of beam. Melting surface layer for electron beam irradiation with $P \geq P_o$ are taking place. Analogous results have been received for Nb_3Ge /5/, being irradiation by such electron beam. Carbon ion beam power density inis insufficient for heat effect and for amorphyzation /6/ .

Thas, the results are the following:

- the threshold of electron beam power density for surface melting at kinetic energy 100-300keV are $\geq 10^7 W/cm^2$ for Y-Ba-Cu-O and $\geq 4 \cdot 10^6 W/cm^2$ for Bi-Ca-Sr-Cu-O;
- surface melting considerably impruved the surface structure of high T_c superconductors by the decrease of micro-roughing;
- surface melting after irradiation provokes the increase of the J_c;
- integral resistance of Y-Ba-Cu-O, Bi-Ca-Sr-Cu-O decreases in 2-2,5 time for electron beam irradiated and in 1,25-1,8 time for ion beam irradiated samples;
- amorphyzation doesn't take place and stoichiometry is conserved after irradiation. This effect open a possibility for the use the properties study of high T_c superconductors.

References.

1. Korenev S.A., Valentovich D., Lushchikov V.I. JINR,rapid comm. N4(30) -88, p.83, JINR, Dubna, 1988.
2. Korenev S.A., Valentovich D. Proc. of 1st conf. on Modification of Constructive Materials by Charged Particles Beams. P.II,p.4,1988, Tomsk.
3. Korenev S.A., e.a. Preprint JINR N13-88-292, Dubna, 1988.
4. Korenev S.A. Proc. of 3rd seminar on Plasma Electronics. p.83, kharkov, 1988.
5. Vavra I., Korenev S.A. Proc. of 2nd Intern. Conf. on Electron Beam Technology. P.III, p.641,BAS, Sofia, 1988.
6. Boiko V.I., Evstigneev V.V. Physics of high current charged particle beams interaction with malter. Energoatomizdat, Moscow, 1988.

ENHANCEMENT OF T_c Y-Ba-Cu-O THIN FILMS AFTER PROTON IRRADIATION

A.A. Nikonov, G.V. Sotnikov, A.S. Tokarev*

Effect of proton irradiation with E=3.9 MeV on superconducting (SC) properties of $YBa_2Cu_3O_7$ films in pure oxygen atmosphere were investigated. T_c dependence on φt was studied. Non-monotonous behavior of T_c from fluence was found. At low fluencies the rate of T_c decreasing is less than corresponding value obtained after the irradiation in vacuum. Subsequent bombardment of the films resulted in T_c increase. T_c rise phenomenon contradicts earlier experimental results of ion irradiation of high-T_c superconducting films.

INTRODUCTION*

All the experiments with superconducting ceramics aimed at introducing radiation damages into crystalline structure by charged-particle bombardment were carried out in vacuum. This paper presents the results of temperature dependence measurements of ac magnetic susceptibility for $YBa_2Cu_3O_7$ films subjected to proton bombardment with E=3.9 MeV in pure oxygen environment under atmospheric pressure.

EXPERIMENT

Samples of the $YBa_2Cu_3O_7$ films were prepared by a method of laser sputtering on a the $SrTiO_3$ sublayer. Conditions of preparing the samples are depicted in more detail in [1]. We have got two series of samples which herebelow will be called #X, #Y, #Z and #R1, #R2 correspondingly.

The irradiation was carried out on Van Graaf's tandem accelerator. Protons, being accelerated to energy 4.5 MeV were passed through aluminium foil into a chamber used for exposure, which contained pure oxygen. Energy of protons after leaving the foil made up 3.9 MeV.

A certain rate of gas flowing through the exposure chamber was set up to rule out the effects connected with a chemical interaction of the forming ozone and the sample.

RESULTS

The first experiments carried out with the samples #X, #Y, #Z show [2] that the rate of T_c decreasing is noticeably lower than that in the case of vacuum irradiation. The sample #X had only 4% decrease of T_c^{ons} (determined by appearence of diamagnetizm [1]) after dose $0.68 \cdot 10^{17}$ p/cm^2. Mostly unexpected was the result obtained with sample #Z. After the same fluence its transition temperature dropped by 6 K. But the following bombardment with a fluence $1.37 \cdot 10^{17}$p/cm^2 brought about the T_c rise by 4.5 K. Because T_c rise phenomenon contradicts earlier experimental results of ion irradiation of high-T_c superconducting films, in the following checking experiments we used samples which had far more narrow transition curves (#R1, #R2). In the first experiments the samples #R1 and #R2 were irradiated at the temperatures 25° and 33° up to fluencies $1.12 \cdot 10^{17}$ and $1.67 \cdot 10^{17}$p/cm^2 respectively.

Fig.1 contains the results of temperature measurements of a superconducting transition defined according to the peak position of imaginary part of susceptibility T_c^{peak} and according to T_c^{ons} depending on the fluence for sample #R1. Fig.2 contains similar results for sample #R2. Since the rate of changing T_c^{ons} and T_c^{peak} for samples #R1 and #R2 up to the abovementioned fluence values proved to be approximately equal within the error of measurement

* Kurchatov Institute of Atomic Energy, Moscow, USSR.

limits, further exposure was continued at greater difference of temperature: 80°C for sample

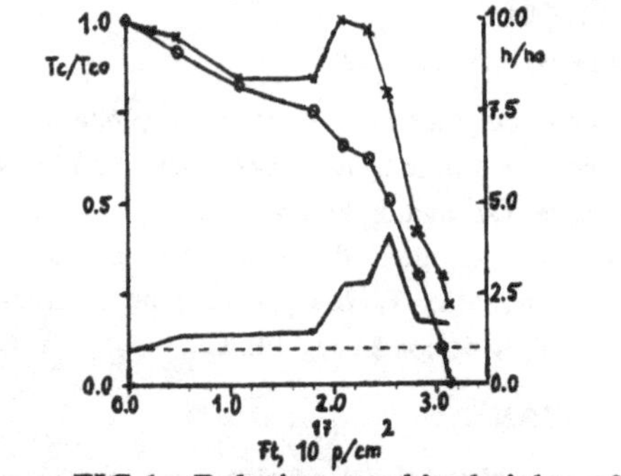

FIG.1. Relative peak's height of x"(T) vs fluence (right scale) and superconducting transition temperatures determined by the onset of transition (x) and by x"(T) peak's position (o) vs fluence for sample #R1 (left scale).

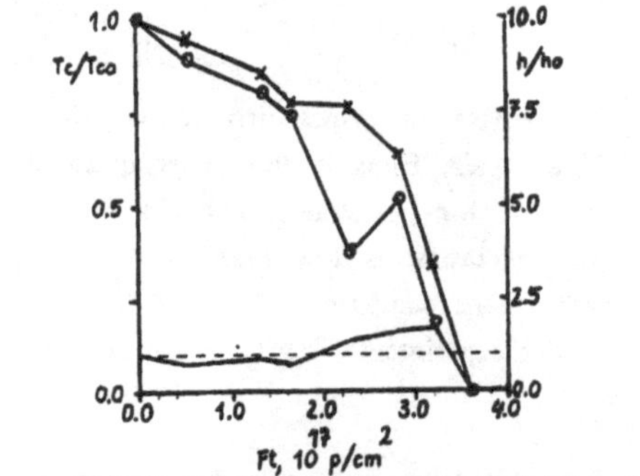

FIG.2. Relative peak's height of x"(T) vs fluence (right scale) and superconducting transition temperatures determined by the onset of transition (x) and by x"(T) peak's position (o) vs fluence for sample #R2 (left scale).

#R1 and 120°C for sample #R2. As can be seen from the given charts, alongside with the generally accepted regularities the results for both samples have their distinguishing features. Presence of two areas by a fluence for which mean degradation rates T_c^{on} and T_c^{peak} differ noticeably is a common feature in the nature of the obtained dependences. **Table** tabulates the mean values of T_c reduction rate for each of the areas and for both methods used for definition of this value. Area 1 lies in interval $0\text{-}2\cdot10^{17}p/cm^2$ and area 2 meets the value of fluence above $2.5\cdot10^{17}$ p/cm^2.

Table

Degradation mean rate	Area1 #R1	Area1 #R2	Area2 #R1	Area2 #R2
$<dT_c^{on}/d(\varphi t)>$ K/($10^{16}p/cm^2$)	1.0	0.95	8.35	6.2
$<dT_c^{peak}/d(\varphi t)>$ K/($10^{16}p/cm^2$)	1.3	1.32	6.22	6.2

Similar to that was observed for sample #Z, we revealed in the area of fluencies $2.0\text{-}2.5\cdot10^{17}p/cm^2$ for sample #R2 a significant increase of T_c^{peak}, which greatly exceeds a possible experiment error, it made up 13 K. (Fig.2). Sample #R1 (Fig.3) demonstrates its abnormal behavior in approximately the same area of fluencies. However, as distinguished from sample #R2 (Fig.4) here we observed T_c^{ons} increase up to the value which the sample possessed before irradiation.(curve 5).

CONSIDERATION

Analyzing the dose dependences of a superconducting transition temperature represented in Fig.1 and 2 we may see that after irradiation in an oxygen medium superconducting properties vanish at fluence $3.2\text{-}3.7\cdot10^{17}p/cm^2$, i.e. approximately at the same values at which they vanish in case of vacuum irradiation [1]. At the same time in the first case T_c degradation rate does not remain unchanged, and at low fluencies it proves to be lower than for vacuum irradiation. Since the fault-formation process should not depend on the medium in which the sample stays, then a possible cause of the degradation rate moderation may be any competing process which either leads to moderation of the defects accumulation or creates a mechanism compensating to some extent Tc degradation, down to a certain level at the expense of the fault formation.

Additional introduction of the oxygen atoms from outside may be one of the reasons

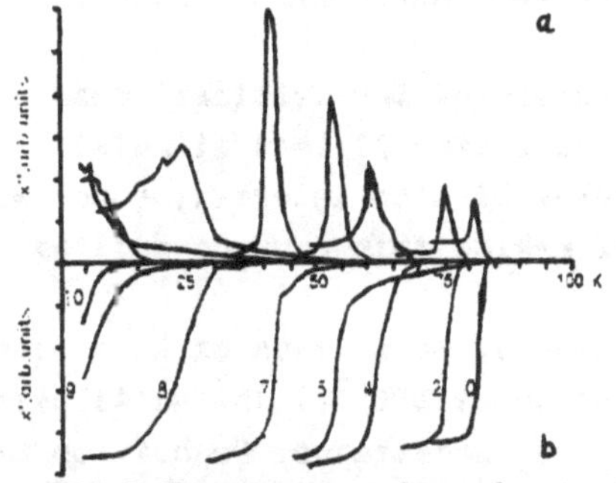

FIG.3. Temperature dependence of a)imaginary x" and b)real x' parts of ac susceptibility of #R1 sample for different fluencies:0- 0;2- $4.18\cdot10^{16}$;4- $1.82\cdot10^{17}$;5- $2.09\cdot10^{17}$;7- $2.57\cdot10^{17}$;8- $2.84\cdot10^{17}$;9-$3.05\cdot10^{17}$;10- $3.13\cdot10^{17}$p/cm^2.

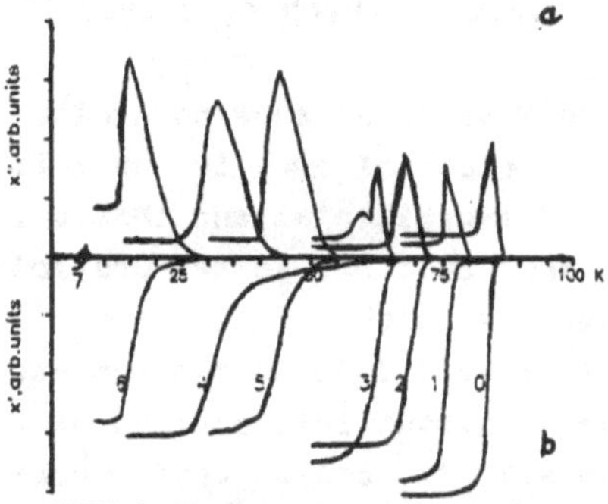

FIG.4. Temperature dependence of a)imaginary x" and b)real x' parts of ac susceptibility of #R2 sample for different fluencies:0- 0;1- $6.93\cdot10^{16}$; 2- $1.36\cdot10^{17}$;3- $1.67\cdot10^{17}$;4- $2.31\cdot10^{17}$; 5- $2.83\cdot10^{17}$;6- $3.19\cdot10^{17}$;7- $3.61\cdot10^{17}$p/cm^2.

which leads to complete or partial "curing" of the defects. This additional introduction may occur both at the expense of direct driving in of the oxygen atoms after they attained the energy upon colliding with the bombarding protons and at the expense of radiation enhancement diffusion and reduction of the surface barrier.

CONCLUSION

As distinguished from the results obtained under vacuum conditions, the data on the irradiation in oxygen show the following:

1. There are two areas by fluence where degradation rates of the superconducting transition temperature much differ between themselves, area of small and area of large fluencies.

2. There is also the third by fluence area 1.5-$2.8\cdot10^{17}$p/cm^2 where alternation of the superconducting properties with a gradual increase of the radiation dose attains quite an unstable character.

3. Upon reaching certain values of a fluence lying in the interval of 2-$2.8\cdot10^{17}$p/cm^2, a temperature increase of a superconducting transition defined both by a peak on an imaginary part of ac susceptibility x" and by the onset on real part of the ac susceptibility is observed.

4. Noted increase of T_c may result in a penetration of the additional amount of oxygen from a gaseous media into a film under the bombardment by the protons and its subsequent redistribution in a crystal lattice according to vacant positions, stimulated by a radiation-accelerated diffusion.

REFERENCES

(1) Nikonov A.A., Sotnikov G.V., Tokarev A.S.: IEEE Trans.on Magn. 2 (1989) 2579.

(2) Nikonov A.A., Sotnikov G.V., Tokarev A.S., Superconductivity: Physics, Chemistry, Technics, 1 (1989) 7.

PREPARATION OF SUPERCONDUCTING THIN FILMS BY ION BEAM MIXING

B.Rauschenbach[1)], W.Erfurth[2)], J.Heydenreich[2)] and K.Hohmuth[1)]

The discovery of superconductors with values of the critical temperature of the order of 90 K in the Y-Ba-Cu oxide system /1/ has stimulated a lot of experimental and theoretical works. Of high interest, especially for applications, is the possibility of making thin superconducting films with high T_c.
Multilayered Y-Ba-Cu samples were prepared by deposition of alternate layers without interruption of vacuum at about 400 °C. The multilayered metals Y, Ba, and Cu were deposited on ZrO_2-substrates. Deposition was carried out in a partial pressure of an argon-oxygen mixture of about 10^{-2} Pa. The thickness of a single Cu-layer was 20 nm, of a Ba-layer 75 nm, and of a Y-layer 15 nm. Because ion beam mixing was induced by 330 keV Xe-ions or 150 keV O-ions, we have chosen a total thickness to about 0.7 µm and 0.3 µm for Xe-ion and O-ion irradiation, respectively. The ion fluences were between $1*10^{14}$ and $1*10^{16}$ ions/cm² and the temperature during irradiation was about 400 °C. After ion mixing, the mixed samples were annealed between 600 and 800 °C up to 2 h under a continuous flow of oxygen (experimental details see Ref. /2/).
After ion induced mixing all samples were amorphous independ on ion species. With a subsequent low-temperature annealing at temperatures between 700 °C and 750 °C in oxygen for times longer than 20 min, the mixed layers recrystallize, became metallic again and were generally superconducting at low temperatures. The micrograph (Fig. 1) shows a electron diffraction pattern of a fine-grained layer after annealing at 700 °C. After 25 min, the electron diffraction pattern demonstrates the orthorhombic structure of superconducting phase. Fig. 2 depicts the temperature dependence of the resistance over the range between the temperature of liquid nitrogen (LNT) and the room temperature (RT) for two typical samples. The RT-resistance for Xe-ion mixed layers is higher than for O-ion mixed layers. Resistance measurements indicate a superconducting onset temperature of 110...120 K, a midpoint temperature of 89...92 K, and a complete zero resistance at 79...82 K for Y-Ba-Cu oxide layers mixed ions. As can be seen, the using of oxygen ions is favourable.

----

1) Zentralinstitut für Kernforschung Rossendorf
DDR-8051 Dresden, PSF 19
2) Institut für Festkörperphysik und Elektronenmikroskopie
DDR-4010 Halle/S., PSF 250

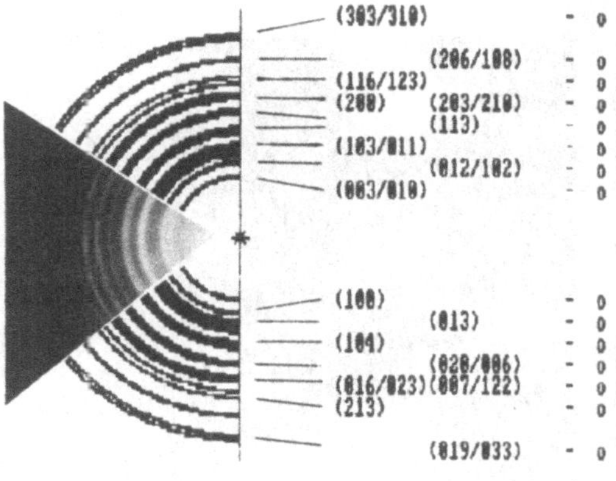

Fig. 1:
Identified ED pattern of multilayer Y-Ba-Cu oxide thin film after mixing and annealing (o = peak of the orthorhombic phase)

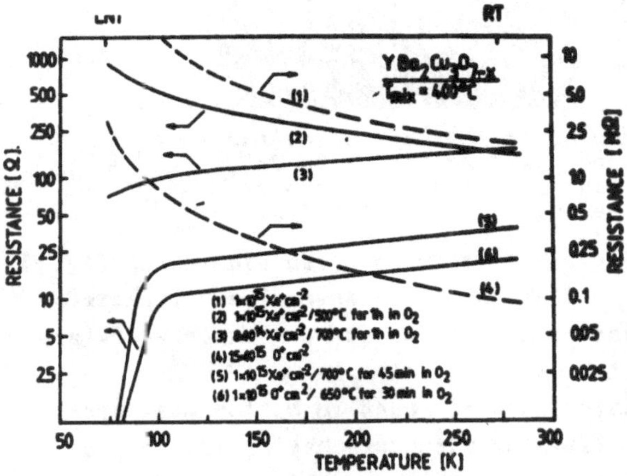

Fig. 2:
Temperature dependence of resistance of multilayered Y-Ba-Cu oxide films in the temperature range between liquid nitrogen and room temperature after mixing

The investigation reveals a topography with distinct appearance after each step of the manufacturing of superconducting thin films. After mixing the topography is very smooth and nearly free from planar defects. After annealing in flowing oxygen at 650 °C for 30 min, two typs of grains (smooth and rough grains) with very distinct appearance may be distinguished.

The topography of the grains seems to be connect with the composition. It is clearly demonstrated using energy dispersive X-ray microanalysis that the local concentration of Y, Ba, and Cu in smooth and rough grains is different and that the content of these elements in the smooth grains are close to the 1-2-3 composition.

The determination of the lattice parameters and the real space group is rather difficult, because the possibility of twinning cannot be completly excluded. Very small differences between the dimension of the lattice parameters a and b also make the situation very complicate. Lattice parameters were calculated from the distance between two appropriate spots in the electron diffraction patterns and by performing least-

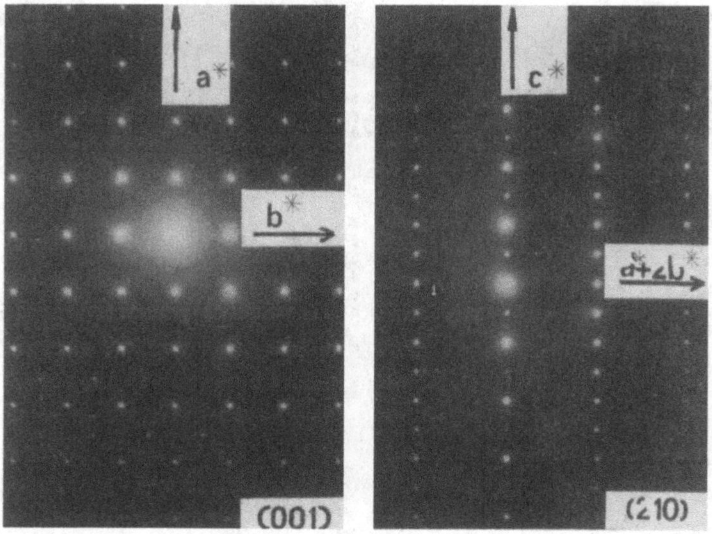

Fig. 3: SAD patterns corresponding to (001) plane and (210) plane of superconducting thin Y-Ba-Cu oxide films after mixing with oxygen ions at 400 °C and post-annealing at 650 °C for 1 h.

squares analysis (refinement). Two SAD patterns are represented in Fig. 3, corresponding to (001) and (210) reciprocal net planes. The calculated lattice parameters of superconducting thin films at room temperature are a=0.383 nm, b=0.389 nm, and c=1.165 nm.

A lamellar domain structure of this phase is found to be the most prominent defect /2/. The alternate lamellar structure having the average spacing of about 0.1 μm and is seen inside of many grains. By comparison with electron diffraction patterns, the assigned plane of this lamellar structure is the (001) plane and the diffraction vector g=110, i.e. parallel to the lamellar direction. Therefore, the edge-on plane separating of the lamellar structure is $(1\bar{1}0)$. The lamellar structure may be considered as alternate layers of matrix and twin with the edge-on boundary being the twin boundary.

In conclusion, we have shown that superconducting thin films can be produced by ion beam induced mixing in a relatively short-time and low-temperature process.

References

/1/ J.G.Bednorz and K.A.Müller: Z. Phys. B 64 (1986) 189

/2/ B.Rauschenbach and K.Hohmuth: Z. Phys. B 74 (1989) 155 and Nucl. Instr. Meth. B 39 (1989) 648

FORMATION OF HIGH-TEMPERATURE SUPERCONDUCTING Bi-Sr-Ca-Cu-O FILMS BY ION-BEAM SPIKE SPUTTERING

A.V.Suvorov, V.E.Chelnokov, G.M.Gusinsky, M.E.Leshchenko, V.N.Makarov, D.D.Prokofiev, V.V.Tretiakov, T.B.Zchukova (a)

Superconducting thin films of Bi-Sr-Ca-Cu-O were formed by ion-beam spike sputtering under vacuum 10^{-6} Torr with subsequent annealing in oxygen atmosphere [1]. The target of nominal composition 1112 was mounted at 45° to the ion beam. The average distance between the target (100 mm in diameter) and the substrate holder was 50 mm. 40-keV Ar^+ or O_2^+ ions were used for sputtering. The films 0.3 - 0.9 μm thick were deposited onto SiC and MgO (100) substrates held either at 430°C or at room temperature. The deposition rate was about 20 Å/min for Ar^+ and 4 Å/min for O_2^+. Before annealing, the films were characterized using Rutherford backscattering spectroscopy (RBS) and x-ray microanalysis (XMA).

The films were annealed at different temperatures: 850, 870, 880 and 890°C for 1 - 16 hours. The critical current density j_c and the resistivity temperature dependence $\rho(T)$ were measured in annealed films using standard d.c. four-probe technique. X-ray diffraction spectra were also obtained for the films.

Fig.1 shows the RBS spectra for as-deposited films produced at the different substrate temperatures.

Table 1 gives the RBS and XMA data on element content in as-deposited films. Bi is normalized to 1. It can be seen that RBS and XMA data on Ca differ from each other considerably. The XMA seems to be more reliable, because of certain difficulties in RBS concentration analysis of the elements with small atomic radius. Comparing the element composition of the target and of the films, we have estimated the mass-transfer coefficients for different sputtering conditions. Using these coefficients, it is possible to correct the element ratio and sputtering conditions with the goal of film content optimization close to the required stoichiometry.

Fig.2 shows the $\rho(T)$ characteristics as a function of the annealing temperature. The j_c(4.2K) and T_c(R=0) for the curves 1,2,3, 4 are 812.0 , 50.0 , 1.6 , 0.84 A/cm^2 and 66.7 , 62.6 , 34.7 , 13.0 K, respectively. It was found that annealing at 880 - 890°C gives the best superconducting properties in case of the annealing duration of 1 hour.

Higher annealing temperature yielded the semiconductor-like behavior of the $\rho(T)$ curves in normal state and no T_c(R=0) above 4.2 K. X-ray

(a) A.F.Ioffe Physico-Technical Institute, Academy of Sciences of the USSR, Politekhnicheskaya 26, SU-194021 Leningrad, USSR.

diffraction patterns indicated the majority of the 2212 phase (c=30.7Å) with c axis oriented normal to the surface. The reflections corresponding to the high-temperature 2223 phase (c=37 Å) were not revealed, whereas curve I in Fig.2 indicates the presence of this phase in some films.

Table I

Sput. ions	Substr.	Substr. temper.	RBS data Bi:Sr:Ca:Cu	XMA data Bi:Sr:Ca:Cu	Mass-transfer coefficients Bi:Sr:Ca:Cu
Ar^+	MgO	room	I:I.3:2.6:2.4	I:I.4:I.2:2.6	0.8:I.I:I.0:I.o
Ar^+	SiC	room	I:I.2:2.4:I.9	I:I.5:I.3:2.5	0.8:I.2:I.0:I.0
Ar^+	MgO	430°C	I:4.I:6.8:7.3	I:4.0:3.4:8.0	0.3:I.2:I.0:I.2
O_2^+	SiC	430°C	I:I.5:5.I:3.5	I:I.7:2.I:4.I	0.6:I.0:I.I:I.2

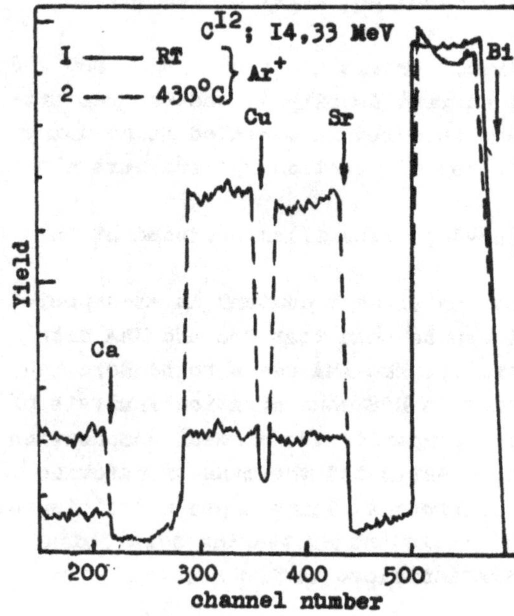

Fig.I. RBS spectra of as-deposited films at different substrate temperature.

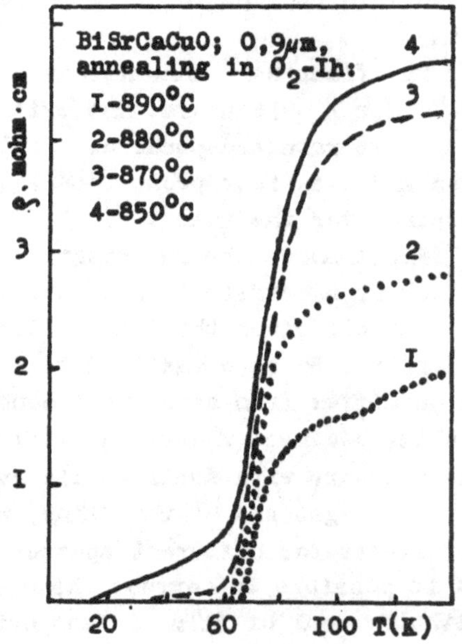

Fig.2. Resistivity vs. temperature at different annealing temperature.

In our experiments, the highest $T_c(R=0)=72.0$ K was revealed in the film,produced by Ar^+ sputtering at 430°C substrate temperature, after annealing at 850°C for I6 hours. The best $j_c(4.2K)=I.5\cdot IO^3$ A/cm^2 was achieved after I hour annealing at 890°C of the film produced by Ar^+ sputtering at room temperature.

The films formed on SiC substrates and annealed in the same conditions indicated nonsuperconductors. It is apparently caused by strong interraction between the substrate and the film at applied annealing temperature that was confirmed by RBS data.

The ion-beam spike sputtering technology used for the superconductor film producing allows to control the deposition process in situ and to combine the sputtering procedure with implantation of some impurities. During experiments, high reproducibility and homogenity were gained without change of the target. The substrate holder area of very good (within 5% for each element) composition homogenity was 60 mm in diameter. From process to process, the element content deviation was not more than 2%. In addition, the half year storage in air has not affected the properties of the films.

It is clear that annealing conditions require further improvement. Special attention has to be devoted to the substrate surface quality. The T_c end point has not exceeded even the liquid -nitrogen temperature in our experiments, in spite of majority of the 2212 phase with $T_c \approx 80-85$ K. Moreover, the values of critical current density and room temperature resistivity were not constant in different parts of the same film.

(I) K.K.Bourdelle, V.N.Makarov, A.V.Suvorov, N.G.Chechenin: Tesisi dokladov XVIII Vsesojuznogo soveshchaniya po fisike vzaimodeystviya zariazhennikh chstits s kristallami, Moskva, MGU, 1988, 166.

PROPERTIES OF SUPERCONDUCTING FILMS Bi-Sr-Ca-Cu-O IMPLANTED BY OXYGEN IONS

A.V.Suvorov, V.E.Chelnokov, M.E.Leshchenko, V.N.Makarov, D.D.Prokofiev, and T.B.Zchukova (a)

Superconducting Bi-Sr-Ca-Cu-O films about 0.9 μm thick were prepared by ion-beam spike sputtering and post-annealing in flowing oxygen. The polycrystalline films were mainly consisted of 22I2 phase. The 5-I0 μm grains were well oriented with c axis perpendicular to the substrate surface. The details of the film formation are described in Ref.I. We also have used non-annealed film for implantation with oxygen to high doses.

The films were irradiated at room temperature with 90-keV O^+ ions in the $I \cdot IO^{I2} - IO^{I8}$ cm^{-2} dose range. The ion beam current density was 0.I $\mu A/cm^2$. Critical current density j_c and resistivity temperature dependence $\rho(T)$ in the films were investigated as a function of dose. X-ray diffraction patterns were also measured for one film before and after oxygen implantation. To recover the superconducting properties, the relatively low-temperature annealing at 600-6I0°C for 0.5 or 5 hours was carried out.

As shown in Figs. Ia) and b), the $\rho(T)$ characteristic changes with increasing dosage. Particularly, the T_c end point decreases up to liquid-helium temperature. Such dose dependence is more distinct in the films with lower initial critical current density. Moreover, no changes in $\rho(T)$ characteristic and j_c (4.2K) value were observed until certain critical dose ($I \cdot IO^{I2}$ cm^{-2} in case presented in Fig. Ia)). The value of this critical dose depends on superconducting properties of the film before irradiation. This behavior of resistivity temperature dependence coinsides with the results reported earlier for the Y-Ba-Cu-O films [2,3,4]. Table I shows the changes of the resistivity at room temperature ρ(293K), as well as the critical value of j at 4.2K in the film (its $\rho(T)$ characteristics are ploted in Fig. Ia)) with increasing irradiation dosage. The pronounced decreasing of j_c at 4.2K is observed. It should be noted for some films that decreasing of the T_c end point was accompanied by semiconductor-like behavior of the resistivity in normal state. In addition, the step at nearly II0 K on the $\rho(T)$ curves observed for high-temperature treated films was not affected by ion irradiation.

Taking into consideration the fact that polycrystalline films consist of grains weakly connected with each other, the results of expe-

(a) A.F.Ioffe Physico-Technical Institute, Academy of Sciences of the USSR, Politekhnicheskaya 26, SU-I9402I Leningrad, USSR.

Table I

Implant. dose, O^+ (cm^{-2})	Room temperature resistivity (mOhm · cm)	Critical current density at 4.2K ($A \cdot cm^{-2}$)	T_c end point (K)
before imp.	3.3	36	59.0
IO^{I2}	3.3	36	59.0
IO^{I3}	3.9	I8	52.6
IO^{I4}	29.0	<6 · IO^{-3}	<4.2
IO^{I4} and annealing 600°C,0.5h.	II.0	not measured	22.6

riments can be interpreted as follows. Since relatively low doses of the ion implantation changed for the worse dramaticaly superconducting properties or the films, such as j_c(4.2K) and T_c end point, it may be proposed that the grain boundaries were mainly affected. The maximum current density in superconducting state is probably limited by the grain boundary state. More slow changing of the T_c onset and mid-point temperatures of superconducting transition confirms this proposition, because they characterize the bulk properties of the material.

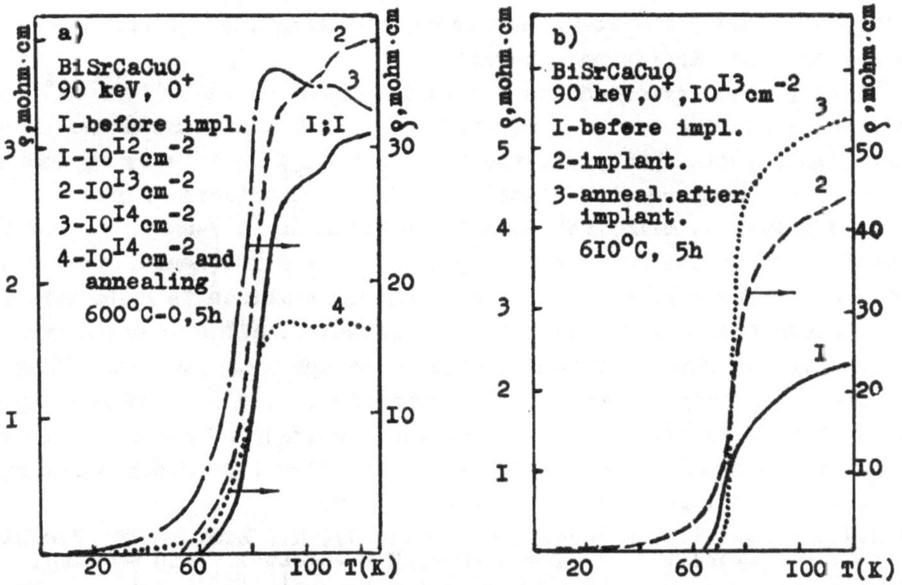

Figs. Ia) and b). Resistivity vs. temperature at different ion doses for two of the films investigated.

After disappearance of superconducting transition at high doses (curve 3 in Fig.Ia) and curve 2 in Fig.Ib)), the films were annealed at relatively low temperatures (600-610°C) for 0.5 or 5 hours. Such temperature was applied to prevent the recrystallization of the film material or any changes of the grain boundary geometry. The results are illustrated by curves 4 and 3 in Figs. Ia) and Ib), respectively. It is clear that enough long annealing promote the recovering of superconducting transition properties, including j_c and ρ(293K). For information, see the Table 2, concerning to the Fig.Ib). However, the irradiation-

Table 2

	Nonimplanted	O^+ impl. $10^{13}cm^{-2}$	Annealing after impl. 610°C,5h.
ρ(293K) (mOhm·cm)	3.4	38.0	6.0
T_c(R=0) (K)	64.5	<4.2	66.0
j_c(54.5K)(A/cm^2)	4.I		3.0

induced semiconductor behavior of the normal state resistivity was not changed. It seems to be an influence of the amorphous layer created by ion bombardment. The films were examined in the optical transmission microscope. The irradiated area had become transparent that can be explaind by formation of the amorphous layer [2]. Not essential disordering of bulk crystal structure was confirmed by comparison of x-ray diffraction peak intensities measured before and after irradiation at $1 \cdot 10^{16}$ cm^{-2} dose. The low temperature annealing was not capable to re crystallize the disordered material.

We have also performed the oxygen implantation to $5 \cdot 10^{17}$ cm^{-2} dose with post-annealing at the same conditions with unimplanted one. Preliminary implantation had rised the T_c end and T_{mid} points by IO and 3K, respectively, compare with films obtaned by conventional manner.

In conclusion, relatively low irradiation doses seem to induce the point defects in the grain boundary area. As a consequence, strong degradation of the current transmission characteristics is observed. The exact mechanisms involved are not identified. Possible loss of oxygen may take place, which can be restored by enough long time annealing at sufficient low temperatures. The experiments emphasize that grain boundaries play an important role in superconductivity of Bi-Sr-Ca-Cu-O polycrystalline films. The O^+ion implantation effects on grain boundaries as Josephson junctions will be published later.

(I) A.V.Suvorov, V.E.Chelnokov, G.M.Gusinsky, M.E.Leshchenko, V.N.Makarov, D.D.Prokofiev, V.V.Tretiakov, T.B.Zchukova, this meeting.
(2) G.J.Clark, A.D.Marwick, R.H.Koch and R.B.Laibowitz: Appl. Phys. Lett. 5I(2),1987,pp.139-14I.
(3) G.J.Clark, F.K.LeGoues, A.D.Marwick, R.B.Laibowitz and R.Koch: Appl. Phys. Lett. 5I(18),1987,pp.1462-1464.
(4) S.Matsui, Y.Ochiai, J.Fujita, T.Yoshitake and Y.Kubo: J. Appl. Phys. 64(2), 1988, pp.936-938.

FORMATION OF THIN SUPERCONDUCTING FILMS BY Cu^+ IMPLANTATION

V. Yu. Petukhov, L. S. Grigorjev, E. F. Kukovitsky, I. B. Khaibullin[1)]

Superconducting thin films have been successfully fabricated by magnetron sputtering [1], laser deposition [2], molecular beam epitaxy [3], ion beam mixing [4] and others, but an elaboration of new methods is still in progress. Ion beam synthesis, i.e. formation of compounds and alloys at a high dose implantation is of a great interest because it has some advantages [5]. Ion synthesized buried silicides, SOI structures etc. is being used in an electronic device technology. However, except for a communication of M. Nastasi et al. [6] and our recent report [7] no other works of successful applications of ion implantation for forming high temperature superconducting films have been published. In this report we present first results of studies in which we have prepared thin films of Er-Ba-Cu-O superconductor by Cu implantation into a base material.

The base material of compound $Er_2Ba_4O_7$ was synthesized in air using Er-oxide and Ba-hydroxide at 1100°C. A selection of the material was conditioned both by suitable phase diagrams of double systems R_2O_3 - BaO and a sufficient stability of $R_2Ba_4O_7$ in a wet air. From this point of view for R some 4f-elements such as Er, Ho, Eu etc. appear to be very convenient [8]. A material formed was pressed into pellets with consequent annealing in air at 1000°C. To identify the phase composition of obtained targets the X-ray diffraction was used. Cu implantation into the received material was performed at ion energy of 40 or 90 keV with doses available to form superconducting thin films. Ion current density was varied in the range of 10 to 40 $\mu A/cm^2$. Sample temperature could be increased by few hundred degrees during implantation depending on ion current density. The temperature of superconducting transition (T_c) was determined from the appearance of superconducting impedance in ultrahigh frequency (UHF) measurements using EPR radiospectrometer "Bruker BER-418" at temperature scanning from room to helium.

In Fig. 1 the dependence of UHF impedance versus temperature for a sample implanted with 90 keV Cu^+ ions is presented. Unlike the report of M. Nastasi [2] in our case superconductivity appears in "as implanted" sample without additional thermal treatments.

---

1) Kazan Phys.-Techn. Institute of the Academy of Sciences of the USSR 420029, Sybirsky trakt 10/7, Kazan, USSR

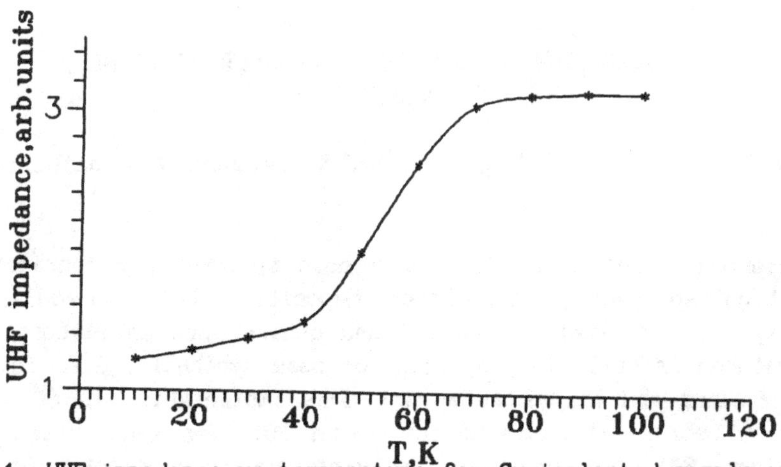

Fig. 1. UHF impedance vs temperature for Cu-implanted sample with ion energy of 90 keV

Measurements show, that T_c depends on the ion energy and the ion current density. For example at energies 40 and 90 keV T_c are 25-30K and 45-50K, respectively. Likewise T_c rises with the increase of current density. Moreover, superconductivity appears only at the dose higher than some threshold according to the stoichiometry content of the known "1-2-3" superconductor in ion synthesized layer. Since superconducting layer is formed in the depth of the target it is covered by a thin film of the basic material that is insulating.

Surface morphology of implanted layer differs from that before implantation. For example,some microcracks appeared on the surface, not allowing surface current flows. Both this fact and a presence of thin film of basic material on the sample surface were a reason of unsuccessful 4-probe measurements in our experiments.

X-ray glancing angle measurements demonstrated multiple phases in synthesized layers. Multiple phase content and radiation damages both lead to appearance of sufficiently long tail in Fig. 1.

In summary, we have shown that the implantation of Cu ions into a base material of $Er_2Ba_4O_7$ is capable of producing superconductors with sufficiently high superconducting transition temperature.

References:

[1]. Lee, S. J., et al. Appl. Phys. Lett. 51 (1987) 1194
[2]. Dijkkamp, D., et al. Appl. Phys. Lett. 51 (1987) 619
[3]. Webb, C., et al. Appl. Phys. Lett. 51 (1987) 1191
[4]. Rauschenbach, B., K. Hohmuth Z. Phys. B.-Condensed Matt. 74 (1989) 155
[5]. Akimchenko, I. P., V. V. Krasnopevtsev Proc. First USA-USSR Seminar on Ion Impl. Albany, USA (1977) 31
[6]. M. Nastasi et al. Appl. Phys. Lett. 52 (1988) 1729
[7]. Grigorjev, L. S. et al. Тезисы докл. 1 Всес. Совещания по ВТСП, Харьков, 1988, т. 3, с. 68-69
[8]. Antipov, et al. Sov. J. Inorg. Chem. 29 (1984) 1624

SUPERCONDUCTING $YBa_2Cu_3O_x$ FILMS ON SILICON PREPARED BY LPVD

M. Panzner, H. Müller, G. Keßler, W. Pompe, S. Völlmar, H. Vinzelberg, K. Fischer, B. Schöneich, C. Ritschel[1]

Introduction

Superconducting $YBa_2Cu_3O_x$ thin films can be prepared by means of any conventional thin film technique, including the laser pulse vapour deposition (LPVD). An advantage of the latter is given by the fact that it does not need UHV. Several authors (4,6,10,11) succeeded in depositing them immediately on silicon substrates. This was achieved with UV-laser pulses (193nm or 248nm, 10-30ns) at high deposition rates on heated substrates in the presence of O_2 such that annealing was not necessary.

This paper concerns with LPVD of the above mentioned films on Si by means of a XeCl excimer laser (type 1701 IOFAN Moscow, 308nm). Its pulse length of 20ns is suitable for stoichiometric deposition, as shown earlier (8). The deposition rate however, was limited by the low pulse energy, and narrow focussing was necessary to achieve the suitable energy density. It was found by means of SIMS that there was undesirable interaction with the substrate (7). Therefore we used ZrO_2 for a diffusion barrier. The favourable effect of diffusion barrier is known from other works (10,11,12) The essential deposition parameters are known from numerous publications. Nevertheless every particular experimental setup requires optimization.

Experiments and results

A scheme of experimental setup is shown in fig. 1. The essential parameters were realized as follows:

wave length	308 nm	repetition rate	50 Hz
pulse length	20 ns	deposition time	20 min
pulse energy	34 mJ	oxygen pressure	40 Pa
energy density	1 J/cm^2	substrate temperature	700°C

The ZrO_2 film was deposited by LPVD from a Y-stabilized ZrO_2 target. Immediately afterwards the superconducting film was deposited from ceramic $YBa_2Cu_3O_x$ targets (tetragonal or orthorhombic). After deposition the samples were left in the chamber for 2 hours at 450°C and 40 kPa O_2.

1) Zentralinstitut für Festkörperphysik und Werkstofforschung, AdW der DDR, Helmholtzstr. 20, Dresden, 8027

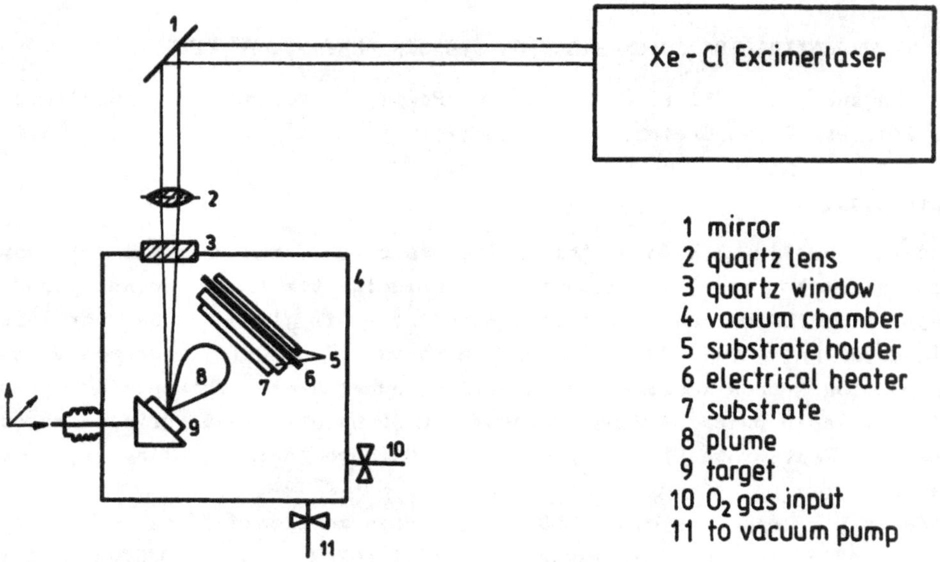

fig.1 Scheme of the experimental arrangement

The metals ratio measured by electron probe microanalysis (EPMA) shows compositions near that of the known 1-2-3-phase. The films show a rough surface with /um-particles. The refractive index was found by ellipsometry as n = 1,12 - 0.3j, wich is compatible with the values found by (3,5). However the accuracy of the measurements was low because of the surface roughness. The resistivity versus temperature curve of our in-situ films differ from those obtained with $SrTiO_3$ substrates (1,2,9,10), indicating deviations from optimal structure (fig. 2). It is the aim of further work to obtain textured films on Si at lower temperatures.

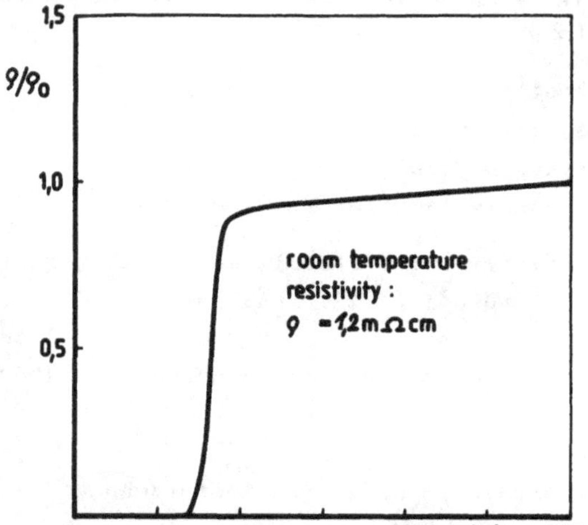

fig. 2 Electrical resistivity of $YBa_2Cu_3O_x$ film in dependence on temperature

References

(1) Frühlingsdorf, J., W. Zander and B. Stritzker: Solid State Comm. 67 (1988) 1071

(2) Golovashkin, A.I., E.V. Ekimov, S.J. Krasnosvobodzev and E.V. Pechen: Physika C 153-155 (1988) 1455

(3) Humlicek, J., M. Garriga, M. Cordona, B. Gegenheimer and E. Schönherr: Solid State Comm. 66 (1988) 1071

(4) Kehlenbeck, M. and H.U. Krebs: Frühjahrstagung der Deutschen Physikalischen Gesellschaft in Münster, 1989, Abstract p. 53

(5) Kelly, M.K., P. Barboux, J.-M. Tarascon, D.E. Aspnes, W.A. Bonner and P.A. Morris: Phys. Rev. B38 (1988) 870

(6) Koren, G., E. Polturak, B. Fisher, D. Cohen and G. Kimel: Appl. Phys. Lett. 53 (1988) 2330

(7) Oswald, S., R. Voigtmann, M. Panzner and G. Keßler: Unveröffentlichte Ergebnisse

(8) Richter, A. and G. Keßler: Solid State Comm. 70 (1989) 1147

(9) Roas, B., B. Hensel, G. Saemann-Ischenko and L. Schultz: Appl. Phys. Lett. 54 (1989) 1051

(10) Venkatesan, T., E.W. Chase, X.D. Wu, C.C. Chang and F.K.Shokoohi: Appl.Phys.Lett. 53 (1988) 243

(11) Witanachchi, S., S. Patel, H.S. Kwok and D.T. Shaw: Appl.Phys. Lett. 54 (1988) 578

(12) Wu, X.D., A. Inam, M.S. Hedge, B. Wilkens, C.C. Chang, D.M. Hwang, Nazar, T. Venkatesan, S. Miura, S. Matsubara, Y. Miyasaka and N. Shohota: Appl. Phys. Lett. 54 (1989) 754

COMPUTATION OF THE DISTRIBUTION OF RADIATION AND TEMPERATURE DURING RAPID THERMAL PROCESSING

Rolf-Reiner Bartl*)

Equipments for rapid thermal processing (RTP) are applied in an increasing rate for the production of VLSI semiconductor devices, for instance for

- annealing and activation of dotants after ion implantation into silicon and other semiconductors
- reflow of passivation layers
- oxydation and nitridation
- silicide formation

The most used RTP technique is radiation with intensive halogen lamps. For its application in a technological process an uniform temperature distribution across the whole wafer must be guaranted. Therefore the presented optimizing computations for a RTP equipment contain informations of the distribution of radiation intensity and temperature.

1. Twodimensional modelling of radiation intensity in the target plane of a RTP equipment

Modelling the radiation of tungsten halogen lamps, we regard some line emitters of the lenght 2*L parallel to the x-axis. The intensity, striking to any elemental area of the target plane, is given by

$$E(x,y) = \Phi/8\pi L \sum_{yo} \int_{xo} a / (a^2 + (xo-x)^2 + (yo-y)^{3/2})\, dyo \qquad (1)$$

with the distance a between source and target plane. The integral of the x-dependence can be solved analytically and yields for n emitters at the position yi

$$E = \frac{\Phi a}{8\pi L} \sum_{i=1}^{n} \frac{1}{a^2+(yi-y)^2} \left(\frac{L-x}{\sqrt{a^2+(yi-y)^2+(L-x)^2}} - \frac{-L-x}{\sqrt{a^2+(yi-y)^2+(L+x)^2}} \right) \qquad (2)$$

Usually in a halogen-furnace emitters and target are sourrounded by reflecting walls. Now we summ the intensity of direct incident and reflected rays of first and second order. The program allows the variation of one or a group of emitter intensities Φ to test the effect of power graduation or breakdown of a lamp. Following formula summarizes the computations:

$$E = \sum_{j=1}^{m} \frac{a1}{8\pi L} \sum_{i=1}^{n} \frac{\Phi(yi)}{a1^2+ y1^2} \left(\frac{x1}{\sqrt{a1^2+ y1^2+ x1^2}} - \frac{x2}{\sqrt{a1^2+ y1^2+ x2^2}} \right) \qquad (3)$$

with a1 = f(a,Z); x1, x2 = f(L,x) and y1 = f(Y,y,yi) describing the modified limits of integration and summation.
The parameters Z, L and Y describe the reflector geometry.
To reduce the duration of calculation, the Fortran program can be restricted to the computing of the profiles E(x,0) and E(0,y).

---

*) VEB Mikroelektronik "Karl Marx" Erfurt, Forschungszentrum
Rudolfstraße 47, Erfurt 5010, GDR

The figures 1.a and 1.b show the intensity distribution in the target plane of RTP laboratory system which is used in VEB Mikroelektronik. The ratio between the local intensity and the intensity in the centre of sample chamber is plotted as a function of x and y. In the centre the intensity has a maximum and decreases parabolically both parallel (1.a) and rectangular (1.b) to the lamps. The curves can be approximated by polynoms with dominating second- and fourth-order-terms. Evidently the superposition of the lamp intensities does not effect periodical intensity variation in the target plane. On the other hand the intensity distribution immediate near (15 mm) the emitters shows ripples, see fig. 2.a.

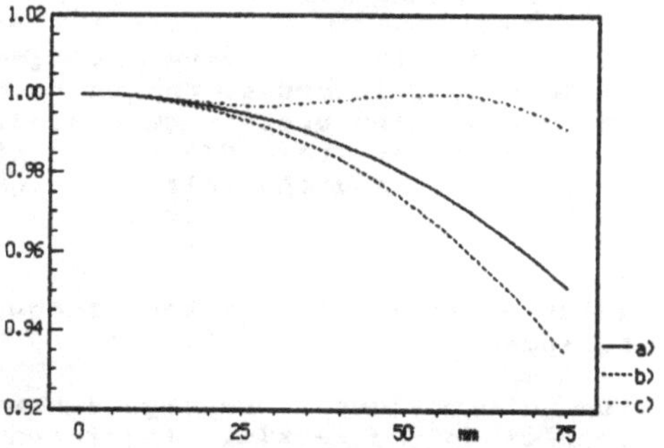

fig.1: relative intensity in the target plane
a)x-dir b)y-dir c)y optim.

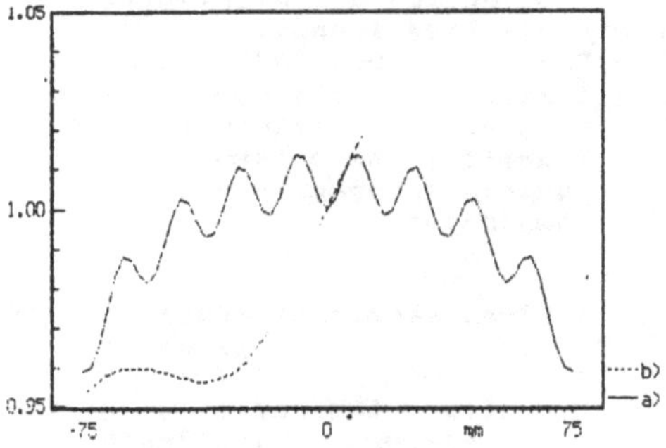

fig.2: relative intensity (y-dir)
a) near the lamps
b) if a lamp breaks down

The intensity in y-direction can be homogenized by an aimed graduation of the lamp power. Fig. 1.c shows the distribution, if 4 of 12 lamps work with another power. Up to 62.5 mm from the target centre (that means along a 5-inch-wafer) the intensity variation is less then 0.3 %.
Breaks down one lamp, the intensity distribution is asymmetrically. Fig. 2.b shows that effect with the fault lamp 27 mm from centre.

2. Modelling of temperature distribution in a semiconductor wafer

The radiated wafer lies thermally isolated on quartz filaments in the centre of the chamber.
Starting from the universal equation of thermal conduction for the computation, following premises were made:

1) During halogen lamp annealing the wafer is after reaching required temperature for some seconds in thermal equlibrium. Thus vanishes the time dependence $\partial T/\partial t = 0$.
2) At the used high temperatures T the thermal conductivity coefficient K is only lightly dependent on the temperature and therefore assumped as constantly.
3) The wafer is cylindrically symmetric: $\partial T/\partial \psi = 0$.
4) Because its thickness is much smaller then its diameter, the perpendicular temperature gradient is not regarded: $\partial T/\partial z = 0$.

With that the thermal conductivity equation for every volume element reads in differential form as follows:

$$d^2/dr^2\, T(r) + 1/r * d/dr\, T(r) = c1*T(r)^4 - c0 \qquad (4)$$

with.

$$c0 = (2*I*(1-R) + 4*\varepsilon*\sigma*Tu^4) / D*K$$

$$c1 = 4*\varepsilon*\sigma / D*K$$

I : the irradiated intensity per unit area
R : the reflectivity coefficient
σ : the Stefan-Boltzmann-constant
ε : the emissivity coefficient
Tu : the ambient temperature
D : the thickness of the wafer

The irradiation from the the wafer edge r=R gives the boundary condition

$$K*\partial T/\partial r\Big|_{r=R} + \sigma*\varepsilon*(T(R)^4 - Tu^4) = 0 \qquad (5)$$

The thermal conduction equation with inhomogenous T^4 term and radiation boundary condition is not solvable analytically and must be calculated numerically. Here the finite differences method (FDM) is used: The differential quotients were approximated by difference quotients using Taylor's method (with a grid spacing h) and provide the equation:

$$Ti = (Ti-1 + Ti+1 + (Ti+1 - Ti-1)/2*(i-1) - (c1*Ti^4 - C)h)/2 \qquad (6)$$

that is calculated by iteration. At the wafer edge the formula

$$Ti\Big|_{i=R/h} = -c1*D*h*(Ti^4 - Tu^4)/4 + Ti-1 \qquad (7)$$

submit. The calculations are realized as Fortran program system. The equation appears convergent, 10 to 20 iterations are needed. Fig. 3 shows the temperature distribution of silicon samples with the diameter 3", 4" and 5", using the parabolic intensity distribution calculated in chapter 1. The curves demonstrate, that the temperature profile is determined by two overlaying effects:

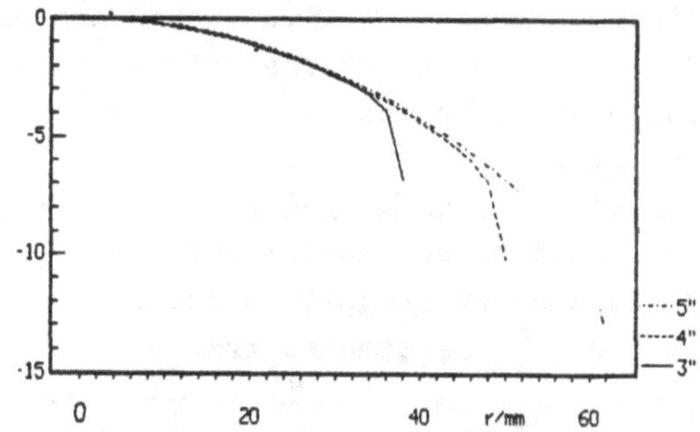

fig.3 radial temperature distribution (T-Tm /K) of wafers during RTA

1) A temperature gradient builds up by the parabolic irradiation.
2) The outside part of the wafer is additional cooled by the lateral irradiation from the target edge.

Both effects increase with the wafer dimension. For instance by a 4 inch wafer (with 1000 °C in the centre) the difference between centre and edge is 11 Kelvin.

3. Conclusions

For the application of RTP in a technological process a homogeneous intensity distribution in the target plane is necassary. Rectangular to the lamps this can be reached software-controlled by lamp power selection. Constructing a new target chamber, this demand cam also be realised by different distances between the lamps. For homogenity in both axis two 90 degree turned lamp arrays are recommendable.

DEFECT PROFILES AND COMPOSITION DISTURBANCES INDUCED BY ION IMPLANTATION

BURENKOV A.F.[1)] and KONOPLJANIK G.G.[1)]

The forward transport equation for the flux Φ_i of the atoms of i-type moving in a compound target material may be writen:

$$\frac{\vec{V}}{V}\frac{\partial\Phi_i(\vec{V},\vec{z})}{\partial\vec{z}} = \sum_k N_k\left[\int d\sigma'_{ik}\Phi'_i(\vec{V}',\vec{z}) - \int d\sigma_{ik}\Phi_i(\vec{V},\vec{z})\right] + N_i\sum_k\int d\sigma''_{ki}\Phi_k(\vec{V}'',\vec{z}) \quad (1)$$

where N_k is the density of the target atoms of k-type; $d\sigma_{ik}$- is the cross section for a scattering of i-atom on k-atom corresponding to $\vec{V}' \rightarrow \vec{V}$ transition for i-atom, $d\sigma_{ik}$ describes transitions of i-atoms from $\vec{V}$ to other states and $d\sigma''_{ki}$ means cross section for a process in which a i-type recoil atom is created as a result of k-atom on i-atom scattering. The sum over k comprises all the components of the target.

Equation (1) describes both the transport of incident ions and that of recoil atoms. When recoil atoms are considered the last term is non-zero otherwise only two first terms are needed.

The system of equations (1) is solved in a multigroup-approximation for a distribution function depending on energy and angle of atomic movement.

The distributions of particles over energy and angular variables is represented by occupation numbers F_l^m where l and m denotes the energy and angular groups. In the multigroup approximation Eq.(1) expresses how F_l^m is changed when the depth X is changed:

$$\frac{\Delta F_l^m(X)}{\Delta X} = -\sigma_l F_l^m(X) + \sum_{j=l+1}^{NE} P_{lj}\sum_{n=1}^{NA} W_{lj}^{mn} F_j^n(X) + G(X) + S(l,m) \quad (2)$$

where σ_l is the probability per unit length that the particles leave the i-energy group. The second term describes increase of the particle number in the (l,m) subgroup; P_{lj} is the probability of transition from j to l energy groups; W_{lj}^{mn} is transition probability from n to m angular subgroup corresponding to a transition in energy groups from j to l ; G(X) is the term describing the generation of recoil atoms and the last term S(l,m) accounts for the stopping of moving particles and includes both the electronic and the low energy transfer elastic stopping. Thus the three terms of equation (2) describe the elastic collisions with a significant change of energy and angle and the last one describes inelastic stopping and elastic stopping with energy transfer restricted by the half width of the energy group $\Delta E/2$. The way of the taking the stopping into account is

1)Institute of Applied Physics of the Belorussian V.I.Lenin State University, Minsk, 220080 USSR presented in /1/.

The coefficients σ_{l}, P_{lj} and W_{lj}^{mn} are taken in the analytical form for the elastic cross section from /2/. The proposed numerical algorithm of the forward transport equation modeling is not sensitive to the size of the depth step ΔX, and to the size of energetic and angular group. Thus, the step ΔX can be in the order of 1/20 to 1/50 of the mean projected range. As a result the recuired computing time is reduced by approximately 10 to 100 times as compared to the algorithm of /3/. A calculation of ion ranges and of atomic transport in cascade during ion irradiation of one-component target recuires 1 to 2 minutes of CPU time on a ES-1036 computer.

From many applications of the FTE we consider here calculations of the damage profiles and stoichiometric disturbances induced by ion irradiation. A typical example of the calculation of such kind is presented by the fig.1 , where the range profile of the phosphorus ions together with the vacancy profile and the netto-defect profile (3) in silicon is depicted. The size of the depth step ΔX=20 nm.
Due to anisotropy of the recoil atom generation at lower depths ($X \leq 0.91 R_p$) dominate defects of vacancy-type and at greater depths the defects of interstitial type.

To show the trends in the netto-defect profiles let us introduce the depth Xo where the vacancy- and interstitial-type defects compensate each other and the mean projected range of the recoil atoms R_r as the difference between the mean depth of the vacancies and of interstitial location. In the Table these quantities related to Rp and ΔRp of the range profile are presented for the case of boron and phosphorus implantation in silicon.

T A B L E

	E, KeV	10	100	1000
Ion	X_o/R_p	0.641	0.854	0.959
B	$R_r/\Delta R_p$	0.128	0.0512	0.0389
Ion	X_o/R_p	0.730	0.815	0.904
P	$R_r/\Delta R_p$	0.379	0.271	0.157

An example of a calculation of the atomic transport in a more complicated case of layered target is shown on the Fig.2.
Due to different mass of In- and of P-atoms they are in a different manner carried away by the atomic cascade and as a result there are more In-atoms closer to the surface and more P-atoms at larger depth.
This effect is the stronger the heavier are the ions which initiate the atomic cascade.

A peculiarity of this example is a rather strong penetration of the silicon recoil atoms into InP substrat.

To sum up, the forward transport equation approach gives an effective way to study the ballistic atomic processes during ion irradiation of solid targets.

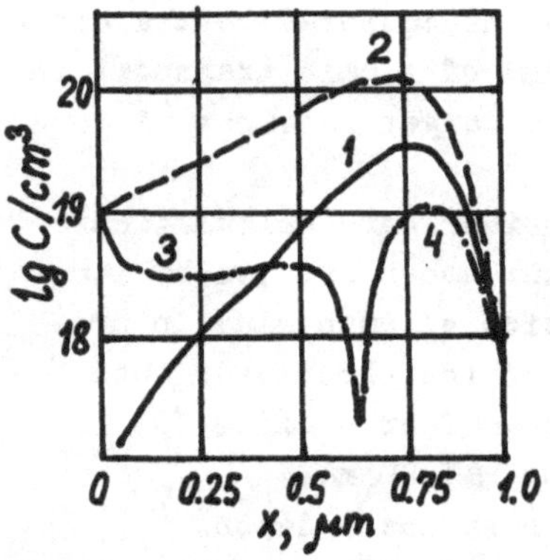

Fig.1. Range profile (1), vacancy concentration (2) and the netto-concentration of vacancies (3) and of interstitials (4) for the case of 500 KeV (10^{15} cm^{-2}) P→Si implantation.

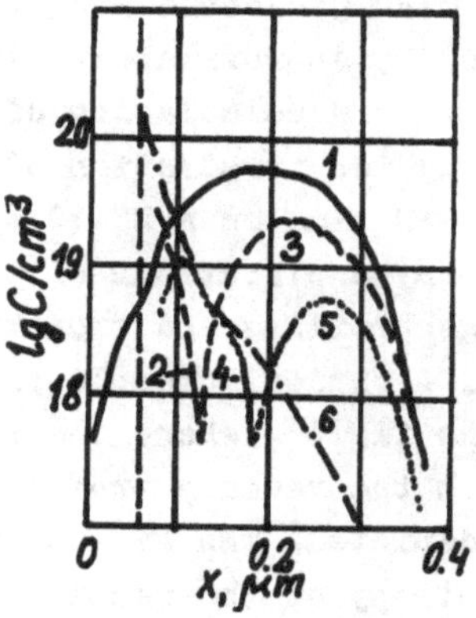

Fig.2. Range profile (1), netto-concentration of vacancies(2) and interstitials (3), the over-stoichiometric excess of In (4) and of P (5) and the Si-recoils concentration (6) for P^{+}-ions implantation into InP covered by 600 nm thick Si-layer (E=150 KeV , D=10^{15} cm^{-2})

REFERENCES

(1) Буренков А.Ф., Ф.Ф.Комаров и М.М.Темкин : Поверхность 4 (1989) 15-21.

(2) Буренков А.Ф., Ф.Ф.Комаров, М.А.Кумахов и М.М.Тёмкин: Пространственые распределения энергии, выделенной в каскаде. Энергоатомиздат, 1985.

(3) Christel L.A. and J.F.Gibbons : J.Appl.Phys. 52(1981) 5050-5055.

ENERGY BALANCE IN NONEQUILIBRIUM GRADIENT-ZONE CRYSTALLIZATION IN A SEMICONDUCTOR

B.S. EVSEEV [1)]

Energy balance as a practical aspect of nonequilibrium gradient-zone crystallization (NGZC) in a semiconductor is studied.

The NGZC is based on concurrent development of the following physical processes /1-3/:

- absorption of "seeding" laser radiation $I_1(\hbar\omega > E_g)$ at a certain depth from the sample surface and formation of a seeding liquid phase region (SLPR);
- absorption of "heating" laser radiation $I_0(\hbar\omega < E_g)$ penetrating the sample on its other side by a part of the SLPR, and formation of a mobile liquid phase domain (LPD);
- excitation of electric field by ion and electron probes along optical wave vectors directed to meet each other;
- transport of the LPD in the electric field through the solid-phase region (SPR).

The succession of absorption of the laser pulses determines the starting conveyer regime under which the LPD is transferred by a conveyer from the energy flux $I_1(\hbar\omega > E_g)$ to the energy flux $I_0(\hbar\omega < E_g)$. The energy flux I_0 supports the LPD during its transport at the temperature T_L exceeding the melting temperature T_m. Fig. 1 presents the phase structure, spatial orientation of the pulsed radiations and the initial condition for temperature distribution in the SPR-LPD-SPR structure at the moment of the LPD formation ($t = t_{L1}$). In the course of the LPD transport into the sample a SPR-LPD-SPR dynamic structure is formed.

A generalized heat conduction equation defining the variation in temperature for the structure is:

$$\frac{\partial T}{\partial t} = a\frac{\partial^2 T}{\partial x^2} + V_L\frac{\partial T}{\partial x} + \frac{T}{t_L}, \tag{1}$$

where T is the temperature, V_L is the LPD transport velocity, t_L is the longest time interval typical of the examined thermal process, $t_L = a/V_L^2$; a is the heat conductivity; t, x are the current values of time and coordinate, respectively.

In a partial derivative equation a complete change of the temperature of the structure (on the left side of the equation) is due to heat conduction (the first term on the right is of diffusional nature), transport of the LPD heat at the velocity V_L (the second term is of drift

1) Institute of Electronics; BSSR Academy of Sciences, 22 Logoiski Trakt, Minsk-90, 220841, USSR

nature) and transformation of a heating radiation energy into heat (the third term is connected with the generation of heat in LPD at the velocity $1/t_L$).

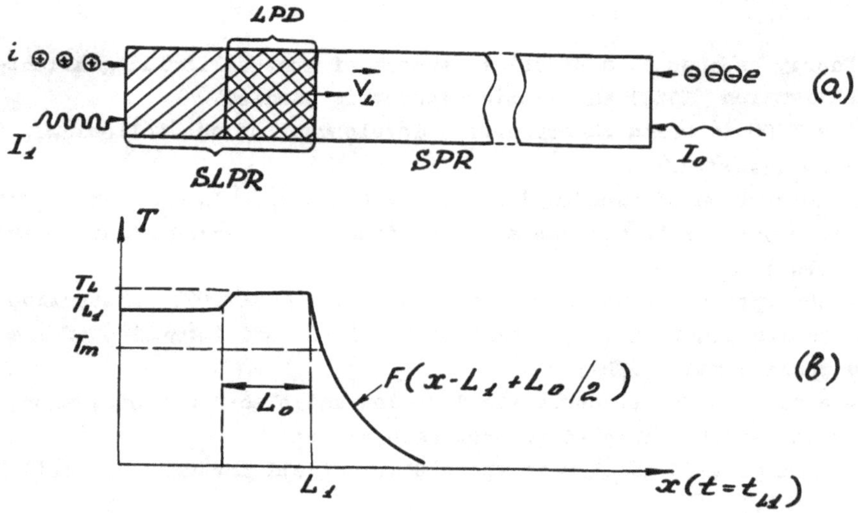

Fig.1 (a) Phase structure and spatial orientation of the pulsed radiations: i is an ion probe; e is an electron probe; V_L is the LPD transport velocity; (б) Initial condition for temperature distribution in the structure at the moment of the LPD formation: L_1 is the initial SLPR extent; L_o is the LPD extent; F is the function defining spatial blurring of the heat front; T_{L1} is the SLPR temperature

The complete energy balance for NGZC under conveyer regime is defined in a unique fashion by equation (1) within range of definition. Start and finish boundary conditions, as defined by nonlinear equations of the following form:

$$\frac{dT}{dt} = \frac{T_c(T) \mp T}{t_L}, \qquad (2)$$

where the negative sign refers to the start condition and the positive sign to the finish condition, $T_c(T)$ is the characteristic start or finish temperature, respectively. $T_c(T)$ depends on heat technological and electrophysical parameters of the semiconductor in solid and liquid phase and energy parameters of the laser pulses. Boundary temperatures were calculated by Laplace transform method of successive approximations. Relative error of calculations (less than 1%) allows only one or two approximations. The initial condition for temperature is specified for the moment of complete transmission of the LPD by the conveyer from the seeding to the heating laser pulse (Fig.1).

The ultimate energy parameters of radiations, determining the process of the LPD formation and transport for NGZC were estimated. The minimum value of energy density is the critical energy parameter for the seeding laser pulse

$$I_1 t_{L1} > \frac{b}{V_1}(\Delta T_m + \eta_L h_L),\ \eta_L = \frac{L_0 + L_1}{a} V_L,\ h_L = \frac{a H_L}{b}, \quad (3)$$

$$H_L = H\Delta\rho + \frac{b}{a}\Delta T_L,\ \Delta T_m = T_m - T_S,\ \Delta T_L = T_L - T_S,\ \Delta\rho = \rho_L - \rho_S$$

where t_{L1} is the seeding pulse duration, H is the latent heat of melting, ρ_L, ρ_S are the semiconductor densities in liquid and solid phases, respectively, b is the thermal conductivity, T_L is the LPD temperature, T_S is the ambient temperature.

For the heating radiation the condition

$$I_0 > H_L V_L \quad (4)$$

defines the minimum density of the energy flux of the laser radiation necessary for the LPD transport to begin. The condition defining the LPD transport through the entire sample is more strict than (4) and is given by inequality

$$I_0 > \frac{b\Delta T_m}{L_b} + H_L V_L \quad (5)$$

for exponential finish pulses, or

$$I_0 > \frac{b\Delta T_m}{L_0} + H_L V_L \quad (6)$$

for the single finish pulse, where L_b is the single pulse width.

The specific estimation for silicon gives the following results:

$$I_1 t_{L1} \approx 0.8\ J/cm^2,\quad H_L V_L \approx 5\cdot 10^5\ W/cm^2$$

$$b\Delta T_m / L_b \approx 1.4\cdot 10^6\ W/cm^2,\ b\Delta T_m / L_0 \approx 7\cdot 10^6\ W/cm^2$$

The solution of the generalized heat conduction equation within the range of definition under given initial and boundary conditions is obtained by successive integration method.

The presented theory allows calculation of specific NGZC regimes and estimation of optimal conditions for the examined technological process for Ge, Si semiconductors, A^3B^5 compounds melting like semiconductor-metal structure.

REFERENCES

(1) Evseev B.S.: Proc. of Second Internat. Conf. on Electron Beam Technologies. EBT 88, Varna (1988) 575-579.
(2) Evseev B.S.: Zh.Tekhn.Fiz. 58 (1988) 1543-1545
(3) Evseev B.S.: Phys.Stat.Sol. (a) 109 (1988) K31-K34

A VACANCY DIFFUSION MODEL OF ION BEAM INDUCED EPITAXIAL CRYSTALLIZATION

V. Heera[1)]

Both, the challenge of future microelectronic technology and the scientific interest in understanding the fundamentals of solid phase epitaxy (SPE) have stimulated the studies on ion beam induced epitaxial crystallization (IBIEC) of silicon. The main feature of IBIEC is that SPE of silicon takes place at an amorphous/crystalline (a/c) interface under ion irradiation at temperatures as low as 200 - 400 °C /1/. In contrast, thermal induced SPE needs temperatures higher than 450 °C.

Extensive experimental investigations have been carried out to clarify the IBIEC-mechanism /2-4/. It was unambiguously demonstrated that the IBIEC-rate scales linearly with the nuclear energy deposited at or near the a/c-interface. In other words, there is a strong evidence that beam-induced vacancy-interstitial-pair generation governs the IBIEC-process. Moreover, the small thermal activation energy of about 0.3 eV found for IBIEC agrees with the migration energy of neutral vacancies in silicon determined from life time measurements. Thus, it was assumed that vacancies induce SPE when arriving at the a/c-interface.

Findings concerning electron beam induced epitaxial crystallization (EBIEC) /5,6/ appear to confirm the dominant role of athermal vacancy generation and subsequent vacancy diffusion towards the a/c-interface for particle induced SPE. Moreover, the EBIEC investigations have indicated, that defects coming from both the amorphous and the crystalline side should contribute to SPE to the same degree. Earlier results on the channeling dependence of IBIEC /7/ can be well understood on the basis of this finding.

First quantitative modelling of EBIEC has been performed by Lulli and coworkers /6/. However, this vacancy diffusion model, which considers first order vacancy annihilation, fails to describe temperature and dose rate dependence /8/ of IBIEC. Hence, a modified vacancy diffusion model has been discussed /9/, which includes vacancy annihilation by divacancy formation.

Another model, which is based on intracascade defect evolution has been recently published /10/.

The vacancy diffusion model of IBIEC is based on the following assumptions:

(i) Ions generate vacancies at the crystalline side as well as at the amorphous one with a rate of $G = Z \cdot j$ $[cm^{-3}s^{-1}]$ where Z $[cm^{-1}]$ is the number of vacancies per unit length generated by a single ion and j $[cm^{-2}s^{-1}]$ is the ion flux. Via the Kinchin-Pease-formula Z depends linearly on the nuclear deposited energy.

(ii) The vacancies diffuse from both the amorphous and the crystalline side towards the a/c-interface/s and the surface, which act as total sinks. The vacancies have a thermal activated diffusivity D $[cm^2s^{-1}] = D_o \exp(-E_m/kT)$ with E_m the vacancy migration energy.

1) Zentralinstitut für Kernforschung Rossendorf, AdW der DDR, PSF 19, DDR-8051 Dresden

(iii) There is a vacancy annihilation in the bulk by divacancy formation given by a rate $A\ [cm^{-3}s^{-1}] = D\ L\ n^2$. L is the critical distance between two vacancies for divacancy formation and n is the vacancy density.

(iv) Every vacancy arriving at the a/c-interface crystallizes a microscopical amorphous volume c.

According to these assumptions the IBIEC-rate r (crystalline layer regrowth per ion dose $[nm/\ 10^{16}\ cm^{-2}]$) is a sum of the amorphous and the crystalline side contribution. The one-side-component r_i (i = a or c) is given by

$$r_i = c\ j_{v,i}(0)/j \tag{1}$$

where the steady state vacancy flux at the a/c-interface (x = 0) coming from the i-side $j_{v,i}(0) = D_i\ dn_i/dx\ _{x=0}$ can be obtained from the balance equation for the vacancy distribution $n_i(x)$ in the i-layer of thickness d_i:

$$D\ d^2n/dx^2 + Z\ j - L\ D\ n^2 = 0; \quad n_i(0) = n_i(d_i) = 0 \tag{2}$$

The nonlinear differential equation (2) can be solved numerically /9/. However, analytical expressions are obtained in the limit of thin and thick layer, respectively:

$$r_i = (1/2)\ c\ Z\ d_i \qquad \text{for} \qquad d_i \ll d_c \tag{3}$$

$$r_i = \sqrt{4/3}\ c \sqrt[4]{Z^3\ L\ D_i/j} \qquad \text{for} \qquad d_i \gg d_c \tag{4}$$

$$d_c = \sqrt{8}\ \sqrt[4]{D_i/(Z\ L\ j)}. \tag{5}$$

According to these results there should be striking differences between the IBIEC-behavior of thin and thick layers, namely independence of temperature (via D) and ion flux j and dependence of layer thickness for thin layers and reversed behavior for thick layers.

Unfortunately, the regrowth behavior of thin layers under ion irradiation has not studied yet. EBIEC results /5,6/ obtained for thin buried amorphous layers correspond very well with theoretical predictions in regard to linear thickness dependence and dose rate independence. From these experiments the critical thickness d_c has been estimated to be in the magnitude of order of some tens of nanometers. However, it should be noted that according to formula (5) the critical thickness may vary in dependence of experimental conditions.

On the other hand, IBIEC experiments on thick amorphous surface layers have shown that the IBIEC rate decreases with increasing ion flux /8/ as expected from formula (4). Furthermore, it has been demonstrated /9/ that the flux dependence theoretically derived fits the experimental results except the case when concurrent amorphization processes become dominant /11/.

Also, the energy dependence of IBIEC reflected from formulas (3) and (4) is in good accordance with experimental results. Only a weak deviation from the linear behavior in the nuclear energy deposition is predicted from formula (4) for thick layers.

According to formula (4) there is no one-to-one correspondence between the activation energy of crystallization and defect migration as commonly suggested. The IBIEC rate should thermally activated for thick layers with an activation energy of $E_{IBIEC} = E_m/4$. Using the generally accepted value of vacancy migration $E_m = 0{,}33$ eV we obtain $E_{IBIEC} = 0{,}08$ eV which is clearly in contradiction with IBIEC-results. In order to overcome this discrepancy we can suppose that **either** an additional thermal activated interface reaction rate exists or that the vacancy migration energy differs from the value determined from vacancy life time measurements. We favor the second assumption since an additional interface reaction rate leads to a more pronounced deviation from the linear scaling with the nuclear deposited energy. We attain an agreement between theory and experiment when we choose $E_m = 1{,}2$ eV. Indeed, there are some indications from dopant diffusion experiments and theoretical calculations that the vacancy migration energy should be in the region of 1...2 eV. Additionally, a migration energy of 1.2 eV could explain some results concerning the competition between crystallization and amorphization under ion irradiation in a straightforward manner /10/.

In conclusion, the vacancy diffusion model fits the IBIEC-experiments provided that (i) the vacancy migration energy is 1.2 eV and (ii) the vacancies annihilate in pairs.

References

/1/ J. Linnros, B. Svensson and G. Holmen
Phys. Rev. B30 (1984) 3629

/2/ J. Linnros, G. Holmen and B. Svensson
Phys. Rev. B32 (1985) 1482

/3/ J.S. Williams, W.L. Brown, R.G. Elliman, R.V. Knoell, D.M. Maher and T.E. Seidel
Mater. Res. Soc. Symp. Proc. 51 (1986) 319

/4/ R.G. Elliman, J.S. Williams, D.M. Maher and W.L. Brown
Mater. Res. Soc. Symp. Proc. 51 (1986) 319

/5/ G. Lulli, P.G. Merli and M.V. Antisari
Phys. Rev. B36 (1987) 8038

/6/ G. Lulli, P.G. Merli and M.V. Antisari
Mater. Res. Soc. Symp. Proc. 100 (1988) 375

/7/ J. Linnros and G. Holmen
J. Appl. Phys. 59 (1986) 1513

/8/ J. Linnros and G. Holmen
J. Appl. Phys. 62 (1987) 4737

/9/ V. Heera
Phys. Stat. Sol. A 114 (1989)

/10/ K.A. Jackson
J. Mater. Res. 3 (1988) 1218

/11/ J. Linnros, R.G. Elliman and W.L. Brown
J. Mater. Res. 3 (1988) 1208

PECULIAR FEATURES OF DEFECT PRODUCTION DUE TO HIGH ENERGY ION IMPLANTATION

L.A.Kazak, F.G.Neshov, V.E.Semyannikov, and A.R.Urmanov[1)]

With the view of further studying earlier discovered /1-5/ features in defect production due to high energy ion implantation (HEII) at depth t substantially smaller than the mean projective range R_p we irradiated Si and W single crystals with nitrogen ions of initial energy E_o ranging from 2 to 16 MeV. Irradiation was carried out for a range of doses D from 10^{15} to 10^{18} cm^{-2}. The sample temperature during irradiation did not exceed 50°C. After irradiation samples were studied using the Rutherford backscattering technique (RBS) in combination with the channeling effect. Qualitative estimates for the degree of lattice distortion at different depths were obtained using the ratio of the aligned backscattering energy spectrum to the random one, χ_{min}. Protons with E_o = 0.6 MeV were used. The RBS probing depths were too smaller R_p.

Figs.1-3 show functions $\chi_{min}(D)$ for different fixed t (indicated in the figures in microns) for Si irradiated by nitrogen ions with E_o = 1.7 MeV (Fig.1) and E_o = 16 MeV (Fig.2) and for W (E_o = 5.6 MeV in Fig.3). The arrows point to the values of $\chi_{min}(0)$ (non-irradiated crystal) for the same depths t.

The following results appear to be quite unusual: i) the presence of the non-monotonic and decreasing functions $\chi_{min}(D)$ for various depths in Fig.2 and 3; ii) the substantially greater relative increase in the amount of defects in W as compared with Si; iii) the excess of quantity $\chi_{min}(0)$ over $\chi_{min}(D)$ for $D \gtrsim 5 \cdot 10^{16}$ cm^{-2} in Fig.2. These results are in striking disagreement with the generally accepted mechanism of defect production due to direct transfer of the energy from ions to the nuclear subsystem of solids (nuclear stopping).

It is clear that the causes of these peculiarities in defect production due to HEII should be looked for in non-equilibrium local excitations of the electronic subsystem in solids brought about by fast ions. Indeed, all the observed relationships in defect formation due to HEII may be explained if we note that, as shown in Ref.6, relaxation of electronic excitations to atoms located outside lattice sites (intersitial atoms, atoms appearing in extended defects, etc.) proceeds more effectively than to atoms occupying regular positions. This may lead to a considerable increase in the mobility of defect-forming atoms and vacancies. In virtue of the specific profile of the energy density generated by ions of energies under consideration into the electronic subsystem (electronic energy losses), one will observe a mobility gradient for above mentioned atoms directed towards the target surface. Thus, it may

1) Ural Polytechnical Institute, SU-620002, Sverdlovsk. USSR

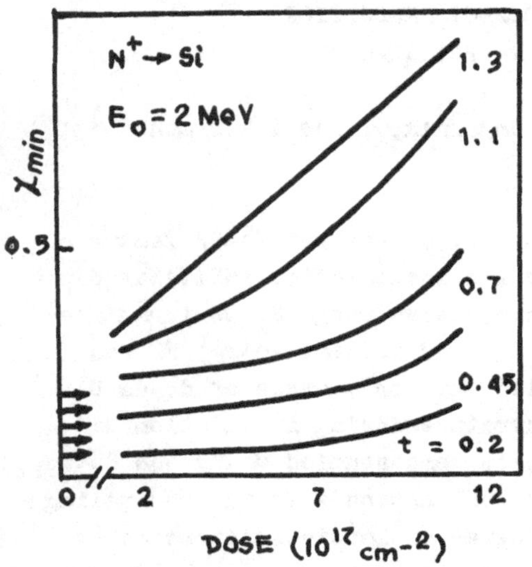

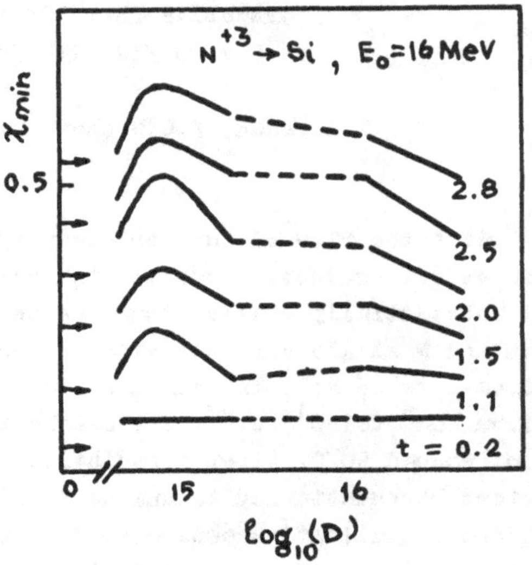

Fig.1. Function $\chi_{min}(D)$ for various depths after irradiation of Si with N ions with E_0 = 2 MeV, R_p = 2 μm.

Fig.2. The same as in Fig.1. but E_0 = 16 MeV, R_p = 12 μm.

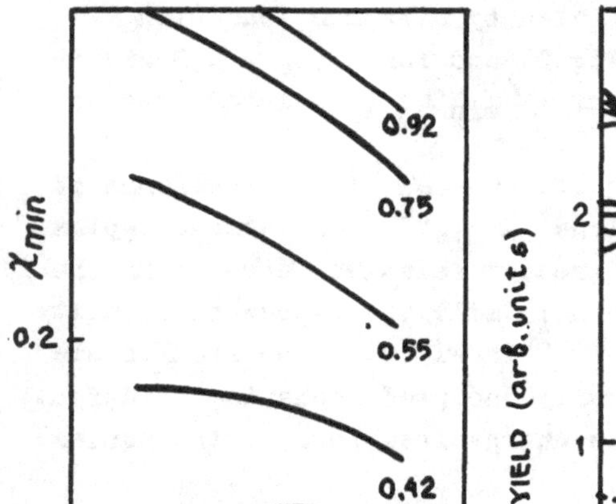

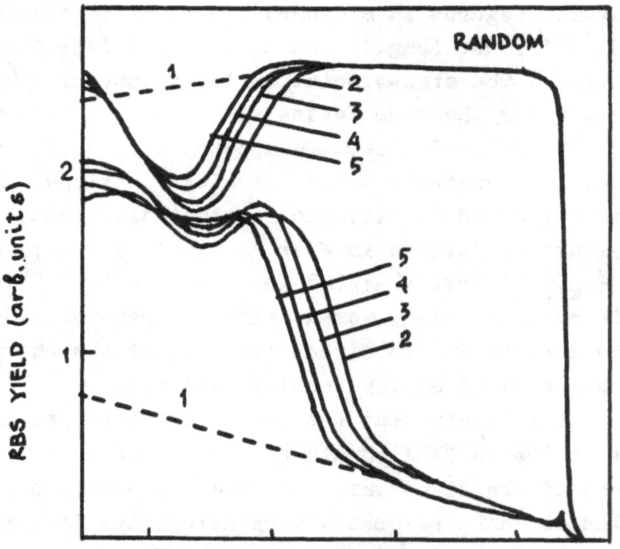

Fig.3. Function $_{min}$(D) for various depths after irradiation of W with N ions with E_0 = 5.8 MeV, R_p = 2.8 μm.

Fig.4. RBS energy spectra for protons of E_0 = 0.6 MeV for aligned (axis 100) and random incidence of protons on target. Curves 1 - non-irradiated Si; 2 - Si crystal having Si_3N_4 layer; 3 - the same as for curves 2, but after irradiation with N ions of E_0 = 16 MeV at random with D = $5.8 \cdot 10^{14}$ cm^{-2}; 4 - the same as for curves 3, but D = $1.9 \cdot 10^{15}$ cm^{-2}; 5 - aligned irradiation (⟨100⟩ axis) with D = $1.6 \cdot 10^{15}$ cm^{-2}.

be assumed that defects forming in the subsurface region of a crystal at HEII are due to their radiation-stimulated diffusion from depth to the surface. Evidently, for the described mechanism to be sufficiently effective, multiple overlap of the irradiated area with cross-sections of fast ion tracks is needed. A single overlap of ion tracks requires, in accordance with the estimates from /6/, doses not greater than $D_s = \pi^{-1}10^{14}$ cm^{-2}.

On the other hand, increased mobility of defect-forming atoms leads to fast relaxation of defects, i.e. to their annealing. Which of these two mechanisms - defect production due to the "drawing out" of defects from specimen depth or their annealing - will dominate and at what doses depends substantially on the depth from which defects are drawn to the surface, i.e. in the long run, on quantity R_p. Apparently, the greater R_p, the higher probability that defects relax during migration. This is what explains such a low amount of defects in Si in Fig.2 ($R_p = 12\,\mu$m) as compared with Fig.1 ($R_p = 2\,\mu$m) and Fig.3 ($R_p = 2.8\,\mu$m).

To provide experimental support for the proposed mechanism of defect production at low depths due to HEII, we irradiated Si single crystals having a layer of compound Si_3N_4 at a depth of $t = 2\,\mu$m /4/ with nitrogen ions with $E_0 = 16$ MeV. It is obvious that this layer is a barrier in the way of defect migrating from depth to the surface, which should lead to a considerable decrease in defect production at depths smaller than $2\,\mu$m as compared with ordinary Si. Then irradiated specimens were studied using RBS. Fig.4 shows RBS spectra measured both at random and with proton beam aligned with the $\langle 100 \rangle$ axis. The figure shows that at depths $t < 2\,\mu$m no defect production occurs, but the lattice structure of Si recovers to the original state at a dose of $2 \cdot 10^{15}$ cm^{-2}. When oriented irradiation of targets ($\langle 100 \rangle$ axis) with the same ions is used, annealing of defects proceeds even more efficiently than in the case of irradiation at random, which, apparently, should be attributed to the fact, that channeled particles enter the region of considerable damadges of the lattice with energy somewhat higher than that of ions moving at random. The random spectra of Fig.4 show that the effect of HEII leads to the "dissipation" of the layer, which is not achived by high-temperature annealing (1250°C) of however-long duration /4/.

Thus, the experiments performed support the validity of the new mechanism of defect production due to HEII at small depths proposed by us.

REFERENCES

1. Ascheron C., et al. phys. stat. sol. (a), 97(1986)15.
2. Bugrov N.V., and S.A.Karamyan. Izv. AN SSSR, ser. fiz. 50(1986)1009.
3. Didyk A.Yu., et al. JINR Rapid Communication. 5/31/-88(1988)36-41.
4. Gorodischenski A.L., et al. Materials of XVIII All-Union Conf. on Phys. of Interaction of Charged Particles with Crystals. (1989)143.
5. Karamyan S.A., A.V.Rykhlyuk, and V.N.Bugrov. JINR Rapid Communication. 5/25/-87(1987)18-24.
6. Martynenko Yu.V., and Yu.N.Yavlinski. Atomnaya Energia 62(1987)80-83.

COMPUTER SIMULATION OF ION MOTION IN MULTILAYERED HETEROSTRUCTURES

K.Lenkeit and A.Pirrwitz [1)]

1. Introduction

Progress in epitaxial growth by the development of the molecular beam epitaxy (MBE) and the metalorganic chemical vapour deposition (MOCVD) allows the preparation of thin (up to monolayers) multilayered microstructures. Especially, strained-layer superlattices (SLSs), grown from lattice-mismatched materials, are promising materials for optoelectronic devices /1/. The RBS- and channeling analysis of such multilayered heteroepitaxial structures require computer simulation programs for the quantitative interpretation of the experimental results.

In the present paper, results of a simulation program for channeling studies in SLSs are presented. The axial-to-planar channeling transition /2,3/ of ions in strained-layer superlattices and the influence of the steering effect /4,5/ on kink angle measurements were investigated.

2. Calculation model

For studying the channeling of fast ions in SLSs a program was developed, which can calculate individual ion trajectories under the conditions of alternately direction change of off-normal crystal planes and axes. The basic concept of these simulations is described in /2/. The ion trajectories were computed in the continuum string model using the Moliére approximation to the Thomas-Fermi potential. The radius vector lies in the transverse plane in a representative area without potential symmetry inside.

Then the ion trajectory is calculated by solving the ion motion equation iteratively without taking into account the ion stopping and scattering processes. Each integration step is preceeded by an estimation of the mean forces according to /6/. The alternately tilting of the off-normal planes and axes at each interface in the SLS was taken into account by corresponding transformations of coordinates for the ion velocity vector

[1)] Humboldt-University Berlin, Dep. of Phys., Invalidenstr. 110, 1040, GDR

The trajectories for all particles with given initial coordinates were calculated up to a defined depth or the dechanneling case. A projectile will be considered as dechanneled if its distance to a string is smaller than the critical distance. The number of dechanneled particles N_D is calculated as a function of incident angles and depth.

3. Results

The case of ion incidence under small angles α to the [110] axis parallel to the (110) plane for 3 MeV He ions in an $In_xGa_{1-x}As/GaAs$ (30x(10nm/10nm)) SLS was simulated to investigate the axial-to-planar channeling transition in SLSs. The [100] was assumed as the normal direction and the kink angle Ψ_k for the [110] axis amounts to 0.71°.

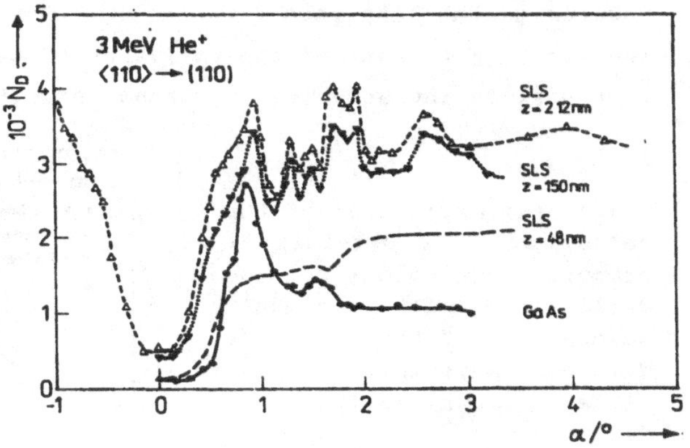

Fig. 1
Computer simulation results for angular scans $N_D(\alpha)$ through the [110] axis in the [110]-(110) axial-to-planar channeling transition case for a $In_xGa_{1-x}As/GaAs$ SLS and for GaAs crystal at various depths z.

The calculated angular scans $N_D(\alpha)$ in various depths z for the SLS and a GaAs reference crystal are presented in Fig. 1. The maxima at angles α >1.2° are caused by the resonance dechanneling effect. In this case the wavelength of the transverse ion motion amounts to a multiple of the distance between the atomic strings in the plane /2,3/. As seen, an increase of the resonance dechanneling effect for axial-to-planar channeling transition in the SLS can be observed. This is of special interest for the application of the resonance dechanneling effect to measure the effective charge of ions inside the solid /7/. The higher dechanneling level under (110) planar channeling conditions (α >2°) for the SLS can be explained by the influence of the alternately tilting of the off-normal (110) planes.

The tetragonal distortion of strained-layer superlattices can be determined by measuring the angular shift $\Delta\alpha$ between angular scans for the first layer and either the angular scan averaged over several layers or angular scan for the second layer. For this method the steering of channeled ions by the top layer with thickness d_t must be taken into account /4,5/. We have carried out computer simulation studies of the steering effect, because this effect is the most important limitation of strain measurements by RBS/channeling technique.

In Fig. 2 the computed energy dependence of the ratio $2\Delta\alpha/\Psi_k$ for superlattices with various thicknesses of the top layer d_t is presented. The computer simulation results show that in the case of thin top layers (8.16 and 14.14 nm) the steering effect leads to a pronounced reduction in the angular shift for energies $E_o < 0.6$ MeV. Furthermore the calculations show that for thicker top layers (30 and 42 nm) the angular shift is reduced to a large degree already at an ion energy of 1.6 MeV. Similar results were obtained by the calculations of the angular dependence of the normalized nuclear encounter probability according to /8/. The orientation dependence of the steering effect and the influence of the axial-to-planar transition will be investigated in further computer simulation studies.

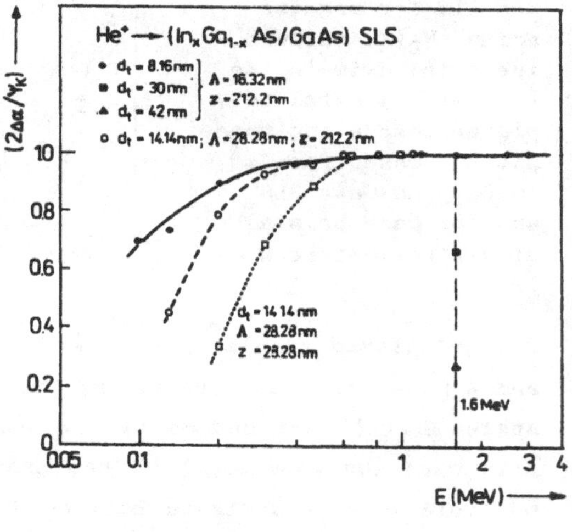

Fig. 2
The energy dependence of the ratio $2\Delta\alpha/\Psi_k$ obtained by computer simulations for superlattices with various thicknesses of the top layer d_t, the superlattice periodes Λ and analyzing depths z.

References

(1) Osbourn,G.C.: J. Vac. Sci. Technol. 3 (1985) 826
(2) Bulgakov,Yu.V. and V.I.Shulga: Rad. Effects 28 (1976) 15
(3) Lenkeit,K. and R.Wedell: phys. stat. sol.(b) 98 (1980) 235
(4) Picraux,S.T., W.K.Chu, W.R.Allen and J.A.Ellison: Nucl. Instr. & Methods B 15 (1986) 306
(5) Hashimoto,S., Y.-Q.Feng, W.M.Gibson, L.Y.Schowalter and B.D.Hunt: Nucl. Instr. & Methods B 13 (1986) 45
(6) Harrison,D.E., W.L.Gay and H.M.Effron: J. Mat. Phys. 10 (1969) 1179
(7) Bulgakov,Yu.V., K.Lenkeit and R.Wedell: phys.stat.sol. (b) (1983) K201
(8) Barrett, J.H.: Phys. Rev. B3 (1971) 1527

THE ROLE OF ABSOLUTE AND CONSTITUTIONAL SUPERCOOLING IN ZONE-MELTING-RECRYSTALLIZATION OF SILICON FILMS

A.B.Limanov, L.V.Musatova +)

Structure of silicon films grown on amorphous substrates by zone-melting-recrystallization (ZMR) depends on the shape of the cellular crystallization front. It is known that the cellular growth is a result of melt supercooling which can be caused by the features of heating and heat sink (absolute supercooling) or by impurity segregation (constitutional supercooling). During ZMR the both types supercooling can arise. Earlier [1,2], the role of the absolute supercooling has been studied for the case where scanning beam in ZMR process had energy densities > 10^4 W/cm^2. In this paper, the model of a linear- shaped ZMR with energy densities ~10^2 W/cm^2 is considered. Following to the approach developed in paper [3], interface stability at constitutuional supercooling is analysed, and critical conditions for the interface stability are calculated.

Considered three - layer - structure (Fig.1) consists of silicon substrate of thickness d_s, oxide film of d_{SiO_2} and silicon film of d_f. This structure is held on point supports and heated to the temperature T_o. A focused in the strip energetic beam creates a line-shaped (normal to the figure plane) melting zone in the polycrystalline silicon film. Here x_{cr} and x_m are coordinates of the crystallization and melting fronts. The zone moves with velocity V along the X direction. The energy irradiation profile is Gaussian : $I(x) = A\exp(-x^2/2a^2)$, and the heat source in silicon film is $P_i(x) = I(x)(1-R_i)$, where $R_s = 0.4$ and $R_l = 0.7$ for solid and liquid phases, respectively.

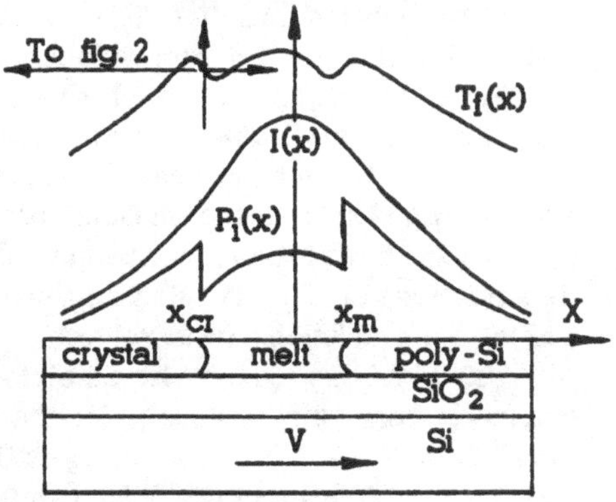

Fig. 1. The process scheme.

We introduce: $T_s(x)$ - the difference between the substrate (silicon wafer) temperature and the preheating temperature T_o, and $T_f^{1,2}(y)$ - the difference between the silicon film temperature and the silicon substrate temperature $T_s(x_{cr,m})$ where $Y^{1,2} = X + x_{cr,m}$. We suppose that all the absorbed in Si film energy is transfered to the substrate across the SiO_2 layer, spreads along axis X and is dissipated from the two substrate surfaces by radiation. In this case, the silicon film and the substrate are heated by source $P_i(x)$. The function of the heat losses from substrate surfaces is defined by the difference of the absolute radiative flow and the flow in preheated state at T_o: $Q_i(x) = (r_s + r_i)HT_s(x) + (r_i - r_s)gT_o^4$. Here we use the linearized approximation of Stefan-Boltzman's equation for $T > T_o$. The coefficient H is determined from the equation for radiation flows at melting point: $H(T_m - T_o) = g(T_m^4 - T_o^4)$, where g - Stefan-Bolthsman's constant, r_i - emission coefficient. The heat loss function for the silicon film is determined from the heat equation for the oxide layer, using the linear approximation for substrate temperature:

$$q_i^{1,2}(y) = (T_f - b^{1,2}y)h + r_i g T_m^4, \text{ where } h = K_{SiO_2}/d_{SiO_2},\ b^{1,2} = dT_s/dx]_{x=x_{cr,m}}.$$

+) 117333, Moscow, Leninsky pr., 59, USSR, Institute of Crystallography USSR Academy of Sciences.

Using the expressions for $P_i(x)$, $Q_i(x)$ and $q_i^{1,2}$ we can write the equation (1) for $T_s(x)$ and equations (2) for $T_f^{1,2}(y)$, where D_i and K_i - the thermal diffusivity and thermal conductivity coefficients, respectively; parameter $b^{1,2}$ is determined from equation (1); $P_i^{1,2} = P_i(x_{cr,m})$ and $a_i^{1,2} = dP_i(x)/dx]_{x=x_{cr,m}}$ are paramites for the heat source linear approximation:

$$dT^2/dx^2+(V/D_s)dT/dx-((r_s+r_l)HT+(r_l-r_s)gT_o^4)/(K_s d_s)+A\exp(-x^2/2a^2)(1-R_l)/(K_s d_s)=0, \quad (1)$$

$$d^2T/dy^2+(V/D_l)dT/dx-((T-b^{1,2}y)h+r_l gT_m^4)/(K_l d_f)+(P_l^{1,2}+a_l^{1,2}y)/(K_l d_f)=0. \quad (2)$$

The solution of Eq.(1) is obtained on $(-\infty,+\infty)$ for boundary conditions: $T_s(\pm\infty)=0$; the continuity of the temperature and heat flow at the points, corresponding to the $P_i(x)$ profile changes. Eqs. (2) are solved in vicinity of crystallization (melting) front, where the approximations for $T_s(x)$ and $P_i(x)$ are correct. Here we used the equality of heat profiles in the silicon film and substrate near the fronts; the temperature equality and latent heat generation (illumination) on the phase boundary ($y=0$).

The expressions for $T_s(x)$ and $T_f^{1,2}(y)$ were calculated for different values of x_{cr} and x_m. From the series of the obtained solutions we selected the values which satisfy the equality of the temperature at crystallization and melting fronts to the melting point T_m. In such a way the problem, concerning the mutual dependence of the initial parameters of the task and it's solutions, was solved.

In Fig.2 are given temperature profiles obtained from the solution for the following parameters: $T=1200°C$, $a=400$, $d_s=350$, $d_f=0.5$, and $d_{SiO_2}=1.0$ microns. $V=0$ for (a), 0.1 for (b), and 0.2 cm/s for (c). The point (0,0) corresponds to x_{cr} and T_m. Three profiles in one picture are different in values of the radiation power and, hence, correspond to different values of melting zone widths: "1" - 200, "2" - 400, "3" - 600 microns.

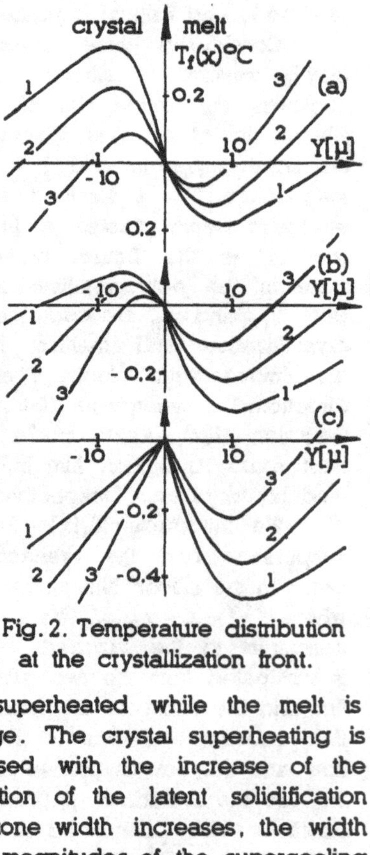

Fig. 2. Temperature distribution at the crystallization front.

Analysing the results of the modelling it can be concluded that an inversion of the sign of the temperature gradient in the silicon film takes place. At $V \le 0.1$ cm/s the crystal is superheated while the melt is supercooled due to the effect of reflectivity change. The crystal superheating is removed and the melt supercooling is increased with the increase of the growth velocity. The effect is due to the evolution of the latent solidification heat on the crystallization front. As the melting zone width increases, the width of the supercooled (superheated) areas and the magnitudes of the supercooling (superheating) decrease. It can be explaned by the temperature gradient in the substrate at x_{cr} (b^1) increas.

To obtain critical conditions a small periodic perturbation of the solidification interface was assumed and the equations for the temperature and impurity concentration fields were solved. The solution determines the sign of perturbation velocity and conditions of it`s development. The equations can be written as:

$$\partial^2 C/\partial x^2 + \partial^2 C/\partial y^2 + (V/D_{im})\partial C/\partial x = 0,$$
$$\partial^2 T_i/\partial x^2 + \partial^2 T_i/\partial y^2 + (V/D_i)\partial T_i/\partial x - q_i^1(x)T_i/(K_i d_f) + P_i(x)/(K_i d_f) = 0, \qquad (3)$$

where D_{im} - coefficient diffusivity for impurity in the melt. According to paper [3], it was shown that crystallization front is stable if the critical condition is realized:

$$(K_s T_s'' - K_l T_l'') - T_m \Gamma \omega^2 (K_s S_s - K_l S_l) - (K_s S_s T_s' - K_l S_l T_l') - mB(K_s S_s - K_l S_l) < 0, \qquad (4)$$

where ' and " indicate first and second derivative on coordinates at the unperturbated crystallization front, Γ - the surface tension coefficient, ω - the perturbation friequency, m - the slope of the liquidus curve of the phase diagram, B - the coefficient, determined from the condition of the continuity of temperature and flow of material at perturbed phase boundary:

$$B = \frac{C_o[T_m \Gamma \omega^2 (K_s S_s - K_l S_l) - (K_s T_s'' - K_l T_l'') + (K_s S_s T_s' - K_l S_l T_l') - \rho L D_{im}(C'' - C' S_{im})/(1-k)}{\rho L V - C_o m(K_s S_s - K_l S_l) + \rho L D_{im} S_{im}/(1-k)},$$

here: C_o - impurity concentration at the moving crystallization front, ρ - density and L - latent heat of silicon, k - segregation coefficient, $S_s = S_s^+$, $S_l = S_l^-$,

$$S_i^{\pm} = -(V/2D) \pm ((V/2D_i)^2 + h/(K_i d_f) + \omega^2)^{1/2}, \quad S_{im} = -(V/2D_{im}) - ((V/2D_{im})^2 + \omega^2))^{1/2}.$$

Impurity distribution in melt was obtained from the condition of the equality of impurity concentrations in melt and in crystal that takes place if melt is not mixed. The values for the impurity concentrations were selected following [4].

The first three terms of the condition (4) are analogous to the results of papers [1,2] for the case of the pure absolute supercooling. Calculations of these terms show that the crystallization front is stable for all regimes shown on Fig.2. Taking into account the constitutional supercooling (the calculation of the forth term) results in the unstability of the crystallization front for $V \geq 0{,}1$ cm/s. Therefore, the cellular growth of the films during ZMR is a result of combined action of absolute and constitutional supercoolings. We suppose that the cells at crystallization front develop on a region with an inversion of the temperature gradient sign. The cell size is determined by the width of this region rather than impurity concentration field, because $1/S_{im}$ is several times greater than $1/S_i$.

In conclusion it should be noted that the dendritic (lamellae) growth was observed *in situ* [5,6] under conditions (at $V < 0{,}1$ cm/s , see Fig. 2 a) of the crystal superheating, however, our calculations indicate that the front is stable. Probably, a system of lamellae at crystallization front is developed due to local melting of the superheated crystal near crystallization front. It is important that in real process the crystallization front temperature $T_\varphi < T_m$ due to impurity contamination of the melt and capillary curvature (see phase boundary on Fig.1) of the crystal - melt interface. Therefore, near phase interface the superheated above T_φ silicon film is not superheated above T_m and exists in solid state, if it's (lamellae) width dont exceeds the size of the region with temperature $T_\varphi < T < T_m$.

(1) Jackson K.A., Kurtze D.A.: J. Crystal Growth 71 (1985) 385

(2) Grigoropoulos C.P., Buckholz R.H., Domoto G.A.: J. Appl. Phys. 59 (1986) 454

(3) Mullins W.W, Sekerka R.F.: J. Appl. Phys. 35 (1964) 444

(4) Geis et al.: Mater. Res. Symp. Proc. 35 (1985) 575

(5) Givargizov E.I., Limanov A.B.: The Crystal Growth, Nauka, Moscow 15 (1986) 5
: Growth of Crystals, Consultants Bureau, New York 15 (1988) 3

(6) Givargizov E.I., Limanov A.B.: Microelectron. Engineering 8 (1988) 273

TIME EVOLUTION OF THE ELECTRON GAS IN A
LASER-INDUCED PLASMA - A THEORETICAL STUDY

R. Müller and W. Stiller[1]

Introduction

In this paper it is intended to present a theoretical study of the interaction of CO_2-laser radiation with metals. In particular the process of formation of a laser-induced plasma is under consideration; especially iron and aluminium are of interest.

In the intensity region considered, namely $I = 10^7 \dots 10^9$ W cm^{-2}, it is necessary to produce a metallic vapour for the formation of a plasma. The coupling of the laser with the metal leads to an acceleration of free electrons. Such charge particles exist both in the metallic solid and in the vapour formed in front of the surface of the solid (initial density of electrons greater than 10^{13} cm^{-3}). In this way the absorption of laser irradiation is guaranteed. If the laser intensity I is greater than a fixed threshold I_v the laser energy absorbed from the metallic surface exceeds the value caused by heat conduction. Therefore a phase transition does occur. In the vapour streaming off from the surface additional laser irradiation can be absorbed. Inelastic collisions of sufficiently accelerated free electrons, i.e. collisions between energy-rich free electrons and atoms of the neutral gas produce ionization events leading to an electron avalanche multiplication. The recombination (three-body collisions) limits the process to a stationary plasma regime.

This paper is stimulated by Beyer (2) and is - to some extent - a generalization of his results.

Coupled system of rate equations

The threshold intensity I_v for vaporization is dependent on the irradiation time t.

The density of the metallic vapour, in which the plasma is formed, is determined in the same way as Afanas'ev and Krokhin (1) did (quasi-stationary vaporization process). According to losses of heat conduction the quantity I has to be replaced by $I-I_v$.

The initial density of the free electrons in the metallic vapour is determined as a function of the ionization energy, temperature, and pressure and is nearly equal to the density of free electrons in the metallic solid.

[1] Central Institute of Isotope and Radiation Research Leipzig,
Academy of Sciences of the GDR,
Permoserstr. 15, Leipzig, DDR-7050, GDR

The value of the ionization energy has to be fixed to a reduced one in comparison with a neutral vapour atmosphere, because it is sufficient to remove an electron so far from the rest atom, that it can not be differed from the other plasma electrons. This decrease is inversely proportional to the Debye radius. In addition the laser field as an external field reduces the ionization energy (prop. $I^{1/4}$).

The energy density of the plasma is determined by the densities and temperatures of the ions and electrons, respectively. Because the laser radiation above all interacts with the electrons, for a description of the plasma absorption, however, only the energy density of the electrons is of importance. With the quantities

n_e, electron density and

$\bar{\varepsilon}$, average electron energy (Maxwellian distributed)

the change of energy density of free electron gas can be described by eq. (1):

$$\frac{d(n_e\bar{\varepsilon})}{dt} = \alpha I - P_c - P_a - P_i - P_{diff} + R_{ee}(E_i - E_a) - R_{en}\bar{\varepsilon} \qquad (1)$$

(αI, laser absorption; P_c, elastic collisions; P_a, excitation collisions; P_i, ionization collisions; P_{diff}, electron diffusion; R_{ee}, three-body recombination ion - electron - electron; R_{en}, three-body recombination ion - electron - neutral gas species). The coefficients denote the following quantities: the P's, changes of the energy density, and the R's, the corresponding rates according to the pattern P = R times energy; E_i, ionization energy; E_a, excitation energy.

From the balance of species for the plasma electrons it can be found

$$\frac{dn_e}{dt} = R_i - R_{diff} - R_{ee} - R_{en}. \qquad (2)$$

Using eq. (2) and the relation $\frac{d(n_e\bar{\varepsilon})}{dt} = n_e \frac{d\bar{\varepsilon}}{dt} + \bar{\varepsilon}\frac{dn_e}{dt}$ it can be followed from eq. (1) for the change of the average electron energy:

$$n_e \frac{d\bar{\varepsilon}}{dt} = \alpha I - P_c - P_a - R_i(E_i + \bar{\varepsilon}) + R_{ee}(E_i - E_a + \bar{\varepsilon}). \qquad (3)$$

Results and discussion

Eqs. (2)+(3) constitute a coupled system of two ordinary nonlinear first-order differential equations. By a numerical solution of the system a survey at the time evolution of the plasma can be obtained:

- In a starting phase the electron energy $\bar{\varepsilon}$ is increasing up to a critical value and remains constant then.
- From this time onward the ionization rate dominates in the system of equations; the balance of species is becoming positive leading to an avalanche-like electron multiplication.

With increasing electron density n_e the electron energy is growing, too, based on the e-i collision frequency and the three-body recombination i-e-e. On the other hand, the three-body recombination limits the electron multiplication. Thus a stationary plasma state is established.

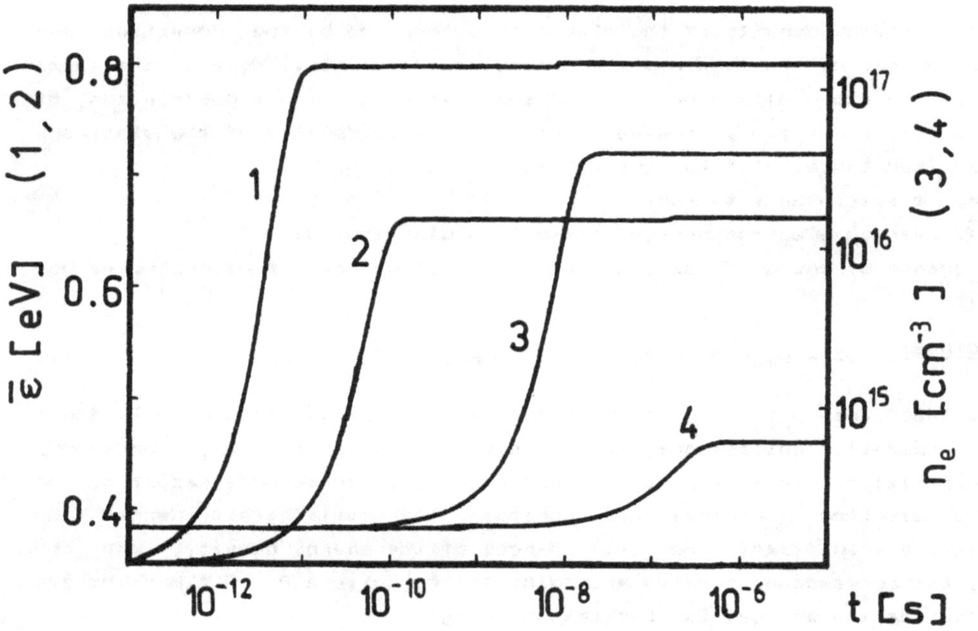

Fig. 1 Average electron energy ε and electron density n_e as a function of time for laser intensities of $I = 1 \cdot 10^7$ W cm^{-2} (curves 2,4) and $I = 5 \cdot 10^7$ W cm^{-2} (curves 1,3) (aluminium, r_F = 400 μm)

The dependence both of the plasma evolution and of the final stationary plasma state on the laser intensity can be seen in Fig. 1. The influence of the focus radius to this evolution is comparatively small. A comparison with Beyer (2) shows lower values for the maximum electron density (within the range of $n_e > 10^{-3}$ $n_{e\ Beyer}$). Parallel to this the second ascent of the average electron energy $\bar{\varepsilon}$ is not so pronounced. A deeper analysis of the origin of the difference is still open. Agreement could be found for the fact, that with increasing intensity (and therefore increasing n) the second ascent of $\bar{\varepsilon}$ is also enlarging.

References

(1) Afanas'ev, Yu. V. and O. N. Krokhin: Zh. Eksp. Teor. Fiz. 52 (1966) 966-975

(2) Beyer, E.: Einfluß des laserinduzierten Plasmas beim Schweißen mit CO_2-Lasern, Deutscher Verlag für Schweißtechnik, Düsseldorf 1985

THE INFLUENCE OF THE LIGHT POLARIZATION ON RESONANT LASER-INDUCED DIFFUSION IN CRYSTALS

V.N.Pavlovich[1]

The effect of diffusion acceleration of light interstitials in semiconductors with irradiation by resonant laser light was suggested in (2),(3). The effect consists in excitation of high excited vibration states by resonant light and subsequent activationless transition in the neighbour equilibrium site. This effect was observed experimentally (1) for the case of oxygen diffusion in silicon. The simple considerations about the necessity of dipole vibration momentum to be parallel to electric field vector allow to make an assumption about the strong influence of light polarization on the possibility of observation of this effect. In this paper, the way of influence of light polarization on longrange diffusion of impurity atoms is considered in details.

The diffusion coefficient is defined by the formulae

$$D \approx P(l_m) D_{l_m} \quad ,$$

where $P(l_m)$ is the population of the nearest to barrier top level, D_{l_m} is the diffusion coefficient on the level l_m: $D_{l_m} = D_o^{l_m} \exp(-E_p/kT)$, where E_p is the part of the activation energy connected with the lattice deformation transfer during the diffusion transition. The pre-exponential factor D_o^{lm} is defined by the transition matrix element from the level l_m of the initial site to the level l_m' of the neighbour site:

$$D_o^{l_m} \sim |\langle l_m | V_{\varkappa'}(r) | l_m' \rangle|^2$$

Here $V_{\varkappa'}(r)$ is the diffusive atom potential energy in the neighbour site. The last formulae can be evaluated using the model of three-dimentional isotropic harmonic oscillator. It is easy to show that diffusion transition in the direction OX, for example, occurs from the state $| l_m \rangle = \Psi_{l_m}(x) \Psi_o(y) \Psi_o(z)$. The matrix transition elements from other states with the same energy are negligible.

Let us consider the excitation of impurity atom vibration by electromagnetic field. The operator of perturbation is proportional to $(\vec{u} \cdot \vec{p})$, where $\vec{u}$ is the light polarization vector, and $\vec{p}$ is the momentum of the particle. The probability of excitation of the vibrational states is defined by the matrix element $\langle l | (\vec{u} \cdot \vec{r}) | l' \rangle$. If $\vec{u} \parallel OX$, then $\langle l | (\vec{u} \cdot \vec{r}) | l' \rangle = \langle l | x | l \pm 1 \rangle$. Thus, the light with polarization vector parallel to OX excites the state $\Psi_l(x) \Psi_o(y) \Psi_o(z)$ which just gives the contribution to diffusion transition matrix element. The light which is polarized in the direction OY (or OZ) leads to excitation of the states of type $\Psi_o(x) \Psi_l(y) \Psi_o(z)$ (or $\Psi_o(x) \Psi_o(y) \Psi_l(z)$) which does not give the contribution to diffusion transition matrix element in the direct-

---

1) Insitute for Nuclear Research, Kiev, USSR

ion OX. Thus, if one can approximately describe the vibration of impurity atom using the model of three-dimentional isotropic harmonic oscillator, then in order to stimulate the longrange diffusion in the given direction (which is defined usually by the direction of gradient of concentration) the light polarization vector must be parallel to this given direction. Allowing for anharmonicity, this conclusion practically is not changed.

However, the real situation differs from that considered above. The anisotropy of vibration is strong sufficiently, and the frequencies of vibrations are strongly different. Therefore, the resonant light is able to excite the vibration only in a given direction, in a given elementary cell, and to stimulate the diffusion jump in this (or in the vicinity of this) direction. If the direction of vibration with a given frequency in the neighbour cell coincides with this given direction, and if the direction of concentration gradient also coincides with this given direction, then the light which is polarized in this direction is able to induce the longrange stimulated diffusion due to concentration gradient.

In order to clear up this question, let us consider the real situation of oxygen diffusion in silicon. Usually, one assumes that the oxygen in silicon lattice breaks the Si-Si bond and forms Si-O-Si molecule. These bonds are bended, and the angle between two Si-O bonds is equal to 162°. Three modes of vibration of such molecule are identified. The antisymmetric mode ν_3 due to vibration of oxygen atom along the direction ⟨111⟩ with dipole momentum parallel to ⟨111⟩ induces the light absorption in 9 μm area (1136 sm^{-1}). One can assume that ν_3 mode is the diffusion mode of oxygen atom, i.e. the multiquantum excitation of this mode leads to the transition across the barrier to the neighbour Si-Si bond which makes the angle 109°29' with the initial Si-Si bond.

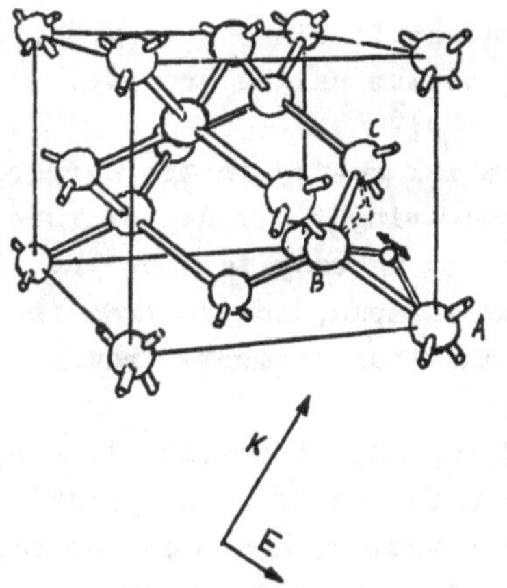

Fig.1. The interstitial position and ν_3 vibration of oxygen atom in silicon lattice.

Let us assume that the concentration gradient is directed along ⟨111⟩ direction. Irradiating the crystal by the resonant light with vector $\vec{k} \perp$⟨111⟩ and with polarization vector $\vec{u} \parallel$⟨111⟩, one can induce the photostimulated transitions of oxygen atoms, which are localized on the bonds along this direction, to three bonds along three other axes ⟨111⟩ (see Fig.1). At these axes the light absorption is attenuated by the factor cos(70°31'), and the probability of transition to the bonds of initial direction decreases. In the result, the population of the bonds, which make the angle with polarization vector, increases, and the effective diffusion coefficient in the direction <111> is defined by the transition probability which is averaged taking into account the changing populations of different bonds.

Irradiating by the light just in this direction $\vec{k}\perp\langle 111\rangle$, but with the polarization vector $\vec{u}\perp\langle 111\rangle$, the longrange diffusion does not appear, and the result of irradiation consists in increasing populations of the bonds of AB type (see Fig.1).

We can analyse also the different cases of relative orientation of crystal axes, concentration gradient and polarization vector of exciting light. For example, if the concentration gradient is directed along <100> axis, the longrange stimulated diffusion is possible with arbitrary direction of light polarization vector.

As a conclusion, we can say that one can observe the longrange laser-induced diffusion of oxygen in silicon with an arbitrary direction of concentration gradient, if the light polarization vector is not perpendicular to any of <111> directions. In opposite case, one can achieve the repopulations of the bonds of different types, which can be observed by the study of absorption of probe resonant beam at different orientations of crystal axes and polarization vector.

REFERENCES

(1) Artsimovich M.V., A.N.Baranov, E.M.Kudriavtsev et al. The resonant laser-induced diffusion of oxygen in silicon. Preprint FIAN,N164, Moscow,1989.

(2) Lubchenko A.F. and V.N.Pavlovich. Phys.stat.sol.(b)78(1976)K97.

(3) Pavlovich V.N. Phys.stat.sol.(b)116(1983)K9.

RESONANT LASER-INDUCED DIFFUSION OF OXYGEN IN SILICON

M.V.Artsimovich[1],A.N.Baranov[2],E.M.Kudriavtsev[2],I.F.Mogilnik[1], V.N.Pavlovich[1],B.N.Romanuk[3],V.I.Soroka[1],V.V.Tokarevski[1],S.D.Zotov[2]

In the last years the diffusion processes in crystals stimulated by different external forces are widely investigated. The one of the ways of influence on point defect diffusion consists in direct excitation of light impurity vibration in semiconductors and insulators by external resonant electromagnetic field of great intensity. Such irradiation can cause the significant acceleration of diffusion of given impurity (2).

According to quantum theory of diffusion, the diffusion coefficient of light impurities in crystals can be presented in the form

$$D \approx P(l_m)D_{l_m} ,$$

where $P(l_m)$ is the population of the nearest to barrier top vibration level, D_{l_m} is the diffusion coefficient on the level l_m. In equilibrium conditions, the function $P(l)$ is the Boltzmann distribution: $P(l)=C\times \exp(-E_l/kT)$. With irradiation of the crystal by the resonant light of great intensity with a frequency equal to that of local vibration of impurity, the function $P(l)$ can differ significantly.

One can evaluate the intensity of irradiation, which is nacessary for noticeable increasing of the highest level population, using the model of harmonic oscillator anharmonically interacting with phonon bath and with external electromagnetic field (3). This evaluation shows that one can achieve the significant acceleration of impurity diffusion with intensity of irradiation of the order of $10^4 W/sm^2$. The calculation was made with two assumptions. First of them is that the lifetime of the excited oscillator levels is of the order of 10^{-12}s which is the typical value for excited states of local vibration. The second assumption is that the polarization vector of electromagnetic field is paralleled to the dipole moment of impurity. The analysis of the structure of the oxygen vibration in silicon crystals leads to the requirement: The polarization vector of the field must be paralleled to concentration gradient of impurity in order to observe the longrange stimulated diffusion.

In order to verify the above theoretical premices the experiment was performed to observe the laser-induced diffusion of oxygen in silicon. The general scheme of experiment is presented on Fig.1. The ions of oxygen with the energies 100-150 keV were implanted into the silicon plates. After the implantation, the samples were annealed at the temperature 650°C during 1-3 h. As a result of annealing, the radiation def-

1) Institute for Nuclear Research, Kiev, USSR
2) Physical Institute, Moscow, USSR
3) Institute of Semiconductors, Kiev, USSR

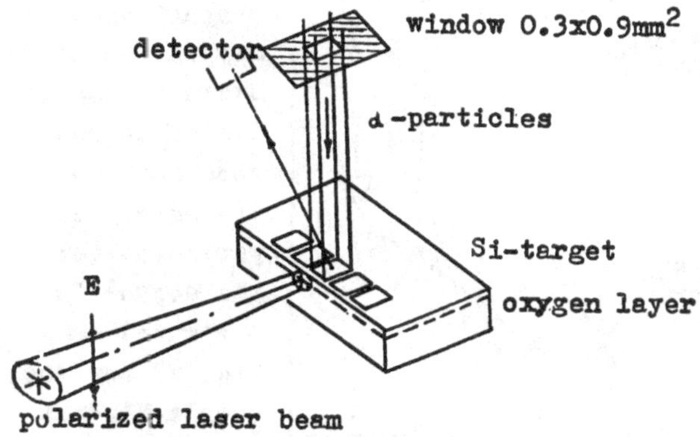

Fig.1. The general scheme of experiment

ects disappear and oxygen atoms stand in interstitial positions forming the bonds Si-O-Si. The atom of oxygen just in this position forms the local vibration which gives rise the sharp absorption band near the 9 μm area (1100 sm^{-1})(1). The state of oxygen after annealing was controlled by the spectra of IR absorption which show that approximately 90% of oxygen atoms are situated in the interstitial positions. Thus, in the samples the interstitial oxygen layer was created at the depth 2000-3000 A.

The such samples were irradiated by IR with wavelength 9.2 μm using tunable CO_2 laser according to the scheme of Fig.1. The power of generation was 1-10 W. After the focusing, the intensity of irradiation in the maximum of the Gauss' beam with a halfwidth 40 μm was 150-500kW/sm^2. The silicon wafer was located in the focusing laser beam according to Fig.1. The electric vector oscillations of the light were perpendicular to the wafer face of implantation and therefore parallel to the gradient of oxygen concentration. During the irradiation, the temperature of the sample surface was measured and was ≤ 60°C.

Several experiments were performed with different conditions of irradiation. We present here the results of two experiments. The sample N1 was irradiated three times. The area of each subsequent irradiation was shifted by 0.1 mm along the implanted layer. The duration of each irradiation was 8 h. On the sample N2, the nine regions were irradiated with distance between them 0.1 mm. Such a geometry was chosen according to a dimension of cross section of the probe beam of α-particles.

In order to investigate the distribution of oxygen in silicon the method of resonant backscattering of α-particles was used in this work which allows to study the distribution of light atoms in the surface layers of heavier matrix. The collimated beam of α-particle with cross section 0.28x0.9 mm^2 was directed to sample perpendicular to implanted surface. The target was scaned by the beam of α-particles with the step 0.31 mm as it is shown on Fig.1. Thus, the irradiated and nonirradiated by laser regions of the target were studied. The depth profile of oxygen atoms was obtained at each region by working of spectra of the resonant backscattering at different energies of incident beam of α-particles.

Fig.2 a),b) shows the results of such measurements. The dependence of related yield of α-particles scattered on the oxygen atoms (which is

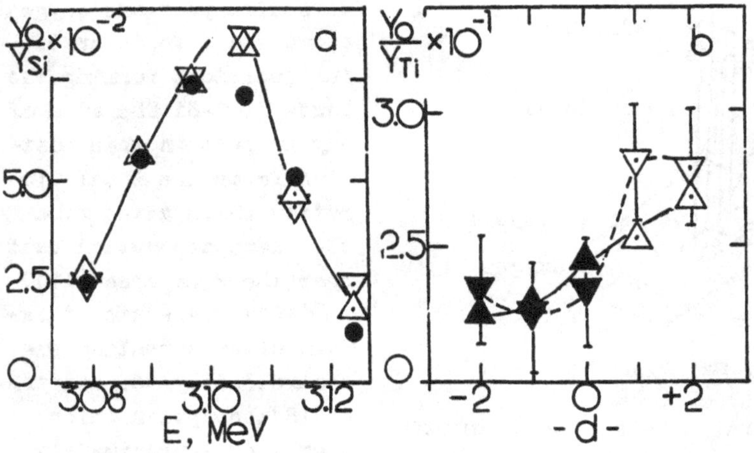

Fig.2.The related yields of scattered -particles.The full symbols stand for irradiated regions.a) energy (depth) dependence of yields in different regions for the sample N1.b) step of scaning dependence of yield for the semple N2 at maxima of oxygen distributions obtained from two (——,- - -) detectors.

proportional to oxygen concentration in condition of narrow resonance) on the energy of α-particles (or on the depth) is presented on Fig.2a) for the sample N1.From this figure one can see the noticeable decrease (exceeding the experimental errors) of oxygen concentration at the maximum of distribution in irradiated region compared to nonirradiated regions. The related yield of scattered α-particles at the energy corresponding to maxima of distributions for different regions of the sample N2 is presented in Fig.2 b).This figure also shows the decrease of the concentration at the maximum of distribution for irradiated regions compared to nonirradiated regions.

The method of treatment of the result was based on a conception of diffusion smearing of Gauss profile of oxygen concentration. One can define the diffusion coefficient from the ratio of oxygen concentrations at the maxima of distributions for irradiated and nonirradiated regions. Taking into account that the probe beam of α-particles averages the yields from irradiated and nonirradiated sections within the region 0.28 mm, we can obtaine for the coefficient of laser-induced diffusion of oxygen the value $D \approx 3\text{-}5\ 10^{-15} sm^2/s$ which corresponds to equilibrium diffusion coefficient at 900-920 K.

In order to show that this diffusion coefficient does not arise due to temperature increase in absorptive layer of oxygen, we made a calculation of temperature distribution in absorptive medium which interacts with Gauss laser beam.This calculation gives that even if the absorption coefficient in oxygen layer is of the order of $10^6 sm^{-1}$,the temperature into the depths of the layer of small thickness exceeds the constant temperature of the surface not more than by 5 K.Therefore, we can made a conclusion that in these experiments we observed the resonant laser-induced diffusion of oxygen in silicon crystals.

REFERENCES

(1) Hrostowski H.J. and R.H.Kaiser. Phys.Rev. 107(1957)966
(2) Lubchenko A.F. and V.N.Pavlovich. Phys.stat.sol.(b)78(1976)K97
(3) Pavlovich V.N. Ukr.Fiz.Zhurn. 29(1984)1430

IMPLANTATION EFFECTS OF LASER EVAPORATED PARTICLES

W. Pompe, M. Bobeth[1] and M. Posselt[2]

Laser pulse vapour deposition (LPVD) is successfully applied for the preparation of nm-scaled multilayers. The formation of such layers is essentially affected by the kinetic energy of the deposited particles. LPVD allows, to some extent, to control the energy spectrum of the particles in the region 1 - 1 000 eV /1/. The reported energy spectra exhibit one or two peaks as for example a peak of neutral atoms at $\approx$ 10 eV and of ions at $\approx$ 100 eV /2/. Since higher energetic particles are able to penetrate the material, in the present paper, thin film formation by LPVD is modelled as simultaneous surface deposition and implantation of particles.

The interaction of energetic particles (20 - 1 000 eV) with the solid has been studied by means of a Monte Carlo simulation based on the TRIM-code by Biersack. As a result the ion range distribution, backscattering coefficient and sputtering yield have been obtained. For example, in connection with the preparation of Ni/C-multilayers, the following data of the mean projected ion range R_p and its standard deviation ΔR_p have been calculated for different particle energies (R_p and ΔR_p in Å, ϱ_C = 2.3 g/cm^3):

	C → Ni		Ni → C		Ni → Ni		C → C	
eV	R_p	ΔR_p	R_p	ΔR_p	R_p	ΔR_p	R_p	ΔR_p
20	2.1	1.0	2.2	0.7	1.5	0.7	-	-
40	2.9	1.6	2.5	0.9	2.0	0.9	-	-
100	5.0	2.9	5.9	1.2	2.2	1.3	6.1	3.2
300	9.9	5.8	10.	2.2	4.8	2.3	14.	7.2
1 000	24.	14.	21.	4.8	9.0	4.8	37.	19.

The obtained backscattering is small except in the case C → Ni with a backscattering coefficient $\alpha \approx 0.2$ for ε = 20 - 1 000 eV.

The film formation has been investigated by adopting a model established in /3/. The evolution of the atomic density of implanted atoms n(z,t) is governed by the equation

$$\partial_t n(z,t) + \partial_z [v_{sp} + v_{dep} + v_{sw}] \, n(z,t) = \nu(z,t) \tag{1}$$

where z denotes the depth measured from the film surface. The velocity

1) Zentralinstitut für Festkörperphysik und Werkstofforschung der AdW der DDR, Dresden, 8027, GDR

2) Zentralinstitut für Kernforschung der AdW der DDR, PF 19, Dresden, 8051, GDR

$v_{sp} < 0$ considers the sputtering effect and v_{dep} corresponds to the film thickness growth due to the surface deposition of low energetic atoms with $\varepsilon < \varepsilon_0$:

$$v_{dep} = (1/n_f) \int_0^{\varepsilon_0} d\varepsilon \, (1 - \alpha(\varepsilon)) \cdot \nu(\varepsilon). \tag{2}$$

n_f is the atomic film density and $\nu(\varepsilon)$ is the time-averaged spectral particle deposition rate. The source term in (1) is given by

$$\nu(z,t) = \theta(t) \int_{\varepsilon_0}^{\infty} d\varepsilon \, u(z,\varepsilon;t) \; (1 - \alpha(\varepsilon)) \cdot \nu(\varepsilon) \tag{3}$$

where $u(z,\varepsilon;t)$ represents the ion range distribution which, in general, depends on the time because of the changing film thickness and chemical composition of the substrate during implantation. The swelling of film and substrate with increasing dose of implanted particles has approximately taken into account by the velocity

$$v_{sw}(z,t) = \int_0^z dz' \, \nu(z',t) \cdot \Omega(z',t) \tag{4}$$

where the atomic volume Ω required to incorporate an atom may differ within the film and the substrate.

As a further example the deposition of carbon on tungsten has been studied numerically. Since the data R_p and ΔR_p do not much differ for the two cases $C \rightarrow W$ and $C \rightarrow C$, the ion range distribution has been approximated by a time-independent Gaussian distribution. The sputtering and backscattering has been neglected. The spectral deposition rate $\nu(\varepsilon)$ has been supposed as Maxwell-distribution of temperature $kT = 15$ eV with a total deposition rate $R = 10^{19}/m^2s$. The resulting evolution of $n(z,t)$ is shown in Fig. 1. The position $z_0(t)$ denotes the original surface of the substrate. The region $z < z_0$ represents the growing carbon film and the region $z > z_0$ corresponds to the arising WC_x interface layer of about 8 Å thickness. The unrealistic peak-shaped C-concentration in tungsten at the beginning is due to the choice of a minimal implantation energy $\varepsilon_0 = 20$ eV. The density of tungsten $n^W(z,t)$ within the interface layer is obtained from the packing condition $n^C \cdot \Omega^C + n^W \cdot \Omega^W = 1$ (cf. Fig. 2).

The model (1) does not include atomic mixing processes by ballistic collisions and by radiation enhanced and thermally activated diffusion. The first two mechanisms play a minor role at low ion energies. The latter one can be excluded in the case of strong chemical bonding at low temperatures. If, however, the carbon concentration in the original tungsten substrate exceeds that of the compound WC, the overstoichiometric, less bonded carbon atoms are probably more mobile and can dif-

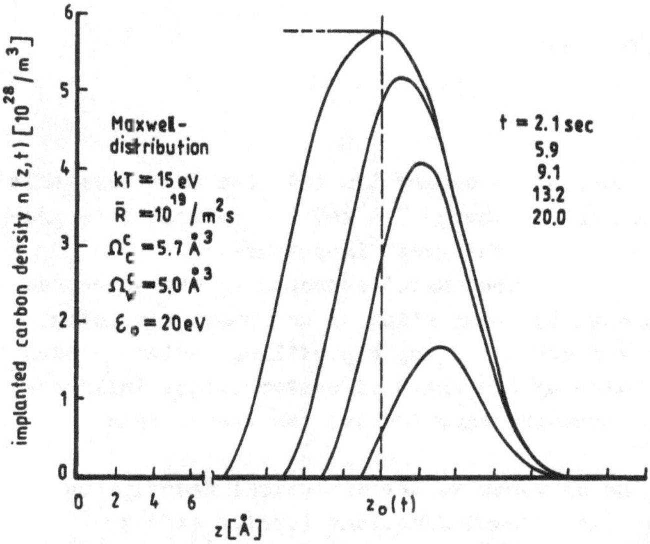

Fig. 1

fuse out of the overstoichiometric region. Following the ideas in /3/, an additional diffusion term in (1), which acts only in this region, would lead to a flattening of the concentration profiles as symbolically indicated by the dotted lines in Fig. 2.

In conclusion, our result suggests that sharp interfaces can be prepared by LPVD using relatively small particle energies obtained e.g. by excimer and Nd-YAG lasers. The formation of a stoichiometric interface compound can eventually increase the thermal stability of the multilayer. Since the calculated ion ranges for the implantation of W -> C and W -> W are smaller than in the case C -> W and C -> C, a sharper interface can be expected for the deposition of tungsten on carbon provided that the particle energy spectrum is the same. This would lead to alternating interface thicknesses in W/C-multilayers, which have been reported for sputter-deposited Mo/Si-multilayers /4/.

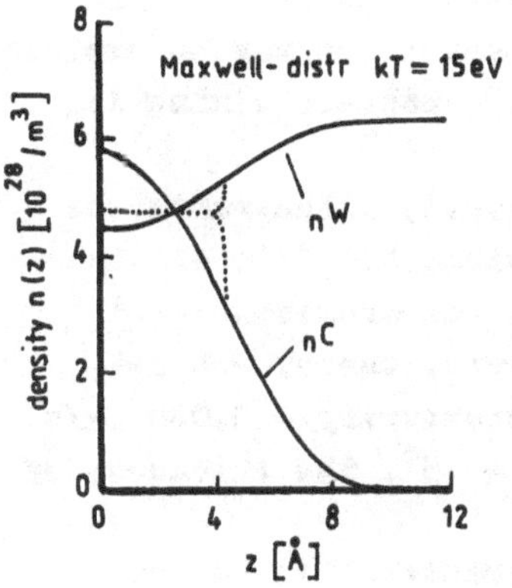

Fig. 2

/1/ S.V. Gaponov and M.D. Strikovsky, "EPM 87", Physical Research vol. 8 (87) p. 140

/2/ A.D. Akhsakhalyan et al., Zh. Tekh. Fiz. 52 (82) 1584

/3/ H.U. Jäger et. al., Thin Solid Films 127 (85) 159

/4/ K. Holloway et al., J. Appl. Phys. 65 (89) 474

AN "INSTRUMENT FUNCTION" OF SECONDARY ION EMISSION

E.V.Popov*

During the investigations of secondary ion emission from crystalls due to initial ion bombardment of energy 3-5 keV the correct interpretation of emission spectrums have the great importance.

The necessary elaboration of the united approach to mass-spectrum interpretation is conditioned by observation of secondary ion going out changeable curves as a function of depth profiling. Actually such a problem is conditioned also by disregard of crater effect influence to secondary ion mass-spectrometry experimental results at some investigations (2).

This paper presents the approach to the analytical description of secondary ion emission "instrument function" (crater effect influence) which allows scientists to interpretate emission spectrums identically to solid body structures.

An "instrument function" is an analytical dependence of distortion between real mass concentration of some element as a function of solid depth profiling and observed emission mass-spectrum.

It should be noted that the appearance of crater effect is more significant during the investigations of monoelement thin films on solid body materials. At the case of implanted impurities a changeable curves of secondary ion emission mass-spectrums may be explained satisfactorily by return processes during the cascade mixing in solids (3).

A typical curve due to crater effect (Fig.1) illustrates the titanium secondary ion going out from titanium thin film on the molybdenum body after the initial argon ion bombardment.

Parametres of initial ion bombardment were: energy - 4 keV, radius of bunch - 1.0 mm, velocity of ion sputtering - 0.008 nm/s, angle of inclination from normal direction - 45°. The thickness of titanium film was 190 nm.

Such curve one may divide into four characteristic regions of initial ion bunch interactions with investigating sample.

First region: ion emission from sample surface (depth smaller than 20 nm). Surface curve top of emission spectrum is conditioned by oxidation of sample surface which results in essential increase of investigating element ionizing coefficient (1).

---

* Urals Polytechnical Institute, SU-620002, Sverdlovsk, USSR

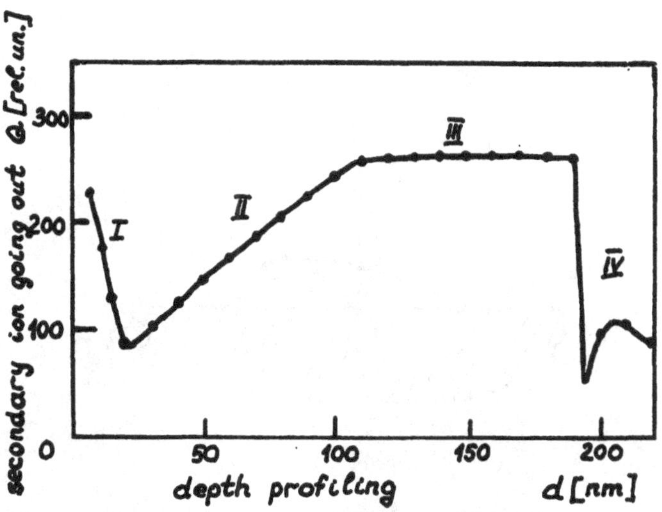

Figure 1. The dependence of the titanium secondary ion emission going out from titanium depth profiling on molybdenum body.

Region II (from 20 nm to 110 nm): the consistent increase of initil ion bunch interaction area at sample surface. At solidity approximation the secondary ion going out Q are proportional the half of ellipsoid of revolution surface area (Fig.2) bending from normal direction under angle φ ($\varphi = 45^{\circ}$):

$$Q(t) = \frac{\pi F}{\sin\varphi} \int_{-r}^{+r} y(x,t) \sqrt{1 + \left(\frac{\partial y}{\partial x}\right)^2}\, dx$$

$$y(x,t) = \ell(t) \sqrt{1 - \frac{x^2}{r^2}}$$

Here: $\ell = \ell(t)$ - the radius of ellipsoid of revolution which is equivalent to the sample depth profiling d at such emission curve region; r - the radius of ion buncn; F - experimental constant.

Because at real situation $\ell \ll r$ one may be obtained:

$$Q(t) = \frac{\pi F r\, \ell(t)}{\sin\varphi} \int_{0}^{\pi} \sin^2\alpha\, d\alpha$$

Therefore, the increase of secondary ion going out at second curve region is proportional to the increase of depth profiling of thin film on body.

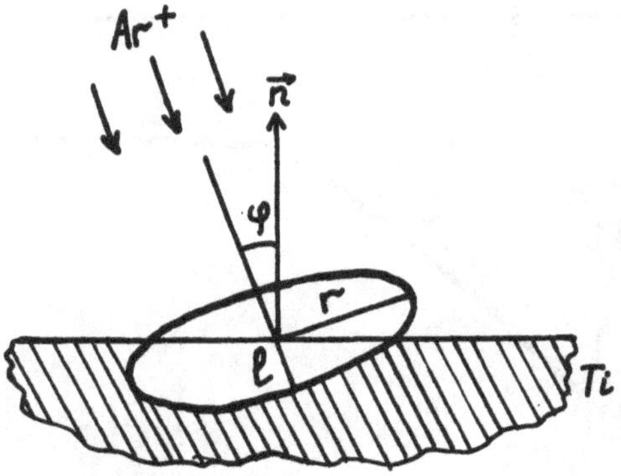

Figure 2. The ellipsoid of revolution which illustrates the initial ion bunch enteraction section with investigating film on body.

Region III (from 110 nm to 190 nm): the consistent velocity of secondary ion going out which is conditioned by established section of initial ion bunch interaction with investigating sample:

$$l = const, \quad l \neq d$$

At this case the radius of ellipsoid of revolution not changes and this ellipsoid only moves deep into the sample.

Region IY: the initial ion bunch after thin film reached the body material. Some quantity of film material at mass-spectrum is conditioned by ion emission from the crater walls due to interaction ellipsoid.

So the approximations of solid film and solid sample and radial symmetry of initial ion bunch interaction area at investigating sample surface are allows us to interpretate the changeable curves of secondary ion emission spectrums due to film depth profiling.

(1) Benninghoven A.: Ztschr. Naturforsch., 5 (1967) 841-843
(2) Kenway-Jackson C.: Vacuum, 3-4 (1984) 479-480
(3) King B.V., I.S.T.Tsong: Nucl. Instr. and Meth., B7/8 (1985) 793-799

ENERGY AND ANGULAR DISTRIBUTIONS OF IONS BACKSCATTERED FROM THE SIDEWALLS DURING THE IMPLANTATION INTO DEEP TRENCHES

M. Posselt, G. Otto[1)]

1. Introduction

Deep trench capacitor technology is often favoured to increase the packing density of VLSI DRAM cells. Ion implantation is one way for doping the sidewalls of these trenches. Since the ions are implanted with a shallow angle to the sidewall surface some specific phenomena appear (1)-(5). After some collisions in the substrate a considerable part of the incident particles is backscattered and may be reimplanted into the opposite sidewalls of the trench. The enhanced sputtering yield can lead to important modifications of the sidewall surface if implantation doses are sufficiently high. In the present work we investigate the backscattering yield and the energy and angular distributions of the backscattered particles by TRIM Monte Carlo simulations (6). Characteristic examples of the trench implantation of As^+ are considered. The histograms obtained for the energy and directional distributions of the reflected ions are fitted to analytical expressions using a least-squares fit method.

2. Results and discussion

If the direction of the incident ions is not identical with the surface perpendicular of the target the energy and angular distribution of the backscattered particles can be determined for two different cases: (i) for a point source and (ii) for a line source of incident ions at the target surface. We consider the second case illustrated in fig. 1. Then, for a given incidence angle θ_0 ($0 \leq \theta_0 \leq \pi/2$) and an implantation energy E_0 the energy and directional distribution of the reflected ions depends on two variables: (i) the energy E of the particles and (ii) the direction of motion characterized by the angle ψ ($0 \leq \psi \leq \pi$). In our TRIM simulations we count the number dY_B of backscattered ions per incident particle.

$$dY_B = y_B(E,\psi)\, dE\, d\psi \qquad (1)$$

$y_B(E,\psi)$ is the energy and angular distribution of the reflected ions or the differential backscattering yield. The total backscattering yield Y_B is obtained from (1) by integration.

We studied the grazing incidence implantation of As^+ into Si for implantation energies of 80, 115 and 150 keV and incidence angles of 4^0, 6^0, 8^0 and 10^0 to the target surface. The values obtained for the

---

1)Academy of Sciences of the GDR, Central Institute for Nuclear Research, Rossendorf, PF 19, DDR-8051 Dresden backscattering yield are given in table 1. In the case of grazing incidence implantation Y_B strongly depends on the incidence angle. For $\theta_0=8^0$ about 35% of the incoming ions are reflected. The backscattering yields of table 1 are nearly independent of the implantation energy. Fig. 2a shows the energy and angular distribution $y_B(E,\psi)$ of the reflected particles for 80 keV As^+ implantation and $\theta_0=8^0$. The figure demonstrates that the forward scattering dominates. The reason of this behaviour is the fact that the reflected ions have to obey the momentum conservation. In the region $3\pi/8 \leq \psi \leq \pi$ only few particles with $E < 0.15\ E_0$ were found by the TRIM simulations. Therefore, this part of the histograms for $y_B(E,\psi)$ is not shown in the figure. The minimum value y_{Bmin} of y_B ($y_{Bmin} \neq 0$) which could be still calculated by our Monte Carlo simulations was about $y_{Bmin} = 5 \cdot 10^{-4}\ keV^{-1}\ rad^{-1}$. We found that there exist a maximum angle $\psi_m(E)$ with

$$\begin{aligned} y_B &\geq y_{Bmin} \quad \text{for } \psi \leq \psi_m(E) \\ y_B &< y_{Bmin} \quad \text{for } \psi > \psi_m(E) \end{aligned} \tag{2}$$

In the region $E > 0.1\ E_0$ $\psi_m(E)$ can be described by

$$\psi_m(E) = \sum_{i=0}^{3} a_i\ (E/E_0)^{3-i} \qquad (\psi_m \text{ in rad}) \tag{3}$$

The 4 coefficients a_i obtained from a least-squares fit procedure are given in table 2. For $0 \leq E \leq E_0$ and $0 \leq \psi \leq \psi_m(E)$ we fitted the histogram of fig. 2a to analytical expressions characterized by 16 parameters b_{ij} (see table 2)

$$y_B(E,\psi) = \sum_{i,j=0}^{3} b_{ij}\ (E/E_0)^{3-i}\ \psi^{3-j} \tag{4}$$

(ψ in rad and y_B in $10^{-3}\ keV^{-1}\ rad^{-1}$)

The results are shown in fig. 2b. The agreement with the histograms can be improved if the number of particle histories followed by TRIM simulations is increased. A better agreement could be also achieved by the choice of a more appropriate fit function. The application of the fit procedures leads to an important reduction of information without the loss of relevant features of the differential backscattering yield $y_B(E,\psi)$: 800 data of the histograms are described by 20 parameters a_i and b_{ij}.

8. Conclusions

A method to calculate the energy and directional distributions of the backscattered particles and the backscattering yield has been applied to the trench implantation of arsenic ions. We have obtained analytical expressions for the energy and angular distributions. Such a representation could be useful in a semi-analytical calculation of the concentration profiles of dopants implanted into a trench structure further results and a more detailed discussion will be published elsewhere.

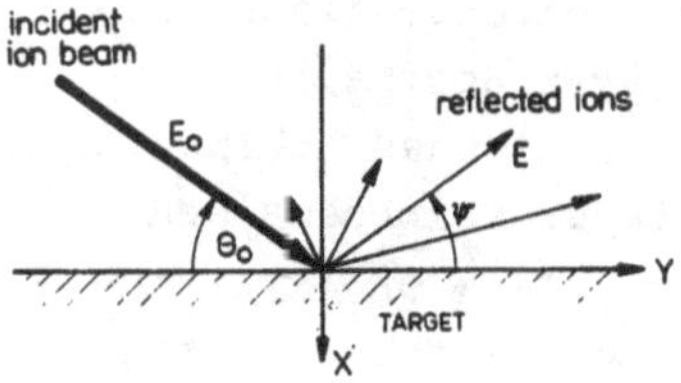

Fig. 1

Table 1: Reflection coefficients

θ_0 /deg. E_0 /keV	4	6	8	10
80	0.5028	0.4310	0.3632	0.3044
115	0.5091	0.4222	0.3492	0.2842
150	0.4964	0.4149	0.3622	0.2905

Table 1

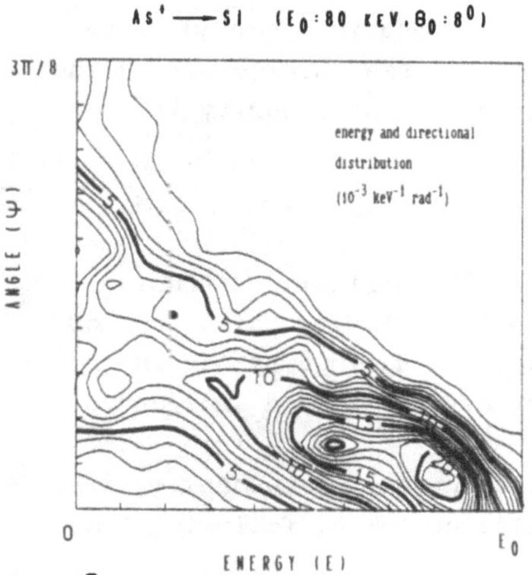

Fig. 2a

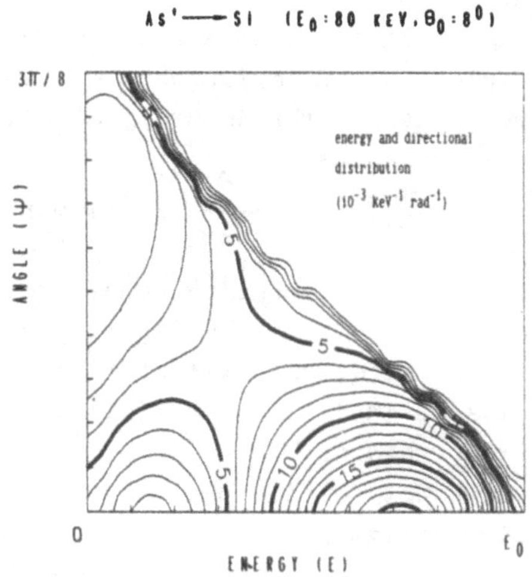

Fig. 2b

Table 2

coefficients a_i:

(i=0,3):	-0.1510E+01	0.2728E+01	-0.2605E+01	0.1403E+01

coefficients b_{ij}:

i: 0	(j=0,3):	0.6951E+02	-0.3886E+03	0.6251E+03	-0.2333E+03
i: 1	(j=0,3):	-0.9278E+02	0.5126E+03	-0.8137E+03	0.3004E+03
i: 2	(j=0,3):	0.1976E+02	-0.1065E+03	0.1686E+03	-0.7139E+02
i: 3	(j=0,3):	0.2374E+01	-0.1196E+02	0.1317E+02	0.4640E+01

Table 2

(1) Fuse,G., H. Unimoto, S. Odanaka, M. Wakabayashi, M. Fukumoto and T. Ohzone: J. Electrochem. Soc. 133 (1986) 996.
(2) Kakoschke,R., H. Binder, S. Röhl, M. Maeßeli, I.W. Rangelow, S. Saler and R. Kassing: Nucl. Instr. and Meth. B21 (1987) 142.
(3) Oura,K., H. Ugawa and T. Hanawa: J. Appl. Phys. 64 (1988) 1795.
(4) Kato,K.: IEEE Trans. on Electron Devices ED-35 (1988) 1820.
(5) Fuse, G., H. Ogawa, K. Tamura, Y. Naito and H. Iwasaki: Appl. Phys. Lett. 54 (1989) 1534.
(6) Ziegler, J.F., J. P. Biersack and U. Littmark: The Stopping and Range of Ions in Solids, Pergamon Press, New York 1985.

THE MECHANISMS OF ATOMIC MIGRATIONS IN IMPLANTED FILMS

E.M. SHPILEVSKY[1)]

The action of ion beam upon bilayer films is accompanied by a release of considerable energies, an increase of the defect concentrations and a nonuniform distribution of the defects over the layer depths, development of temperature and pressure gradients. The magnitudes of the above mentioned changes depend on the ion energy, layer thickness and the thickness ratios, thermal and thermodynamic properties of layer materials. These changes give rise to unsteady-state conditions for mass transfer, which under such conditions can be realized by many mechanisms.

A ballistic mechanism is a transfer of atoms due to a series of collisions both of implanted atom itself and those ones knocked-out by the former against the impurity atoms. The atomic flow of impurity is expressed then as

$$J_f = - \frac{\partial}{\partial x} D_f n + v_f n \qquad (1)$$

where n is the impurity concentration, D_f is the ballistic diffusion coefficient, v_f is the impurity drift velosity including anisotropy of transfer. The ballistic diffusion coefficient is defined by the energy scattered in elastic collisions, the mixing efficiency and independent of temperature.

A variety of the ballistic mechanism is a channeling mechanism, i.e. motion of impurity atoms that acquired collision pulse, following the crystallographic direction along the channel confined by atomic planes. The channeled atom can be expressed by

$$J_\alpha = \alpha v_\alpha n \qquad (2)$$

where α is the coefficient that takes account of the relative amount of channeled atoms; v_α is the mean drift velocity; n is the impurity concentration.

In a high energy cascades ($>$1 eV/atom) temperature peaks and liquid phase microvolumes can appear that allow such a mechanism as atomic motion to be realized in liquid phase.

In the liquid phase mass can be transferred by the vortex mechanism, that is mass transfer by liquid particles, which depends on the flow turbulence and hydrodynamic conditions. The development of vortex motion

1) Byelorussian State University, Leninskii Pr. 4, 220080 Minsk, USSR

leads to a fast transverse mass transfer, gives rise to turbulence and, consequently, to flow mixing. The total mass transfer by atomic and vortex diffusions can be expressed /1/ by

$$J_{L+W} = - \frac{(D_L + D_W)\Delta P}{\rho \gamma v_L} \frac{dn}{dx} \qquad (3)$$

where D_L and D_W are the atomic and vortex diffusion coefficients, ρ, γ are the density and viscosity of liquid phase, ΔP, v_L are the pressure drop and flow velocity. While the temperature peak is being cooled, the impurity can be expelled during crystallization.

Since ion implantation brings about various types of defects and their complexes, mass transfer may be carried out by the mechanisms associated with migration of some defects and complexes (by vacancies, internodes, dislocations, etc.). Then the diffusion flows induced by defect migration are equal to

$$J_i = - D_i \frac{dn_i}{dx} \qquad (4)$$

The heating of samples by the energy transferred by ion flow should be taken into account /2/ as

$$\frac{dT}{dt} = \frac{1}{\rho \cdot c \cdot d} \quad M - 2\varepsilon \sigma (T^4 - T_o^4) \qquad (5)$$

where ρ, c, d are the sample density, heat capacity and thickness, M is the ion energy, ε is the surface emissivity, σ is the Stefan-Boltzmann constant, T_o is the ambient temperature.

The heat diffusion due to a temperature gradient in the ion implanted layer can enhance or inhibit the mass transfer since the "equilibrium" concentrations of the components, including the defects in the regions of higher and lower temperatures, become different. Thus, a particle flow is

$$J_T = -D_i \frac{dn_i}{dx} - \frac{D_T}{T} \frac{dT}{dx} \qquad (6)$$

where D_T is the heat diffusion coefficient, D_i is the diffusion coefficient due to the i-th type defect.

Neglecting the interaction between the flows, the total implantation mass transfer can be expressed by

$$J = \sum (J_f + J_\alpha + J_{L+W} + J_i + J_T) \qquad (7)$$

Annealing of samples following the implantation leads to redistribution of the components and annealing of the defects. In that case mass transfer can follow the mechanisms associated with defect migration. The presence of high concentration of the defects, concentration gradients and fields of mechanical stresses is characteristic of diffusion processes in implanted samples. The diffusion flow of impurity can be expressed here as

$$J_n = - D_i(x,t) \frac{dn_i(x,t)}{dx} \pm n_i(x,t) \frac{\dot{\mathcal{E}}(x,t)}{E} \qquad (8)$$

where $D_i(x,t)$ and $n(x,t)$ are the diffusion coefficients of various types of defects and their concentrations, $\dot{\mathcal{E}}$ is the lattice deformation, E is the Young modulus.

The contribution of each mechanism depends on both the parameters of implanted ions and the properties of sample materials and on the layer thickness.

REFERENCES

1. Kafarov V.V.: The Fundamentals of Mass Transfer, Vysshaya Shkola, Moscow, 1972, 496 pp. (in Russian).
2. Russel H., and I. Ruge: Ion Implantation. Nauka, Moscow, 1989, 360 pp. (in Russian).

MODELING OF THE NITROGEN HIGH DOSE IMPLANTATION INTO SILICON FOR BURIED INSULATING LAYERS

E. Sobeslavsky and W. Skorupa
Zentralinstitut für Kernforschung, DDR-8051 Dresden, PF 19

1. Introduction

Buried layers of silicon nitride serve to isolate the surface silicon layer from the substrate in silicon-on-insulator (SOI) technology. In previous work /1/, /2/, we presented model calculations to find $n^N(z,t)$, the as-implanted nitrogen concentration profile after high dose implantation. In the present paper we are going to briefly discuss how alterations of the model parameters influence the resulting nitrogen profile for representative ion energies. Our model basically involves a step by step solution of the continuity equation. We take into account the change of target volume and the alteration of the incoming ion range profile due to the formation of silicon nitride. A Pearson type IV statistical distribution serves as incoming nitrogen ion range profile. In target depths where the nitrogen to silicon ratio exceeds that of stoichiometric Si_3N_4 we assume a broadening of the concentration profile caused by unbonded nitrogen species; these species have a volume of Ω^N_{exc}. We also consider the sputtering process at the target surface. A dose matching procedure /3/ was used to describe the alteration of the ion range profiles in depths, where the target composition changes from silicon to silicon nitride.

2. Choice of model parameters

The most sensitive parameters of our calculations are the density of the synthesized nitride, $\rho(Si_3N_4)$, and the parameters of the incoming nitrogen range distribution. Contrary to our earlier papers the latter were now extracted from the improved data of Burenkov et al. /4/. For comparison, in the present table we also mention the parameters formerly used. All ranges and stragglings refer to a nitride density of $\rho_{ref}(Si_3N_4)$ = 3.18 $g \cdot cm^{-3}$. As visible, the improved incoming range distributions are moved toward the surface.

	E = 70		150		330 keV	
Silicon, $\rho(Si) = 2.33\ g \cdot cm^{-3}$						
		ref. /5/		ref. /5/		ref. /5/
$R_p(N \rightarrow Si)$, µm	0.177	(0.182)	0.359	(0.377)	0.697	(0.743)
$\Delta R_p(N \rightarrow Si)$, µm	0.057	(0.061)	0.087	(0.095)	0.119	(0.133)
$\gamma(N \rightarrow Si)$	-0.55	(-0.59)	-1.00	(-1.01)	-1.49	(-1.56)
$\beta(N \rightarrow Si)$	3.89	(4.01)	5.97	(6.03)	10.00	(10.57)
Silicon nitride, assumed $\rho(Si_3N_4) = \rho_{ref} = 3.18\ g \cdot cm^{-3}$						
		ref. /5/		ref. /5/		ref. /5/
$R_p(N \rightarrow Si_3N_4)$, µm	0.122	(0.124)	0.246	(0.254)	0.472	(0.492)
$\Delta R_p(N \rightarrow Si_3N_4)$, µm	0.035	(0.037)	0.053	(0.058)	0.071	(0.077)
$\gamma(N \rightarrow Si_3N_4)$	-0.64	(-0.67)	-1.11	(-1.11)	-1.66	(-1.67)
$\beta(N \rightarrow Si_3N_4)$	4.20	(4.34)	6.75	(6.73)	11.79	(11.93)

3. Results

A comparison of the measured and calculated nitrogen profiles show that we must estimate the synthesized nitride density to be $\rho(Si_3N_4) \cong 4\ g \cdot cm^{-3}$. It is much higher than the nitride density of $\rho_{ref}(Si_3N_4)$ $3.18\ g \cdot cm^{-3}$ known from various textbooks. The present finding is in line with our previous results. Interestingly, the density parameter seems to grow with growing implantation energies. The figures show that also the unbonded nitrogen must be in a high density phase.

Also visible (cf. the high density plots, left below) is a systematically moved maximum of the computed nitrogen profiles to the bulk side of the target, although the Tables /4/ used provide substantial lower R_p data than other ones. We are not able to really interpret such displacement, further work has to clear up this.

References

/1/ E. Sobeslavsky et al., Phys. Stat. Sol. (a) 105 (1988) 387

/2/ E. Sobeslavsky et al., Phys. Stat. Sol. (a) 114 (1989) 935

/3/ B.A.J. Amaratunga et al., IEEE Trans. Electron. Devices 32 (1985) 1889

/4/ A.F. Burenkov et al.: Prostranstvennye raspredeleniya energii, vydelenoi v kaskade atomnykh stolknovenii v tverdykh telakh, Energoatomizdat, Moscow, 1985

/5/ A.F. Burenkov et al.: Tablitsy parametrov prostranstvennogo raspredeleniya ionno-implantirovannykh primessei, Izd. V.I. Lenin State University, Minsk 1980

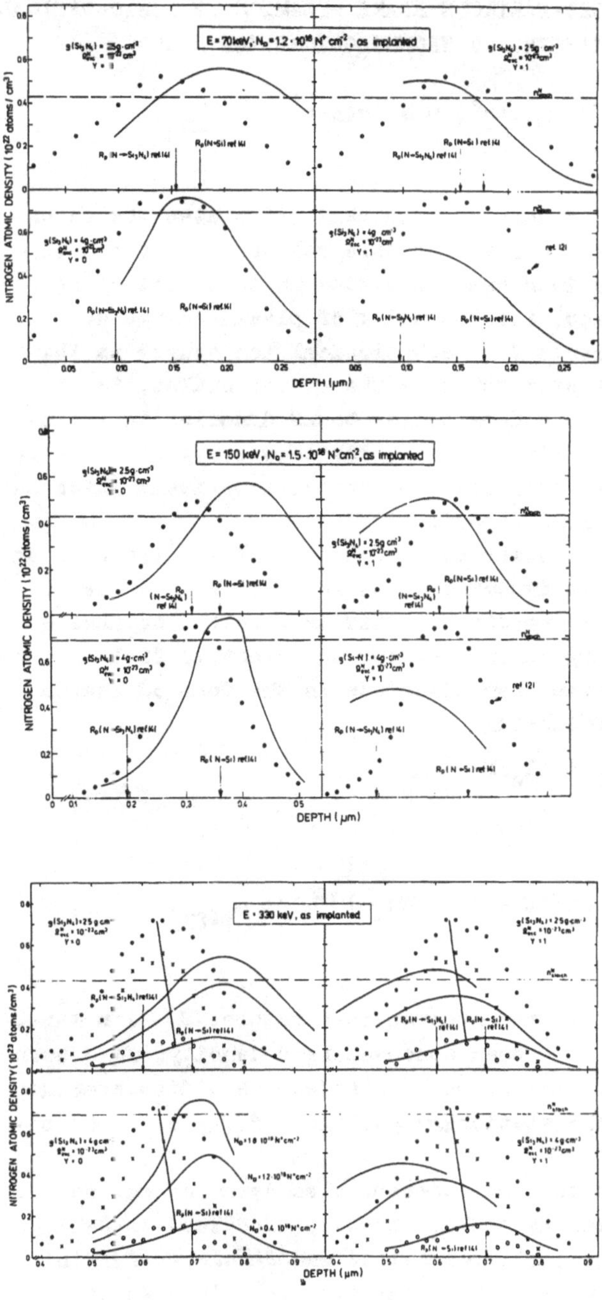

a)
E = 70 keV

b)
E = 150 keV

c)
E = 330 keV

Fig. 1

Implantation of a) 70 keV, b) 150 keV and c) 330 keV N^+-ions into silicon ($\Omega^N_{exc} = 1 \cdot 10^{-23}$ cm^3). Comparison of the as-implanted experimental curves of the nitrogen concentration (...) with theoretical profiles (——) for nitride densities of $\rho(Si_3N_4) = 2.5$ g·cm^{-3} and $\rho(Si_3N_4) = 4$ g·cm^{-3}. The dashed lines denote the stoichiometric nitrogen concentration typical for Si_3N_4 of these densities.

NUMERICAL SIMULATION OF CHARGED-PARTICLE BEAMS DYNAMICS IN IONIC-OPTICAL SYSTEMS WITH PLASMA EMITTERS OF TECHNOLOGICAL ION SOURCES

A.F.Stekolnikov*, O.M.Ivanov*

The use of ion beams for a surface treatment stipulates the necessity of creating effective ion sources for this region. One of the effective method of obtaining of a ions beam consistes in their selecting from a plasma formation surface, i.e. in using of plasma emitters.

One of the main requirements for technological ion source is the ensuring an ion beam with set properties at the source outlet. So the task of an investigation of charged-particles beams dynamics in ionic-optical systems with plasma emitters arises. The complexity of the experimental study of such systems reguires a numerical experiment which has an appropriate advantages of real experiment.

The aim of this work was a numerical simulation of the ion-beam extraction from plasma and of its dynamics in an ionic-optical systems.

A generalised mathematical wording of this task which includes a correct formulation of boundary conditions for the electric field on the moving surface of the plasma-ion beam interface in the form of Stefan condition [1] is reduced to following:

$$\vec{J}\big|_{f\tau} = 0.57\, e \cdot n_i \cdot (\, kT_e/M\,)^{1/2}\, \vec{n}\,,$$
$$d\vec{v}/dt = -(\, e/M\,)\nabla\varphi,$$
$$\nabla^2\varphi = -\rho/\varepsilon_0,$$
$$\varphi\big|_{f\tau} = \varphi_1\,, \quad ds/d\tau = (\,\varepsilon_0/Mn_i\,)^{1/2}\partial\varphi/\partial n\big|_{f\tau}\,,$$
$$\varphi\big|_{f_{el}} = \varphi_2\,;$$

where n_i - ion density, T_e - plasma electron temperature, M - ion mass, φ - electric field potential, ρ - beam space-charged density, $f\tau$ - boundary of the surface of the plasma-ion beam interface, s - displacement of the plasma boundary from the stationary position, f_{el} - electrod boundary.

Mathematical modeling of the charged-particles dynamics was done on a space grid by large-particles method on the base of "Cloud-in-Cell" model [2]. The classical large-particles method was developed for this case taking into account a number of peculiarities in a large-particles motion near the plasma-ion beam boundary.

The algorithm of the numerical solution of the task consisted in setting the geometry of the calculation region, the electrodes potentials, beam current, large-particles parameters and the initial position of the plasma boundary; solving Laplace equation for the electric field

* Minsk Radioengineering Institute, P.Brovka 6, Minsk 220600, USSR

potential and calculating the strength of the electric field; inletting a group of large particles with the ion-sonic velosity from the plasma boundary into the calculation region; solving the set of the equations of the large-particles motion; detemining the grid density of the space charge and solving Poisson equation with the set boundary; defining the new position of the plasma-beam boundary according to Stefan condition and passing to the injection of the large particles and repeating all the calculations. If the displacements of the plasma-beam boundary as well as meanings of the potential and the space-charge density cease to change within the set accuracy calculations are stoped and received results are taken as a task solution.

In order to calculate the output characteristics of the beam from the plasma boundary surface probe particles were injected and the equation of motion for them was solved throughout the whole region within the precalculated self-consistent field.

A numerical simulation of the extraction of different ion beams and its dynamics in three- and four electrode ionic-optical systems at different densities of the ionic current extracted from the plasma surface and the beam energies has been perfomed. Some results of such simulation are presented in Fig. 1 - 3.

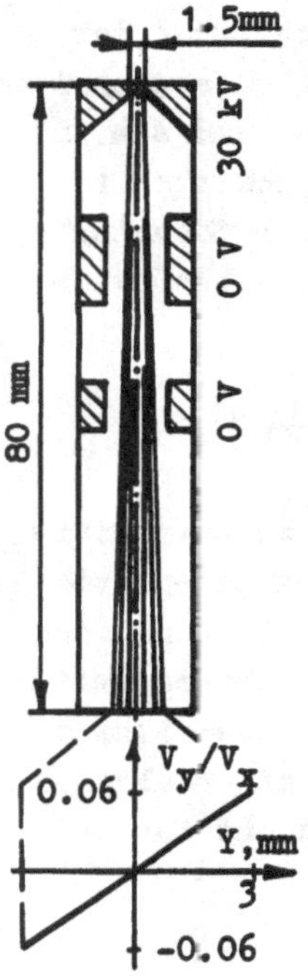

Fig. 1

Boron ion trajectories on slit three electrode ionic-optical system for the current density $j = 10$ mA/cm^2 and the ion energy $E = 30$keV are shown in Fig.1. A laminar motion of separate ion beam components with an average angle of divergence $\alpha \sim 0.03$ radn. is characteristically for this system. Ion beam phase diagram shows actually the lack of aberrations.

Argon ion trajectories in multibeam three electrode ionic-optical system with round holes $j = 5$ mA/cm^2 and $E = 40$ keV are presented in Fig.2a. The motion of separate ion beam components is nonlaminar: the extreme part of the beam suffers the overfocusing; then follows the convergent part of beam; central part of beam

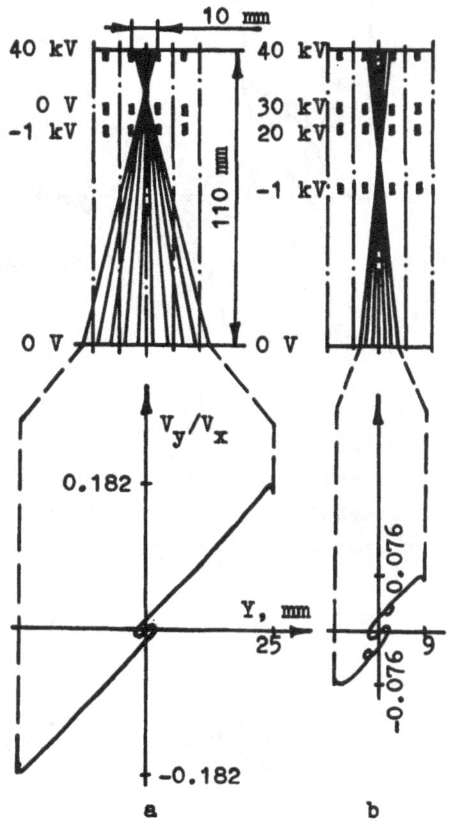

Fig. 2

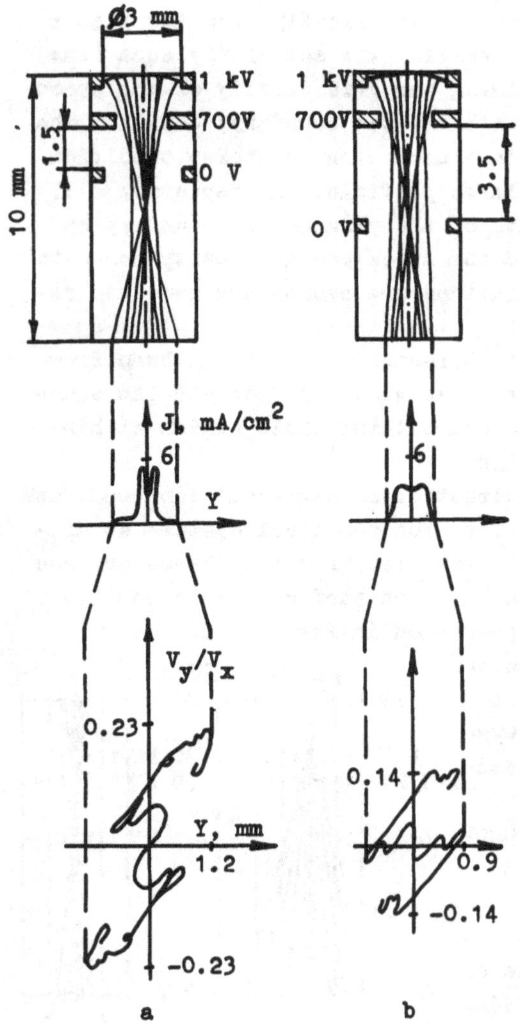

Fig. 3

is weakdivergent. The phase diagram of the beam consisted of 1000 particles in the output section d∼11 cm is also given at this picture. Using the extra fourth electrode as shown in Fig.2b leads to a significant improvement of the beam focusing in view of obtaining both a small crossover and a minimum angle of divergence.

Argon ion trajectories in two different on geometry multibeam three electrod ionic-optical systems with round holes for $j = 1$ mA/cm^2 and $E = 1$ keV as well as the beam phase diagrame and the ion current density distributions in the output section d∼1 cm are presented in Fig.3. Choicing an appropriate geometry as one can see in this figure it is possible to get both a good beam focusing and a uniform spread beams.

Conclusions

The task of the mathematical modeling of charged-particles beams dynamics in ionic-optical systems with plasma emitters was done. The algorithm of the numerical sulution of the task was described. A numerical simulation of the extraction of ion beams and its dynamics was carried out for several ionic-optical systems.

References

(1) Stekolnikov A.F.: Sov.Phys. Tech.Phys. 28 (1983) 1369.
(2) Hockney R.W. and Eastwood J.W.: Computer Simulation Using Particles, McGraw-Hill, New York, 1981.

ON THE MECHANISM OF ION-STIMULATED CRYSTALLIZATION

N.P.Stepina, G.A.Kachurin, S.I.Romanov, V.P.Popov, V.Sh.Aliev

A number of papers has demonstrated the enhancement of amorphous layer crystallization under the influence of ion bombardment. The mechanism of this phenomenon, however, is not yet clear.

The present paper aims to separate the role of diffusion of point defects from the influence of ion bombardment (through the displacement cascades, electron excitation, thermal spikes) directly on the interphase boundary. Experimental scheme is presented in Fig.1. 120 nm thick buried amorphous layers were formed by implantation of $2 \cdot 10^{15} cm^{-2}$ Si^+ ions with E=200 keV in monocrystalline Si samples at the room temperature. The thickness of the top monocrystalline Si layer after heating 460°C, 40 minutes was about 90 nm. The movement of amorphous-crystalline boundary under the influence of Ne^+ ions with E=1.5-100 keV at irradiation temperatures T_i=100-400°C was studied. Fig.1 shows that up to E=30 keV Ne^+ ions lose nearly all their energy before reaching the interphase boundary. Consequently, in these cases all the possible reasons for its displacement should be eliminated, except the sink of point defects.

Figures 2,3 present the Rutherford backscattering spectra (He^+, 1.5 MeV, $\alpha_s \approx 96°$) for the structures under study after Ne^+ ion bombardment with dose of $10^{17} cm^{-2}$. For the case E=1.5 keV and T_i=100°C we have observed the increase of amorphous layer thickness by ~ 30 nm (Fig. 2). So, at the given T_i point defects causing the boundary displacement, were able to get over the distance ~ 90 nm. Bombardment of the targets heated up to T_i=400°C (E=7 keV) caused decrease of the amorphous layer thikness (Fig.3). Simultaneously the disorder in the region of maximum elastic losses of Ne^+ions increased, and that reduced the efficiency of crystallization with the dose increase. Our experiments prove that for ion-beam induced crystallization (IBIC) occurrence the diffusional approach of point defects to the interphase boundary is quite sufficient.

We have analyzed the dependence of the re-crystallized layer thickness on the elastic losses (ν) of Ne^+ ions with different energy, the dose being fixed (Fig.3). The thickness turned to be proportional to ν, despite the variation of the distance between the maximum losses R_d and the interphase boundary. The same proportionality is seen

Institute of Semiconductor Physics, Ac. Sci. USSR, Siberian Branch, 630090 Novosibirsk, USSR

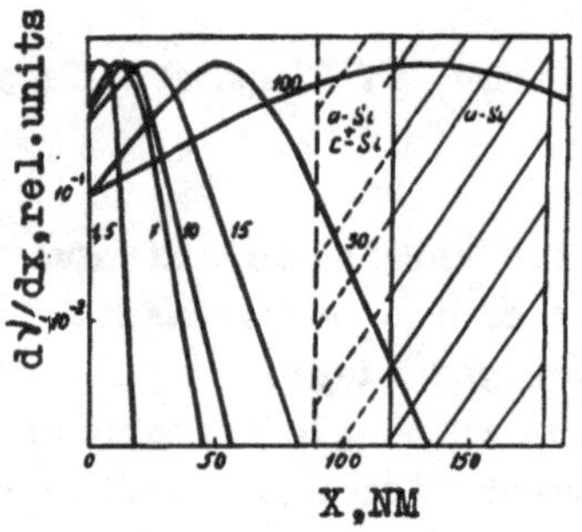

Fig.1 Elastic energy losses of Ne^+ ions relatively to position of buried desordered region in Si. Indicated are the initial ion energies (1.5-100keV), the amorphous layer a-Si and mixed amorphous-crystalline part a-Si + c-Si.

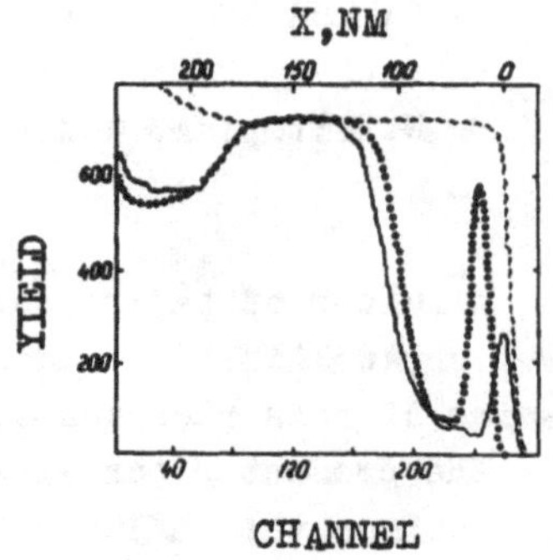

Fig.2. RBS spectra (He^+, 1.5 MeV) for silicon samples
~~~~ - random
—— - aligned before Ne^+ bombardment,
○○○○ - aligned after bombardment with Ne^+, 1.5 keV, $*10^{17} cm^{-2}$, T_i=100°C.

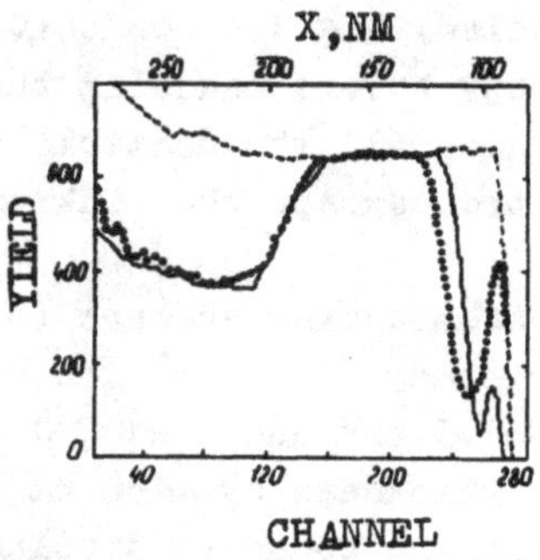

Fig.3. RBS spectra (He^+,1.5 MeV) for silicon samples.
~~~ - random
—— -aligned before Ne^+ bombardment,
○○○○ - aligned after bombardment with Ne^+, 7 keV, 10^{17} cm^{-2}, T_i=400°C.

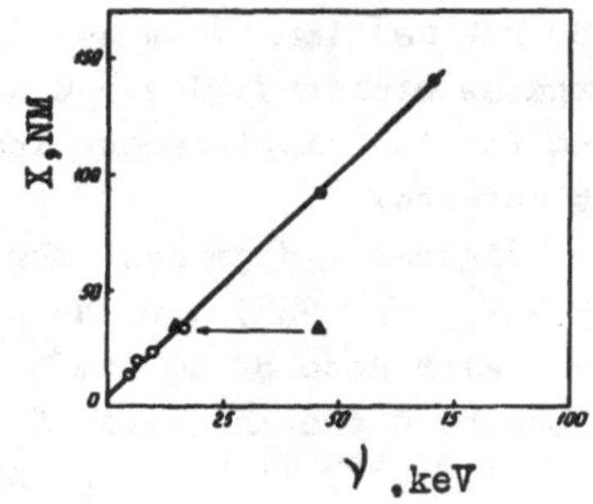

Fig.4. Thicknesses of the recrystallized layers versus elastic energy losses of ions. ○ - this work ●, ◐ - data /1/ for N^+ and Ne^+, respectively, ▲ - this work for Ne^+, 100keV, see text.

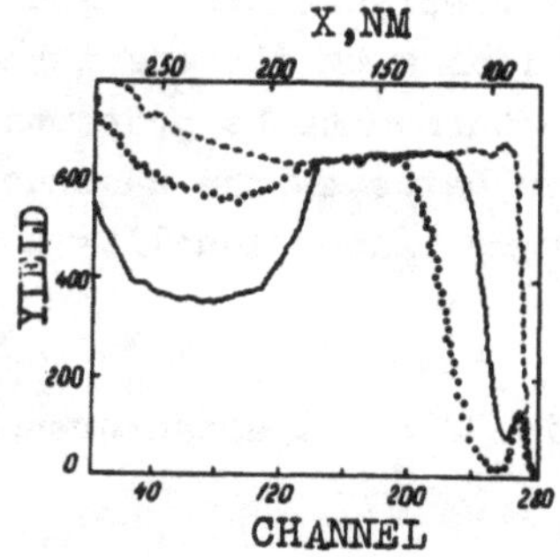

Fig.5. RBS spectra (He^+, 1.5 MeV) for silicon samples.
~~~ - random
—— - aligned before Ne^+ bombardment
○○○○ - aligned after bombardment with Ne^+, 100 keV, 10^{17} cm^{-2}, T_i=400°C.
~~~

also from the data /1/ for light N^+ and Ne^+ ions , however at T_i=300°C (Fig.3). The straight line slope corresponds to 1 atom in crystalline layer for ~ 200-300 atomic displacements. As it was noted above, even during the irradiation by light Ne^+ ions and at T_i=400°C, an intensive accumulation of defects takes place in region R_d, that reduces the efficiency of IBIC with time. We suppose, that the bombardment by heavy ions at the very first stages of irradiation causes the gross structural imperfections, which serve as sinks for mobile point defects. Most of mobile defects in region R_d vanish at the sinks and recombine in dense displacement spikes of heavy ions. In this case, diffusion length of defects strongly decreases and only the defects formed at small distances from interphase boundary take part in IBIC.

Even the authors, considering that the point defect migration is responsible for the IBIC process, do not agree from which region (i.e. amorphous, crystalline or near-boundary one) sink the defects, which stimulate crystallization.

Ne^+ ion-stimulated crystallization with ion energy 100 keV at T_i= 400° C, dose of $10^{17}cm^{-2}$ was performed. The crystallized layer turned to be ~ 30 nm thick, with total elastic losses ~ 40 keV (Fig.4). But Fig.1 shows that Ne^+ ions with E=100 keV lose in the upper crystalline Si layer less than a half of their elastic energy (~ 40%). If only 17 keV , deposited in crystalline layer, is taken into account, then the point falls on the straight line in Fig.3 (arrow indicates the shift).

So, the experimental data obtained in the present paper prove the following. The diffusion approach of non-equilibrium point defects to the amorphous-crystalline boundary is sufficient for the IBIC proceeding. Their diffusion length is demonstrated to be of tens of nanometres even at T_i ~ 100°C.

---

(1) Linnros J., Svensson B., Holmen G.: Phys.Rev.B32 , 1985,2770-2776

DECOMPOSITION OF SOLID SOLUTIONS UNDER ION IRRADIATION

Yu.V. Trushin[1)]

Irradiation of solid solutions by fast ions is able to stimulate their decomposition with falling out a secondary phase as a precipitate. Such a decomposition is referred to as radiation-induced [1] . As shown by the experiments on various composites [2] , as well as the theoretical calculations [3-5] , the radiation-induced decomposition can be a physical process that favours the decrease in the radiation damage of solid solutions, and in particular in the radiation swelling.

Demonstrate this effect using example of a two-component solid solution exposed to hard radiation. During the decomposition the forming precipitation of the second phase passes two basic stages [1] : coherent precipitation nuclei and precipitation. The former stage is of greatest interest from the standpoint of a decreased radiation swelling [3] .

Let the coherent precipitation nuclei be of radius R_p and deformation $\varepsilon_I > 0$, i.e. the volume effect is positive. Ignoring the elastic and shape anisotropy, one assumes the inner stress inside a compressed precipitation nucleus σ_p^I and that in an expanded matrix σ_p^{II} to be constant [6] with $Sp\,\sigma_p^I > 0$, $Sp\,\sigma_p^{II} < 0$, the equilibrium condition

$$v_p\,Sp\,\sigma_p^I + (1 - v_p)\,Sp\,\sigma_p^{II} = 0 \qquad (1)$$

being met over the volume on an average. Here v_p is the volume fraction of precipitation nuclei.

Refs. [3,5] have given a general expression for the swelling rate of a heterogeneous material. Apply it for a case when in the material there are only coherent precipitation nuclei adsorbing vacancies, i.e. the precipitation P is compressed. Then the expression for the swelling rate has the form

$$\dot{S} = (1-v_p)\Omega\left\{g_i-g_v-\frac{\Delta\Omega_v^p-\Delta\Omega_v}{\Omega}\frac{v_p}{1-v_p}g^p\alpha_p^v\bar{R}_p C_p D_v C_v^+\left(1+\frac{\varkappa_p^v}{C_v^+}\right)\left(1-\frac{\Delta\Omega_v^p-\Delta\Omega_v}{\Omega}\right)\right\}+\bar{R}_v C_v\left[\alpha_v^v D_v\cdot C_v^+\left(1+\frac{\varkappa_v^v}{C_v^+}\right)-\alpha_v^i D_i C_i^+\left(1+\frac{\varkappa_v^i}{C_i^+}\right)\right], \qquad (2)$$

where $g_i-g_v = v_p g^p (1 - v_p)^{-1}$.

If in the solid solution without precipitation nuclei ($v_p=0$) there are dislocations ($q = D$), the irradiation creates voids ($q = V$) and interstitial dislocation loops ($q = L_i$), expression (4) takes the form

1) A.F. Ioffe Physico-Technical Institute of the USSR Academy of Sciences, Polytecknicheskaya 26, 194021 Leningrad, USSR

$$\dot{S}_0 = \Omega R_v^c C_v^o \left[\alpha_v^v D_v C_v^+ \left(1 + \frac{\varkappa_{vo}^v}{C_{vo}^+}\right) - \alpha_v^i D_i C_{io}^+ \left(1 + \frac{\varkappa_{vo}^i}{C_{io}^+}\right) \right. \tag{3}$$

where C_{jo}^+ are stationary supersaturations over defects (when $v_p = 0$), C_v^o is an average void concentration, D_j is the diffusion coefficient of defects j; α_{qo}^j , $\varkappa_{qo}^j$ are the coefficients depending on the material.

It is seen from the comparison of S from (2) with S_o from (3) that reducing (or suppression) the swelling can take place in our model only by way of diminishing the density of vacancies in voids. In its turn, this can be gained only by vacancy recombination increase when the interstices are both free and combined into clusters (e.g. q = Q).

To calculate the ratio $\dot{S}/\dot{S}_o$, it is needed, in both cases, to perform a self-consistent calculation procedure for the purpose of eliminating the dependences of the sums of the sink strengths from k_{jo}^2 and k_j^2 . This purpose can be obtained by finding functional linkages of $\bar{R}_q^o C_q^o$ and $\bar{R}_q C_q$ with the supersaturations C_{jo}^+ and C_j^+ . This procedure is rather complicated to be realized analytically. However, for a simpliest assumption system, such a calculation has been made in Ref. [5] for a material with nickel parameters: $\nu = 1/3$, $G = 8\cdot10^{11}$ dn/cm^2, $\rho_D = 10^{10}$ cm^{-2}, $b = 3\cdot10^{-8}$cm, $\Delta\Omega_v/\Omega = 0.23$, $\Delta\Omega_v/\Delta\Omega_v^p = 0.8$, $v_p = 0.1$ at T = 600 K and $g^p \simeq g_i \simeq g_v = 10^{16}$cm^{-3}s^{-1}. The estimate of the ratio of $\dot{S}$ by formula (2), and $\dot{S}_o$ by (3), gives $\dot{S}/\dot{S}_o \simeq 0.18$. In this way, the rate of the decomposed solid solution in which the coherent precipitation nuclei are forming, becomes one fifth as high as that in a material without precipitation nuclei, which is well-consistent with the experimental data [2] .

For a quantitative comparison of the values $\dot{S}_o$ and $\dot{S}$, take that , assuming also that $\varkappa_{qo}^j / C_{jo}^+ \simeq \varkappa_p^v / C_v^+ \to 0$. With allowance for the foregoing, one has

$$\frac{\dot{S}_o}{\dot{S}} = \frac{\bar{R}_v^o C_v^o}{\bar{R}_p C_p}\left(1 - \frac{k_{vo}^2}{k_{io}^2}\right)\frac{C_{vo}^+}{C_v^+}\left[1 + v_p\left(\frac{k_v^2}{\alpha_p^v R_p C_p} - 1\right)\right]^{-1}. \tag{4}$$

The ratio of the sums of sink strengths, k_{vo}^2/k_{io}^2, in a material without precipitation is always less than 1, $C_{vo}^+/C_v^+ \sim 1$. Now one has

$$S_o/S \simeq (\bar{R}_v^o C_v^o/\bar{R}_p C_p)(1 + v_p). \tag{5}$$

Making use of the data for a material with $v_p = 0$ from Ref. [7] ($R_p \simeq 2 \div 3\cdot10^{-7}$cm, $C_p \simeq 10^{16}$cm^{-3}) and selecting for $R_v^o C_v^o$, as in Refs. [7-9] $R_v^o = 8\cdot10^{-7}$cm, $C_v^o = 2\cdot10^{16}$cm^{-3}, respectively, one obtains that $S_o/S \simeq 5$.

REFERENCES

(1) Physical Metallurgy: ed. R.W. Cahn, P. Haasen, North-Holland Physics Publishing. Amsterdam etc., 1983

(2) Zelenskij, V.F., I.M. Neklyudov, B.V. Matvienko et al.: In: Reactor Metallurgy/Proc. Int. Conf., Alushta, 1978, M., CRI Atominform 2(1978) 21-43

(3) Orlov, A.N., A.M. Parshin, Yu.V. Trushin: Sov. Phys. Techn. Phys. 28(1983) 1455-1458

(4) Orlov, A.N., W. Pompe, Yu.V. Trushin: In: Proc. Int. Conf. "Energy Pulse Modification of Semiconductors and Related Materials", Dresden, 1984

(5) Trushin,Yu.V., A.N. Orlov: Sov. Phys. Techn. Phys. 31(1986) 763-767

(6) Pompe, W., Bahr, H.-A., G. Gille et al.: In: Modern Theories and Experimental Evidence, Amsterdam 12(1985) 213-471

(7) Brailsford,A.D.: J. Nucl. Mater. 102(1981) 54-65

(8) Heald, D.T.: Phil. Mag. 31(1975) 555-558

(9) Bullogh, R., T.M. Quigley: J. Nucl. Mater. 104(1981) 1397-1402

PHASE DIAGRAMS OF BINARY ALLOYS UNDER IRRADIATION

A.A.Turkin and A.S.Bakaj *

Frequently, irradiation induces changes in the phase microstructure of alloys, e.g., the dissolution of thermally stable precipitates or, conversely, the precipitation of nonequilibrium phases /1/. These processes cannot be predicted by standard phase diagrams. Obviously, to forecast the behavior of the alloy under irradiation, one needs an analog of the phase diagram. The location of phase boundaries on the irradiation-modified phase diagram is determined by the kinetics of structural-phase transformations. The points of the irradiation-modified phase diagram, where the rate of transformation goes to zero, form the curve of equilibrium which separates the stability regions of different structural-phase states.

In this paper we consider the effect of irradiation on the kinetics of the first-order phase transformation, viz, on the growth of second phase precipitates in a random solid solution of elements A and B (A is the solute). To find the growth rate of the precipitate, we use the model for irradiation-induced segregation described in /2,3/. Due to the coupling between fluxes of point defects (vacancies and interstitials) and fluxes of alloying elements, the growth rate of the precipitate essentially depends on its structural state, i.e., whether the precipitate contains a sink for point defects or not. The boundary of the coherent precipitate is free of defect sites and does not absorb vacancies and interstitials. In the vicinity of the coherent precipitate the defect fluxes are small compared to the solute flux and practically do not influence the growth of the coherent precipitate. The total solute flux to the surface of the coherent precipitate of radius R is given by /4/

$$J_A = 4\pi R\left(\frac{d_{Av} d_{Bv} \bar{c}_v}{d_{Av}\bar{c}_A + d_{Bv}\bar{c}_B} + \frac{\delta_{Bi}\,\delta_{Ai}^R\,\bar{c}_i}{\delta_{Ai}^R c_A^R + \delta_{Bi}^R c_B^R}\right)\left(\bar{c}_A - c_A^R\right), \tag{1}$$

$$\delta_{Ai} = d_{Ai}\zeta\,(c_B + \zeta c_A)^{-1}, \quad \delta_{Bi} = d_{Bi}\,(c_B + \zeta c_A)^{-1},$$

where $C_{v,i}$ are the vacancy and interstitial concentrations; $C_{A,B}$ the concentrations of alloying elements ($C_A + C_B = 1$); d_{Av}, d_{Bv} the diffusivity coefficients of vacancies; d_{Ai}, d_{Bi} the diffusivity coefficients of interstitials; $\zeta = \exp(H^i_{B\to A}/T)$ ($H^i_{B\to A}$ is the energy gained by converting a B-interstitial (BB-dumbbell) into an A-interstitial (AB-dumbell). The superscript R refers to the quantities at the precipitate-matrix interface, the symbols with the bar denote the average values far from the precipitate. It is seen that in the case of the coherent precipitate, irradiation enhances the solute diffusion, whereas the flux dependence upon the solute supersaturation remains the same as that for the unir-

* Institute of Physics & Technology, the Ukrainian Academy of Sciences, 310108 Kharkov, USSR

radiated alloy. As regards the incoherent precipitate, its surface is a good sink for irradiation-induced vacancies and interstitials. The migration of point defects and annihilation at the incoherent interface give rise to an additional mechanism of solute transport toward (or away from) the precipitate. As a result, under irradiation the stability of the incoherent precipitate changes. In ref. /4/ (see also refs. /2,3/) we have defined the kinetically equilibrium concentration of the solute, C_A^{irr}, i.e., the concentration at which the solid solution is in equilibrium with an extended incoherent precipitate. The thermal and irradiation-altered solubilities are connected by

$$\frac{C_A^{irr}}{1-C_A^{irr}} = \frac{C_A^e}{1-C_A^e}\left[1+\left(\frac{\bar{C}_v}{C_v^e}-1\right)\frac{1+\varkappa}{1+(\varkappa-1)C_A^{irr}}\right]^{\frac{1-\varkappa}{1+\varkappa}}, \quad (2)$$

where $\varkappa = \zeta d_{Ai}d_{Bv}/d_{Bi}d_{Av}$, $C_A^e = \lim_{R\to\infty} C_A^R$ is the solute solubility limit, C_v^e is the thermal vacancy concentration. If $\bar{C}_A < C_A^{irr}$, then the incoherent precipitates dissolve under irradiation (the same is true for the precipitates on sinks). Eq.(2) contains the average vacancy concentration. To express it in terms of the displacement rate, K, the dislocation density, ρ, the density, N, and the mean radius, r_v, of voids, we shall use the conservation equations for the defect concentrations

$$K-\alpha_R\bar{D}_i\bar{C}_i\bar{C}_v-(4\pi r_v N+Z\rho)\frac{1+(\lambda-1)C_A^{irr}}{1+(\varkappa-1)C_A^{irr}}d_{Bv}(\bar{C}_v-C_v^e)=0, \quad \lambda\equiv\zeta\frac{d_{Ai}}{d_{Bi}}, \quad (3)$$

$$K-\alpha_R\bar{D}_i\bar{C}_i\bar{C}_v-(4\pi r_v N+Z\rho)\bar{D}_i\bar{C}_i=0, \quad (4)$$

where α_R is the recombination rate constant, $Z=2\pi\left(\ln\frac{R_0}{r_d}\right)^{-1}$ the capture efficiency of a dislocation for point defects, R_0 the average distance between defect sinks, r_d the capture radius. The solution of eqs.(3) and (4) for C_v is written as

$$\bar{C}_v = C_v^e+\frac{1}{2}\left(C_v^e+\frac{k^2}{\alpha_R}\right)\left(\sqrt{1+\frac{4K}{\alpha_R D^*}\left(C_v^e+\frac{k^2}{\alpha_R}\right)^{-2}}-1\right). \quad (5)$$

Here $D^* = d_{Bv}\frac{1+(\lambda-1)C_A^{irr}}{1+(\varkappa-1)C_A^{irr}}$, $k^2=4\pi r_v N+Z\rho$ is the total sink strength.

It is convenient to represent the results of the phase stability studies on the irradiation-modified phase diagram. There are two types of the diagrams. The portions of the diagrams for the Ni-Si alloy ($\varkappa>1$, solute atoms migrate preferentially via interstitials) and the Ni-Al alloy ($\varkappa<1$, solute transport via vacancies dominates) are shown in fig. 1. It is seen that the regions of stability for coherent and incoherent precipitates differ essentially. To plot the curve of C_A^{irr} vs T eqs.(2) and (5) were solved numerically at $K = 10^{-6}$ dpa/s, $k^2 = 10^{10}$ cm^{-2}. The parameters common for the both alloys are the same as for pure Ni. Diffusivity coefficients were taken in the usual form as $d_{Ll} = d_{lo} \times \exp(-H_{Ll}^m/T)$ (L = A,B; l = i,v). To demonstrate the sensitivity of C_A^{irr} to the variation of the activation energies, the calculations were carried out for several values of the parameters $H_\lambda = H_{B\to A}^i - H_{Ai}^m + H_{Bi}^m$, $H_\mu = H_{Bv}^m - H_{Av}^m$. The three-dimensional structural-phase diagram of the

Ni-Si alloy is depicted in fig.2. The section of this diagram by the plane of C_A = const (see fig.3) represents the stability diagram of the undersaturated solid solution. The dependence of the threshold dose rate on temperature and the alloy parameters is given by

$$K_{th}(T) = k^2 d_{Bv} c_v^e W \frac{1+(\lambda-1)C_A}{1+\varkappa}\left[1+\frac{\alpha_R c_v^e}{k^2}\left(W\frac{1+(\varkappa-1)C_A}{1+\varkappa}+1\right)\right], \quad (6)$$

where

$$W = \left[\frac{C_A^e(1-C_A)}{C_A(1-C_A^e)}\right]^{\frac{\varkappa+1}{\varkappa-1}} - 1. \quad (7)$$

The diagrams of alloys with $\varkappa > 1$ (see figs.1a and 3) qualitatively agree with the experimental diagrams of Ni-Si and Al-Zn alloys /5,6/.

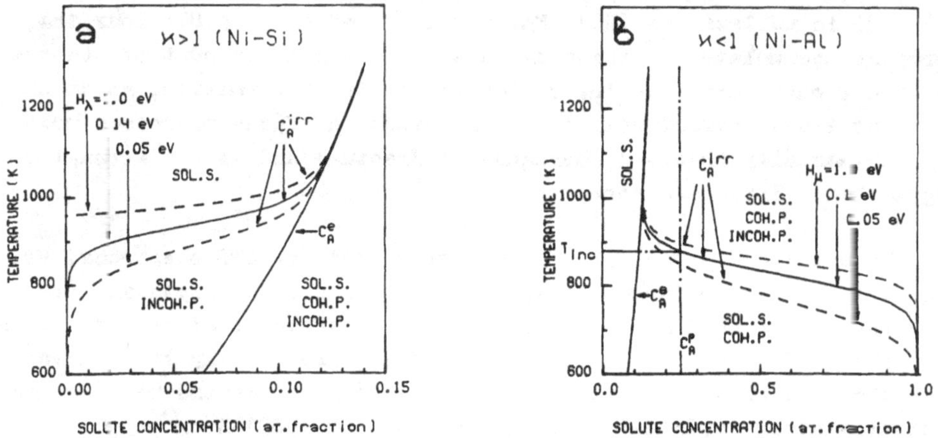

Fig. 1

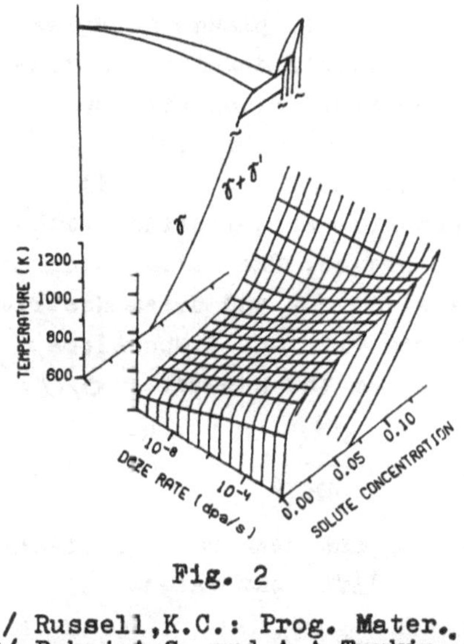

Fig. 2

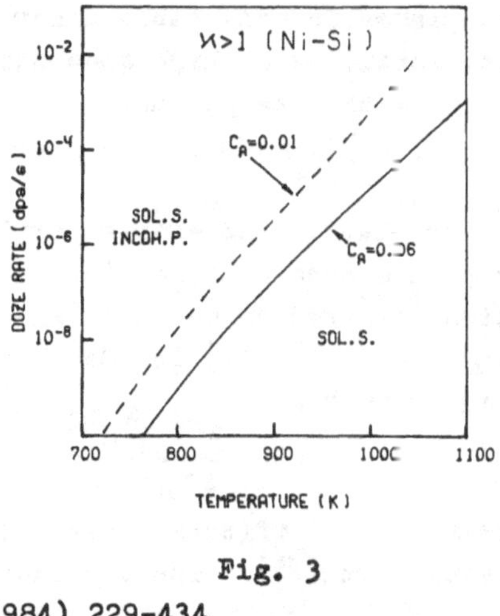

Fig. 3

/1/ Russell,K.C.: Prog. Mater. Sci. 28 (1984) 229-434
/2/ Bakaj,A.S. and A.A.Turkin: Pis'ma Zh. Tech. Fiz. 13 (1987) 1281-1284
/3/ Bakaj,A.S. and A.A.Turkin: J. Nucl. Mater. 152 (1988) 331-333
/4/ Bakaj,A.S. and A.A.Turkin: Radiation-induced modification of the binary alloy phase diagram (in Russian),TsNIIatominform, Moscow 1988
/5/ Barbu,A., G.Martin and A.Chamberod: J.Appl.Phys. 51 (1980) 6192-6196
/6/ Cauvin,R. and G.Martin: Phys. Rev. B23 (1981) 3333-3348

THE ENERGY ACCUMULATION IN SOLIDS IRRADIATED BY HIGH POWER BEAMS OF CHARGED PARTICLES

Valyaev A.N.[1]

Abstract. Under the irradiation of solids by high power pulse beams (HPB) of charged particles the absorbed energy accumulation is possible both in radiated region and non-radiated one. In dielectrics two its mechanisms have been considered: 1) in electronic crystal sub-system (ES); 2) in nuclear one (NS). For metals irradiated by HPB ions the energy is accumulated in the non-radiated region on account of the sharp loss shock wave energy in the region of its front formation, spent on point and linear structural defect's formation. These processes role in experimentally observed phenomena of fracture and metals strengthening is being discussed here.

The brittle fracture of solids, irradiated by HPB electrons, has been observed for semiconductors, metals, rock stones, glasses. These works review has been given in ref [6] The lowest value of this fracture threshold is for ionic crystals, $W_0 \sim$ Erg/cm^3, practically two orders lower than its binding energy. The HPB electron parameters were: maximal particales energy o,25 MeV; current density (50-1500) A/cm^2; pulse duration (2-3o) ns. The quasi-stational electron-hole plasma (EHP) n density under irradiation is $\sim 10^{17}$ cm^{-3}, its non-ideal parameter is $(U/K) \sim 1$, where U and K are potential and kinetic plasma energies. This plasma is non-stable under the positive particle density fluctuation. Particles of high concentration form the simmetrical structural lattice with the parameter 2a. Full EHP energy density is :

$$E(n) = (\mathcal{E}_e + \mathcal{E}_{ok} + \mathcal{E}_{ex} + \mathcal{E}_{corr} + \mathcal{E}_p)\, n \qquad (1)$$

where $\mathcal{E}_e$, $\mathcal{E}_{ok}$, $\mathcal{E}_{ex}$, $\mathcal{E}_{corr}$, $\mathcal{E}_p$ are the electrostatic, zero-kinetic, exchange, correlative and excite particles pair energies. The calculation of $\mathcal{E}_e$ has been made for the Caussian space electrons and holes distribution, centred on cation's and anion's centres with the dispersions δ_e^2 and δ_h^2 [1]. The Madelung const β versus the parameter δ/a is shown in Fig. 1.

$$\mathcal{E}_{ok} = \frac{3^{5/3}\hbar^2 n^{2/3}}{40\pi^{2/3}\mu} \qquad (2) \qquad\qquad \mathcal{E}_{ex} = -\frac{3^{1/3} e^2 n^{1/3}}{4\pi^{1/3}} \qquad (3)$$

where μ is the effective electron-hole mass. $\mathcal{E}_{corr}$ has been used in extrapolated form [7] . The dependence $E(n)$ for different values $\beta(\delta)$

---

[1] USSR, Ust-Kamenogorsk, Institute of Civil Engineering and Roads

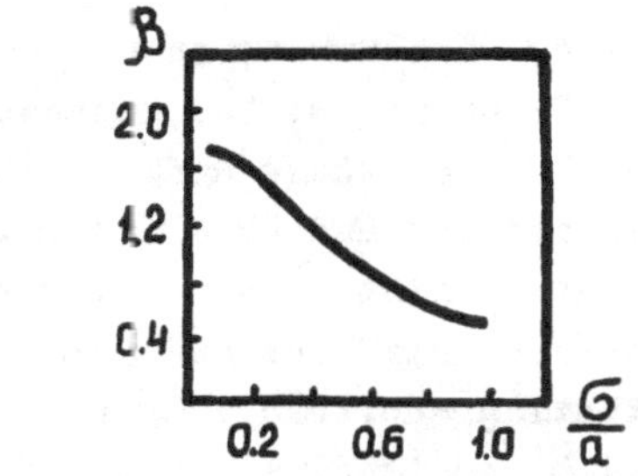

Fig 1. The Madeling const β versus the parameter σ ($\sigma = \sigma_e = \sigma_h$).

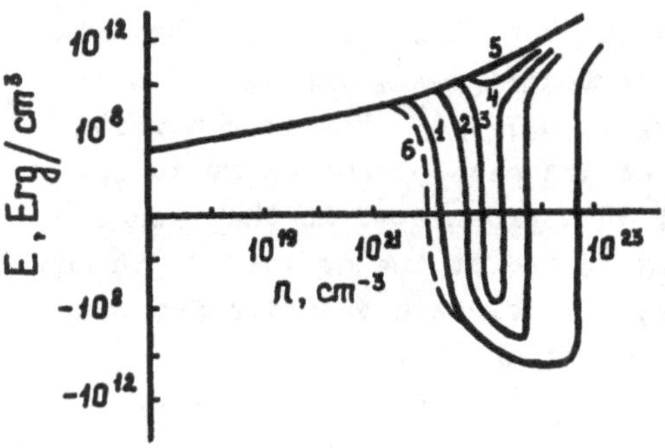

Fig 2. Full EHP energy density versus the particle concentration for different values β and μ (1- β = 1.765; μ = 1; 2-1.765; 3-1.55; 4-1.50; μ = 0.5; 6- μ =1; β = 1.765 with correlative energy).

and μ is shown in Fig. 2. The critical value β^* has been found. For $\beta > \beta^*$ $E(n)$ has the deep minimum, indicating the possible EHP condensation in electron-hole condensational phase (EHCP) with the equilibrium concentration $n_e \sim 10^{22} cm^{-3}$ and energy density $\bar{\rho}_d \sim 10^{10}$ Erg/cm^3, closed to binding energy.

During the electron beams passage through the EHP plasma, the plasma instabilities creations are performed. These instabilities cause the local fluctuations of particles density and play the artificial seed's role of the phase transition and EHCP replenishment. The existence stability according to Lindeman's melting criterion for E. Vigner electron's crystal and A. Abrikosov electron-ions one is caused by the excess of electric and exchange interactional particles energy under these zero-kinetic and thermal ones. The critical EHCP existence temperature according to L. Keldyish estimate for exitons condensation is $\sim 0{,}1\mathcal{E}_B \sim 1000\,^\circ$K, ($\mathcal{E}_B$ is the binding energy for EHCP particles pair in order to several eV), therefore EHCP existence is possible at room's temperature. Considering that full EHP converts into EHCP spherical form drops, the upper EHCP parameters estimates have been made [2]:

$$\bar{a}_d \approx R_h = V_h \tau_h \quad (4) \qquad \bar{n}_d = \left(\tfrac{4}{3}\pi R_h^3\right)^{-1} \quad (5)$$

$$\bar{\mathcal{E}}_d = W_0/\bar{n}_d \quad (6) \qquad \bar{r}_d = (3\bar{\mathcal{E}}_d/4\pi\bar{\rho}_d)^{1/3} \quad (7)$$

where $\bar{n}_d$ is drops concentration; $\bar{a}_d$ is the middle distance between drops; $\bar{\mathcal{E}}_d$ is drops energy with radius $\bar{r}_d$; $\tau_h \sim 10^{-11}$s is the autolocalised hole's time; $V_h \sim 10^6$ cm/s is the thermolised hole's velocity. For KCl crystal the following estimates have been made: $\bar{a}_d \sim 10^{-5}$cm; $\bar{n}_d \sim 10^{15}$cm^{-3}; $\bar{\mathcal{E}}_d \sim 10^{-7}$ Erg; $\bar{r}_d \sim 10^{-6}$cm. The energy transition from ES to NS is going on for the time $\sim 10^{-9}$s, such as the reaction of autolocalised hole recombinate with thermolised electron flowing with high output of

Frenkel's pairs defects (FP).

According to the second mechanism, the energy accumulation occurs in FP localisation result in NS macroregions (FP drops) at the expense of: 1) the activation energy reduction of new FP hear those appeared before; 2) FP interaction among themselves dislocations and FP clasters. The drop is non-stable, any FP recombination gives rise to thermal wave, initiating shortrange FP recombination. From randomly blusted drops thermal and accoustic waves are spread, initiating shortrange drop blusts till in the last analysis crystal fracture with longitudinal sound's velocity is obtained. This processes modelation has been shown in 3 . The comparison of the calculated and experimental time characteristics made it possible to determinate the middle lifetime of the FP drop till its spontaneous blust $\sim 1{,}4 \ 10^{-2}$s.

In ref [4,9] the HPB ions irradiated influence on the mechanical properties of metals has been investigated. Together with the known microhardness increasing ΔH_0 in irradiated region $(\sim 1 \mu m)$, the second peak of the microhardness increasing $\Delta H_2 = (0{,}6 - 0{,}7)\Delta H_0$ has been noticed at the depths $\sim (40 - 60)\mu m$. The gas dynamics plasma spray in irradiated region causes the shock wave (SW), propagated in the non-irradiated region. The SW front (SWF) is formed at the defined depth and it has the finite width $\sim(0{,}1 - 10)\mu m$[8]. The kinetic equation for the FP concentration in the SWF formations region is [5]:

$$\frac{dn}{dt} = A\exp\{-[E_a - B\rho(x,t)]/kT\} - K_r n^2 - K_d n \qquad (8)$$

where E_a is the activations energy; K_r is the FP recombinations coefficient; K_d is the interstitials flow efficience to dislocations; $A, B = const$ $\rho(x,t)$ is the SW amplitude in the point x at the time moment t . The $\rho(x,t)$ calculation has been made by the grid method. The N_d interstitials amount flowed together to dislocations in the SWF region has been estimated $\sim(10^{19} - 10^{20})$ cm^{-3}. Considering the linear dependence ΔH_2 from N_d , it has been obtained $\Delta H_2 = \alpha N_d$, where α = const.

References.

(1)Vaisburd D.I., Valyaev A.N., Balyichev I.N. Phycika (USSR) 3 (1973)159.
(2) Valyaev A.N. Synopsis of dissertation thesis (1978), USSR,Sverdlovsk.
(3) Valyaev A.N. Lappa A.V., Phycika Tverdogo Tela, (USSR) 18 (1977),3534
(4) Valyaev A.N., Plotnikov S.V., Nazarov U.K. Kojshibaev R.G., Thesis of Reports, USSR Sympozium on High Current Electronic (Tomsk) (1986) 210.
(5) Valyaev A.N., Pogrebniak, A.D., Remnev G.E., S.V.Plotnikov, Itern. Conf of Ion Implantation in Semiconductors and Other Materials (1988), Lublin, Poland, 163.
(6) Gering G.I., Valyaev A.N., Phycika (USSR) 5 (1974) 157.
(7) Gordon R.G., Kim Y.S., J. Chem. Phys. 56 (1972) 3122.
(8) Mogilevsky M.A., Phys. Reports 97, no 6 (1983) 357.
(9) Pogrebniak A.D., Remnev G.E., Chistyakov S.A., Ligachev A.E., Phycika (USSR) 1 (1987) 52.

NUCLEATION AT VARYING TEMPERATURES

D.Wolff, H.D.Geiler[1)]

1. Introduction

Temperature fields generated by short laser pulses vary in a nanosecond or picosecond time scale. Therefore, solidification scenarios in strongly undercooled molten systems have to take into the consideration the real time dependent nucleation process [1]. Starting with the classical hierarchy of kinetic cluster reactions [2], the problem of nucleation in undercooled molten silicon is tracted, which plays an important role in nucleation controlled explosive crystallization [3].

2. Basic equations of nucleation

The problem will be handeled using the well known Szillard model [2], which describes the time dependent cluster concentration $z_i(t)$ of a cluster containing i monomers by a hierarchy of construction and destruction reactions. Assuming monomer exchange only and a constant reservoire of monomers $z_1 = 1/V_1$ (V_1 - atomic volume), which is a good approximation for the description of nucleation events, the following set of master equations is used [4]:

$$\frac{d}{dt} z_i = g^+_{i-1} z_{i-1} - (g^+_i + g^-_i) z_i + g^-_{i+1} z_{i+1} \quad , \ i=2 \ldots i_m-1$$

$$\frac{d}{dt} z_{i_m} = g^+_{i_m-1} z_{i_m-1} - (g^+_{i_m} + g^-_{i_m}) z_{i_m} \qquad (1)$$

This hierarchy of cluster equations is broken at a maximum i_m, which is determined by the condition that all monomers up to this time incorporated into the cluster distribution are contained in clusters with size $i \le i_m$. This means $z_{i+1} = 0$. This condition must be proved in each time step. The reaction rates are taken from the theory of absolute reaction rates taking into account the classical thermodynamic consideration of surface tension σ of the nucleus [4]:

$$g^\pm_i = i_o \Gamma_o \exp\{-E/(kT) - \mu^\pm_i/(kT)\} \qquad (2)$$

with the difference of the chemical potential

$$\mu^\pm_i = \left\{ \begin{array}{ll} 2\sigma V_1(1/r(i) - 1/r^*) & \text{for sign + and } i<i^* \\ 0 & \text{for sign - and } i \le i^* \\ 0 & \text{for sign + and } i \ge i^* \\ -2\sigma V_1(1/r(i) - 1/r^*) & \text{for sign - and } i>i^* \end{array} \right\}$$

using the critical radius r^*

---

1) Friedrich-Schiller-University, dept. physics
Max-Wien-Platz 1, DDR-6900 Jena, GDR

$r^* = 2\sigma / L\, T/(T_m - T)$, $r(i) = \sqrt[3]{V_1 i 3/(4\pi)}$

(i_o - number of nucleation sites at the cluster surface, Γ_o - jumping frequency, L - latent heat and T_m - melting temperature).

The linear system (1) is solved numerically taking into account the temperature function generated by a laser pulse.

3. The temperature pulse

The laser pulse is fitted by a quadratic function:

$$I(t) = \begin{cases} I_m(1-(1-\sqrt{2}\, t/\tau)^2) & \text{for } 0 \leq t \leq \sqrt{2}\,\tau \\ 0 & \text{for } \sqrt{2}\,\tau < t \end{cases}$$

To get a realistic temperature pulse homogeneous energy deposition into a volume is assumed including heat exchange via linear heat transfer governed by the heat loss parameter α with the surrounding which is kept at a temperature T_o. The solution of the heat balance equation

$$\frac{\partial}{\partial t} T = -\alpha(T-T_o) + I(t)/(\varrho c_p)$$

provides

$$T(t) = \begin{cases} -R[t^2/a^2-2(t/(\alpha a)+1)+2(1/(\alpha a)+1)/(\alpha a)]+ \\ \qquad +\exp(-\alpha t)2R(1/(\alpha a)+1)/(\alpha a)+T_o & \text{for } t \leq 2a \\ -R\exp(-\alpha(t-2a))2(1/(\alpha a)-1)/(\alpha a)+ \\ \qquad +\exp(-\alpha t)2R(1/(\alpha a)+1)/(\alpha a)+T_o & \text{for } t > 2a \end{cases} \qquad (3)$$

with $R=I_m/(\varrho c_p \alpha)$, $a=\tau/\sqrt{2}$.

I_m is adjusted to reach a given maximum temperature T_{max}. This T-pulse represents the heating and cooling governed by τ and α, respectively.

4. Results of nucleation in strongly undercooled molten silicon

The model system of amorphous silicon is considered, which is molten at $T=T_{ma}=1350$ K. In this moment the laser pulse is applied to heat the melt. Fig.1 shows the temperature pulse and the resulting concentration of all overcritical nuclei $N^* = \sum_{i=i^*}^{i_m} z_i$.

The temperature pulse is generated by a laser pulse of 5 ps duration and different cooling rates. The concentration of all overcritical nuclei N^* grows rapidly in the moment of melting because of the tiny critical radius and the vanishing back reactions. With increasing temperature the rate of N^* decreases and up to a critical temperature range the total amount of N^* decreases because of the increasing critical radius r^*. Hence, a lot of the generated overcritical cluster became undercritical. This process is reversed if the cooling takes place and the temperature decreases. The temporal behaviour of N^* now is determined by the competition of the decreasing r^* and the increasing influence of the back reactions of the rate $g_i^- z_i$. Shorter heating which causes a narrower cluster distribution therefore influences the nucleation in the cooling periode by a higher rate of N^* because of the diminished influence of the back reactions. Then the rate during the cooling periode approaches to the rate of the virgine melt in the heating periode. So the cross over of the curves in fig.1 can be understood. This effect vanishes if r^* reaches this region of the cluster distribution with stabilized back reactions. The nucleation process is stopped at $T=T_{ma}$ by quenching. The results clearly demonstrate that it is impossible to suppress the nucleation by shorter temperature pulses.

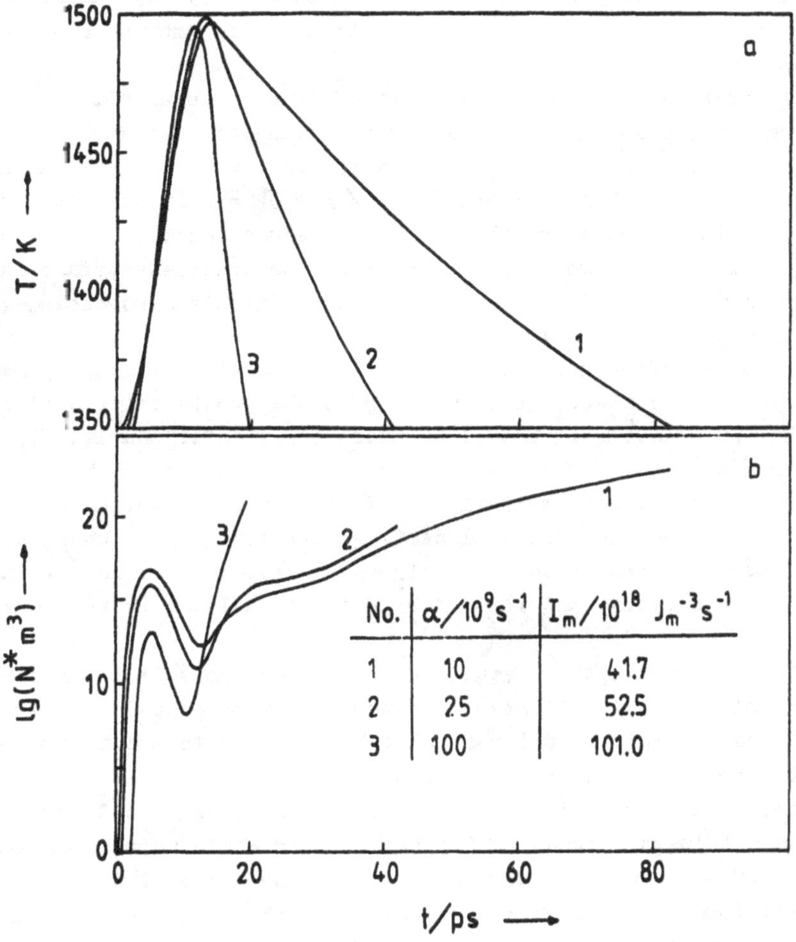

Fig.1: time dependent nucleation in melting amorphous Si heated by a 5 ps laser pulse for different heating and cooling rates
(a) temperature versus time
(b) concentration of all overcritical clusters N^* versus time

References:

[1] R.F.Wood, G.A.Geist: Phys.Rev. B34,2606(1986)
[2] D.Kashchiev: Cryst.Res.Technol. 19,1413(1984)
[3] H.D.Geiler, E.Glaser, G.Andrä, D.Wolff, G.Götz: Appl.Surf.Sci. 36,632(1989)
[4] D.Wolff, H.D.Geiler, G.Götz : phys.stat.sol.(a) 112,727(1989)

MODELLING OF PHASE TRANSFORMATIONS IN AMORPHOUS SILICON BY NANOSECOND PULSED-LASER IRRADIATION

S.P. Zhvavyi and O.L. Sadovskaya[1)]

As shown in /4/, for short (< 45 ns) and long (> 45 ns) nanosecond laser pulses processes of explosive crystallization in amorphous silicon exhibit a marked difference.

In this paper, we compare the results of computational simulation of the phase transitions in nanosecond laser annealing of amorphous silicon (a-Si) on monocrystalline (c-Si) substrate for two values of the radiation pulse width τ = 32 ns /1/ and 70 ns /3/ (λ = 0.69 μm). The statement of the problem is based on the following assumptions:

1. At the initial stage of annealing when the specimen surface layer is heated up to $T \gg T_a$ = 1433 K, a-Si melts and a highly supercooled (metastable) silicon melt (sl-Si) is formed.

2. Immediately after melting of a-Si, a fine-grained (FG) polycrystalline silicon (p-Si) starts to form owing to the growth of crystal phase nuclei already existing in the sl-Si phase ("intrinsic" nuclei)/2,5/.

3. Further heating of the specimen results in melting of the FG p-Si and formation of the silicon melt (l-Si). If the rate of heating of the metastable melt exceeds the critical heating rate for growing nucleus, melting will occur at temperature $T_m = T_{mo}(1 - A/r)$, where r is the radius of the growing nucleus, and T_{mo} = 1683 K is the melting temperature of c-Si.

4. If the melting depth is equal to or exceeds the a-Si layer thickness, cooling of the l-Si will result in the formation of c-Si. Otherwise the layer of coarse-grained (CG) p-Si will be formed with its thickness determined by the l-Si penetration depth.

In mathematical formulation of the problem, melting of a-Si and c-Si was described within the bounds of Stephan problem while melting of a-Si and formation of p-Si and c-Si were analyzed on the basis of the mechanism of nuclei formation and growth described by the Kolmogorov's equation. It was assumed that melting of a-Si and crystallization of l-Si occur as a result of homogeneous nucleation via two-dimensional (melting) and three-dimensional (crystallization) mechanisms, and the formation of FG-p-Si proceeds through three-dimensional growth of "intrinsic" nuclei in the sl-Si.

It follows from our calculations for τ =32 ns that by the time a-Si melting begins overheating of the near-surface layer reaches $\Delta T = T - T_a \sim 20$ K. The absorption of the latent heat of phase transition during a-Si melting relieves the overheating and the heating rate is sub-

[1)]Institute of Electronics, BSSR Academy of Sciences,
22 Logoiski Trakt, Minsk-90, 220841, USSR

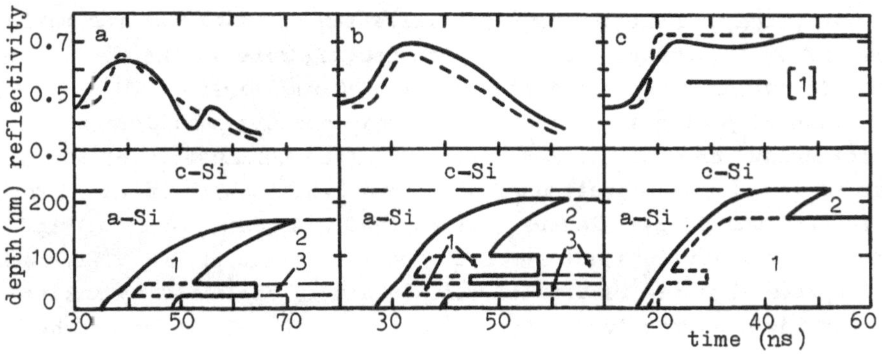

Fig. 1. Time dependences of the reflectivity ($\lambda = 0.633\ \mu m$) and penetration depth of the melt for the energy densities E = 0.19 (a), 0.27 (b), and 0.7 J/cm^2 (c) and τ = 32 ns, 1 - l-Si, 2 - FG p-Si 3 - CG p-Si.

stantially retarded. At the same time, FG p-Si starts to grow on the "intrinsic" nuclei existing in the sl-Si volume. As a result of the crystallization heat evolution in the near-surface region of sl-Si, the rate of heating of the deeper sl-Si layers becomes higher than the critical heating rate which supports the growth of FG p-Si grains of radius r. As a result of this, underneath a thin near-surface layer where growth of FG p-Si grains continues, a region is formed in which the crystalline phase nuclei dissolve, l-Si forms, and the conditions for formation of FG p-Si disappear. However, absorption of heat during melting of a-Si and FG p-Si grains results once again in a decrease in the rate of sl-Si heating, and the motion of the l-Si/a-Si interface ceases (Fig. 1a).

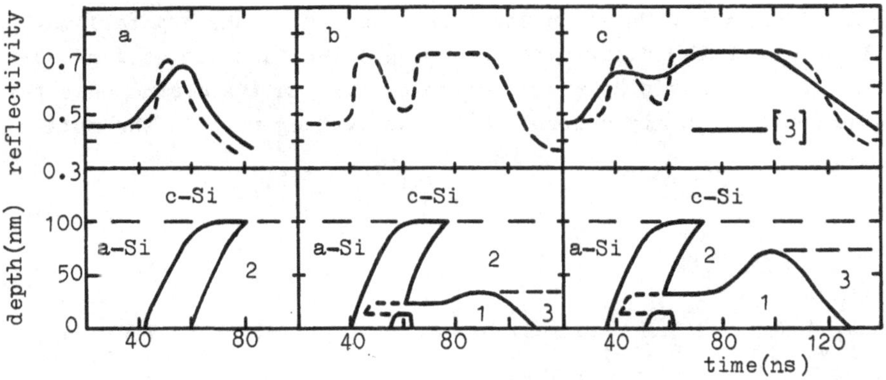

Fig. 2. Time dependences of the reflectivity ($\lambda = 1.06\ \mu m$) and penetration depth of the melt for the energy densities E = 0.8 (a), 1.0 (b), and 1.2 J/cm^2 (c) and τ = 70 ns.

At high energy densities (see Fig. 1b) by the time the l-Si/sl-Si interface motion stops, the influence of the surface FG p-Si layer on further annealing process becomes increasingly important. As the crystallized fraction of silicon is increased, the sample becomes more "antireflecting" and evolution of heat during crystallization of sl-Si becomes maximum. Therefore, after a short-time decay of the heating rate in the sl-Si which causes the formation of intermediate thin (~ 10 nm) layer of FG p-Si, melting of the FG p-Si grains occurs once again and the l-Si formed starts to move following the front of a-Si melting. After termination of the laser pulse action, crystallization of sl-Si and l-Si results in the formation of FG p-Si and CG p-Si layers, respectively. As canbe seen in Fig. 1a,b, the reflectivity does not reach its maximum value of R = 0.72. Such behaviour agrees with the results of ref. 1 and is explained by "antireflection" of the specimen due to formation of the FG p-Si layer on the surface.

As seen in Fig. 2, for τ = 70 ns, an increase in the FG p-Si fraction in the near-surface region also results in decreasing R. In this case, however, the formation of FG p-Si layer on the surface results in that the main heat source caused by absorption of laser radiation in the liquid phase and the evolution of the latent heat during crystallization of sl-Si moves from the surface to the specimen depth. Therefore, T_m is firstly reached at the depth of the mixed layer (FG p-Si + sl-Si) and, here, the FG p-Si layer begins to melt and the melt front moves from the depth to the surface to meet the laser beam. When the melt front appears on the surface, the reflectivity rises abruptly to its maximum value and retains this value until the annealing process stops.

In conclusion, by taking into account the intermediate crystallization of sl-Si on the "intrinsic" nuclei and the dependence of the melting temperature of growing FG p-Si grains on their sizes, it is possible not only to explain the experimental data available but, also, to show that for τ = 70 ns and τ = 32 ns the dynamics of phase transitions are substantially different. Thus, in the first case it is the crystallization of the sl-Si on the "intrinsic" nuclei and the "antireflecting" action of the FG p-Si that play the major role while in the second case the annealing process is mainly controlled by the heating rate in the metastable melt.

References

(1) Bruines, J.J.P., R.P.M. van Hal, H.M.J. Boots, A. Polman, F.W. Saris: Appl. Phys. Lett. 49 (1986) 1160

(2) Ivlev, G.D., V.L. Malevich: Phys. Stat. Sol. (a) 103 (1987) K87

(3) Ivlev, G.D.: Zhurn. Tekh. Fiz. Pisma. 8 (1982) 468

(4) Peercy, P.S., J.Y. Tsao, S.R. Stiffler, M.O. Thompson: Appl. Phys. Lett. 52 (1988) 203

(5) Zhvavyi, S.P., O.L. Sadovskaya: Zhurn. Tekh. Fiz. Pisma. 13 (1987) 1171

ELASTIC WAVE GENERATION IN THE CONTINUUM BY A TUBULAR BEAM OF CHARGED PARTICLES

I.V.Bar'yakhtar, Yu.V.Filimonov, Eh.V.Lysenko, Yu.V.Tkach, V.T.Uvarov[1)]

In the interaction of a high-current charged beam with the surface of the medium, a significant portion of energy is transformed into elastic waves. The present report is concerned with the problem of elastic wave generation by a pulsed tubular beam of charged particles. The peculiar character of the problem stems from the elastic wave source geometry specified by the beam.

We shall describe the excitation of longitudinal elastic waves in the medium using the following wave equation for the velocity ($\vec{v}=\vec{\nabla}\varphi$) potential φ

$$\vec{\nabla}^2\varphi - \frac{1}{c^2}\frac{\partial^2\varphi}{\partial t^2} = m(t)\,\delta(z)\,\delta(x^2+y^2-R_0^2) \qquad (1)$$

Here m(t) is the volume source of the mass, the δ-functions define the geometry of the problem, namely, the elastic wave source as a thin ring of radius R_0 in the plane z=0, c is the velocity of longitudinal elastic waves in an unbounded medium. However, if the depth of charged particle penetration into the medium is small as compared with the length λ of the emitted wave, it is clearly evident , in view of the symmetry of the problem relative to the z=0 plane, that Eq. (1) describes the generation of elastic waves in a half-space. In this case we neglect the surface effects, i.e., we analyse the solution in the $z \gg \lambda$ region.

For the liquid medium, with a knowledge of the expression for the potential φ (r,t) it is easy to derive the formula for the pressure: $P(r,t) = -\rho_0 \frac{\partial\varphi}{\partial t}$, where ρ_0 is the density of the medium (see Figs. 1-3 which are given for the symmetric pulse, $\tau_{in} = \tau_{dec}$).

For solids, the components of the stress tensor σ_{ik} are given by

$$\sigma_{rr} = -2\mu\left(\Phi_{zz} + \frac{1}{r}\Phi_r\right) \;;\quad \sigma_{\varphi\varphi} = -2\mu\left(\Phi_{zz} + \Phi_{rr}\right) + \rho_0\Phi_{tt}$$

$$\sigma_{zz} = -2\mu\left(\frac{1}{r}\Phi_r + \Phi_{rr}\right) - \rho_0\Phi_{tt} \;;\quad \sigma_{rz} = 2\mu\Phi_{rz} \;;\quad \sigma_{z\varphi} = \sigma_{r\varphi} = 0$$

Here $\Phi(\vec{r},t)$ is the shift potential ($u = \nabla\Phi$) related to the velocity potential φ (r,t) by

$$\Phi(\vec{r},t) = \int_0^t \varphi(\vec{r},t')\,dt'$$

We choose m(t) in the form typical of the thermal mechanism of elastic wave generation by a pulsed beam of charged particles

$$m(t) = \frac{m_0}{\tau_{in}}\,t\left[\theta(t)-\theta(t-\tau_{in})\right] + \frac{m_0}{\tau_{dec}}\left[\theta(t-\tau_{in})-\theta(t-T)\right](T-t) \qquad (2)$$

---

1) Kharkov Institute of Physics and Technology, Ukrainian SSR Academy of Sciences, 310108 Kharkov, USSR

where τ_{in} is the time of the pulse rise from zero to m_0, τ_{dec} is the pulse decay time, $T = \tau_{in} + \tau_{dec}$.

The form of the directivity diagram depends on the parameter $\gamma = R_0/cT$. If $\gamma \ll 1$, then the emission occurs similarly to that of the point source case. If $\gamma \gg 1$, then the directivity diagram has the form of a lobe directed along the z axis. In this case, the angle α, in which the main energy portion is emitted, may be estimated by the formula:

$$\alpha(\gamma) \approx 2 \arcsin(1/2\pi\gamma) \tag{3}$$

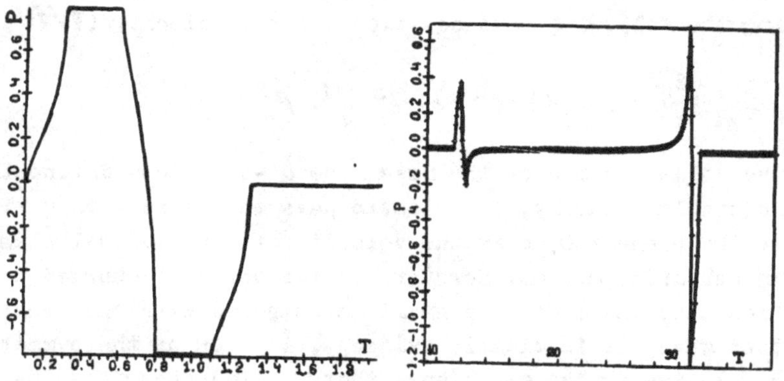

a) b)

Fig. 1. Pulse P(t) shapes in liquid. a: $\gamma \ll 1$ or $\gamma \gg 1$, if $r/z \ll 1$. b: $\gamma \gg 1$, $r/z \gg 1$.

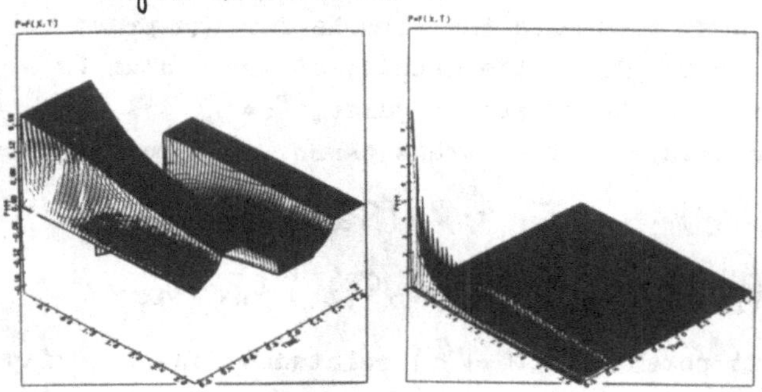

a) b)

Fig. 2. Pulse P(t) shapes in liquid for fixed z. a: $\gamma \ll 1$, b: $\gamma \gg 1$.

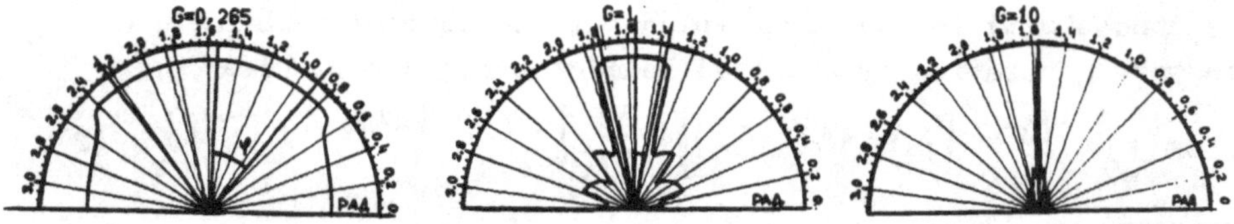

a) b) c)

Fig. 3. Directivity diagrams. a: $\gamma \ll 1$, $R=10^2 R_0$, b: $\gamma \approx 1$, $R=50 R_0$, c: $\gamma \gg 1$, $R=5 R_0$.

NUMERICAL CALCULATION OF THERMAL PROCESSES IN LASER PROCESSING OF SEMICONDUCTOR MATERIALS

Bobitski Ya.V.[1], Gotra Z.Yu.[1], Demkovich I.V.[2], Semerak M.M.[2]

In this paper the thermal processes in the semiconductor materials under the pulsed laser processing are analyzed by the numerical methods Particular attention is paid to consideration of dependence of the optical and thermophysical semiconductor characteristics upon the temperature and also to their possible melting with further crystallisation. Various numerical calculation schemes are considered. The used explicit differential schemes to solve the set tasks are rather simple but require the short-time itteration resulting in long-time machine calculations(4). Proceeding from this we developed a program for the numerical solution of the quasilinear equation of heat conductance by an inexplicit scheme (3) permitting greatly to reduce calculation time.

Whem dimentions of an area to be exposed are significantly larger than the effective depth of heat penetration in a sample, then

$$d \gg (2\lambda\tau/c\rho)^{1/2} , \tag{1}$$

where d - diameter of an area to be exposed; τ - time interval to be analyzed; λ, ρ, c - heat conductivity factor, density and heat capacity of material, respectively. Supposing the uniform distribution of the laser pulse power over a cross-section, heating process during the laser processing is described by the one-dimensional equation of heat conductivity

$$c(t,x)\rho(t,x)\frac{\partial t}{\partial \tau} = q(x\tau) + \frac{\partial}{\partial x}\left(\lambda(t,x)\frac{\partial t}{\partial x}\right), \tag{2}$$

provided the

$$t(x,0)=t_0, \quad t(l,\tau)=t_0, \qquad -\frac{\partial t(0,\tau)}{\partial x}=0, \tag{3}$$

where t - temperature; x - distance from the surface to the sample depth; $q(x,\tau)$ - density of heat sources; t_0 - initial temperature; l - sample thickness.

The heat source in the equation (2) is expressed as

$$q(x,\tau) = \alpha(x)(1-R)J(\tau)\exp\left[-\int_0^x \alpha(\xi)d\xi\right] , \tag{4}$$

where R - reflectivity and ξ - interration variable. Pulse waweform

1) Lvov Polytechnical Institute

2) Institute for Applied Problems of Physics and Mechanics under the Ukr.SSR Academy of Sciences, Lvov

was assumed to be triangular

$$\mathcal{J}(\tau) = \mathcal{J}_{max} \begin{cases} \tau \cdot \tau_{max}, & 0 \leq \tau \leq \tau_{max} \\ (\tau_u - \tau)(\tau_u - \tau_{max}), & \tau_{max} < \tau \leq \tau_u \\ 0, & \tau > \tau_u \end{cases} , \quad (5)$$

where τ_u - pulse length; τ_{max} - time interval when power reaches the maximum value, and maximum density of the laser pulse power is related with energy density by ratio $\mathcal{J}_{max} = E \cdot 2\tau^{-1}$.

Using the integration-interpolation method we describe the law of heat conservation over the section $x_{i-1/2} \leq x \leq x_i + 1/2$ for time $\Delta\tau = \tau_{j+1} - \tau_j$

Numerical calculation of thermal action of the ruby laser radiation pulse of τ_u = 100 ns (τ_{max} = 50 ns) on monocrystal germanium sample at t_o = 300 K serves as an example for the specific use of the method described above. The sample was aussumed to be homogeneous, i.e. relationship between the sample properties and depth was determined only by their temperature dependences at different temperatures for the layers and their phase states.

Temperature dependences in the surface layer of 50 nm thickness on time for a laser pulse with energy density E = 0,5 J/cm^2 are given in Fig.1. Curve 1 has been obtained by solving the equation (2) using the inexplicit scheme (Δx = 50 nm, $\Delta\tau$ = 2 ns) by the finite differences method with independent characteristics of material and without accounting for the phase transitions. Values of the material parameters at t_o = 300 K have

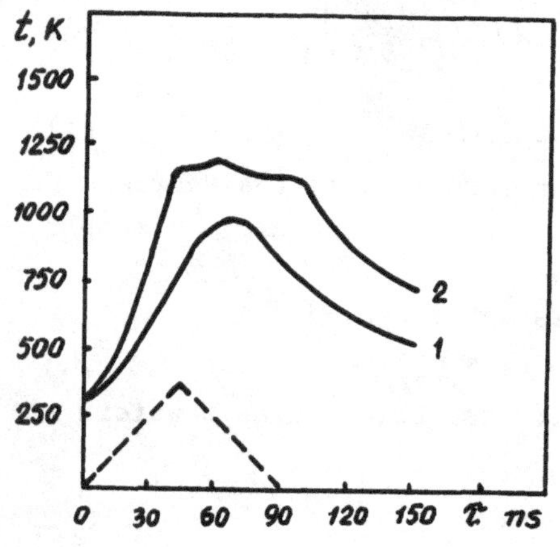

Fig.1

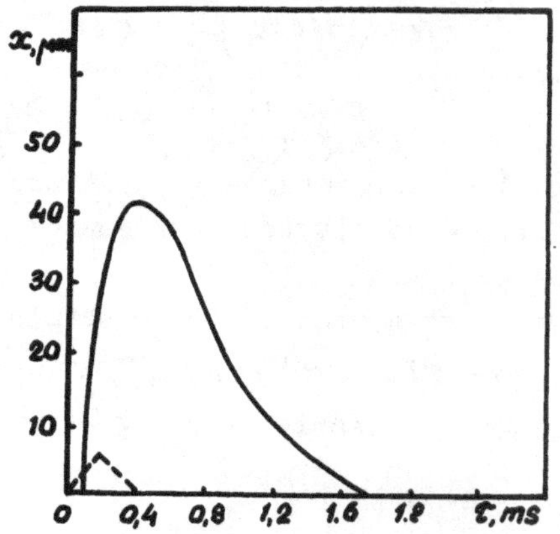

Fig.2

been used. The numerical calculation of the analytical equation (2) solution with similar conditions gives the results coincident with curve Curve 2 represents the calculation results for the same energy density with accounting for phase changes and temperature dependences of the material properties. Calculation according to the explicit scheme (Δx = 50 nm, $\Delta \tau$ = 0,1 ns) gives the same results. Decrease of the heat conductivity and increase in the absorption coefficient with the temperature rised lead to accumulation of heat in the sample surface layer resulting in its melting. Threshold power dencity of the pulse required to melt the germanium surface equals 0,23 J/cm^2 which coincides with an experimental value of 0,2 J/cm^2 obtained with the shorter pulse (τ_u = 66 ns)(1) Results of the numerical calculation allow to determine the recrystallization rate of the melted material which is reduced from 350 cm/s at E = 0,5 J/cm^2 to 210 cm/s at E = 1,5 J/cm^2.

Calculation results with the other pulse lengts and initial temperatures showed that the recrystallization rate insignificantly depends upon the pulse length and decreases significantly with the initial temperature rise.

Calculations of the thermal effect on a silicon monocrystal exposed by the nonunimodel laser pulse having a length of τ_u = 0,4 μs (τ_{max} = 0,2 ms) at t_0 = 300 K also have been made. There have been used the following temperature dependences within the temperature range from t_0 = 300 K to t_M = 1700 K (2).

Fig.2 shows the phase boundaries position depending upon the time for the pulse with power density of 200 J/cm^2. The maximum melting depth amounts to 42 μm, recrystallization rate equals 3,5 cm/s. The maximum temperature of material surface reaches 2600 K while calculation using the material parameters at t_0 = 300 K and without their temperature variations gives only the 1100 K value. The phase change parameters adequately coincide with evaluations of the experimental study for impurity distribution in silicon.

References

(1) Auston D.H., Golovchenko I.A., Simons A.L. and others: APPl.Phys. Let.54 (1979) 777-781

(2) Meyer, I.R., M.R.Kruer and F.G.Bartoli: J.Appl.Phys. 51 (1980) 5513-5518.

(3) Samarski, A.A.: Theory of differential schemes, Nauka, Moskow 1977

(4) Semerak, M.M., Z.Gotra, I.Demkovich, S.Osered'ko, Ya.Bobitski: Engineering Physical Journal, 2 (1984) 326

A MODEL OF DOPAND REDISTRIBUTION UNDER LASER RECRYSTALLIZATION

Bobitski Ya.V., Gotra Z.Yu., Fan Tam, Demkovich I.V.[1)]

In this paper dopant redistribution has been modelled taking into account physical processes on the atomic level. The new dopant redistribution model is based on a thermodynamic description of the laser liquid phase recrystallization in combination with an atomic and kinetic description of processes on the phase interface.

The metastable adsorbed phase is formed in the condicions of fast melt cooling, limited atom mobility, short solidification time. It then changes into the stable solid phase by the formation and subsequent growth of the latter's centres, which are two-dimentional in the case of a layer by layer crystal growth mechanism. The extent of the metastable phase then is a value of the order of magnitude of the atomic layer.

The adsorbed atoms (adatoms) in the metastable phase can migrate over the growth surface colliding with others and under favourable conditions they can form clusters of adatoms - two-dimensional growth nuclei. The latter are more stable than separate adatoms as regards the inverse transition into the liquid phase, because they are bound with each other by interaction forces characteristic of nucleation energy. As soon as they reach a certain critical size, they no longer disintegrate adatoms but grow on trapping the surrounding adatoms within a neighbourhood $r_\alpha = (D_\alpha \tau_\alpha)^{1/2}$ in radius, where D_α - is the adatoms surface diffusion coefficient, τ_α - is the adatoms average lifetime. These neighbourhoods will be trapping zones. As the time τ_α passes, the adatoms out of trapping zones get desorbed, i.e. change into liquid phase. The inverse transition into metastable phase can be accomplished only after recrystallization of the absorbed layer considered is over. This is connected with a significant disbalance of laser recrystallization, wheh the fluctuations of the thermodynamical parameters of the balanced stable liquid phase near the metastable phase interface are insufficient for spontaneously transferrihg it into metastable phase.

The degree of filling up a growth surface with critical nuclei and their trapping zones is approximately equal to

$$F(t) = N(t) D_\alpha \tau_\alpha$$

Lvov Polytechnical Institute, Lvov, USSR
1)Institute for Applied Problems of Physics and Mechanics under the Ukr.SSR Academy of Sciences, Lvov

where N(t) - is the surface concentration of critical crystal nuclei — at moment t from the beginning of recrystallization of the subsequent atomic layer considered.

It is clear that trapping zones are depleted of adatoms, that is why further nucleation can take place only on that area of the growth surface, which is free of them. Hence the following nucleation differencial equation can be put down

$$\frac{dN}{dt} = J[1 - F(t)] ,$$

where J - is the stationary frequency of forming crystal nuclei of a critical size.

Upon solving it one can obtain

$$F(t) = 1 - \exp(-JD_\alpha \tau_\alpha t) ,$$

$$N(t) = \frac{1}{D_\alpha \tau_\alpha}[1 - \exp(-JD_\alpha \tau_\alpha t)] .$$

According to the presented model of laser recrystallization and segregation a new approach to calculating dopant redistribution profiles has been suggested. It is based on solving a onedimentional eguation of dopant diffusion in the liquid phase during the time interval of the ordered formation of an atomic layer consisting of two subintervals $[0, \tau_\alpha^-], [\tau_\alpha^+, \tau]$. Having the same boundary conditions and differing in initial conditions. It is considered that phase interfaces are fixed they move stepwise at a distance of the atomic layer thickness at the moment of its complete formation.

Dopant atoms in subintervals $[0, \tau_\alpha^-], [\tau_\alpha^+, \tau]$ do not cross the interface between the metastable phase and the liquid one. Subsequently the dopant concentration gradient at that interface, is equal to zero. At the moment τ_α in the diffusive dopant distribution there appears a segregational component of the distribution of the desorbed dopant atoms. It has been suggested that while moving stepwisely in the direction of the liquid phase over a random distance x of the order of magnitude a_0 the desorbed dopand atoms are distributed according to the normal probability distribution of a random quantity x with an expectation a_0 and a standard deviation σ, equal to the amplitude of the thermal oscillation of atoms at the crystallization temperature (Fig.1)

$$C_{des} = \frac{C_0}{\sigma\sqrt{2\pi}} \exp\left(\frac{x - a_0}{\sigma}\right)^2 .$$

The constant C_0 is determined from the corresponding normalization condition. Let us put down a set of equations of the initial and boundary conditions, that describe dopant distribution during a time interval $[(n-1)\tau, n\tau]$, when the phase interfaces are situated as on Fig.1.

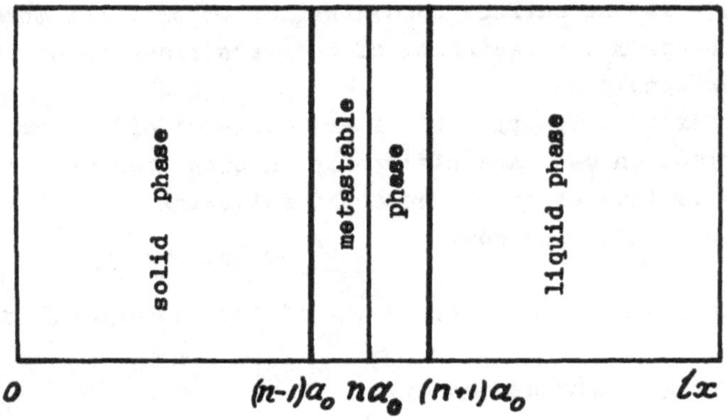

Fig. 1

Then $(n-1)\tau \leq t \leq (n-1)\tau + \tau_\alpha^-$, $na_0 \leq x \leq l$, one obtains

$$D_{\alpha l}\frac{\partial^2 c}{\partial x^2} = \frac{\partial c}{\partial t}; \quad \frac{\partial c}{\partial x}\Big|_{x=na_0} = 0; \quad D_{\alpha l}\frac{\partial c}{\partial x}\Big|_{x=l} = hc(l,t);$$

where $D_{\alpha l}$ - is the dopant diffusion coefficient in the liquid phase; h - is the rate of dopant evapouration from the sample surface; l - is the molten layer thickness; $f(x)$- is the dopant distribution in the liquid phase after the recrystallization of the (n-1)-th atomic layer.

The constant C_0 is determined from the normalization condition

$$\frac{C_0}{\sigma\sqrt{2\pi}}\int_{na_0}^{\infty} exp\left[\frac{x+(n+1)\,a_0}{\sigma}\right]dx = (1-h)\,c\left(na_0,(n-1)\tau + \tau_\alpha^-\right)$$

The approach suggested is seen to differ from the others in that the segregation is intensified in the inicial condition of the diffusion equation.

On the whole, formulating the mathematical problem of dopand redistribution taking into account its segregation at the phase interface makes it possible to analitically solve the diffusion equation in particular by the Green function technique as is (1).

References

(1) Tam, H.S., C.K. Ong: J.Phys.Solidi Status, 14 (1983) 2799-2807

S. Wagner, Th. Otto, F. Neubert (Technische Universität
Karl-Marx-Stadt, GDR)

MIS-Transistors in thin polycrystalline silicon and applications in liquid crystal displays and three dimensional integrated circuits

MIS-transistors in thin silicon films provide numerous advantages related to application parameters as for instance velocity, packing density, radiation hardness and parallel data processing. If quasi-monocrystalline silicon layers produced with some kind of recrystallization are used the main determining material parameters approximately reach to the values of bulk disks, with the minority life time as an exception. During the recrystallization process however danger exists to deteriorate underlying structures or substrates by the applied high temperatures. Beyond this there are restrictions relating to the available chip area. As an alternate approach polysilicon thin film devices passivated with hydrogenium offer possibilities to construct integrated circuits with electronic elements not as good as possible but as good as necessary. Three dimensional integrated circuits with some active semiconducting layers interconnected by numerous contacts can be realized in principle. But care must be taken to avoid high temperatures in any step of the technological process. Three dimensional modular circuits offer new possibilities in data processing principles.

Investigations have been carried out for analysing specific electrically efficacious parameters of polysilicon grain boundaries, surface states and radiation induced defects in the semiconductor-insulator system after hydrogenium implantation.

Polysilicon layers enclosed in insulating layers on both sides have been investigated regarding concentration and energetical dependance of grain boundary traps and traps at the semiconductor-insulator interface using quasistatic capacity-voltage measurements. For large grain material this resulted in approximately $10^{11}/cm^2$ (eV) rechargable surface states in the middle of the forbidden gap and rapidly growing densities near the conduction and valence band, up to $10^{14}/cm^2$ (eV) for the interface. After implantation of hydrogenium (100 keV, 10^{16} cm^{-2}) traps near the conduction band were diminuished and at the middle of the gap enhanced to approximately $10^{12}/cm^2$ (eV). So the sum of all rechargable states and the interface charge became about five times lower.

It is difficult to distinct how much surface states at the interface and at the grain boundaries participate on the effective trap density measured with the capacity-voltage methode but in combination with measurements of current characteristics in the subthreshold region of MIS field effect transistors this becomes possible in principle.

In investigations with a large area insulated gate-electrode of variable voltage effecting the polysilicon layer the grain boundary states may be regarded to be of uniform density all over the grain volume with an effective value defined by grain boundary state density divided by the grain diameter. In this model no potential barriers exist at grain boundaries which are situated perpendicular to the semiconductor-insulator interface. The grain boundary states are fixing an important part of the majority carriers generated by the dopand atoms making them unmoveable. If the effective density of majority-traps lying above the fermi level is larger than the dopand concentration of the grain it is completely depleted. The layer seems to be undoped and the resistivity is high. One can define an effective majority density like the difference of dopand concentration and effective grain boundary state density. If the dopand concentration is dominating the resistivity falls down and the Fermi level is positioned near the limit of the band gap corresponding to the effective majority-density.

In a more detailed view the existence of potential barriers have to be regarded. While in the upper model of effective majority concentrations the charging degree of the traps depends on the position of the effective Fermi level, i.e. on the effective majority deosity in a logarithmic manner, in reality the degree of charging is smaller and the voltage deviation at the grain boundaries is larger. Therefore the energy dependance of the grain boundary states plays a more important role. Midgab traps eralier change over the Fermi level and loose their electrical charge. States of inversal type are charged at the grain boundaries, with opposite sign. States at the inverse bound of the gap get more importance. Experimental results are in agreement with this interpretation.

MIS-field effect transistors have been investigated experimentally. Test structures have been prepared and their electrical characteristics before and after implantation of hydrogenium have been compared. The dependance of the drain current on the gate and drain voltages exhibited drastic differences to analogue characteristics in single-crystalline silicon. These differences can be explained by the recharching process of the traps in qualitative and quantitative order.

With growing gate voltage and constant drain voltage the inversion channel begins to builed up at the source junction if the threshold voltage is reached. This threshold voltage corresponds to the effective majority concentration, i.e. the doping concentration diminuished by the equivalent grain boundary majority trap density outside the channel. But grain boundaries and their potential barriers touching the channel or lying inside the channel are recharged by inversion charges comming from the source region. The potential barriers are compensated and traps charged with majority carriers are neutralized. The inverse traps are charged with opposite sign and a new opposite potential barrier grows up, lowering the effective mobility of the inversion carriers. The charge fixed in the inverse traps play a contrary role to the free movable carriers in the inversion channel by occupying a part of the dielectric flux of the gate electrode and weakening the charge density of the channel. The peaks of the potential barriers inside the channel are slowly further growing with gate voltage but not in such a drastic manner as when entering the channel. Finaly the channel reaches the drain junction if gate voltage is further enlarged and all grain boundaries inside the channel have been recharged. This is the transition to the active transistor regime with an other, i.e. higher treshold voltage corresponding to the sum of the densities of dopands and equivalent inversaly charged grain boundary states. The threshold voltage difference follows from the square roots of the dopand concentrations diminuished by the majority traps or enlarged by the minority traps, respectively. If gate voltage is kept constant the threshold voltage also is a fixed value.

Implantation of hydrogenium caused radiation damages in the gate oxide layer, resulting in additional positive interface charges and altering threshold voltages by minus 1.5 Volts.

Transistors were examined to suitability for controling pixels on large screens of liquid crystal displays. Each cell has a poly-silicon transistor or inverter. The transistors exhibite better quality parameters than with hydrogenated amorphous silicon as regards the on/off property of channel current or the electron mobility or the long time stability. In polysilicon complete integrated on-glass substrate realization is possible of all peripherical circuits for controling signals and voltages of lines and columns for more than 10^5 pixels and conventional television picture frequency. Gate delay times of 10 ns have been measured.

Chips with many stacked polysilicon layers with active devices and circuits, separated by thin insulating layers and interconnected with numerous contacts offer new fields of still more sofisticated data handling. Examples can be demonstrated for parallel data processing, picture recognition, damage tolerant memories and realization of circuits similar to neuronal networks.

A MIXED BALLISTIC AND THERMODYNAMIC DESCRIPTION OF ION-BEAM MIXING

ROGER KELLY[1)]

Bombardment-induced mixing has been traditionally regarded as a wholly ballistic process. It is therefore of considerable interest that Cheng et al. have established, in studies of 20 different systems, that for ΔH_m < 0 the extent of mixing scales as $(\Delta H_m \phi/E_c)^{1/2}$, where ΔH_m is the heat of mixing, ϕ is the fluence (dose), and E_c is the cohesive energy. Likewise Peiner and Kopitzki established that for $\Delta H_m > 0$ the extent of bombardment-induced solubility decreased with ΔH_m. We explore mixing in terms of the extended diffusion equation $\partial \alpha_i/\partial \phi = - v\partial \alpha_i/\partial x + D\partial^2 \alpha_i/\partial x^2$, where α is atom fraction, D is the diffusion coefficient (here treated as purely ballistic), and v is the drift velocity (here treated as purely thermodynamic). This equation is easily solved for a mixing geometry provided v is approximated as concentration-independent. It is shown that, for $\Delta H_m < 0$, mixing is complete and the extent scales as $\Delta H_m \phi/E_c$. For $\Delta H_m > 0$ demixing occurs in the sense that the mixing approaches a steady state which depends on the value of ΔH_m. The results of both Cheng et al. and of Peiner and Kopitzki are thus understandable in a general sense even if our arguments fail to be exact.

1. INTRODUCTION

We have recently considered the role of chemical driving forces in particle-surface interactions [1-4]. Situations where this is important include the following bombardment-induced phenomena: sputtering in general [1], segregation [2], mixing [3], demixing [5], precipitate growth, and decomposition [3]. To this group of closely related bombardment-induced effects we add one more: the sputtering of condensed gases and other systems where the system is able to reach the thermodynamic critical tremperature (T_{tc}) during the thermal-spike phase of each ion impact [4].

1.1 Collisional sputtering in general

The role of chemical driving forces in sputtering is well-known. Provided the mechanism involves binary collisions rather than thermal spikes, it consists in the fact that the surface binding energy, U, enters inversely [6] in the yield, Y:

$$Y \propto 1/U. \tag{1}$$

The chemical driving force seen in eq. (1) is fairly strong, as U is typically of order 2 to 9 eV (table 1). For a discussion of U for binary alloys and for oxides we refer to [1].

1.2 Bombardment-induced segregation ("BIS"), mixing, demixing, and precipitate growth

With BIS [2], what is unexpected it that the effect can occur at all given the small numerical value of the driving force, namely the heat of segregation, ΔH^{seg} = 0.06-0.52 eV (table 1). Likewise mixing [3] is now recognized not to be totally ballistic as historically assumed [e.g. 7] but to be in part chemically driven by the heat of mixing, ΔH_m [8-11], typically with absolute values of 1.3 eV or less (table 1). Demixing and precipitate growth were not considered in [2,3], in part because very little was known about them. It is therefore important that demixing was explicitly described by Averback et al. [5], while there is a recent study by Marton et al. [12] in which demixing can be inferred to have occurred even if the authors opted for a different interpretation. A further example that relates closely to demixing is the observation [13] that multilayer systems with $\Delta H_m > 0$ showed incomplete mixing and, moreover, that the effect varied systematically (fig. 1 [13]). Concerning precipitate growth there is an unusual study by Johnson and Moulton [14] in which N implantation into Nb at 5 K gave good crystalline δ-NbN in spite of the limited mobility that might be assumed for such a temperature.

The problem with understanding these effects is that the chemical (i.e. thermodynamic) driving force is unusually small, much smaller than the U seen in eq. (1). Bearing this in mind, it was proposed in [3] that "chemical guidance" as in fig. 2 occurs: each basically ballistic relocation is envisaged as being terminated by a chemical step which could range from a relaxation of the disordered lattice (seen in simulation [15]), to a formal interstitial or interstitialcy diffusion [16,17], and even to vacancy diffusion [5]. For example, it was shown by simulation that the interstitial population at the end of the collisional phase was much higher than the terminal interstitial population [18].

1.3 Bombardment-induced decomposition

We have used the term "decomposition" to refer to the characteristic compositional changes found with oxides and oxysalts [3]. These were historically identified with electron diffraction [e.g. 19], but are now commonly studied with X-ray photoelectron spectroscopy (XPS). The latter presents the possibility of confirming valence states which do not

[1)]IBM Research Division, T.J. Watson Research Center, Yorktown Heights, NY 10598, U.S.A.

TABLE 1
Examples of chemical energy differences which can act as driving forces for changes in bombarded solids. The heat of vaporization, ΔH^{vap}, is similar to the surface binding energy, U, which guides collisional sputtering (eq. (1)). It is also equivalent, provided the system vaporizes as atoms, to the cohesive energy, E_c. The heat of segregation, ΔH^{seg}, leads to BIS, while the heat of mixing, ΔH_m, contributes to bombardment-induced mixing and demixing. The various heats, with ΔH_m applying to systems with 1:1 proportions, are taken mostly from [2]. The entries AES, FIM, ISS, and XPS are the usual acronyms referring to the methods of surface analysis.

System	ΔH^{vap} (eV) (similar to U or E_c)	ΔH^{seg} (eV) (experimental from Arrhenius plots)	ΔH^{seg} (eV) (experimental from individual data points)	ΔH_m (eV)
Ag-Au	2.94; 3.82		0.04-0.07 ISS	-0.048
Ag-Ni	2.94; 4.46	...	...	+0.23 [35]
Ag-Pd	2.94; 3.90	0.09-0.13 AES	0.02-0.04; 0.098 AES	-0.052
Al-Fe	3.41; 4.31	...	...	-0.26
Au-Cu	3.82; 3.49	0.13 AES; 0.13 ISS 0.04-0.10 ISS	0.03-0.06 AES;	-0.053
Au-Ni	3.82; 4.46	0.52 AES; 0.45 ISS	~0.4 AES[(a)]	+0.078
Au-Pd	3.82; 3.90	... 0.09-0.11 ISS	0.05-0.13 AES;	-0.081
Bi-Cu	2.17; 3.49	...	...	+0.28 [35]
Cr-Mo	4.12; 6.82	...	~0.02 AES[(b)]	+0.075
Cu-Nb	3.49; 7.48	...	...	(positive) [35]
Cu-Ni	3.49; 4.46	0.42 ISS	0.21 FIM; 0.2-0.3 ISS	+0.018
Cu-Pd	3.49; 3.90	0.059 AES	0.02-0.07 AES	-0.111
Cu-Pt	3.49; 5.85	... 0.09-0.2 ISS	0.1-0.2 AES, XPS;	-0.115
Mg-Al	1.52; 3.41	...	0.1 AES	-0.034
Mo-W	6.82; 8.80	...	~0.4 AES	+0.021
Ni-Co	4.46; 4.44	0.069; 0.18 AES	0.04-0.09 AES	0.000
Ni-Mo	4.46; 6.82	...		-0.011
Ni-Pt	4.46; 5.85	0.15; 0.25 AES; 0.11-0.24 ISS		-0.096
Pd-Ni	3.90; 4.46	0.31 AES	0.06 AES	-0.006
Pt-Ti	5.85; 4.87	...		-1.26 [35]

(a) This information is all for dilute Au. For concentrated Au, ΔH^{seg} is much lower, ~0.07.
(b) The information on the surface composition is not fully self-consistent.

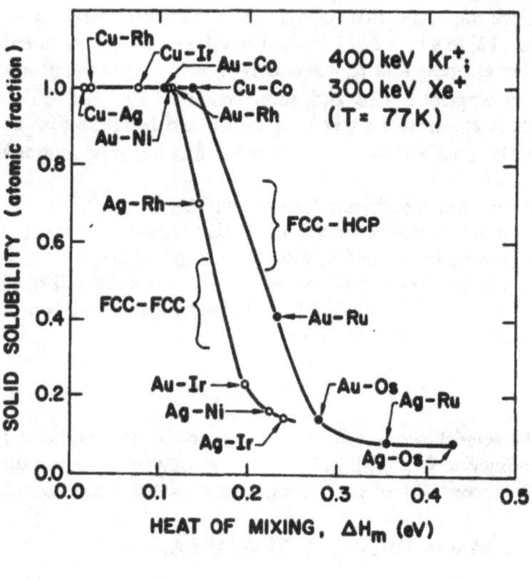

Fig. 1. Solid solubilities obtained by bombarding multilayer targets with 400 keV Kr^+ or 300 keV Xe^+ at 77 K. All systems have positive ΔH_m, the values used for the figure being those of Miedema [35]. The results are mainly due to Peiner and Kopitzki [13], but Cu-Ag, Au-Co, Au-Ni, Cu-Co, and Ag-Ni are due to Mayer et al. [36]. We interpret the existence of solubility less than complete as due to the steady state that occurs in mixing when ΔH_m obeys $\Delta H_m > 0$. See eq. (7).

give stable bulk compounds or well-defined diffraction patterns. The typical result here is that decomposition tends to occur only if the product involves an energy increase of less than about 0.7 eV per atom [3].

The model proposed to explain decomposition [3] differs from the "chemical guidance" discussed above. In chemical guidance the system responds so as lower the potential energy. Decomposition was envisaged in terms similar to those proposed to describe bombardment-induced amorphization [20]. The idea was that defect accumulation causes the potential energy of a system to increase until the amorphous state is favored with respect to the defective crystalline state.

1.4 Condensed-gas sputtering

The sputtering of condensed gases and related systems has been long known to involve exceedingly high yields and to give energy spectra suggesting particle energies much lower than what is necessary to surmount the surface binding energy. It has been recently proposed [4,21] that the important process with such sputtering is not collisional but based on gas flow. It is envisaged that each ion impact leads to an atomically small region of target subject to the usual sequence of first a collision cascade and then an elevated temperature ("thermal spike"). If the latter obeys $T > T_{tc}$, where T_{tc} is the thermodynamic critical temperature, then the material should flow like a gas, undergoing a one-dimensional unsteady adiabatic expansion [4]. The final result is that yields should be much higher than those for collisional sputtering and the energy spectra will tend to correlate with T_{tc} rather than ΔH^{vap}.

What we intend to do in the following is to treat mainly mixing and demixing, while recognizing that the mathematical formalism, although it fails to describe the act of segregation, does at least describe the development of the profile characteristic of BIS.

2. A TIME-DEPENDENT SOLUTION OF THE MIXING PROBLEM

Bombardment-induced mixing was traditionally [e.g. 7] regarded as ballistic, a stand that is at first sight reasonable given the numerical smallness of ΔH_m (table 1). Nevertheless, there was a discrepancy in that there was more mixing than could be explained in purely ballistic terms. This led to a model combining ballistic motion with interstitial diffusion [16,17]. This model did not, let it be noted, contain a chemical driving force.

A particularly important step in the understanding of mixing was therefore the demonstration by Cheng et al. [8-11] of the existence of chemical effects, namely a correlation with ΔH_m. As seen in fig. 3 [11] it takes the form of an approximate linearity between the quadratic mixing rate, $d(4Dt)/d\phi$, and the ratio $\Delta H_m/E_c$. This correlation leads to the conclusion that the purely ballistic effect, seen for $\Delta H_m = 0$, is less important by a factor of about 3 than the purely chemical effect. Cheng et al. [8-11] therefore argued that mixing takes place in an intermediate phase of the cascade when the average particle energy simulates a temperature of 1-2 eV. Mixing was thus identified with a thermal-spike type of process like that proposed for thermal sputtering [e.g. 22,23] before the alternative gas-flow model [4,21] was developed.

The main problem with this picture is, in our opinion, the assumption that thermal spikes play a role. For example there is the general objection that the smallest ΔH_m values seen in table 1 and fig. 3 (~0.05 eV) violate the condition

$$\text{(driving force)} \gtrsim kT$$

by more than an order of magnitude. This problem would arise also if BIS were attributed to a thermal spike.

It is therefore appropriate to propose an alternative in which chemical energy differences could be important to mixing without being tied to the concept of a thermal spike, a point of view apparently first taken by Averback et al. [5]. The traditional view is accepted that mixing is <u>initiated</u> by a ballistic event [17,24,25]. Consider a ballistic mixing process expressed in the "diffusion limit" [25]. Then, if ϕ is the incident particle fluence (dose), z is the relocation distance, and $d\sigma_i(z)$ is the relocation cross section of component i, we have

$$\partial\alpha_i/\partial\phi = -\frac{\partial\alpha_i}{\partial x}\int_{-\infty}^{\infty} d\sigma_i(z)\,z + \frac{\partial^2\alpha_i}{2\partial x^2}\int_{-\infty}^{\infty} d\sigma_i(z)\,z^2 \equiv -v_i\frac{\partial\alpha_i}{\partial x} + D_i\frac{\partial^2\alpha_i}{\partial x^2}, \qquad (2)$$

Here, in our usual notation, α_i is the atom fraction of component i, v_i is the drift velocity (thenceforth written v), D_i is the bombardment-induced diffusion coefficient (thenceforth written D), and additional terms to conserve sites [25] have been omitted. Consider fig. 2. We identify D with the ballistic part of each relocation and note that, if a given atom is displaced n times in a bombardment and each time executes random ballistic motion with average length $m_i\lambda$, then the total mean square distance moved along a given axis is

$$\langle z^2\rangle = nm_i^2\lambda^2\,\frac{\int_0^{\pi}\cos^2\theta\,\sin\theta\,d\theta}{\int_0^{\pi}\sin\theta\,d\theta} = (1/3)nm_i^2\lambda^2. \qquad (3a)$$

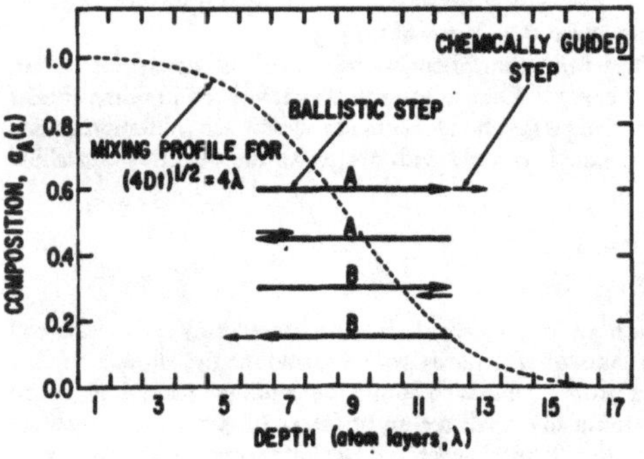

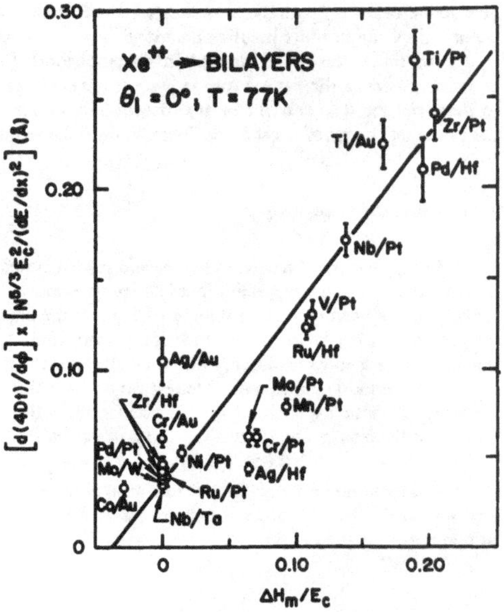

Fig. 2. Sketch of how chemical guidance could lead to bombardment-induced mixing. It is envisaged that the ballistic trajectories tend to be followed by chemically guided steps. The steps are here assumed to involve interstitials, i.e. are located at the end of the ballistic relocations, but vacancy motion is also possible [5]. The ballistic events are random and therefore lead to a diffusion coefficient, D, whereas the guided steps are directed and lead to a drift velocity, v. For a sufficiently large number of jumps per atom, n, or for a large enough driving force, m_2, drift (eq. (3b)) will always dominate diffusion (eq. (3a)). Also shown is the mixing profile given by $(1/2)\mathrm{erfc}[(x - x_o)/(4Dt)^{1/2}]$, with $x_o = 9\lambda$ and $(4Dt)^{1/2} = 4\lambda$. This relation is equivalent to eq. (6a) with $v = 0$.

Fig. 3. Correlation suggesting that the extent of ion-beam mixing of metallic bilayers using 600 keV Xe^{++} at 77 K scales as $(\Delta H_m \phi/E_c)^{1/2}$ as in eq. (12b). ΔH_m is the heat of mixing, E_c is the cohesive energy for an equimolar system, $d(4Dt)/d\phi$ is the experimentally observed quadratic mixing rate, ϕ is the fluence (dose), and N is the average number density (cm^{-3}) of the target. Due to Cheng et al. [11].

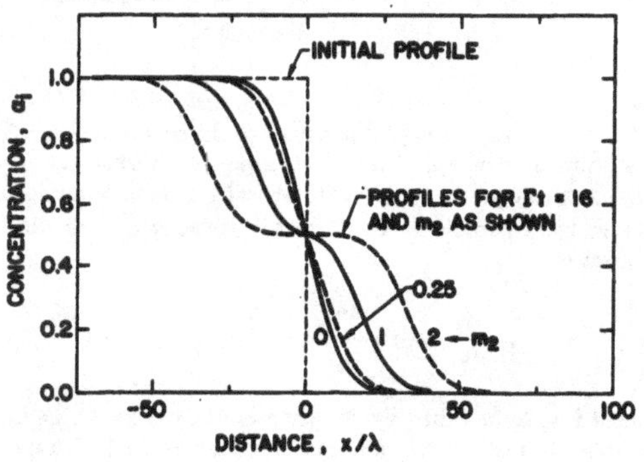

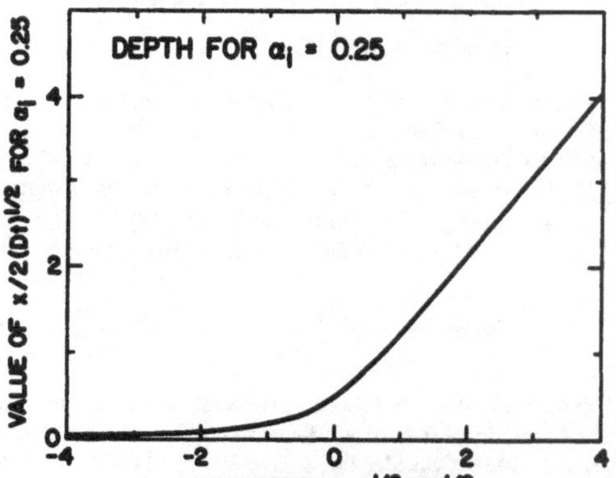

Fig. 4. Mixing profiles appropriate to a bilayer system obtained with eq. (6a) for $\Gamma t = 16$, $m_1 = 4$, and values of m_2 from 0 to 2. As m_2 increases there is a marked increase in the extent of mixing. Note that the drift velocity is related to m_2 by eq. (8a), namely $v = \Gamma m_2 \lambda$, and also that $m_2\lambda$ is the average length of the chemically guided steps which are assumed to follow each random ballistic event. The relation between the two types of steps is seen also in fig. 2.

Fig. 5. For $a_i = 1/4$ and for designated values of the parameter $vt^{1/2}/2D^{1/2}$, eq. (6a) yields values of $x/2(Dt)^{1/2}$. The result, as plotted here, effectively shows the mixing rate as a function of the drift velocity and agrees in spirit, eq. (12a) (even if not in detail, eq. (12b)) with the experimental results of fig. 3.

Here λ is the mean atomic spacing, typically 0.24-0.26 nm. Likewise, we identify v with low-energy terminal steps involving interstitials (as in fig. 2) or vacancies [5] which have an average length $m_2\lambda$. They lead to a "v" and not a "D" because they tend to have a unique direction governed by the chemical driving force. Because of this "v" the same atom executes an additional motion the mean square value of which along a given axis is

$$<z^2> = (nm_2\lambda)^2. \quad (3b)$$

It follows that a minor chemical perturbation, i.e. what we call chemically guided steps, of a basically ballistic process could dominate as far as mixing is concerned. The condition for this to be so is just eq. (3b) $\gg$ eq. (3a), an inequality which should be always valid if n or m_2 is large enough.

As far as the problems at hand are concerned, we would note that v takes on the following values:

(1) $v > 0$ for mixing. Here v is a real thermodynamic drift velocity.

(2) $v < 0$ for demixing. Here v is again thermodynamic.

(3) $v < 0$ for describing the profile characteristic of BIS. Here v describes surface recession rather than a thermodynamic effect.

Eq. (2) is readily expressed as a Laplace transform:

$$s\tilde{\alpha}_i - \alpha_i(t = +0) = -v\frac{d\tilde{\alpha}_i}{dx} + D\frac{d^2\tilde{\alpha}_i}{dx^2}, \quad (4)$$

where the system is assumed to be binary with $\alpha_i + \alpha_j = 1$, $\tilde{\alpha}_i$ is the transform of α_i, s is the transform variable, and $\alpha_i(t = +0)$ is zero. Provided v can be taken as concentration-independent, the solution of eq. (4) is very simple:

$$\tilde{\alpha}_i = \frac{1}{2s}\exp(-hx), \quad (5a)$$

where $x = 0$ is the position of the interface, $\alpha_i = \frac{1}{2}$ for $x = 0$ (cf. fig. 2), and h stands for

$$h = -\frac{v}{2D} + \left(\frac{v^2}{4D^2} + \frac{s}{D}\right)^{1/2} \quad (5b)$$

Eqs. (5) can be inverted explicitly [26]:

$$\alpha_i = \frac{1}{4}\operatorname{erfc}\left(\frac{x}{2(Dt)^{1/2}} - \frac{vt^{1/2}}{2D^{1/2}}\right) + \frac{1}{4}\exp\left(\frac{vx}{D}\right)\operatorname{erfc}\left(\frac{x}{2(Dt)^{1/2}} + \frac{vt^{1/2}}{2D^{1/2}}\right); \quad (6a)$$

$x \geq 0$.

For $x \leq 0$ we have, by symmetry,

$$\alpha_i(x = -n) = 1 - \alpha_i(x = +n). \quad (6b)$$

We emphasize at this point that the assumption that v is concentration-independent is not always correct. Thus, according to an argument based on assuming the system to be a regular solution (sect. 3), v scales as

$$v \propto \alpha_j(\partial\alpha_i/\partial x),$$

where, although α_j can be approximated as unity with little error, the problem of $\partial\alpha_i/\partial x$ is more fundamental.

Eq. (6a) is easily restated in steady-state, "∞", form. For mixing, i.e. $v > 0$, the result is

$$\alpha_i^{\infty} \sim \frac{1}{2},$$

corresponding to the effect tending to go to completion. For demixing, i.e. $v < 0$, the result is

$$\alpha_i^{\infty} \sim \frac{1}{2}\exp(-|v|x/D), \quad (7)$$

showing that the system goes to a steady state which is intermediate between the mixed and unmixed conditions.

It is convenient to express v and D in atomistic terms:

$$v = \Gamma m_2\lambda; \quad (8a)$$

$$D = \frac{1}{6}\Gamma m_1^2\lambda^2, \quad (8b)$$

where Γ is the displacement frequency per atom (s^{-1}) and m_1 and m_2 have been defined in connection with eqs. (3a) and (3b). Littmark [27] suggests that m_1 is an ill-defined but atomically small quantity and we will take it as $m_1 = 4$ in what follows. We are unable to estimate the size of m_2 and will therefore treat it as a parameter.

3. EVALUATION OF v

Johnson [28], followed by others [5,8-11], proposed that chemical driving forces in mixing could be taken into account by the standard analysis named for Darken [29]. Briefly, the argument is as follows. In thermally activated diffusion the flux for non-ideal systems scales not as the concentration gradient but as the chemical potential ("μ_i") gradient [29]:

$$\Omega J_i = -\alpha_i B_i(\partial\mu_i/\partial x), \tag{9}$$

where $\Omega = \lambda^3$ is the atomic volume and B_i is the mobility, i.e. the velocity per unit force. We recognize the standard relation between μ_i and the activity coefficient, γ_i:

$$\mu_i = \mu_i^\bullet + RT\ell n\gamma_i\alpha_i,$$

as well as the relation for γ_i of a regular solution [30],

$$RT\ell n\gamma_i = \alpha_j^2 h_m, \tag{10}$$

where $h_m = \Delta H_m/\alpha_i\alpha_j$ is constant for a given system. For most real systems, ΔH_m is parabolic [31] such that eq. (10) is a surprisingly good approximation. We are now in a position to expand eq. (9):

$$\begin{aligned}\Omega J_i &= -\alpha_i B_i RT(\partial\ell n\alpha_i/\partial x) + 2\alpha_i\alpha_j h_m B_i(\partial\alpha_i/\partial x)\\ &\equiv -D(\partial\alpha_i/\partial x) + v\alpha_i,\end{aligned} \tag{11a}$$

where we have arbitrarily chosen to identify the extra term with a "v" rather than as a correction [28] to a "D". Specifically, our "v" is given by

$$v = (2\alpha_j h_m D/RT)(\partial\alpha_i/\partial x); \tag{11b}$$

$$v > 0 \text{ for } h_m < 0; \quad v < 0 \text{ for } h_m > 0.$$

An important argument is made by van Rossum and Cheng [32] (and supported by simulation [18]) concerning the role of the cohesive energy, E_c, in displacement production. They noted that although stable Frenkel pairs are formed in a quantity which scales as E_d^{-1}, E_d being the displacement threshold, mixing will come about from both stable and unstable Frenkel pairs and the relevant scaling is therefore E_c^{-1}. The quantity D in eq. (11b), being ballistic, therefore also scales as E_c^{-1} and we find altogether

$$v \propto \Delta H_m/E_c. \tag{11c}$$

We note that eq. (11c) was obtained without assuming thermal-spike behavior and that this constitutes a major difference from [9-11,28,32].

4. DISCUSSION

Eq. (6a) is shown in fig. 4 for $\Gamma t = 16$, $m_1 = 4$, and values of m_2 from 0 to 2. We note a striking increase in the extent of mixing as m_2 increases. Also, there is a characteristic change in the profile shape such that it acquires a step-like nature. Such profiles were found recently by Choupyatova et al. [33] for bilayers of Al on Fe, a system which has an unusually large, negative ΔH_m (table 1).

Mixing is shown in a different way in fig. 5, which gives values of $x/2(Dt)^{1/2}$ for $\alpha_i = 0.25$ (this being a measure of the mixing rate) as a function of $vt^{1/2}/2D^{1/2}$. Approximate linearity is found between $x/2(Dt)^{1/2}$ and $vt^{1/2}/2D^{1/2}$. Since v scales as $\Delta H_m/E_c$ (eq. (11c)) and, in addition, we have $t \propto \phi$, it follows that the linearity seen in fig. 5 is equivalent to

$$\text{(linear extent of mixing)} \propto \Delta H_m\phi/E_c. \tag{12a}$$

Fig. 5 thus agrees in spirit, but not in detail, with what is observed by Cheng et al., namely (see fig. 3)

$$\text{(linear extent of mixing)} \propto (\Delta H_m\phi/E_c)^{1/2}. \tag{12b}$$

Fig. 6 shows how a multilayer target should evolve at steady state for both positive and negative v. It was assumed in constructing fig. 6 that eq. (7) could be used even when the mixing from consecutive layers overlapped. While fig. 6a corresponds to the situation of fig. 3, i.e. $\Delta H_m \leq 0$, figs. 6b and 6c should apply to the situation of fig. 1, i.e. $\Delta H_m \geq 0$. We note that, for $m_2 = 0.1$, there is extensive mixing in spite of the positive ΔH_m, while, for $m_2 = 1$, there is a strong resistance to mixing, i.e. demixing could be said to occur. This difference is well borne out in fig. 1.

Demixing can be seen in another way in terms of the mean atom fraction, $<\alpha_i>$, to be expected for $v < 0$:

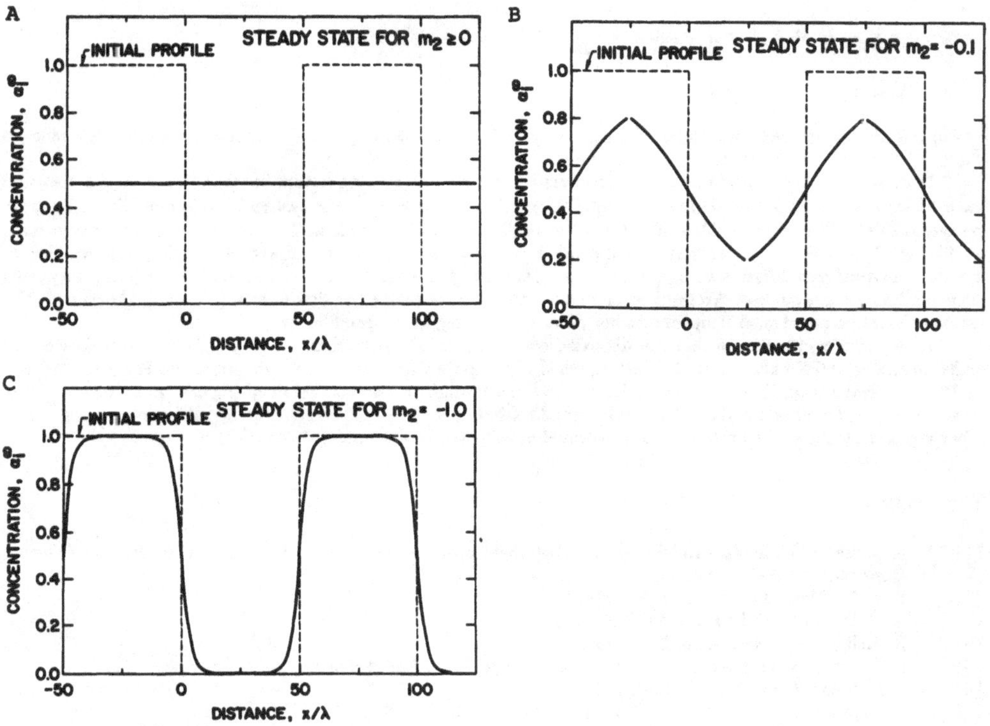

Fig. 6. Steady-state mixing profiles appropriate to a multilayer system as obtained with eq. (7). The width of each layer (2ℓ) in the assumed initial profile is that for the specimens of Peiner and Kopitzki [13], namely about 50 λ. (6a) shows the universal final condition, $\alpha_i = 0.5$, that occurs for $m_2 \geq 0$. (6b) shows the steady-state profile that occurs for $m_2 = -0.1$ provided eq. (7) can be used in a situation where the equation is not fully valid owing to an overlapping of the mixing from consecutive layers. (6c) shows the profile that occurs for $m_2 = -1.0$. On passing from (a) to (b) to (c) we see a drastic decrease in the extent of mixing.

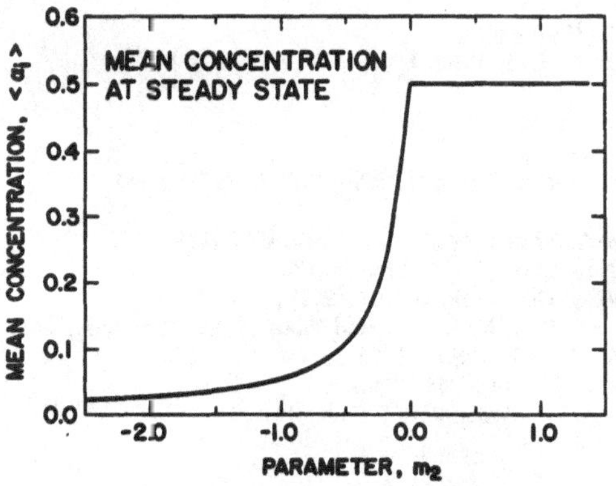

Fig. 7. Referring to the steady-state mixing profiles of fig. 6, it follows that a measure of the mixing is the area from 0 to ℓ with $\ell = 25\lambda$. This area, given by eq. (13), is shown as a function of m_2. We regard this figure as being in the spirit of the experimental results of fig. 1.

$$< \alpha_i > = \frac{1}{\ell}\int_0^{\ell}\frac{1}{3}\exp(-ax)dx = \frac{1}{2a\ell}[1 - \exp(-a\ell)]; \qquad (13)$$

$$a = |v|/D = 3|m_2|/8\lambda,$$

where ℓ is the half-thickness of a layer. Taking $\ell = 25\ \lambda$ [13] we obtain fig. 7 and note it to be reasonably similar to fig. 1.

Marton et al. [12] prepared multilayer specimens of Ag-Ni with each Ag section having a thickness of 3-4 nm and each Ni section a thickness of ~50 nm, the total thickness being ~320 nm. The system is again one with $\Delta H_m > 0$, namely 0.23 eV. They then profiled with 4 keV Ar^+ and got a diffuse profile at T = 299 K but unexpectedly sharp profiles at T = 400 to 603 K. Their interpretation was that, at the lowest temperature, mixing occurred due to bombardment-induced diffusion involving a complex. At the higher temperatures the complex broke up and the mixing therefore became less marked. We propose as an alternative that demixing like that of refs. [5] and [13] occurred. The process should be active at all temperatures but more so at the higher temperatures.

A very similar situation to that just discussed was encountered by Averback et al. [5]. Here Cu-Bi showed "95 units" of mixing at 6 K but none at all at 295 K, while with Cu-Nb there were "7 units" of mixing at 6 K against "3 units" at 295 K. Their explanation is closely similar to the stand taken here, namely that demixing can occur for $\Delta H_m > 0$. Tsaur et al. [34] found ~16% Ag in Ni by mixing at 83 K but only ~8% at 523 K, the equilibrium amount being < 2%. They argued in terms of the nucleation rate but could equally well have considered demixing.

REFERENCES

[1] R. Kelly, in: Materials Modification by High-fluence Ion Beams, eds. R. Kelly and M.F. da Silva (Kluwer, Dordrecht, Netherlands) p. 305.
[2] R. Kelly, Nucl. Instr. Meth. B39 (1989) 43.
[3] R. Kelly, Mat. Sci. Eng. A115 (1989) 11.
[4] R. Kelly, Nucl. Instr. Meth. B (in press).
[5] R.S. Averback, D. Peak, and L.J. Thompson, Appl. Phys. A39 (1986) 59.
[6] P. Sigmund, Phys. Rev. 184 (1969) 383.
[7] P. Sigmund and A. Gras-Marti, Nucl. Instr. Meth. 182/183 (1981) 23.
[8] Y-T. Cheng, M. van Rossum, M-A. Nicolet, and W. L. Johnson, Appl. Phys. Lett. 45 (1984) 185.
[9] M. van Rossum, Y-T. Cheng, M-A. Nicolet, and W.L. Johnson, Appl. Phys. Lett. 46 (1985) 610.
[10] T.W. Workman, Y.T. Cheng, W.L. Johnson, and M-A. Nicolet, Appl. Phys. Lett. 50 (1987) 1485.
[11] Y-T. Cheng, T. W. Workman, M-A. Nicolet, and W. L. Johnson, Mat. Res. Soc. Symp. Proc. 74 (1987) 419.
[12] D. Marton, J. Fine, and G.P. Chambers, Phys. Rev. Lett. 61 (1988) 2697.
[13] E. Peiner and K. Kopitzki, Nucl. Instr. Meth. B34 (1988) 173.
[14] J.P. Johnson and W.G. Moulton, Rad. Effects (in press).
[15] D.E. Harrison and R.P. Webb, Nucl. Instr. Meth. 218 (1983) 727.
[16] F. Besenbacher, J. Bøttiger, S.K. Nielsen, and H.J. Whitlow, Appl. Phys. A29 (1982) 141.
[17] P. Sigmund, Appl. Phys. A30 (1983) 43.
[18] W.E. King and R. Benedek, J. Nucl. Mat. 117 (1983) 26.
[19] D.K. Murti and R. Kelly, Thin Sol. Films 33 (1976) 149.
[20] H.J. Fecht and W.L. Johnson, Nature 334 (1988) 50.
[21] H.M. Urbassek and J. Michl, Nucl. Instr. Meth. B22 (1987) 480.
[22] M. Szymonski, R.S. Bhattacharya, H. Overeijnder, and A.E. de Vries, J. Phys. D11 (1978) 751.
[23] M. Szymonski and A. Poradzisz, Appl. Phys. A28 (1982) 175.
[24] P. Sigmund, A. Oliva, and G. Falcone, Nucl. Instr. Meth. 194 (1982) 541.
[25] A. Oliva, R. Kelly, and G. Falcone, Surf. Sci. 166 (1986) 403.
[26] F. Oberhettinger and L. Badii, Tables of LaPlace Transforms (Springer, Berlin, F.R.G., 1973) p.264.
[27] U. Littmark, Nucl. Instr. Meth. B7/8 (1985) 684.
[28] W.L. Johnson, Y.T. Cheng, M. van Rossum, and M-A. Nicolet, Nucl. Instr. Meth. B7/8 (1985) 657.
[29] P.G. Shewmon, Diffusion in Solids (McGraw-Hill, New York, U.S.A., 1963) p. 122.
[30] R.A. Swalin, Thermodynamics of Solids, 2nd ed. (Wiley, New York, U.S.A., 1972) p. 147.
[31] R. Hultgren, P.D. Desai, D.T. Hawkins, M. Gleiser, and K.K. Kelley, Selected Values of the Thermodynamic Properties of Binary Alloys (Am. Soc. for Metals, Metals Park, U.S.A., 1973).
[32] M. van Rossum and Y-T. Cheng, Defects and Diffusion Forum 57/58 (1988) 1.
[33] L.P. Choupyatova, I.A. Komarovsky, I.G. Murzin, and T.E. Videnskaya (this book).
[34] B.Y. Tsaur and J.W. Mayer, Appl. Phys. Lett. 37 (1980) 389.
[35] A.R. Miedema, Philips Tech. Rev. 36 (1976) 217.
[36] J.W. Mayer, B.Y. Tsaur, S.S. Lau, and L-S. Hung, Nucl. Instr. Meth. 182/183 (1981) 1.

Physical Research

Vol. 1: Harald Böttger/Valerij V. Bryksin,
HOPPING CONDUCTION IN SOLIDS.
1985 - 398 pages - 107 figures - 1 table

Vol. 2: Heinz-Eberhard Albrecht,
LASER - DOPPLER - STRÖMUNGSMESSUNG (in German).
1986 - 202 pages - 136 figures - 14 tables

Vol. 3: Eckehard W. Mielke,
GEOMETRODYNAMICS OF GAUGE FIELDS.
1987 - 242 pages - 11 figures

Vol. 4: Klaus Baumgärtel/Konrad Sauer,
TOPICS ON NONLINEAR WAVE-PLASMA-INTERACTION.
1987 - 222 pages - 111 figures

Vol. 5: Friedhelm Bechstedt/Rolf Enderlein,
SEMICONDUCTOR SURFACES AND INTERFACES.
1988 - 441 pages - 149 figures - 51 tables

Vol. 6: Gerhard Götz/Konrad Gärtner (Eds.),
HIGH ENERGY ION BEAM ANALYSIS OF SOLIDS.
1988 - 376 pages - 178 figures - 33 tables

Vol. 7: Isaak M. Tsidilkovski,
GAPLESS SEMICONDUCTORS - A NEW CLASS OF MATERIALS.
1988 - 134 pages - 36 figures

Vol. 8: Klaus Hennig (Ed.),
EPM 87.
Energy Pulse and Particle Beam Modification of Materials
- International Conference held September 7 - 11, 1987,
Dresden, G. D. R.
1988 - 594 pages - 470 figures - 44 tables

Vol. 9: Wolfgang Kreher/Wolfgang Pompe,
INTERNAL STRESSES IN HETEROGENEOUS SOLIDS.
1989 - 225 pages - 49 figures

Vol. 10: Wolfgang Osten/Ryszard J. Pryputniewicz/Graeme T. Reid/
Hans Rottenkolber (Eds.),
FRINGE '89.
Proceedings of the I. International Workshop on Automatic
Processing of Fringe Patterns held in Berlin (GDR),
April 25 - 28, 1989.
1989 - 168 pages - 90 figures - 1 table

Vol. 11: Hubert Pfeiffer/Thomas Klupsch/Werner Haubenreißer,
MICROSCOPIC THEORY OF CRYSTAL GROWTH.
1989 - 401 pages - 73 figures

Vol. 12: Dietrich Schulze (Ed.),
AMORPHOUS STRUCTURES - METHODS AND RESULTS.
1990 - 400 pages - 150 figures

<u>in preparation:</u>

Vol. 14: Ulrich Messerschmidt/Fritz Appel/Johannes Heydenreich/
Volker Schmidt (Eds.),
ELECTRON MICROSCOPY IN PLASTICITY AND FRACTURE RESEARCH
OF MATERIALS.
Proceedings of the International Symposium held at Holzhau
near Dresden, October 8 - 13, 1989.
1990 - 600 pages - 400 figures

Vol. 15: Gottfried Frankowski/Niels Abramson/Zoltan Füzessy (Eds.),
APPLICATION OF METROLOGICAL LASER METHODS IN MACHINES AND
SYSTEMS.
1990 - 350 pages - 250 figures

Vol. 16: Joachim Kunze,
NITROGEN AND CARBON IN IRON AND LOW ALLOYED STEEL.
1990 - 250 pages - 60 figures - 50 tables

FSC
www.fsc.org
MIX
Papier | Fördert gute Waldnutzung
FSC® C083411

Kolibri 360 GmbH
Ferdinand-Jühlke-Straße 7
99095 Erfurt, Deutschland
produktsicherheit@kolibri360.de